Polarized Light

OPTICAL ENGINEERING

Founding Editor

Brian J. Thompson

University of Rochester
Rochester, New York

1. Electron and Ion Microscopy and Microanalysis: Principles and Applications, *Lawrence E. Murr*
2. Acousto-Optic Signal Processing: Theory and Implementation, *edited by Nor man J. Berg and John N. Lee*
3. Electro-Optic and Acousto-Optic Scanning and Deflection, *Milton Gottlieb, Clive L. M. Ireland, and John Martin Ley*
4. Single-Mode Fiber Optics: Principles and Applications, *Luc B. Jeunhomme*
5. Pulse Code Formats for Fiber Optical Data Communication: Basic Principles and Applications, *David J. Morris*
6. Optical Materials: An Introduction to Selection and Application, *Solomon Musikant*
7. Infrared Methods for Gaseous Measurements: Theory and Practice, *edited by Joda Wormhoudt*
8. Laser Beam Scanning: Opto-Mechanical Devices, Systems, and Data Storage Optics, *edited by Gerald F. Marshall*
9. Opto-Mechanical Systems Design, *Paul R. Yoder, Jr.*
10. Optical Fiber Splices and Connectors: Theory and Methods, *Calvin M. Miller with Stephen C. Mettler and Ian A. White*
11. Laser Spectroscopy and Its Applications, *edited by Leon J. Radziemski, Richard W. Solarz, and Jeffrey A. Paisner*
12. Infrared Optoelectronics: Devices and Applications, *William Nunley and J. Scott Bechtel*
13. Integrated Optical Circuits and Components: Design and Applications, *edited by Lynn D. Hutcheson*
14. Handbook of Molecular Lasers, *edited by Peter K. Cheo*
15. Handbook of Optical Fibers and Cables, *Hiroshi Murata*
16. Acousto-Optics, *Adrian Korpel*
17. Procedures in Applied Optics, *John Strong*
18. Handbook of Solid-State Lasers, *edited by Peter K. Cheo*
19. Optical Computing: Digital and Symbolic, *edited by Raymond Arrathoon*
20. Laser Applications in Physical Chemistry, *edited by D. K. Evans*
21. Laser-Induced Plasmas and Applications, *edited by Leon J. Radziemski and David A. Cremers*
22. Infrared Technology Fundamentals, *Irving J. Spiro and Monroe Schlessinger*
23. Single-Mode Fiber Optics: Principles and Applications, Second Edition, Re vised and Expanded, *Luc B. Jeunhomme*
24. Image Analysis Applications, *edited by Rangachar Kasturi and Mohan M. Trivedi*
25. Photoconductivity: Art, Science, and Technology, *N. V. Joshi*
26. Principles of Optical Circuit Engineering, *Mark A. Mentzer*
27. Lens Design, *Milton Laikin*
28. Optical Components, Systems, and Measurement Techniques, *Rajpal S. Sirohi and M. P. Kothiyal*

29. Electron and Ion Microscopy and Microanalysis: Principles and Applications, Second Edition, Revised and Expanded, *Lawrence E. Murr*
30. Handbook of Infrared Optical Materials, *edited by Paul Klocek*
31. Optical Scanning, *edited by Gerald F. Marshall*
32. Polymers for Lightwave and Integrated Optics: Technology and Applications, *edited by Lawrence A. Hornak*
33. Electro-Optical Displays, *edited by Mohammad A. Karim*
34. Mathematical Morphology in Image Processing, *edited by Edward R. Dougherty*
35. Opto-Mechanical Systems Design: Second Edition, Revised and Expanded, *Paul R. Yoder, Jr.*
36. Polarized Light: Fundamentals and Applications, *Edward Collett*
37. Rare Earth Doped Fiber Lasers and Amplifiers, *edited by Michel J. F. Digonnet*
38. Speckle Metrology, *edited by Rajpal S. Sirohi*
39. Organic Photoreceptors for Imaging Systems, *Paul M. Borsenberger and David S. Weiss*
40. Photonic Switching and Interconnects, *edited by Abdellatif Marrakchi*
41. Design and Fabrication of Acousto-Optic Devices, *edited by Akis P. Goutzoulis and Dennis R. Pape*
42. Digital Image Processing Methods, *edited by Edward R. Dougherty*
43. Visual Science and Engineering: Models and Applications, *edited by D. H. Kelly*
44. Handbook of Lens Design, *Daniel Malacara and Zacarias Malacara*
45. Photonic Devices and Systems, *edited by Robert G. Hunsberger*
46. Infrared Technology Fundamentals: Second Edition, Revised and Expanded, *edited by Monroe Schlessinger*
47. Spatial Light Modulator Technology: Materials, Devices, and Applications, *edited by Uzi Efron*
48. Lens Design: Second Edition, Revised and Expanded, *Milton Laikin*
49. Thin Films for Optical Systems, *edited by Francoise R. Flory*
50. Tunable Laser Applications, *edited by F. J. Duarte*
51. Acousto-Optic Signal Processing: Theory and Implementation, Second Edition, *edited by Norman J. Berg and John M. Pellegrino*
52. Handbook of Nonlinear Optics, *Richard L. Sutherland*
53. Handbook of Optical Fibers and Cables: Second Edition, *Hiroshi Murata*
54. Optical Storage and Retrieval: Memory, Neural Networks, and Fractals, *edited by Francis T. S. Yu and Suganda Jutamulia*
55. Devices for Optoelectronics, *Wallace B. Leigh*
56. Practical Design and Production of Optical Thin Films, *Ronald R. Willey*
57. Acousto-Optics: Second Edition, *Adrian Korpel*
58. Diffraction Gratings and Applications, *Erwin G. Loewen and Evgeny Popov*
59. Organic Photoreceptors for Xerography, *Paul M. Borsenberger and David S. Weiss*
60. Characterization Techniques and Tabulations for Organic Nonlinear Optical Materials, *edited by Mark G. Kuzyk and Carl W. Dirk*
61. Interferogram Analysis for Optical Testing, *Daniel Malacara, Manuel Servin, and Zacarias Malacara*
62. Computational Modeling of Vision: The Role of Combination, *William R. Uttal, Ramakrishna Kakarala, Spiram Dayanand, Thomas Shepherd, Jagadeesh Kalki, Charles F. Lunskis, Jr., and Ning Liu*
63. Microoptics Technology: Fabrication and Applications of Lens Arrays and Devices, *Nicholas Borrelli*
64. Visual Information Representation, Communication, and Image Processing, *edited by Chang Wen Chen and Ya-Qin Zhang*
65. Optical Methods of Measurement, *Rajpal S. Sirohi and F. S. Chau*
66. Integrated Optical Circuits and Components: Design and Applications, *edited by Edmond J. Murphy*

67. Adaptive Optics Engineering Handbook, *edited by Robert K. Tyson*
68. Entropy and Information Optics, *Francis T. S. Yu*
69. Computational Methods for Electromagnetic and Optical Systems, *John M. Jarem and Partha P. Banerjee*
70. Laser Beam Shaping, *Fred M. Dickey and Scott C. Holswade*
71. Rare-Earth-Doped Fiber Lasers and Amplifiers: Second Edition, Revised and Expanded, *edited by Michel J. F. Digonnet*
72. Lens Design: Third Edition, Revised and Expanded, *Milton Laikin*
73. Handbook of Optical Engineering, *edited by Daniel Malacara and Brian J. Thompson*
74. Handbook of Imaging Materials: Second Edition, Revised and Expanded, *edited by Arthur S. Diamond and David S. Weiss*
75. Handbook of Image Quality: Characterization and Prediction, *Brian W. Keelan*
76. Fiber Optic Sensors, *edited by Francis T. S. Yu and Shizhuo Yin*
77. Optical Switching/Networking and Computing for Multimedia Systems, *edited by Mohsen Guizani and Abdella Battou*
78. Image Recognition and Classification: Algorithms, Systems, and Applications, *edited by Bahram Javidi*
79. Practical Design and Production of Optical Thin Films: Second Edition, Revised and Expanded, *Ronald R. Willey*
80. Ultrafast Lasers: Technology and Applications, *edited by Martin E. Fermann, Almantas Galvanauskas, and Gregg Sucha*
81. Light Propagation in Periodic Media: Differential Theory and Design, *Michel Nevière and Evgeny Popov*
82. Handbook of Nonlinear Optics, Second Edition, Revised and Expanded, *Richard L. Sutherland*
83. Polarized Light: Second Edition, Revised and Expanded, *Dennis Goldstein*

Additional Volumes in Preparation

Optical Remote Sensing: Science and Technology, *Walter Egan*

Nonlinear Optics: Theory, Numerical Modeling, and Applications, *Partha P. Banerjee*

Polarized Light

Second Edition, Revised and Expanded

Dennis Goldstein
Air Force Research Laboratory
Eglin Air Force Base, Florida, U.S.A.

MARCEL DEKKER

MARCEL DEKKER, INC. NEW YORK • BASEL

The first edition was published as *Polarized Light: Fundamentals and Applications*, Edward Collett (Marcel Dekker, Inc., 1993).

ISBN: 0-8247-4053-X

This book is printed on acid-free paper.

Headquarters
Marcel Dekker, Inc.
270 Madison Avenue, New York, NY 10016
tel: 212-696-9000; fax: 212-685-4540

Eastern Hemisphere Distribution
Marcel Dekker AG
Hutgasse 4, Postfach 812, CH-4001 Basel, Switzerland
tel: 41-61-261-8482; fax: 41-61-261-8896

World Wide Web
http://www.dekker.com

The publisher offers discounts on this book when ordered in bulk quantities. For more information, write to Special Sales/Professional Marketing at the headquarters address above.

Current printing (last digit):
10 9 8 7 6 5 4 3 2 1

PRINTED IN THE UNITED STATES OF AMERICA

Preface to the Second Edition

Where there is light, there is polarized light. It is in fact difficult to find a source of light that is completely randomly polarized. As soon as light interacts with anything, whether through reflection, transmission, or scattering, there is opportunity for polarization to be induced. As pointed out in the first sentence of the Preface to the First Edition, polarization is a fundamental characteristic of the transverse wave that is light. More than ever, it is a characteristic that must be addressed in modern optical systems and applications.

Since 1993 when the first edition of this text appeared, there have been many new developments in the measurement and application of polarized light. This revised edition includes revisions and corrections of the original text and substantive new material. Most of the original figures have been redone. Chapter 8 has been expanded to include the derivation of the Fresnel equations with plots of the magnitude and phase of the reflection coefficients. Also included in Part I is a chapter with in-depth discussion of the mathematics and meaning of the Mueller matrix. In this chapter, there is a discussion of physical realizability and elimination of error sources with eigenvector techniques, and a discussion of Mueller matrix decomposition. The Lu–Chipman decomposition has shown that Mueller matrices are separable, so that a general Mueller matrix may be decomposed into a set of product matrices, each dependent on only one of the quantities of diattenuation, retardance, or depolarization. A chapter on devices and components has been added to Part III, Applications. Those interested in use or measurement of polarized light should have knowledge of available devices and components that serve as polarizers and retarders for various wavelength regions and for various conditions of achromaticity. Chapters on Stokes polarimetry and Mueller matrix polarimetry have been inserted in Part III. These polarimetric techniques are essential to an understanding of measurement of polarized light and characterization of optical elements.

Appendixes have been added with summaries of the Jones and Stokes vectors for various states of polarized light, and with summaries of Jones and Mueller matrices for various optical elements. An appendix has been included that gives the relations between the Jones and Mueller matrix elements. Finally, a comprehensive bibliography has been included.

Ed Collett collected a wonderful set of topics for students of polarized light for the first edition of this book, and he provided a resource that did not exist before. It is my hope that the revisions and additions contained in this second edition will make this text even more useful and thorough. I express my gratitude to the following colleagues and friends for their critical comments during the creation of this work: Russell A. Chipman of the University of Arizona, Robert R. Kallman of the University of North Texas, J. Scott Tyo of the University of New Mexico, and E.E. (Gene) Youngblood and Lynn L. Diebler of the Air Force Research Laboratory. David Goetsch of Okaloosa-Walton Community College provided wise counsel. Finally, I express gratitude to my wife, Carole, and daughters, Dianne and Laura, for their presence and support.

Dennis Goldstein

Preface to the First Edition

Light is characterized by its intensity, wavelength, and polarization. Remarkably, in spite of the importance of polarized light, no book is devoted just to this subject. Nearly every book on optics contains several chapters on polarized light. However, if one tries to obtain a deeper understanding of the subject, one quickly discovers that it is almost always necessary to go to the original papers in the literature. The objective of this book therefore is to provide a single source that describes the fundamental behavior of polarized light and its interaction with matter. The book is designed to be used by scientists and engineers working in the fields of physics, optics, opto-electronics, chemistry, biology, and mechanical and electrical engineering as well as advanced undergraduate and graduate students.

There are two well-known books on polarized light. The first is W. A. Shurcliff's *Polarized Light*, an excellent introductory and reference book on the subject. The other book, also excellent, is *Ellipsometry and Polarized Light* by R. M. A. Azzam and N. M. Bashara. It is very advanced and is directed to those working in the field of ellipsometry. While it contains much information on polarized light, its approach to the subject is very different. Ellipsometry is important, however, and an introductory discussion is included here in the final chapter.

This book is divided into three parts. One can begin the study of polarized light with Maxwell's equations. However, one soon discovers that in optics, unlike the field of microwave physics, Maxwell's equations are not readily apparent; this was why in the nineteenth century Fresnel's elastic equations were only slowly displaced by Maxwell's equations. Much of the subject of polarized light can be studied and understood almost independently of Maxwell's equations. This is the approach taken in Part I. We begin with the wave equation and quickly move on to the polarization ellipse. At this point the observable concept of the optical field is introduced, and in succeeding chapters we discover that much new information is revealed on the nature as well as the description of polarized light and its interaction with polarizing elements. Ultimately, however, it becomes necessary to describe the *source* of the radiation field and polarized light. At this point no further progress can be made without Maxwell's equations. Therefore, in Part II of this book, Maxwell's equations are introduced and then used to describe the emission of polarized radiation by accelerating electrons. In turn, the emitted radiation is then formulated in terms of the Stokes vector and Mueller matrices and applied to the description of unpolarized light, the Zeeman effect, synchrotron radiation, scattering, and the Faraday effect. In particular, we shall see that the Stokes vector takes on a very interesting role in describing spectral lines. In Part III, a number of important applications of polarized light are presented, namely, propagation in anisotropic media (crystals), opto-isolators, electro-optical modulation, reflection from metals, and a final introductory chapter on ellipsometry.

The creation of this book could have happened only with the support of my family. I wish to express my gratitude to my children Ronald Edward and Gregory Scott, and especially to my wife, Marilyn, for their continuous support, encouragement and interest. Without it, this book would have never been completed.

Edward Collett

Contents

Preface to the Second Edition		*iii*
Preface to the First Edition		*v*
A Historical Note Edward Collett		*xiii*

PART I: THE CLASSICAL OPTICAL FIELD

Chapter 1	Introduction	1
	References	2
Chapter 2	The Wave Equation in Classical Optics	3
	2.1 Introduction	3
	2.2 The Wave Equation	3
	2.3 Young's Interference Experiment	11
	2.4 Reflection and Transmission of a Wave at an Interface	15
	References	19
Chapter 3	The Polarization Ellipse	21
	3.1 Introduction	21
	3.2 The Instantaneous Optical Field and the Polarization Ellipse	22
	3.3 Specialized (Degenerate) Forms of the Polarization Ellipse	25
	3.4 Elliptical Parameters of the Polarization Ellipse	27
	References	30
Chapter 4	The Stokes Polarization Parameters	31
	4.1 Introduction	31
	4.2 Derivation of the Stokes Polarization Parameters	32
	4.3 The Stokes Vector	37
	4.4 Classical Measurement of the Stokes Polarization Parameters	43

	4.5 Stokes Parameters for Unpolarized and Partially Polarized Light	47
	4.6 Additional Properties of the Stokes Polarization Parameters	49
	4.7 Stokes Parameters and Wolf's Coherency Matrix	59
	References	63
Chapter 5	The Mueller Matrices for Polarizing Components	65
	5.1 Introduction	65
	5.2 The Mueller Matrix of a Polarizer	67
	5.3 The Mueller Matrix of a Retarder	72
	5.4 The Mueller Matrix of a Rotator	75
	5.5 Mueller Matrices for Rotated Polarizing Components	77
	5.6 Generation of Elliptically Polarized Light	83
	References	86
Chapter 6	Methods of Measuring the Stokes Polarization Parameters	87
	6.1 Introduction	87
	6.2 Classical Measurement Method: The Quarter-Wave Retarder Polarizer Method	87
	6.3 Measurement of the Stokes Parameters Using a Circular Polarizer	91
	6.4 The Null-Intensity Method	95
	6.5 Fourier Analysis Using a Rotating Quarter-Wave Retarder	98
	6.6 The Method of Kent and Lawson	102
	6.7 Simple Tests to Determine the State of Polarization of an Optical Beam	108
	References	115
Chapter 7	The Measurement of the Characteristics of Polarizing Elements	117
	7.1 Introduction	117
	7.2 Measurement of Attenuation Coefficients of a Polarizer (Diattenuator)	117
	7.3 Measurement of Phase Shift of a Retarder	124
	7.4 Measurement of Rotation Angle of a Rotator	130
	Reference	132
Chapter 8	Mueller Matrices for Reflection and Transmission	133
	8.1 Introduction	133
	8.2 Fresnel's Equations for Reflection and Transmission	134
	8.3 Mueller Matrices for Reflection and Transmission at an Air–Dielectric Interface	145
	8.4 Special Forms for the Mueller Matrices for Reflection and Transmission	152
	References	163

Chapter 9 The Mathematics of the Mueller Matrix 165
9.1 Introduction 165
9.2 Constraints on the Mueller Matrix 167
9.3 Eigenvector and Eigenvalue Analysis 168
9.4 Example of Eigenvector Analysis 171
9.5 The Lu–Chipman Decomposition 175
9.6 Summary 185
References 185

Chapter 10 The Mueller Matrices for Dielectric Plates 187
10.1 Introduction 187
10.2 The Diagonal Mueller Matrix and the *ABCD* Polarization Matrix 187
10.3 Mueller Matrices for Single and Multiple Dielectric Plates 195
References 209

Chapter 11 The Jones Matrix Calculus 211
11.1 Introduction 211
11.2 The Jones Vector 212
11.3 Jones Matrices for the Polarizer, Retarder, and Rotator 216
11.4 Applications of the Jones Vector and Jones Matrices 221
11.5 Jones Matrices for Homogeneous Elliptical Polarizers and Retarders 231
References 239

Chapter 12 The Poincaré Sphere 241
12.1 Introduction 241
12.2 Theory of the Poincaré Sphere 242
12.3 Projection of the Complex Plane onto a Sphere 258
12.4 Applications of the Poincaré Sphere 265
References 273

Chapter 13 The Interference Laws of Fresnel and Arago 275
13.1 Introduction 275
13.2 Mathematical Statements for Unpolarized Light 276
13.3 Young's Interference Experiment with Unpolarized Light 278
13.4 The First Experiment: First and Second Interference Laws 282
13.5 The Second Experiment: Third Interference Law 287
13.6 The Third Experiment: Fourth Interference Law 289
13.7 The Herschel–Stokes Experiment 292
13.8 Summary of the Fresnel–Arago Interference Laws 294
References 296

PART II: THE CLASSICAL AND QUANTUM THEORY OF RADIATION BY ACCELERATING CHARGES

Chapter 14 Introduction to the Classical and Quantum Theory of Radiation by Accelerating Charges 297
References 298

Chapter 15 Maxwell's Equations for the Electromagnetic Field 301
References 306

Chapter 16 The Classical Radiation Field 307
16.1 Field Components of the Radiation Field 307
16.2 Relation Between the Unit Vector in Spherical Coordinates and Cartesian Coordinates 309
16.3 Relation Between the Poynting Vector and the Stokes Parameters 312
References 317

Chapter 17 Radiation Emitted by Accelerating Charges 319
17.1 Stokes Vector for a Linearly Oscillating Charge 319
17.2 Stokes Vector for an Ensemble of Randomly Oriented Oscillating Charges 322
17.3 Stokes Vector for a Charge Rotating in a Circle 325
17.4 Stokes Vector for a Charge Moving in an Ellipse 328
References 329

Chapter 18 The Radiation of an Accelerating Charge in the Electromagnetic Field 331
18.1 Motion of a Charge in an Electromagnetic Field 331
18.2 Stokes Vectors for Radiation Emitted by Accelerating Charges 345
References 350

Chapter 19 The Classical Zeeman Effect 351
19.1 Historical Introduction 351
19.2 Motion of a Bound Charge in a Constant Magnetic Field 352
19.3 Stokes Vector for the Zeeman Effect 361
References 366

Chapter 20 Further Applications of the Classical Radiation Theory 369
20.1 Relativistic Radiation and the Stokes Vector for a Linear Oscillator 369
20.2 Relativistic Motion of a Charge Moving in a Circle: Synchrotron Radiation 376
20.3 Čerenkov Effect 382
20.4 Thomson and Rayleigh Scattering 393
References 401

Chapter 21 The Stokes Parameters and Mueller Matrices for Optical Activity and Faraday Rotation 403
21.1 Introduction 403
21.2 Optical Activity 405
21.3 Faraday Rotation in a Transparent Medium 411
21.4 Faraday Rotation in a Plasma 414
References 416

Chapter 22 The Stokes Parameters for Quantum Systems 417
22.1 Introduction 417
22.2 Relation Between Stokes Polarization Parameters and Quantum Mechanical Density Matrix 419
22.3 Note on Perrin's Introduction of Stokes Parameters, Density Matrix, and Linearity of the Mueller Matrix Elements 427
22.4 Radiation Equations for Quantum Mechanical Systems 432
22.5 Stokes Vectors for Quantum Mechanical Systems 435
References 442

Part III: APPLICATIONS

Chapter 23 Introduction 445

Chapter 24 Crystal Optics 447
24.1 Introduction 447
24.2 Review of Concepts from Electromagnetism 448
24.3 Crystalline Materials and Their Properties 450
24.4 Crystals 451
24.5 Application of Electric Fields: Induced Birefringence and Polarization Modulation 463
24.6 Magneto-optics 470
24.7 Liquid Crystals 471
24.8 Modulation of Light 475
24.9 Concluding Remarks 476
References 477

Chapter 25 Optics of Metals 479
25.1 Introduction 479
25.2 Maxwell's Equations for Absorbing Media 480
25.3 Principal Angle of Incidence Measurement of Refractive Index and Extinction Coefficient of Optically Absorbing Materials 489
25.4 Measurement of Refractive Index and Extinction Coefficient at an Incident Angle of 45° 496
References 509

Chapter 26 Polarization Optical Elements 511
26.1 Introduction 511
26.2 Polarizers 511

	26.3 Retarders	522
	26.4 Rotators	529
	26.5 Depolarizers	531
	References	532
Chapter 27	Stokes Polarimetry	533
	27.1 Introduction	533
	27.2 Rotating Element Polarimetry	534
	27.3 Oscillating Element Polarimetry	537
	27.4 Phase Modulation Polarimetry	543
	27.5 Techniques in Simultaneous Measurement of Stokes Vector Elements	545
	27.6 Optimization of Polarimeters	554
	References	556
Chapter 28	Mueller Matrix Polarimetry	559
	28.1 Introduction	559
	28.2 Dual Rotating-Retarder Polarimetry	563
	28.3 Other Mueller Matrix Polarimetry Methods	578
	References	583
Chapter 29	Ellipsometry	585
	29.1 Introduction	585
	29.2 Fundamental Equation of Classical Ellipsometry	586
	29.3 Classical Measurement of the Ellipsometric Parameters Psi (ψ) and Delta (Δ)	588
	29.4 Solution of the Fundamental Equation of Ellipsometry	597
	29.5 Further Developments in Ellipsometry: The Mueller Matrix Representation of ψ and Δ	616
	References	623

Appendix A: Jones and Stokes Vectors 625
Appendix B: Jones and Mueller Matrices 627
Appendix C: Relationships Between the Jones and Mueller Matrix Elements 631
Appendix D: Vector Representation of the Optical Field: Application to Optical Activity 633
Bibliography 645
Index 647

A Historical Note

At the midpoint of the nineteenth century the wave theory of light developed by Augustin Jean Fresnel (1788–1827) and his successors was a complete triumph. The wave theory completely explained the major optical phenomena of interference, diffraction, and polarization. Furthermore, Fresnel had successfully applied the wave theory to the problem of the propagation and polarization of light in anisotropic media, that is, crystals. A further experiment was carried out in 1851 by Armand Hypolite Louis Fizeau (1819–1896), who showed that the speed of light was less in an optically dense medium than in a vacuum, a result predicted by the wave theory. The corpuscular theory, on the other hand, had predicted that in an optically dense medium the speed of light would be greater than in a vacuum. Thus, in practically all respects Fresnel's wave theory of light appeared to be triumphant.

By the year 1852, however, a crisis of quite significant proportions was slowly simmering in optics. The crisis, ironically, had been brought on by Fresnel himself 35 years earlier. In the year 1817 Fresnel, with the able assistance of his colleague Dominique François Arago (1786–1853), undertook a series of experiments to determine the influence of polarized light on the interference experiments of Thomas Young (1773–1829). At the beginning of these experiments Fresnel and Arago held the view that light vibrations were longitudinal. At the end of their experiments they were unable to understand their results on the basis of longitudinal vibrations. Arago communicated the puzzling results to Young, who then suggested that the experiments could be understood if the light vibrations were transverse, consisted of only two orthogonal components, and there was no longitudinal component. Indeed, this did make some, but not all, of the results comprehensible. At the conclusion of their experiments Fresnel and Arago summarized their results in a series of statements that have come down to us as the four interference laws of Fresnel and Arago.

All physical laws are described in terms of verbal statements from which mathematical statements can then be written (e.g., Kepler's laws of planetary motion and Newton's laws of motion). Fresnel understood this very well. Upon completing his experiments, he turned to the problem of developing the mathematical statements for the four interference laws. Fresnel's wave theory was an amplitude description of light and was completely successful in describing completely polarized light, that is, elliptically polarized light and its degenerate states, linearly and circularly polarized light. However, the Fresnel–Arago experiments were carried out not with completely polarized light but with another state of polarized light called unpolarized light. In order to describe the Fresnel–Arago experiments it would be necessary for Fresnel to provide the mathematical statements for unpolarized light, but much to his surprise, on the basis of his amplitude formulation of light, he was unable to write the mathematical statements for unpolarized light! And he never succeeded. With his untimely death in 1827 the task of describing unpolarized light (or for that matter any state of polarized light within the framework of classical optics) along with providing the mathematical statements of the Fresnel–Arago interference laws passed to others. For many years his successors were no more successful than he had been.

By 1852, 35 years had elapsed since the enunciation of the Fresnel–Arago laws and there was still no satisfactory description of unpolarized light or the interference laws. It appeared that unpolarized light, as well as so-called partially polarized light, could not be described within the framework of the wave theory of light, which would be a crisis indeed.

The year 1852 is a watershed in optics because in that year Sir George Gabriel Stokes (1819–1903) published two remarkable papers in optics. The first appeared with the very bland title "On the Composition and Resolution of Streams of Polarized Light from Different Sources," a title that appears to be far removed from the Fresnel–Arago interference laws; the paper itself does not appear to have attracted much attention. It is now, however, considered to be one of the great papers of classical optics. After careful reading of his paper, one discovers that it provides the mathematical formulation for describing any state of polarized light and, most importantly, the mathematical statements for unpolarized light: the mathematical statements for the Fresnel–Arago interference laws could now be written. Stokes had been able to show, finally, that unpolarized light and partially polarized light could be described within the framework of the wave theory of light.

Stokes was successful where all others had failed because he developed a highly novel approach for describing unpolarized and partially polarized light. He abandoned the fruitless attempts of his predecessors to describe unpolarized light in terms of amplitudes and, instead, resorted to an experimental definition of unpolarized light. In other words, he was led to a formulation of polarized light in terms of measured quantities, that is, intensities (observables). This was a completely unique point of view for the nineteenth century. The idea of observables was not to reappear again in physics until the advent of quantum mechanics in 1925 by Werner Heisenberg (1901–1976) and later in optics with the observable formulation of the optical field in 1954 by Emil Wolf (1922–   ).

Stokes showed that his intensity formulation of polarized light could be used to describe not only unpolarized and partially polarized light but completely polarized light as well. Thus, his formulation was applicable to any state of polarized light. His entire paper is devoted to describing in all the detail of mid-nineteenth-century algebra the properties of various combinations of polarized and unpolarized light. Near the end of his paper Stokes introduced his discovery that four parameters, now known as the Stokes polarization parameters, could characterize any state of polarized light. Unlike the amplitude formulation of the optical field, his parameters were directly accessible to measurement. Furthermore, he then used these parameters to obtain a correct mathematical statement for unpolarized light. The stage had now been set to write the mathematical statements for the Fresnel–Arago interference laws.

At the end of Stokes' paper he turns, at long last, to his first application, the long awaited mathematical statements for the Fresnel–Arago interference laws. In his paper he states, "Let us now apply the principles and formulae which have just been established to a few examples. And first let us take one of the fundamental experiments by which MM. Arago and Fresnel established the laws of interference of polarized light, or *rather an analogous experiment mentioned by Sir John Herschel.*" Thus, with these few words Stokes abandoned his attempts to provide the mathematical statements for the Fresnel–Arago laws. At this point Stokes knew that to apply his formulation to the formulation of the Fresnel–Arago interference laws was a considerable undertaking. It was sufficient for Stokes to know that his mathematical formulation of polarized light would explain them. Within several more pages, primarily devoted to correcting several experiments misunderstood by his colleagues, he concluded his paper.

This sudden termination is remarkable in view of its author's extraordinary effort to develop the mathematical machinery to describe polarized light, culminating in the Stokes polarization parameters. One must ask why he brought his paper to such a rapid conclusion. In my opinion, and this shall require further historical research, the answer lies in the paper that immediately follows Stokes' polarization paper, published only two months later. Its title was, "On the Change of the Refrangibility of Light."

In the beginning of this Historical Note it was pointed out that by 1852 there was a crisis in optics over the inability to find a suitable mathematical description for unpolarized light and the Fresnel–Arago interference laws. This crisis was finally overcome with the publication of Stokes' paper on polarized light in 1852. But this next paper by Stokes dealt with a new problem of very disconcerting proportions. It was the first in a series of papers that would lead, 75 years later, to quantum mechanics. The subject of this second paper is a topic that has become known as the fluorescence of solutions. It is a monumental paper and was published in two parts. The first is a 20-page abstract! The second is the paper itself, which consists of nearly 150 pages. After reading this paper it is easy to understand why Stokes had concluded his paper on the Fresnel–Arago interference laws. He was deeply immersed in numerous experiments exploring the peculiar phenomenon of fluorescence. After an enormous amount of experimental effort Stokes was able to enunciate his now famous law of fluorescence, namely, that the wavelength of the emitted fluorescent radiation was greater than the excitation wavelength; he also found that the fluorescence radiation appeared to be unpolarized. Stokes was never able to find the reason for this peculiar behavior of fluorescence or the basis of his law. He would spend the next 50 years searching for the reason for his empirical law until his death in 1903. Ironically, in 1905, two years after Stokes' death, a young physicist by the name of Albert Einstein (1879–1955) published a paper entitled "On a Heuristic Point of View Concerning the Generation and Conversion of Light" and showed that Stokes' law of fluorescence could be easily explained and understood on the basis of the quantum hypothesis of Max Planck (1858–1947). It is now clear that Stokes never had the slightest chance of explaining the phenomenon of fluorescence within the framework of classical optics. Thus, having helped to remove one of the last barriers to the acceptance of the wave theory of light, Stokes' investigations on the nature of light had led him to the discovery of the first law ever associated with the quantum phenomenon. Unknowingly, Stokes had stumbled onto the quantum nature of light. Thirty-five years later, in 1888, a similar chain of events was repeated when Heinrich Hertz (1857–1894), while verifying the electromagnetic field theory of James Clerk Maxwell (1831–1879), the ultimate proof of the truth of the classical wave theory of light, also discovered a new and unexplainable phenomenon, the photoelectric effect. We now know that this too can be understood only in terms of the quantum theory. Science is filled with ironies.

Within two months of the publication in March 1852 of his paper on polarized light, in which the formulation of classical optics appeared to be complete, with the May 1852 publication of his paper on fluorescence, Stokes went from complete triumph to complete dismay. He would constantly return to the subject of fluorescence for the remainder of his life, always trying but never succeeding in understanding the origin of his law of fluorescence.

Stoke's great paper on polarization was practically forgotten because by the mid-nineteenth century classical optics was believed to be complete and physicists had turned their attention to the investigation of the electromagnetic field and the statistical mechanics of molecules. His paper was buried in the scientific literature for nearly a century. Its importance was finally recognized with its "discovery" in the 1940s by the Nobel laureate Subrahmanya Chandrasekhar (1910–   ), who used the Stokes parameters to include the effects of polarized light in the equations of radiative transfer.

In this book we shall see that the Stokes polarization parameters provide a rich and powerful tool for investigating and understanding polarized light and its interaction with matter. The use of these parameters provides a mathematical formulation of polarized light whose power is far greater than was ever imagined by their originator and serves as a tribute to his genius.

Edward Collett

REFERENCES

Papers

1. Stokes, G. G. Trans. Camb. Phil. Soc. 9, 399, 1852. Reprinted in *Mathematical and Physical Papers*, Vol. 3, p. 233, Cambridge University Press, London, 1901.
2. Stokes, G. G. Proc. of the Royal Soc. b, 195, 1852. Reprinted in *Mathematical and Physical Papers*, Vol. 3, p. 259, Cambridge University Press, London, 1901.
3. Einstein, A. *Ann. Phys. 17*, 132, 1905.
4. Heisenberg, W. *Zs. f. Phys. 33*, 879, 1925.
5. Wolf, E. *Nuovo Cimento, 12*, 884, 1954.
6. Collett, E. *Amer. J. Phys., 39*, 1483, 1971.
7. Mulligan, J. *Physics Today, 42*, 50, March, 1989.

Books

1. Fresnel, A. J. *L'Oeuvres Completes*, Henri de Senarmont, Emile Verdet et Leonor Fresnel, Vol. I. Paris, 1866.
2. Born, M. and Wolf, E. *Principles of Optics*, 3rd ed. Pergamon Press, Inc., New York, 1965.
3. Whittaker, E. *A History of the Theories of Aether and Electricity*, Vol. I. Philosophical Society, New York, 1951.
4. S. Chandrasekhar, *Radiative Transfer*, Dover Publications, pp. 24–34, New York, 1960.

Polarized Light

1

Introduction

The polarization light is one of its fundamental properties, the others being its intensity, frequency, and coherence. In this book the nature of polarized light and its numerous applications are described. Aside from its remarkable properties, the study of polarized light has led to a deeper understanding of the nature of light itself. The investigations of polarized light began with the discovery by Erasmus Bartholinus (1625–1698) in 1669 of the phenomenon of double refraction in calcite crystals (calspar). This was followed by the work of Christian Huygens (1629–1695), who interpreted double refraction by assuming that in the calspar crystal there is, in addition to a primary spherical wave, a secondary ellipsoidal wave. In the course of his investigations in 1690, Huygens also made a fundamental discovery on polarization, namely, each of the two rays arising from refraction by calcite can be extinguished by passing it through a second calcite crystal if the latter crystal is rotated about the direction of the ray. Isaac Newton (1642–1727) interpreted these phenomena by assuming that the rays have "sides." Indeed, this "transversality" appeared to him to be a serious objection to the acceptance of the wave theory. In Newton's time, scientists, from their work on the propagation of sound, were familiar only with longitudinal waves; it was believed that light "waves," if they existed, were similar to sound waves.

During the eighteenth century the corpuscular theory of light supported by Newton held sway. However, in 1801 Thomas Young (1773–1829) gave new life to the wave theory when he enunciated his principle of interference and applied it to the explanation of the colors of thin films. In addition, Young carried out a rather spectacular and extraordinarily simple experiment to demonstrate the interference of light, namely, the two-pinhole interference experiment. However, because Young's views were largely qualitative, they did not gain immediate acceptance.

In 1808, Etienne-Louis Malus (1775–1812), an officer in the French army was in the Palais de Luxembourg in Paris, where he made a remarkable discovery. He observed the reflection of the sun from a windowpane through a calspar crystal and found that the two images obtained by double refraction were extinguished alternately as he rotated the calcite crystal. Malus reported this result but offered no explanation. Several years later, in 1812, Sir David Brewster (1781–1868) also investigated the behavior of light reflected from glass. He discovered that at a particular angle of incidence (Brewster's angle) the reflected light viewed through a calcite crystal could be extinguished. Further investigations by Brewster revealed that there was a simple relation between what was to be called the Brewster angle and the refractive index of the glass. The importance of this work was further enhanced because it allowed the refractive index of optical glass to be determined by reflection rather than by refraction (transmission). The significance of Brewster's discovery was immediately recognized by his contemporaries, and he received the Gold Medal from the Royal Society in 1815.

While Brewster was actively working in Great Britain, Augustin Jean Fresnel (1788–1827) in France was placing the wave theory on a firm theoretical foundation using the Fresnel–Huygens integral to solve the problem of diffraction. In 1818 he was awarded the prize for the solution of the diffraction problem by the Paris Academy of Science after his friend and colleague, Dominique François Arago (1786–1853), experimentally showed the existence of a small bright spot in the shadow of a small circular disk, a result predicted by Fresnel's theory. The wave theory was further enhanced when it was used to describe the propagation of polarized light through optically active media. As a result of Fresnel's work and others, the wave theory of light gained almost universal acceptance.

The wave equation appears in classical optics as a hypothesis. It was accepted because it led to the understanding and description of the propagation, diffraction, interference, and polarization of light. Furthermore, the calculations made using the wave equations led to results in complete agreement with experiments. A true experimental foundation for the wave equation would have to wait until James Clerk Maxwell's (1831–1879) electrodynamic theory and its experimental confirmation by Heinrich Hertz (1857–1894) in the second half of the nineteenth century. To discuss polarized light, we need to investigate first the wave equation and its properties. We therefore begin our study of polarized light with the wave equation.

REFERENCES

1. Sommerfeld, A., *Lectures on Theoretical Physics*, Vols. I–V, Academic Press, New York, 1952.
2. Whittaker, E., *A History of the Theories of Aether and Electricity*, Vol. 1, Philosophical Society, New York, 1951.
3. Born, M. and Wolf, E., *Principles of Optics*, 3rd ed., Pergamon Press, New York, 1965.

2

The Wave Equation in Classical Optics

2.1 INTRODUCTION

The concept of the interference of waves, developed in mechanics in the eighteenth century, was introduced into optics by Thomas Young at the beginning of the nineteenth century. In the eighteenth century the mathematical physicists Euler, d'Alembert, and Lagrange had developed the wave equation from Newtonian mechanics and investigated its consequences, e.g., propagating and standing waves. It is not always appreciated that Young's "leap of genius" was to take the ideas developed in one field, mechanics, and apply them to the completely different field of optics.

In addition to borrowing the idea of wave interference, Young found that it was also necessary to use another idea from mechanics. He discovered that the superposition of waves was insufficient to describe the phenomenon of optical interference; it, alone, did not lead to the observed interference pattern. To describe the interference pattern he also borrowed the concept of energy from mechanics. This concept had been developed in the eighteenth century, and the relation between the amplitude of a wave and its energy was clearly understood. In short, the mechanical developments of the eighteenth century were crucial to the work of Young and to the development of optics in the first half of the nineteenth century. It is difficult to imagine the rapid progress which took place in optics without these previous developments. In order to have a better understanding of the wave equation and how it arose in mechanics and was then applied to optics, we now derive the wave equation from Newton's laws of motion.

2.2 THE WAVE EQUATION

Consider a homogeneous string of length l fixed at both ends and under tension T_0, as shown in Fig. 2-1. The lateral displacements are assumed to be small compared with l. The angle θ between any small segment of the string and the straight line (dashed) joining the points of support are sufficiently small so that $\sin\theta$ is closely approximated by $\tan\theta$. Similarly, the tension T_0 in the string is assumed to be unaltered by the small lateral displacements; the motion is restricted to the xy plane.

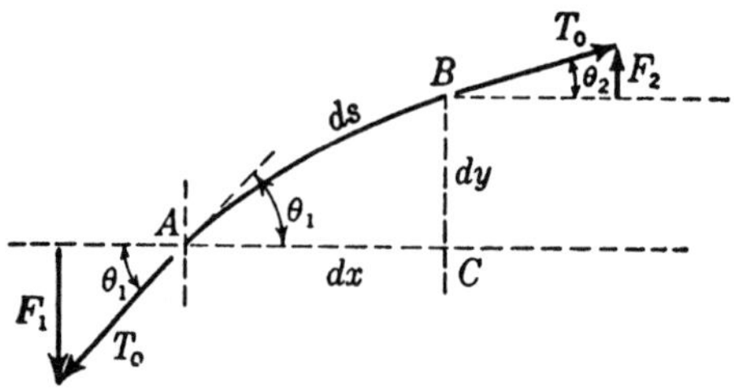

Figure 2-1 Derivation of the wave equation. Motion of a string under tension.

The differential equation of motion is obtained by considering a small element ds of the string and is shown exaggerated as the segment $\overline{AB}$ in Fig. 2-1. The y component of the force acting on ds consists of F_1 and F_2. If θ_1 and θ_2 are small, then

$$F_1 = T_0 \sin\theta_1 \simeq T_0 \tan\theta_1 = T_0\left(\frac{\partial y}{\partial x}\right)_A \tag{2-1a}$$

$$F_2 = T_0 \sin\theta_2 \simeq T_0 \tan\theta_2 = T_0\left(\frac{\partial y}{\partial x}\right)_B \tag{2-1b}$$

where the derivatives are partials because y depends on time t as well as on the distance x. The subscripts signify that the derivatives are to be evaluated at points A and B, respectively. Then, by Taylor's expansion theorem,

$$\left(\frac{\partial y}{\partial x}\right)_A = \frac{\partial y}{\partial x} - \left[\frac{\partial}{\partial x}\frac{\partial y}{\partial x}\right]\frac{dx}{2} = \frac{\partial y}{\partial x} - \frac{\partial^2 y}{\partial x^2}\frac{dx}{2} \tag{2-2a}$$

$$\left(\frac{\partial y}{\partial x}\right)_B = \frac{\partial y}{\partial x} + \left[\frac{\partial}{\partial x}\frac{\partial y}{\partial x}\right]\frac{dx}{2} = \frac{\partial y}{\partial x} + \frac{\partial^2 y}{\partial x^2}\frac{dx}{2} \tag{2-2b}$$

in which the derivatives without subscripts are evaluated at the midpoint of ds. The resultant force in the y direction is

$$F_2 - F_1 = T_0\left(\frac{\partial^2 y}{\partial x^2}\right)dx \tag{2-3}$$

If ρ is the mass per unit length of the string, the inertial reaction (force) of the element ds is $\rho ds(\partial^2 y/\partial t^2)$. For small displacements, ds can be written as $ds \simeq dx$. The equation of motion is then obtained by equating the inertial reaction to the applied force (2-3), so we have

$$\frac{\partial^2 y}{\partial t^2} = \frac{T_0}{\rho}\frac{\partial^2 y}{\partial x^2} \tag{2-4}$$

Equation (2-4) is the wave equation in one dimension. In optics $y(x, t)$ is equated with the "optical disturbance" $u(x, t)$. Also, the ratio of the tension to the density in the string T/ρ is found to be related to the velocity of propagation v by the equation:

$$v^2 = \frac{T_0}{\rho} \tag{2-5}$$

The form of (2-5) is easily derived by a dimensional analysis of (2-4). Equation (2-4) can then be written as

$$\frac{\partial^2 u(x,t)}{\partial x^2} = \frac{1}{v^2}\frac{\partial^2 u(x,t)}{\partial t^2} \tag{2-6}$$

in which form it appears in optics. Equation (2-6) describes the propagation of an optical disturbance $u(x,t)$ in a direction x at a time t. For a wave propagating in three dimensions it is easy to show that the wave equation is

$$\frac{\partial^2 u(r,t)}{\partial x^2} + \frac{\partial^2 u(r,t)}{\partial y^2} + \frac{\partial^2 u(r,t)}{\partial z^2} = \frac{1}{v^2}\frac{\partial^2 u(r,t)}{\partial t^2} \tag{2-7}$$

where $r = (x^2 + y^2 + z^2)^{1/2}$. Equation (2-7) can be written as

$$\nabla^2 u(r,t) = \frac{1}{v^2}\frac{\partial^2 u(r,t)}{\partial t^2} \tag{2-8}$$

where ∇^2 is the Laplacian operator,

$$\nabla^2 \equiv \frac{\partial^2}{\partial x^2} + \frac{\partial^2}{\partial y^2} + \frac{\partial^2}{\partial z^2} \tag{2-9}$$

Because of the fundamental importance of the wave equation in both mechanics and optics, it has been thoroughly investigated. Equation (2-7) shall now be solved in several ways. Each method of solution yields useful insights.

2.2.1 Plane Wave Solution

Let $\mathbf{r}(x,y,z)$ be a position vector of a point P in space, $\mathbf{s}(s_x, s_y, s_z)$ a unit vector in a fixed direction. Any solution of (2-7) of the form:

$$u = u(\mathbf{s}\cdot\mathbf{r}, t) \tag{2-10}$$

is said to represent a plane-wave solution, since at each instant of time u is constant over each of the planes,

$$\mathbf{s}\cdot\mathbf{r} = \text{constant} \tag{2-11}$$

Equation (2-11) is the vector equation of a plane; a further discussion of plane waves and (2-11) will be given later.

Figure 2-2 shows a Cartesian coordinate sytem Ox, Oy, Oz. We now choose a new set of Cartesian axes, $O\xi$, $O\zeta$, $O\eta$, with $O\zeta$ in the direction $\mathbf{s}\cdot\mathbf{r} = \zeta$. Then $\partial/\partial x = (\partial\zeta/\partial x)\cdot\partial/\partial\zeta$, etc., so

$$s_x x + s_y y + s_z z = \zeta \tag{2-12a}$$

and we can write

$$\frac{\partial}{\partial x} = s_x\frac{\partial}{\partial \zeta} \qquad \frac{\partial}{\partial y} = s_y\frac{\partial}{\partial \zeta} \qquad \frac{\partial}{\partial z} = s_z\frac{\partial}{\partial \zeta} \tag{2-12b}$$

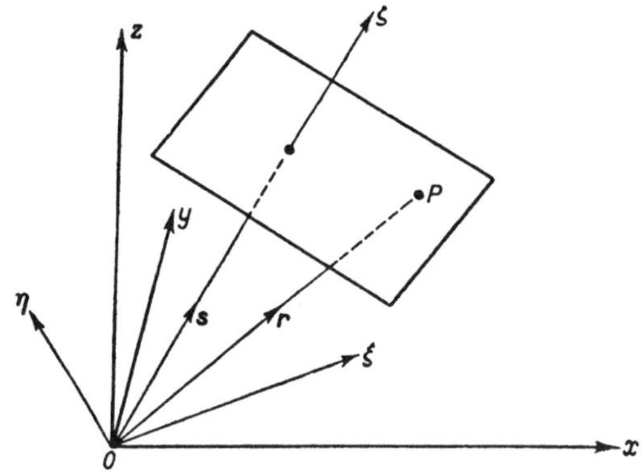

Figure 2-2 Propagation of plane waves.

Since $s_x^2 + s_y^2 + s_z^2 = 1$, we easily find that

$$\nabla^2 u = \frac{\partial^2 u}{\partial \zeta^2} \tag{2-13}$$

so that (2-8) becomes

$$\frac{\partial^2 u}{\partial \zeta^2} - \frac{1}{v^2}\frac{\partial^2 u}{\partial t^2} = 0 \tag{2-14}$$

Thus, the transformation (2-12) reduces the three-dimensional wave equation to a one-dimensional wave equation. Next, we set

$$\zeta - vt = p \qquad \zeta + vt = q \tag{2-15}$$

and substitute (2-15) into (2-14) to find

$$\frac{\partial^2 u}{\partial p \partial q} = 0 \tag{2-16}$$

The solution of (2-16) is

$$u = u_1(p) + u_2(q) \tag{2-17}$$

as a simple differentiation quickly shows. Thus, the general solution of (2-14) is

$$u = u_1(\mathbf{s} \cdot \mathbf{r} - vt) + u_2(\mathbf{s} \cdot \mathbf{r} + vt) \tag{2-18}$$

where u_1 and u_2 are arbitrary functions. The argument of u is unchanged when (ζ, t) is replaced by $(\zeta + v\tau, t + \tau)$, where τ is an arbitrary time. Thus, $u_1(\zeta + v\tau)$ represents a disturbance which is propagated with a velocity v in the negative ζ direction. Similarly, $u_2(\zeta - v\tau)$ represents a disturbance which is propagated with a velocity v in the positive ζ direction.

2.2.2 Spherical Waves

Next, we consider solutions representing spherical waves, i.e.,

$$u = (r, t) \tag{2-19}$$

where $r = |r| = (x^2 + y^2 + z^2)^{1/2}$. Using the relations

$$\frac{\partial}{\partial x} = \frac{\partial r}{\partial x}\frac{\partial}{\partial r} = \frac{x}{r}\frac{\partial x}{\partial r}, \text{ etc.} \tag{2-20}$$

one finds after a straightforward calculation that

$$\nabla^2(u) = \frac{1}{r}\frac{\partial^2(ru)}{\partial r^2} \tag{2-21}$$

The wave equation (2-8) then becomes

$$\frac{\partial^2(ru)}{\partial r^2} - \frac{1}{v^2}\frac{\partial^2(ru)}{\partial t^2} = 0 \tag{2-22}$$

Following (2-14) the solution of (2-22) is

$$u(r, t) = \frac{u_1(r - vt)}{r} + \frac{u_2(r + vt)}{r} \tag{2-23}$$

where u_1 and u_2 are, again, arbitrary functions. The first term in (2-23) represents a spherical wave diverging from the origin, and the second term is a spherical wave converging toward the origin; the velocity of propagation being v in both cases.

2.2.3 Fourier Transform Method

The method for solving the wave equation requires a considerable amount of insight and experience. It would be desirable to have a formal method for solving partial differential equations of this type. This can be done by the use of Fourier transforms.

Let us again consider the one-dimensional wave equation:

$$\frac{\partial^2 u(\zeta, t)}{\partial \zeta^2} = \frac{1}{v^2}\frac{\partial^2 u(\zeta, t)}{\partial t^2} \tag{2-24}$$

The Fourier transform pair for $u(\zeta, t)$ is defined in the time domain, t, to be

$$u(\zeta, t) = \frac{1}{2\pi}\int_{-\infty}^{\infty} \mathbf{u}(\zeta, \omega)e^{i\omega t}\, d\omega \tag{2-25a}$$

and

$$\mathbf{u}(\zeta, \omega) = \int_{-\infty}^{\infty} u(\zeta, t)e^{-i\omega t}\, dt \tag{2-25b}$$

We can then write

$$\frac{\partial^2 u(\zeta, t)}{\partial \zeta^2} = \frac{1}{2\pi}\int_{-\infty}^{\infty} \frac{\partial^2 \mathbf{u}(\zeta, \omega)e^{i\omega t}}{\partial \zeta^2} d\omega$$

$$\frac{\partial^2 u(\zeta, t)}{\partial t^2} = \frac{1}{2\pi}\int_{-\infty}^{\infty} \mathbf{u}(\zeta, \omega)(-\omega^2)e^{i\omega t}\, d\omega \tag{2-26}$$

so (2-24) is transformed to

$$\frac{\partial^2 \mathbf{u}(\zeta, \omega)}{\partial \zeta^2} = \frac{-\omega^2 \mathbf{u}(\zeta, \omega)}{v^2} \tag{2-27}$$

Equation (2-27) is recognized immediately as the equation of a harmonic oscillator whose solution is

$$\mathbf{u}(\zeta, \omega) = A(\omega)e^{ik\zeta} + B(\omega)e^{-ik\zeta} \tag{2-28}$$

where $k = \omega/v$. We note that the "constants" of integration, $A(\omega)$ and $B(\omega)$, must be written as functions of ω because the partial differentiation in (2-24) is with respect to ζ. The reader can easily check that (2-28) is the correct solution by differentiating it according to (2-27). The solution of (2-24) can then be found by substituting $\mathbf{u}(\zeta, \omega)$ in (2-28) into the Fourier transform $u(\zeta, t)$ in (2-25a)

$$u(\zeta, t) = \frac{1}{2\pi}\int_{-\infty}^{\infty} [A(\omega)e^{ik\zeta} + B(\omega)e^{-ik\zeta}]e^{i\omega t}\, d\omega \tag{2-29}$$

or

$$= \frac{1}{2\pi}\int_{-\infty}^{\infty} A(\omega)e^{i\omega(t+\zeta/v)}\, d\omega + \frac{1}{2\pi}\int_{-\infty}^{\infty} B(\omega)e^{i\omega(t-\zeta/v)}\, d\omega \tag{2-30}$$

From the definition of the Fourier transform, Eq. (2-25), we then see that

$$u(\zeta, t) = u_1\left(t + \frac{\zeta}{v}\right) + u_2\left(t - \frac{\zeta}{v}\right) \tag{2-31}$$

which is equivalent to the solution (2-18).

Fourier transforms are used throughout physics and provide a powerful method for solving partial differential equations. Finally, the Fourier transform pair shows that the simplest sinusoidal solution of the wave equation is

$$u(\zeta, t) = A\sin(\omega t + k\zeta) + B\sin(\omega t - k\zeta) \tag{2-32}$$

where A and B are constants. The reader can easily check that (2-32) is the solution of the wave equation (2-24).

2.2.4 Mathematical Representation of the Harmonic Oscillator Equation

Before we end the discussion of the wave equation, it is also useful to discuss, further, the harmonic oscillator equation. From mechanics the differential equation of the harmonic oscillator motion is

$$m\frac{d^2x}{dt^2} = -kx \tag{2-33a}$$

or

$$\frac{d^2x}{dt^2} = -\frac{k}{m}x = -\omega_0^2 x \tag{2-33b}$$

where m is the mass of the oscillator, k is the force constant of the spring, and $\omega_0 = 2\pi f$ is the angular frequency where f is the frequency in cycles per second.

Equation (2-33b) can be solved by multiplying both sides of the equation by $dx/dt = v$ (v = velocity):

$$v\frac{dv}{dt} = -\omega_0^2 x\frac{dx}{dt} \tag{2-34a}$$

or

$$vdv = -\omega_0^2 xdx \tag{2-34b}$$

Integrating both sides of (2-34b) yields

$$\frac{v^2}{2} = -\frac{\omega_0^2}{2}x^2 + A^2 \tag{2-35a}$$

where A^2 is the constant of integration. Solving for v, we have

$$v = \frac{dx}{dt} = (A^2 - \omega_0^2 x^2)^{1/2} \tag{2-35b}$$

which can be written as

$$\frac{dx}{(A^2 - \omega_0^2 x^2)^{1/2}} = dt \tag{2-36}$$

The solution of (2-36) is well known from integral calculus and is

$$x = a\sin(\omega_0 t + \delta) \tag{2-37}$$

where a and δ are constants of integration. Equation (2-37) can be rewritten in another form by using the trigonometric expansion:

$$\sin(\omega_0 t + \delta) = \sin(\omega_0 t)\cos\delta + \cos(\omega_0 t)\sin\delta \tag{2-38}$$

so

$$x(t) = A\sin\omega_0 t + B\cos\omega_0 t \tag{2-39}$$

where

$$A = a\cos\delta \qquad B = a\sin\delta \tag{2-40}$$

Another form for (2-39) is to express $\cos\omega_0 t$ and $\sin\omega_0 t$ in terms of exponents; that is,

$$\cos\omega_0 t = \frac{e^{i\omega_0 t} + e^{-i\omega_0 t}}{2} \tag{2-41a}$$

$$\sin\omega_0 t = \frac{e^{i\omega_0 t} - e^{-i\omega_0 t}}{2i} \tag{2-41b}$$

Substituting (2-41a) and (2-41b) into (2-39) and grouping terms leads to

$$x(t) = Ce^{i\omega_0 t} + De^{-i\omega_0 t} \tag{2-42a}$$

where

$$C = \frac{A - iB}{2} \qquad D = \frac{A + iB}{2} \tag{2-42b}$$

and C and D are complex constants. Thus, we see that the solution of the harmonic oscillator can be written in terms of purely real quantities or complex quantities.

The form of (2-35a) is of particular interest. The differential equation (2-33a) clearly describes the amplitude motion of the harmonic oscillator. Let us retain the original form of (2-33a) and multiply through by $dx/dt = v$, so we can write

$$mv\frac{dv}{dt} = -kx\frac{dx}{dt} \tag{2-43}$$

We now integrate both sides of (2-43), and we are led to

$$\frac{mv^2}{2} = \frac{-kx^2}{2} + C \tag{2-44}$$

where C is a constant of integration. Thus, by merely carrying out a formal integration we are led to a new form for describing the motion of the harmonic oscillator. At the beginning of the eighteenth century the meaning of (2-44) was not clear. Only slowly did physicists come to realize that (2-44) describes the motion of the harmonic oscillator in a completely new way, namely the description of motion in terms of energy. The terms $mv^2/2$ and $-kx^2/2$ correspond to the kinetic energy and the potential energy for the harmonic oscillator, respectively. Thus, early on in the development of physics a connection was made between the amplitude and energy for oscillatory motion. The energy of the wave could be obtained by merely squaring the amplitude. This point is introduced because of its bearing on Young's interference experiment, specifically, and on optics, generally. The fact that a relation exists between the amplitude of the harmonic oscillator and its energy was taken directly over from mechanics into optics and was critical for Young's interference experiment. In optics, however, the energy would become known as the intensity.

2.2.5 A Note on the Equation of a Plane

The equation of a plane was stated in (2-11) to be

$$\mathbf{s} \cdot \mathbf{r} = \text{constant} \tag{2-11}$$

We can show that (2-11) does indeed describe a plane by referring to Fig. 2-2. Inspecting the figure, we see that $\mathbf{r}$ is a vector with its origin at the origin of the coordinates, so,

$$\mathbf{r} = x\mathbf{i} + y\mathbf{j} + z\mathbf{k} \tag{2-45}$$

and $\mathbf{i}$, $\mathbf{j}$, and $\mathbf{k}$ are unit vectors. Similarly, from Fig. 2-2 we see that

$$\mathbf{s} = s_x\mathbf{i} + s_y\mathbf{j} + s_z\mathbf{k} \tag{2-46}$$

Suppose we now have a vector $\mathbf{r}_0$ along $\mathbf{s}$ and the plane is perpendicular to $\mathbf{s}$. Then $\overline{OP}$ is the vector $\mathbf{r} - \mathbf{r}_0$ and is perpendicular to $\mathbf{s}$. Hence, the equation of the plane is

$$\mathbf{s} \cdot (\mathbf{r} - \mathbf{r}_0) = 0 \tag{2-47}$$

or

$$\mathbf{s} \cdot \mathbf{r} = \zeta \tag{2-48}$$

where $\zeta = \mathbf{s} \cdot \mathbf{r}_0$ is a constant. Thus, the name *plane-wave solutions* arises from the fact that the wave front is characterized by a plane of infinite extent.

2.3 YOUNG'S INTERFERENCE EXPERIMENT

In the previous section we saw that the developments in mechanics in the eighteenth century led to the mathematical formulation of the wave equation and the concept of energy.

Around the year 1800, Thomas Young performed a simple, but remarkable, optical experiment known as the two-pinhole interference experiment. He showed that this experiment could be understood in terms of waves; the experiment gave the first clear-cut support for the wave theory of light. In order to understand the pattern that he observed, he adopted the ideas developed in mechanics and applied them to optics, an extremely novel and radical approach. Until the advent of Young's work, very little progress had been made in optics since the researches of Newton (the corpuscular theory of light) and Huygens (the wave theory of light). The simple fact was that by the year 1800, aside from Snell's law of refraction and the few things learned about polarization, there was no theoretical basis on which to proceed. Young's work provided the first critical step in the development and acceptance of the wave theory of light.

The experiment carried out by Young is shown in Fig. 2-3. A source of light, σ, is placed behind two pinholes s_1 and s_2, which are equidistant from σ. The pinholes then act as secondary monochromatic sources that are in phase, and the beams from them are superposed on the screen Σ at an arbitrary point P. Remarkably, when the screen is then observed, one does not see a uniform distribution of light. Instead, a distinct pattern consisting of bright bands alternating with dark bands is observed. In order to explain this behavior, Young assumed that each of the pinholes, s_1 and s_2, emitted waves of the form:

$$u_1 = u_{01} \sin(\omega t - kl_1) \tag{2-49a}$$

$$u_2 = u_{02} \sin(\omega t - kl_2) \tag{2-49b}$$

where pinholes s_1 and s_2 are in the source plane A, and are distances l_1 and l_2 from a point $P(x, y)$ in the plane of observation Σ. The pattern is observed on the plane Oxy normal to the perpendicular bisector of $\overline{s_1 s_2}$ and with the x axis parallel to $\overline{s_1 s_2}$. The separation of the pinholes is d, and a is the distance between the line joining the

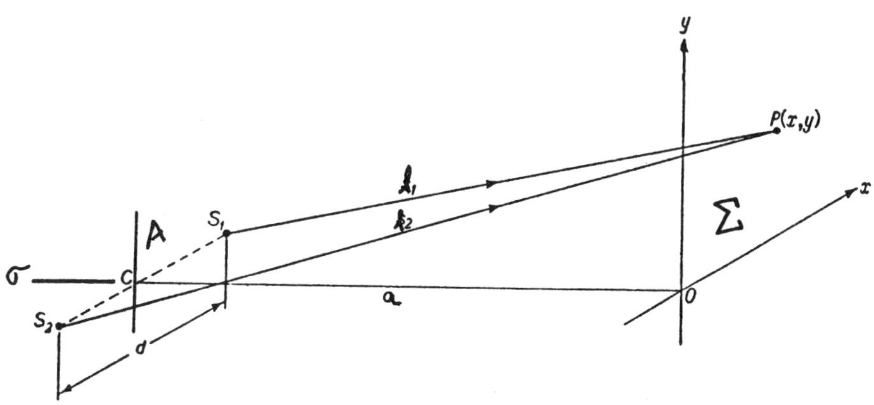

Figure 2-3 Young's interference experiment.

pinholes and the plane of observation Σ. For the point $P(x, y)$ on the screen, Fig. 2-3 shows that

$$l_1^2 = a^2 + y^2 + \left(x - \frac{d}{2}\right)^2 \tag{2-50a}$$

$$l_2^2 = a^2 + y^2 + \left(x + \frac{d}{2}\right)^2 \tag{2-50b}$$

Thus,

$$l_2^2 - l_1^2 = 2xd \tag{2-51}$$

Equation (2-51) can be written as

$$(l_2 - l_1)(l_1 + l_2) = 2xd \tag{2-52}$$

Now if x and y are small compared to a, then $l_1 + l_2 \simeq 2a$. Thus,

$$l_2 - l_1 = \Delta l = \frac{xd}{a} \tag{2-53}$$

At this point we now return to the wave theory. The secondary sources s_1 and s_2 are assumed to be equal, so $u_{01} = u_{02} = u_0$. In addition, the assumption is made that the optical disturbances u_1 and u_2 can be superposed at $P(x, y)$ (the principle of coherent superposition), so

$$\begin{aligned} u(t) &= u_1 + u_2 \\ &= u_0[\sin(\omega t - kl_1) + \sin(\omega t - kl_2)] \end{aligned} \tag{2-54}$$

A serious problem now arises. While (2-54) certainly describes an interference behavior, the parameter of time enters in the term ωt. In the experiment the observed pattern does not vary over time, so the time factor cannot enter the final result. This suggests that we average the amplitude $u(t)$ over the time of observation T. The time average of $u(t)$ written as $u(t)$, is then defined to be

$$\langle u(t) \rangle = \lim_{T\to\infty} \frac{\int_0^T u(t)\, dt}{\int_0^T dt} \tag{2-55a}$$

$$= \lim_{T\to\infty} \frac{1}{T}\int_0^T u(t)\, dt \tag{2-55b}$$

Substituting (2-54) into (2-55) yields

$$\langle u(t) \rangle = \lim_{T\to\infty} \frac{u_0}{T}\int_0^T [\sin(\omega t - kl_1) + \sin(\omega t - kl_2)]\, dt \tag{2-56}$$

Using the trigonometric identity:

$$\sin(\omega t - kl) = \sin(\omega t)\cos(kl) - \cos(\omega t)\sin(kl) \tag{2-57}$$

and averaging over one cycle in (2-56) yields

$$\langle u(t) \rangle = 0 \tag{2-58}$$

This is not observed. That is, the time average of the amplitude is calculated to be zero, but observation shows that the pattern exhibits nonzero intensities. At this point we must abandon the idea that the interference phenomenon can be explained only in terms of amplitudes $u(t)$. Another idea must now be borrowed from mechanics. Namely, the optical disturbance must be described in terms of squared quantities, analogous to energy, $u^2(t)$. But this, too, contains a time factor. Again, a time average is introduced, and a new quantity, I, in optics called the intensity, is defined:

$$I = \langle u^2(t) \rangle = \lim_{T \to \infty} \frac{1}{T} \int_0^T u^2(t)\, dt \tag{2-59}$$

Substituting $u^2(t) = (u_0 \sin(\omega t - kl))^2$ into (2-59) and averaging over one cycle yields

$$\begin{aligned} I = \langle u^2(t) \rangle &= \lim_{T \to \infty} \frac{1}{T} \int_0^T u_0^2 \sin^2(\omega t - kl)\, dt \\ &= \frac{u_0^2}{2} = I_0 \end{aligned} \tag{2-60}$$

Thus, the intensity is constant over time; this behavior is observed.

The time average of $u^2(t)$ is now applied to the superposed amplitudes (2-54). Squaring $u^2(t)$ yields

$$u^2(t) = u_0^2[\sin^2(\omega t - kl_1) + \sin^2(\omega t - kl_2) + 2\sin(\omega t - kl_1)\sin(\omega t - kl_2)] \tag{2-61}$$

The last term is called the interference. Equation (2-61) can be rewritten with the help of the well-known trigonometric identity:

$$2\sin(\omega t - kl_1)\sin(\omega t - kl_2) = \cos(k[l_2 - l_1]) - \cos(2\omega t - k[l_1 + l_2]) \tag{2-62}$$

Thus, (2-61) can be written as

$$\begin{aligned} u^2(t) = u_0^2[&\sin^2(\omega t - kl_1) + \sin^2(\omega t - kl_2) \\ &+ \cos(k[l_2 - l_1]) - \cos(2\omega t - k[l_1 + l_2])] \end{aligned} \tag{2-63}$$

Substituting (2-63) into (2-59), we obtain the intensity on the screen to be

$$I = \langle u^2(t) \rangle = 2I_0[1 + \cos k(l_2 - l_1)] = 4I_0 \cos^2\left[\frac{k(l_2 - l_1)}{2}\right] \tag{2-64a}$$

or

$$I = 4I_0 \cos^2 \frac{kxd}{2a} \tag{2-64b}$$

where, from (2-53)

$$l_2 - l_1 = \Delta l = \frac{xd}{a} \tag{2-53}$$

Equation (2-64b) is Young's famous interference formula. We note that from (2-60) we would expect the intensity from a single source to be $u_0^2/2 = I_0$, so the intensity from two independent optical sources would be $2I$. Equation (2-64a) [or (2-64b)] shows a remarkable result, namely, when the intensity is observed from a single source in which the beam is divided, the observed intensity varies between 0 and $4I_0$; the intensity can be double or even zero from that expected from two independent optical sources! We see from (2-64b) that there will be maximum intensities ($4I_0$) at

$$x = \frac{a\lambda n}{d} \qquad n = 0, \pm 1, \pm 2, \ldots \tag{2-65a}$$

and minimum intensities (null) at

$$x = \frac{a\lambda}{d}\left(\frac{2n+1}{2}\right) \qquad n = 0, \pm 1, \pm 2, \ldots \tag{2-65b}$$

Thus, in the vicinity of O on the plane Σ an interference pattern consisting of bright and dark bands is aligned parallel to the OY axis (at right angles to the line $\overline{s_1 s_2}$ joining the two sources).

Young's experiment is of great importance because it was the first step in establishing the wave theory of light and was the first theory to provide an explanation of the observed interference pattern. It also provides a method, albeit one of low precision, of measuring the wavelength of light by measuring d, a, and the fringe spacing according to (2-65a) or (2-65b). The separation Δx between the central bright line and the first bright line is, from (2-65a),

$$\Delta x = x_1 - x_0 = \frac{a\lambda}{d} \tag{2-66}$$

The expected separation on the observing screen can be found by assuming the following values:

$$\begin{aligned} &a = 100\,\text{cm} \qquad d = 0.1\,\text{cm} \\ &\lambda = 5 \times 10^{-5}\text{cm} \qquad \Delta x = 0.05\,\text{cm} = 0.5\,\text{mm} \end{aligned} \tag{2-67}$$

The resolution of the human eye at a distance of 25 cm is, approximately, of the same order of magnitude, so the fringes can be observed with the naked eye.

Young's interference gave the first real support for the wave theory. However, aside from the important optical concepts introduced here to explain the interference pattern, there is another reason for discussing Young's interference experiment. Around 1818, Fresnel and Arago repeated his experiments with polarized light to determine the effects, if any, on the interference phenomenon. The results were surprising to understand in their entirety. To explain these experiments it was necessary to understand the nature and properties of polarized light. Before we turn to the subject of polarized light, however, we discuss another topic of importance, namely, the reflection and transmission of a wave at an interface separating two different media.

2.4 REFLECTION AND TRANSMISSION OF A WAVE AT AN INTERFACE

The wave theory and the wave equation allow us to treat an important problem, namely, the reflection and transmission of wave at an interface between two different media. Specifically, in optics, light is found to be partially reflected and partially transmitted at the boundary of two media characterized by different refractive indices. The treatment of this problem was first carried out in mechanics, however, and shows how the science of mechanics paved the way for the introduction of the wave equation into optics.

Two media can be characterized by their ability to support two different velocities v_1 and v_2. In Fig. 2-4 we show an incident wave coming from the left which is partially transmitted and reflected at the interface (boundary).

We saw earlier that the solution of the wave equation in complex form is

$$u(x) = Ae^{-ikx} + Be^{+ikx} \tag{2-68}$$

where $k = \omega/v$. The time factor $\exp(i\omega t)$ has been suppressed. The term Ae^{-ikx} describes propagation to the right, and the term Be^{+ikx} describes propagation to the left. The fields to the left and right of the interface (boundary) can be described by a superposition of waves propagating to the right and left, that is,

$$u_1(x) = Ae^{-ik_1x} + Be^{+ik_1x} \qquad x < 0 \tag{2-69a}$$

$$u_2(x) = Ce^{-ik_2x} + De^{+ik_2x} \qquad x > 0 \tag{2-69b}$$

where $k_1 = \omega/v_1$ and $k_2 = \omega/v_2$.

We must now evaluate A, B, C, and D. To do this, we assume that at the interface the fields are continuous—that is,

$$u_1(x)\,|_{x=0} = u_2(x)\,|_{x=0} \tag{2-70}$$

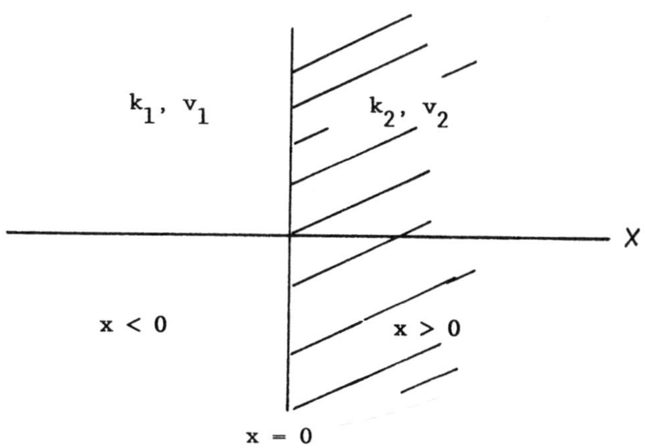

Figure 2-4 Reflection and transmission of a wave at the interface between two media.

and that the slopes of $u_1(x)$ and $u_2(x)$ are continuous at the interface—that is, the derivatives of $u_1(x)$ and $u_2(x)$, so

$$\left.\frac{\partial u_1(x)}{\partial x}\right|_{x=0} = \left.\frac{\partial u_2(x)}{\partial x}\right|_{x=0} \tag{2-71}$$

We also assume that there is no source of waves in the medium to the right of the interface, i.e, $D = 0$. This means that the wave which propagates to the left on the left side of the interface is due only to reflection of the incident wave.

With $D = 0$, and applying the boundary conditions in (2-70) and (2-71) to (2-69a) and (2-69b) we easily find

$$A + B = C \tag{2-72a}$$

$$k_1 A - k_1 B = k_2 C \tag{2-72b}$$

We solve for B and C in terms of the amplitude of the incident wave, A, and find

$$B = \left(\frac{k_1 - k_2}{k_1 + k_2}\right) A \tag{2-73a}$$

$$C = \left(\frac{2k_1}{k_1 + k_2}\right) A \tag{2-73b}$$

The B term is associated with the reflected wave in (2-69a). If $k_1 = k_2$, i.e., the two media are the same, then (2-73a) and (2-73b) show that $B = 0$ and $C = A$; that is, there is no reflected wave, and we have complete transmission as expected.

We can write (2-69a) as the sum of an incident wave $u_i(x)$ and a reflected wave $u_r(x)$:

$$u_1(x) = u_i(x) + u_r(x) \tag{2-74a}$$

and we can write (2-69b) as a transmitted wave:

$$u_2(x) = u_t(x) \tag{2-74b}$$

The energies corresponding to $u_i(x)$, $u_r(x)$, and $u_t(x)$, are then the squares of these quantities. We can use complex quantities to bypass the formal time-averaging procedure and define the energies of these waves to be

$$\varepsilon_i = u_i(x) u_i^*(x) \tag{2-75a}$$

$$\varepsilon_r = u_r(x) u_r^*(x) \tag{2-75b}$$

$$\varepsilon_t = u_t(x) u_t^*(x) \tag{2-75c}$$

The principle of conservation of energy requires that

$$\varepsilon_i = \varepsilon_r + \varepsilon_t \tag{2-76}$$

The fields $u_i(x)$, $u_r(x)$, and $u_t(x)$ from (2-69a) and (2-69b) are

$$u_i(x) = Ae^{-ik_1 x} \tag{2-77a}$$

$$u_r(x) = Be^{+ik_1 x} \tag{2-77b}$$

$$u_t(x) = Ce^{-ik_2 x} \tag{2-77c}$$

The energies corresponding to (2-77) are then substituted in (2-76), and we find

$$A^2 = B^2 + C^2 \tag{2-78a}$$

or

$$\left(\frac{B}{A}\right)^2 + \left(\frac{C}{A}\right)^2 = 1 \tag{2-78b}$$

The quantities $(B/A)^2$ and $(C/A)^2$ are the normalized reflection and transmission coefficients, which we write as R and T, respectively. Thus, (2-78b) becomes

$$R + T = 1 \tag{2-79a}$$

where

$$R = \left(\frac{k_1 - k_2}{k_1 + k_2}\right)^2 \tag{2-79b}$$

$$T = \left(\frac{2k_1}{k_1 + k_2}\right)^2 \tag{2-79c}$$

from (2-73a) and (2-73b). Equation (2-79b) and (2-79c) can be seen to satisfy the conservation condition (2-79a).

The coefficients B and C show an interesting behavior, which is as follows. From (2-73a) and (2-73b) we write

$$\frac{B}{A} = \frac{1 - k_2/k_1}{1 + k_2/k_1} \tag{2-80a}$$

$$\frac{C}{A} = \frac{2}{1 + k_2/k_1} \tag{2-80b}$$

where

$$\frac{k_2}{k_1} = \frac{\omega/v_2}{\omega/v_1} = \frac{v_1}{v_2} \tag{2-80c}$$

Now if $v_2 = 0$, that is, there is no propagation in the second medium, (2-80c) becomes

$$\lim_{v_2 \to 0} \frac{k_2}{k_1} = \frac{v_1}{v_2} = \infty \tag{2-81}$$

With this limiting value, (2-81), we see that (2-80a) and (2-80b) become

$$\frac{B}{A} = -1 = e^{i\pi} \tag{2-82a}$$

$$\frac{C}{A} = 0 \tag{2-82b}$$

Equation (2-82a) shows that there is a 180° (π rad) phase reversal upon total reflection. Thus, the reflected wave is completely out of phase with the incident wave, and we have total cancellation. This behavior is described by the term *standing waves*. We now derive the equation which specifically shows that the resultant wave does not propagate.

The field to the left of the interface is given by (2-69a) and is

$$u_1(x,t) = e^{i\omega t}(Ae^{-ik_1x} + Be^{ik_1x}) \qquad x < 0 \tag{2-83}$$

where we have reintroduced the (suppressed) time factor $\exp(i\omega t)$. From (2-82a) we can then write

$$u_1(x,t) = Ae^{i\omega t}(e^{-ik_1x} - e^{ik_1x}) \tag{2-84a}$$

$$= Ae^{i(\omega t - k_1x)} - Ae^{i(\omega t + k_1x)} \tag{2-84b}$$

$$= u_-(x,t) - u_+(x,t) \tag{2-84c}$$

where

$$u_-(x,t) = Ae^{i(\omega t - k_1x)} \tag{2-84d}$$

$$u_+(x,t) = Ae^{i(\omega t + k_1x)} \tag{2-84e}$$

The phase velocity v_p of a wave can be defined in terms of amplitude as

$$v_p = -\frac{(\partial u/\partial t)}{(\partial u/\partial x)} \tag{2-85}$$

Applying (2-85) to (2-84d) and (2-84e), respectively, we find that

$$v_p(-) = \frac{\omega}{k_1} \tag{2-86a}$$

$$v_p(+) = \frac{-\omega}{k_1} \tag{2-86b}$$

so the total velocity of the wave is

$$v = v_p(-) + v_p(+) = 0 \tag{2-87}$$

Thus, the resultant velocity of the wave is zero according to (2-87); that is, the wave does not propagate and it appears to be standing in place. The equation for the standing wave is given by (2-84a), which can be written as

$$u_1(x,t) = 2Ae^{i\omega t}\sin(k_1x) \tag{2-88}$$

It is customary to take the real part of (2-88)

$$u(x,t) = 2A\cos(\omega t)\sin(kx) \tag{2-89}$$

where we have dropped the subscript 1. We see that there is no propagator $\omega t - kx$, so (2-89) does not describe propagation.

Thus, we see that the wave equation and wave theory lead to a correct description of the transmission and reflection of a wave at a boundary. While this behavior was first studied in mechanics in the eighteenth century, it was applied with equal success to optics in the following century. It appears that this was first done by Fresnel, who derived the equations for reflection and transmission at an interface between two media characterized by refractive indices n_1 and n_2. Fresnel's equations are derived in Chapter 8.

With this material on the wave equation behind us, we can now turn to the study of one of the most interesting properties of light, its polarization.

REFERENCES

Books

1. Born, M. and Wolf, E., *Principles of Optics*, 3rd ed., Pergamon Press, New York, 1965.
2. Whittaker, E., *A History of the Theories of Aether and Electricity*, Vol. I, Philosophical Society, New York, 1951.
3. Sommerfeld, A., *Lectures on Theoretical Physics*, Vols. I–V, Academic Press, New York, 1952.
4. Towne, D. H., *Wave Phenomena*, Addison-Wesley, Reading, MA, 1964.
5. Becker, R. A., *Introduction to Theoretical Mechanics*, McGraw-Hill, New York, 1954.

3

The Polarization Ellipse

3.1 INTRODUCTION

Christian Huygens was the first to suggest that light was not a scalar quantity based on his work on the propagation of light through crystals; it appeared that light had "sides" in the words of Newton. This vectorial nature of light is called *polarization*. If we follow mechanics and equate an optical medium to an isotropic elastic medium, it should be capable of supporting three independent oscillations (optical disturbances): $u_x(r, t)$, $u_y(r, t)$, and $u_z(r, t)$. Correspondingly, three independent wave equations are then required to describe the propagation of the optical disturbance, namely,

$$\nabla^2 u_i(r, t) = \frac{1}{v^2}\frac{\partial^2 u_i(r, t)}{\partial t^2} \qquad i = x, y, z \tag{3-1}$$

where v is the velocity of propagation of the oscillation and $\mathbf{r} = \mathbf{r}(x, y, z)$. In a Cartesian system the components $u_x(\mathbf{r}, t)$ and $u_y(\mathbf{r}, t)$ are said to be the transverse components, and the component $u_z(\mathbf{r}, t)$ is said to be the longitudinal component when the propagation is in the z direction. Thus, according to (3-1) the optical field components should be

$$u_x(\mathbf{r}, t) = u_{0x} \cos(\omega t - \mathbf{k} \cdot \mathbf{r} + \delta_x) \tag{3-2a}$$

$$u_y(\mathbf{r}, t) = u_{0y} \cos(\omega t - \mathbf{k} \cdot \mathbf{r} + \delta_y) \tag{3-2b}$$

$$u_z(\mathbf{r}, t) = u_{0z} \cos(\omega t - \mathbf{k} \cdot \mathbf{r} + \delta_z) \tag{3-2c}$$

In 1818 Fresnel and Arago carried out a series of fundamental investigations on Young's interference experiment using polarized light. After a considerable amount of experimentation they were forced to conclude that the longitudinal component (3-2c) did not exist. That is, light consisted only of the transverse components (3-2a) and (3-2b). If we take the direction of propagation to be in the z direction, then the optical field in free space must be described only by

$$u_x(z, t) = u_{0x} \cos(\omega t - \mathrm{k}z + \delta_x) \tag{3-3a}$$

$$u_y(z, t) = u_{0y} \cos(\omega t - \mathrm{k}z + \delta_y) \tag{3-3b}$$

where u_{0x} and u_{0y} are the maximum amplitudes and δ_x and δ_y are arbitrary phases. There is no reason, a priori, for the existence of only transverse components on the basis of an elastic medium (the "ether" in optics). It was considered to be a defect in Fresnel's theory. Nevertheless, in spite of this (3-3a) and (3-3b) were found to describe satisfactorily the phenomenon of interference using polarized light.

The "defect" in Fresnel's theory was overcome by the development of a new theory, which we now call Maxwell's electrodynamic theory and his equations. One of the immediate results of solving his equations was that in free space only transverse components arose; there was no longitudinal component. This was one of the first triumphs of Maxwell's theory. Nevertheless, Maxwell's theory took nearly 40 years to be accepted in optics due, in large part, to the fact that up to the end of the nineteenth century it led to practically nothing that could not be explained or understood by Fresnel's theory.

Equations (3-3a) and (3-3b) are spoken of as the polarized or polarization components of the optical field. In this chapter we consider the consequences of these equations. The results are very interesting and lead to a surprising number of revelations about the nature of light.

3.2 THE INSTANTANEOUS OPTICAL FIELD AND THE POLARIZATION ELLIPSE

In previous sections we pointed out that the experiments of Fresnel and Arago led to the discovery that light consisted only of two transverse components. The components were perpendicular to each other and could be chosen for convenience to be propagating in the z direction. The waves are said to be "instantaneous" in the sense that the time duration for the wave to go through one complete cycle is only 10^{-15} sec at optical frequencies. In this chapter we find the equation that arises when the propagator is eliminated between the transverse components. In order to do this we show in Fig. 3-1 the transverse optical field propagating in the z direction.

The transverse components are represented by

$$E_x(z,t) = E_{0x}\cos(\tau + \delta_x) \tag{3-4a}$$

$$E_y(z,t) = E_{0y}\cos(\tau + \delta_y) \tag{3-4b}$$

where $\tau = \omega t - \kappa z$ is the propagator. The subscripts x and y refer to the components in the x and y directions, E_{0x} and E_{0y} are the maximum amplitudes, and δ_x and δ_y are the phases, respectively. As the field propagates, $E_x(z,t)$ and $E_y(z,t)$ give rise to a resultant vector. This vector describes a locus of points in space, and the curve generated by those points will now be derived. In order to do this (3-4a) and (3-4b) are written as

$$\frac{E_x}{E_{0x}} = \cos\tau\cos\delta_x - \sin\tau\sin\delta_x \tag{3-5a}$$

$$\frac{E_y}{E_{0y}} = \cos\tau\cos\delta_y - \sin\tau\sin\delta_y \tag{3-5b}$$

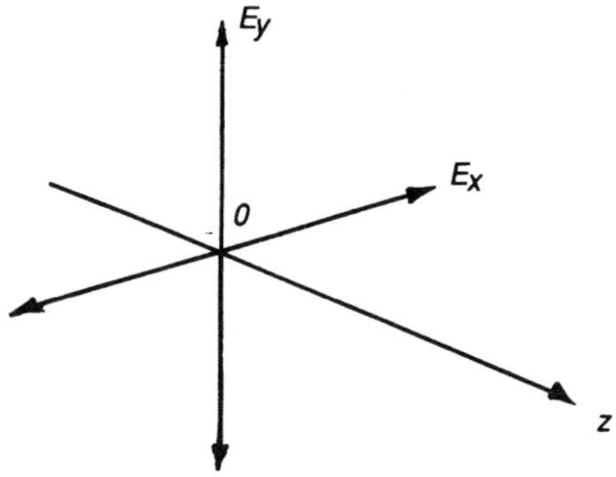

Figure 3-1 Propagation of the transverse optical field.

Hence,

$$\frac{E_x}{E_{0x}}\sin\delta_y - \frac{E_y}{E_{0y}}\sin\delta_x = \cos\tau\sin(\delta_y - \delta_x) \tag{3-6a}$$

$$\frac{E_x}{E_{0x}}\cos\delta_y - \frac{E_y}{E_{0y}}\cos\delta_x = \sin\tau\sin(\delta_y - \delta_x) \tag{3-6b}$$

Squaring (3-6a) and (3-6b) and adding gives

$$\frac{E_x^2}{E_{0x}^2} + \frac{E_y^2}{E_{0y}^2} - 2\frac{E_x}{E_{0x}}\frac{E_y}{E_{0y}}\cos\delta = \sin^2\delta \tag{3-7a}$$

where

$$\delta = \delta_y - \delta_x \tag{3-7b}$$

Equation (3-7a) is recognized as the equation of an ellipse and shows that at any instant of time the locus of points described by the optical field as it propagates is an ellipse. This behavior is spoken of as *optical polarization*, and (3-7a) is called the *polarization ellipse*. In Fig. 3-2 the ellipse is shown inscribed within a rectangle whose sides are parallel to the coordinate axes and whose lengths are $2E_{0x}$ and $2E_{0y}$.

We now determine the points where the ellipse is tangent to the sides of the rectangle. We write (3-7a) as

$$E_{0x}^2E_y^2 - (2E_{0x}E_{0y}E_x\cos\delta)E_y + E_{0y}^2(E_x^2 - E_{0x}^2\sin^2\delta) = 0 \tag{3-8}$$

The solution of this quadratic equation (3-8) is

$$E_y = \frac{E_{0y}E_x\cos\delta}{E_{0x}} \pm \frac{E_{0y}\sin\delta}{E_{0x}}(E_{0x}^2 - E_x^2)^{1/2} \tag{3-9}$$

At the top and bottom of the ellipse where it is tangent to the rectangle the slope is 0. We now differentiate (3-9), set $E_y' = dE_y/dE_x = 0$, and find that

$$E_x = \pm E_{0x}\cos\delta \tag{3-10a}$$

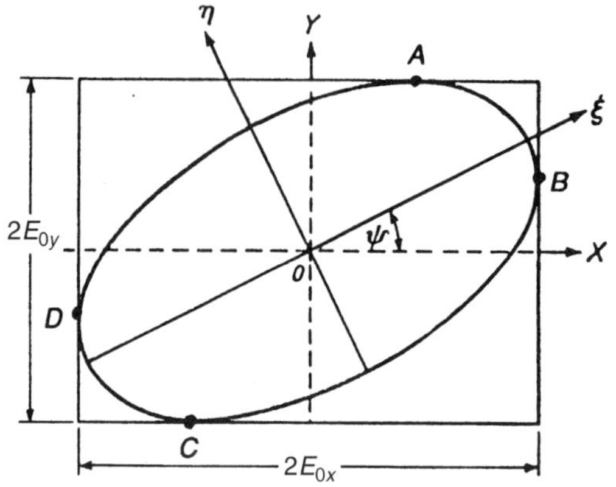

Figure 3-2 An elliptically polarized wave and the polarization ellipse.

Substituting (3-10a) into (3-9), the corresponding values of E_y are found to be

$$E_y = \pm E_{0y} \tag{3-10b}$$

Similarly, by considering (3-9) where the slope is $E'_y = \infty$ on the sides of the rectangle, the tangent points are

$$E_x = \pm E_{0x} \tag{3-11a}$$

$$E_y = \pm E_{0y} \cos\delta \tag{3-11b}$$

Equations (3-10) and (3-11) show that the maximum length of the sides of the ellipse are $E_x = \pm E_{0x}$ and $E_y = \pm E_{0y}$. The ellipse is tangent to the sides of the rectangle at $(\pm E_{0x}, \pm E_{0y}\cos\delta)$ and $(\pm E_{0x}\cos\delta, \pm E_{0y})$. We also see that (3-10) and (3-11) show that the extrema of E_x and E_y are $\pm E_{0x}$ and $\pm E_{0y}$, respectively.

In Fig. 3-2 the ellipse is shown touching the rectangle at point A, B, C, and D, the coordinates of which are

$$A: \quad +E_{0x}\cos\delta, \ +E_{0y} \tag{3-12a}$$

$$B: \quad +E_{0x}, \ +E_{0y}\cos\delta \tag{3-12b}$$

$$C: \quad -E_{0x}\cos\delta, \ -E_{0y} \tag{3-12c}$$

$$D: \quad -E_{0x}, \ -E_{0y}\cos\delta \tag{3-12d}$$

The presence of the "cross term" in (3-7a) shows that the polarization ellipse is, in general, rotated, and this behavior is shown in Fig. 3-2 where the ellipse is shown rotated through an angle ψ. More will be said about this later.

It is also of interest to determine the maximum and minimum areas of the polarization ellipse which can be inscribed within the rectangle. We see that along the x axis the ellipse is tangent at the extrema $x = -E_{0x}$ and $x = +E_{0x}$. The area of the ellipse above the x axis is given by

$$A = \int_{-E_{0x}}^{+E_{0x}} E_y \, dx \tag{3-13}$$

Substituting (3-9) into (3-13) and evaluating the integrals, we find that the area of the polarization ellipse is

$$A = \pi E_{0x} E_{0y} \sin \delta \tag{3-14}$$

Thus, the area of the polarization ellipse depends on the lengths of the major and minor axes, E_{0x} and E_{0y}, and the phase shift δ between the orthogonal transverse components. We see that for $\delta = \pi/2$ the area is $\pi E_{0x} E_{0y}$, whereas for $\delta = 0$ the area is zero. The significance of these results will soon become apparent.

In general, completely polarized light is elliptically polarized. However, there are certain degenerate forms of the polarization ellipse which are continually encountered in the study of polarized light. Because of the importance of these special degenerate forms we now discuss them as special cases in the following section. These are the cases where either E_{0x} or E_{0y} is zero or E_{0x} and E_{0y} are equal and/or where $\delta = 0$, $\pi/2$, or π radians.

3.3 SPECIALIZED (DEGENERATE) FORMS OF THE POLARIZATION ELLIPSE

The polarization ellipse (3-7a) degenerates to special forms for certain values of E_{0x}, E_{0y}, and δ. We now consider these special forms.

1. $E_{0y} = 0$. In this case $E_y(z, t)$ is zero and (3-4) becomes

$$E_x(z, t) = E_{0x} \cos(\tau + \delta_x) \tag{3-15a}$$

$$E_y(z, t) = 0 \tag{3-15b}$$

In this case there is an oscillation only in the x direction. The light is then said to be linearly polarized in the x direction, and we call this *linear horizontally polarized light*. Similarly, if $E_{0x} = 0$ and $E_y(z, t) \neq 0$, then we have a linear oscillation along the y axis, and we speak of *linear vertically polarized light*.

2. $\delta = 0$ or π. Equation (3-7a) reduces to

$$\frac{E_x^2}{E_{0x}^2} + \frac{E_y^2}{E_{0y}^2} \pm 2 \frac{E_x}{E_{0x}} \frac{E_y}{E_{0y}} = 0 \tag{3-16}$$

Equation (3-16) can be written as

$$\left(\frac{E_x}{E_{0x}} \pm \frac{E_y}{E_{0y}} \right)^2 = 0 \tag{3-17}$$

whence

$$E_y = \pm \left(\frac{E_{0y}}{E_{0x}} \right) E_x \tag{3-18}$$

Equation (3-18) is recognized as the equation of a straight line with slope $\pm(E_{0y}/E_{0x})$ and zero intercept. Thus, we say that we have linearly polarized light with slope $\pm(E_{0y}/E_{0x})$. The value $\delta = 0$ yields a negative slope, and the value $\delta = \pi$ a positive slope. If $E_{0x} = E_{0y}$, then we see that

$$E_y = \pm E_x \tag{3-19}$$

The positive value is said to represent *linear* +45° *polarized light*, and the negative value is said to represent *linear* −45° *polarized light*.

3. $\delta = \pi/2$ or $3\pi/2$. The polarization ellipse reduces to

$$\frac{E_x^2}{E_{0x}^2} + \frac{E_y^2}{E_{0y}^2} = 1 \tag{3-20}$$

This is the standard equation of an ellipse. Note that $\delta = \pi/2$ or $\delta = 3\pi/2$ yields the identical polarization ellipse.

4. $E_{0x} = E_{0y} = E_0$ and $\delta = \pi/2$ or $\delta = 3\pi/2$. The polarization ellipse now reduces to

$$\frac{E_x^2}{E_0^2} + \frac{E_y^2}{E_0^2} = 1 \tag{3-21}$$

Equation (3-21) describes the equation of a circle. Thus, for this condition the light is said to be right or left circularly polarized ($\delta = \pi/2$ and $3\pi/2$, respectively). Again, we note that (3-21) shows that it alone cannot determine if the value of δ is $\pi/2$ or $3\pi/2$.

Finally, in the previous section we showed that the area of the polarization ellipse was

$$A = \pi E_{0x} E_{0y} \sin\delta \tag{3-22}$$

We see that for $\delta = 0$ or π the area of the polarization ellipse is zero, which is to be expected for linearly polarized light. For $\delta = \pi/2$ or $3\pi/2$ the area of the ellipse is a maximum; that is, $\pi E_{0x} E_{0y}$. It is important to note that even if the phase shift between the orthogonal components is $\pi/2$ or $3\pi/2$, the light is, in general, elliptically polarized. Furthermore, the polarization ellipse shows that it is in the standard form as given by (3-20).

For the more restrictive condition where the orthogonal amplitudes are equal so that $E_{0x} = E_{0y} = E_0$ and, when $\delta = \pi/2$ or $3\pi/2$, (3-22) becomes

$$A = \pi E_0^2 \tag{3-23}$$

which is, of course, the area of a circle.

The previous special forms of the polarization ellipse are spoken of as being degenerate states. We can summarize these results by saying that the degenerate states of the polarization ellipse are (1) linear horizontally or vertically polarized light, (2) linear +45° or −45° polarized light, and (3) right or left circularly polarized light.

Aside from the fact that these degenerate states appear quite naturally as special cases of the polarization ellipse, there is a fundamental reason for their importance: they are relatively easy to create in an optical laboratory and can be used to create "null-intensity" conditions. Polarization instruments, which may be based on null-intensity conditions, enable very accurate measurements to be made.

3.4 ELLIPTICAL PARAMETERS OF THE POLARIZATION ELLIPSE

The polarization ellipse has the form:

$$\frac{E_x^2}{E_{0x}^2} + \frac{E_y^2}{E_{0y}^2} - 2\frac{E_x}{E_{0x}}\frac{E_y}{E_{0y}}\cos\delta = \sin^2\delta \tag{3-7a}$$

where $\delta = \delta_y - \delta_x$. In general, the axes of the ellipse are not in the Ox and Oy directions. In (3-7a) the presence of the "product" term E_xE_y shows that it is actually a rotated ellipse; in the standard form of an ellipse the product term is not present. In this section we find the mathematical relations between the parameters of the polarization ellipse, E_{0x}, E_{0y}, and δ and the angle of rotation ψ, and another important parameter, χ, the ellipticity angle.

In Fig. 3-3 we show the rotated ellipse. Let Ox and Oy be the initial, unrotated, axes, and let Ox' and Oy' be a new set of axes along the rotated ellipse. Furthermore, let $\psi (0 \leq \psi \leq \pi)$ be the angle between Ox and the direction Ox' of the major axis.

The components E'_x and E'_y are

$$E'_x = E_x \cos\psi + E_y \sin\psi \tag{3-24a}$$

$$E'_y = -E_x \sin\psi + E_y \cos\psi \tag{3-24b}$$

If $2a$ and $2b$ $(a \geq b)$ are the lengths of the major and minor axes, respectively, then the equation of the ellipse in terms of Ox' and Oy' can be written as

$$E'_x = a\cos(\tau + \delta') \tag{3-25a}$$

$$E'_y = \pm b\sin(\tau + \delta') \tag{3-25b}$$

where τ is the propagator and δ' is an arbitrary phase. The $\pm$ sign describes the two possible senses in which the end point of the field vector can describe the ellipse.

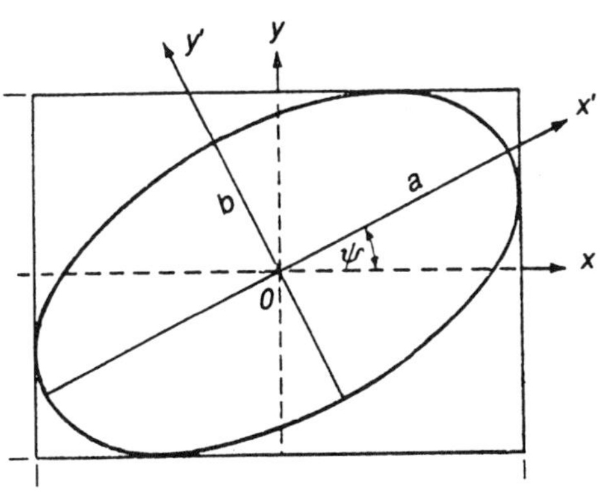

Figure 3-3 The rotated polarization ellipse.

The form of (3-25) is chosen because it is easy to see that it leads to the standard form of the ellipse, namely,

$$\frac{E_x'^2}{a^2} + \frac{E_y'^2}{b^2} = 1 \tag{3-26}$$

We can relate a and b in (3-25) to the parameters E_{0x} and E_{0y} in (3-7a) by recalling that the original equations for the optical field are

$$\frac{E_x}{E_{0x}} = \cos(\tau + \delta_x) \tag{3-27a}$$

$$\frac{E_y}{E_{0y}} = \cos(\tau + \delta_y) \tag{3-27b}$$

We then substitute (3-25) and (3-27) into (3-24), expand the terms, and write

$$\begin{aligned} a(\cos\tau\cos\delta' - \sin\tau\sin\delta') &= E_{0x}(\cos\tau\cos\delta_x - \sin\tau\sin\delta_x)\cos\psi \\ &\quad + E_{0y}(\cos\tau\cos\delta_y - \sin\tau\sin\delta_y)\sin\psi \end{aligned} \tag{3-28a}$$

$$\begin{aligned} \pm b(\sin\tau\cos\delta' + \cos\tau\sin\delta') &= -E_{0x}(\cos\tau\cos\delta_x - \sin\tau\sin\delta_x)\sin\psi \\ &\quad + E_{0y}(\cos\tau\cos\delta_y - \sin\tau\sin\delta_y)\cos\psi \end{aligned} \tag{3-28b}$$

Equating the coefficients of $\cos\tau$ and $\sin\tau$ leads to the following equations:

$$a\cos\delta' = E_{0x}\cos\delta_x\cos\psi + E_{0y}\cos\delta_y\sin\psi \tag{3-29a}$$

$$a\sin\delta' = E_{0x}\sin\delta_x\cos\psi + E_{0y}\sin\delta_y\sin\psi \tag{3-29b}$$

$$\pm b\cos\delta' = E_{0x}\sin\delta_x\sin\psi - E_{0y}\sin\delta_y\cos\psi \tag{3-29c}$$

$$\pm b\sin\delta' = E_{0x}\cos\delta_x\sin\psi - E_{0y}\cos\delta_y\cos\psi \tag{3-29d}$$

Squaring and adding (3-29a) and (3-29b) and using $\delta = \delta_y - \delta_x$, we find that

$$a^2 = E_{0x}^2\cos^2\psi + E_{0y}^2\sin^2\psi + 2E_{0x}E_{0y}\cos\psi\sin\psi\cos\delta \tag{3-30a}$$

Similarly, from (3-29c) and (3-29d) we find that

$$b^2 = E_{0x}^2\sin^2\psi + E_{0y}^2\cos^2\psi - 2E_{0x}E_{0y}\cos\psi\sin\psi\cos\delta \tag{3-30b}$$

Hence,

$$a^2 + b^2 = E_{0x}^2 + E_{0y}^2 \tag{3-31}$$

Next, we multiply (3-29a) by (3-29c), (3-29b) by (3-29d), and add. This gives

$$\pm ab = E_{0x}E_{0y}\sin\delta \tag{3-32}$$

Further, dividing (3-29d) by (3-29a) and (3-29c) by (3-29b) leads to

$$(E_{0x}^2 - E_{0y}^2)\sin 2\psi = 2E_{0x}E_{0y}\cos\delta\cos 2\psi \tag{3-33a}$$

or

$$\tan 2\psi = \frac{2E_{0x}E_{0y}\cos\delta}{E_{0x}^2 - E_{0y}^2} \tag{3-33b}$$

which relates the angle of rotation ψ to E_{0x}, E_{0y}, and δ.

We note that, in terms of the phase δ, ψ is equal to zero only for $\delta = 90^\circ$ or 270°. Similarly, in terms of amplitude, only if E_{0x} or E_{0y} is equal to zero is ψ equal to zero.

An alternative method for determining ψ is to transform (3-7a) directly to (3-26). To show this we write (3-24a) and (3-24b) as

$$E_x = E'_x \cos\psi - E'_y \sin\psi \tag{3-34a}$$

$$E_y = E'_x \sin\psi + E'_y \cos\psi \tag{3-34b}$$

Equation (3-34) can be obtained from (3-24) by solving for E_x and E_y or, equivalently, replacing ψ by $-\psi$, E_x by E'_x, and E_y by E'_y. On substituting (3-34a) and (3-34b) into (3-7a), the cross term is seen to vanish only for the condition given by (3-33).

It is useful to introduce an auxiliary angle $\alpha (0 \le \alpha \le \pi/2)$ for the polarization ellipse defined by

$$\tan\alpha = \frac{E_{0y}}{E_{0x}} \tag{3-35}$$

Then (3-33) is easily shown by using (3-34) to reduce to

$$\tan 2\psi = \frac{2E_{0x}E_{0y}}{E_{0x}^2 E_{0y}^2}\cos\delta = \frac{2\tan\alpha}{1-\tan^2\alpha}\cos\delta \tag{3-36}$$

which then yields

$$\tan 2\psi = (\tan 2\alpha)\cos\delta \tag{3-37}$$

We see that for $\delta = 0$ or π the angle of rotation is

$$\psi = \pm\alpha \tag{3-38}$$

For $\delta = \pi/2$ or $3\pi/2$ we have $\psi = 0$, so the angle of rotation is also zero.

Another important parameter of interest is the angle of ellipticity, χ. This is defined by

$$\tan\chi = \frac{\pm b}{a} \qquad -\frac{\pi}{4} \le \chi \le \frac{\pi}{4} \tag{3-39}$$

We see that for linearly polarized light $b = 0$, so $\chi = 0$. Similarly, for circularly polarized light $b = a$, so $\chi = \pm\pi/4$. Thus, (3-39) describes the extremes of the ellipticity of the polarization ellipse.

Using (3-31), (3-32), and (3-35), we easily find that

$$\frac{\pm 2ab}{a^2+b^2} = \frac{2E_{0x}E_{0y}}{E_{0x}^2 + E_{0y}^2}\sin\delta = (\sin 2\alpha)\sin\delta \tag{3-40}$$

Next, using (3-39) we easily see that the left-hand side of (3-40) reduces to $\sin 2\chi$, so we can write (3-40) as

$$\sin 2\chi = (\sin 2\alpha)\sin\delta \tag{3-41}$$

which is the relation between the ellipticity of the polarization ellipse and the parameters E_{0x}, E_{0y}, and δ of the polarization ellipse.

We note that only for $\delta = \pi/2$ or $3\pi/2$ does (3-41) reduce to

$$\chi = \pm\alpha \tag{3-42}$$

which is to be expected.

The results that we have obtained here will be used again, so it is useful to summarize them. The elliptical parameters E_{0x}, E_{0y}, and δ of the polarization ellipse are related to the orientation angle ψ and ellipticity angle χ by the following equations:

$$\tan 2\psi = (\tan 2\alpha)\cos\delta \qquad 0 \le \psi \le \pi \tag{3-43a}$$

$$\sin 2\chi = (\sin 2\alpha)\sin\delta \qquad -\frac{\pi}{4} < \chi \le \frac{\pi}{4} \tag{3-43b}$$

where $0 \le \alpha \le \pi/2$ and

$$a^2 + b^2 = E_{0x}^2 + E_{0y}^2 \tag{3-43c}$$

$$\tan\alpha = \frac{E_{0y}}{E_{0x}} \tag{3-43d}$$

$$\tan\chi = \frac{\pm b}{a} \tag{3-43e}$$

We emphasize that the polarization ellipse can be described either in terms of the orientation and ellipticity angles ψ and χ on the left-hand sides of (3-43a) and (3-43b) or the major and minor axes E_{0x} and E_{0y} and the phase shift δ on the right-hand sides of (3-43a) and (3-43b).

Finally, a few words must be said on the terminology of polarization. Two cases of polarization are distinguished according to the sense in which the end point of the field vector describes the ellipse. It seems natural to call the polarization right-handed or left-handed according to whether the rotation of **E** and the direction of propagation form a right-handed or left-handed screw. The traditional terminology, however, is just the opposite and is based on the apparent behavior of **E** when viewed face on by the observer. In this book we shall conform to the traditional, that is, customary usage. Thus, the polarization is *right-handed* when to an observer looking in the direction from which the light is coming, the end point of the electric vector would appear to describe the ellipse in the *clockwise* sense. If we consider the value of (3-4) for two time instants separated by a quarter of a period, we see that in this case $\sin\delta > 0$, or by (3-43), $0 < \chi \le \pi/4$. For *left-handed* polarization the opposite is the case; i.e., to an observer looking in the direction from which the light is propagated, the electric vector would appear to describe the ellipse *counterclockwise*; in this case $\sin\delta < 0$, so that $-\pi/4 \le \chi < 0$.

REFERENCES

1. Born, M. and Wolf, E., *Principles of Optics*, 3rd ed., Pergamon Press, New York, 1965.
2. Wood, R. W., *Physical Optics*, 3rd ed., Optical Society of America, Washington, D.C., 1988.
3. Strong, J., *Concepts of Classical Optics*, W. H. Freeman and Company, San Francisco, 1958.
4. Jenkins, F. S. and White, H. E., *Fundamentals of Optics*, McGraw-Hill, New York, 1957.
5. Stone, J. M., *Radiation and Optics*, McGraw-Hill, New York, 1963.

4

The Stokes Polarization Parameters

4.1 INTRODUCTION

In Chapter 3 we saw that the elimination of the propagator between the transverse components of the optical field led to the polarization ellipse. Analysis of the ellipse showed that for special cases it led to forms which can be interpreted as linearly polarized light and circularly polarized light. This description of light in terms of the polarization ellipse is very useful because it enables us to describe by means of a single equation various states of polarized light. However, this representation is inadequate for several reasons. As the beam of light propagates through space, we find that in a plane transverse to the direction of propagation the light vector traces out an ellipse or some special form of an ellipse, such as a circle or a straight line in a time interval of the order 10^{-15} sec. This period of time is clearly too short to allow us to follow the tracing of the ellipse. This fact, therefore, immediately prevents us from ever observing the polarization ellipse. Another limitation is that the polarization ellipse is only applicable to describing light that is completely polarized. It cannot be used to describe either unpolarized light or partially polarized light. This is a particularly serious limitation because, in nature, light is very often unpolarized or partially polarized. Thus, the polarization ellipse is an idealization of the true behavior of light; it is only correct at any given instant of time. These limitations force us to consider an alternative description of polarized light in which only observed or measured quantities enter. We are, therefore, in the same situation as when we dealt with the wave equation and its solutions, neither of which can be observed. We must again turn to using average values of the optical field which in the present case requires that we represent polarized light in terms of observables.

In 1852, Sir George Gabriel Stokes (1819–1903) discovered that the polarization behavior could be represented in terms of observables. He found that any state of polarized light could be completely described by four measurable quantities now known as the Stokes polarization parameters. The first parameter expresses the total intensity of the optical field. The remaining three parameters describe the polarization state. Stokes was led to his formulation in order to provide a suitable mathematical description of the Fresnel–Arago interference laws (1818). These laws were based on experiments carried out with an unpolarized light source, a quantity which Fresnel and his successors were never able to characterize mathematically. Stokes succeeded where others had failed because he abandoned the attempts to describe unpolarized light in terms of amplitude. He resorted to an experimental definition, namely, unpolarized light is light whose intensity is unaffected when a polarizer is rotated or by the presence of a retarder of any retardance value. Stokes also showed that his parameters could be applied not only to unpolarized light but to partially polarized and completely polarized light as well. Unfortunately, Stokes' paper was forgotten for nearly a century. Its importance was finally brought to the attention of the scientific community by the Nobel laureate S. Chandrasekhar in 1947, who used the Stokes parameters to formulate the radiative transfer equations for the scattering of partially polarized light. The Stokes parameters have been a prominent part of the optical literature on polarized light ever since.

We saw earlier that the amplitude of the optical field cannot be observed. However, the quantity that can be observed is the intensity, which is derived by taking a time average of the square of the amplitude. This suggests that if we take a time average of the unobserved polarization ellipse we will be led to the observables of the polarization ellipse. When this is done, as we shall show shortly, we obtain four parameters, which are exactly the Stokes parameters. Thus, the Stokes parameters are a logical consequence of the wave theory. Furthermore, the Stokes parameters give a complete description of any polarization state of light. Most important, the Stokes parameters are exactly those quantities that are measured. Aside from this important formulation, however, when the Stokes parameters are used to describe physical phenomena, e.g., the Zeeman effect, one is led to a very interesting representation. Originally, the Stokes parameters were used only to describe the measured intensity and polarization state of the optical field. But by forming the Stokes parameters in terms of a column matrix, the so-called Stokes vector, we are led to a formulation in which we obtain not only measurables but also observables, which can be seen in a spectroscope. As a result, we shall see that the formalism of the Stokes parameters is far more versatile than originally envisioned and possesses a greater usefulness than is commonly known.

4.2 DERIVATION OF THE STOKES POLARIZATION PARAMETERS

We consider a pair of plane waves that are orthogonal to each other at a point in space, conveniently taken to be $z = 0$, and not necessarily monochromatic, to be represented by the equations:

$$E_x(t) = E_{0x}(t)\cos[\omega t + \delta_x(t)] \tag{4-1a}$$

$$E_y(t) = E_{0y}(t)\cos[\omega t + \delta_y(t)] \tag{4-1b}$$

where $E_{0x}(t)$ and $E_{0y}(t)$ are the instantaneous amplitudes, ω is the instantaneous angular frequency, and $\delta_x(t)$ and $\delta_y(t)$ are the instantaneous phase factors. At all times the amplitudes and phase factors fluctuate slowly compared to the rapid vibrations of the cosinusoids. The explicit removal of the term ωt between (4-1a) and (4-1b) yields the familiar polarization ellipse, which is valid, in general, only at a given instant of time:

$$\frac{E_x^2(t)}{E_{0x}^2(t)} + \frac{E_y^2(t)}{E_{0y}^2(t)} - \frac{2E_x(t)E_y(t)}{E_{0x}(t)E_{0y}(t)}\cos\delta(t) = \sin^2\delta(t) \tag{4-2}$$

where $\delta(t) = \delta_y(t) - \delta_x(t)$.

For monochromatic radiation, the amplitudes and phases are constant for all time, so (4-2) reduces to

$$\frac{E_x^2(t)}{E_{0x}^2} + \frac{E_y^2(t)}{E_{0y}^2} - \frac{2E_x(t)E_y(t)}{E_{0x}E_{0y}}\cos\delta = \sin^2\delta \tag{4-3}$$

While E_{0x}, E_{0y}, and δ are constants, E_x and E_y continue to be implicitly dependent on time, as we see from (4-1a) and (4-1b). Hence, we have written $E_x(t)$ and $E_y(t)$ in (4-3). In order to represent (4-3) in terms of the observables of the optical field, we must take an average over the time of observation. Because this is a long period of time relative to the time for a single oscillation, this can be taken to be infinite. However, in view of the periodicity of $E_x(t)$ and $E_y(t)$, we need average (4-3) only over a single period of oscillation. The time average is represented by the symbol $\langle\cdots\rangle$, and so we write (4-3) as

$$\frac{\langle E_x^2(t)\rangle}{E_{0x}^2} + \frac{\langle E_y^2(t)\rangle}{E_{0y}^2} - \frac{2\langle E_x(t)E_y(t)\rangle}{E_{0x}E_{0y}}\cos\delta = \sin^2\delta \tag{4-4a}$$

where

$$\langle E_i(t)E_j(t)\rangle = \lim_{T\to\infty}\frac{1}{T}\int_0^T E_i(t)E_j(t)\,dt \qquad i, j = x, y \tag{4-4b}$$

Multiplying (4-4a) by $4E_{0x}^2E_{0y}^2$, we see that

$$4E_{0y}^2\langle E_x^2(t)\rangle + 4E_{0x}^2\langle E_y^2(t)\rangle - 8E_{0x}E_{0y}\langle E_x(t)E_y(t)\rangle\cos\delta = (2E_{0x}E_{0y}\sin\delta)^2 \tag{4-5}$$

From (4-1a) and (4-1b), we then find that the average values of (4-5) using (4-4b) are

$$\langle E_x^2(t)\rangle = \frac{1}{2}E_{0x}^2 \tag{4-6a}$$

$$\langle E_y^2(t)\rangle = \frac{1}{2}E_{0y}^2 \tag{4-6b}$$

$$\langle E_x(t)E_y(t)\rangle = \frac{1}{2}E_{0x}E_{0y}\cos\delta \tag{4-6c}$$

Substituting (4-6a), (4-6b), and (4-6c) into (4-5) yields

$$2E_{0x}^2E_{0y}^2 + 2E_{0x}^2E_{0y}^2 - (2E_{0x}E_{0y}\cos\delta)^2 = (2E_{0x}E_{0y}\sin\delta)^2 \tag{4-7}$$

Since we wish to express the final result in terms of intensity this suggests that we add and subtract the quantity $E_{0x}^4 + E_{0y}^4$ to the left-hand side of (4-7); doing this leads to perfect squares. Upon doing this and grouping terms, we are led to the following equation:

$$(E_{0x}^2 + E_{0y}^2)^2 - (E_{0x}^2 - E_{0y}^2)^2 - (2E_{0x}E_{0y}\cos\delta)^2 = (2E_{0x}E_{0y}\sin\delta)^2 \tag{4-8}$$

We now write the quantities inside the parentheses as

$$S_0 = E_{0x}^2 + E_{0y}^2 \tag{4-9a}$$

$$S_1 = E_{0x}^2 - E_{0y}^2 \tag{4-9b}$$

$$S_2 = 2E_{0x}E_{0y}\cos\delta \tag{4-9c}$$

$$S_3 = 2E_{0x}E_{0y}\sin\delta \tag{4-9d}$$

and then express (4-8) as

$$S_0^2 = S_1^2 + S_2^2 + S_3^2 \tag{4-10}$$

The four equations given by (4-9) are the Stokes polarization parameters for a plane wave. They were introduced into optics by Sir George Gabriel Stokes in 1852. We see that the Stokes parameters are real quantities, and they are simply the observables of the polarization ellipse and, hence, the optical field. The first Stokes parameter S_0 is the total intensity of the light. The parameter S_1 describes the amount of linear horizontal or vertical polarization, the parameter S_2 describes the amount of linear +45° or −45° polarization, and the parameter S_3 describes the amount of right or left circular polarization contained within the beam; this correspondence will be shown shortly. We note that the four Stokes parameters are expressed in terms of intensities, and we again emphasize that the Stokes parameters are *real* quantities.

If we now have partially polarized light, then we see that the relations given by (4-9) continue to be valid for very short time intervals, since the amplitudes and phases fluctuate slowly. Using Schwarz's inequality, one can show that for any state of polarized light the Stokes parameters always satisfy the relation:

$$S_0^2 \geq S_1^2 + S_2^2 + S_3^2 \tag{4-11}$$

The equality sign applies when we have completely polarized light, and the inequality sign when we have partially polarized light or unpolarized light.

In Chapter 3, we saw that orientation angle ψ of the polarization ellipse was given by

$$\tan 2\psi = \frac{2E_{0x}E_{0y}\cos\delta}{E_{0x}^2 - E_{0y}^2} \tag{3-33b}$$

Inspecting (4-9) we see that if we divide (4-9c) by (4-9b), ψ can be expressed in terms of the Stokes parameters:

$$\tan 2\psi = \frac{S_2}{S_1} \tag{4-12}$$

Similarly, from (3-40) and (3-41) in Chapter 3 the ellipticity angle χ was given by

$$\sin 2\chi = \frac{2E_{0x}E_{0y}\sin\delta}{E_{0x}^2 + E_{0y}^2} \tag{4-13}$$

Again, inspecting (4-9) and dividing (4-9d) by (4-9a), we can see that χ can be expressed in terms of the Stokes parameters:

$$\sin 2\chi = \frac{S_3}{S_0} \tag{4-14}$$

The Stokes parameters enable us to describe the degree of polarization P for any state of polarization. By definition,

$$P = \frac{I_{\text{pol}}}{I_{\text{tot}}} = \frac{(S_1^2 + S_2^2 + S_3^2)^{1/2}}{S_0} \qquad 0 \le P \le 1 \tag{4-15}$$

where I_{pol} is the intensity of the sum of the polarization components and I_{tot} is the total intensity of the beam. The value of $P = 1$ corresponds to completely polarized light, $P = 0$ corresponds to unpolarized light, and $0 < P < 1$ corresponds to partially polarized light.

To obtain the Stokes parameters of an optical beam, one must always take a time average of the polarization ellipse. However, the time-averaging process can be formally bypassed by representing the (real) optical amplitudes, (4-1a) and (4-1b), in terms of complex amplitudes:

$$E_x(t) = E_{0x}\exp[i(\omega t + \delta_x)] = E_x\exp(i\omega t) \tag{4-16a}$$

$$E_y(t) = E_{0y}\exp[i(\omega t + \delta_y)] = E_y\exp(i\omega t) \tag{4-16b}$$

where

$$E_x = E_{0x}\exp(i\delta_x) \tag{4-16c}$$

and

$$E_y = E_{0y}\exp(i\delta_y) \tag{4-16d}$$

are complex amplitudes. The Stokes parameters for a plane wave are now obtained from the formulas:

$$S_0 = E_xE_x^* + E_yE_y^* \tag{4-17a}$$

$$S_1 = E_xE_x^* - E_yE_y^* \tag{4-17b}$$

$$S_2 = E_xE_y^* + E_yE_x^* \tag{4-17c}$$

$$S_3 = i(E_xE_y^* - E_yE_x^*) \tag{4-17d}$$

We shall use (4-17), the complex representation, henceforth, as the defining equations for the Stokes parameters. Substituting (4-16c) and (4-16d) into (4-17) gives

$$S_0 = E_{0x}^2 + E_{0y}^2 \tag{4-9a}$$

$$S_1 = E_{0x}^2 - E_{0y}^2 \tag{4-9b}$$

$$S_2 = 2E_{0x}E_{0y}\cos\delta \tag{4-9c}$$

$$S_3 = 2E_{0x}E_{0y}\sin\delta \tag{4-9d}$$

which are the Stokes parameters obtained formally from the polarization ellipse.

As examples of the representation of polarized light in terms of the Stokes parameters, we consider (1) linear horizontal and linear vertical polarized light, (2) linear +45° and linear −45° polarized light, and (3) right and left circularly polarized light.

4.2.1 Linear Horizontally Polarized Light (LHP)

For this case $E_{0y} = 0$. Then, from (4-9) we have

$$S_0 = E_{0x}^2 \tag{4-18a}$$

$$S_1 = E_{0x}^2 \tag{4-18b}$$

$$S_2 = 0 \tag{4-18c}$$

$$S_3 = 0 \tag{4-18d}$$

4.2.2 Linear Vertically Polarized Light (LVP)

For this case $E_{0x} = 0$. From (4-9) we have

$$S_0 = E_{0y}^2 \tag{4-19a}$$

$$S_1 = -E_{0y}^2 \tag{4-19b}$$

$$S_2 = 0 \tag{4-19c}$$

$$S_3 = 0 \tag{4-19d}$$

4.2.3 Linear +45° Polarized Light (*L* + 45)

The conditions to obtain $L + 45$ polarized light are $E_{0x} = E_{0y} = E_0$ and $\delta = 0°$. Using these conditions and the definition of the Stokes parameters (4-9), we find that

$$S_0 = 2E_0^2 \tag{4-20a}$$

$$S_1 = 0 \tag{4-20b}$$

$$S_2 = 2E_0^2 \tag{4-20c}$$

$$S_3 = 0 \tag{4-20d}$$

4.2.4 Linear −45° Polarized Light (*L* − 45)

The conditions on the amplitude are the same as for $L + 45$ light, but the phase difference is $\delta = 180°$. Then, from (4-9) we see that the Stokes parameters are

$$S_0 = 2E_0^2 \tag{4-21a}$$

$$S_1 = 0 \tag{4-21b}$$

$$S_2 = -2E_0^2 \tag{4-21c}$$

$$S_3 = 0 \tag{4-21d}$$

4.2.5 Right Circularly Polarized Light (RCP)

The conditions to obtain RCP light are $E_{0x} = E_{0y} = E_0$ and $\delta = 90°$. From (4-9) the Stokes parameters are then

$$S_0 = 2E_0^2 \tag{4-22a}$$

$$S_1 = 0 \tag{4-22b}$$

$$S_2 = 0 \tag{4-22c}$$

$$S_3 = 2E_0^2 \tag{4-22d}$$

4.2.6 Left Circularly Polarized Light (LCP)

For LCP light the amplitudes are again equal, but the phase shift between the orthogonal, transverse components is $\delta = -90°$. The Stokes parameters from (4-9) are then

$$S_0 = 2E_0^2 \tag{4-23a}$$

$$S_1 = 0 \tag{4-23b}$$

$$S_2 = 0 \tag{4-23c}$$

$$S_3 = -2E_0^2 \tag{4-23d}$$

Finally, the Stokes parameters for elliptically polarized light are, of course, given by (4–9).

Inspection of the four Stokes parameters suggests that they can be arranged in the form of a column matrix. This column matrix is called the Stokes vector. This step, while simple, provides a formal method for treating numerous complicated problems involving polarized light. We now discuss the Stokes vector.

4.3 THE STOKES VECTOR

The four Stokes parameters can be arranged in a column matrix and written as

$$S = \begin{pmatrix} S_0 \\ S_1 \\ S_2 \\ S_3 \end{pmatrix} \tag{4-24}$$

The column matrix (4-24) is called the Stokes vector. Mathematically, it is not a vector, but through custom it is called a vector. Equation (4-24) should correctly be called the Stokes column matrix. The Stokes vector for elliptically polarized light is then written from (4-9) as

$$S = \begin{pmatrix} E_{0x}^2 + E_{0y}^2 \\ E_{0x}^2 - E_{0y}^2 \\ 2E_{0x}E_{0y}\cos\delta \\ 2E_{0x}E_{0y}\sin\delta \end{pmatrix} \tag{4-25}$$

Equation (4-25) is also called the Stokes vector for a plane wave.

The Stokes vectors for linearly and circularly polarized light are readily found from (4-25). We now derive these Stokes vectors.

4.3.1 Linear Horizontally Polarized Light (LHP)

For this case $E_{0y} = 0$, and we find from (4-25) that

$$S = I_0 \begin{pmatrix} 1 \\ 1 \\ 0 \\ 0 \end{pmatrix} \tag{4-26}$$

where $I_0 = E_{0x}^2$ is the total intensity.

4.3.2 Linear Vertically Polarized Light (LVP)

For this case $E_{0x} = 0$, and we find that (4-25) reduces to

$$S = I_0 \begin{pmatrix} 1 \\ -1 \\ 0 \\ 0 \end{pmatrix} \tag{4-27}$$

where, again, I_0 is the total intensity.

4.3.3 Linear +45° Polarized Light ($L + 45$)

In this case $E_{0x} = E_{0y} = E_0$ and $\delta = 0$, so (4-25) becomes

$$S = I_0 \begin{pmatrix} 1 \\ 0 \\ 1 \\ 0 \end{pmatrix} \tag{4-28}$$

where $I_0 = 2E_0^2$.

4.3.4 Linear −45° Polarized Light ($L - 45$)

Again, $E_{0x} = E_{0y} = E_0$, but now $\delta = 180°$. Then (4-25) becomes

$$S = I_0 \begin{pmatrix} 1 \\ 0 \\ -1 \\ 0 \end{pmatrix} \tag{4-29}$$

and $I_0 = 2E_0^2$.

4.3.5 Right Circularly Polarized Light (RCP)

In this case $E_{0x} = E_{0y} = E_0$ and $\delta = 90^\circ$. Then (4-25) becomes

$$S = I_0 \begin{pmatrix} 1 \\ 0 \\ 0 \\ 1 \end{pmatrix} \tag{4-30}$$

and $I_0 = 2E_0^2$.

4.3.6 Left Circularly Polarized Light (LCP)

Again, we have $E_{0x} = E_{0y}$, but now the phase shift δ between the orthogonal amplitudes is $\delta = -90^\circ$. Equation (4-25) then reduces to

$$S = I_0 \begin{pmatrix} 1 \\ 0 \\ 0 \\ -1 \end{pmatrix} \tag{4-31}$$

and $I_0 = 2E_0^2$.

We also see from (4-25) that if $\delta = 0^\circ$ or 180°, then (4-25) reduces to

$$S = \begin{pmatrix} E_{0x}^2 + E_{0y}^2 \\ E_{0x}^2 - E_{0y}^2 \\ \pm 2E_{0x}E_{0y} \\ 0 \end{pmatrix} \tag{4-32}$$

We recall that the ellipticity angle χ and the orientation angle ψ for the polarization ellipse are given, respectively, by

$$\sin 2\chi = \frac{S_3}{S_0} \qquad \frac{-\pi}{4} \leq \chi \leq \frac{\pi}{4} \tag{4-33a}$$

$$\tan 2\psi = \frac{S_2}{S_1} \qquad 0 \leq \psi < \pi \tag{4-33b}$$

We see that S_3 is zero, so the ellipticity angle χ is zero and, hence, (4-32) is the Stokes vector for linearly polarized light. The orientation angle according to (4-33b) is

$$\tan 2\psi = \frac{\pm 2E_{0x}E_{0y}}{E_{0x}^2 - E_{0y}^2} \tag{4-34}$$

The form of (4-32) is a useful representation for linearly polarized light. Another useful representation can be made by expressing the amplitudes E_{0x} and E_{0y} in terms of an angle. To show this, we first rewrite the total intensity S_0 as

$$S_0 = E_{0x}^2 + E_{0y}^2 = E_0^2 \tag{4-35}$$

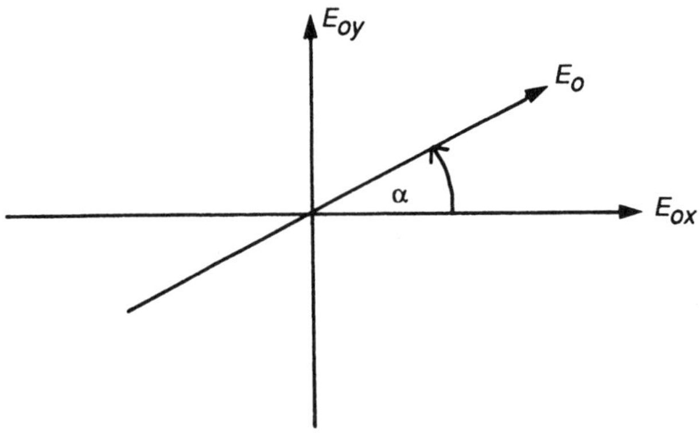

Figure 4-1 Resolution of the optical field components.

Equation (4-35) suggests Fig. 4-1. From Fig. 4-1 we see that

$$E_{0x} = E_0 \cos\alpha \tag{4-36a}$$

$$E_{0y} = E_0 \sin\alpha \qquad 0 \le \alpha \le \frac{\pi}{2} \tag{4-36b}$$

The angle α is called the auxiliary angle; it is identical to the auxiliary angle used to represent the orientation angle and ellipticity equations summarized earlier. Substituting (4-36) into (4-32) leads to the following Stokes vector for linearly polarized light:

$$S = I_0 \begin{pmatrix} 1 \\ \cos 2\alpha \\ \sin 2\alpha \\ 0 \end{pmatrix} \tag{4-37}$$

where $I_0 = E_0^2$ is the total intensity. Equation (4-36) can also be used to represent the Stokes vector for elliptically polarized light, (4-25). Substituting (4-36) into (4-25) gives

$$S = I_0 \begin{pmatrix} 1 \\ \cos 2\alpha \\ \sin 2\alpha \cos\delta \\ \sin 2\alpha \sin\delta \end{pmatrix} \tag{4-38}$$

It is customary to write the Stokes vector in normalized form by setting $I_0 = 1$. Thus, (4-38) is written as

$$S = \begin{pmatrix} 1 \\ \cos 2\alpha \\ \sin 2\alpha \cos\delta \\ \sin 2\alpha \sin\delta \end{pmatrix} \tag{4-39}$$

The orientation angle ψ and the ellipticity angle χ of the polarization ellipse are given by (4-33a) and (4-33b). Substituting S_1, S_2, and S_3 into (4-39) into (4-33a) and (4-33b) gives

$$\tan 2\psi = \tan 2\alpha \cos \delta \tag{4-40a}$$

$$\sin 2\chi = \sin 2\alpha \sin \delta \tag{4-40b}$$

which are identical to the relations we found earlier.

The use of the auxiliary angle α enables us to express the orientation and ellipticity in terms of α and δ. Expressing (4-39) in this manner shows that there are two unique polarization states. For $\alpha = 45°$, (4-39) reduces to

$$S = \begin{pmatrix} 1 \\ 0 \\ \cos \delta \\ \sin \delta \end{pmatrix} \tag{4-41}$$

Thus, the polarization ellipse is expressed only in terms of the phase shift δ between the orthogonal amplitudes. The orientation angle ψ is seen to be always 45°. The ellipticity angle, (4-40b) however, is

$$\sin 2\chi = \sin \delta \tag{4-42}$$

so $\chi = \delta/2$. The Stokes vector (4-41) expresses that the polarization ellipse is rotated 45° from the horizontal axis and that the polarization state of the light can vary from linearly polarized ($\delta = 0, 180°$) to circularly polarized ($\delta = 90°, 270°$).

Another unique polarization state occurs when $\delta = 90°$ or 270°. For this condition (4-39) reduces to

$$S = \begin{pmatrix} 1 \\ \cos 2\alpha \\ 0 \\ \pm \sin 2\alpha \end{pmatrix} \tag{4-43}$$

We see that we now have a Stokes vector and a polarization ellipse, which depends only on the auxiliary angle α. From (4-40a) the orientation angle ψ is always zero. However, (4-40b) and (4-43) show that the ellipticity angle χ is now given by

$$\sin 2\chi = \pm \sin 2\alpha \tag{4-44}$$

so $\chi = \pm\, \alpha$. In general, (4-46) shows that we will have elliptically polarized light. For $\alpha = +45°$and $-45°$ we obtain right and left circularly polarized light. Similarly, for $\alpha = 0°$ and 90° we obtain linear horizontally and vertically polarized light.

The Stokes vector can also be expressed in terms of S_0, ψ, and χ. To show this we write (4-33a) and (4-33b) as

$$S_3 = S_0 \sin 2\chi \tag{4-45a}$$

$$S_2 = S_1 \tan 2\psi \tag{4-45b}$$

In Section 4.2 we found that

$$S_0^2 = S_1^2 + S_2^2 + S_3^2 \tag{4-10}$$

Substituting (4-45a) and (4-45b) into (4-10), we find that

$$S_1 = S_0 \cos 2\chi \cos 2\psi \tag{4-46a}$$

$$S_2 = S_0 \cos 2\chi \sin 2\psi \tag{4-46b}$$

$$S_3 = S_0 \sin 2\chi \tag{4-46c}$$

Arranging (4-46) in the form of a Stokes vector, we have

$$S = S_0 \begin{pmatrix} 1 \\ \cos 2\chi \cos 2\psi \\ \cos 2\chi \sin 2\psi \\ \sin 2\chi \end{pmatrix} \tag{4-47}$$

The Stokes parameters (4-46) are almost identical in form to the well-known equations relating Cartesian coordinates to spherical coordinates. We recall that the spherical coordinates r, θ, and ϕ are related to the Cartesian coordinates x, y, and z by

$$x = r \sin\theta \cos\phi \tag{4-48a}$$

$$y = r \sin\theta \sin\phi \tag{4-48b}$$

$$z = r \cos\theta \tag{4-48c}$$

Comparing (4-48) with (4-46), we see that the equations are identical if the angles are related by

$$\theta = 90^\circ - 2\chi \tag{4-49a}$$

$$\phi = 2\psi \tag{4-49b}$$

In Fig. 4-2 we have drawn a sphere whose center is also at the center of the Cartesian coordinate system. We see that expressing the polarization state of an optical beam in terms of χ and ψ allows us to describe its ellipticity and orientation on a sphere; the radius of the sphere is taken to be unity. The representation of the polarization state on a sphere was first introduced by Henri Poincaré in 1892 and is, appropriately, called the Poincaré sphere. However, at that time, Poincaré introduced the sphere in an entirely different way, namely, by representing the polarization equations in a complex plane and then projecting the plane on to a sphere, a so-called stereographic projection. In this way he was led to (4-46). He does not appear to have known that (4-46) were directly related to the Stokes parameters. Because the Poincaré sphere is of historical interest and is still used to describe the polarization state of light, we shall discuss it in detail later. It is especially useful for describing the change in polarized light when it interacts with polarizing elements.

The discussion in this chapter shows that the Stokes parameters and the Stokes vector can be used to describe an optical beam which is completely polarized. We have, at first sight, only provided an alternative description of completely polarized light. All of the equations derived here are based on the polarization ellipse given in Chapter 3, that is, the amplitude formulation. However, we have pointed out that the Stokes parameters can also be used to describe unpolarized and partially polarized light, quantities which cannot be described within an amplitude

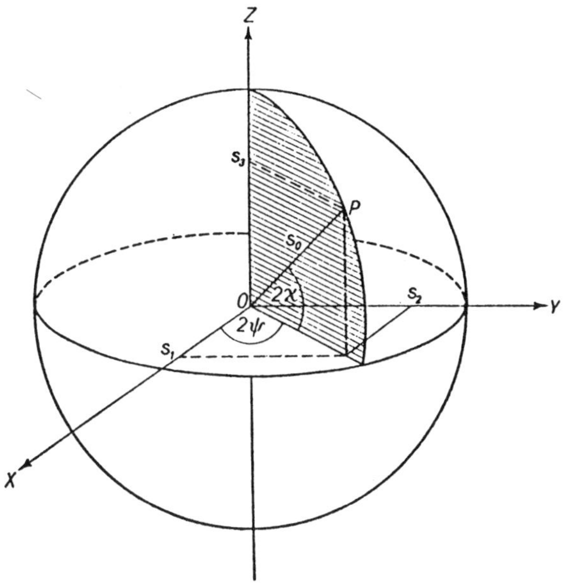

Figure 4-2 The Poincaré representation of polarized light on a sphere.

formulation of the optical field. In order to extend the Stokes parameters to unpolarized and partially polarized light, we must now consider the classical measurement of the Stokes polarization parameters.

4.4 CLASSICAL MEASUREMENT OF THE STOKES POLARIZATION PARAMETERS

The Stokes polarization parameters are immediately useful because, as we shall now see, they are directly accessible to measurement. This is due to the fact that they are an intensity formulation of the polarization state of an optical beam. In this section we shall describe the measurement of the Stokes polarization parameters. This is done by allowing an optical beam to pass through two optical elements known as a retarder and a polarizer. Specifically, the incident field is described in terms of its components, and the field emerging from the polarizing elements is then used to determine the intensity of the emerging beam. Later, we shall carry out this same problem by using a more formal but powerful approach known as the Mueller matrix formalism. In the following chapter we shall also see how this measurement method enables us to determine the Stokes parameters for unpolarized and partially polarized light.

We begin by referring to Fig. 4-3, which shows an monochromatic optical beam incident on a polarizing element called a retarder. This polarizing element is then followed by another polarizing element called a polarizer. The components of the incident beam are

$$E_x(t) = E_{0x}e^{i\delta_x}e^{i\omega t} \tag{4-50a}$$

$$E_y(t) = E_{0y}e^{i\delta_y}e^{i\omega t} \tag{4-50b}$$

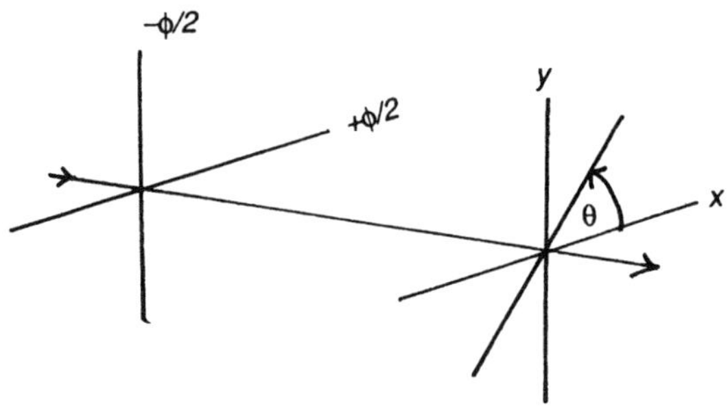

Figure 4-3 Measurement of the Stokes polarization parameters.

In Section 4.2 we saw that the Stokes parameters for a plane wave written in complex notation could be obtained from

$$S_0 = E_x E_x^* + E_y E_y^* \tag{4-17a}$$

$$S_1 = E_x E_x^* - E_y E_y^* \tag{4-17b}$$

$$S_2 = E_x E_y^* + E_y E_x^* \tag{4-17c}$$

$$S_3 = i(E_x E_y^* - E_y E_x^*) \tag{4-17d}$$

where $i = \sqrt{-1}$ and the asterisk represents the complex conjugate.

In order to measure the Stokes parameters, the incident field propagates through a phase-shifting element which has the property that the phase of the x component (E_x) is advanced by $\phi/2$ and the phase of the y component E_y is retarded by $\phi/2$, written as $-\phi/2$. The components E'_x and E'_y emerging from the phase-shifting element component are then

$$E'_x = E_x e^{i\phi/2} \tag{4-51a}$$

$$E'_y = E_y e^{-i\phi/2} \tag{4-51b}$$

In optics, a polarization element that produces this phase shift is called a retarder; it will be discussed in more detail later.

Next, the field described by (4-51) is incident on a component which is called a polarizer. It has the property that the optical field is transmitted only along an axis known as the transmission axis. Ideally, if the transmission axis of the polarizer is at an angle θ only the components of E'_x and E'_y in this direction can be transmitted perfectly; there is complete attenuation at any other angle. A polarizing element which behaves in this manner is called a polarizer. This behavior is described in Fig. 4-4. The component of E'_x along the transmission axis is $E'_x \cos\theta$. Similarly, the component of E'_y is $E'_y \sin\theta$. The field transmitted along the transmission axis is the sum of these components so the total field E emerging from the polarizer is

$$E = E'_x \cos\theta + E'_y \sin\theta \tag{4-52}$$

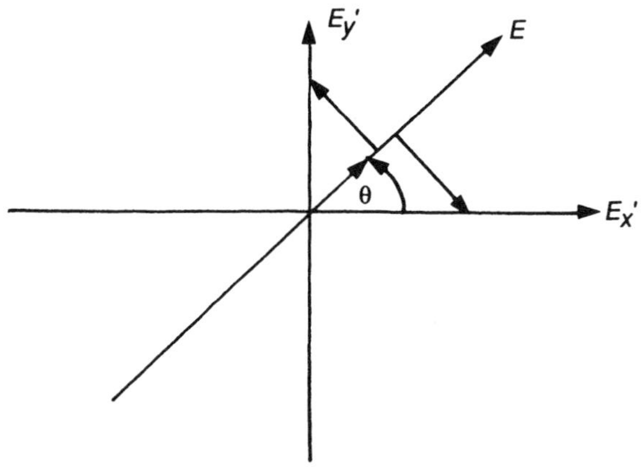

Figure 4-4 Resolution of the optical field components by a polarizer.

Substituting (4-51) into (4-52), the field emerging from the polarizer is

$$E = E_x e^{i\phi/2} \cos\theta + E_y e^{-i\phi/2} \sin\theta \tag{4-53}$$

The intensity of the beam is defined by

$$I = E \cdot E^* \tag{4-54}$$

Taking the complex conjugate of (4-53) and forming the product in accordance with (4-54), the intensity of the emerging beam is

$$\begin{aligned} I(\theta, \phi) = {} & E_x E_x^* \cos^2\theta + E_y E_y^* \sin^2\theta \\ & + E_x^* E_y e^{-i\phi} \sin\theta\cos\theta + E_x E_y^* e^{i\phi} \sin\theta\cos\theta \end{aligned} \tag{4-55}$$

Equation (4-55) can be rewritten by using the well-known trigonometric half-angle formulas:

$$\cos^2\theta = \frac{1+\cos 2\theta}{2} \tag{4-56a}$$

$$\sin^2\theta = \frac{1-\cos 2\theta}{2} \tag{4-56b}$$

$$\sin\theta\cos\theta = \frac{\sin 2\theta}{2} \tag{4-56c}$$

Using (4-56) in (4-55) and grouping terms, we find that the intensity $I(\theta, \phi)$ becomes

$$\begin{aligned} I(\theta, \phi) = {} & \frac{1}{2}[(E_x E_x^* + E_y E_y^*) + (E_x E_x^* - E_y E_y^*) \cos 2\theta \\ & + (E_x E_y^* + E_y E_x^*) \cos\phi \sin 2\theta + i(E_x E_y^* - E_y E_x^*) \sin\phi \sin 2\theta] \end{aligned} \tag{4-57}$$

The terms within parentheses are exactly the Stokes parameters given in (4-17). It was first derived by Stokes and is the manner in which the Stokes parameters were first introduced in the optical literature. Replacing the terms in (4-57) by the definitions of the Stokes parameters given in (4-17), we arrive at

$$I(\theta, \phi) = \frac{1}{2}[S_0 + S_1 \cos 2\theta + S_2 \cos\phi \sin 2\theta + S_3 \sin\phi \sin 2\theta] \tag{4-58}$$

Equation (4-58) is Stokes' famous intensity formula for measuring the four Stokes parameters. Thus, we see that the Stokes parameters are directly accessible to measurement; that is, they are observable quantities.

The first three Stokes parameters are measured by removing the retarder ($\phi = 0°$) and rotating the transmission axis of the polarizer to the angles $\theta = 0°$, $+45°$, and $+90°$, respectively. The final parameter, S_3, is measured by reinserting a so-called quarter-wave retarder ($\phi = 90°$) into the optical path and setting the transmission axis of the polarizer to $\theta = 45°$. The intensities are then found from (4-58) to be

$$I(0°, 0°) = \frac{1}{2}[S_0 + S_1] \tag{4-59a}$$

$$I(45°, 0°) = \frac{1}{2}[S_0 + S_2] \tag{4-59b}$$

$$I(90°, 0°) = \frac{1}{2}[S_0 - S_1] \tag{4-59c}$$

$$I(45°, 90°) = \frac{1}{2}[S_0 + S_3] \tag{4-59d}$$

Solving (4-59) for the Stokes parameters, we have

$$S_0 = I(0°, 0°) + I(90°, 0°) \tag{4-60a}$$

$$S_1 = I(0°, 0°) - I(90°, 0°) \tag{4-60b}$$

$$S_2 = 2I(45°, 0°) - I(0°, 0°) - I(90°, 0°) \tag{4-60c}$$

$$S_3 = 2I(45°, 90°) - I(0°, 0°) - I(90°, 0°) \tag{4-60d}$$

Equation (4-60) is really quite remarkable. In order to measure the Stokes parameters it is necessary to measure the intensity at four angles. We must remember, however, that in 1852 there were no devices to measure the intensity *quantitatively*. The intensities can be measured quantitatively only with an optical detector. But when Stokes introduced the Stokes parameters, such detectors did not exist. The only optical detector was the human eye (retina), a detector capable of measuring only the null or greater-than-null state of light, and so the above method for measuring the Stokes parameters could not be used! Stokes did not introduce the Stokes parameters to describe the optical field in terms of observables as is sometimes stated. The reason for his derivation of (4-58) was not to measure the Stokes polarization parameters but to provide the solution to an entirely different problem, namely, a mathematical statement for unpolarized light. We shall soon see that (4-58) is perfect for doing this. It is possible to measure all four Stokes parameters using the human eye, however, by using a null-intensity technique. This method is described in Section 6.4.

Unfortunately, after Stokes solved this problem and published his great paper on the Stokes parameters and the nature of polarized light, he never returned to this subject again. By the end of his researches on this subject he had turned his attention to the problem of the fluorescence of solutions. This problem would become the major focus of his attention for the rest of his life. Aside from Lord Rayleigh in England and Emil Verdet in France, the importance of Stokes' paper and the Stokes parameters was not fully recognized, and the paper was, practically, forgotten for nearly a century by the optical community. Fortunately, however, Emil Verdet did understand the significance of Stokes' paper and wrote a number of subsequent papers on the Stokes polarization parameters. He thus began a tradition in France of studying the Stokes parameters. The Stokes polarization parameters did not really appear in the English-speaking world again until they were "rediscovered" by S. Chandrasekhar in the late 1940s when he was writing his monumental papers on radiative transfer. Previous to Chandrasekhar no one had included optical polarization in the equations of radiative transfer. In order to introduce polarization into his equations, he eventually found Stokes' original paper. He immediately recognized that because the Stokes parameters were an intensity formulation of optical polarization they could be introduced into radiative equations. It was only after the publication of Chandrasekhar's papers that the Stokes parameters reemerged. They have remained in the optical literature ever since.

We now describe Stokes' formulation for unpolarized light.

4.5 STOKES PARAMETERS FOR UNPOLARIZED AND PARTIALLY POLARIZED LIGHT

The intensity $I(\theta, \phi)$ of a beam of light emerging from the retarder/polarizer combination was seen in the previous section to be

$$I(\theta, \phi) = \frac{1}{2}[S_0 + S_1 \cos 2\theta + S_2 \sin 2\theta \cos\phi + S_3 \sin 2\theta \sin\phi] \tag{4-58}$$

where S_0, S_1, S_2, and S_3 are the Stokes parameters of the incident beam, θ is the rotation angle of the transmission axis of the polarizer, and ϕ is the phase shift of the retarder. By setting θ to 0°, 45°, or 90° and ϕ to 0° or 90°, with the proper pairings of angles, all four Stokes parameters can then be measured. However, it was not Stokes' intention to merely cast the polarization of the optical field in terms of the intensity rather than the amplitude. Rather, he was interested in finding a suitable mathematical description for unpolarized light. Stokes, unlike his predecessors and his contemporaries, recognized that it was impossible to describe unpolarized light in terms of amplitudes. Consequently, he abandoned the amplitude approach and sought a description based on the observed intensity.

To describe unpolarized light using (4-58), Stokes observed that unpolarized light had a very unique property, namely, its intensity was unaffected by (1) rotation of a linear polarizer (when a polarizer is used to analyze the state of polarization, it is called an analyzer) or (2) the presence of a retarder. Thus, for unpolarized light the only way the observed intensity $I(\theta, \phi)$ could be independent of θ, ϕ was for (4-58) to satisfy

$$I(\theta, \phi) = \frac{1}{2} S_0 \tag{4-61a}$$

and

$$S_1 = S_2 = S_3 = 0 \tag{4-61b}$$

Equations (4-61a) and (4-61b) are the mathematical statements for unpolarized light. Thus, Stokes had finally provided a correct mathematical statement. From a conceptual point of view S_1, S_2, and S_3 describe the polarizing behavior of the optical field. Since there is no polarization, (4-61a) and (4-61b) must be the correct mathematical statements for unpolarized light. Later, we shall show how (4-61) is used to formulate the interference laws of Fresnel and Arago.

In this way Stokes discovered an entirely different way to describe the polarization state of light. His formulation could be used to describe completely polarized light and completely unpolarized light as well. Furthermore, Stokes had been led to a formulation of the optical field in terms of measurable quantities (observables), the Stokes parameters. This was a unique point of view for nineteenth-century optical physics. The representation of radiation phenomena in terms of observables would not reappear again in physics until 1925 with the discovery of the laws of quantum mechanics by Werner Heisenberg.

The Stokes parameters described in (4-58) arise from an experimental configuration. Consequently, they were associated for a long time with the experimental measurement of the polarization of the optical field. Thus, a study of classical optics shows that polarization was conceptually understood with the nonobservable polarization ellipse, whereas the measurement was made in terms of intensities, the Stokes parameters. In other words, there were two distinct ways to describe the polarization of the optical field.

We have seen, however, that the Stokes parameters are actually a consequence of the wave theory and arise naturally from the polarization ellipse. It is only necessary to transform the nonobservable polarization ellipse to the observed intensity domain, whereupon we are led directly to the Stokes parameters. Thus, the Stokes polarization parameters must be considered as part of the conceptual foundations of the wave theory.

For a completely polarized beam of light we saw that

$$S_0^2 = S_1^2 + S_2^2 + S_3^2 \tag{4-10}$$

and we have just seen that for unpolarized light

$$S_0^2 > 0, \qquad S_1 = S_2 = S_3 = 0 \tag{4-62}$$

Equations (4-10) and (4-62) represent extreme states of polarization. Clearly, there must be an intermediate polarization state. This intermediate state is called partially polarized light. Thus, (4-10) can be used to describe all three polarization conditions by writing it as

$$S_0^2 \geq S_1^2 + S_2^2 + S_3^2 \tag{4-11}$$

For perfectly polarized light "$\geq$" is replaced by "$=$"; for unpolarized light "$\geq$" is replaced by "$>$" with $S_1 = S_2 = S_3 = 0$; and for partially polarized light "$\geq$" is replaced by "$>$."

An important quantity which describes these various polarization conditions is the degree of polarization P. This quantity can be expressed in terms of the

Stokes parameters. To derive P we decompose the optical field into unpolarized and polarized portions, which are mutually independent. Then, and this will be proved later, the Stokes parameters of a combination of independent waves are the sums of the respective Stokes parameters of the separate waves. The four Stokes parameters, S_0, S_1, S_2, and S_3 of the beam are represented by S. The total intensity of the beam is then S_0. We subtract the polarized intensity $(S_1^2 + S_2^2 + S_3^2)^{1/2}$ from the total intensity S_0 and we obtain the unpolarized intensity. Thus, we have

$$S^{(u)} = S_0 - \sqrt{S_1^2 + S_2^2 + S_3^2}, 0, 0, 0 \tag{4-63a}$$

and

$$S^{(p)} = \sqrt{S_1^2 + S_2^2 + S_3^2}, S_1, S_2, S_3 \tag{4-63b}$$

where $S^{(u)}$ represents the unpolarized part and $S^{(p)}$ represents the polarized part. The degree of polarization P is then defined to be

$$P = \frac{I_{\text{pol}}}{I_{\text{tot}}} = \frac{\sqrt{S_1^2 + S_2^2 + S_3^2}}{S_0} \qquad 0 \leq P \leq 1 \tag{4-64}$$

Thus, $P = 0$ indicates that the light is unpolarized, $P = 1$ that the light is (completely) polarized, and $0 < P < 1$ that the light is partially polarized.

The use of the Stokes parameters to describe polarized light rather than the amplitude formulation enables us to deal directly with the quantities measured in an optical experiment. Thus, we carry out the analysis in the amplitude domain and then transform the amplitude results to the Stokes parameters, using the defining equations. When this is done, we can easily relate the experimental results to the theoretical results. Furthermore, when we obtain the Stokes parameters, or rather the Stokes vector, we shall see that we are led to a description of radiation in which the Stokes parameters not only describe the measured quantities but can also be used to truly describe the observed spectral lines in a spectroscope. In other words, we shall arrive at observables in the strictest sense of the word.

4.6 ADDITIONAL PROPERTIES OF THE STOKES POLARIZATION PARAMETERS

Before we proceed to apply the Stokes parameters to a number of problems of interest, we wish to discuss a few of their additional properties. We saw earlier that the Stokes parameters could be used to describe any state of polarized light. In particular, we saw how unpolarized light and completely polarized light could both be written in terms of a Stokes vector. The question remains as to how we can represent partially polarized light in terms of the Stokes parameters and the Stokes vector. To answer this question, we must establish a fundamental property of the Stokes parameters, the property of additivity whereby the Stokes parameters of two completely independent beams can be added. This property is another way of describing the principle of incoherent superposition. We now prove this property of additivity.

We recall that the Stokes parameters for an optical beam can be represented in terms of complex amplitudes by

$$S_0 = E_x E_x^* + E_y E_y^* \tag{4-17a}$$

$$S_1 = E_x E_x^* - E_y E_y^* \tag{4-17b}$$

$$S_2 = E_x E_y^* + E_y E_x^* \tag{4-17c}$$

$$S_3 = i(E_x E_y^* - E_y E_x^*) \tag{4-17d}$$

Consider now that we have two optical beams each of which is characterized by its own set of Stokes parameters represented as $S^{(1)}$ and $S^{(2)}$:

$$S_0^{(1)} = E_{1x} E_{1x}^* + E_{1y} E_{1y}^* \tag{4-65a}$$

$$S_1^{(1)} = E_{1x} E_{1x}^* - E_{1y} E_{1y}^* \tag{4-65b}$$

$$S_2^{(1)} = E_{1x} E_{1y}^* + E_{1y} E_{1x}^* \tag{4-65c}$$

$$S_3^{(1)} = i(E_{1x} E_{1y}^* - E_{1y} E_{1x}^*) \tag{4-65d}$$

and

$$S_0^{(2)} = E_{2x} E_{2x}^* + E_{2y} E_{2y}^* \tag{4-66a}$$

$$S_1^{(2)} = E_{2x} E_{2x}^* - E_{2y} E_{2y}^* \tag{4-66b}$$

$$S_2^{(2)} = E_{2x} E_{2y}^* + E_{2y} E_{2x}^* \tag{4-66c}$$

$$S_3^{(2)} = i(E_{2x} E_{2y}^* - E_{2y} E_{2x}^*) \tag{4-66d}$$

The superscripts and subscripts 1 and 2 refer to the first and second beams, respectively. These two beams are now superposed. Then by the principle of superposition for amplitudes the total field in the x and y direction is

$$E_x = E_{1x} + E_{2x} \tag{4-67a}$$

$$E_y = E_{1y} + E_{2y} \tag{4-67b}$$

We now form products of (4-67a) and (4-67b) according to (4-17):

$$\begin{aligned} E_x E_x^* &= (E_{1x} + E_{2x})(E_{1x} + E_{2x})^* \\ &= E_{1x} E_{1x}^* + E_{1x} E_{2x}^* + E_{2x} E_{1x}^* + E_{2x} E_{2x}^* \end{aligned} \tag{4-68a}$$

$$\begin{aligned} E_y E_y^* &= (E_{1y} + E_{2y})(E_{1y} + E_{2y})^* \\ &= E_{1y} E_{ly}^* + E_{1y} E_{2y}^* + E_{2y} E_{1y}^* + E_{2y} E_{2y}^* \end{aligned} \tag{4-68b}$$

$$\begin{aligned} E_x E_y^* &= (E_{1x} + E_{2x})(E_{1y} + E_{2y})^* \\ &= E_{1x} E_{1y}^* + E_{1x} E_{2y}^* + E_{2x} E_{1y}^* + E_{2x} E_{2y}^* \end{aligned} \tag{4-68c}$$

$$\begin{aligned} E_y E_x^* &= (E_{1y} + E_{2y})(E_{1x} + E_{2x})^* \\ &= E_{1y} E_{1x}^* + E_{2y} E_{1x}^* + E_{1y} E_{2x}^* + E_{2y} E_{2x}^* \end{aligned} \tag{4-68d}$$

Let us now assume that the two beams are completely independent of each other with respect to their amplitudes and phase. We describe the degree of independence by writing an overbar which signifies a time average over the product of E_x and E_y, that is, $\overline{E_x E_x^*}$, $\overline{E_y E_y^*}$, etc., so

$$\overline{E_i E_j^*} \quad i, j = x, y \tag{4-69}$$

Since the two beams are completely independent, we express this behavior by

$$\overline{E_{1i} E_{2j}^*} = \overline{E_{2i} E_{1j}^*} = 0 \qquad i \neq j \tag{4-70a}$$

$$\overline{E_{1i} E_{1j}^*} \neq 0 \qquad i, j = x, y \tag{4-70b}$$

$$\overline{E_{2i} E_{2j}^*} \neq 0 \qquad i, j = x, y \tag{4-70c}$$

The value of zero in (4-70a) indicates complete independence. On the other hand, the nonzero value in (4-70b) and (4-70c) means that there is some degree of dependence. Operating on (4-68a) through (4-68b) with an overbar and using the conditions expressed by (4-70), we find that

$$\overline{E_x E_x^*} = \overline{E_{1x} E_{1x}^*} + \overline{E_{2x} E_{2x}^*} \tag{4-71a}$$

$$\overline{E_y E_y^*} = \overline{E_{1y} E_{1y}^*} + \overline{E_{2y} E_{2y}^*} \tag{4-71b}$$

$$\overline{E_x E_y^*} = \overline{E_{1x} E_{1y}^*} + \overline{E_{2x} E_{2y}^*} \tag{4-71c}$$

$$\overline{E_y E_x^*} = \overline{E_{1y} E_{1x}^*} + \overline{E_{2y} E_{2x}^*} \tag{4-71d}$$

We now form the Stokes parameters according to (4-17), drop the overbar because the noncorrelated terms have been eliminated, and group terms. The result is

$$S_0 = E_x E_x^* + E_y E_y^* = (E_{1x} E_{1x}^* + E_{1y} E_{1y}^*) + (E_{2x} E_{2x}^* + E_{2y} E_{2y}^*) \tag{4-72a}$$

$$S_1 = E_x E_x^* - E_y E_y^* = (E_{1x} E_{1x}^* - E_{1y} E_{1y}^*) + (E_{2x} E_{2x}^* - E_{2y} E_{2y}^*) \tag{4-72b}$$

$$S_2 = E_x E_y^* + E_y E_x^* = (E_{1x} E_{1y}^* + E_{1y} E_{1x}^*) + (E_{2x} E_{2y}^* + E_{2y} E_{2x}^*) \tag{4-72c}$$

$$S_3 = i(E_x E_y^* - E_y E_x^*) = i(E_{1x} E_{1y}^* - E_{1y} E_{1x}^*) + i(E_{2x} E_{2y}^* - E_{2y} E_{2x}^*) \tag{4-72d}$$

From (4-65) and (4-66) we see that we can then write (4-72) as

$$S_0 = S_0^{(1)} + S_0^{(2)} \tag{4-73a}$$

$$S_1 = S_1^{(1)} + S_1^{(2)} \tag{4-73b}$$

$$S_2 = S_2^{(1)} + S_2^{(2)} \tag{4-73c}$$

$$S_3 = S_3^{(1)} + S_3^{(2)} \tag{4-73d}$$

Thus, the Stokes parameters of two completely independent optical beams can be added and represented by the Stokes parameters of the combined beams. We can write (4-73) in terms of Stokes vectors, i.e.,

$$\begin{pmatrix} S_0 \\ S_1 \\ S_2 \\ S_3 \end{pmatrix} = \begin{pmatrix} S_0^{(1)} \\ S_1^{(1)} \\ S_2^{(1)} \\ S_3^{(1)} \end{pmatrix} + \begin{pmatrix} S_0^{(2)} \\ S_1^{(2)} \\ S_2^{(2)} \\ S_3^{(2)} \end{pmatrix} \tag{4-74}$$

or simply

$$S = S^{(1)} + S^{(2)} \tag{4-75}$$

so the Stokes vectors, $S^{(i)}, i = 1, 2$, are also additive.

As a first application of this result, (4-74), we recall that the Stokes vector for unpolarized light is

$$S = I_0 \begin{pmatrix} 1 \\ 0 \\ 0 \\ 0 \end{pmatrix} \tag{4-76}$$

We also saw that the Stokes vector could be written in terms of the orientation angle ψ and the ellipticity χ as

$$S = I_0 \begin{pmatrix} 1 \\ \cos 2\chi \cos 2\psi \\ \cos 2\chi \sin 2\psi \\ \sin 2\chi \end{pmatrix} \tag{4-47}$$

Thus, for a beam of light (which may be a result of combining two beams), we see from (4-74) that we can write (4-76), using (4-47), as

$$I_0 \begin{pmatrix} 1 \\ 0 \\ 0 \\ 0 \end{pmatrix} = \frac{I_0}{2} \begin{pmatrix} 1 \\ \cos 2\chi \cos 2\psi \\ \cos 2\chi \sin 2\psi \\ \sin 2\chi \end{pmatrix} + \frac{I_0}{2} \begin{pmatrix} 1 \\ -\cos 2\chi \cos 2\psi \\ -\cos 2\chi \sin 2\psi \\ -\sin 2\chi \end{pmatrix} \tag{4-77}$$

We can also express (4-74) in terms of two beams of equal intensity $I_0/2$ using the form in (4-47) as

$$I_0 \begin{pmatrix} 1 \\ 0 \\ 0 \\ 0 \end{pmatrix} = \frac{I_0}{2} \begin{pmatrix} 1 \\ \cos 2\chi_1 \cos 2\psi_1 \\ \cos 2\chi_1 \sin 2\psi_1 \\ \sin 2\chi_1 \end{pmatrix} + \frac{I_0}{2} \begin{pmatrix} 1 \\ \cos 2\chi_2 \cos 2\psi_2 \\ \cos 2\chi_2 \sin 2\psi_2 \\ \sin 2\chi_2 \end{pmatrix} \tag{4-78}$$

Comparing the Stokes parameters in the second column in (4-78) with (4-77), we see that

$$\cos 2\chi_2 \cos 2\psi_2 = -\cos 2\chi_1 \cos 2\psi_1 \tag{4-79a}$$

$$\cos 2\chi_2 \sin 2\psi_2 = -\cos 2\chi_1 \sin 2\psi_1 \tag{4-79b}$$

$$\sin 2\chi_2 = -\sin 2\chi_1 \tag{4-79c}$$

Equation (4-79c) is only true if

$$\chi_2 = -\chi_1 \tag{4-80}$$

Thus, the ellipticity of beam 2 is the negative of that of beam 1. We now substitute (4-80) into (4-79a) and (4-79b) and we have

$$\cos 2\psi_2 = -\cos 2\psi_1 \tag{4-81a}$$

$$\sin 2\psi_2 = -\sin 2\psi_1 \tag{4-81b}$$

Equations (4-81a) and (4-81b) can only be satisfied if

$$2\psi_1 = 2\psi_2 \pm \pi \tag{4-82a}$$

or

$$\psi_2 = \psi_1 \pm \frac{\pi}{2} \tag{4-82b}$$

Thus, the polarization ellipse for the second beam is oriented 90° ($\pi/2$) from the first beam. The conditions

$$\chi_2 = -\chi_1 \tag{4-80b}$$

$$\psi_2 = \psi_1 \pm \frac{\pi}{2} \tag{4-82b}$$

are said to describe two polarization ellipses of orthogonal polarization. Thus, unpolarized light is a superposition or mixture of two beams of equal intensity and orthogonal polarization. As special cases of (4-77) we see that unpolarized light can be decomposed into (independent) beams of linear and circular polarized light; that is,

$$I_0\begin{pmatrix}1\\0\\0\\0\end{pmatrix} = \frac{I_0}{2}\begin{pmatrix}1\\1\\0\\0\end{pmatrix} + \frac{I_0}{2}\begin{pmatrix}1\\-1\\0\\0\end{pmatrix} \tag{4-83a}$$

$$I_0\begin{pmatrix}1\\0\\0\\0\end{pmatrix} = \frac{I_0}{2}\begin{pmatrix}1\\0\\1\\0\end{pmatrix} + \frac{I_0}{2}\begin{pmatrix}1\\0\\-1\\0\end{pmatrix} \tag{4-83b}$$

$$I_0\begin{pmatrix}1\\0\\0\\0\end{pmatrix}=\frac{I_0}{2}\begin{pmatrix}1\\0\\0\\1\end{pmatrix}+\frac{I_0}{2}\begin{pmatrix}1\\0\\0\\-1\end{pmatrix} \tag{4-83c}$$

Of course, the intensity of each beam is half the intensity of the unpolarized beam.

We now return to our original problem of representing partially polarized light in terms of the Stokes vector. Recall that the degree of polarization P is defined by

$$P=\frac{\sqrt{S_1^2+S_2^2+S_3^2}}{S_0}\qquad 0\le P\le 1 \tag{4-84}$$

This equation suggests that partially polarized light can be represented by a superposition of unpolarized light and completely polarized light by using (4-74). A little thought shows that if we have a beam of partially polarized light, which we can write as

$$S=\begin{pmatrix}S_0\\S_1\\S_2\\S_3\end{pmatrix} \tag{4-85}$$

Equation (4-85) can be written as

$$S=\begin{pmatrix}S_0\\S_1\\S_2\\S_3\end{pmatrix}=(1-P)\begin{pmatrix}S_0\\0\\0\\0\end{pmatrix}+P\begin{pmatrix}S_0\\S_1\\S_2\\S_3\end{pmatrix}\qquad 0\le P\le 1 \tag{4-86}$$

The first Stokes vector on the right-hand side of (4-86) represents unpolarized light, and the second Stokes vector represents completely polarized light. For $P=0$, unpolarized light, (4-86) reduces to

$$S=\begin{pmatrix}S_0\\0\\0\\0\end{pmatrix} \tag{4-87a}$$

and for $P=1$, completely polarized light, (4-86) reduces to

$$S=\begin{pmatrix}S_0\\S_1\\S_2\\S_3\end{pmatrix} \tag{4-87b}$$

We note that S_0 on the left-hand side of (4-86) always satifies

$$S_0 \geq \sqrt{S_1^2 + S_2^2 + S_3^2} \tag{4-88a}$$

whereas S_0 in the Stokes vector associated with P on the right-hand side of (4-86) always satisfies

$$S_0 = \sqrt{S_1^2 + S_2^2 + S_3^2} \tag{4-88b}$$

Another representation of partially polarized light in terms of P is the decomposition of a beam into two completely polarized beams of orthogonal polarizations, namely,

$$\begin{pmatrix} S_0 \\ S_1 \\ S_2 \\ S_3 \end{pmatrix} = \frac{1+P}{2P}\begin{pmatrix} PS_0 \\ S_1 \\ S_2 \\ S_3 \end{pmatrix} + \frac{1-P}{2P}\begin{pmatrix} PS_0 \\ -S_1 \\ -S_2 \\ -S_3 \end{pmatrix} \qquad 0 < P \leq 1 \tag{4-89a}$$

where

$$PS_0 = \sqrt{S_1^2 + S_2^2 + S_3^2} \tag{4-89b}$$

Thus, partially polarized light can also be decomposed into two orthogonally polarized beams.

While we have restricted this discussion to two beams, it is easy to see that we could have described the optical field in terms of n beams, that is, extended (4-75) to

$$\begin{aligned} S &= S^{(1)} + S^{(2)} + S^{(3)} + \cdots + S^{(n)} \\ &= \sum_{i=1}^{n} S^{(i)} \qquad i = 1, \ldots, n \end{aligned} \tag{4-90}$$

We have not done this for the simple reason that, in practice, dealing with two beams is sufficient. Nevertheless, the reader should be aware that the additivity law can be extended to n beams. Lastly, we note that for partially polarized light the intensities of the two beams are given by

$$S_0^{(1)} = \frac{1}{2}S_0 + \frac{1}{2}\sqrt{S_1^2 + S_2^2 + S_3^2} \tag{4-91a}$$

$$S_0^{(2)} = \frac{1}{2}S_0 - \frac{1}{2}\sqrt{S_1^2 + S_2^2 + S_3^2} \tag{4-91b}$$

Only for unpolarized light are the intensities of the two beams equal. This is also shown by (4-89a).

It is of interest to express the parameters of the polarization ellipse in terms of the Stokes parameters. To do this, we recall that

$$S_0 = E_{0x}^2 + E_{0y}^2 = I_0 \tag{4-92a}$$

$$S_1 = E_{0x}^2 - E_{0y}^2 = I_0 \cos 2\alpha \tag{4-92b}$$

$$S_2 = 2E_{0x}E_{0y} \cos \delta = I_0 \sin 2\alpha \cos \delta \tag{4-92c}$$

$$S_3 = 2E_{0x}E_{0y} \sin \delta = I_0 \sin 2\alpha \sin \delta \tag{4-92d}$$

We can then write (4-92) as

$$E_{0x}^2 = \frac{S_0 + S_1}{2} \tag{4-93a}$$

$$E_{0y}^2 = \frac{S_0 - S_1}{2} \tag{4-93b}$$

$$\cos\delta = \frac{S_2}{2E_{0x}E_{0y}} \tag{4-93c}$$

$$\sin\delta = \frac{S_3}{2E_{0x}E_{0y}} \tag{4-93d}$$

We recall that the instantaneous polarization ellipse is

$$\frac{E_x^2}{E_{0x}^2} + \frac{E_y^2}{E_{0y}^2} - \frac{2E_xE_y}{E_{0x}E_{0y}}\cos\delta = \sin^2\delta \tag{4-94}$$

Substituting (4-93) into the appropriate terms in (4-94) gives

$$\frac{2E_x^2}{S_0 + S_1} + \frac{2E_y^2}{S_0 - S_1} - \frac{4S_2E_xE_y}{S_0^2 - S_1^2} = \frac{S_3^2}{S_0^2 - S_1^2} \tag{4-95}$$

where we have used $E_{0x}^2E_{0y}^2 = (S_0^2 - S_1^2)/4$ from (4-93a) and (4-93b). Multiplying through (4-95) by $(S_0^2 - S_1^2)/S_3^2$ then yields

$$\frac{2(S_0 - S_1)E_x^2}{S_3^2} + \frac{2(S_0 + S_1)E_y^2}{S_3^2} - \frac{4S_2E_xE_y}{S_3^2} = 1 \tag{4-96}$$

We now write (4-96) as

$$Ax^2 - 2Cxy + 2By^2 = 1 \tag{4-97a}$$

where

$$A = \frac{2(S_0 - S_1)}{S_3^2} \tag{4-97b}$$

$$B = \frac{2(S_0 + S_1)}{S_3^2} \tag{4-97c}$$

$$C = \frac{2S_2}{S_3^2} \tag{4-97d}$$

and for convenience we have set $x = E_x$ and $y = E_y$.

We can now find the orientation and ratio of the axes in terms of the Stokes parameters (4-97). To do this we first express x and y in polar coordinates:

$$x = \rho\cos\phi \tag{4-98a}$$

$$y = \rho\sin\phi \tag{4-98b}$$

Substituting (4-98a) and (4-98b) into (4-97) we find

$$A\rho^2 \cos^2 \phi - 2C\rho^2 \sin \phi \cos \phi + B\rho^2 \sin^2 \phi = 1 \tag{4-99}$$

Using the half-angle formulas for $\cos^2 \phi$ and $\sin^2 \phi$, (4-99) then becomes

$$\frac{A\rho^2(1 + \cos 2\phi)}{2} - C\rho^2 \sin 2\phi + \frac{B\rho^2(1 - \cos 2\phi)}{2} = 1 \tag{4-100}$$

We now introduce the parameter L defined in terms of ρ as

$$L = \frac{2}{\rho^2} \tag{4-101}$$

substitute (4-101) into (4-100), and write

$$L = (A + B) - 2C \sin 2\phi + (A - B) \cos 2\phi \tag{4-102}$$

The major and minor axes of the ellipse correspond to maximum and minimum values of ρ, respectively, whereas L is a minimum and maximum. The angle ϕ where this maximum and minimum occur can be found in the usual way by setting $dL/d\phi = 0$ and solving for ϕ. We, therefore, have

$$\frac{dL}{d\phi} = -4C \cos 2\phi - 2(A - B) \sin 2\phi = 0 \tag{4-103}$$

and

$$\frac{\sin 2\phi}{\cos 2\phi} = \tan 2\phi = \frac{-2C}{A - B} \tag{4-104}$$

Solving for ϕ, we find that

$$\phi = \frac{-1}{2} \tan^{-1} \frac{2C}{A - B} \tag{4-105}$$

To find the corresponding maximum and minimum values of L in (4-102), we must express $\sin 2\phi$ and $\cos 2\phi$ in terms of A, B, and C. We can find unique expressions for $\sin 2\phi$ and $\cos 2\phi$ from (4-104) by constructing the right triangle in Fig. 4-5. We see from the right triangle that (4-104) is satisfied by

$$\sin 2\phi = \frac{-2C}{\sqrt{(-2C)^2 + (A - B)^2}} \tag{4-106a}$$

$$\cos 2\phi = \frac{A - B}{\sqrt{(-2C)^2 + (A - B)^2}} \tag{4-106b}$$

or

$$\sin 2\phi = \frac{2C}{\sqrt{(2C)^2 + (-(A - B))^2}} \tag{4-106c}$$

$$\cos 2\phi = \frac{-(A - B)}{\sqrt{(2C)^2 + (-(A - B))^2}} \tag{4-106d}$$

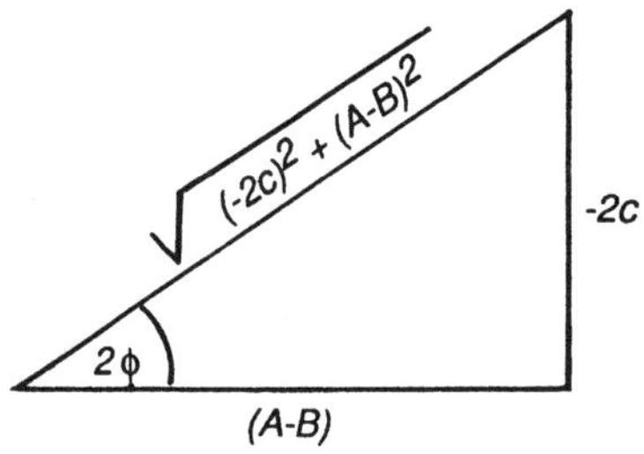

Figure 4-5 Right triangle corresponding to Eq. (4-104).

Substituting (4-106a) and (4-106b) into (4-102) yields

$$L_{\max} = (A + B) + \sqrt{(-2C)^2 + (A - B)^2} \tag{4-107a}$$

and, similarly, substituting (4-106c) and (4-106d) into (4-102) yields

$$L_{\min} = (A + B) - \sqrt{(2C)^2 + (-(A - B))^2} \tag{4-107b}$$

We have written "max" and "min" on L in (4-107a) and (4-107b) to indicate that these are the maximum and minimum values of L. We also note that (4-106a) and (4-106c) are related by

$$\sin 2\phi_1 = -\sin 2\phi_2 \tag{4-108a}$$

and (4-106b) and (4-106d) by

$$\cos 2\phi_1 = -\cos 2\phi_2 \tag{4-108b}$$

We see that (4-106a) and (4-106b) are satisfied by setting

$$\phi_2 = \phi_1 + \frac{\pi}{2} \tag{4-109}$$

Thus, the maximum and minimum lengths, that is, the major and minor axes, are at ϕ_1 and $\phi_1 + 90^\circ$, respectively, which is exactly what we would expect. We thus see from (4-101) that

$$\rho^2_{\min} = \frac{2}{L_{\max}} \tag{4-110a}$$

$$\rho^2_{\max} = \frac{2}{L_{\min}} \tag{4-110b}$$

The ratio of the square of the lengths of the major axis to the minor axis is defined to be

$$R = \frac{\rho^2_{\max}}{\rho^2_{\min}} \tag{4-111}$$

so from (4-107a) and (4-107b) we have

$$R = \frac{(A+B) - \sqrt{(2C)^2 + (A-B)^2}}{(A+B) + \sqrt{(2C^2 + (A-B)^2}} \tag{4-112}$$

We can now express (4-112) in terms of the Stokes parameters from (4-97b), (4-97c) and (4-97d) and we find that (4-112) becomes

$$R = \frac{S_0 - \sqrt{S_1^2 + S_2^2}}{S_0 + \sqrt{S_1^2 + S_2^2}} \tag{4-113}$$

Thus, we have found the relation between the length of the major and minor axes of the polarization ellipse and the Stokes parameters. This can be expressed directly by using (4-110) and (4-97) or as a ratio R given by (4-113).

Not surprisingly there are other interesting relations between the Stokes parameters and the parameters of the polarization ellipse. These relations are fundamental to the development of the Poincaré sphere, so we shall discuss them in Chapter 12.

4.7 STOKES PARAMETERS AND WOLF'S COHERENCY MATRIX

We have demonstrated that the state of polarization is specified completely by the four Stokes parameters S_0, S_1, S_2, and S_3. There is another representation in which the polarization is described by a 2×2 matrix known as Wolf's coherency matrix. Furthermore, there is a direct relationship between the elements of the coherency matrix and the Stokes parameters. This relationship, as well as the required mathematical background, is thoroughly discussed in the text by Born and Wolf. For the sake of completeness, however, we briefly discuss the coherency matrix as it relates to the Stokes parameters.

Consider an optical field consisting of the components:

$$E_x(t) = E_{0x}(t)e^{-i(\omega t + \delta_x)} \tag{4-114a}$$

$$E_y(t) = E_{0y}(t)e^{-i(\omega t + \delta_y)} \tag{4-114b}$$

If we take the real part of these expressions, i.e., let

$$E_x(t) = \mathrm{Re}[E_{0x}(t)e^{-i(\omega t + \delta_x)}] \tag{4-115a}$$

$$E_y(t) = \mathrm{Re}[E_{0y}(t)e^{-i(\omega t + \delta_y)}] \tag{4-115b}$$

then these are equivalent to (4-1).

The element J_{ij} of the coherency matrix J are defined to be

$$J_{ij} = \langle E_i E_j^* \rangle = \lim_{T\to\infty} \frac{1}{2T} \int_{-T}^{T} E_i E_j^* \, dt \qquad (i, j = x, y) \tag{4-116}$$

It follows that

$$J_{xy} = J_{yx}^* \tag{4-117}$$

and so the coherency matrix in Hermitian. The coherency matrix is defined to be the array:

$$\boldsymbol{J} = \begin{pmatrix} J_{xx} & J_{xy} \\ J_{yx} & J_{yy} \end{pmatrix} = \begin{pmatrix} \langle E_x E_x^* \rangle & \langle E_x E_y^* \rangle \\ \langle E_y E_x^* \rangle & \langle E_y E_y^* \rangle \end{pmatrix} \tag{4-118}$$

The trace of this matrix, i.e.,

$$\mathrm{Tr}\boldsymbol{J} = J_{xx} + J_{yy} = \langle E_x E_x^* \rangle + \langle E_y E_y^* \rangle \tag{4-119}$$

is equal to the total intensity of the light.

There is a direct connection between the Stokes parameters and the elements of the coherency matrix. The Stokes parameters for a quasi-monochromatic wave are defined to be [see (4-17)]

$$S_0 = \langle E_x E_x^* \rangle + \langle E_y E_y^* \rangle \tag{4-120a}$$

$$S_1 = \langle E_x E_x^* \rangle - \langle E_y E_y^* \rangle \tag{4-120b}$$

$$S_2 = \langle E_x E_y^* \rangle + \langle E_y E_x^* \rangle \tag{4-120c}$$

$$S_3 = i(\langle E_x E_y^* \rangle - \langle E_y E_x^* \rangle) \tag{4-120d}$$

where the angular brackets are the time averages. We see immediately from (4-117) and (4-120) that

$$S_0 = J_{xx} + J_{yy} \tag{4-121a}$$

$$S_1 = J_{xx} - J_{yy} \tag{4-121b}$$

$$S_2 = J_{xy} + J_{yx} \tag{4-121c}$$

$$S_3 = i(J_{xy} - J_{yx}) \tag{4-121d}$$

Equations (4-121) show that the Stokes parameters and the elements of the coherency matrix are linearly related. A specification of the wave in terms of the coherency matrix is in all respects equivalent to its specification in terms of the Stokes parameters.

There is a very simple way of describing the degree of polarization using the coherency matrix. From Schwarz's inequality we have

$$\int A_i A_i^*\, dt \int A_j A_j^*\, dt \geq \int A_i A_j^*\, dt \int A_i A_j^*\, dt \qquad i,j = x,y \tag{4-122}$$

From the definition given by (4-117) it follows that

$$J_{xx} J_{yy} \geq J_{xy} J_{yx} \tag{4-123}$$

or, using (4-118),

$$J_{xx} J_{yy} - J_{yx} J_{xy} \geq 0 \tag{4-124}$$

The equality sign clearly refers to completely polarized light, and the $>$ sign to partially polarized light. Furthermore, we see from (4-119) that (4-124) is the determinant of (4-119) so

$$\det \boldsymbol{J} = 0 \qquad \text{complete polarization} \tag{4-125a}$$

$$\det \boldsymbol{J} > 0 \qquad \text{partial polarization} \tag{4-125b}$$

One can readily determine the coherency matrices for various states of polarized light, using (4-121). We easily find for unpolarized light that

$$\boldsymbol{J} = \frac{S_0}{2}\begin{pmatrix} 1 & 0 \\ 0 & 1 \end{pmatrix} \tag{4-126a}$$

for linearly horizontally polarized light

$$\boldsymbol{J} = S_0\begin{pmatrix} 1 & 0 \\ 0 & 0 \end{pmatrix} \tag{4-126b}$$

and for right circularly polarized light

$$\boldsymbol{J} = \frac{S_0}{2}\begin{pmatrix} 1 & -i \\ i & 1 \end{pmatrix} \tag{4-126c}$$

The degree of polarization is readily found to be

$$P = \sqrt{1 - \frac{4 \det \boldsymbol{J}}{(\mathrm{Tr}\, \boldsymbol{J})^2}} \tag{4-127}$$

where $\mathrm{Tr}\, \boldsymbol{J}$ is the trace of the matrix $\boldsymbol{J}$ and is defined as the sum of the diagonal elements; that is

$$\mathrm{Tr}\, \boldsymbol{J} = J_{xx} + J_{yy} \tag{4-128}$$

The coherency matrix elements can also be introduced by considering the measurement of the polarization state of an optical beam. We recall that the intensity of a beam emerging from a retarder/polarizer combination is

$$\begin{aligned} I(\theta, \phi) = {} & E_x E_x^* \cos^2\theta + E_y E_y^* \sin^2\theta \\ & + E_x^* E_y e^{-i\phi} \sin\theta\cos\theta + E_x E_y^* e^{i\phi} \sin\theta\cos\theta \end{aligned} \tag{4-55}$$

The Stokes parameters were then found by expressing the sinusoidal terms in terms of the half-angle trigonometric formulas. If we had a quasi-monochromatic wave, then we could time-average the quadratic field terms and express (4-55) as

$$\begin{aligned} I(\theta, \phi) = {} & \langle E_x E_x^* \rangle \cos^2\theta + \langle E_y E_y^* \rangle \sin^2\theta \\ & + \langle E_x^* E_y \rangle e^{-i\phi} \sin\theta\cos\theta + \langle E_x E_y^* \rangle e^{i\phi} \sin\theta\cos\theta \end{aligned} \tag{4-129a}$$

or

$$\begin{aligned} I(\theta, \phi) = {} & J_{xx} \cos^2\theta + J_{yy} \sin^2\theta \\ & + J_{xy} e^{-i\phi} \sin\theta\cos\theta + J_{yx} e^{i\phi} \sin\theta\cos\theta \end{aligned} \tag{4-129b}$$

where the J_{ij} are defined to be

$$J_{ij} = \langle E_i E_j^* \rangle \tag{4-129c}$$

which are the coherency matrix elements.

Finally, there is a remarkable relation between the Stokes parameters and the coherency matrix. We first note from (4-121) that

$$J_{xx} = \frac{S_0 + S_1}{2} \tag{4-130a}$$

$$J_{yy} = \frac{S_0 - S_1}{2} \tag{4-130b}$$

$$J_{xy} = \frac{S_2 - iS_3}{2} \tag{4-130c}$$

$$J_{yx} = \frac{S_2 + iS_3}{2} \tag{4-130d}$$

so we can express (4-130) in matrix form as

$$\boldsymbol{J} = \begin{pmatrix} J_{xx} & J_{xy} \\ J_{yx} & J_{yy} \end{pmatrix} = \frac{1}{2}\begin{pmatrix} S_0 + S_1 & S_2 - iS_3 \\ S_2 + iS_3 & S_0 - S_1 \end{pmatrix} \tag{4-131}$$

One can easily decompose (4-131) into 2 × 2 matrices such that

$$\boldsymbol{J} = \frac{1}{2}\sum_{i=0}^{3} \sigma_i S_i \tag{4-132a}$$

where

$$\sigma_0 = \begin{pmatrix} 1 & 0 \\ 0 & 1 \end{pmatrix} \tag{4-132b}$$

$$\sigma_1 = \begin{pmatrix} 0 & 1 \\ 1 & 0 \end{pmatrix} \tag{4-132c}$$

$$\sigma_2 = \begin{pmatrix} 0 & -i \\ i & 0 \end{pmatrix} \tag{4-132d}$$

$$\sigma_3 = \begin{pmatrix} 1 & 0 \\ 0 & -1 \end{pmatrix} \tag{4-132e}$$

The remarkable fact about this decomposition is that σ_1, σ_2, the σ_3 are the three Pauli spin matrices of quantum mechanics with the addition of the identity matrix, σ_0. This connection between the coherency matrix, the Stokes parameters, and the Pauli spin matrices appears to have been first pointed out by U. Fano in 1954. What is even more surprising about the appearance of the Pauli spin matrices is that they were introduced into quantum mechanics by Pauli in order to describe the behavior of the spin of the electron, a particle. Indeed, in quantum mechanics the wave function that describes a pure state of polarization can be expanded in a complete set of orthonormal eigenfunctions; it has the same form for electromagnetic radiation and particles of spin 1/2 (the electron).

The coherency matrix is treated in an elegant manner by Born and Wolf and the reader is referred to their text for further information on this subject.

REFERENCES

Papers

1. Stokes, G. G., *Trans. Camb. Phil. Soc.*, *9*, 399 (1852). Reprinted in *Mathematical and Physical Papers*, Vol. 3, Cambridge University Press, London, 1901, p. 233.
2. Wolf, E., *Nuovo Cimento*, *12*, 884 (1954).
3. Walker, M. J., *Am. J. Phys.*, *22*, 170 (1954).
4. McMaster, W. H., *Rev. Mod. Phys.*, *33*, 1 (1961).
5. Collett, E., *Am. J. Phys.*, *36*, 713 (1968).
6. Fano, U., *Phys. Rev.*, *93*, 121 (1954).
7. Wolf, E., *Nuovo Cimento*, *13*, 1165 (1959).

Books

1. Born, M. and Wolf, E., *Principles of Optics*, 3rd ed., Pergamon Press, New York, 1965.
2. Shurcliff, W., *Polarized Light*, Harvard University Press, Cambridge, MA, 1962.

5

The Mueller Matrices for Polarizing Components

5.1 INTRODUCTION

In the previous chapters we have concerned ourselves with the fundamental properties of polarized light. In this chapter we now turn our attention to the study of the interaction of polarized light with elements which can change its state of polarization and see that the matrix representation of the Stokes parameters leads to a very powerful mathematical tool for treating this interaction. In Fig. 5-1 we show an incident beam interacting with a polarizing element and the emerging beam. In Fig. 5-1 the incident beam is characterized by its Stokes parameters S_i, where $i = 0, 1, 2, 3$. The incident polarized beam interacts with the polarizing medium, and the emerging beam is characterized by a new set of Stokes parameters S'_1, where, again, $i = 0, 1, 2, 3$. We now assume that S'_1 can be expressed as a linear combination of the four Stokes parameters of the incident beam by the relations:

$$S'_0 = m_{00}S_0 + m_{01}S_1 + m_{02}S_2 + m_{03}S_3 \tag{5-1a}$$

$$S'_1 = m_{10}S_0 + m_{11}S_1 + m_{12}S_2 + m_{13}S_3 \tag{5-1b}$$

$$S'_2 = m_{20}S_0 + m_{21}S_1 + m_{22}S_2 + m_{23}S_3 \tag{5-1c}$$

$$S'_3 = m_{30}S_0 + m_{31}S_1 + m_{32}S_2 + m_{33}S_3 \tag{5-1d}$$

In matrix form (5-1) is written as

$$\begin{pmatrix} S'_0 \\ S'_1 \\ S'_2 \\ S'_3 \end{pmatrix} = \begin{pmatrix} m_{00} & m_{01} & m_{02} & m_{03} \\ m_{10} & m_{11} & m_{12} & m_{13} \\ m_{20} & m_{21} & m_{22} & m_{23} \\ m_{30} & m_{31} & m_{32} & m_{33} \end{pmatrix} \begin{pmatrix} S_0 \\ S_1 \\ S_2 \\ S_3 \end{pmatrix} \tag{5-2}$$

or

$$S' = M \cdot S \tag{5-3}$$

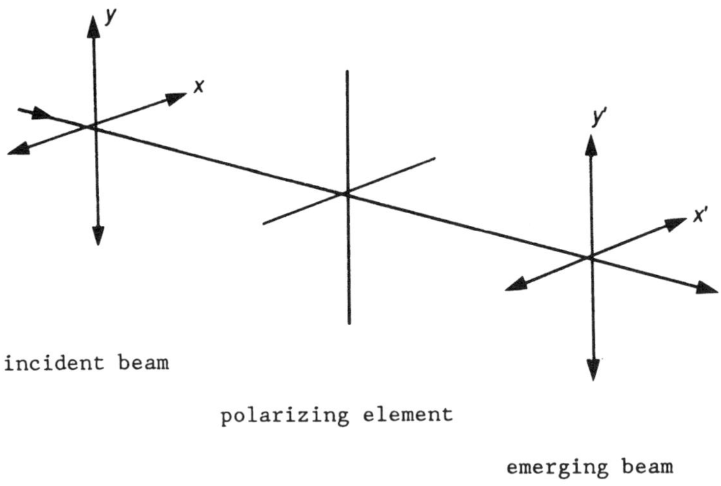

Figure 5-1 Interaction of a polarized beam with a polarizing element.

where S and S' are the Stokes vectors and M is the 4×4 matrix known as the Mueller matrix. It was introduced by Hans Mueller during the early 1940s. While Mueller appears to have based his 4×4 matrix on a paper by F. Perrin and a still earlier paper by P. Soleillet, his name is firmly attached to it in the optical literature. Mueller's important contribution was that he, apparently, was the first to describe polarizing components in terms of his Mueller matrices. Remarkably, Mueller never published his work on his matrices. Their appearance in the optical literature was due to others, such as N.G. Park III, a graduate student of Mueller's who published Mueller's ideas along with his own contributions and others shortly after the end of the Second World War.

When an optical beam interacts with matter its polarization state is almost always changed. In fact, this appears to be the rule rather than the exception. The polarization state can be changed by (1) changing the amplitudes, (2) changing the phase, (3) changing the direction of the orthogonal field components, or (4) transferring energy from polarized states to the unpolarized state. An optical element that changes the orthogonal amplitudes unequally is called a *polarizer* or *diattenuator*. Similarly, an optical device that introduces a phase shift between the orthogonal components is called a *retarder*; other names used for the same device are wave plate, compensator, or phase shifter. If the optical device rotates the orthogonal components of the beam through an angle θ as it propagates through the element, it is called a *rotator*. Finally, if energy in polarized states goes to the unpolarized state, the element is a *depolarizer*. These effects are easily understood by writing the transverse field components for a plane wave:

$$E_x(z,t) = E_{0x}\cos(\omega t - \kappa z + \delta_x) \tag{5-4a}$$

$$E_y(z,t) = E_{0y}\cos(\omega t - \kappa z + \delta_y) \tag{5-4b}$$

Equation (4) can be changed by varying the amplitudes, E_{0x} or E_{0y}, or the phase, δ_x or δ_y and, finally, the direction of $E_x(z,t)$ and $E_y(z,t)$. The corresponding devices for causing these changes are the polarizer, retarder, and rotator. The use of the names *polarizer* and *retarder* arose, historically, before the behavior of these polarizing elements was fully understood. The preferable names would be diattenuator for a polarizer and phase shifter for the retarder. All three polarizing elements, polarizer, retarder, and rotator, change the polarization state of an optical beam.

In the following sections we derive the Mueller matrices for these polarizing elements. We then apply the Mueller matrix formalism to a number of problems of interest and see its great utility.

5.2 THE MUELLER MATRIX OF A POLARIZER

A polarizer is an optical element that attenuates the orthogonal components of an optical beam unequally; that is, a polarizer is an anisotropic attenuator; the two orthogonal transmission axes are designated p_x and p_y. Recently, it has also been called a diattenuator, a more accurate and descriptive term. A polarizer is sometimes described also by the terms generator and analyzer to refer to its use and position in the optical system. If a polarizer is used to create polarized light, we call it a generator. If it is used to analyze polarized light, it is called an analyzer. If the orthogonal components of the incident beam are attenuated equally, then the polarizer becomes a neutral density filter. We now derive the Mueller matrix for a polarizer.

In Fig. 5-2 a polarized beam is shown incident on a polarizer along with the emerging beam. The components of the incident beam are represented by E_x and E_y. After the beam emerges from the polarizer the components are E'_x and E'_y, and they are parallel to the original axes. The fields are related by

$$E'_x = p_x E_x \qquad 0 \le p_x \le 1 \tag{5-5a}$$

$$E'_y = p_y E_y \qquad 0 \le p_y \le 1 \tag{5-5b}$$

The factors p_x and p_y are the amplitude attenuation coefficients along orthogonal transmission axes. For no attenuation or perfect transmission along an orthogonal axis $p_x(p_y) = 1$, whereas for complete attenuation $p_x(p_y) = 0$. If one of the axes has an absorption coefficient which is zero so that there is no transmission along this axis, the polarizer is said to have only a single transmission axis.

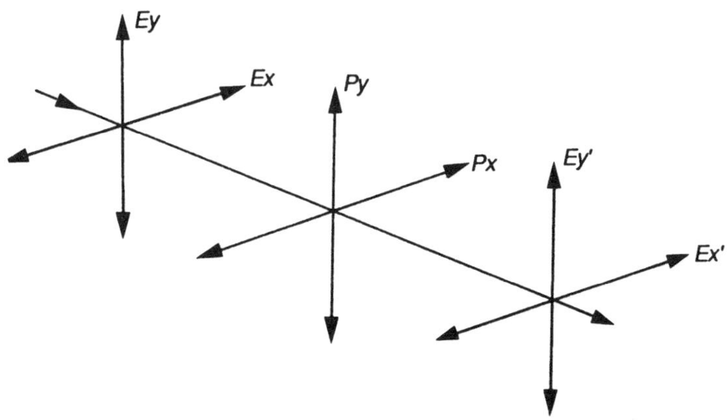

Figure 5-2 The Mueller matrix of a polarizer with attenuation coefficients p_x and p_y.

The Stokes polarization parameters of the incident and emerging beams are, respectively,

$$S_0 = E_x E_x^* + E_y E_y^* \tag{5-6a}$$

$$S_1 = E_x E_x^* - E_y E_y^* \tag{5-6b}$$

$$S_2 = E_x E_y^* + E_y E_x^* \tag{5-6c}$$

$$S_3 = i(E_x E_y^* - E_y E_x^*) \tag{5-6d}$$

and

$$S_0' = E_x' E_x'^* + E_y' E_y'^* \tag{5-7a}$$

$$S_1' = E_x' E_x'^* - E_y' E_y'^* \tag{5-7b}$$

$$S_2' = E_x' E_y'^* + E_y' E_x'^* \tag{5-7c}$$

$$S_3' = i(E_x' E_y'^* - E_y' E_x'^*) \tag{5-7d}$$

Substituting (5-5) into (5-7) and using (5-6), we then find

$$\begin{pmatrix} S_0' \\ S_1' \\ S_2' \\ S_3' \end{pmatrix} = \frac{1}{2} \begin{pmatrix} p_x^2 + p_y^2 & p_x^2 - p_y^2 & 0 & 0 \\ p_x^2 - p_y^2 & p_x^2 + p_y^2 & 0 & 0 \\ 0 & 0 & 2p_x p_y & 0 \\ 0 & 0 & 0 & 2p_x p_y \end{pmatrix} \begin{pmatrix} S_0 \\ S_1 \\ S_2 \\ S_3 \end{pmatrix} \tag{5-8}$$

The 4×4 matrix in (5-8) is written by itself as

$$M = \frac{1}{2} \begin{pmatrix} p_x^2 + p_y^2 & p_x^2 - p_y^2 & 0 & 0 \\ p_x^2 - p_y^2 & p_x^2 + p_y^2 & 0 & 0 \\ 0 & 0 & 2p_x p_y & 0 \\ 0 & 0 & 0 & 2p_x p_y \end{pmatrix} \qquad 0 \le p_{x,y} \le 1 \tag{5-9}$$

Equation (5-9) is the Mueller matrix for a polarizer with amplitude attenuation coefficients p_x and p_y. In general, the existence of the m_{33} term shows that the polarization of the emerging beam of light will be elliptically polarized.

For a neutral density filter $p_x = p_y - p$ and (5-9) becomes

$$M = p^2 \begin{pmatrix} 1 & 0 & 0 & 0 \\ 0 & 1 & 0 & 0 \\ 0 & 0 & 1 & 0 \\ 0 & 0 & 0 & 1 \end{pmatrix} \tag{5-10}$$

which is a unit diagonal matrix. Equation (5-10) shows that the polarization state is not changed by a neutral density filter, but the intensity of the incident beam is reduced by a factor of p^2. This is the expected behavior of a neutral density filter, since it only affects the magnitude the intensity and not the polarization state. According to (5-10), the emerging intensity I' is then

$$I' = p^2 I \tag{5-11}$$

where I is the intensity if the incident beam.

Equation (5-9) is the Mueller matrix for a polarizer which is described by unequal attenuations along the p_x and p_y axes. An *ideal* linear polarizer is one which has transmission along only one axis and no transmission along the orthogonal axis. This behavior can be described by first setting, say, $p_y = 0$. Then (5-9) reduces to

$$M = \frac{p_x^2}{2}\begin{pmatrix} 1 & 1 & 0 & 0 \\ 1 & 1 & 0 & 0 \\ 0 & 0 & 0 & 0 \\ 0 & 0 & 0 & 0 \end{pmatrix} \tag{5-12}$$

Equation (5-12) is the Mueller matrix for an ideal linear polarizer which polarizes only along the x axis. It is most often called a linear horizontal polarizer, arbitrarily assigning the horizontal to the x direction. It would be a perfect linear polarizer if the transmission factor p_x was unity ($p_x = 1$). Thus, the Mueller matrix for an ideal perfect linear polarizer with its transmission axis in the x direction is

$$M = \frac{1}{2}\begin{pmatrix} 1 & 1 & 0 & 0 \\ 1 & 1 & 0 & 0 \\ 0 & 0 & 0 & 0 \\ 0 & 0 & 0 & 0 \end{pmatrix} \tag{5-13}$$

If the original beam is completely unpolarized, the maximum intensity of the emerging beam which can be obtained with a perfect ideal polarizer is only 50% of the original intensity. It is the price we pay for obtaining perfectly polarized light. If the original beam is perfectly horizontally polarized, there is no change in intensity. This element is called a linear polarizer because it affects a linearly polarized beam in a unique manner as we shall soon see.

In general, all linear polarizers are described by (5-9). There is only one known natural material that comes close to approaching the perfect ideal polarizer described by (5-13), and this is calcite. A synthetic material known as Polaroid is also used as a polarizer. Its performance is not as good as calcite, but its cost is very low in comparison with that of natural calcite polarizers, e.g., a Glan–Thompson prism. Nevertheless, there are a few types of Polaroid which perform extremely well as "ideal" polarizers. We shall discuss the topic of calcite and Polaroid polarizers in Chapter 26.

If an ideal perfect linear polarizer is used in which the role of the transmission axes is reversed from that of our linear horizontal polarizer, that is, $p_x = 0$ and $p_y = 1$, then (5-9) reduces to

$$M = \frac{1}{2}\begin{pmatrix} 1 & -1 & 0 & 0 \\ -1 & 1 & 0 & 0 \\ 0 & 0 & 0 & 0 \\ 0 & 0 & 0 & 0 \end{pmatrix} \tag{5-14}$$

which is the Mueller matrix for a linear vertical polarizer.

Finally, it is convenient to rewrite the Mueller matrix, of a general linear polarizer, (5-9), in terms of trigonometric functions. This can be done by setting

$$p_x^2 + p_y^2 = p^2 \tag{5-15a}$$

and

$$p_x = p\cos\gamma \qquad p_y = p\sin\gamma \tag{5-15b}$$

Substituting (5-15) into (5-9) yields

$$M = \frac{p^2}{2}\begin{pmatrix} 1 & \cos 2\gamma & 0 & 0 \\ \cos 2\gamma & 1 & 0 & 0 \\ 0 & 0 & \sin 2\gamma & 0 \\ 0 & 0 & 0 & \sin 2\gamma \end{pmatrix} \tag{5-16}$$

where $0 \leq \gamma \leq 90^\circ$. For an ideal perfect linear polarizer $p = 1$. For a linear horizontal polarizer $\gamma = 0$, and for a linear vertical polarizer $\gamma = 90^\circ$. The usefulness of the trigonometric form of the Mueller matrix, (5-16), will appear later.

The reason for calling (5-13) and (5-14) linear polarizers is due to the following result. Suppose we have an incident beam of arbitrary intensity and polarization so that its Stokes vector is

$$S = \begin{pmatrix} S_0 \\ S_1 \\ S_2 \\ S_3 \end{pmatrix} \tag{5-17}$$

We now matrix multiply (5-17) by (5-13) or (5-14), and we can write

$$\begin{pmatrix} S_0' \\ S_1' \\ S_2' \\ S_3' \end{pmatrix} = \frac{1}{2}\begin{pmatrix} 1 & \pm 1 & 0 & 0 \\ \pm 1 & 1 & 0 & 0 \\ 0 & 0 & 0 & 0 \\ 0 & 0 & 0 & 0 \end{pmatrix}\begin{pmatrix} S_0 \\ S_1 \\ S_2 \\ S_3 \end{pmatrix} \tag{5-18}$$

Carrying out the matrix multiplication in (5-18), we find that

$$\begin{pmatrix} S_0' \\ S_1' \\ S_2' \\ S_3' \end{pmatrix} = \frac{1}{2}(S_0 \pm S_1)\begin{pmatrix} 1 \\ \pm 1 \\ 0 \\ 0 \end{pmatrix} \tag{5-19}$$

Inspecting (5-19), we see that the Stokes vector of the emerging beam is always *linearly* horizontally (+) or vertically (−) polarized. Thus an ideal linear polarizer always creates linearly polarized light regardless of the polarization state of the incident beam; however, note that because the factor $2p_xp_y$ in (5-9) is never zero, in practice there is no known perfect linear polarizer and all polarizers create elliptically polarized light. While the ellipticity may be small and, in fact, negligible, there is always some present.

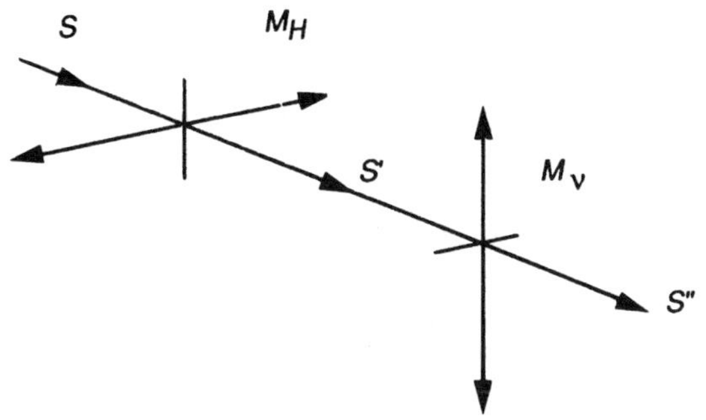

Figure 5-3 Testing for a linear polarizer.

The above behavior of linear polarizers allows us to develop a test to determine if a polarizing element is actually a linear polarizer. The test to determine if we have a linear polarizer is shown in Fig. 5-3. In the test we assume that we have a linear polarizer and set its axis in the horizontal (H) direction. We then take another polarizer and set its axis in the vertical (V) direction as shown in the figure. The Stokes vector of the incident beam is S, and the Stokes vector of the beam emerging from the first polarizer (horizontal) is

$$S' = M_H S \tag{5-20}$$

Next, the S' beam propagates to the second polarizer (vertical), and the Stokes vector S'' of the emerging beam is now

$$S'' = M_V S' = M_V M_H S = MS \tag{5-21}$$

where we have used (5-20). We see that M is the Mueller matrix of the combined vertical and linear polarizer:

$$M = M_V M_H \tag{5-22}$$

where M_H and M_V are given by (5-13) and (5-14), respectively. These results, (5-21) and (5-22), show that we can relate the Stokes vector of the emerging beam to the incident beam by merely multiplying the Mueller matrix of each component and finding the resulting Mueller matrix. In general, the matrices do not commute.

We now carry out the multiplication in (5-22) and write, using (5-13) and (5-14),

$$M = \frac{1}{4}\begin{pmatrix} 1 & -1 & 0 & 0 \\ -1 & 1 & 0 & 0 \\ 0 & 0 & 0 & 0 \\ 0 & 0 & 0 & 0 \end{pmatrix}\begin{pmatrix} 1 & 1 & 0 & 0 \\ 1 & 1 & 0 & 0 \\ 0 & 0 & 0 & 0 \\ 0 & 0 & 0 & 0 \end{pmatrix} = \begin{pmatrix} 0 & 0 & 0 & 0 \\ 0 & 0 & 0 & 0 \\ 0 & 0 & 0 & 0 \\ 0 & 0 & 0 & 0 \end{pmatrix} \tag{5-23}$$

Thus, we obtain a null Mueller matrix and, hence, a null output intensity regardless of the polarization state of the incident beam. The appearance of a null Mueller matrix (or intensity) occurs only when the linear polarizers are in the *crossed polarizer* configuration. Furthermore, the null Mueller matrix always arises whenever the polarizers are crossed, regardless of the angle of the transmission axis of the first polarizer.

5.3 THE MUELLER MATRIX OF A RETARDER

A retarder is a polarizing element which changes the phase of the optical beam. Strictly speaking, its correct name is phase shifter. However, historical usage has led to the alternative names retarder, wave plate, and compensator. Retarders introduce a phase shift of ϕ between the orthogonal components of the incident field. This can be thought of as being accomplished by causing a phase shift of $+\phi/2$ along the x axis and a phase shift of $-\phi/2$ along the y axis. These axes of the retarder are referred to as the *fast* and *slow* axes, respectively. In Fig. 5-4 we show the incident and emerging beam and the retarder. The components of the emerging beam are related to the incident beam by

$$E'_x(z,t) = e^{+i\phi/2}E_x(z,t) \tag{5-24a}$$

$$E'_y(z,t) = e^{-i\phi/2}E_y(z,t) \tag{5-24b}$$

Referring again to the definition of the Stokes parameters (5-6) and (5-7) and substituting (5-24a) and (5-24b) into these equations, we find that

$$S'_0 = S_0 \tag{5-25a}$$

$$S'_1 = S_1 \tag{5-25b}$$

$$S'_2 = S_2\cos\phi + S_3\sin\phi \tag{5-25c}$$

$$S'_3 = -S_2\sin\phi + S_3\cos\phi \tag{5-25d}$$

Equation (5-25) can be written in matrix form as

$$\begin{pmatrix} S'_0 \\ S'_1 \\ S'_2 \\ S'_3 \end{pmatrix} = \begin{pmatrix} 1 & 0 & 0 & 0 \\ 0 & 1 & 0 & 0 \\ 0 & 0 & \cos\phi & \sin\phi \\ 0 & 0 & -\sin\phi & \cos\phi \end{pmatrix}\begin{pmatrix} S_0 \\ S_1 \\ S_2 \\ S_3 \end{pmatrix} \tag{5-26}$$

Note that for an ideal phase shifter (retarder) there is no loss in intensity; that is, $S'_0 = S_0$.

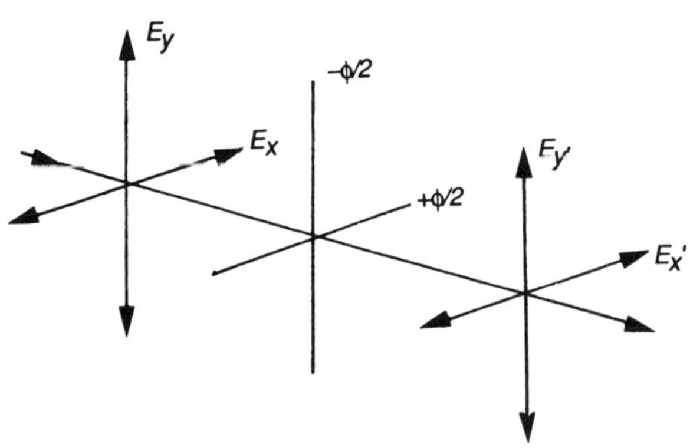

Figure 5-4 Propagation of a polarized beam through a retarder.

The Mueller matrix for a retarder with a phase shift ϕ is, from (5-26),

$$M = \begin{pmatrix} 1 & 0 & 0 & 0 \\ 0 & 1 & 0 & 0 \\ 0 & 0 & \cos\phi & \sin\phi \\ 0 & 0 & -\sin\phi & \cos\phi \end{pmatrix} \tag{5-27}$$

There are two special cases of (5-27) which appear often in polarizing optics. These are the cases for quarter-wave retarders ($\phi = 90^\circ$, i.e., the phase of one component of the light is delayed with respect to the orthogonal component by one quarter wave) and half-wave retarders ($\phi = 180^\circ$, i.e., the phase of one component of the light is delayed with respect to the orthogonal component by one half wave), respectively. Obviously, a retarder is naturally dependent on wavelength, although there are achromatic retarders that are slowly dependent on wavelength. We will discuss these topics in more detail in Chapter 26. For a quarter-wave retarder (5-27) becomes

$$M = \begin{pmatrix} 1 & 0 & 0 & 0 \\ 0 & 1 & 0 & 0 \\ 0 & 0 & 0 & 1 \\ 0 & 0 & -1 & 0 \end{pmatrix} \tag{5-28}$$

The quarter-wave retarder has the property that it transforms a linearly polarized beam with its axis at $+45^\circ$ or -45° to the fast axis of the retarder into a right or left circularly polarized beam, respectively. To show this property, consider the Stokes vector for a linearly polarized $\pm 45^\circ$ beam:

$$S = I_0 \begin{pmatrix} 1 \\ 0 \\ \pm 1 \\ 0 \end{pmatrix} \tag{5-29}$$

Multiplying (5-29) by (5-28) yields

$$S' = I_0 \begin{pmatrix} 1 \\ 0 \\ 0 \\ \mp 1 \end{pmatrix} \tag{5-30}$$

which is the Stokes vector for left (right) circularly polarized light. The transformation of linearly polarized light to circularly polarized light is an important application of quarter-wave retarders. However, circularly polarized light is obtained only if the incident linearly polarized light is oriented at $\pm 45^\circ$.

On the other hand, if the incident light is right (left) circularly polarized light, then multiplying (5-30) by (5-28) yields

$$S' = I_0 \begin{pmatrix} 1 \\ 0 \\ \mp 1 \\ 0 \end{pmatrix} \tag{5-31}$$

which is the Stokes vector for linear -45° or $+45^\circ$ polarized light. The quarter-wave retarder can be used to transform linearly polarized light to circularly polarized light or circularly polarized light to linearly polarized light.

The other important type of wave retarder is the half-wave retarder ($\phi = 180^\circ$). For this condition (5-27) reduces to

$$M = \begin{pmatrix} 1 & 0 & 0 & 0 \\ 0 & 1 & 0 & 0 \\ 0 & 0 & -1 & 0 \\ 0 & 0 & 0 & -1 \end{pmatrix} \tag{5-32}$$

A half-wave retarder is characterized by a diagonal matrix. The terms $m_{22} = m_{33} = -1$ reverse the ellipticity and orientation of the polarization state of the incident beam. To show this formally, we have initially

$$S = \begin{pmatrix} S_0 \\ S_1 \\ S_2 \\ S_3 \end{pmatrix} \tag{5-17}$$

We also saw previously that the orientation angle ψ and the ellipticity angle χ are given in terms of the Stokes parameters:

$$\tan 2\psi = \frac{S_2}{S_1} \tag{4-12}$$

$$\sin 2\chi = \frac{S_3}{S_0} \tag{4-14}$$

Multiplying (5-17) by (5-32) gives

$$S' = \begin{pmatrix} S'_0 \\ S'_1 \\ S'_2 \\ S'_3 \end{pmatrix} = \begin{pmatrix} S_0 \\ S_1 \\ -S_2 \\ -S_3 \end{pmatrix} \tag{5-33}$$

where

$$\tan 2\psi' = \frac{S'_2}{S'_1} \tag{5-34a}$$

$$\sin 2\chi' = \frac{S'_3}{S'_0} \tag{5-34b}$$

Substituting (5-33) into (5-34) yields

$$\tan 2\psi' = \frac{-S_2}{S_1} = -\tan 2\psi \tag{5-35a}$$

$$\sin 2\chi' = \frac{-S_3}{S_0} = -\sin 2\chi \tag{5-35b}$$

Hence,

$$\psi' = 90^\circ - \psi \tag{5-36a}$$

$$\chi' = 90^\circ + \chi \tag{5-36b}$$

Half-wave retarders also possess the property that they can rotate the polarization ellipse. This important property shall be discussed in Section 5.5.

5.4 THE MUELLER MATRIX OF A ROTATOR

The final way to change the polarization state of an optical field is to allow a beam to propagate through a polarizing element that rotates the orthogonal field components $E_x(z,t)$ and $E_y(z,t)$ through an angle θ. In order to derive the Mueller matrix for rotation, we consider Fig. 5-5. The angle θ describes the rotation of E_x to E'_x and of E_y to E'_y. Similarly, the angle β is the angle between E and E_x. In the figure the point P is described in the E'_x, E'_y coordinate system by

$$E'_x = E\cos(\beta - \theta) \tag{5-37a}$$

$$E'_y = E\sin(\beta - \theta) \tag{5-37b}$$

In the E_x, E_y coordinate system we have

$$E_x = E\cos\beta \tag{5-38a}$$

$$E_y = E\sin\beta \tag{5-38b}$$

Expanding the trigonometric functions in (5-37) gives

$$E'_x = E(\cos\beta\cos\theta + \sin\beta\sin\theta) \tag{5-39a}$$
$$E'_y = E(\sin\beta\cos\theta - \sin\theta\cos\beta) \tag{5-39b}$$

Collecting terms in (5-39) using (5-38) then gives

$$E'_x = E_x\cos\theta + E_y\sin\theta \tag{5-40a}$$
$$E'_y = -E_x\sin\theta + E_y\cos\theta \tag{5-40b}$$

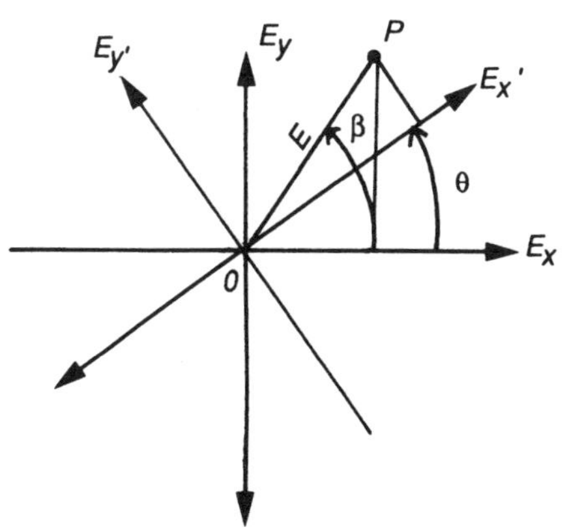

Figure 5-5 Rotation of the optical field components by a rotator.

Equations (5-40a) and (5-40b) are the amplitude equations for rotation. In order to find the Mueller matrix we form the Stokes parameters for (5-40) as before and find the Mueller matrix for rotation:

$$M(2\theta) = \begin{pmatrix} 1 & 0 & 0 & 0 \\ 0 & \cos 2\theta & \sin 2\theta & 0 \\ 0 & -\sin 2\theta & \cos 2\theta & 0 \\ 0 & 0 & 0 & 1 \end{pmatrix} \tag{5-41}$$

We note that a physical rotation of θ leads to the appearance of 2θ in (5-41) rather than θ because we are working in the intensity domain; in the amplitude domain we would expect just θ.

Rotators are primarily used to change the orientation angle of the polarization ellipse. To see this behavior, suppose the orientation angle of an incident beam is ψ. Recall that

$$\tan 2\psi = \frac{S_2}{S_1} \tag{4-12}$$

For the emerging beam we have a similar expression with the variables in (4-12) replaced with primed variables. Using (5-41) we see that the orientation angle ψ' is then

$$\tan 2\psi' = \frac{-S_1 \sin 2\theta + S_2 \cos 2\theta}{S_1 \cos 2\theta + S_2 \sin 2\theta} \tag{5-42}$$

Equation (4-12) is now written as

$$S_2 = S_1 \tan 2\psi \tag{5-43}$$

Substituting (5-43) into (5-42), we readily find that

$$\tan 2\psi' = \tan(2\psi - 2\theta) \tag{5-44}$$

so

$$\psi' = \psi - \theta \tag{5-45}$$

Equation (5-45) shows that a rotator merely rotates the polarization ellipse of the incident beam; the ellipticity remains unchanged. The sign is negative in (5-45) because the rotation is clockwise. If the rotation is counterclockwise, that is, θ is replaced by $-\theta$ in (5-41), then we find

$$\psi' = \psi + \theta \tag{5-46}$$

In the derivation of the Mueller matrices for a polarizer, retarder, and rotator, we have assumed that the axes of these devices are aligned along the E_x and E_y (or x, y axes), respectively. In practice, we find that the polarization elements are often rotated. Consequently, it is also necessary for us to know the form of the Mueller matrices for the rotated polarizing elements. We now consider this problem.

5.5 MUELLER MATRICES FOR ROTATED POLARIZING COMPONENTS

To derive the Mueller matrix for rotated polarizing components, we refer to Fig. 5-6. The axes of the polarizing component are seen to be rotated through an angle θ to the x' and y' axes. We must, therefore, also consider the components of the incident beam along the x' and y' axes. In terms of the Stokes vector of the incident beam, S, we then have

$$S' = M_R(2\theta)S \tag{5-47}$$

where $M_R(2\theta)$ is the Mueller matrix for rotation (5-41) and S' is the Stokes vector of the beam whose axes are along x' and y'.

The S' beam now interacts with the polarizing element characterized by its Mueller matrix M. The Stokes vector S'' of the beam emerging from the rotated polarizing component is

$$S'' = MS' = MM_R(2\theta)S \tag{5-48}$$

where we have used (5-47). Finally, we must take the components of the emerging beam along the original x and y axes as seen in Fig. 5-6. This can be described by a counterclockwise rotation of S'' through $-\theta$ and back to the original x, y axes, so

$$\begin{aligned} S''' &= M_R(-2\theta)S'' \\ &= [M_R(-2\theta)MM_R(2\theta)]S \end{aligned} \tag{5-49}$$

where $M_R(-2\theta)$ is, again, the Mueller matrix for rotation and S''' is the Stokes vector of the emerging beam. Equation (5-49) can be written as

$$S''' = M(2\theta)S \tag{5-50}$$

where

$$M(2\theta) = M_R(-2\theta)MM_R(2\theta) \tag{5-51}$$

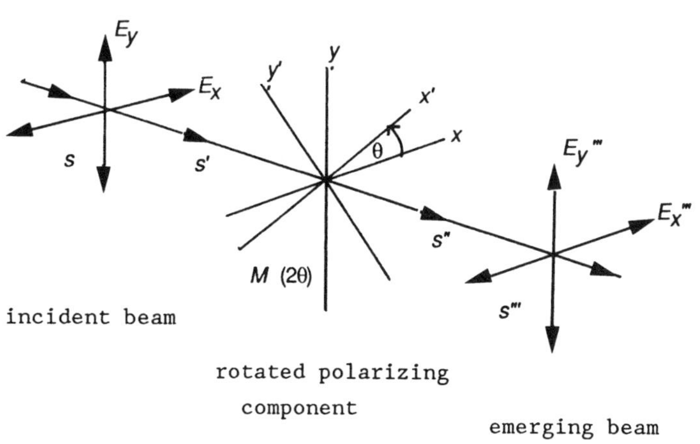

Figure 5-6 Derivation of the Mueller matrix for rotated polarizing components.

Equation (5-51) is the Mueller matrix of a rotated polarizing component. We recall that the Mueller matrix for rotation $M_R(2\theta)$ is given by

$$M_R(2\theta) = \begin{pmatrix} 1 & 0 & 0 & 0 \\ 0 & \cos 2\theta & \sin 2\theta & 0 \\ 0 & -\sin 2\theta & \cos 2\theta & 0 \\ 0 & 0 & 0 & 1 \end{pmatrix} \tag{5-52}$$

The rotated Mueller matrix expressed by (5-51) appears often in the treatment of polarized light. Of particular interest are the Mueller matrices for a rotated polarizer and a rotated retarder. The Mueller matrix for a rotated "rotator" is also interesting, but in a different way. We recall that a rotator rotates the polarization ellipse by an amount θ. If the rotator is now rotated through an angle α, then one discovers, using (5-51), that $M(2\theta) = M_R(2\theta)$; that is, the rotator is unaffected by a mechanical rotation. Thus, the polarization ellipse cannot be rotated by rotating a rotator! The rotation comes about only by the intrinsic behavior of the rotator. It is possible, however, to rotate the polarization ellipse mechanically by rotating a half-wave plate, as we shall soon demonstrate.

The Mueller matrix for a rotated polarizer is most conveniently found by expressing the Mueller matrix of a polarizer in angular form, namely,

$$M = \frac{p^2}{2} \begin{pmatrix} 1 & \cos 2\gamma & 0 & 0 \\ \cos 2\gamma & 1 & 0 & 0 \\ 0 & 0 & \sin 2\gamma & 0 \\ 0 & 0 & 0 & \sin 2\gamma \end{pmatrix} \tag{5-16}$$

Carrying out the matrix multiplication according to (5-51) and using (5-52), the Mueller matrix for a rotated polarizer is

$$M = \frac{1}{2} \begin{pmatrix} 1 & \cos 2\gamma \cos 2\theta & \cos 2\gamma \sin 2\theta & 0 \\ \cos 2\gamma \cos 2\theta & \cos^2 2\theta + \sin 2\gamma \sin^2 2\theta & (1 - \sin 2\gamma) \sin 2\theta \cos 2\theta & 0 \\ \cos 2\gamma \sin 2\theta & (1 - \sin 2\gamma) \sin 2\theta \cos 2\theta & \sin^2 2\theta + \sin 2\gamma \cos^2 2\theta & 0 \\ 0 & 0 & 0 & \sin 2\gamma \end{pmatrix} \tag{5-53}$$

In (5-53) we have set p^2 to unity. We note that $\gamma = 0°$, $45°$, and $90°$ correspond to a linear horizontal polarizer, a neutral density filter, and a linear vertical polarizer, respectively.

The most common form of (5-53) is the Mueller matrix for an ideal linear horizontal polarizer ($\gamma = 0°$). For this value (5-53) reduces to

$$M_P(2\theta) = \frac{1}{2} \begin{pmatrix} 1 & \cos 2\theta & \sin 2\theta & 0 \\ \cos 2\theta & \cos^2 2\theta & \sin 2\theta \cos 2\theta & 0 \\ \sin 2\theta & \sin 2\theta \cos 2\theta & \sin^2 2\theta & 0 \\ 0 & 0 & 0 & 0 \end{pmatrix} \tag{5-54}$$

In (5-54) we have written $M_P(2\theta)$ to indicate that this is the Mueller matrix for a rotated ideal linear polarizer. The form of (5-54) can be checked immediately by setting $\theta = 0$ (no rotation). Upon doing this, we obtain the Mueller matrix of a linear horizontal polarizer:

$$M_P(0^\circ) = \frac{1}{2}\begin{pmatrix} 1 & 1 & 0 & 0 \\ 1 & 1 & 0 & 0 \\ 0 & 0 & 0 & 0 \\ 0 & 0 & 0 & 0 \end{pmatrix} \tag{5-55}$$

One can readily see that for $\theta = 45^\circ$ and 90° (5-54) reduces to the Mueller matrix for an ideal linear $+45^\circ$ and vertical polarizer, respectively. The Mueller matrix for a rotated ideal linear polarizer, (5-54), appears often in the generation and analysis of polarized light.

Next, we turn to determining the Mueller matrix for a retarder or wave plate. We recall that the Mueller matrix for a retarder with phase shift ϕ is given by

$$M_c = \begin{pmatrix} 1 & 0 & 0 & 0 \\ 0 & 1 & 0 & 0 \\ 0 & 0 & \cos\phi & \sin\phi \\ 0 & 0 & -\sin\phi & \cos\phi \end{pmatrix} \tag{5-56}$$

Somtimes the term compensator is used in place of retarder, and so we have used the subscript "c."

From (5-51) the Mueller matrix for the rotated retarder (5-56) is found to be

$$M_c(\phi, 2\theta) = \begin{pmatrix} 1 & 0 & 0 & 0 \\ 0 & \cos^2 2\theta + \cos\phi \sin^2 2\theta & (1-\cos\phi)\sin 2\theta \cos 2\theta & -\sin\phi \sin 2\theta \\ 0 & (1-\cos\phi)\sin 2\theta \cos 2\theta & \sin^2 2\theta + \cos\phi \cos^2 2\theta & \sin\phi \cos 2\theta \\ 0 & \sin\phi \sin 2\theta & -\sin\phi \cos 2\theta & \cos\phi \end{pmatrix} \tag{5-57}$$

For $\theta = 0^\circ$, (5-57) reduces to (5-56) as expected. There is a particularly interesting form of (5-57) for a phase shift of $\phi = 180^\circ$, a so-called half-wave retarder. For $\phi = 180^\circ$ (5-57) reduces to

$$M_c(180^\circ, 4\theta) = \begin{pmatrix} 1 & 0 & 0 & 0 \\ 0 & \cos 4\theta & \sin 4\theta & 0 \\ 0 & \sin 4\theta & -\cos 4\theta & 0 \\ 0 & 0 & 0 & -1 \end{pmatrix} \tag{5-58}$$

Equation (5-58) looks very similar to the Mueller matrix for rotation $M_R(2\theta)$, (5-52), which we write simply as M_R:

$$M_R = \begin{pmatrix} 1 & 0 & 0 & 0 \\ 1 & \cos 2\theta & \sin 2\theta & 0 \\ 0 & -\sin 2\theta & \cos 2\theta & 0 \\ 0 & 0 & 0 & 1 \end{pmatrix} \tag{5-59}$$

However, (5-58) differs from (5-59) in some essential ways. The first is the ellipticity. The Stokes vector of an incident beam is, as usual,

$$S = \begin{pmatrix} S_0 \\ S_1 \\ S_2 \\ S_3 \end{pmatrix} \tag{5-17}$$

Multiplying (5-17) by (5-59) yields the Stokes vector S':

$$S' = \begin{pmatrix} S_0 \\ S_1 \cos 2\theta + S_2 \sin 2\theta \\ -S_1 \sin 2\theta + S_2 \cos 2\theta \\ S_3 \end{pmatrix} \tag{5-60}$$

The ellipticity angle χ' is

$$\sin 2\chi' = \frac{S_3'}{S_0'} = \frac{S_3}{S_0} = \sin 2\chi \tag{5-61}$$

Thus, the ellipticity is not changed under true rotation. Multiplying (5-17) by (5-58), however, yields a Stokes vector S' resulting from a half-wave retarder:

$$S' = \begin{pmatrix} S_0 \\ S_1 \cos 4\theta + S_2 \sin 4\theta \\ S_1 \sin 4\theta - S_2 \cos 4\theta \\ -S_3 \end{pmatrix} \tag{5-62}$$

The ellipticity angle χ' is now

$$\sin 2\chi' = \frac{S_3'}{S_0'} = \frac{-S_3}{S_0} = -\sin 2\chi \tag{5-63}$$

Thus,

$$\chi' = \chi + 90^\circ \tag{5-64}$$

so the ellipticity angle χ of the incident beam is advanced 90° by using a rotated half-wave retarder.

The next difference is for the orientation angle ψ'. For a rotator, (5-59), the orientation angle associated with the incident beam, ψ, is given by the equation:

$$\tan 2\psi = \frac{S_2}{S_1} \tag{5-65}$$

so we immediately find from (5-65) and (5-60) that

$$\tan 2\psi' = \frac{S_2'}{S_1'} = \frac{\sin 2\psi \cos 2\theta - \sin 2\theta \cos 2\psi}{\cos 2\psi \cos 2\theta + \sin 2\psi \sin 2\theta} = \frac{\sin(2\psi - 2\theta)}{\cos(2\psi - 2\theta)} \tag{5-66}$$

whence

$$\psi' = \psi - \theta \tag{5-67}$$

Equation (5-67) shows that a mechanical rotation in θ increases ψ by the same amount and in the same direction (by definition, a clockwise rotation of θ increases). On the other hand, for a half-wave retarder the orientation angle ψ' is given by the equation, using (5-17) and (5-62),

$$\tan 2\psi' = \frac{\cos 2\psi \sin 4\theta - \sin 2\psi \cos 4\theta}{\cos 2\psi \cos 4\theta + \sin 2\psi \sin 4\theta} = \frac{\sin(4\theta - 2\psi)}{\cos(4\theta - 2\psi)} \tag{5-68}$$

so

$$\psi' = 2\theta - \psi \tag{5-69a}$$

or

$$\psi' = -(\psi - 2\theta) \tag{5-69b}$$

Comparing (5-69b) with (5-67), we see that rotating the half-wave retarder clockwise causes ψ' to rotate counterclockwise by an amount twice that of a rotator. Because the rotation of a half-wave retarder is opposite to a true rotator, it is called a *pseudorotator*. When a mechanical rotation of θ is made using a half-wave retarder the polarization ellipse is rotated by 2θ and in a direction opposite to the direction of the mechanical rotation. For a true mechanical rotation of θ the polarization ellipse is rotated by an amount θ and in the same direction as the rotation.

This discussion of rotation of half-wave retarders is more than academic, however. Very often manufacturers sell half-wave retarders as polarization rotators. Strictly speaking, this belief is quite correct. However, one must realize that the use of a half-wave retarder rather than a true rotator requires a mechanical mount with twice the resolution. That is, if we use a rotator in a mount with, say $2'$ of resolution, then in order to obtain the same resolution with a half-wave retarder a mechanical mount with $1'$ of resolution is required. The simple fact is that *doubling* the resolution of a mechanical mount can be very expensive in comparison with using a true rotator. The cost for doubling the resolution of a mechanical mount can easily double, whereas the cost increase between a quartz rotator and a half-wave retarder is usually much less. In general, if the objective is to rotate the polarization ellipse by a known fixed amount, it is better to use a rotator rather than a half-wave retarder.

A half-wave retarder is very useful as a rotator. Half-wave retarders can also be used to "reverse" the polarization state. In order to illustrate this behavior, consider that we have an incident beam which is right or left circularly polarized. Its Stokes vector is

$$S = I_0 \begin{pmatrix} 1 \\ 0 \\ 0 \\ \pm 1 \end{pmatrix} \tag{5-70}$$

Multiplying (5-70) by (5-58) and setting $\theta = 0^\circ$ yields

$$S' = I_0 \begin{pmatrix} 1 \\ 0 \\ 0 \\ \mp 1 \end{pmatrix} \tag{5-71}$$

We see that we again obtain circularly polarized light but opposite to its original state; that is, right circularly polarized light is transformed to left circularly polarized light, and vice versa. Similarly, if we have incident linear $+45^\circ$ polarized light, the emerging beam is linear -45° polarized light. It is this property of reversing the ellipticity and the orientation, manifested by the negative sign in m_{22} and m_{33}, that also makes half-wave plates very useful.

Finally, we consider the Mueller matrix of a rotated quarter-wave retarder. We set $\phi = 90^\circ$ in (5-58) and we have

$$M_c(90^\circ, 2\theta) = \begin{pmatrix} 1 & 0 & 0 & 0 \\ 0 & \cos^2 2\theta & \sin 2\theta \cos 2\theta & -\sin 2\theta \\ 0 & \sin 2\theta \cos 2\theta & \sin^2 2\theta & \cos 2\theta \\ 0 & \sin 2\theta & -\cos 2\theta & 0 \end{pmatrix} \tag{5-72}$$

Consider that we have an incident linearly horizontally polarized beam, so its Stokes vector is ($I_0 = 1$)

$$S = \begin{pmatrix} 1 \\ 1 \\ 0 \\ 0 \end{pmatrix} \tag{5-73}$$

We multiply (5-73) by (5-72), and we find that the Stokes vector S' is

$$S' = \begin{pmatrix} 1 \\ \cos^2 2\theta \\ \sin 2\theta \cos 2\theta \\ \sin 2\theta \end{pmatrix} \tag{5-74}$$

We see immediately from (5-74) that the orientation angle ψ' and the ellipticity angle χ' of the emerging beam are given by

$$\tan 2\psi' = \tan 2\theta \tag{5-75a}$$

$$\sin 2\chi' = \sin 2\theta \tag{5-75b}$$

Thus, the rotated quarter-wave plate has the property that it can be used to generate any desired orientation and ellipticity starting with an incident linearly horizontally polarized beam. However, we can only select one of these parameters; we have no control over the other parameter. We also note that if we initially have right or left circularly polarized light the Stokes vector of the output beam is

$$S' = \begin{pmatrix} 1 \\ \mp \sin 2\theta \\ \pm \cos 2\theta \\ 0 \end{pmatrix} \tag{5-76}$$

which is the Stokes vector for linearly polarized light. While it is well known that a quarter-wave retarder can be used to create linearly polarized light, (5-76) shows that an additional variation is possible by rotating the retarder, namely, the orientation can be controlled.

Equation (5-76) shows that we can generate any desired orientation or ellipticity of a beam, but not both. This leads to the question of how we can generate an elliptically polarized beam of any desired orientation and ellipticity regardless of the polarization state of an incident beam.

5.6 GENERATION OF ELLIPTICALLY POLARIZED LIGHT

In the previous section we derived the Mueller matrices for a rotated polarizer and a rotated retarder. We now apply these matrices to the generation of an elliptically polarized beam of any desired orientation and ellipticity. In order to do this we refer to Fig. 5-7. In the figure we show an incident beam of arbitrary polarization. The beam propagates first through an ideal polarizer rotated through an angle θ and then through a retarder, with its fast axis along the x axis. The Stokes vector of the incident beam is

$$S = \begin{pmatrix} S_0 \\ S_1 \\ S_2 \\ S_3 \end{pmatrix} \tag{5-17}$$

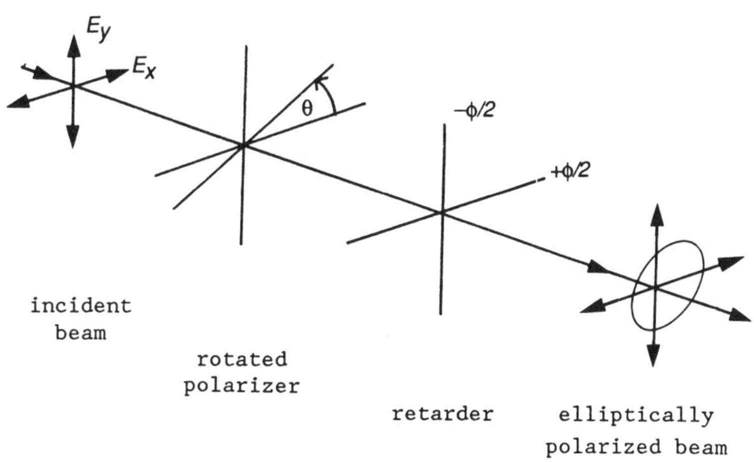

Figure 5-7 The generation of elliptically polarized light.

It is important that we consider the optical source to be arbitrarily polarized. At first sight, for example, we might wish to use unpolarized light or linearly polarized light. However, unpolarized light is surprisingly difficult to generate, and the requirement to generate ideal linearly polarized light calls for an excellent linear polarizer. We can avoid this problem if we consider that the incident beam is of unknown but arbitrary polarization. Our objective is to create an elliptically polarized beam of any desired ellipticity and orientation and which is totally independent of the polarization state of the incident beam.

The Mueller matrix of a rotated ideal linear polarizer is

$$M_P(2\theta) = \frac{1}{2}\begin{pmatrix} 1 & \cos 2\theta & \sin 2\theta & 0 \\ \cos 2\theta & \cos^2 2\theta & \sin 2\theta \cos 2\theta & 0 \\ \sin 2\theta & \sin 2\theta \cos 2\theta & \sin^2 2\theta & 0 \\ 0 & 0 & 0 & 0 \end{pmatrix} \tag{5-54}$$

Multiplying (5-17) by (5-54) yields

$$S' = \frac{1}{2}(S_0 + S_1 \cos 2\theta + S_2 \sin 2\theta)\begin{pmatrix} 1 \\ \cos 2\theta \\ \sin 2\theta \\ 0 \end{pmatrix} \tag{5-77}$$

The Mueller matrix of the retarder (nonrotated) is

$$M_c = \begin{pmatrix} 1 & 0 & 0 & 0 \\ 0 & 1 & 0 & 0 \\ 0 & 0 & \cos\phi & \sin\phi \\ 0 & 0 & -\sin\phi & \cos\phi \end{pmatrix} \tag{5-56}$$

Multiplying (5-77) by (5-56) then gives the Stokes vector of the beam emerging from the retarder:

$$S'' = I(\theta)\begin{pmatrix} 1 \\ \cos 2\theta \\ \cos\phi \sin 2\theta \\ -\sin\phi \sin 2\theta \end{pmatrix} \tag{5-78a}$$

where

$$I(\theta) = \frac{1}{2}(S_0 + S_1 \cos 2\theta + S_2 \sin 2\theta) \tag{5-78b}$$

Equation (5-78a) is the Stokes vector of an elliptically polarized beam. We immediately find from (5-78a) that the orientation angle ψ (we drop the double prime) is

$$\tan 2\psi = \cos\phi \tan 2\theta \tag{5-79a}$$

and the ellipticity angle χ is

$$\sin 2\chi = -\sin\phi \sin 2\theta \tag{5-79b}$$

We must now determine the θ and ϕ which will generate the desired values of ψ and χ. We divide (5-79a) by $\tan 2\theta$ and (5-79b) by $\sin 2\theta$, square the equations, and add. The result is

$$\cos 2\theta = \pm \cos 2\chi \cos 2\psi \tag{5-80}$$

To determine the required phase shift ϕ, we divide (5-79b) by (5-79a):

$$\frac{\sin 2\chi}{\tan 2\psi} = -\tan\phi \cos 2\theta \tag{5-81}$$

Solving for $\tan\phi$ and using (5-80), we easily find that

$$\tan\phi = \mp \frac{\tan 2\chi}{\sin 2\psi} \tag{5-82}$$

Thus, (5-80) and (5-82) are the equations for the angles θ and ϕ to which the polarizer and the retarder must be set in order to obtain the desired ellipticity and orientation angles χ and ψ.

We have thus shown that using only a rotated ideal linear polarizer and a retarder we can generate any state of elliptically polarized light. There is a final interesting fact about (5-80) and (5-82). We write (5-80) and (5-82) as a pair in the form

$$\cos 2\theta = \pm \cos 2\chi \cos 2\psi \tag{5-80}$$

$$\tan 2\chi = \mp \sin 2\psi \tan\phi \tag{5-83}$$

Equations (5-80) and (5-83) are recognized as equations arising from spherical trigonometry for a right spherical triangle. In Fig. 5-8 we have drawn a right spherical triangle. The angle 2ψ (the orientation of the polarization ellipse) is plotted on the equator, and the angle 2χ (the ellipticity of the polarization ellipse) is plotted on the longitude. If a great circle is drawn from point A to point B, the length of the arc $\overline{AB}$ is given by (5-80) and corresponds to 2θ as shown in the figure. Similarly, the

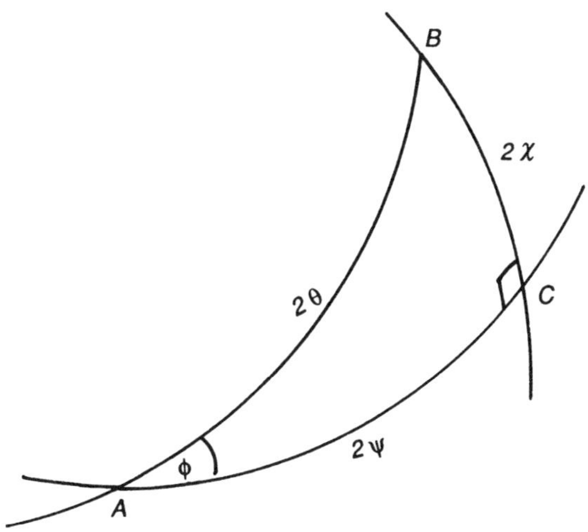

Figure 5-8 A right spherical triangle drawn on the surface of a sphere.

phase ϕ is the angle between the arc $\overline{AB}$ and the equator; its value is given by (5-83). We see from Fig. 5-8 that we can easily determine θ and ϕ by (1) measuring the length of the arc $\overline{AB}$ and (2) measuring the angle between the arc $\overline{AB}$ and the equator on a sphere.

The polarization equations (5-80) and (5-83) are intimately associated with spherical trigonometry and a sphere. Furthermore, we recall from Section 4.3 that when the Stokes parameters were expressed in terms of the orientation angle and the ellipticity angle they led directly to the Poincaré sphere. In fact, (5-80) and (5-83) describe a spherical triangle which plots directly on to the Poincaré sphere. Thus, we see that even at this early stage in our study of polarized light there is a strong connection between the equations of polarized light and its representation on a sphere. In fact, one of the most remarkable properties of polarized light is that there is such a close relation between these equations and the equations of spherical trigonometry. In Chapter 12, on the Poincaré sphere, these relations will be discussed in depth. In order to provide the reader with background material on right spherical triangles a brief discussion of the fundamentals of spherical trigonometry is presented at the end of Section 12.2.

REFERENCES

Papers

1. Soleillet, P., Ann. *Phys., 12* (10), 23 (1929).
2. Perrin, F., *J. Chem. Phys., 10*, 415 (1942).
3. Mueller, H., *J. Opt. Soc. Am., 37*, 110 (1947).
4. Parke, N. G., III, *Statistical Optics*. II: *Mueller Phenomenological Algebra*, RLE TR-119, Research Laboratory of Elect. at M.I.T. (1949).
5. McMaster, W. H., *Am. J. Phys., 22*, 351 (1954).
6. Walker, M. J., *J. Phys., 22*, 170 (1954).
7. McMaster, W. H., *Rev. Mod. Phys., 33*, 8 (1961).
8. Collett, *E., Am. J. Phys., 36*, 713 (1968).
9. Collett, *E., Am. J. Phys., 39*, 517 (1971).

Books

1. Shurcliff, W. A., *Polarized Light*, Harvard University Press, Cambridge, MA, 1962.
2. Gerrard, A. and Burch, J. M., *Introduction to Matrix Methods in Optics*, Wiley, London, 1975.

6

Methods of Measuring the Stokes Polarization Parameters

6.1 INTRODUCTION

We now turn our attention to the important problem of measuring the Stokes polarization parameters. In Chapter 7 we shall also discuss the measurement of the Mueller matrices. The first method for measuring the Stokes parameters is due to Stokes and is probably the best known method; this method was discussed in Section 4.4. There are other methods for measuring the Stokes parameters. However, we have refrained from discussing these methods until we had introduced the Mueller matrices for a polarizer, a retarder, and a rotator. The Mueller matrix and Stokes vector formalism allows us to treat all of these measurement problems in a very simple and direct manner. While, of course, the problems could have been treated using the amplitude formulation, the use of the Mueller matrix formalism greatly simplifies the analysis.

In theory, the measurement of the Stokes parameters should be quite simple. However, in practice there are difficulties. This is due, primarily, to the fact that while the measurement of S_0, S_1, and S_2 is quite straightforward, the measurement of S_3 is more difficult. In fact, as we pointed out, before the advent of optical detectors it was not even possible to measure the Stokes parameters using Stokes' measurement method (Section 4.4). It is possible, however, to measure the Stokes parameter using the eye as a detector by using a so-called null method; this is discussed in Section 6.4. In this chapter we discuss Stokes' method along with other methods, which includes the circular polarizer method, the null-intensity method, the Fourier analysis method, and the method of Kent and Lawson.

6.2 CLASSICAL MEASUREMENT METHOD: THE QUARTER-WAVE RETARDER POLARIZER METHOD

The Mueller matrices for the polarizer (diattenuator), retarder (phase shifter), and rotator can now be used to analyze various methods for measuring the Stokes parameters. A number of methods are known. We first consider the application of the Mueller matrices to the classical measurement of the Stokes polarization parameters using a quarter-wave retarder and a polarizer. This is the same problem that was treated in Section 4.4; it is the problem originally considered by Stokes (1852). The result is identical, of course, with that obtained by Stokes. However, the advantage of using the Mueller matrices is that a formal method can be used to treat not only this type of problem but other polarization problems as well.

The Stokes parameters can be measured as shown in Fig. 6-1. An optical beam is characterized by its four Stokes parameters S_0, S_1, S_2, and S_3. The Stokes vector of this beam is represented by

$$S = \begin{pmatrix} S_0 \\ S_1 \\ S_2 \\ S_3 \end{pmatrix} \tag{6-1}$$

The Mueller matrix of a retarder with its fast axis at 0° is

$$M = \begin{pmatrix} 1 & 0 & 0 & 0 \\ 0 & 1 & 0 & 0 \\ 0 & 0 & \cos\phi & \sin\phi \\ 0 & 0 & -\sin\phi & \cos\phi \end{pmatrix} \tag{6-2}$$

The Stokes vector S' of the beam emerging from the retarder is obtained by multiplication of (6-2) and (6-1), so

$$S' = \begin{pmatrix} S_0 \\ S_1 \\ S_2\cos\phi + S_3\sin\phi \\ -S_2\sin\phi + S_3\cos\phi \end{pmatrix} \tag{6-3}$$

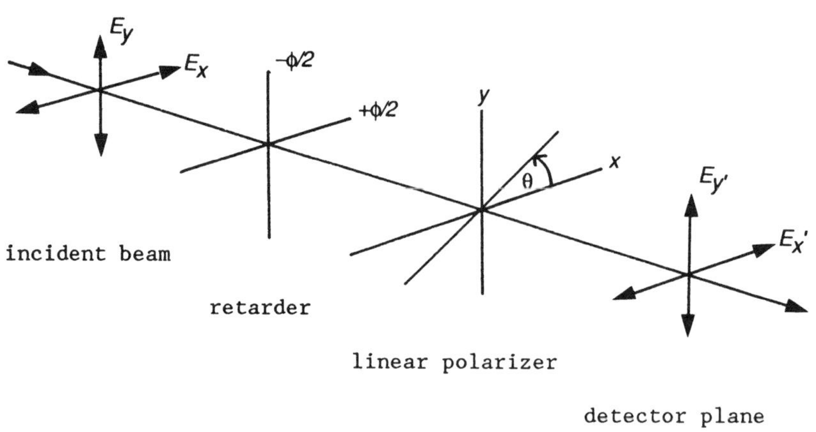

Figure 6-1 Classical measurement of the Stokes parameters.

The Mueller matrix of an ideal linear polarizer with its transmission axis set at an angle θ is

$$M = \frac{1}{2}\begin{pmatrix} 1 & \cos 2\theta & \sin 2\theta & 0 \\ \cos 2\theta & \cos^2 2\theta & \sin 2\theta \cos 2\theta & 0 \\ \sin 2\theta & \sin 2\theta \cos 2\theta & \sin^2 2\theta & 0 \\ 0 & 0 & 0 & 0 \end{pmatrix} \tag{6-4}$$

The Stokes vector S'' of the beam emerging from the linear polarizer is found by multiplication of (6-3) by (6-4). However, we are only interested in the intensity I'', which is the first Stokes parameter S_0'' of the beam incident on the optical detector shown in Fig. 6-1. Multiplying the first row of (6-4) with (6-3), we then find the intensity of the beam emerging from the quarter-wave retarder–polarizer combination to be

$$I(\theta, \phi) = \frac{1}{2}[S_0 + S_1 \cos 2\theta + S_2 \sin 2\theta \cos \phi + S_3 \sin 2\theta \sin \phi] \tag{6-5}$$

Equation (6-5) is Stokes' famous intensity relation for the Stokes parameters. The Stokes parameters are then found from the following conditions on θ and ϕ:

$$S_0 = I(0°, 0°) + I(90°, 0°) \tag{6-6a}$$

$$S_1 = I(0°, 0°) - I(90°, 0°) \tag{6-6b}$$

$$S_2 = 2I(45°, 0°) - S_0 \tag{6-6c}$$

$$S_3 = 2I(45°, 90°) - S_0 \tag{6-6d}$$

In practice, S_0, S_1, and S_2 are easily measured by removing the quarter-wave retarder ($\phi = 90°$) from the optical train. In order to measure S_3, however, the retarder must be reinserted into the optical train with the linear polarizer set at $\theta = 45°$. This immediately raises a problem because the retarder absorbs some optical energy. In order to obtain an accurate measurement of the Stokes parameters the absorption factor must be introduced, *ab initio*, into the Mueller matrix for the retarder. The absorption factor which we write as p must be determined from a separate measurement and will then appear in (6-5) and (6-6). We can easily derive the Mueller matrix for an absorbing retarder as follows.

The field components E_x and E_y of a beam emerging from an absorbing retarder in terms of the incident field components E_x and E_y are

$$E_x' = E_x e^{+i\phi/2} e^{-\alpha_x} \tag{6-7a}$$

$$E_y' = E_y e^{-i\phi/2} e^{-\alpha_y} \tag{6-7b}$$

where α_x and α_y are the absorption coefficients. We can also express the exponential absorption factors in (6-7) as

$$p_x = e^{-\alpha_x} \tag{6-8a}$$

$$p_y = e^{-\alpha_y} \tag{6-8b}$$

Using (6-7) and (6-8) in the defining equations for the Stokes parameters, we find the Mueller matrix for an anisotropic absorbing retarder:

$$M = \frac{1}{2}\begin{pmatrix} p_x^2 + p_y^2 & p_x^2 - p_y^2 & 0 & 0 \\ p_x^2 - p_y^2 & p_x^2 + p_y^2 & 0 & 0 \\ 0 & 0 & 2p_x p_y \cos\phi & 2p_x p_y \sin\phi \\ 0 & 0 & -2p_x p_y \sin\phi & 2p_x p_y \cos\phi \end{pmatrix} \tag{6-9}$$

Thus, we see that an absorbing retarder behaves simultaneously as a polarizer and a retarder. If we use the angular representation for the polarizer behavior, Section 5.2, equation (5-15b), then we can write (6-9) as

$$M = \frac{p^2}{2}\begin{pmatrix} 1 & \cos 2\gamma & 0 & 0 \\ \cos 2\gamma & 1 & 0 & 0 \\ 0 & 0 & \sin 2\gamma\cos\phi & \sin 2\gamma\sin\phi \\ 0 & 0 & -\sin 2\gamma\sin\phi & \sin 2\gamma\cos\phi \end{pmatrix} \tag{6-10}$$

where $p_x^2 + p_y^2 = p^2$. We note that for $\gamma = 45^\circ$ we have an isotropic retarder; that is, the absorption is equal along both axes. If p^2 is also unity, then (6-9) reduces to an ideal phase retarder.

The intensity of the emerging beam $I(\theta, \phi)$ is obtained by multiplying (6-1) by (6-10) and then by (6-4), and the result is

$$\begin{aligned} I(\theta, \phi) = \frac{p^2}{2}[&(1 + \cos 2\theta\cos 2\gamma)S_0 + (\cos 2\gamma + \cos 2\theta)S_1 \\ &+ (\sin 2\gamma\cos\phi\sin 2\theta)S_2 + (\sin 2\gamma\sin\phi\sin 2\theta)S_3] \end{aligned} \tag{6-11}$$

If we were now to make all four intensity measurements with a quarter-wave retarder in the optical train, then (6-11) would reduce for each of the four combinations of θ and $\phi = 90^\circ$ to

$$S_0 = \frac{1}{p^2}[I(0^\circ, 0^\circ) + I(90^\circ, 0^\circ)] \tag{6-12a}$$

$$S_1 = \frac{1}{p^2}[I(0^\circ, 0^\circ) - I(90^\circ, 0^\circ)] \tag{6-12b}$$

$$S_2 = \frac{2}{p^2}I(45^\circ, 0^\circ) - S_0 \tag{6-12c}$$

$$S_3 = \frac{2}{p^2}I(45^\circ, 90^\circ) - S_0 \tag{6-12d}$$

Thus, each of the intensities in (6-12) are reduced by p^2, and this has no effect on the final value of the Stokes parameters with respect to each other. Furthermore, if we are interested in the ellipticity and the orientation, then we take ratios of the Stokes parameters S_3/S_0 and S_2/S_1 and the absorption factor p^2 cancels out. However, this is not exactly the way the measurement is made. Usually, the first three intensity measurements are made *without* the retarder present, so the first three parameters are measured according to (6-6). The last measurement is done *with* a quarter-wave retarder in the optical train, (6-12d), so the equations are

$$S_0 = I(0^\circ, 0^\circ) + I(90^\circ, 0^\circ) \tag{6-13a}$$

$$S_1 = I(0^\circ, 0^\circ) - I(90^\circ, 0^\circ) \tag{6-13b}$$

$$S_2 = 2I(45^\circ, 0^\circ) - S_0 \tag{6-13c}$$

$$S_3 = \frac{2}{p^2} I(45^\circ, 90^\circ) - S_0 \tag{6-13d}$$

Thus, (6-13d) shows that the absorption factor p^2 enters in the measurement of the fourth Stokes parameters S_3. It is therefore necessary to measure the absorption factor p^2. The easiest way to do this is to place a linear polarizer between an optical source and a detector and measure the intensity; this is called I_0. Next, the retarder with its fast axis in the horizontal x direction is inserted between the linear polarizer and the detector. The intensity is then measured with the polarizer generating linear horizontally and linear vertically polarized light [see (6-11)]. Dividing each of these measured intensities by I_0 and adding the results gives p^2. Thus, we see that the measurement of the first three Stokes parameters is very simple, but the measurement of the fourth parameter S_3 requires a considerable amount of additional effort.

It would therefore be preferable if a method could be devised whereby the absorption measurement could be eliminated. A method for doing this can be devised, and we now consider this method.

6.3 MEASUREMENT OF THE STOKES PARAMETERS USING A CIRCULAR POLARIZER

The problem of absorption by a retarder can be completely overcome by using a *single* polarizing element, namely, a circular polarizer; this is described below. The beam is allowed to enter one side of the circular polarizer, whereby the first three parameters can be measured. The circular polarizer is then flipped 180°, and the final Stokes parameter is measured. A circular polarizer is made by cementing a quarter-wave retarder to a linear polarizer with its axis at 45° to the fast axis of the retarder. This ensures that the retarder and polarizer axes are always fixed with respect to each other. Furthermore, because the same optical path is used in all four measurements, the problem of absorption vanishes; the four intensities are reduced by the same amount.

The construction of a circular polarizer is illustrated in Fig. 6-2.

The Mueller matrix for the polarizer–retarder combination is

$$M = \frac{1}{2}\begin{pmatrix} 1 & 0 & 0 & 0 \\ 0 & 1 & 0 & 0 \\ 0 & 0 & 0 & 1 \\ 0 & 0 & -1 & 0 \end{pmatrix}\begin{pmatrix} 1 & 0 & 1 & 0 \\ 0 & 0 & 0 & 0 \\ 1 & 0 & 1 & 0 \\ 0 & 0 & 0 & 0 \end{pmatrix} \tag{6-14a}$$

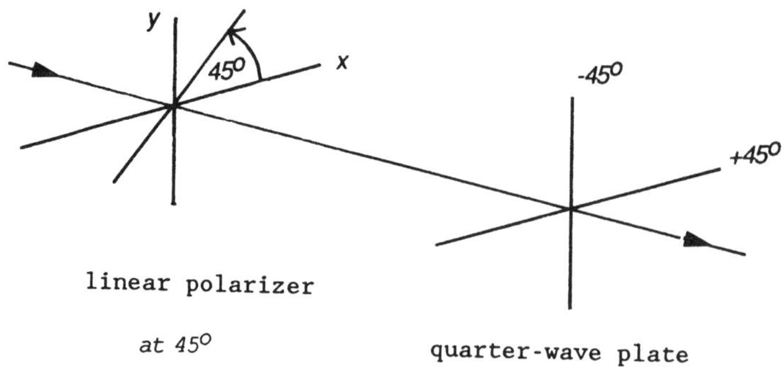

Figure 6-2 Construction of a circular polarizer using a linear polarizer and a quarter-wave retarder.

and thus

$$M = \frac{1}{2}\begin{pmatrix} 1 & 0 & 1 & 0 \\ 0 & 0 & 0 & 0 \\ 0 & 0 & 0 & 0 \\ -1 & 0 & -1 & 0 \end{pmatrix} \tag{6-14b}$$

Equation (6-14b) is the Mueller matrix of a circular polarizer. The reason for calling (6-14b) a circular polarizer is that regardless of the polarization state of the incident beam the emerging beam is always circularly polarized. This is easily shown by assuming that the Stokes vector of an incident beam is

$$S = \begin{pmatrix} S_0 \\ S_1 \\ S_2 \\ S_3 \end{pmatrix} \tag{6-1}$$

Multiplication of (6-1) by (6-14b) then yields

$$S' = \frac{1}{2}(S_0 + S_2)\begin{pmatrix} 1 \\ 0 \\ 0 \\ -1 \end{pmatrix} \tag{6-15}$$

which is the Stokes vector for left circularly polarized light (LCP). Thus, regardless of the polarization state of the incident beam, the output beam is *always* left circularly polarized. Hence, the name *circular* polarizer. Equation (6-14b) defines a circular polarizer.

Next, consider that the quarter-wave retarder–polarizer combination is "flipped"; that is, the linear polarizer now follows the quarter–wave retarder. The Mueller matrix for this combination is obtained with the Mueller matrices in (6-14a) interchanged; we note that the axis of the linear polarizer when it is flipped causes a sign change in the Mueller matrix (see Fig. 6-2). Then

$$M = \frac{1}{2}\begin{pmatrix} 1 & 0 & -1 & 0 \\ 0 & 0 & 0 & 0 \\ -1 & 0 & 1 & 0 \\ 0 & 0 & 0 & 0 \end{pmatrix}\begin{pmatrix} 1 & 0 & 0 & 0 \\ 0 & 1 & 0 & 0 \\ 0 & 0 & 0 & 1 \\ 0 & 0 & -1 & 0 \end{pmatrix} \tag{6-16a}$$

so

$$M = \frac{1}{2}\begin{pmatrix} 1 & 0 & 0 & -1 \\ 0 & 0 & 0 & 0 \\ -1 & 0 & 0 & 1 \\ 0 & 0 & 0 & 0 \end{pmatrix} \tag{6-16b}$$

Equation (6-16b) is the matrix of a linear polarizer. That (6-16b) is a linear polarizer can be easily seen by multiplying (6-1) by (6-16b):

$$S' = \frac{1}{2}(S_0 - S_3)\begin{pmatrix} 1 \\ 0 \\ -1 \\ 0 \end{pmatrix} \tag{6-17}$$

which is the Stokes vector for linear -45° polarized light. Regardless of the polarization state of the incident beam, the final beam is always linear $+45^\circ$ polarized. It is of interest to note that in the case of the "circular" side of the polarizer configuration, (6-15), the intensity varies only with the linear component, S_2, in the incident beam. On the other hand, for the "linear" side of the polarizer, (6-17), the intensity varies only with S_3, the circular component in the incident beam.

The circular polarizer is now placed in a rotatable mount. We saw earlier that the Mueller matrix for a rotated polarizing component, M, is given by the relation:

$$M(2\theta) = M_R(-2\theta) M M_R(2\theta) \tag{5-51}$$

where $M_R(2\theta)$ is the rotation Mueller matrix:

$$M_R(2\theta) = \begin{pmatrix} 1 & 0 & 0 & 0 \\ 0 & \cos 2\theta & \sin 2\theta & 0 \\ 0 & -\sin 2\theta & \cos 2\theta & 0 \\ 0 & 0 & 0 & 1 \end{pmatrix} \tag{5-52}$$

and $M(2\theta)$ is the Mueller matrix of the rotated polarizing element. The Mueller matrix for the circular polarizer with its axis rotated through an angle θ is then found by substituting (6-14b) into (5-51). The result is

$$M_C(2\theta) = \frac{1}{2}\begin{pmatrix} 1 & -\sin 2\theta & \cos 2\theta & 0 \\ 0 & 0 & 0 & 0 \\ 0 & 0 & 0 & 0 \\ 1 & \sin 2\theta & -\cos 2\theta & 0 \end{pmatrix} \tag{6-18}$$

where the subscript C refers to the fact that (6-18) describes the circular side of the polarizer combination. We see immediately that the Stokes vector emerging from the beam of the rotated circular polarizer is, using (6-18) and (6-1),

$$S_C = \frac{1}{2}(S_0 - S_1 \sin 2\theta + S_2 \cos 2\theta)\begin{pmatrix} 1 \\ 0 \\ 0 \\ -1 \end{pmatrix} \tag{6-19}$$

Thus, as the circular polarizer is rotated, the intensity varies but the polarization state remains unchanged, i.e., circular. We note again that the total intensity depends on S_0 and on the linear components, S_1 and S_2, in the incident beam.

The Mueller matrix when the circular polarizer is flipped to its linear side is, from (6-16b) and (5-51),

$$M_L(2\theta) = \frac{1}{2}\begin{pmatrix} 1 & 0 & 0 & -1 \\ \sin 2\theta & 0 & 0 & -\sin 2\theta \\ -\cos 2\theta & 0 & 0 & \cos 2\theta \\ 0 & 0 & 0 & 0 \end{pmatrix} \tag{6-20}$$

where the subscript L refers to the fact that (6-20) describes the linear side of the polarizer combination. The Stokes vector of the beam emerging from the rotated linear side of the polarizer, multiplying, (6-20) and (6-1), is

$$S_L = \frac{1}{2}(S_0 - S_3)\begin{pmatrix} 1 \\ \sin 2\theta \\ -\cos 2\theta \\ 0 \end{pmatrix} \tag{6-21}$$

Under a rotation of the circular polarizer on the linear side, (6-21) shows that the polarization is always linear. The total intensity is constant and depends on S_0 and the circular component S_3 in the incident beam.

The intensities detected on the circular and linear sides are, respectively, from (6-19) and (6-21),

$$I_C(\theta) = \frac{1}{2}(S_0 - S_1 \sin 2\theta + S_2 \cos 2\theta) \tag{6-22a}$$

$$I_L(\theta) = \frac{1}{2}(S_0 + S_3) \tag{6-22b}$$

The intensity on the linear side, (6-22b), is seen to be independent of the rotation angle of the polarizer. This fact allows a simple check when the measurement is being made. If the circular polarizer is rotated and the intensity does not vary, then one knows the measurement is being made on I_L, the linear side.

In order to obtain the Stokes parameters, we first use the circular side of the polarizing element and rotate it to $\theta = 0^\circ$, 45°, and 90°, and then flip it to the linear side. The measured intensities are then

$$I_C(0^\circ) = \frac{1}{2}(S_0 + S_2) \tag{6-23a}$$

$$I_C(45^\circ) = \frac{1}{2}(S_0 - S_1) \tag{6-23b}$$

$$I_C(90^\circ) = \frac{1}{2}(S_0 - S_2) \tag{6-23c}$$

$$I_L(0^\circ) = \frac{1}{2}(S_0 - S_3) \tag{6-23d}$$

The I_L value is conveniently taken to be $\theta = 0^\circ$. Solving (6-23) for the Stokes parameters yields

$$S_0 = I_C(0^\circ) + I_C(90^\circ) \tag{6-24a}$$

$$S_1 = S_0 - 2I_C(45^\circ) \tag{6-24b}$$

$$S_2 = I_C(0^\circ) - I_C(90^\circ) \tag{6-24c}$$

$$S_3 = S_0 - 2I_L(0^\circ) \tag{6-24d}$$

Equation (6-24) is similar to the classical equations for measuring the Stokes parameters, (6-6), but the intensity combinations are distinctly different. The use of a circular polarizer to measure the Stokes parameters is simple and accurate because (1) only a single rotating mount is used, (2) the polarizing beam propagates through the same optical path so that the problem of absorption losses can be ignored, and (3) the axes of the wave plate and polarizer are permanently fixed with respect to each other.

6.4 THE NULL-INTENSITY METHOD

In previous sections the Stokes parameters were expressed in terms of measured intensities. These measurement methods, however, are suitable only for use with quantitative detectors. We pointed out earlier that before the advent of solid-state detectors and photomultipliers the only available detector was the human eye. It can only measure the presence of light or no light (a null intensity). It is possible, as we shall now show, to measure the Stokes parameters from the condition of a null-intensity state. This can be done by using a variable retarder (phase shifter) followed by a linear polarizer in a rotatable mount. Devices are manufactured which can change the phase between the orthogonal components of an optical beam. They are called Babinet–Soleil compensators, and they are usually placed in a rotatable mount. Following the compensator is a linear polarizer, which is also placed in a rotatable mount. This arrangement can be used to obtain a null intensity. In order to carry out the analysis, the reader is referred to Fig. 6-3.

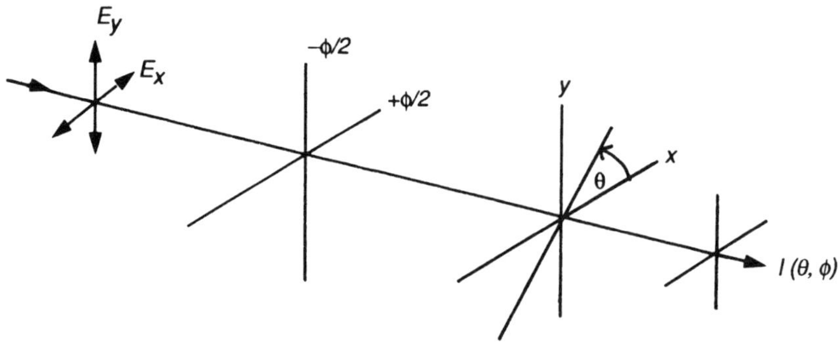

Figure 6-3 Null intensity measurement of the Stokes parameters.

The Stokes vector of the incident beam to be measured is

$$S = \begin{pmatrix} S_0 \\ S_1 \\ S_2 \\ S_3 \end{pmatrix} \tag{6-1}$$

The analysis is simplified considerably if the α, δ form of the Stokes vector derived in Section 4.3 is used:

$$S = I_0 \begin{pmatrix} 1 \\ \cos 2\alpha \\ \sin 2\alpha \cos \delta \\ \sin 2\alpha \sin \delta \end{pmatrix} \tag{4-38}$$

The axis of the Babinet–Soleil compensator is set at 0°. The Stokes vector of the beam emerging from the compensator is found by multiplying the matrix of the nonrotated compensator (Section 5.3, equation (5-27)) with (4-40):

$$S' = I_0 \begin{pmatrix} 1 & 0 & 0 & 0 \\ 0 & 1 & 0 & 0 \\ 0 & 0 & \cos\phi & \sin\phi \\ 0 & 0 & -\sin\phi & \cos\phi \end{pmatrix} \begin{pmatrix} 1 \\ \cos 2\alpha \\ \sin 2\alpha \cos \delta \\ \sin 2\alpha \sin \delta \end{pmatrix} \tag{6-25}$$

Carrying out the matrix multiplication in (6-25) and using the well-known trigonometric sum formulas, we readily find

$$S' = I_0 \begin{pmatrix} 1 \\ \cos 2\alpha \\ \sin 2\alpha \cos(\delta - \phi) \\ \sin 2\alpha \sin(\delta - \phi) \end{pmatrix} \tag{6-26}$$

Two important observations on (6-26) can be made. The first is that (6-26) can be transformed to linearly polarized light if S'_3 can be made to be equal to zero. This can be done by setting $\delta - \phi$ to 0°. If we then analyze S' with a linear polarizer, we see that a null intensity can be obtained by rotating the polarizer; at the null setting we can then determine α. This method is the procedure that is almost always used to obtain a null intensity. The null-intensity method works because δ in (6-25) is simply transformed to $\delta - \phi$ in (6-26) after the beam propagates through the compensator (retarder). For the moment we shall retain the form of (6-26) and not set $\delta - \phi$ to 0°. The function of the Babinet–Soleil compensator in this case is to transform elliptically polarized light to linearly polarized light.

Next, the beam represented by (6-26) is incident on a linear polarizer with its transmission axis at an angle θ. The Stokes vector S'' of the beam emerging from the rotated polarizer is now

$$S'' = \frac{I_0}{2}\begin{pmatrix} 1 & \cos 2\theta & \sin 2\theta & 0 \\ \cos 2\theta & \cos^2 2\theta & \sin 2\theta \cos 2\theta & 0 \\ \sin 2\theta & \sin 2\theta \cos 2\theta & \sin^2 2\theta & 0 \\ 0 & 0 & 0 & 0 \end{pmatrix}\begin{pmatrix} 1 \\ \cos 2\alpha \\ \sin 2\alpha \cos(\delta - \phi) \\ \sin 2\alpha \sin(\delta - \phi) \end{pmatrix} \tag{6-27}$$

where we have used the Mueller matrix of a rotated linear polarizer, Equation (5-54) Section 5.5. We are interested only in the intensity of the beam emerging from the rotated polarizer; that is, $S''_0 = I(\theta, \phi)$. Carrying out the matrix multiplication with the first row in the Mueller matrix and the Stokes vector in (6-27) yields

$$I(\theta, \phi) = \frac{I_0}{2}[1 + \cos 2\theta \cos 2\alpha + \sin 2\theta \sin 2\alpha \cos(\delta - \phi)] \tag{6-28}$$

We now set $\delta - \phi = 0$ in (6-28) and find

$$I(\theta, \delta) = \frac{I_0}{2}[1 + \cos 2\theta \cos 2\alpha + \sin 2\theta \sin 2\alpha] \tag{6-29a}$$

which reduces to

$$I(\theta, \delta) = \frac{I_0}{2}[1 + \cos 2(\theta - \alpha)] \tag{6-29b}$$

The linear polarizer is rotated until a null intensity is observed. At this angle $\theta - \alpha = \pi/2$, and we have

$$I\left(\alpha + \frac{\pi}{2}, \delta\right) = 0 \tag{6-30}$$

The angles δ and α associated with the Stokes vector of the incident beam are thus found from the conditions:

$$\delta = \phi \tag{6-31a}$$

$$\alpha = \theta - \frac{\pi}{2} \tag{6-31b}$$

Equations (6-31a) and (6-31b) are the required relations between α and δ of the Stokes vector (6-26) and ϕ and θ, the phase setting on the Babinet–Soleil compensator and the angle of rotation of the linear polarizer, respectively.

From the values obtained for α and δ we can determine the corresponding values for the orientation angle ψ and the ellipticity χ of the incident beam. We saw in (4-40) (Section 4.3) that ψ and χ could be expressed in terms of α and δ, namely,

$$\tan 2\psi = \tan 2\alpha \cos \delta \tag{4-40a}$$

$$\sin 2\chi = \sin 2\alpha \sin \delta \tag{4-40b}$$

Substituting (6-31) into (4-40), we see that ψ and χ can be expressed in the terms of the measured values of θ and ϕ:

$$\tan 2\psi = \tan 2\theta \cos \phi \tag{6-32a}$$

$$\sin 2\chi = -\sin 2\theta \sin \phi \tag{6-32b}$$

Remarkably, (6-32) is identical to (4-40) in form. It is only necessary to take the measured values of θ and ϕ and insert them into (6-32) to obtain ψ and χ. Equations (4-40a) and (4-40b) can be solved in turn for α and δ following the derivation given in Section 5.6, and we have

$$\cos 2\alpha = \pm \cos 2\chi \cos 2\psi \tag{6-33a}$$

$$\tan \delta = \frac{\tan 2\chi}{\sin 2\psi} \tag{6-33b}$$

The procedure to find the null-intensity angles θ and ϕ is first to set the Babinet–Soleil compensator with its fast axis to 0° and its phase angle to 0°. The phase is then adjusted until the intensity is observed to be a minimum. At this point in the measurement the intensity will not necessarily be zero, only a minimum, as we see from (6-29b),

$$I(\theta, \delta) = \frac{I_0}{2}[1 + \cos 2(\theta - \alpha)] \tag{6-29b}$$

Next, the linear polarizer is rotated through an angle θ until a null intensity is observed; the setting at which this angle occurs is then measured. In theory this completes the measurement. In practice, however, one finds that a small adjustment in phase of the compensator and rotation angle of the linear polarizer are almost always necessary to obtain a null intensity. Substituting the observed angular settings on the compensator and the polarizer into (6-32) and (6-33), we then find the Stokes vector (4-38) of the incident beam. We note that (4-38) is a normalized representation of the Stokes vector if I_0 is set to unity.

6.5 FOURIER ANALYSIS USING A ROTATING QUARTER-WAVE RETARDER

Another method for measuring the Stokes parameters is to allow a beam to propagate through a rotating quarter-wave retarder followed by a linear horizontal polarizer; the retarder rotates at an angular frequency of ω. This arrangement is shown in Fig. 6-4.

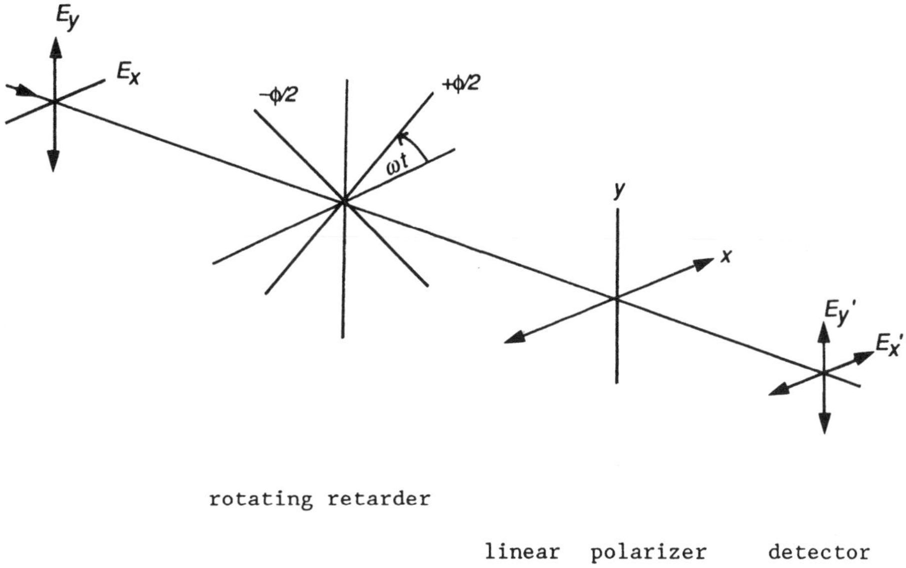

Figure 6-4 Measurement of the Stokes parameters using a rotating quarter-wave retarder and a linear polarizer.

The Stokes vector of the incident beam to be measured is

$$S = \begin{pmatrix} S_0 \\ S_1 \\ S_2 \\ S_3 \end{pmatrix} \tag{6-1}$$

The Mueller matrix of the rotated quarter-wave retarder (Section 5.5) is

$$M = \begin{pmatrix} 1 & 0 & 0 & 0 \\ 0 & \cos^2 2\theta & \sin 2\theta \cos 2\theta & -\sin 2\theta \\ 0 & \sin 2\theta \cos 2\theta & \sin^2 2\theta & \cos 2\theta \\ 0 & \sin 2\theta & -\cos 2\theta & 0 \end{pmatrix} \tag{5-72}$$

and for a rotating retarder we consider $\theta = \omega t$. Multiplying (6-1) by (5-72) yields

$$S' = \begin{pmatrix} S_0 \\ S_1 \cos^2 2\theta + S_2 \sin 2\theta \cos 2\theta - S_3 \sin 2\theta \\ S_1 \sin 2\theta \cos 2\theta + S_2 \sin^2 2\theta + S_3 \cos 2\theta \\ S_1 \sin 2\theta - S_2 \cos 2\theta \end{pmatrix} \tag{6-34}$$

The Mueller matrix of the linear horizontal polarizer is

$$M = \frac{1}{2}\begin{pmatrix} 1 & 1 & 0 & 0 \\ 1 & 1 & 0 & 0 \\ 0 & 0 & 0 & 0 \\ 0 & 0 & 0 & 0 \end{pmatrix} \tag{5-13}$$

The Stokes vector of the beam emerging from the rotating quarter-wave retarder–horizontal polarizer combination is then found from (6-34) and (5-13) to be

$$S' = \frac{1}{2}(S_0 + S_1 \cos^2 2\theta + S_2 \sin 2\theta \cos 2\theta - S_3 \sin 2\theta) \begin{pmatrix} 1 \\ 1 \\ 0 \\ 0 \end{pmatrix} \tag{6-35}$$

The intensity $S'_0 = I(\theta)$ is

$$I(\theta) = \frac{1}{2}(S_0 + S_1 \cos^2 2\theta + S_2 \sin 2\theta \cos 2\theta - S_3 \sin 2\theta) \tag{6-36}$$

Equation (6-36) can be rewritten by using the trigonometric half-angle formulas:

$$I(\theta) = \frac{1}{2}\left[\left(S_0 + \frac{S_1}{2}\right) + \frac{S_1}{2}\cos 4\theta + \frac{S_2}{2}\sin 4\theta - S_3 \sin 2\theta\right] \tag{6-37}$$

Replacing θ with ωt, (6-37) can be written as

$$I(\omega t) = \frac{1}{2}[A - B\sin 2\omega t + C\cos 4\omega t + D\sin 4\omega t] \tag{6-38a}$$

where

$$A = S_0 + \frac{S_1}{2} \tag{6-38b}$$

$$B = S_3 \tag{6-38c}$$

$$C = \frac{S_1}{2} \tag{6-38d}$$

$$D = \frac{S_2}{2} \tag{6-38e}$$

Equation (6-38) describes a truncated Fourier series. It shows that we have a d.c. term (A), a double frequency term (B), and two quadruple frequency terms (C and D). The coefficients are found by carrying out a Fourier analysis of (6-38). We easily find that ($\theta = \omega t$)

$$A = \frac{1}{\pi}\int_0^{2\pi} I(\theta)\, d\theta \tag{6-39a}$$

$$B = \frac{2}{\pi}\int_0^{2\pi} I(\theta)\sin 2\theta\, d\theta \tag{6-39b}$$

$$C = \frac{2}{\pi}\int_0^{2\pi} I(\theta)\cos 4\theta\, d\theta \tag{6-39c}$$

$$D = \frac{2}{\pi}\int_0^{2\pi} I(\theta)\sin 4\theta\, d\theta \tag{6-39d}$$

Solving (6-38) for the Stokes parameters gives

$$S_0 = A - C \tag{6-40a}$$

$$S_1 = 2C \tag{6-40b}$$

$$S_2 = 2D \tag{6-40c}$$

$$S_3 = B \tag{6-40d}$$

In practice, the quarter-wave retarder is placed in a fixed mount which can be rotated and driven by a stepper motor through N steps. Equation (6-38a) then becomes, with $\omega t = n\theta_j$ (θ_j is the step size),

$$I_n(\theta_j) = \frac{1}{2}[A - B\sin 2n\theta_j + C\cos 4n\theta_j + D\sin 4n\theta_j] \tag{6-41a}$$

and

$$A = \frac{2}{N}\sum_{n=1}^{N} I(n\theta_j) \tag{6-41b}$$

$$B = \frac{4}{N}\sum_{n=1}^{N} I(n\theta_j)\sin 2n\theta_j \tag{6-41c}$$

$$C = \frac{4}{N}\sum_{n=1}^{N} I(n\theta_j)\cos 4n\theta_j \tag{6-41d}$$

$$D = \frac{4}{N}\sum_{n=1}^{N} I(n\theta_j)\sin 4n\theta_j \tag{6-41e}$$

As an example of (6-41), consider the rotation of a quarter-wave retarder that makes a complete rotation in 16 steps, so $N = 16$. Then the step size is $\theta_j = 2\pi/N = 2\pi/16 = \pi/8$. Equation (6-41) is then written as

$$A = \frac{1}{8}\sum_{n=1}^{16} I\left(n\frac{\pi}{8}\right) \tag{6-42a}$$

$$B = \frac{1}{4}\sum_{n=1}^{16} I\left(n\frac{\pi}{8}\right)\sin\left(n\frac{\pi}{4}\right) \tag{6-42b}$$

$$C = \frac{1}{4}\sum_{n=1}^{16} I\left(n\frac{\pi}{8}\right)\cos\left(n\frac{\pi}{2}\right) \tag{6-42c}$$

$$D = \frac{1}{4}\sum_{n=1}^{16} I\left(n\frac{\pi}{8}\right)\sin\left(n\frac{\pi}{2}\right) \tag{6-42d}$$

Thus, the data array consists of 16 measured intensities I_1 through I_{16}. We have written each intensity value as $I(n\pi/8)$ to indicate that the intensity is measured at intervals of $\pi/8$; we observe that when $n = 16$ we have $I(2\pi)$ as expected. At each step the intensity is stored to form (6-42a), multiplied by $\sin(n\pi/4)$ to form B, $\cos(n\pi/2)$ to form C, and $\sin(n\pi/2)$ to form D. The sums are then performed according to (6-42), and we obtain A, B, C, and D. The Stokes parameters are then found from (6-40) using these values.

6.6 THE METHOD OF KENT AND LAWSON

In Section 6.4 we saw that the null-intensity condition could be used to determine the Stokes parameters and, hence, the polarization state of an optical beam. The null-intensity method remained the only practical way to measure the polarization state of an optical beam before the advent of photodetectors. It is fortunate that the eye is so sensitive to light and can easily detect its presence or absence. Had this not been the case, the progress made in polarized light would surely not have been as rapid as it was. One can obviously use a photodetector as well as the eye, using the null-intensity method described in Section 6.4. However, the existence of photodetectors allows one to consider an extremely interesting and novel method for determining the polarization state of an optical beam.

In 1937, C. V. Kent and J. Lawson proposed a new method for measuring the ellipticity and orientation of a polarized optical beam using a Babinet–Soleil compensator and a photomultiplier tube (PMT). They noted that it was obvious that a photomultiplier could simply replace the human eye as a detector, and used to determine the null condition. However, Kent and Lawson went beyond this and made several important observations. The first was that the use of the PMT could obviously overcome the problem of eye fatigue. They also noted that, in terms of sensitivity (at least in 1937) for weak illuminations, determining the null intensity was as difficult with a PMT as with the human eye. They observed that the PMT really operated best with full illumination. In fact, because the incident light at a particular wavelength is usually much greater than the laboratory illumination the measurement could be done with the room lights on. They now noted that this property of the PMT could be exploited fully if the incident optical beam whose polarization was to be determined was transformed not to linearly polarized light but to circularly polarized light. By then analyzing the beam with a rotating linear polarizer, a constant intensity would be obtained when the condition of circularly polarized light was obtained or, as they said, "no modulation." From this condition of "no modulation" the ellipticity and orientation angles of the incident beam could then be determined. Interestingly, they detected the circularly polarized light by converting the optical signal to an audio signal and then used a headphone set to determine the constant-intensity condition.

It is worthwhile to study this method because it enables us to see how photodetectors provide an alternative method for measuring the Stokes parameters and how they can be used to their optimum, that is, in the measurement of polarized light at high intensities. The measurement is described by the experimental configuration in Fig. 6-5. The Stokes vector of the incident elliptically polarized beam to be measured is represented by

$$S = \begin{pmatrix} S_0 \\ S_1 \\ S_2 \\ S_3 \end{pmatrix} \tag{6-1}$$

The primary use of a Babinet–Soleil compensator is to create an arbitrary state of elliptically polarized light. This is accomplished by changing the phase

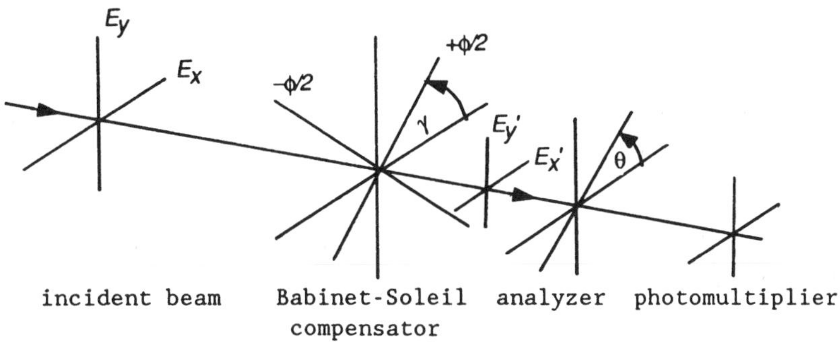

Figure 6-5 Measurement of the ellipticity and orientation of an elliptically polarized beam using a compensator and a photodetector.

and orientation of the incident beam. We recall from Section 5.5 that the Mueller matrix for a rotated retarder is

$$M_C(\phi, 2\gamma) = \begin{pmatrix} 1 & 0 & 0 & 0 \\ 0 & \cos^2 2\gamma + \cos\phi \sin^2 2\gamma & (1 - \cos\phi)\sin 2\gamma \cos 2\gamma & -\sin\phi \sin 2\gamma \\ 0 & (1 - \cos\phi)\sin 2\gamma \cos 2\gamma & \sin^2 2\gamma + \cos\phi \cos^2 2\gamma & \sin\phi \cos 2\gamma \\ 0 & \sin\phi \sin 2\gamma & -\sin\phi \cos 2\gamma & \cos\phi \end{pmatrix} \tag{6-43}$$

where γ is the angle that the fast axis makes with the horizontal x axis and ϕ is the phase shift.

The beam emerging from the Babinet–Soleil compensator is then found by multiplying (6-1) by (6-43):

$$S' = \begin{pmatrix} S_0 \\ S_1(\cos^2 2\gamma + \cos\phi \sin^2 2\gamma) + S_2(1 - \cos\phi)\sin 2\gamma \cos 2\gamma - S_3 \sin\phi \sin 2\gamma \\ S_1(1 - \cos\phi)\sin 2\gamma \cos 2\gamma + S_2(\sin^2 2\gamma + \cos\phi \cos^2 2\gamma) + S_3 \sin\phi \cos 2\gamma \\ S_1 \sin\phi \sin 2\gamma - S_2 \sin\phi \cos 2\gamma + S_3 \cos\phi \end{pmatrix} \tag{6-44}$$

For the moment let us assume that we have elliptically polarized light incident on a rotating ideal linear polarizer. The Stokes vector of the beam incident on the rotating linear polarizer is represented by

$$S = \begin{pmatrix} 1 \\ \cos 2\alpha \\ \sin 2\alpha \cos\delta \\ \sin 2\alpha \sin\delta \end{pmatrix} \tag{4-39}$$

The Mueller matrix of the rotating linear polarizer is

$$M = \frac{1}{2}\begin{pmatrix} 1 & \cos 2\theta & \sin 2\theta & 0 \\ \cos 2\theta & \cos^2 2\theta & \sin 2\theta \cos 2\theta & 0 \\ \sin 2\theta & \sin 2\theta \cos 2\theta & \sin^2 2\theta & 0 \\ 0 & 0 & 0 & 0 \end{pmatrix} \tag{6-4}$$

The Stokes vector of the beam emerging from the rotating analyzer is found by multiplying (6-1) by (6-4)

$$S' = \frac{1}{2}[1 + \cos 2\alpha \cos 2\theta + \sin 2\alpha \cos \delta \sin 2\theta]\begin{pmatrix} 1 \\ \cos 2\theta \\ \sin 2\theta \\ 0 \end{pmatrix} \tag{6-45}$$

Thus, as the analyzer is rotated we see that the intensity is modulated. If the intensity is to be independent of the rotation angle θ, then we must have

$$\cos 2\alpha = 0 \tag{6-46a}$$

$$\sin 2\alpha \cos \delta = 0 \tag{6-46b}$$

We immediately see that (6-46a) and (6-46b) are satisfied if $2\alpha = 90^\circ$ (or 270°) and $\delta = 90^\circ$. Substituting these values in (4-39), we have

$$S = \begin{pmatrix} 1 \\ 0 \\ 0 \\ 1 \end{pmatrix} \tag{6-47}$$

which is the Stokes vector for right circularly polarized light.

In order to obtain circularly polarized light, the Stokes parameters in (6-44) must satisfy the conditions:

$$S_0' = S_0 \tag{6-48a}$$

$$S_1' = S_1(\cos^2 2\gamma + \cos\phi \sin^2 2\gamma) + S_2(1 - \cos\phi)\sin 2\gamma \cos 2\gamma - S_3 \sin\phi \sin 2\gamma = 0 \tag{6-48b}$$

$$S_2' = S_1(1 - \cos\phi)\sin 2\gamma \cos 2\gamma + S_2(\sin^2 2\gamma + \cos\phi \cos^2 2\gamma) + S_3(\sin\phi \cos 2\gamma) = 0 \tag{6-48c}$$

$$S_3' = S_1(\sin\phi \sin 2\gamma) - S_2(\sin\phi \cos 2\gamma) + S_3 \cos\phi \tag{6-48d}$$

We must now solve these equations for S_1, S_2, and S_3 in terms of γ and ϕ (S_0 is unaffected by the wave plate). While it is straightforward to solve (6-48), the algebra is surprisingly tedious and complicated. Fortunately, the problem can be solved in another way, because we know the transformation equation for describing a rotated compensator.

To solve this problem, we take the following approach. According to Fig. 6-5, the Stokes vector of the beam S' emerging from the compensator is related to the Stokes vector of the incident beam S by the equation:

$$S' = M_C(2\gamma)S \tag{6-49}$$

where $M_C(2\gamma)$ is given by (6-43) above. We recall that $M_C(2\gamma)$ is the rotated Mueller matrix for a retarder, so (6-49) can also be written as

$$S' = [M(-2\gamma)M_C M(2\gamma)]S \tag{6-50a}$$

where

$$M(2\gamma) = \begin{pmatrix} 1 & 0 & 0 & 0 \\ 0 & \cos 2\gamma & \sin 2\gamma & 0 \\ 0 & -\sin 2\gamma & \cos 2\gamma & 0 \\ 0 & 0 & 0 & 1 \end{pmatrix} \tag{6-50b}$$

and

$$M_C = \begin{pmatrix} 1 & 0 & 0 & 0 \\ 0 & 1 & 0 & 0 \\ 0 & 0 & \cos\phi & \sin\phi \\ 0 & 0 & -\sin\phi & \cos\phi \end{pmatrix} \tag{6-50c}$$

We now demand that our resultant Stokes vector represents right circularly polarized light and write (6-50a) as

$$S' = M(-2\gamma)M_C M(2\gamma)\begin{pmatrix} S_0 \\ S_1 \\ S_2 \\ S_3 \end{pmatrix} = \begin{pmatrix} 1 \\ 0 \\ 0 \\ 1 \end{pmatrix} \tag{6-51}$$

While we could immediately invert (6-51) to find the Stokes vector of the incident beam, it is simplest to find S in steps. Multiplying both sides of (6-51) by $M(2\gamma)$, we have

$$M_C M(2\gamma)\begin{pmatrix} S_0 \\ S_1 \\ S_2 \\ S_3 \end{pmatrix} = M(2\gamma)\begin{pmatrix} 1 \\ 0 \\ 0 \\ 1 \end{pmatrix} = \begin{pmatrix} 1 \\ 0 \\ 0 \\ 1 \end{pmatrix} \tag{6-52}$$

Next, we multiply (6-52) by M_C^{-1} to find

$$M(2\gamma)\begin{pmatrix} S_0 \\ S_1 \\ S_2 \\ S_3 \end{pmatrix} = M_C^{-1}\begin{pmatrix} 1 \\ 0 \\ 0 \\ 1 \end{pmatrix} = \begin{pmatrix} 1 \\ 0 \\ -\sin\phi \\ \cos\phi \end{pmatrix} \tag{6-53}$$

Finally, (6-53) is multiplied by $M(-2\gamma)$, and we have

$$\begin{pmatrix} S_0 \\ S_1 \\ S_2 \\ S_3 \end{pmatrix} = M(-2\gamma)\begin{pmatrix} 1 \\ 0 \\ -\sin\phi \\ \cos\phi \end{pmatrix} = \begin{pmatrix} 1 \\ -\sin 2\gamma \sin\phi \\ \cos 2\gamma \sin\phi \\ \cos\phi \end{pmatrix} \tag{6-54}$$

We can check to see if (6-54) is correct. We know that if $\phi = 0^\circ$, that is, the retarder is not present, then the only way S' can be right circularly polarized is if the incident beam S is right circularly polarized. Substituting $\phi = 0^\circ$ into (6-54), we find

$$S = \begin{pmatrix} 1 \\ 0 \\ 0 \\ 1 \end{pmatrix} \tag{6-55}$$

which is the Stokes vector for right circularly polarized light.

The numerical value of the Stokes parameters can be determined directly from (6-54). However, we can also express the Stokes parameters in terms of α and δ in (4-39) or in terms of the orientation and ellipticity angles ψ and χ (Section 4-3). Thus, we can equate (4-39) to (6-54) and write

$$\begin{pmatrix} S_0 \\ S_1 \\ S_2 \\ S_3 \end{pmatrix} = \begin{pmatrix} 1 \\ \cos 2\alpha \\ \sin 2\alpha \cos\delta \\ \sin 2\alpha \sin\delta \end{pmatrix} = \begin{pmatrix} 1 \\ -\sin 2\gamma \sin\phi \\ \cos 2\gamma \sin\phi \\ \cos\phi \end{pmatrix} \tag{6-56}$$

or, in terms of the orientation and ellipticity angles,

$$\begin{pmatrix} S_0 \\ S_1 \\ S_2 \\ S_3 \end{pmatrix} = \begin{pmatrix} 1 \\ \cos 2\chi \cos 2\psi \\ \cos 2\chi \sin 2\psi \\ \sin 2\chi \end{pmatrix} = \begin{pmatrix} 1 \\ -\sin 2\gamma \sin\phi \\ \cos 2\gamma \sin\phi \\ \cos\phi \end{pmatrix} \tag{6-57}$$

We now solve for S in terms of the measured values of γ and ϕ. Let us first consider (6-56) and equate the matrix elements:

$$\cos 2\alpha = \pm \sin 2\gamma \sin\phi \tag{6-58a}$$

$$\sin 2\alpha \cos\delta = \cos 2\gamma \sin\phi \tag{6-58b}$$

$$\sin 2\alpha \sin\delta = \pm \cos\phi \tag{6-58c}$$

In (6-58) we have written $\pm$ to include left circularly polarized light. We divide (6-58b) by (6-58c) and find

$$\cot\delta = \pm \cos 2\gamma \tan\phi \tag{6-59a}$$

Similarly, we divide (6-58b) by (6-58a) and find

$$\cos\delta = \pm \cot 2\gamma \cot 2\alpha \tag{6-59b}$$

We can group the results by renumbering (6-58a) and (6-59) and write

$$\cos 2\alpha = \pm \sin 2\gamma \sin\phi \tag{6-60a}$$

$$\cot\delta = \pm \cos 2\gamma \tan\phi \tag{6-60b}$$

$$\cos\delta = \pm \cot 2\gamma \cot 2\alpha \tag{6-60c}$$

Equations (6-60) are the equations of Kent and Lawson.

Thus, by measuring γ and ϕ, the angular rotation and phase shift of the Babinet–Soleil compensator, respectively, we can determine the azimuth α and phase δ of the incident beam. We also pointed out that we can use γ and ϕ to determine the ellipticity χ and orientation ψ of the incident beam from (6-57). Equating terms in (6-57) we have

$$\cos 2\chi \cos 2\psi = \pm \sin 2\gamma \sin\phi \tag{6-61a}$$

$$\cos 2\chi \sin 2\psi = \cos 2\gamma \sin\phi \tag{6-61b}$$

$$\sin 2\chi = \pm \cos\phi \tag{6-61c}$$

Dividing (6-61b) by (6-61a), we find

$$\tan 2\psi = \pm \cot 2\gamma \tag{6-62}$$

Squaring (6-61a) and (6-61b), adding, and taking the square root gives

$$\cos 2\chi = \sin\phi \tag{6-63}$$

Dividing (6-61c) by (6-63) then gives

$$\tan 2\chi = \pm \cot\phi \tag{6-64}$$

We renumber (6-62) and (6-63) as the pair:

$$\tan 2\psi = \pm \cot 2\gamma \tag{6-65a}$$

$$\tan 2\chi = \pm \cot\phi \tag{6-65b}$$

We can rewrite (6-65a) and (6-65b) as

$$\tan 2\psi = \pm \tan(90^\circ - 2\gamma) \tag{6-66a}$$

$$\tan 2\chi = \pm \tan(90^\circ - \phi) \tag{6-66b}$$

so

$$\psi = 45^\circ - \gamma \tag{6-67a}$$

$$\chi = 45^\circ - \frac{\phi}{2} \tag{6-67b}$$

We can check (6-67a) and (6-67b). We know that a linear $+45^\circ$ polarized beam of light is transformed to right circularly polarized light if we send it through a quarter-wave retarder. In terms of the incident beam, $\psi = 45^\circ$ and $\chi = 0^\circ$. Substituting these values in (6-67a) and (6-67b), respectively, we find that $\gamma = 0^\circ$ and $\phi = 90^\circ$ for the retarder. This is exactly what we would expect using a quarter-wave retarder with its fast axis in the x direction.

While nulling techniques for determining the elliptical parameters are very common, we see that the method of Kent and Lawson provides a very interesting alternative. We emphasize that nulling techniques were developed long before the appearance of photodetectors. Nulling techniques continue to be used because they are extremely sensitive and require, in principle, only an analyzer. Nevertheless, the method of Kent and Lawson has a number of advantages, foremost of which is that it can be used in ambient light and with high optical intensities. The method of Kent and Lawson requires the use of a Babinet–Soleil compensator and a rotatable polarizer. However, the novelty and potential of the method and its full exploitation of the quantitative nature of photodetectors should not be overlooked.

6.7 SIMPLE TESTS TO DETERMINE THE STATE OF POLARIZATION OF AN OPTICAL BEAM

In the laboratory one often has to determine if an optical beam is unpolarized, partially polarized, or completely polarized. If it is completely polarized, then we must determine if it is elliptically polarized or linearly or circularly polarized. In this section we consider this problem. Stokes' method for determining the Stokes parameters is a very simple and direct way of carrying out these tests (Section 4.4).

We recall that the polarization state can be measured using a linear polarizer and a quarter-wave retarder. If a polarizer made of calcite is used, then it transmits satisfactorily from 0.2 μm to 2.0 μm, more than adequate for visual work and into the near infrared. Quarter-wave retarders, on the other hand, are designed to transmit at a single wavelength, e.g., He–Ne laser radiation at 0.6328 μm. Therefore, the quarter-wave retarder should be matched to the wavelength of the polarizing radiation. In Fig. 6-6 we show the experimental configuration for determining the state of polarization. We emphasize that we are not trying to determine the Stokes parameters quantitatively but merely determining the polarization state of the light.

We recall from Section 6.2 that the intensity $I(\theta, \phi)$ of the beam emerging from the retarder–polarizer combination shown in Fig. 6-6 is

$$I(\theta, \phi) = \frac{1}{2}[S_0 + S_1 \cos 2\theta + S_2 \cos\phi \sin 2\theta + S_3 \sin\phi \sin 2\theta] \tag{6-5}$$

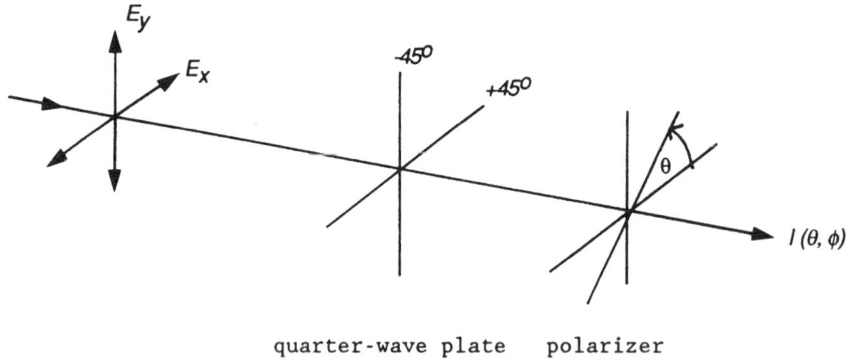

Figure 6-6 Experimental configuration to determine the state of polarization of an optical beam.

where θ is the angle of rotation of the polarizer and ϕ is the phase shift of the retarder. In our tests we shall set ϕ to 0° (no retarder in the optical train) or 90° (a quarter-wave retarder in the optical train). The respective intensities according to (6-5) are then

$$I(\theta, 0^\circ) = \frac{1}{2}[S_0 + S_1 \cos 2\theta + S_2 \sin 2\theta] \tag{6-68a}$$

$$I(\theta, 90^\circ) = \frac{1}{2}[S_0 + S_1 \cos 2\theta + S_3 \sin 2\theta] \tag{6-68b}$$

The first test we wish to perform is to determine if the light is unpolarized or completely polarized. In order to determine if it is unpolarized, the retarder is removed ($\phi = 0^\circ$), so we use (6-68a). The polarizer is now rotated through 180°. If the intensity remains constant throughout the rotation, then we must have

$$S_1 = S_2 = 0 \qquad \text{and} \qquad S_0 \neq 0 \tag{6-69}$$

If the intensity varies so (6-69) is not satisfied, then we know that we do not have unpolarized light. If, however, the intensity remains constant, then we are still not certain if we have unpolarized light because the parameter S_3 may be present.
We must, therefore, test for its presence. The retarder is now reintroduced into the optical train, and we use (6-68b):

$$I(\theta, 90^\circ) = \frac{1}{2}[S_0 + S_1 \cos 2\theta + S_3 \sin 2\theta] \tag{6-68b}$$

The polarizer is now rotated. If the intensity remains constant, then

$$S_1 = S_3 = 0 \qquad \text{and} \qquad S_0 \neq 0 \tag{6-70}$$

Thus, from (6-69) and (6-70) we see that (6-5) becomes

$$I(\theta, \phi) = \frac{1}{2} S_0 \tag{6-71}$$

which is the condition for unpolarized light.

If neither (6-69) or (6-70) is satisfied, we then assume that the light is elliptically polarized; the case of partially polarized light is excluded for the moment. Before we test for elliptically polarized light, however, we test for linear or circular polarization. In order to test for linearly polarized light, the retarder is removed from the optical train and so the intensity is again given by (6-68a):

$$I(\theta, 0^\circ) = \frac{1}{2}[S_0 + S_1 \cos 2\theta + S_2 \sin 2\theta] \tag{6-68a}$$

We recall that the Stokes vector for elliptically polarized light is

$$S = I_0 \begin{pmatrix} 1 \\ \cos 2\alpha \\ \sin 2\alpha \cos \delta \\ \sin 2\alpha \sin \delta \end{pmatrix} \tag{4-38}$$

Substituting S_1 and S_2 in (4-38) into (6-68a) gives

$$I(\theta, 0^\circ) = \frac{1}{2}[1 + \cos 2\alpha \cos 2\theta + \sin 2\alpha \cos \delta \sin 2\theta] \tag{6-72}$$

The polarizer is again rotated. If we obtain a null intensity, then we know that we have linearly polarized light because (6-68a) can only become a null if $\delta = 0^\circ$ or 180°, a condition for linearly polarized light. For this condition we can write (6-72) as

$$I(\theta, 0^\circ) = \frac{1}{2}[1 + \cos(2\alpha - 2\theta)] \tag{6-73}$$

which can only be zero if the incident beam is linearly polarized light. However, if we do not obtain a null intensity, we can have elliptically polarized light or circularly polarized light. To test for these possibilities, the quarter-wave retarder is reintroduced into the optical train so that the intensity is again given by (6-68b):

$$I(\theta, 90^\circ) = \frac{1}{2}[S_0 + S_1 \cos 2\theta + S_3 \sin 2\theta] \tag{6-68b}$$

Now, if we have circularly polarized light, then S_1 must be zero so (6-68b) will become

$$I(\theta, 90^\circ) = \frac{1}{2}[S_0 + S_3 \sin 2\theta] \tag{6-74}$$

The polarizer is again rotated. If a null intensity is obtained, then we must have circularly polarized light. If, on the other hand, a null intensity is not obtained, then we must have a condition described by (6-68b), which is elliptically polarized light.

To summarize, if a null intensity is not obtained with either the polarizer by itself or with the combination of the polarizer and the quarter-wave retarder, then we must have elliptically polarized light.

Thus, by using a polarizer–quarter-wave retarder combination, we can test for the polarization states. The only state remaining is partially polarized light. If none of these tests described above is successful, we then assume that the incident beam is partially polarized.

To be completely confident of the tests, it is best to use a high-quality calcite polarizer and a quartz quarter-wave retarder. It is, of course, possible to make these tests with Polaroid and mica quarter-wave retarders. However, these materials are not as good, in general, as calcite and quartz and there is less confidence in the results. See Chapter 26 for information on these elements.

If we are certain that the light is elliptically polarized, then we can consider (6-5) further. Equation (6-5) is

$$I(\theta, \phi) = \frac{1}{2}[S_0 + S_1 \cos 2\theta + S_2 \cos \phi \sin 2\theta + S_3 \sin \phi \sin 2\theta] \tag{6-5}$$

We can express (6-5) as

$$I(\theta, \phi) = \frac{1}{2}[S_0 + S_1 \cos 2\theta + (S_2 \cos \phi + S_3 \sin \phi) \sin 2\theta] \tag{6-75}$$

or

$$I(\theta, \phi) = [A + B\cos 2\theta + C\sin 2\theta] \tag{6-76a}$$

where

$$A = \frac{S_0}{2} \tag{6-76b}$$

$$B = \frac{S_1}{2} \tag{6-76c}$$

$$C = \frac{S_2\cos\phi + S_3\sin\phi}{2} \tag{6-76d}$$

For an elliptically polarized beam given by (4-38), I_0 is normalized to 1, and we write

$$S = \begin{pmatrix} 1 \\ \cos 2\alpha \\ \sin 2\alpha\cos\delta \\ \sin 2\alpha\sin\delta \end{pmatrix} \tag{4-39}$$

so from (6-76) we see that

$$A = \frac{1}{2} \tag{6-77a}$$

$$B = \frac{\cos 2\alpha}{2} \tag{6-77b}$$

$$C = \frac{\cos(\phi - \delta)\sin 2\alpha}{2} \tag{6-77c}$$

The intensity (6-76a) can then be written as

$$I = \frac{1}{2}[1 + \cos 2\alpha\cos 2\theta + \sin 2\alpha\cos(\phi - \delta)\sin 2\theta] \tag{6-77d}$$

We now find the maximum and minimum intensities of (6-77d) by differentiating (6-77d) with respect to θ and setting $dI(\theta)/d\theta = 0$. The angles where the maximum and minimum intensities occur are then found to be

$$\tan 2\theta = \frac{C}{B} = \frac{-C}{-B} \tag{6-78}$$

Substituting (6-78) into (6-76a), the corresponding maximum and minimum intensities are, respectively,

$$I(\max) = A + \sqrt{B^2 + C^2} \tag{6-79a}$$

$$I(\min) = A - \sqrt{B^2 + C^2} \tag{6-79b}$$

From (6-69) we see that we can then write (6-79) as

$$I(\max, \min) = \frac{1}{2}\left[1 \pm \sqrt{\cos^2 2\alpha + \sin^2 2\alpha\cos^2(\phi - \delta)}\right] \tag{6-80}$$

Let us now remove the retarder from the optical train so that $\phi = 0^\circ$; we then have only a linear polarizer which can be rotated through θ. Equation (6-80) then reduces to

$$I(\max, \min) = \frac{1}{2}\left[1 \pm \sqrt{\cos^2 2\alpha + \sin^2 2\alpha \cos^2 \delta}\,\right] \tag{6-81}$$

For linearly polarized light $\delta = 0^\circ$ or 180°, so (4-39) becomes

$$S = \begin{pmatrix} 1 \\ \cos 2\alpha \\ \pm\sin 2\alpha \\ 0 \end{pmatrix} \tag{6-82}$$

and (6-81) becomes

$$I(\max, \min) = \frac{1}{2}[1 \pm 1] = 1, 0 \tag{6-83}$$

Thus, linearly polarized light always gives a maximum intensity of unity and a minimum intensity of zero (null).

Next, if we have circularly polarized light, $\delta = 90^\circ$ or 270° and $\alpha = 45^\circ$, as is readily shown by inspecting (4-39). For this condition (6-81) reduces to

$$I(\max, \min) = \frac{1}{2}[1 \pm 0] = \frac{1}{2} \tag{6-84}$$

so the intensity is always constant and reduced to 1/2. We also see that if we have *only* the condition $\delta = 90^\circ$ or 270°, then (4-39) becomes

$$S = \begin{pmatrix} 1 \\ \cos 2\alpha \\ 0 \\ \pm\sin 2\alpha \end{pmatrix} \tag{6-85}$$

which is the Stokes vector of an ellipse in a standard form, i.e., unrotated. The corresponding intensity is, from (6-80),

$$I(\max, \min) = \frac{1}{2}[1 \pm \cos 2\alpha] \tag{6-86}$$

Similarly, if $\alpha = \pm 45^\circ$ and δ is not equal to either 90° or 270°, then (4-39) becomes

$$S = \begin{pmatrix} 1 \\ 0 \\ \cos\delta \\ \sin\delta \end{pmatrix} \tag{6-87}$$

and (6-81) reduces to

$$I(\max, \min) = \frac{1}{2}[1 \pm \cos\delta] \tag{6-88}$$

This final analysis confirms the earlier results given in the first part of this chapter. We see that if we rotate a linear polarizer and we observe a

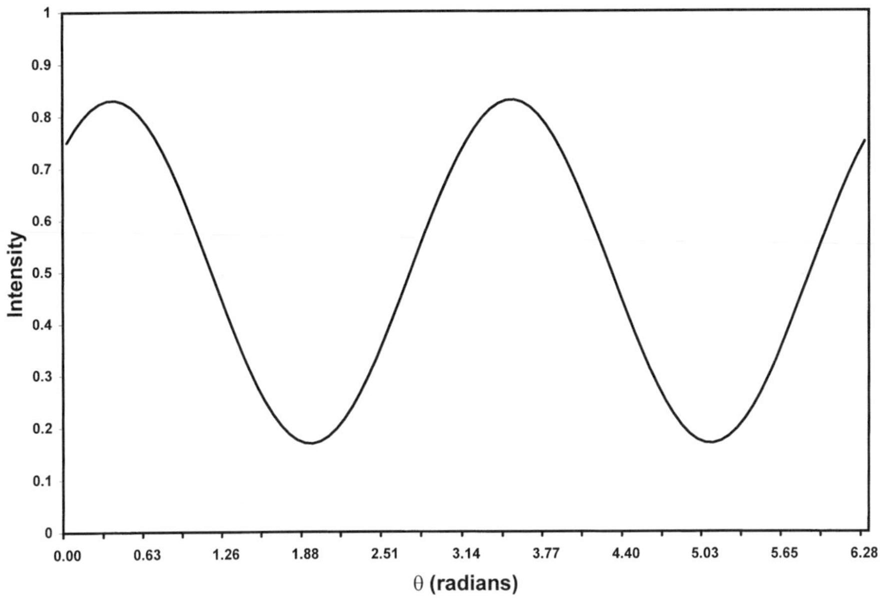

Figure 6-7 Intensity plot of an elliptically polarized beam for $\alpha = \pi/6$ and $\delta = \pi/3$.

null intensity at two angles over a single rotation, we have linearly polarized light; if we observe a constant intensity, we have circularly polarized light; and if we observe maximum and minimum (non-null) intensities, we have elliptically polarized light.

In Figs. 6-7 and 6-8 we have plotted the intensity as a function of the rotation angle of the analyzer. Specifically, in Fig. 6-7 we show the intensity for the condition where the parameters of the incident beam described by (4-39) are $\alpha = \pi/6$ (30°) and $\delta = \pi/3$ (60°); the compensator is not in the wave train, so $\phi = 0$. According to (4-39), the Stokes vector is

$$S = \begin{pmatrix} 1 \\ 1/2 \\ \sqrt{3}/4 \\ 3/4 \end{pmatrix} \tag{6-89}$$

The intensity expected for (6-89) is seen from (6-77d) to be

$$I(\theta) = \frac{1}{2}\left[1 + \frac{1}{2}\cos 2\theta + \frac{\sqrt{3}}{4}\sin 2\theta\right] \tag{6-90}$$

The plot of (6-90) is given in Fig. 6-7.

We see from (6-89) that the square root of the sum of the squares S_1, S_2, and S_3 is equal to unity as expected. Inspecting Fig. 6-8, we see that there is a maximum intensity and a minimum intensity. However, because there is no null intensity we know that the light is elliptically polarized, which agrees, of course, with (6-89).

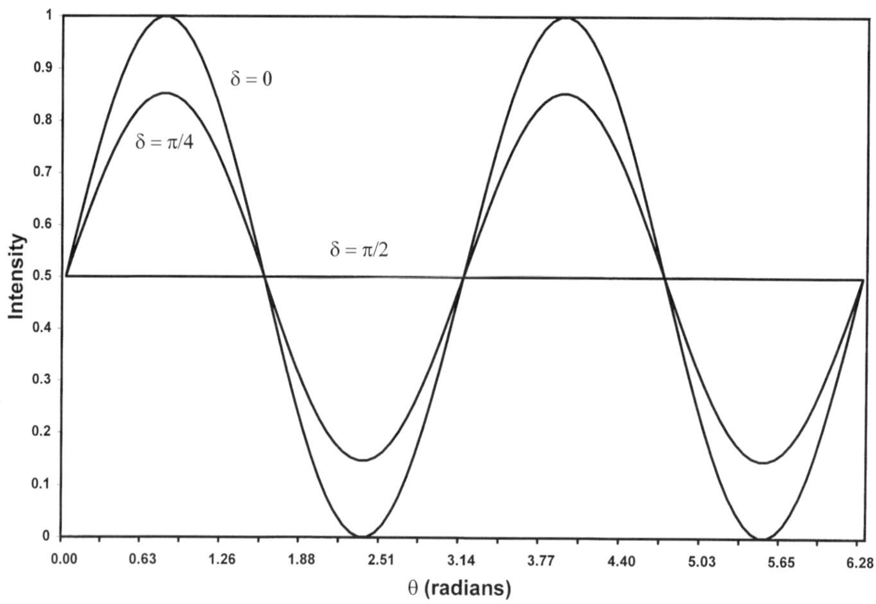

Figure 6-8 Plot of the intensity for a linearly polarized beam, an elliptically polarized beam, and a circularly polarized beam.

In Fig. 6-8 we consider an elliptically polarized beam such that $\alpha = \pi/4$ and we have arbitrary phase δ. This beam is described by the Stokes vector given by (6-87):

$$S = \begin{pmatrix} 1 \\ 0 \\ \cos\delta \\ \sin\delta \end{pmatrix} \tag{6-87}$$

The corresponding intensity for (6-87), according to (6-77d), is

$$I = \frac{1}{2}[1 + \cos\delta \sin 2\theta] \tag{6-91}$$

We now consider (6-87) for $\delta = 0$, $\pi/4$, and $\pi/2$. The Stokes vectors corresponding to these conditions are, respectively,

$$S(0) = \begin{pmatrix} 1 \\ 0 \\ 1 \\ 0 \end{pmatrix} \qquad s\left(\frac{\pi}{4}\right) = \begin{pmatrix} 1 \\ 0 \\ \frac{1}{\sqrt{2}} \\ \frac{1}{\sqrt{2}} \end{pmatrix} \qquad S\left(\frac{\pi}{2}\right) = \begin{pmatrix} 1 \\ 0 \\ 0 \\ 1 \end{pmatrix} \tag{6-92}$$

The Stokes vectors in (6-92) correspond to linear +45° polarized light, elliptically polarized light, and right circularly polarized light. Inspection of Fig. 6-8 shows the corresponding plot for the intensities given by (6-91) for each of the

Stokes vectors in (6-92). The linearly polarized beam gives a null intensity, the elliptically polarized beam gives maximum and minimum intensities, and the circularly polarized beam yields a constant intensity of 0.5.

REFERENCES

Papers

1. Brown, T. B., *Am. J. Phys.*, *26*, 183 (1958).
2. Collett, E., *Opt. Commun.*, *52*, 77 (1984).
3. Sekera, Z., *Adv. Geophys.*, *3*, 43 (1956).
4. Budde, W., *Appl. Opt.*, *1*, 201 (1962).
5. Kent, C. V. and Lawson, J., *J. Opt. Soc. Am.*, *27*, 117 (1937).

Books

1. Born, M. and Wolf, E., *Principles of Optics*, 3rd ed., Pergamon Press, New York, 1965.
2. Azzam, R. M. A, and Hyde, W. L., Eds, Proceedings of SPIE, Vol. 88, *Polarized Light*, SPIE, Bellingham WA, Jan 1976.
3. Clarke, D. and Grainger, J. F., *Polarized Light and Optical Measurement*, Pergamon Press, Oxford, 1971.
4. Shurcliff, W. A., *Polarized Light*, Harvard University Press, Cambridge, MA, 1962.
5. Strong, J., *Concepts of Classical Optics*, Freeman, San Francisco, 1959.
6. Jenkins, F. S. and White, H. E., *Fundamentals of Optics*, McGraw-Hill, New York, 1957.
7. Stone, J. M., *Radiation and Optics*, McGraw-Hill, New York, 1963.

7

The Measurement of the Characteristics of Polarizing Elements

7.1 INTRODUCTION

In the previous chapter we described a number of methods for measuring and characterizing polarized light in terms of the Stokes polarization parameters. We now turn our attention to measuring the characteristics of the three major optical polarizing elements, namely, the polarizer (diattenuator), retarder, and rotator. For a polarizer it is necessary to measure the attenuation coefficients of the orthogonal axes, for a retarder the relative phase shift, and for a rotator the angle of rotation. It is of practical importance to make these measurements. Before proceeding with any experiment in which polarizing elements are to be used, it is good practice to determine if they are performing according to their specifications. This characterization is also necessary because over time polarizing components change: e.g., the optical coatings deteriorate, and in the case of Polaroid the material becomes discolored. In addition, one finds that, in spite of one's best laboratory controls, quarter-wave and half-wave retarders, which operate at different wavelengths, become mixed up. Finally, the quality control of manufacturers of polarizing components is not perfect, and imperfect components are sold.

The characteristics of all three types of polarizing elements can be determined by using a pair of high-quality calcite polarizers that are placed in high-resolution angular mounts; the polarizing element being tested is placed between these two polarizers. A practical angular resolution is 0.1° ($6'$ of arc) or less. High-quality calcite polarizers and mounts are expensive, but in a laboratory where polarizing components are used continually their cost is well justified.

7.2 MEASUREMENT OF ATTENUATION COEFFICIENTS OF A POLARIZER (DIATTENUATOR)

A linear polarizer is characterized by its attenuation coefficients p_x and p_y along its orthogonal x and y axes. We now describe the experimental procedure for

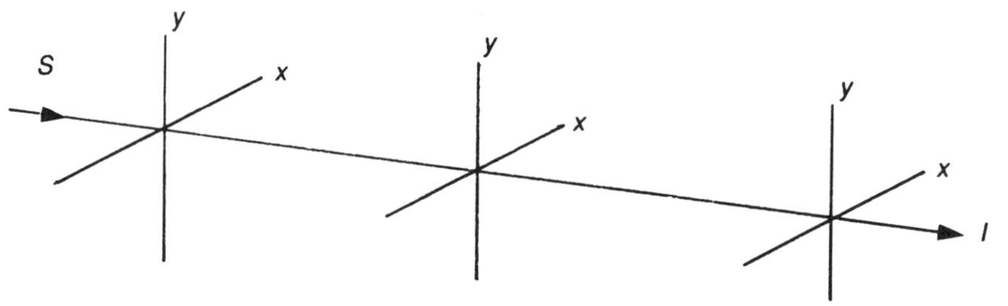

Figure 7-1 Experimental configuration to measure the attenuation coefficients p_x and p_y of a polarizer (diattenuator).

measuring these coefficients. The measurement configuration is shown in Fig. 7-1. In the experiment the polarizer to be tested is inserted between the two polarizers as shown. The reason for using two polarizers is that the same configuration can also be used to test retarders and rotators. Thus, we can have a single, permanent, test configuration for measuring all three types of polarizing components.

The Mueller matrix of a polarizer (diattenuator) with its axes along the x and y directions is

$$M_p = \frac{1}{2}\begin{pmatrix} p_x^2+p_y^2 & p_x^2-p_y^2 & 0 & 0 \\ p_x^2-p_y^2 & p_x^2+p_y^2 & 0 & 0 \\ 0 & 0 & 2p_xp_y & 0 \\ 0 & 0 & 0 & 2p_xp_y \end{pmatrix} \qquad 0 \le p_{x,y} \le 1 \tag{7-1}$$

It is convenient to rewrite (7-1) as

$$M_p = \begin{pmatrix} A & B & 0 & 0 \\ B & A & 0 & 0 \\ 0 & 0 & C & 0 \\ 0 & 0 & 0 & C \end{pmatrix} \tag{7-2a}$$

where

$$A = \frac{1}{2}(p_x^2 + p_y^2) \tag{7-2b}$$

$$B = \frac{1}{2}(p_x^2 - p_y^2) \tag{7-2c}$$

$$C = \frac{1}{2}(2p_xp_y) \tag{7-2d}$$

In practice, while we are interested only in determining p_x^2 and p_y^2, it is useful to measure p_xp_y as well, because a polarizer satisfies the relation:

$$A^2 = B^2 + C^2 \tag{7-3}$$

as the reader can easily show from (7-2). Equation (7-3) serves as a useful check on the measurements. The optical source emits a beam characterized by a Stokes vector

$$S = \begin{pmatrix} S_0 \\ S_1 \\ S_2 \\ S_3 \end{pmatrix} \tag{7-4}$$

In the measurement the first polarizer, which is often called *the generating polarizer*, is set to $+45^\circ$. The Stokes vector of the beam emerging from the generating polarizer is then

$$S = I_0 \begin{pmatrix} 1 \\ 0 \\ 1 \\ 0 \end{pmatrix} \tag{7-5}$$

where $I_0 = (1/2)(S_0 + S_2)$ is the intensity of the emerging beam. The Stokes vector of the beam emerging from the test polarizer is found to be, after multiplying (7-2a) and (7-5),

$$S' = I_0 \begin{pmatrix} A \\ B \\ C \\ 0 \end{pmatrix} \tag{7-6}$$

The polarizer before the optical detector is often called *the analyzing polarizer* or simply *the analyzer*. The analyzer is mounted so that it can be rotated to an angle α. The Mueller matrix of the rotated analyzer is (see Chap. 5)

$$M_A = \frac{1}{2}\begin{pmatrix} 1 & \cos 2\alpha & \sin 2\alpha & 0 \\ \cos 2\alpha & \cos^2 2\alpha & \sin 2\alpha \cos 2\alpha & 0 \\ \sin 2\alpha & \sin 2\alpha \cos 2\alpha & \sin^2 2\alpha & 0 \\ 0 & 0 & 0 & 0 \end{pmatrix} \tag{7-7}$$

The Stokes vector of the beam incident on the optical detector is then seen from multiplying (7-6) by (7-7) to be

$$S' = \frac{I_0}{2}(A + B\cos 2\alpha + C\sin 2\alpha)\begin{pmatrix} 1 \\ \cos 2\alpha \\ \sin 2\alpha \\ 0 \end{pmatrix} \tag{7-8}$$

and the intensity of the beam is

$$I(\alpha) = \frac{I_0}{2}(A + B\cos 2\alpha + C\sin 2\alpha) \tag{7-9}$$

First method:

By rotating the analyzer to $\alpha = 0°$, $45°$, and $90°$, (7-9) yields the following equations:

$$I(0°) = \frac{I_0}{2}(A + B) \tag{7-10a}$$

$$I(45°) = \frac{I_0}{2}(A + C) \tag{7-10b}$$

$$I(90°) = \frac{I_0}{2}(A - B) \tag{7-10c}$$

Solving for A, B, and C, we then find that

$$A = \frac{I(0°) + I(90°)}{I_0} \tag{7-11a}$$

$$B = \frac{I(0°) - I(90°)}{I_0} \tag{7-11b}$$

$$C = \frac{2I(45°) - I(0°) - I(90°)}{I_0} \tag{7-11c}$$

which are the desired relations. From (7-2) we also see that

$$p_x^2 = A + B \tag{7-12a}$$

$$p_y^2 = A - B \tag{7-12b}$$

so that we can write (7-10a) and (7-10c) as

$$p_x^2 = \frac{2I(0°)}{I_0} \tag{7-13a}$$

$$p_y^2 = \frac{2I(90°)}{I_0} \tag{7-13b}$$

Thus, it is only necessary to measure $I(0°)$ and $I(90°)$, the intensities in the x and y directions, respectively, to obtain p_x^2 and p_y^2. The intensity I_0 of the beam emerging from the generating polarizer is measured without the polarizer under test and the analyzer in the optical train.

It is not necessary to measure C. Nevertheless, experience shows that the additional measurement of $I(45°)$ enables one to use (7-3) as a check on the measurements.

In order to determine p_x^2 and p_y^2 in (7-13) it is necessary to know I_0. However, a relative measurement of p_y^2/p_x^2 is just as useful. We divide (7-12b) by (7-12a) and we obtain

$$\frac{p_y^2}{p_x^2} = \frac{I(90°)}{I(0°)} \tag{7-14}$$

We see that this type of measurement does not require a knowledge of I_0. Thus, measuring $I(0°)$ and $I(90°)$ and forming the ratio yields the relative value of the absorption coefficients of the polarizer.

In order to obtain A, B, and C and then p_x^2 and p_y^2 in the method described above, an optical detector is required. However, the magnitude of p_x^2 and p_y^2 can also be obtained using a null-intensity method. To show this we write (7-3) again

$$A^2 = B^2 + C^2 \tag{7-3}$$

This suggests that we can write

$$B = A\cos\gamma \tag{7-15a}$$

$$C = A\sin\gamma \tag{7-15b}$$

Substituting (7-15a) and (7-15b) into (7-9), we then have

$$I(\alpha) = \frac{I_0 A}{2}[1 + \cos(2\alpha - \gamma)] \tag{7-16a}$$

and

$$\tan\gamma = \frac{C}{B} \tag{7-16b}$$

where (7-16b) has been obtained by dividing (7-15a) by (7-15b).

We see that $I(\alpha)$ leads to a null intensity at

$$\alpha_{\text{null}} = 90^\circ + \frac{\gamma}{2} \tag{7-17}$$

where α_{null} is the angle at which the null is observed. Substituting (7-17) into (7-16b) then yields

$$\frac{C}{B} = \tan 2\alpha_{\text{null}} \tag{7-18}$$

Thus by measuring γ from the null-intensity condition, we can find B/A and C/A from (7-15a) and (7-15b), respectively. For convenience we set $A = 1$. Then we see from (7-12) that

$$p_x^2 = 1 + B \tag{7-19a}$$

$$p_y^2 = 1 - B \tag{7-19b}$$

The ratio C/B in (7-18) can also be used to determine the ratio p_y/p_x, which we can then square to form p_y^2/p_x^2. From (7-2)

$$B = \frac{1}{2}(p_x^2 - p_y^2) \tag{7-2c}$$

$$C = \frac{1}{2}(2p_x p_y) \tag{7-2d}$$

Substituting (7-2b) and (7-2c) into (7-18) gives

$$\tan 2\alpha_{\text{null}} = \frac{2p_x p_y}{p_x^2 - p_y^2} \tag{7-20}$$

The form of (7-20) suggests that we set

$$p_x = p\cos\beta \qquad p_y = p\sin\beta \tag{7-21a}$$

so

$$\tan 2\alpha_{\text{null}} = \frac{\sin 2\beta}{\cos 2\beta} = \tan 2\beta \tag{7-21b}$$

and

$$\beta = \alpha_{\text{null}} \tag{7-21c}$$

This leads immediately to

$$\frac{p_y}{p_x} = \tan\beta = \tan(\alpha_{\text{null}}) \tag{7-22a}$$

or, using (7-17)

$$\frac{p_y^2}{p_x^2} = \cot^2\left(\frac{\gamma}{2}\right) \tag{7-22b}$$

Thus, the shift in the intensity, (7-16a) enables us to determine p_y^2/p_x^2 directly from γ. We always assume that $p_y^2/p_x^2 \le 1$. A neutral density filter is described by $p_x^2 = p_y^2$ so the range on p_y^2/p_x^2 limits γ to

$$90^\circ \le \gamma \le 180^\circ \tag{7-22c}$$

For $p_y^2/p_x^2 = 0$, an ideal polarizer, $\gamma = 180^\circ$, whereas for $p_y^2/p_x^2 = 1$, a neutral density filter $\gamma = 90^\circ$ as shown by (7-22b). We see that the closer the value of γ is to 180°, the better is the polarizer. As an example, for commercial Polaroid HN22 at 0.550 μm $p_y^2/p_x^2 = 2 \times 10^{-6}/0.48 = 4.2 \times 10^{-6}$ so from (7-22b) we see that $\gamma = 179.77^\circ$ and $\alpha_{\text{null}} = 179.88^\circ$, respectively; the nearness of γ to 180° shows that it is an excellent polarizing material.

Second method:

The parameters A, B, and C can also be obtained by Fourier-analyzing (7-9), assuming that the analyzing polarizer can be continuously rotated over a half or full cycle. Recall that Eq. (7-9) is

$$I(\alpha) = \frac{I_0}{2}(A + B\cos 2\alpha + C\sin 2\alpha) \tag{7-9}$$

From the point of view of Fourier analysis A describes a d.c. term, and B and C describe second-harmonic terms. It is only necessary to integrate over half a cycle, that is, from 0° to π, in order to determine A, B, and C. We easily find that

$$A = \frac{2}{\pi I_0}\int_0^\pi I(\alpha)\,d\alpha \tag{7-23a}$$

$$B = \frac{4}{\pi I_0}\int_0^\pi I(\alpha)\cos 2\alpha\,d\alpha \tag{7-23b}$$

$$C = \frac{4}{\pi I_0}\int_0^\pi I(\alpha)\sin 2\alpha\,d\alpha \tag{7-23c}$$

Throughout this analysis we have assumed that the axes of the polarizer being measured lie along the x and y directions. If this is not the case, then the polarizer under test should be rotated to its x and y axes in order to make the measurement. The simplest way to determine rotation angle θ is to remove the polarizer under test and rotate the generating polarizer to 0° and the analyzing polarizer to 90°.

Third method:

Finally, another method to determine A, B, and C is to place the test polarizer in a rotatable mount between polarizers in which the axes of both are in the y direction. The test polarizer is then rotated until a minimum intensity is observed from which A, B, and C can be found. The Stokes vector emerging from the y generating polarizer is

$$S = \frac{I_0}{2}\begin{pmatrix} 1 \\ -1 \\ 0 \\ 0 \end{pmatrix} \tag{7-24}$$

The Mueller matrix of the rotated test polarizer (7-2a) is

$$M = \begin{pmatrix} A & B\cos 2\theta & B\sin 2\theta & 0 \\ B\cos 2\theta & A\cos^2 2\theta + C\sin^2 2\theta & (A-C)\sin 2\theta\cos 2\theta & 0 \\ B\sin 2\theta & (A-C)\sin 2\theta\cos 2\theta & A\sin^2 2\theta + C\cos^2 2\theta & 0 \\ 0 & 0 & 0 & 0 \end{pmatrix} \tag{7-25}$$

The intensity of the beam emerging from the y analyzing polarizer is

$$I(\theta) = \frac{I_0}{4}[(A+C) - 2B\cos 2\theta + (A-C)\cos^2 2\theta] \tag{7-26}$$

Equation (7-26) can be solved for its maximum and minimum values by differentiating $I(\theta)$ with respect to θ and setting $dI(\theta)/d\theta = 0$. We then find

$$\sin 2\theta[B - (A-C)\cos 2\theta] = 0 \tag{7-27}$$

The solutions of (7-27) are

$$\sin 2\theta = 0 \tag{7-28a}$$

and

$$\cos 2\theta = \frac{B}{A-C} \tag{7-28b}$$

For (7-28a) we have $\theta = 0^\circ$ and 90°. The corresponding values of the intensities are then, from (7-26)

$$I(0^\circ) = \frac{I_0}{2}[A - B] \tag{7-29a}$$

$$I(90^\circ) = \frac{I_0}{2}[A + B] \tag{7-29b}$$

The second solution (7-28b), on substitution into (7-26), leads to $I(\theta) = 0$. Thus, the minimum intensity is given by (7-29a) and the maximum intensity by (7-29b). Because both the generating and analyzing polarizers are in the y direction, this is exactly what one would expect. We also note in passing that at $\theta = 45^\circ$, (7-26) reduces to

$$I(45^\circ) = \frac{I_0}{4}[A + C] \tag{7-29c}$$

We can again divide (7-29) through by I_0 and then solve (7-29) for A, B, and C.

We see that several methods can be used to determine the absorption coefficients of the orthogonal axes of a polarizer. In the first method we generate a linear $+45^\circ$ polarized beam and then rotate the analyzer to obtain A, B, and C of the polarizer being tested. This method requires a quantitative optical detector. However, if an optical detector is not available, it is still possible to determine A, B, and C by using the null-intensity method; rotating the analyzer until a null is observed leads to A, B, and C. On the other hand, if the analyzer can be mounted in a rotatable mount, which can be stepped (electronically), then a Fourier analysis of the signal can be made and we can again find A, B, and C. Finally, if the transmission axes of the generating and analyzing polarizers are parallel to one another, conveniently chosen to be in the y direction, and the test polarizer is rotated, then we can also determine A, B, and C by rotating the test polarizer to 0°, 45°, and 90°.

7.3 MEASUREMENT OF PHASE SHIFT OF A RETARDER

There are numerous occasions when it is important to know the phase shift of a retarder. The most common types of retarders are quarter-wave and half-wave retarders. These two types are most often used to create circularly polarized light and to rotate or reverse the polarization ellipse, respectively.

Two methods can be used for measuring the phase shift using two linear polarizers following the experimental configuration given in the previous section.

First method:

In the first method a retarder is placed between the two linear polarizers mounted in the "crossed" position. Let us set the transmission axes of the first and second polarizers to be in the x and y directions, respectively. By rotating the retarder, the direction (angle) of the fast axis is rotated and, as we shall soon see, the phase can be found. The second method is very similar to the first except that the fast axis of the retarder is rotated to 45°. In this position the phase can also be found. We now consider both methods.

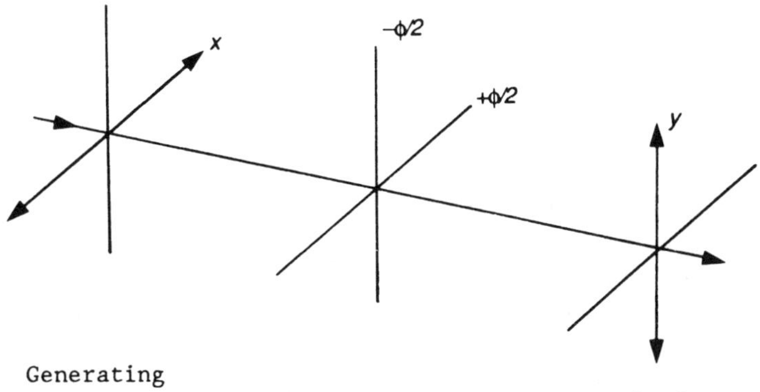

Figure 7-2 Closed polarizer method to measure the phase of a retarder.

For the first method we refer to Fig. 7-2. It is understood that the correct wavelength must be used; that is, if the retarder is specified for, say 6328 Å, then the optical source should emit this wavelength. In the visible domain calcite polarizers are, as usual, best. However, high-quality Polaroid is also satisfactory, but its optical bandpass is much more restricted. In Fig. 7-2 the transmission axes of the polarizers (or diattenuators) are in the x (horizontal) and y (vertical) directions, respectively. The Mueller matrix for the retarder rotated through an angle θ is

$$M(\phi,\theta)\begin{pmatrix} 1 & 0 & 0 & 0 \\ 0 & \cos^2 2\theta + \cos\phi \sin^2 2\theta & (1-\cos\phi)\sin 2\theta\cos 2\theta & -\sin\phi\sin 2\theta \\ 0 & (1-\cos\phi)\sin 2\theta\cos 2\theta & \sin^2 2\theta + \cos\phi\cos^2 2\theta & \sin\phi\cos 2\theta \\ 0 & \sin\phi\sin 2\theta & -\sin\phi\cos 2\theta & \cos\phi \end{pmatrix} \tag{7-30}$$

where the phase shift ϕ is to be determined. The Mueller matrix for an ideal linear polarizer is

$$M_{x,y} = \frac{1}{2}\begin{pmatrix} 1 & \pm 1 & 0 & 0 \\ \pm 1 & 1 & 0 & 0 \\ 0 & 0 & 0 & 0 \\ 0 & 0 & 0 & 0 \end{pmatrix} \tag{7-31}$$

where the plus sign corresponds to a horizontal polarizer and the minus sign to a vertical polarizer. The Mueller matrix for Fig. 7-2 is then

$$M = M_y M(\phi,\theta) M_x \tag{7-32}$$

Carrying out the matrix multiplication in (7-32) using (7-30) and (7-31) then yields

$$M = \frac{(1-\cos\phi)(1-\cos 4\theta)}{8}\begin{pmatrix} 1 & 1 & 0 & 0 \\ -1 & -1 & 0 & 0 \\ 0 & 0 & 0 & 0 \\ 0 & 0 & 0 & 0 \end{pmatrix} \tag{7-33}$$

Equation (7-33) shows that the polarizing train behaves as a pseudopolarizer. The intensity of the optical beam on the detector is then

$$I(\theta, \phi) = I_0 \frac{(1 - \cos\phi)(1 - \cos 4\theta)}{4} \tag{7-34}$$

where I_0 is the intensity of the optical source.

Equation (7-34) immediately allows us to determine the direction of the fast axis of the retarder. When the retarder is inserted between the crossed polarizers, the intensity on the detector should be zero, according to (7-34), at $\theta = 0°$. If it is not zero, the retarder should be rotated until a null intensity is observed. After this angle has been found, the retarder is rotated 45° according to (7-34) to obtain the maximum intensity. In order to determine ϕ, it is necessary to know I_0. The easiest way to do this is to rotate the x polarizer (the first polarizer) to the y position and remove the retarder; both linear polarizers are then in the y direction. The intensity I_D on the detector is then (let us assume that unpolarized light enters the first polarizer)

$$I_D = \frac{I_0}{2} \tag{7-35}$$

so (7-34) can be written as

$$I(\theta, \phi) = I_D \frac{(1 - \cos\phi)(1 - \cos 4\theta)}{2} \tag{7-36}$$

The retarder is now reinserted into the polarizing train. The maximum intensity, $I(\theta, \phi)$, takes place when the retarder is rotated to $\theta = 45°$. At this angle (7-36) is solved for ϕ, and we have

$$\phi = \cos^{-1}\left[1 - \frac{I(45°, \phi)}{I_D}\right] \tag{7-37}$$

The disadvantage of using the crossed-polarizer method is that it requires that we know the intensity of the beam, I_0, entering the polarizing train. This problem can be overcome by another method, namely, rotating the analyzing polarizer and fixing the retarder at 45°. We now consider this second method.

Second method:

The experimental configuration is identical to the first method except that the analyzer can be rotated through an angle α. The Stokes vector of the beam emerging from the generating polarizer is (again let us assume that unpolarized light enters the generating polarizer)

$$S = \frac{I_0}{2}\begin{pmatrix} 1 \\ 1 \\ 0 \\ 0 \end{pmatrix} \tag{7-38}$$

Multiplication of (7-38) by (7-30) yields

$$S' = \frac{I_0}{2}\begin{pmatrix} 1 \\ \cos^2 2\theta + \cos\phi \sin^2 2\theta \\ (1-\cos\phi)\sin 2\theta \cos 2\theta \\ \sin\phi \sin 2\theta \end{pmatrix} \tag{7-39}$$

We assume that the fast axis of the retarder is at $\theta = 0^\circ$. If it is not, the retarder should be adjusted to $\theta = 0^\circ$ by using the crossed-polarizer method described in the first method; we note that at $\theta = 0^\circ$, (7-39) reduces to

$$S' = \frac{I_0}{2}\begin{pmatrix} 1 \\ 1 \\ 0 \\ 0 \end{pmatrix} \tag{7-40}$$

so that the analyzing polarizer should give a null intensity when it is in the y direction. Assuming that the retarder's fast axis is now properly adjusted, we rotate the retarder counterclockwise to $\theta = 45^\circ$. Then (7-39) reduces to

$$S' = \frac{I_0}{2}\begin{pmatrix} 1 \\ \cos\phi \\ 0 \\ \sin\phi \end{pmatrix} \tag{7-41}$$

This is a Stokes vector for elliptically polarized light. The conditions $\phi = 90^\circ$ and 180° correspond to right circularly polarized and linear vertically polarized light, respectively. We note that the linear vertically polarized state arises because for $\phi = 180^\circ$ the retarder behaves as a pseudorotator. The Mueller matrix of the analyzing polarizer is

$$M(\phi) = \frac{1}{2}\begin{pmatrix} 1 & \cos 2\alpha & \sin 2\alpha & 0 \\ \cos 2\alpha & \cos^2 2\alpha & \sin 2\alpha \cos 2\alpha & 0 \\ \sin 2\alpha & \sin 2\alpha \cos 2\alpha & \sin^2 2\alpha & 0 \\ 0 & 0 & 0 & 0 \end{pmatrix} \tag{7-42}$$

The Stokes vector of the beam emerging from the analyzer is then

$$S = \frac{I_0}{4}(1 + \cos\phi \cos 2\alpha)\begin{pmatrix} 1 \\ \cos 2\alpha \\ \sin 2\alpha \\ 0 \end{pmatrix} \tag{7-43}$$

so the intensity is

$$I(\alpha, \phi) = \frac{I_0}{4}(1 + \cos\phi \cos 2\alpha) \tag{7-44}$$

In order to find ϕ, (7-44) is evaluated at $\alpha = 0^\circ$ and 90°, and

$$I(0^\circ, \phi) = \frac{I_0}{4}(1 + \cos\phi) \tag{7-45a}$$

$$I(90^\circ, \phi) = \frac{I_0}{4}(1 - \cos\phi) \tag{7-45b}$$

Equation (7-45a) is divided by (7-45b) and solved for $\cos\phi$:

$$\cos\phi = \frac{I(0^\circ, \phi) - I(90^\circ, \phi)}{I(0^\circ, \phi) + I(90^\circ, \phi)} \tag{7-46}$$

We note that in this method the source intensity need not be known.

We can also determine the direction of the fast axis of the retarder in a "dynamic" fashion. The intensity of the beam emerging from the analyzer when it is in the y position is (see (7-39) and (7-42))

$$I_y = \frac{I_0}{4}[1 - (\cos^2 2\theta + \cos\phi \sin^2 2\theta)] \tag{7-47a}$$

where θ is the angle of the fast axis measured from the horizontal x axis. We now see that when the analyzer is in the x position:

$$I_x = \frac{I_0}{4}[1 + (\cos^2 2\theta + \cos\phi \sin^2 2\theta)] \tag{7-47b}$$

Adding (7-47a) and (7-47b) yields

$$I_x + I_y = \frac{I_0}{2} \tag{7-48a}$$

Next, subtracting (7-47a) from (7-47b) yields

$$I_x - I_y = \frac{I_0}{2}(\cos^2 2\theta + \cos\phi \sin^2 2\theta) \tag{7-48b}$$

We see that when $\theta = 0$ the sum and difference intensities (7-48) are equal. Thus, one can measure I_x and I_y continuously as the retarder is rotated and the analyzer is flipped between the horizontal and vertical directions until (7-48a) equals (7-48b). When this occurs, the amount of rotation that has taken place determines the magnitude of the rotation angle of the fast axis from the x axis.

Third method:

Finally, if a compensator is available, the phase shift can be measured as follows. Figure 7-3 shows the measurement method. The compensator is placed between the retarder under test and the analyzer. The transmission axes of the generating and analyzing polarizers are set at $+45^\circ$ and $+135^\circ$, that is, in the crossed position.

The Stokes vector of the beam incident on the test retarder is

$$S = \frac{I_0}{2}\begin{pmatrix} 1 \\ 0 \\ 1 \\ 0 \end{pmatrix} \tag{7-49}$$

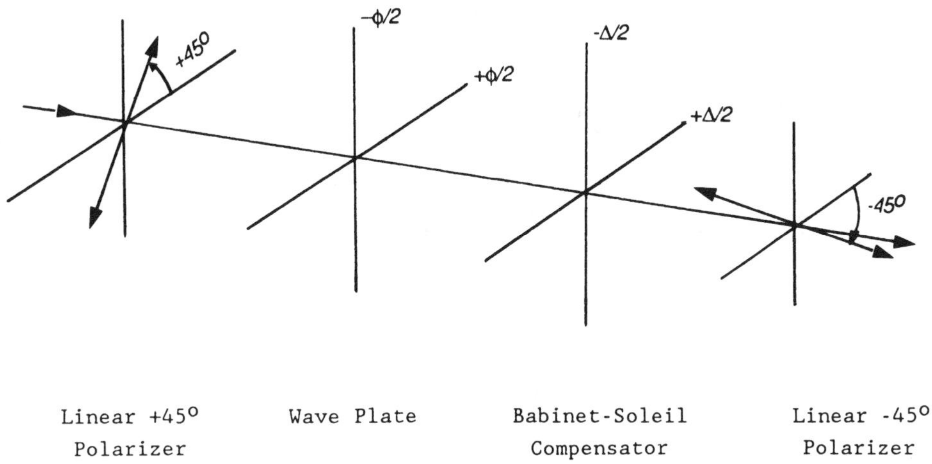

Figure 7-3 Measurement of the phase shift of a wave plate using a Babinet–Soleil compensator.

The Mueller matrix of the test retarder is

$$M = \begin{pmatrix} 1 & 0 & 0 & 0 \\ 0 & 1 & 0 & 0 \\ 0 & 0 & \cos\phi & \sin\phi \\ 0 & 0 & -\sin\phi & \cos\phi \end{pmatrix} \tag{7-50}$$

Multiplying (7-49) by (7-50) yields

$$S = \frac{I_0}{2}\begin{pmatrix} 1 \\ 0 \\ \cos\phi \\ -\sin\phi \end{pmatrix} \tag{7-51}$$

The Mueller matrix of the Babinet–Soleil compensator is

$$M = \begin{pmatrix} 1 & 0 & 0 & 0 \\ 0 & 1 & 0 & 0 \\ 0 & 0 & \cos\Delta & \sin\Delta \\ 0 & 0 & -\sin\Delta & \cos\Delta \end{pmatrix} \tag{7-52}$$

Multiplying (7-51) by (7-52) yields the Stokes vector of the beam incident on the linear -45° polarizer:

$$S = \frac{I_0}{2}\begin{pmatrix} 1 \\ 0 \\ \cos(\Delta+\phi) \\ -\sin(\Delta+\phi) \end{pmatrix} \tag{7-53}$$

Finally, the Mueller matrix for the ideal linear polarizer with its transmission axis at $-45^\circ(+135^\circ)$ is

$$M = \frac{1}{2}\begin{pmatrix} 1 & 0 & -1 & 0 \\ 0 & 0 & 0 & 0 \\ -1 & 0 & 1 & 0 \\ 0 & 0 & 0 & 0 \end{pmatrix} \tag{7-54}$$

Multiplying (7-53) by the first row of (7-54) gives the intensity on the detector, namely,

$$I(\Delta + \phi) = \frac{I_0}{4}[1 - \cos(\Delta + \phi)] \tag{7-55}$$

We see that a null intensity is found at

$$\Delta = 360^\circ - \phi \tag{7-56}$$

from which we then find ϕ.

There are still other methods to determine the phase of the retarder, and the techniques developed here can provide a useful starting point. However, the methods described here should suffice for most problems.

7.4 MEASUREMENT OF ROTATION ANGLE OF A ROTATOR

The final type of polarizing element that we wish to characterize is a rotator. The Mueller matrix of a rotator is

$$M = \begin{pmatrix} 1 & 0 & 0 & 0 \\ 0 & \cos 2\theta & \sin 2\theta & 0 \\ 0 & -\sin 2\theta & \cos 2\theta & 0 \\ 0 & 0 & 0 & 1 \end{pmatrix} \tag{7-57}$$

First method:

The angle θ can be determined by inserting the rotator between a pair of polarizers in which the generating polarizer is fixed in the y position and the analyzing polarizer can be rotated. This configuration is shown in Fig. 7-4.

The Stokes vector of the beam incident on the rotator is

$$S = \frac{I_0}{2}\begin{pmatrix} 1 \\ -1 \\ 0 \\ 0 \end{pmatrix} \tag{7-58}$$

The Stokes vector of the beam incident on the analyzer is then found by multiplying (7-58) by (7-57)

$$S' = \frac{I_0}{2}\begin{pmatrix} 1 \\ -\cos 2\theta \\ \sin 2\theta \\ 0 \end{pmatrix} \tag{7-59}$$

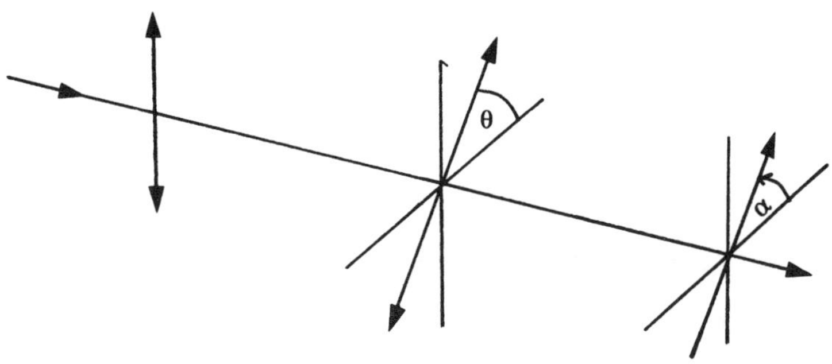

Figure 7-4 Measurement of the rotation angle θ of a rotator.

The Mueller matrix of the analyzer is

$$M = \frac{1}{2}\begin{pmatrix} 1 & \cos 2\alpha & \sin 2\alpha & 0 \\ \cos 2\alpha & \cos^2 2\alpha & \sin 2\alpha \cos 2\alpha & 0 \\ \sin 2\alpha & \sin 2\alpha \cos 2\alpha & \sin^2 2\alpha & 0 \\ 0 & 0 & 0 & 0 \end{pmatrix} \tag{7-60}$$

The intensity of the beam emerging from the analyzer is then seen from the product of (7-60) and (7-59) to be

$$I(\alpha) = \frac{I_0}{4}[1 - \cos(2\alpha + 2\theta)] \tag{7-61}$$

The analyzer is rotated and, according to (7-61), a null intensity will be observed at

$$\alpha = 180^\circ - \theta \tag{7-62a}$$

or, simply,

$$\theta = 180^\circ - \alpha \tag{7-62b}$$

Second method:

Another method for determining the angle θ is to rotate the generating polarizer sequentially to 0°, 45°, 90°, and 135°. The rotator and the analyzing polarizer are fixed with their axes in the horizontal direction. The intensities of the beam emerging from the analyzing polarizer for these four angles are then

$$I(0^\circ) = \frac{I_0}{4}(1 + \cos 2\theta) \tag{7-63a}$$

$$I(45^\circ) = \frac{I_0}{4}(1 + \sin 2\theta) \tag{7-63b}$$

$$I(90^\circ) = \frac{I_0}{4}(1 - \cos 2\theta) \tag{7-63c}$$

$$I(135^\circ) = \frac{I_0}{4}(1 - \sin 2\theta) \tag{7-63d}$$

Subtracting (7-63c) from (7-63a) and (7-63d) from (7-63b) yields

$$\left(\frac{I_0}{2}\right)\cos 2\theta = I(0^\circ) - I(90^\circ) \tag{7-64a}$$

$$\left(\frac{I_0}{2}\right)\sin \theta = I(45^\circ) - I(135^\circ) \tag{7-64b}$$

Dividing (7-64b) by (7-64a) then yields the angle of rotation θ:

$$\theta = \tan^{-1}[(I(45^\circ) - I(135^\circ))/(I(0^\circ) - I(90^\circ))] \tag{7-65}$$

In the null-intensity method an optical detector is not required, whereas in this second method a photodetector is needed. However, one soon discovers that even a null measurement can be improved by several orders of magnitude below the sensitivity of the eye by using an optical detector–amplifier combination.

Finally, as with the measurement of retarders, other configurations can be considered. However, the two methods described here should, again, suffice for most problems.

REFERENCE

Book

1. Clark, D. and Grainger, J. F., *Polarized Light and Optical Measurement*, Pergamon Press, Oxford, 1971.

8

Mueller Matrices for Reflection and Transmission

8.1 INTRODUCTION

In previous chapters the Mueller matrices were introduced in a very formal manner. The Mueller matrices were derived for a polarizer, retarder, and rotator in terms of their fundamental behavior; their relation to actual physical problems was not emphasized. In this chapter we apply the Mueller matrix formulation to a number of problems of great interest and importance in the physics of polarized light. One of the major reasons for discussing the Stokes parameters and the Mueller matrices in these earlier chapters is that they provide us with an excellent tool for treating many physical problems in a much simpler way than is usually done in optical textbooks. In fact, one quickly discovers that many of these problems are sufficiently complex that they preclude any but the simplest to be considered without the application of the Stokes parameters and the Mueller matrix formalism.

One of the earliest problems encountered in the study of optics is the behavior of light that is reflected and transmitted at an air–glass interface. Around 1808, E. Malus discovered, quite by accident, that unpolarized light became polarized when it was reflected from glass. Further investigations were made shortly afterward by D. Brewster, who was led to enunciate his famous law relating the polarization of the reflected light and the refractive index of the glass to the incident angle now known as the Brewster angle; the practical importance of this discovery was immediately recognized by Brewster's contemporaries. The study of the interaction of light with material media and its reflection and transmission as well as its polarization is a topic of great importance.

The interaction of light beams with dielectric surfaces and its subsequent reflection and transmission is expressed mathematically by a set of equations known as Fresnel's equations for reflection and transmission. Fresnel's equations can be derived from Maxwell's equations. We shall derive Fresnel's equations in the next Section.

In practice, if one attempts to apply Fresnel's equations to any but the simplest problems, one quickly finds that the algebraic manipulation is very involved. This complexity accounts for the omission of many important derivations in numerous textbooks. Furthermore, the cases that are treated are usually restricted to, say, incident linearly polarized light. If one is dealing with a different state of polarized light, e.g., circularly polarized or unpolarized light, one must usually begin the problem anew. We see that the Stokes parameters and the Mueller matrix are ideal to handle this task.

The problems of complexity and polarization can be readily treated by expressing Fresnel's equations in the form of Stokes vectors and Mueller matrices. This formulation of Fresnel's equations and its application to a number of interesting problems is the basic aim of the present chapter. As we shall see, both reflection and refraction (transmission) lead to Mueller matrices that correspond to polarizers for materials characterized by a real refractive index n. Furthermore, for total internal reflection (TIR) at the critical angle the Mueller matrix for refraction reduces to a null Mueller matrix, whereas the Mueller matrix for reflection becomes the Mueller matrix for a phase shifter (retarder).

The Mueller matrices for reflection and refraction are quite complicated. However, there are three angles for which the Mueller matrices reduce to very simple forms. These are for (1) normal incidence, (2) the Brewster angle, and (3) an incident angle of 45°. All three reduced matrix forms suggest interesting ways to measure the refractive index η of the dielectric material. These methods will be discussed in detail.

In practice, however, we must deal not only with a single air–dielectric interface but also with a dielectric medium of finite thickness, that is, dielectric plates. Thus, we must consider the reflection and transmission of light at multiple surfaces. In order to treat these more complicated problems, we must multiply the Mueller matrices. We quickly discover, however, that the matrix multiplication requires a considerable amount of effort because of the presence of the off-diagonal terms in the Mueller matrices. This suggests that we first transform the Mueller matrices to a diagonal representation; matrix multiplication of diagonal matrices leads to another diagonal matrix. Therefore, in the final chapters of this part of the book, we introduce the diagonalized Mueller matrices and treat the problem of transmission through a single dielectric plate and through several dielectric plates. This last problem is of particular importance, because at present it is one of the major ways to create polarized light in the infrared spectrum.

8.2 FRESNEL'S EQUATIONS FOR REFLECTION AND TRANSMISSION

In this section we derive Fresnel's equations. Although this material can be found in many texts, it is useful and instructive to reproduce it here because it is so intimately tied to the polarization of light. Understanding the behavior of both the amplitude and phase of the components of light is essential to designing polarization components or analyzing optical system performance. We start with a review of concepts from electromagnetism.

8.2.1 Definitions

Recall from electromagnetism that:

$\vec{\mathbf{E}}$ is the electric field
$\vec{\mathbf{B}}$ is the magnetic induction
$\vec{\mathbf{D}}$ is the electric displacement
$\vec{\mathbf{H}}$ is the magnetic field
ε_0 is the permittivity of free space
ε is the permittivity
μ_0 is the permeability of free space
μ is the permeability

$$\varepsilon_r = \frac{\varepsilon}{\varepsilon_0} = (1+\chi) \tag{8-1a}$$

where ε_r is the relative permittivity or dielectric constant and χ is the electric susceptibility,

$$\mu_r = \frac{\mu}{\mu_0} = (1+\chi_m) \tag{8-1b}$$

and where μ_r is the relative permeability and χ_m is the magnetic susceptibility.

Thus,

$$\varepsilon = \varepsilon_0\varepsilon_r = \varepsilon_0(1+\chi) \tag{8-1c}$$

and

$$\mu = \mu_0\mu_r = \mu_0(1+\chi_m) \tag{8-1d}$$

Recall that (we use rationalized MKSA units here):

$$\vec{\mathbf{B}} = \mu\vec{\mathbf{H}} \tag{8-1e}$$

and

$$\vec{\mathbf{D}} = \varepsilon\vec{\mathbf{E}} \tag{8-1f}$$

Maxwell's equations, where there are no free charges or currents, are

$$\vec{\nabla}\cdot\vec{\mathbf{D}} = 0 \tag{8-2a}$$

$$\vec{\nabla}\cdot\vec{\mathbf{B}} = 0 \tag{8-2b}$$

$$\vec{\nabla}\times\vec{\mathbf{E}} = -\frac{\partial\vec{\mathbf{B}}}{\partial t} \tag{8-2c}$$

$$\vec{\nabla}\times\vec{\mathbf{H}} = \frac{\partial\vec{\mathbf{D}}}{\partial t} \tag{8-2d}$$

8.2.2 Boundary Conditions

In order to complete our review of concepts from electromagnetism, we must recall the boundary conditions for the electric and magnetic field components. The integral form of Maxwell's first equation, (8.2a), is

$$\oint\!\!\oint \vec{\mathbf{D}} \cdot d\vec{\mathbf{A}} = 0 \tag{8-3}$$

This equation implies that, at the interface, the normal components on either side of the interface are equal, i.e.,

$$D_{n1} = D_{n2} \tag{8-4}$$

The integral form of Maxwell's second equation, (8.2b), is

$$\oint\!\!\oint \vec{\mathbf{B}} \cdot d\vec{\mathbf{A}} = 0 \tag{8-5}$$

which implies again that the normal components on either side of the interface are equal, i.e.,

$$B_{n1} = B_{n2} \tag{8-6}$$

Invoking Ampere's law, we have

$$\oint \vec{\mathbf{H}} \cdot d\vec{\mathbf{l}} = \mathbf{I} \tag{8-7}$$

which implies

$$H_{t1} = H_{t2} \tag{8-8}$$

i.e., the tangential component of H is continuous across the interface.
Lastly,

$$\oint \vec{\mathbf{E}} \cdot d\vec{\mathbf{l}} = \iint \bar{\nabla} \times \vec{\mathbf{E}} \cdot dA = 0 \tag{8-9}$$

which implies

$$E_{t1} = E_{t2} \tag{8-10}$$

i.e., the tangential component of E is continuous across the interface.

8.2.3 Derivation of the Fresnel Equations

We now have all the tools we need derive Fresnel's equations. Suppose we have a light beam intersecting an interface between two linear isotropic media. Part of the incident beam is reflected and part is refracted. The plane in which this interaction takes place is called the plane of incidence, and the polarization of light is defined by the direction of the electric field vector. There are two situations that can occur. The electric field vector can either be perpendicular to the plane of incidence or parallel to the plane of incidence. We consider the perpendicular case first.

Case 1: $\vec{\mathbf{E}}$ is Perpendicular to the Plane of Incidence

This is the "s" polarization (from the German "senkrecht" for perpendicular) or σ polarization. This is also known as transverse electric, or TE, polarization (refer to Fig. 8-1). Light travels from a medium with (real) index n_1 and encounters an interface with a linear isotropic medium that has index n_2. The angles of incidence (or reflection) and refraction are θ_i and θ_r, respectively.

In Fig. 8-1, the y axis points into the plane of the paper consistent with the usual Cartesian coordinate system, and the electric field vectors point out of the plane of the paper, consistent with the requirements of the cross product and the direction of energy flow. The electric field vector for the incident field is represented using the symbol $\vec{\mathbf{E}}$, whereas the fields for the reflected and transmitted components are represented by $\vec{\mathbf{R}}$ and $\vec{\mathbf{T}}$, respectively. Using Maxwell's third equation (8.2c) we can write

$$\vec{\mathbf{k}} \times \vec{\mathbf{E}} = \omega \vec{\mathbf{B}} \tag{8-11}$$

We can write this last equation as

$$\vec{\mathbf{H}} = \frac{\vec{\mathbf{k}}_n \times \vec{\mathbf{E}}}{\omega \mu_0} \tag{8-12}$$

where $\vec{\mathbf{k}}_n$ is the wave vector in the medium, and $\vec{\mathbf{k}}_n$ is

$$\vec{\mathbf{k}}_n = \omega \sqrt{\mu_0 \varepsilon}\, \hat{a}_n \tag{8-13}$$

where $\hat{a}_n$ is a unit vector in the direction of the wave vector.

Now we can write

$$\vec{\mathbf{H}} = \omega \sqrt{\mu_0 \varepsilon}\, \frac{\hat{a}_n \times \vec{\mathbf{E}}}{\omega \mu_0} = \frac{\hat{a}_n \times \vec{\mathbf{E}}}{\sqrt{\mu_0 / \varepsilon}} \tag{8-14}$$

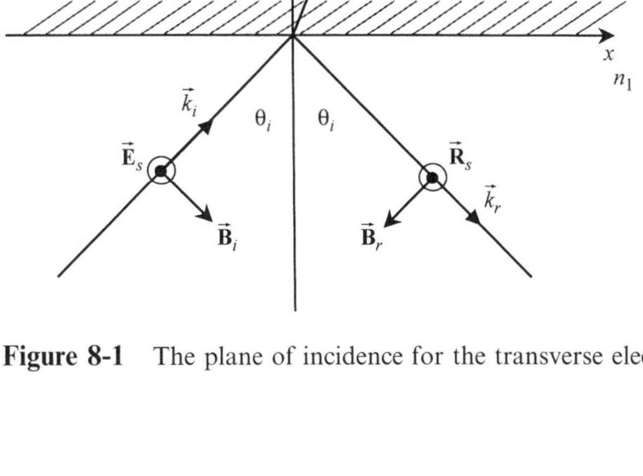

Figure 8-1 The plane of incidence for the transverse electric case.

or

$$\vec{\mathbf{H}} = \frac{\hat{a}_n \times \vec{\mathbf{E}}}{\eta} \tag{8-15}$$

where

$$\eta = \sqrt{\frac{\mu_0}{\varepsilon_0 \varepsilon_r}} = \frac{\eta_0}{n} \qquad \left(\eta_0 = \sqrt{\frac{\mu_0}{\varepsilon_0}} \quad \text{and} \quad n = \sqrt{\varepsilon_r}\right) \tag{8-16}$$

where n is the refractive index and we have made the assumption that $\mu_r \approx 1$. This is the case for most dielectric materials of interest.

The unit vectors in the directions of the incident, reflected, and transmitted wave vectors are

$$\hat{a}_i = \sin\theta_i \hat{a}_x + \cos\theta_i \hat{a}_z \tag{8-17a}$$

$$\hat{a}_r = \sin\theta_i \hat{a}_x - \cos\theta_i \hat{a}_z \tag{8-17b}$$

$$\hat{a}_t = \sin\theta_t \hat{a}_x + \cos\theta_t \hat{a}_z \tag{8-17c}$$

The magnetic field in each region is given by

$$\vec{\mathbf{H}}_i = \frac{\hat{a}_i \times \vec{\mathbf{E}}_s}{\eta_1} \qquad \vec{\mathbf{H}}_r = \frac{\hat{a}_r \times \vec{\mathbf{R}}_s}{\eta_1} \qquad \vec{\mathbf{H}}_t = \frac{\hat{a}_t \times \vec{\mathbf{T}}_s}{\eta_2} \tag{8-18}$$

and the electric field vectors tangential to the interface are

$$\vec{\mathbf{E}}_s = -\mathbf{E}_s \hat{a}_y \qquad \vec{\mathbf{R}}_s = -R_s \hat{a}_y \qquad \vec{\mathbf{T}}_s = -T_s \hat{a}_y \tag{8-19}$$

We can now write the magnetic field components as

$$\vec{\mathbf{H}}_i = \left[\frac{-E_s \sin\theta_i \hat{a}_z}{\eta_1} + \frac{E_s \cos\theta_i \hat{a}_x}{\eta_1}\right] \tag{8-20a}$$

$$\vec{\mathbf{H}}_r = \left[\frac{-R_s \sin\theta_i \hat{a}_z}{\eta_1} - \frac{R_s \cos\theta_i \hat{a}_x}{\eta_1}\right] \tag{8-20b}$$

$$\vec{\mathbf{H}}_t = \left[\frac{-T_s \sin\theta_r \hat{a}_z}{\eta_2} + \frac{T_s \cos\theta_r \hat{a}_x}{\eta_2}\right] \tag{8-20c}$$

We know the tangential component of $\vec{\mathbf{H}}$ is continuous, and we can find the tangential component by taking the dot product of each $\vec{\mathbf{H}}$ with $\hat{a}_x$. We have, for the tangential components:

$$H_i^{\tan} + H_r^{\tan} = H_t^{\tan} \tag{8-21a}$$

or

$$\frac{E_s \cos\theta_i}{\eta_1} - \frac{R_s \cos\theta_i}{\eta_1} = \frac{T_s \cos\theta_r}{\eta_2} = \frac{(E_s + R_s)\cos\theta_r}{\eta_2} \tag{8-21b}$$

using the fact that the tangential component of E is continuous, i.e., $E_s + R_s = T_s$.

We rearrange (8.21b) to obtain

$$E_s[\eta_2 \cos\theta_i - \eta_1 \cos\theta_r] = R_s[\eta_2 \cos\theta_i + \eta_1 \cos\theta_r] \tag{8-21c}$$

and now Fresnel's equation for the reflection amplitude is

$$R_s = \frac{\eta_2 \cos\theta_i - \eta_1 \cos\theta_r}{\eta_2 \cos\theta_i + \eta_1 \cos\theta_r} E_s \tag{8-21d}$$

Using the relation in (8-16) for each material region, we can express the reflection amplitude in terms of the refractive index and the angles as

$$R_s = \frac{n_1 \cos\theta_i - n_2 \cos\theta_r}{n_1 \cos\theta_i + n_2 \cos\theta_r} E_s \tag{8-22a}$$

This last equation can be written, using Snell's law, $n_1 \sin\theta_i = n_2 \sin\theta_r$, to eliminate the dependence on the index:

$$R_s = -\frac{\sin(\theta_i - \theta_r)}{\sin(\theta_i + \theta_r)} E_s \tag{8-22b}$$

An expression for Fresnel's equation for the transmission amplitude can be similarly derived and is

$$T_s = \frac{2n_1 \cos\theta_i}{n_1 \cos\theta_i + n_2 \cos\theta_r} E_s \tag{8-23a}$$

or

$$T_s = \frac{2\sin\theta_r \cos\theta_i}{\sin(\theta_i + \theta_r)} E_s \tag{8-23b}$$

Case 2: $\vec{\mathbf{E}}$ is Parallel to the Plane of Incidence

This is the "p" polarization (from the German "parallel" for parallel) or π polarization. This is also known as transverse magnetic, or TM, polarization (refer to Fig. 8-2). The derivation for the parallel reflection amplitude and transmission amplitude proceeds in a manner similar to the perpendicular case, and Fresnel's equations for the TM case are

$$R_p = \frac{n_2 \cos\theta_i - n_1 \cos\theta_r}{n_2 \cos\theta_i + n_1 \cos\theta_r} E_p \tag{8-24a}$$

or

$$R_p = \frac{\tan(\theta_i - \theta_r)}{\tan(\theta_i + \theta_r)} E_p \tag{8-24b}$$

and

$$T_p \equiv \frac{2n_1 \cos\theta_i}{n_2 \cos\theta_i + n_1 \cos\theta_r} E_p \tag{8-25a}$$

or

$$T_p = \frac{2\sin\theta_r \cos\theta_i}{\sin(\theta_i + \theta_r)\cos(\theta_i - \theta_r)} E_p \tag{8-25b}$$

Figures 8-1 and 8-2 have been drawn as if light goes from a lower index medium to a higher index medium. This reflection condition is called an external reflection.

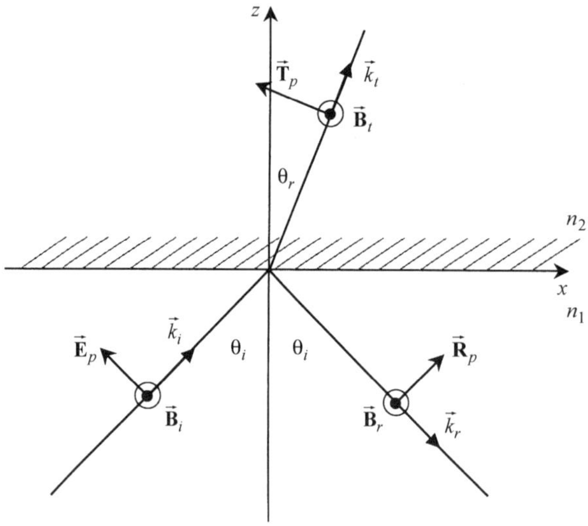

Figure 8-2 The plane of incidence for the transverse magnetic case.

Fresnel's equations also apply if the light is in a higher index medium and encounters an interface with a lower index medium, a condition known as an internal reflection.

Before we show graphs of the reflection coefficients, there are two special angles we should consider. These are Brewster's angle and the critical angle.

First, consider what happens to the amplitude reflection coefficient in (8-24b) when $\theta_i + \theta_r$ sums to 90°. The amplitude reflection coefficient vanishes for light polarized parallel to the plane of incidence. The incidence angle for which this occurs is called Brewster's angle. From Snell's law, we can relate Brewster's angle to the refractive indices of the media by a very simple expression, i.e.,

$$\theta_{i_B} = \tan^{-1}\frac{n_2}{n_1} \tag{8-26}$$

The other angle of importance is the critical angle. When we have an internal reflection, we can see from Snell's law that the transmitted light bends to ever larger angles as the incidence angle increases, and at some point the transmitted light leaves the higher index medium at a grazing angle. This is shown in Fig. 8-3. The incidence angle at which this occurs is the critical angle. From Snell's law, $n_2 \sin\theta_i = n_1 \sin\theta_r$ [writing the indices in reverse order to emphasize the light progression from high (n_2) to low (n_1) index], when $\theta_r = 90°$,

$$\sin\theta_i = \frac{n_1}{n_2} \tag{8-27a}$$

or

$$\theta_c = \sin^{-1}\frac{n_1}{n_2} \tag{8-27b}$$

where θ_c is the critical angle. For any incidence angle greater than the critical angle, there is no refracted ray and we have TIR.

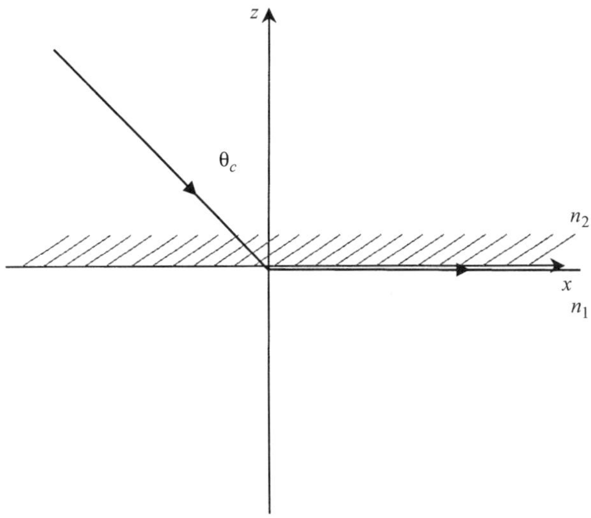

Figure 8-3 The critical angle where the refracted light exists the surface at grazing incidence.

The amplitude reflection coefficients, i.e.,

$$r_s \equiv \frac{R_s}{E_s} \tag{8-28a}$$

and

$$r_p \equiv \frac{R_p}{E_p} \tag{8-28b}$$

and their absolute values for external reflection for $n_1 = 1$ (air) and $n_2 = 1.5$ (a typical value for glass in the visible spectrum) are plotted in Fig. 8.4. Both the incident and reflected light has a phase associated with it, and there may be a net phase change upon reflection. The phase changes for external reflection are plotted in Fig. 8.5. The amplitude reflection coefficients and their absolute values for the same indices for internal reflection are plotted in Fig. 8-6. The phase changes for internal reflection are plotted in Fig. 8-7. An important observation to make here is that the reflection remains total beyond the critical angle, but the phase change is a continuously changing function of incidence angle. The phase changes beyond the critical angle, i.e., when the incidence angle is greater than the critical angle, are given by

$$\tan\frac{\varphi_s}{2} = \frac{\sqrt{\sin^2\theta_r - \sin^2\theta_c}}{\cos\theta_r} \tag{8-29a}$$

and

$$\tan\frac{\varphi_p}{2} = \frac{\sqrt{\sin^2\theta_r - \sin^2\theta_c}}{\cos\theta_r \sin^2\theta_c} \tag{8-29b}$$

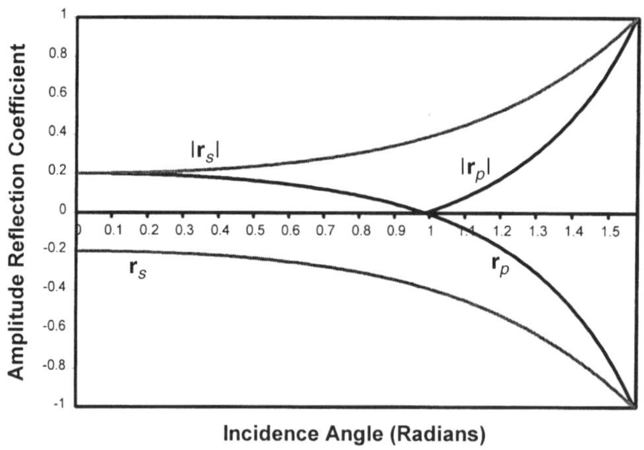

Figure 8-4 Amplitude reflection coefficients and their absolute values versus incidence angle for external reflection for $n_1 = 1$ and $n_2 = 1.5$.

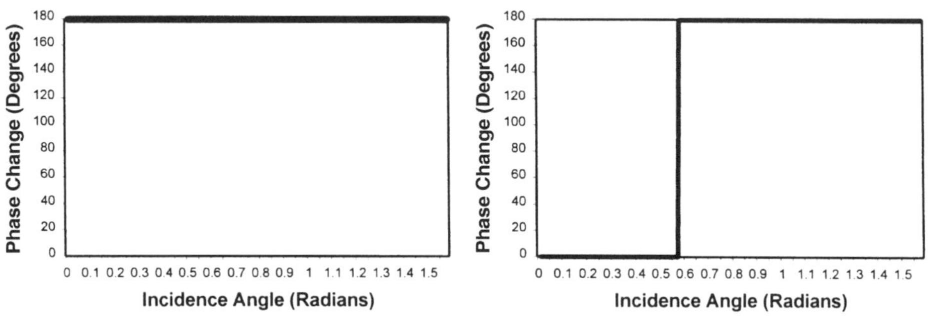

(a) TE case

(b) TM case

Figure 8-5 Phase changes for external reflection versus incidence angle for $n_1 = 1$ and $n_2 = 1.5$.

where φ_s and φ_p are the phase changes for the TE and TM cases, respectively. The reflected intensities, i.e., the square of the absolute value of the amplitude reflection coefficients, $\mathscr{R} = |r^2|$, for external and internal reflection are plotted in Figs. 8-8 and 8-9, respectively.

The results in this section have assumed real indices of refraction for linear, isotropic materials. This may not always be the case, i.e., the materials may be anisotropic and have complex indices of refraction and, in this case, the expressions for the reflection coefficients are not so simple. For example, the amplitude reflection coefficients for internal reflection at an isotropic to anisotropic interface [as would be the case for some applications, e.g., attenuated total reflection (see Deibler)], are

$$r_s = \frac{\sqrt{n_x^2 - k_x^2 + 2in_x k_x - n_1^2 \sin^2\theta} - n_1 \cos\theta}{\sqrt{n_x^2 - k_x^2 + 2in_x k_x - n_1^2 \sin^2\theta} + n_1 \cos\theta} \tag{8-30a}$$

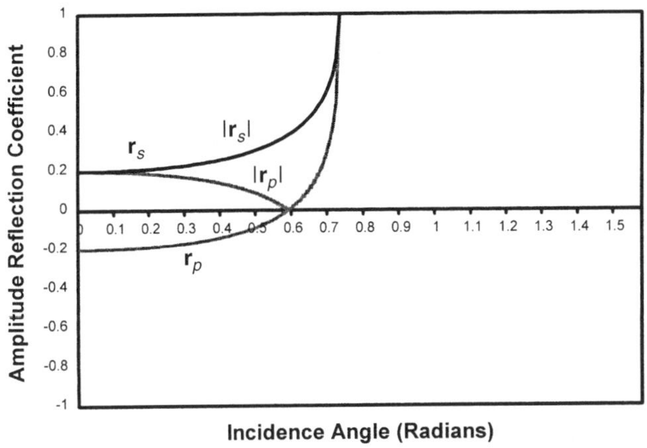

Figure 8-6 Amplitude reflection coefficients and their absolute values versus incidence angle for internal reflection for $n_1 = 1$ and $n_2 = 1.5$.

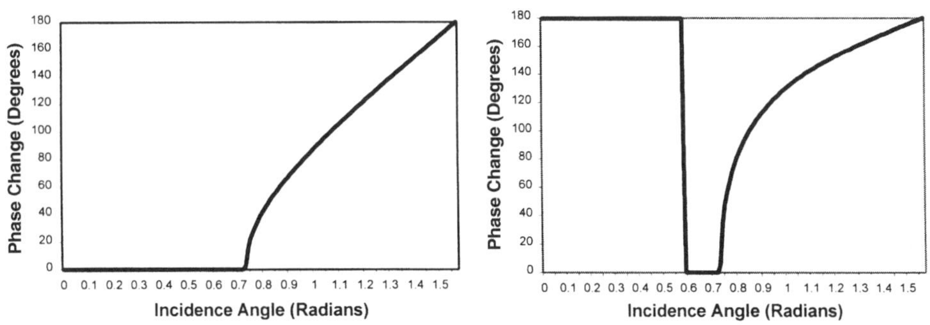

(a) TE case

(b) TM case

Figure 8-7 Phase changes for internal reflection versus incidence angle for $n_2 = 1.5$ and $n_1 = 1$.

and

$$r_p = \frac{n_1\sqrt{n_z^2 - k_z^2 + 2in_zk_z - n_1^2\sin^2\theta} - [n_yn_z - k_yk_z + i(k_yn_z + k_zn_z)]\cos\theta}{n_1\sqrt{n_z^2 - k_z^2 + 2in_zk_z - n_1^2\sin^2\theta} + [n_yn_z - k_yk_z + i(k_yn_z + k_zn_z)]\cos\theta} \tag{8-30b}$$

where n_x, n_y, and n_z are the real parts of the complex indices of the anisotropic material, and k_x, k_y, and k_z are the imaginary parts (in general, materials can have three principal indices). Anisotropic materials and their indices are covered in Chapter 24.

Before we go on to describe the reflection and transmission process in terms of Stokes parameters and Mueller matrices we make note of two important points.

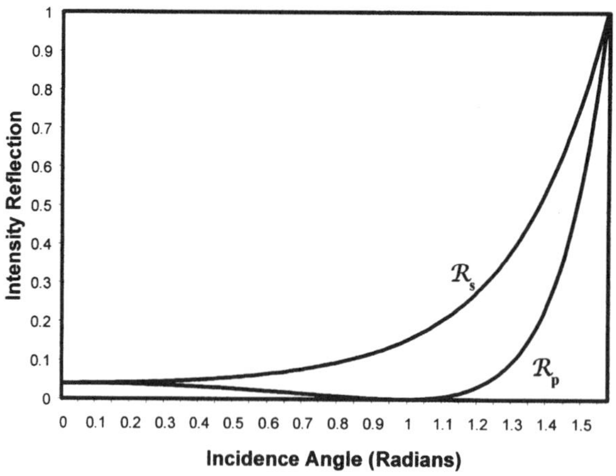

Figure 8-8 Intensity reflection for external reflection versus incidence angle for $n_1 = 1$ and $n_2 = 1.5$.

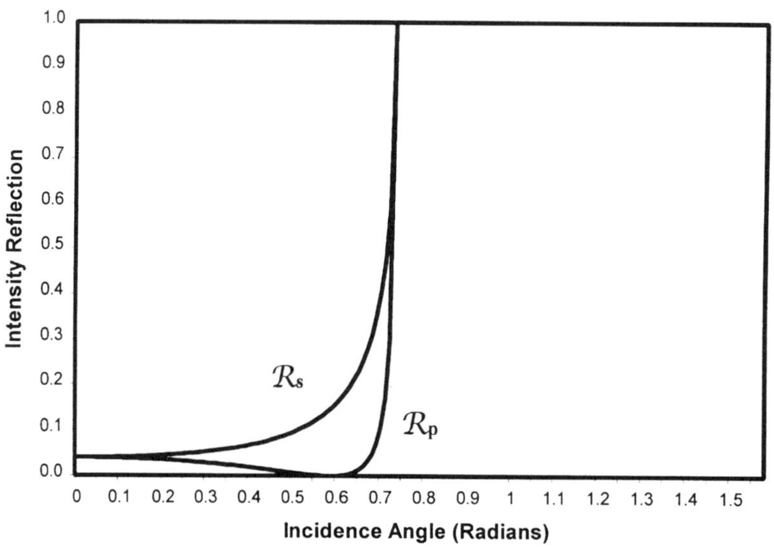

Figure 8-9 Intensity reflection for internal reflection versus incidence angle for $n_2 = 1.5$ and $n_1 = 1$.

First, the Stokes parameters must be defined appropriately for the field within and external to the dielectric medium. The first Stokes parameter represents the total intensity of the radiation and must correspond to a quantity known as the Poynting vector. This vector describes the flow of power of the propagating field components of the electromagnetic field. The Poynting vector is defined to be

$$\vec{\mathbf{S}} = (\vec{\mathbf{E}} \times \vec{\mathbf{H}}) \tag{8-31a}$$

In an isotropic dielectric medium, the time-averaged Poynting vector is

$$\langle\vec{\mathbf{S}}\rangle = \frac{\varepsilon_r}{2}\vec{\mathbf{E}}\cdot\vec{\mathbf{E}}^* \tag{8-31b}$$

Second, the direction of the Poynting vector and the surface normal are different. This requires that the component of the Poynting vector in the direction of the surface normal must be taken. Consequently, a cosine factor must be introduced into the definition of the Stokes parameters.

With these considerations, we will arrive at the correct Mueller matrices for reflection and transmission at a dielectric interface, as we will now show.

8.3 MUELLER MATRICES FOR REFLECTION AND TRANSMISSION AT AN AIR–DIELECTRIC INTERFACE

The Stokes parameters for the incident field in air ($n = 1$) are defined to be

$$S_0 = \cos\theta_i(E_sE_s^* + E_pE_p^*) \tag{8-32a}$$

$$S_1 = \cos\theta_i(E_sE_s^* - E_pE_p^*) \tag{8-32b}$$

$$S_2 = \cos\theta_i(E_sE_p^* + E_pE_s^*) \tag{8-32c}$$

$$S_3 = i\cos\theta_i(E_sE_p^* - E_pE_s^*) \tag{8-32d}$$

where E_s and E_p are the orthogonal components of the incident beam perpendicular and parallel to the plane of incidence, respectively, and the asterisk represents the complex conjugate. The factor i in (8-32d) is $\sqrt{-1}$.

Similarly, the Stokes parameters for the reflected field are

$$S_{0R} = \cos\theta_i(R_sR_s^* + R_pR_p^*) \tag{8-33a}$$

$$S_{1R} = \cos\theta_i(R_sR_s^* - R_pR_p^*) \tag{8-33b}$$

$$S_{2R} = \cos\theta_i(R_sR_p^* + R_pR_s^*) \tag{8-33c}$$

$$S_{3R} = i\cos\theta_i(R_sR_p^* - R_pR_s^*) \tag{8-33d}$$

The subscript R indicates that these are the Stokes parameters associated with the reflected beam. Substituting the values of R_s and R_p from Eqs. (8-22a) and

(8-24a) into (8-33) and using (8-32), the Stokes vector for the reflected beam S_R is found to be related to the Stokes vector of the incident beam S by

$$\begin{pmatrix} S_{0R} \\ S_{1R} \\ S_{2R} \\ S_{3R} \end{pmatrix} = \frac{1}{2}\left(\frac{\tan\theta_-}{\sin\theta_+}\right)^2 \times \begin{pmatrix} \cos^2\theta_- + \cos^2\theta_+ & \cos^2\theta_- - \cos^2\theta_+ & 0 & 0 \\ \cos^2\theta_- - \cos^2\theta_+ & \cos^2\theta_- + \cos^2\theta_+ & 0 & 0 \\ 0 & 0 & -2\cos\theta_+\cos\theta_- & 0 \\ 0 & 0 & 0 & -2\cos\theta_+\cos\theta_- \end{pmatrix} \times \begin{pmatrix} S_0 \\ S_1 \\ S_2 \\ S_3 \end{pmatrix} \tag{8-34}$$

where $\theta_\pm = \theta_i \pm \theta_r$. In the Mueller formalism, the matrix of a polarizer is

$$M = \frac{1}{2}\begin{pmatrix} p_S^2 + p_p^2 & p_S^2 - p_p^2 & 0 & 0 \\ p_s^2 - p_p^2 & p_s^2 + p_p^2 & 0 & 0 \\ 0 & 0 & 2p_s p_p & 0 \\ 0 & 0 & 0 & 2p_s p_p \end{pmatrix} \tag{8-35}$$

Comparing (8-34) with (8-35) we see that the 4 × 4 matrix in (8-34) corresponds to a Mueller matrix of a polarizer; this is to be expected from the form of Fresnel's equations, (8-22) and (8-24), in Section 8.2.

The Stokes parameters for the transmitted field are defined to be

$$S_{0T} = n\cos\theta_r(T_s T_s^* + T_p T_p^*) \tag{8-36a}$$

$$S_{1T} = n\cos\theta_r(T_s T_s^* - T_p T_p^*) \tag{8-36b}$$

$$S_{2T} = n\cos\theta_r(T_s T_p^* + T_p T_s^*) \tag{8-36c}$$

$$S_{3T} = in\cos\theta_r(T_s T_p^* - T_p T_s^*) \tag{8-36d}$$

where the subscript T indicates the Stokes parameters of the transmitted beam, and T_s and T_p are the transmitted field components perpendicular and parallel to the plane of incidence. Substituting the values of T_s and T_p from Eqs. (8-23) and (8-25) into (8-36) and using (8-32), the Stokes vector S_T is found to be

$$\begin{pmatrix} S_{0T} \\ S_{1T} \\ S_{2T} \\ S_{3T} \end{pmatrix} = \frac{\sin 2\theta_i \sin 2\theta_r}{2(\sin\theta_+ \cos\theta_-)^2} \times \begin{pmatrix} \cos^2\theta_- + 1 & \cos^2\theta_- - 1 & 0 & 0 \\ \cos^2\theta_- - 1 & \cos^2\theta_- + 1 & 0 & 0 \\ 0 & 0 & 2\cos\theta_- & 0 \\ 0 & 0 & 0 & 2\cos\theta_- \end{pmatrix} \begin{pmatrix} S_0 \\ S_1 \\ S_2 \\ S_3 \end{pmatrix} \tag{8-37}$$

We see that the 4×4 matrix in (8-37) also corresponds to the Mueller matrix of a polarizer.

It is straightforward to show from (8-34) and (8-37) that the following relation exists:

$$S_0 = S_{0R} + S_{0T} \tag{8-38}$$

Thus, the sum of the reflected intensity and the transmitted intensity is equal to the incident intensity, as expected from the principle of the conservation of energy.

Equation (8-34) shows that incident light which is completely polarized remains completely polarized. In addition to the case of incident light that is completely polarized, (8-34) allows us to consider the interesting case where the incident light is unpolarized. This case corresponds to Malus' discovery. It was very important because up to the time of Malus' discovery the only known way to obtain completely polarized light was to allow unpolarized light to propagate through a calcite crystal. Two beams were observed to emerge, called the ordinary and extraordinary rays, and each was found to be orthogonally linearly polarized.

The Stokes vector for unpolarized light is

$$S = I_0 \begin{pmatrix} 1 \\ 0 \\ 0 \\ 0 \end{pmatrix} \tag{8-39}$$

From (8-34) we then see that (8-39) yields

$$S_R = \begin{pmatrix} S_{0R} \\ S_{1R} \\ S_{2R} \\ S_{3R} \end{pmatrix} = \frac{1}{2}\left(\frac{\tan\theta_-}{\sin\theta_+}\right)^2 \times \begin{pmatrix} \cos^2\theta_- + \cos^2\theta_+ \\ \cos^2\theta_- - \cos^2\theta_+ \\ 0 \\ 0 \end{pmatrix} \tag{8-40}$$

The degree of polarization P is then

$$P = \left|\frac{S_1}{S_0}\right| = \left|\frac{\cos^2\theta_- - \cos^2\theta_+}{\cos^2\theta_- + \cos^2\theta_+}\right| \tag{8-41}$$

In general, because the numerator in (8-41) is less than the denominator, the degree of polarization is less than 1. However, a closer inspection of (8-41) shows that if cos θ_+ is zero, then $P = 1$; that is, the degree of polarization is 100%. This condition occurs at

$$\cos\theta_+ = \cos(\theta_i + \theta_r) = 0 \tag{8-42a}$$

so

$$\theta_i + \theta_r = \frac{\pi}{2} = 90^\circ \tag{8-42b}$$

Thus, when the sum of the incident angle and the refracted angle is 90° the reflected light is completely polarized. We found this earlier in Section 8.2 and this is confirmed by setting cos $\theta_+ = 0$ in (8-40), which then reduces to

$$S_R = \begin{pmatrix} S_{0R} \\ S_{1R} \\ S_{2R} \\ S_{3R} \end{pmatrix} = \frac{1}{2}\cos^2 2\theta_{iB} \begin{pmatrix} 1 \\ 1 \\ 0 \\ 0 \end{pmatrix} \tag{8-43}$$

The Stokes vector in (8-43) shows that the reflected light is linearly horizontally polarized. Because the degree of polarization is 1 (100%) at the angle of incidence which satisfies (8-42b), we have labeled θ_i as θ_{i_B}, Brewster's angle.

In Fig. 8-10 we have plotted (8-41), the degree of polarization P versus the incident angle θ_i, for a material with a refractive index of 1.50. Figure 8-10 shows that as the incident angle is increased P increases, reaches a maximum, and then returns to zero at $\theta_i = 90^\circ$. Thus, P is always less than 1 everywhere except at the

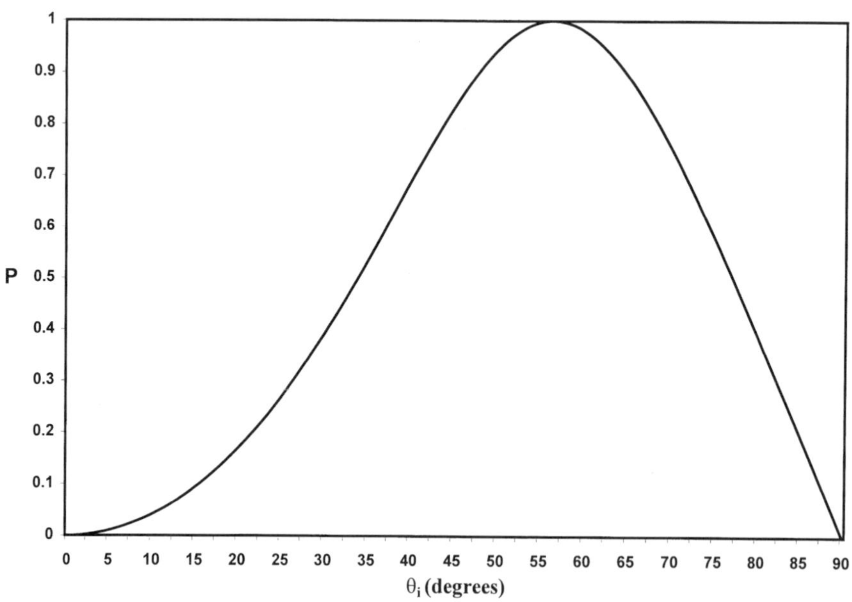

Figure 8-10 Plot of the degree of polarization P versus the incident angle θ_i for incident unpolarized light which is reflected from glass with a refractive index of 1.5.

maximum. The angle at which the maximum takes place is 56.7° (this will be shown shortly) and P is 0.9998 or 1.000 to three significant places. At this particular angle incident unpolarized light becomes completely polarized on being reflected. This angle is known as the polarization or Brewster angle (written θ_{i_B}). We shall see shortly that at the Brewster angle the Mueller matrix for reflection (8-34) simplifies significantly. This discovery by Brewster is very important because it allows one not only to create completely polarized light but partially polarized light as well. This latter fact is very often overlooked. Thus, if we have a perfect unpolarized light source, we can by a single reflection obtain partially polarized light to any degree we wish. In addition to this behavior of unpolarized light an extraordinarily simple mathematical relation emerges between the Brewster angle and the refractive indices of the dielectric materials, i.e., (8-26): this relation was used to obtain the value 56.7°.

With respect to creating partially polarized light, it is of interest to determine the intensity of the reflected light. From (8-40) we see that the intensity I_R of the reflected beam is

$$I_R = \frac{1}{2}\left(\frac{\tan\theta_-}{\sin\theta_+}\right)^2 (\cos^2\theta_- + \cos^2\theta_+) \tag{8-44}$$

In Fig. 8-11 we have plotted the magnitude of the reflected intensity I_R as a function of incident angle θ_i for a dielectric (glass) with a refractive index of 1.5. Figure 8-11 shows that as the incidence angle increases, the reflected intensity increases, particularly at the larger incidence angles. This explains why when the sun is low in the sky the light reflected from the surface of water appears to be quite strong. In fact, at these "low" angles polarizing sunglasses are only partially

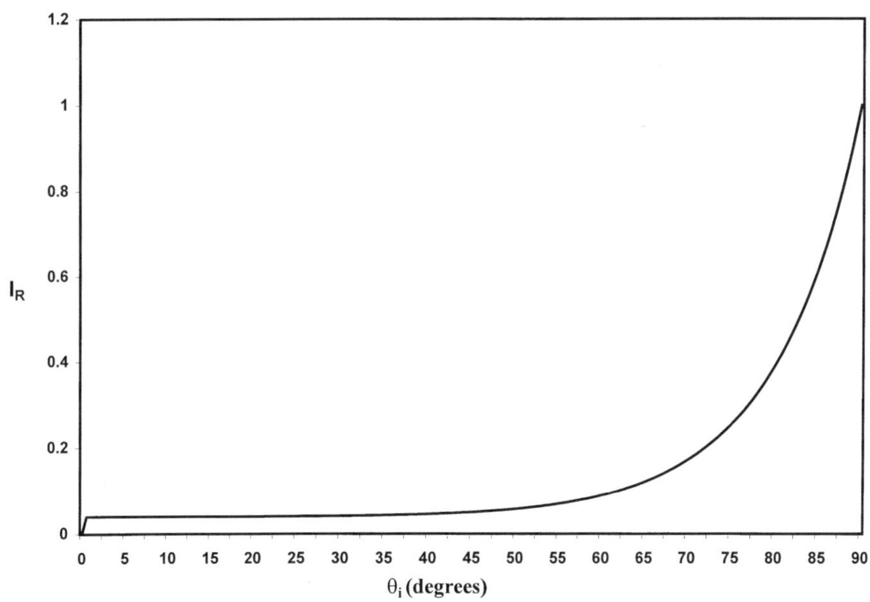

Figure 8-11 Plot of the intensity of a beam reflected by a dielectric of refractive index of 1.5. The incident beam is unpolarized.

effective because the reflected light is not completely polarized. If the incident angle were at the Brewster angle, the sunglasses would be completely effective. The reflected intensity at the Brewster angle θ_{i_B} (56.7°), according to (8-43) is only 7.9%.

In a similar manner (8-37) shows that the θ_{i_B} Stokes vector for the transmitted beam where the incident beam is again unpolarized is

$$\begin{pmatrix} S_{0T} \\ S_{1T} \\ S_{2T} \\ S_{3T} \end{pmatrix} = \frac{\sin 2\theta_i \sin 2\theta_r}{2(\sin\theta_+ \cos\theta_-)^2} \begin{pmatrix} \cos^2\theta_- + 1 \\ \cos^2\theta_- - 1 \\ 0 \\ 0 \end{pmatrix} \tag{8-45}$$

The degree of polarization P of the transmitted beam is

$$P = \left| \frac{\cos^2\theta_- - 1}{\cos^2\theta_- + 1} \right| \tag{8-46}$$

We again see that P is always less than 1. In Fig. 8-12 a plot has been made of the degree of polarization versus the incident angle. The refractive index of the glass is again $n = 1.50$.

The transmitted light remains practically unpolarized for relatively small angles of incidence. However, as the incident angle increases, the degree of polarization increases to a maximum value of 0.385 at 90°. Thus, unlike reflection, one can never obtain completely polarized light ($P = 1$) by the transmission of unpolarized light through a single surface. However, it is possible to increase the degree of polarization by using a dielectric material with a larger refractive index.

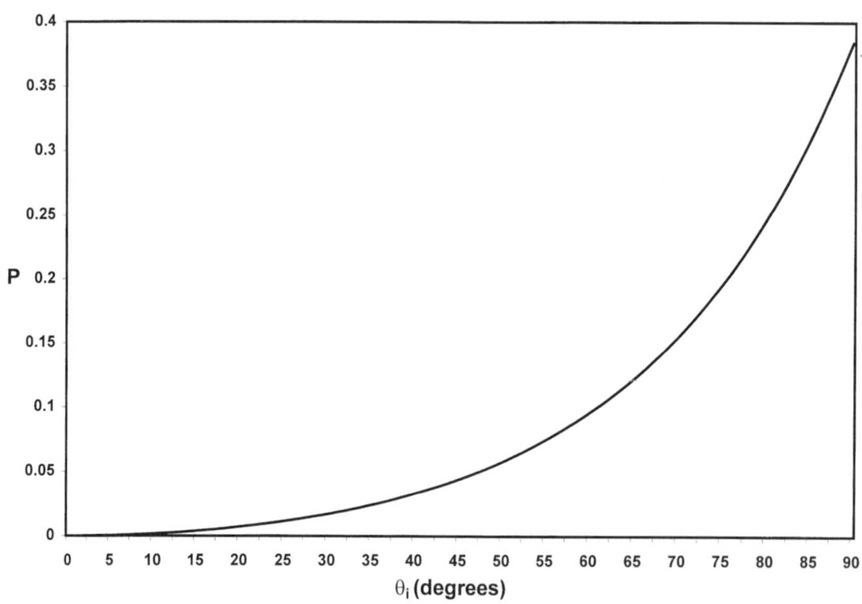

Figure 8-12 Plot of the degree of polarization versus the incident angle for incident unpolarized light transmitted through a single glass surface. The refractive index is again 1.5.

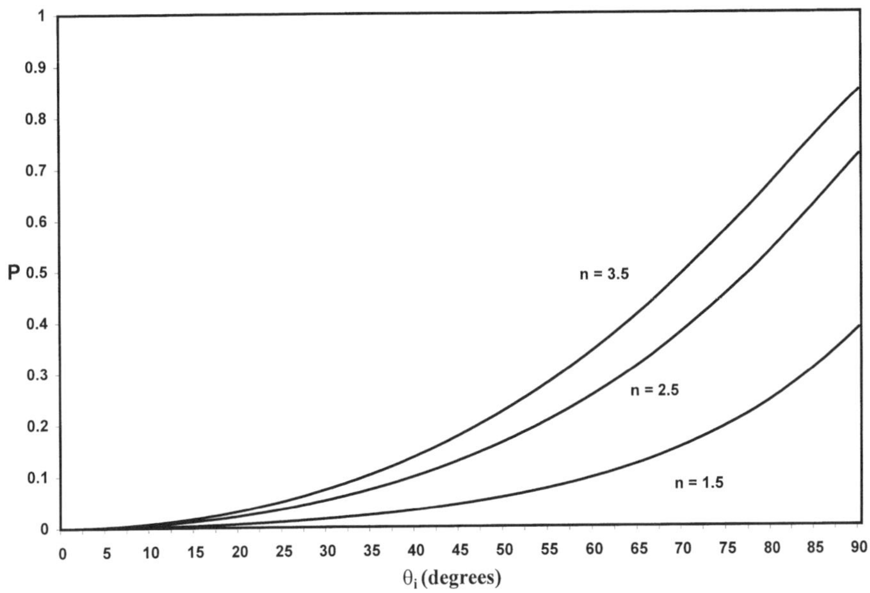

Figure 8-13 Plot of the degree of polarization versus the incident angle for differing refractive indices for an incident unpolarized beam transmitted through a single dielectric surface.

In Fig. 8-13 a plot has been made of the degree of polarization versus incident angle for materials with refractive indices of $n = 1.5$, 2.5, and 3.5. We see that there is a significant increase in the degree of polarization as n increases.

The final question of interest is to determine the intensity of the transmitted beam. From (8-45) we see that the transmitted intensity I_T is

$$I_T = \frac{\sin 2\theta_i \sin 2\theta_r}{2(\sin\theta_+ \cos\theta_-)^2}(\cos^2\theta_- + 1) \tag{8-47}$$

It is also of interest to determine the form of (8-47) at the Brewster angle θ_{i_B}. Using this condition, (8-42b), we easily find that (8-47) reduces to

$$I_{TB} = \frac{1}{2}(1 + \sin^2 2\theta_{i_B}) \tag{8-48}$$

For the Brewster angle of 56.7° ($n = 1.5$) we see that the transmitted intensity is 92.1%. We saw earlier that the corresponding intensity for the reflected beam was 7.9%. Thus, the sum of the reflected intensity and the transmitted intensity is 100%, in agreement with the general case expressed by (8-38), which is always true.

In Fig. 8-14 we have plotted (8-47) as a function of the incident angle for a beam transmitted through a dielectric with a refractive index of $n = 1.5$. We observe that the transmission remains practically constant up to the value of approximately 60°, whereupon the intensity drops rapidly to zero as the incidence angle approaches 90°.

We can extend these results to the important case of dielectric plates and multiple plates. Before we deal with this problem, however, we consider some

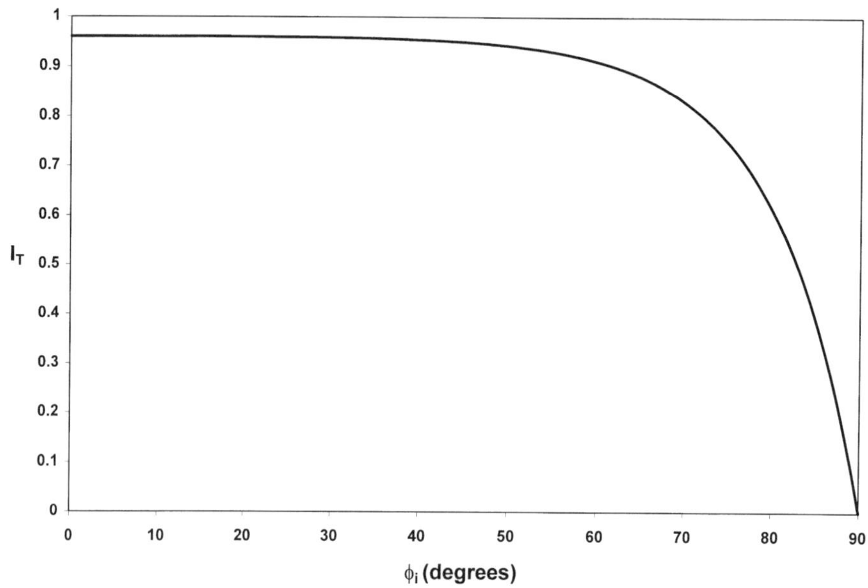

Figure 8-14 The intensity of a beam transmitted through a dielectric with a refractive index of 1.5 as a function of incidence angle. The incident beam is unpolarized.

simplifications in the Mueller matrices (8-34) and (8-37) in the next section. These simplifications occur at normal incidence ($\theta_i = 0^\circ$), at the Brewster angle θ_{i_B}, and at $\theta_i = 45^\circ$.

8.4 SPECIAL FORMS FOR THE MUELLER MATRICES FOR REFLECTION AND TRANSMISSION

There are three cases where the Mueller matrix for reflection by a dielectric surface simplifies. We now consider these three cases. In addition, we also derive the corresponding Mueller matrices for transmission.

8.4.1 Normal Incidence

In order to determine the form of the Mueller matrices at normal incidence for reflection and transmission, (8-34) and (8-37), we first express Snell's law for refraction for small angles. For small angles we have the approximations ($\theta \ll 1$):

$$\cos\theta \simeq 1 \tag{8-49a}$$

$$\sin\theta \simeq \theta \tag{8-49b}$$

Snell's law for refraction for small angles can then be written as

$$\theta_i \simeq n\theta_r \tag{8-50}$$

and we can then write

$$\tan\theta_- \simeq \theta_- = \theta_i - \theta_r \tag{8-51a}$$

$$\sin\theta_+ \simeq \theta_+ = \theta_i + \theta_r \tag{8-51b}$$

$$\cos\theta_+ \simeq 1 \tag{8-51c}$$

$$\cos\theta_- \simeq 1 \tag{8-51d}$$

Using these approximations (8-51), the Mueller matrix (8-34) then reduces to

$$M \simeq \frac{1}{2}\left(\frac{\theta_i - \theta_r}{\theta_i + \theta_r}\right)^2 \begin{pmatrix} 2 & 0 & 0 & 0 \\ 0 & 2 & 0 & 0 \\ 0 & 0 & -2 & 0 \\ 0 & 0 & 0 & -2 \end{pmatrix} \tag{8-52}$$

Substituting Snell's law for small angles (8-50) into (8-52), we then have

$$M_R = \left(\frac{n-1}{n+1}\right)^2 \begin{pmatrix} 1 & 0 & 0 & 0 \\ 0 & 1 & 0 & 0 \\ 0 & 0 & -1 & 0 \\ 0 & 0 & 0 & -1 \end{pmatrix} \tag{8-53}$$

which is the Mueller matrix for reflection at normal incidence. The significance of the negative sign in the matrix elements m_{22} and m_{33} is that on reflection the ellipticity and the orientation of the incident beam are reversed.

In a similar manner we readily determine the corresponding Mueller matrix for transmission at normal incidence. From (8-37) we have for small angles that

$$M = \frac{(2\theta_i)(2\theta_r)}{2(\theta_+)^2} \begin{pmatrix} 2 & 0 & 0 & 0 \\ 0 & 2 & 0 & 0 \\ 0 & 0 & 2 & 0 \\ 0 & 0 & 0 & 2 \end{pmatrix} \tag{8-54}$$

Again, using the small-angle approximation for Snell's law (8-50) we see that (8-54) reduces to

$$M_T = \frac{4n}{(n+1)^2} \begin{pmatrix} 1 & 0 & 0 & 0 \\ 0 & 1 & 0 & 0 \\ 0 & 0 & 1 & 0 \\ 0 & 0 & 0 & 1 \end{pmatrix} \tag{8-55}$$

which is the Mueller matrix for transmission at normal incidence.

The reflected intensity at normal incidence is seen from (8-53) to be

$$I_R = \left(\frac{n-1}{n+1}\right)^2 I_0 \tag{8-56}$$

and from (8-55) the transmitted intensity is

$$I_T = \frac{4n}{(n+1)^2} I_0 \tag{8-57}$$

Adding (8-56) and (8-57) yields

$$I_R + I_T = I_0 \tag{8-58}$$

as expected.

The normal incidence condition indicates that we can determine, in principle, the refractive index of the dielectric medium by reflection, (8-56). At first sight this might appear to be simple. However, in order to use a "normal incidence configuration" the reflected beam must be separated from the incident beam. We can only do this by inserting another optical component in the optical path. Thus, in spite of the seeming simplicity of (8-56), we cannot use it to measure the reflected beam and the refractive index of the dielectric (e.g., glass) directly.

8.4.2 The Brewster Angle

The Mueller matrix for reflection M_R is; from (8-34),

$$M_R = \frac{1}{2}\left(\frac{\tan\theta_-}{\sin\theta_+}\right)^2 \times \begin{pmatrix} \cos^2\theta_- + \cos^2\theta_+ & \cos^2\theta_- - \cos^2\theta_+ & 0 & 0 \\ \cos^2\theta_- - \cos^2\theta_+ & \cos^2\theta_- + \cos^2\theta_+ & 0 & 0 \\ 0 & 0 & -2\cos\theta_+\cos\theta_- & 0 \\ 0 & 0 & 0 & -2\cos\theta_+\cos\theta_- \end{pmatrix} \tag{8-59}$$

Similarly, the Mueller matrix for transmission M_T, from (8-37), is

$$M_T = \frac{\sin 2\theta_i \sin 2\theta_r}{2(\sin\theta_+ \cos\theta_-)^2} \begin{pmatrix} \cos^2\theta_- + 1 & \cos^2\theta_- - 1 & 0 & 0 \\ \cos^2\theta_- - 1 & \cos^2\theta_- + 1 & 0 & 0 \\ 0 & 0 & 2\cos\theta_- & 0 \\ 0 & 0 & 0 & 2\cos\theta_- \end{pmatrix} \tag{8-60}$$

Equation (8-60) has a very interesting simplification for the condition $\theta_+ = \theta_i + \theta_r = 90^\circ$. We write

$$\theta_+ = \theta_{i_B} + \theta_{r_B} = 90^\circ \tag{8-61a}$$

so

$$\theta_{r_B} = 90^\circ - \theta_{i_B} \tag{8-61b}$$

We shall show that this condition defines the Brewster angle. We now also write, using (8-61b)

$$\theta_- = \theta_{i_B} - \theta_{r_B} = 2\theta_{i_B} - 90^\circ \tag{8-62}$$

Substituting (8-62) into (8-59) along with $\theta_+ = 90^\circ$, we see that (8-59) reduces to

$$M_{R_B} = \frac{1}{2}\cos^2 2\theta_{i_B} \begin{pmatrix} 1 & 1 & 0 & 0 \\ 1 & 1 & 0 & 0 \\ 0 & 0 & 0 & 0 \\ 0 & 0 & 0 & 0 \end{pmatrix} \tag{8-63}$$

where we have used the relation:

$$\sin(2\theta_{i_B} - 90^\circ) = -\cos 2\theta_{i_B} \tag{8-64}$$

The result of (8-63) shows that for $\theta_+ = \theta_{i_B} + \theta_{r_B} = 90^\circ$ the Mueller matrix reduces to an ideal linear horizontal polarizer. This angle where the dielectric behaves as an ideal linear polarizer was first discovered by Sir David Brewster in 1812 and is known as the Brewster angle. Equation (8-63) also shows very clearly that at the Brewster angle the reflected beam will be completely polarized in the s direction. This has the immediate practical importance of allowing one to create, as we saw in Section 8.3, a completely linearly polarized beam from either partially or unpolarized light or from elliptically polarized light.

At the interface between a dielectric in air Brewster's relation becomes, from (8.26),

$$\tan\theta_{i_B} = n \tag{8-65}$$

This is a truly remarkable relation because it shows that the refractive index n, which we usually associate with the phenomenon of transmission, can be obtained by a reflection measurement. At the time of Brewster's discovery, using Brewster's angle was the first new method for measuring the refractive index of an optical material since the development of transmission methods in the seventeenth and eighteenth centuries. In fact, the measurement of the refractive index to a useful resolution is surprisingly difficult, in spite of the extraordinarily simple relation given by Snell's law. Relation (8-65) shows that the refractive index of a medium can be determined by a reflection measurement if the Brewster angle can be measured. Furthermore, because a dielectric surface behaves as a perfect linear polarizer at the Brewster angle, the reflected beam will always be linearly polarized regardless of the state of polarization of the incident beam. By then using a polarizer to analyze the reflected beam, we will obtain a null intensity only at the Brewster angle. From this angle the refractive index n can immediately be determined from (8-65).

At the Brewster angle the Mueller matrix for transmission (8-37) is readily seen to reduce to

$$M_{T,B} = \frac{1}{2}\begin{pmatrix} \sin^2 2\theta_{i_B} + 1 & \sin^2 2\theta_{i_B} - 1 & 0 & 0 \\ \sin^2 2\theta_{i_B} - 1 & \sin^2 2\theta_{i_B} + 1 & 0 & 0 \\ 0 & 0 & 2\sin 2\theta_{i_B} & 0 \\ 0 & 0 & 0 & 2\sin 2\theta_{i_B} \end{pmatrix} \tag{8-66}$$

which is a matrix of a polarizer. Thus, at the Brewster angle the Mueller matrix for transmission still behaves as a polarizer.

8.4.3 45° Incidence

The fact the Fresnel's equations simplify at normal incidence and at the Brewster angle is well known. However, there is another angle where Fresnel's equations and the Mueller matrices also simplify, the incidence angle of 45°. Remarkably, the resulting simplification in Fresnel's equations appears to have been first noticed by Humphreys-Owen only around 1960. We now derive the Mueller matrices for reflection and transmission at an incidence angle of 45°. The importance of the Mueller matrix for reflection at this angle of incidence is that it leads to another method for measuring the refractive index of an optical material. This method has a number of advantages over the normal incidence method and the Brewster angle method.

At an incidence angle of $\theta_i = 45^\circ$, Fresnel's equations for R_s and R_p, (8-22b) and (8-24b), reduce to

$$R_s = \left[\frac{\cos\theta_r - \sin\theta_r}{\cos\theta_r + \sin\theta_r}\right] E_s \tag{8-67a}$$

and

$$R_p = \left[\frac{\cos\theta_r - \sin\theta_r}{\cos\theta_r + \sin\theta_r}\right]^2 E_p \tag{8-67b}$$

We see that from (8-67) and the definitions of the amplitude reflection coefficients in (8-28) we have

$$r_s^2 = r_p \tag{8-68}$$

Later, we shall see that a corresponding relation exists between the orthogonal intensities I_s and I_p.

Using the condition that the incidence angle is 45° in (8-33) and using (8-67) we are led to the following Mueller matrix for incident 45° light:

$$M_R(\theta_i = 45^\circ) = \frac{1 - \sin 2\theta_r}{(1 + \sin 2\theta_r)^2} \begin{pmatrix} 1 & \sin 2\theta_r & 0 & 0 \\ \sin 2\theta_r & 1 & 0 & 0 \\ 0 & 0 & -\cos 2\theta_r & 0 \\ 0 & 0 & 0 & -\cos 2\theta_r \end{pmatrix} \tag{8-69}$$

Thus, at +45° incidence the Mueller matrix for reflection also takes on a simplified form. It still retains the form of a polarizer, however, Equation (8-69) now suggests a simple way to determine the refractive index n of an optical material by reflection. First, we irradiate the optical surface with s polarized light with an intensity I_0. Its Stokes vector is

$$S_s = I_0 \begin{pmatrix} 1 \\ 1 \\ 0 \\ 0 \end{pmatrix} \tag{8-70}$$

Multiplication of (8-70) by (8-69) leads to an intensity:

$$I_s = I_0 \frac{1 - \sin 2\theta_r}{1 + \sin 2\theta_r} \tag{8-71}$$

Next, the surface is irradiated with p polarized light so its Stokes vector is

$$S_p = I_0 \begin{pmatrix} 1 \\ -1 \\ 0 \\ 0 \end{pmatrix} \tag{8-72}$$

Multiplication of (8-72) by (8-69) leads to an intensity:

$$I_p = I_0 \left(\frac{1 - \sin 2\theta_r}{1 + \sin 2\theta_r} \right)^2 \tag{8-73}$$

Equations (8-71) and (8-73) for intensity are analogous to (8-67a) and (8-67b) for amplitude. Further, squaring (8-71) and using (8-73) leads to the relation:

$$\left(\frac{I_s}{I_0} \right)^2 = \frac{I_p}{I_0} \tag{8-74}$$

or

$$\frac{I_s^2}{I_p} = I_0 \tag{8-75}$$

Using the intensity reflection coefficients:

$$\mathscr{R}_s = \frac{I_s}{I_0} \tag{8-76a}$$

and

$$\mathscr{R}_p = \frac{I_p}{I_0} \tag{8-76b}$$

we have

$$\mathscr{R}_s^2 = \mathscr{R}_p \tag{8-77}$$

which is the analog of (8-68) in the intensity domain. Equation (8-75) shows that if I_p and I_s of the reflected beam can be measured, then the intensity of the incident beam I_0 can be determined.

Equations (8-71) and (8-73) also allow a unique expression for the refractive index to be found in terms of I_s and I_p. To show this, (8-73) is divided by (8-71), and we have

$$\frac{I_p}{I_s} = \frac{1 - \sin 2\theta_r}{1 + \sin 2\theta_r} \tag{8-78}$$

Solving (8-78) for $\sin 2\theta_r$ then yields

$$\sin 2\theta_r = \frac{I_s - I_p}{I_s + I_p} \tag{8-79}$$

We now write $\sin 2\theta_r$ in (8-79) in terms of the half-angle formula

$$2\sin\theta_r\cos\theta_r = \frac{I_s - I_p}{I_s + I_p} \tag{8-80}$$

Equation (8-80) can be written further as

$$(\sqrt{2}\sin\theta_r)(\sqrt{2}\cos\theta_r) = \frac{(\sqrt{I_s} - \sqrt{I_p})(\sqrt{I_s} + \sqrt{I_p})}{(\sqrt{I_s + I_p})(\sqrt{I_s + I_p})} \tag{8-81}$$

This form suggests that we equate the left- and right-hand sides as

$$\sqrt{2}\sin\theta_r = \frac{\sqrt{I_s} - \sqrt{I_p}}{\sqrt{I_s + I_p}} \tag{8-82a}$$

$$\sqrt{2}\cos\theta_r = \frac{\sqrt{I_s} + \sqrt{I_p}}{\sqrt{I_s + I_p}} \tag{8-82b}$$

We see that this decomposition is satisfactory because (8-82a) also leads to $\theta_r = 0^\circ$ for $I_s = I_p$ as in (8-79). Proceeding further, we have, from Snell's law for an incidence angle of $\theta_i = 45^\circ$,

$$\sqrt{2}\sin\theta_r = \frac{1}{n} \tag{8-83}$$

Equating (8-83) and (8-82a) then yields

$$n = \frac{\sqrt{I_s + I_p}}{\sqrt{I_s} - \sqrt{I_p}} \tag{8-84}$$

Equation (8-84) shows that at an incidence angle of 45° a very simple relation exists between the measured orthogonal intensities I_s and I_p and the refractive index n of an optical material. With the existence of photodetectors this suggests another way to measure the refractive index of an optical material.

Thus, we see that there are several methods for measuring the refractive index. Most importantly, the foregoing analysis enables us to use a single description for determining the behavior of light that is reflected and transmitted by a dielectric surface.

8.4.4 Total Internal Reflection

Fresnel's equations predict correctly the magnitude of the reflected and transmitted intensities of an optical beam. An added success of Fresnel's equations, however, is that they not only describe the behavior of light at an air–dielectric interface for "proper" reflection but, remarkably, for total internal reflection (TIR) as well. The phenomenon of TIR, occurs when light propagates from an optically denser medium into one which is less optically dense. In order to derive the Mueller matrix for TIR, we must first obtain the correct form of Fresnel's equations for TIR.

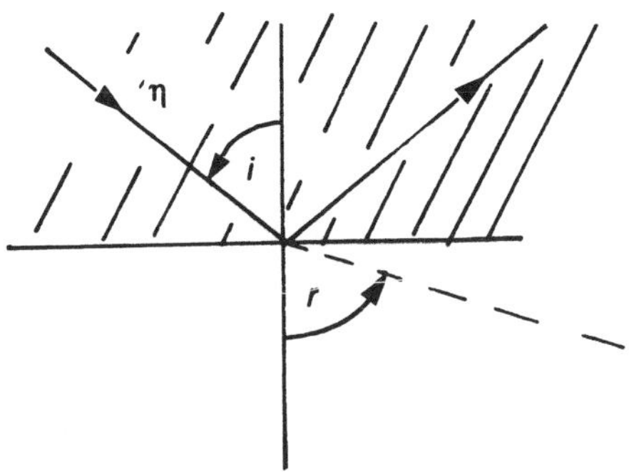

Figure 8-15 Total internal reflection.

In Fig. 8-15, we show an optical beam propagating in an optically denser medium and being reflected at the dielectric air–interface. Snell's law for Fig. 8-15 is now written

$$n \sin \theta_i = \sin \theta_r \tag{8-85}$$

For TIR to occur, the following condition must be satisfied:

$$n \sin \theta_i > 1 \tag{8-86}$$

We recall that Fresnel's reflection equations are

$$R_p = \frac{\tan(\theta_i - \theta_r)}{\tan(\theta_i + \theta_r)} E_p \tag{8-24b}$$

and

$$R_s = -\frac{\sin(\theta_i - \theta_r)}{\sin(\theta_i + \theta_r)} E_s \tag{8-22b}$$

Expanding the trigonometric functions in (8-24b) and (8-22b) gives

$$R_p = \frac{\sin \theta_i \cos \theta_i - \sin \theta_r \cos \theta_r}{\sin \theta_i \cos \theta_i + \sin \theta_r \cos \theta_r} E_p \tag{8-87a}$$

$$R_s = \frac{-\sin \theta_i \cos \theta_r - \sin \theta_r \cos \theta_i}{-\sin \theta_i \cos \theta_r + \sin \theta_r \cos \theta_i} E_s \tag{8-87b}$$

Snell's law (8-85) can be rewritten as

$$\cos \theta_r = i\sqrt{n^2 \sin^2 \theta_i = 1} \quad n \sin \theta_i > 1 \tag{8-88}$$

Substituting (8-88) into (8-87a) and (8-87b) yields

$$R_p = \frac{\cos \theta_i - in\sqrt{n^2 \sin^2 \theta_i - 1}}{\cos \theta_i + in\sqrt{n^2 \sin^2 \theta_i - 1}} E_p \tag{8-89a}$$

$$R_s = \frac{n\cos\theta_i - i\sqrt{n^2\sin^2\theta_i - 1}}{n\cos\theta_i + i\sqrt{n^2\sin^2\theta_i - 1}} E_s \tag{8-89b}$$

Let us consider (8-89a) in further detail. We can express

$$\frac{\cos\theta_i - in\sqrt{n^2\sin^2\theta_i - 1}}{\cos\theta_i + in\sqrt{n^2\sin^2\theta_i - 1}} \tag{8-90}$$

as

$$g = \frac{a - ib}{a + ib} \tag{8-91a}$$

where

$$a = \cos\theta_i \tag{8-91b}$$

and

$$b = n\sqrt{n^2\sin^2\theta_i - 1} \tag{8-91c}$$

The factor g is easily seen to be unimodular; that is, $gg^* = 1$, where the asterisk refers to a complex conjugate. Thus, (8-89a) can be expressed as

$$g = e^{-i\delta p} = \frac{a - ib}{a + ib} \tag{8-92a}$$

and

$$g = \cos\delta_p - i\sin\delta_p \tag{8-92b}$$

where δ_p refers to the phase associated with the parallel component, (8-89a). Equating the real and imaginary parts in (8-92) yields

$$\cos\delta_p = \frac{a^2 - b^2}{a^2 + b^2} \tag{8-93a}$$

$$\sin\delta_p = \frac{2ab}{a^2 + b^2} \tag{8-93b}$$

Dividing (8-93b) by (8-93a) then gives

$$\frac{\sin\delta_p}{\cos\delta_p} = \tan\delta_p = \frac{2ab}{a^2 - b^2} \tag{8-94}$$

Equation (8-94) can be further simplified by noting that sin δ_p and cos δ_p can be written in terms of their half-angle representations; that is,

$$\frac{\sin\delta_p}{\cos\delta_p} = \frac{2\sin(\delta_p/2)\cos(\delta_p/2)}{\cos^2(\delta_p/2) - \sin^2(\delta_p/2)} = \frac{2ab}{a^2 - b^2} \tag{8-95}$$

We arbitrarily set

$$\sin\frac{\delta_p}{2} = b \tag{8-96a}$$

$$\cos\frac{\delta_p}{2} = a \tag{8-96b}$$

Dividing (8-96a) by (8-96b) yields

$$\tan\frac{\delta_p}{2} = \frac{b}{a} \tag{8-97a}$$

and, from (8-91b) and (8-91c),

$$\tan\frac{\delta_p}{2} = \frac{n\sqrt{n^2\sin^2\theta_i - 1}}{\cos\theta_i} \tag{8-97b}$$

In exactly the same manner we find that

$$\tan\frac{\delta_s}{2} = \frac{\sqrt{n^2\sin^2\theta_i - 1}}{n\cos\theta_i} \tag{8-97c}$$

It is straightforward now to show that the following relation between the phases, $\delta = \delta_s - \delta_p$, can be derived from (8-97b) and (8-97c)

$$\tan\frac{\delta}{2} = -\frac{\cos\theta_i\sqrt{n^2\sin^2\theta_i - 1}}{n\sin^2\theta_i} \tag{8-98}$$

Returning to Fresnel's equations (8-89a) and (8-89b), we now see that for TIR they can be written simply as

$$R_p = e^{-i\delta_p}E_p \tag{8-99a}$$

$$R_s = e^{-i\delta_s}E_s \tag{8-99b}$$

From the definition of the Stokes parameters for reflection, we easily find that the Mueller matrix for TIR is

$$M_R = \begin{pmatrix} 1 & 0 & 0 & 0 \\ 0 & 1 & 0 & 0 \\ 0 & 0 & \cos\delta & -\sin\delta \\ 0 & 0 & \sin\delta & \cos\delta \end{pmatrix} \tag{8-100}$$

where $\delta = \delta_s - \delta_p$. Thus, TIR is described by the Mueller matrix for a retarder.

The phenomenon of TIR was first used by Fresnel (around 1820) to create circularly polarized light from linearly polarized light. In order to do this, Fresnel designed and then cut and polished a piece of glass in the form of a rhomb as shown in Fig. 8-16.

For a glass such as BK7, a commonly used optical glass made by Schott, the refractive index n at a wavelength of 6328 Å (He–Ne wavelength) is 1.5151. From (8-98) we see that for an angle of $\theta_i = 55°05'$ the phase shift δ at the first surface is

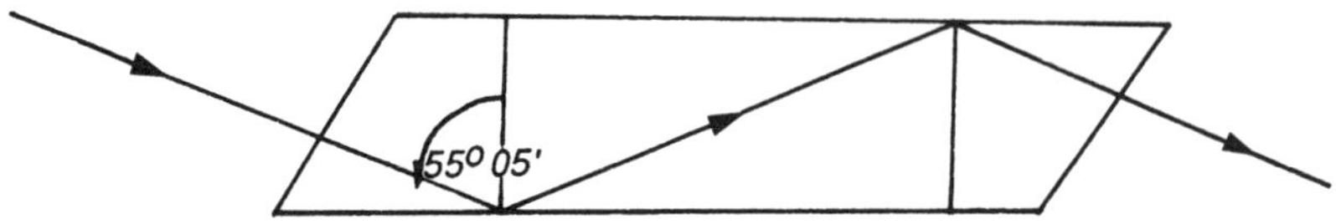

Figure 8-16 The Fresnel rhomb.

$\delta_L = 45.00°$. There is a similar phase shift δ_U at the upper surface for a total phase shift of 90.00°. Formally, we have from (8-100)

$$M = \begin{pmatrix} 1 & 0 & 0 & 0 \\ 0 & 1 & 0 & 0 \\ 0 & 0 & \cos\delta_U & -\sin\delta_U \\ 0 & 0 & \sin\delta_U & \cos\delta_U \end{pmatrix} \begin{pmatrix} 1 & 0 & 0 & 0 \\ 0 & 1 & 0 & 0 \\ 0 & 0 & \cos\delta_L & -\sin\delta_L \\ 0 & 0 & \sin\delta_L & \cos\delta_L \end{pmatrix} \tag{8-101a}$$

which leads to

$$M = \begin{pmatrix} 1 & 0 & 0 & 0 \\ 0 & 1 & 0 & 0 \\ 0 & 0 & \cos(\delta_U+\delta_L) & -\sin(\delta_U+\delta_L) \\ 0 & 0 & \sin(\delta_U+\delta_L) & \cos(\delta_U+\delta_L) \end{pmatrix} \tag{8-101b}$$

For the Fresnel rhomb $\delta = \delta_U + \delta_L = 90°$, so the Mueller matrix is

$$M = \begin{pmatrix} 1 & 0 & 0 & 0 \\ 0 & 1 & 0 & 0 \\ 0 & 0 & 0 & -1 \\ 0 & 0 & 1 & 0 \end{pmatrix} \tag{8-102}$$

If the incident beam is represented by

$$S = \begin{pmatrix} S_0 \\ S_1 \\ S_2 \\ S_3 \end{pmatrix} \tag{8-103}$$

then the Stokes vector of the beam emerging from the Fresnel rhomb is found by multiplication of (8-103) by (8-102) to be

$$S' = \begin{pmatrix} 1 & 0 & 0 & 0 \\ 0 & 1 & 0 & 0 \\ 0 & 0 & 0 & -1 \\ 0 & 0 & 1 & 0 \end{pmatrix} \begin{pmatrix} S_0 \\ S_1 \\ S_2 \\ S_3 \end{pmatrix} = \begin{pmatrix} S_0 \\ S_1 \\ -S_3 \\ S_2 \end{pmatrix} \tag{8-104}$$

If the incident beam is linear +45° polarized light then its Stokes vector is

$$S = I_0 \begin{pmatrix} 1 \\ 0 \\ 1 \\ 0 \end{pmatrix} \tag{8-105}$$

we see that the Stokes vector of the emerging beam is

$$S = I_0 \begin{pmatrix} 1 \\ 0 \\ 0 \\ 1 \end{pmatrix} \tag{8-106}$$

which is, of course, the Stokes vector of right circularly polarized light. Fresnel was the first to design and construct the rhombohedral prism which bears his name. He then used the prism to create circularly polarized light. Before Fresnel did so, *no one* had ever created circularly polarized light! This success was another triumph for his wave theory and his amplitude formulation of polarized light.

The major advantage of casting the problem of reflection and transmission at an optical interface into the formalism of the Mueller matrix calculus and the Stokes parameters is that we then have a single formulation for treating any polarization problem. In particular, very simple forms of the Mueller matrix arise at an incidence angle of 0°, the Brewster angle θ_{i_B}, an incidence angle of 45°, and TIR. However, in practice we usually deal with optical materials of finite thickness. We therefore now extend the results in this chapter toward treating the problem of reflection and transmission by dielectric plates.

REFERENCES

Papers

1. Collett, E., *Am. J. Phys.*, *39*, 517 (1971).
2. Collett, E., *Opt. Commun*, *63*, 217 (1987).
3. Humphreys-Owen, S. P. F., *Proc. Phys. Soc.*, *77*, 949 (1960).

Books

1. Born, M. and Wolf, E., *Principles of Optics*, 3rd ed., Pergamon Press, New York, 1965.
2. Jackson, J. D., *Classical Electrodynamics*, Wiley, New York, 1962.
3. Sommerfeld, A., *Lectures on Theoretical Physics*, Vols. I–V, Academic Press, New York, 1952.
4. Wood, R. W., *Physical Optics*, 3rd ed., Optical Society of America, Washington, DC, 1988.
5. Strong, J. *Concepts of Classical Optics*, Freeman, San Francisco, 1959.
6. Jenkins, F. S. and White, H. E., *Fundamentals of Optics*, McGraw-Hill, New York, 1957.
7. Stone, J. M., *Radiation and Optics*, McGraw-Hill, New York, 1963.
8. Shurcliff, W. A., *Polarized Light*, Harvard University Press, Cambridge, MA, 1962.
9. Hecht, E. and Zajac, A., *Optics*, Addison-Wesley, Reading, MA, 1974.
10. Longhurst, R. S., *Geometrical and Physical Optics*, 2nd Ed., Wiley, New York, 1967.
11. Deibler, L. L., *Infrared Polarimetry Using Attenuated Total Reflection*, Phd dissertation, University of Alabama in Huntsvilla, 2001.

9

The Mathematics of the Mueller Matrix

9.1 INTRODUCTION

Mathematical development to better understand and describe the information contained in the Mueller matrix is given in this chapter. The experimental Mueller matrix can be a complicated function of polarization, depolarization, and noise. How do we separate the specific information we are interested in, e.g., depolarization or retardance, from the measured Mueller matrix? When does an experimental matrix represent a physically realizable polarization element and when does it not? If it does not represent a physically realizable polarization element, how do we extract that information which will give us information about the equivalent physically realizable element? These are the questions we attempt to answer in this chapter.

Two algebraic systems have been developed for the solution of polarization problems in optics, the Jones formalism and the Mueller formalism. The Jones formalism is a natural consequence of the mathematical phase and amplitude description of light. The Mueller formalism comes from experimental considerations of the intensity measurements of polarized light.

R.C. Jones developed the Jones formalism in a series of papers published in the 1940s [1–3] and reprinted in a collection of historically significant papers on polarization [4]. The Jones formalism uses Jones vectors, two element vectors that describe the polarization state of light, and Jones matrices, 2×2 matrices that describe optical elements. The vectors are complex and describe the amplitude and phase of the light, i.e.,

$$\vec{\mathbf{J}}(t) = \begin{pmatrix} \vec{\mathbf{E}}_x(t) \\ \vec{\mathbf{E}}_y(t) \end{pmatrix} \tag{9-1}$$

is a time-dependent Jones vector where $\vec{\mathbf{E}}_x$, $\vec{\mathbf{E}}_y$ are the x and y components of the electric field of light traveling along the z axis. The matrices are also complex and describe the action in both amplitude and phase of optical elements on a light beam.

The Jones matrix is of the form:

$$J = \begin{pmatrix} j_{11} & j_{12} \\ j_{21} & j_{22} \end{pmatrix} \tag{9-2}$$

where the elements $j_{ij} = a_{ij} + ib_{ij}$ are complex. The two elements of the Jones vector are orthogonal and typically represent the horizontal and vertical polarization states. The four elements of the Jones matrix make up the transfer function from the input to the output Jones vector. Since these elements are complex, the Jones matrix contains eight constants and has eight degrees of freedom corresponding to the eight kinds of polarization behavior. A physically realizable polarization element results from any Jones matrix, i.e., there are no physical restrictions on the values of the Jones matrix elements. The Jones formalism is discussed in more detail in Chapter 11.

The Mueller formalism, already discussed in previous chapters but reviewed here, owes its name to Hans Mueller, who built on the work of Stokes [5], Soleillet [6], and Perrin [7] to formalize polarization calculations based on intensity. This work, as Jones', was also done during the 1940s but originally appeared in a now declassified report [8] and in a course of lectures at M.I.T. in 1945–1946. As we have learned, the Mueller formalism uses the Stokes vector to represent the polarization state. The Mueller matrix is a 4×4 matrix of real numbers. There is redundancy built into the Mueller matrix, since only seven of its elements are independent if there is no depolarization in the optical system. In the most general case, the Mueller matrix can have 16 independent elements; however, not every 4×4 Mueller matrix is a physically realizable polarizing element.

For each Jones matrix, there is a corresponding Mueller matrix. On conversion to a Mueller matrix, the Jones matrix phase information is discarded. A matrix with eight pieces of information is transformed to a matrix with seven pieces of information. Transformation equations for converting Jones matrices to Mueller matrices are given in Appendix C. The Mueller matrices can also be generated from equations. The Jones matrix is related to the Mueller matrix by

$$M = A(J \otimes J^*)A^{-1} \tag{9-3}$$

where $\otimes$ denotes the Kronecker product and A is

$$A = \begin{bmatrix} 1 & 0 & 0 & 1 \\ 1 & 0 & 0 & -1 \\ 0 & 1 & 1 & 0 \\ 0 & i & -i & 0 \end{bmatrix} \tag{9-4}$$

The elements of the Mueller matrix can also be obtained from the relation:

$$m_{ij} = \frac{1}{2}\mathrm{Tr}(J\sigma_i J^{\dagger}\sigma_j) \tag{9-5}$$

where $J^{\dagger}$ is the Hermitian conjugate of J and the σ are the set of four 2×2 matrices that comprise the identity matrix and the Pauli matrices (see Section 9.3).

The Jones matrix cannot represent a depolarizer or scatterer. The Mueller matrix can represent depolarizers and scatterers (see, e.g., [9]). Since the Mueller matrix contains information on depolarization, the conversion of Mueller matrices to Jones matrices must discard depolarization information. There is no phase information in a Mueller matrix, and the conversion conserves seven degrees of freedom.

The Mueller formalism has two advantages for experimental work over the Jones formalism. The intensity is represented explicitly in the Mueller formalism, and scattering can be included in the calculations. The Jones formalism is easier to use and more elegant for theoretical studies.

9.2 CONSTRAINTS ON THE MUELLER MATRIX

The issue of constraints on the Mueller matrix has been investigated by a number of researchers, e.g., [10–15]. The fundamental requirement that Mueller matrices must meet in order to be physically realizable is that they map physical incident Stokes vectors into physical resultant Stokes vectors. This recalls our requirement on Stokes vectors that the degree of polarization must always be less than or equal to one, i.e.,

$$P = \frac{(S_1^2 + S_2^2 + S_3^2)^{1/2}}{S_0} \leq 1 \tag{9-6}$$

A well-known constraint on the Mueller matrix is the inequality [16]:

$$\mathrm{Tr}(MM^T) = \sum_{i,j=0}^{3} m_{ij}^2 \leq 4m_{00}^2 \tag{9-7}$$

The equals sign applies for nondepolarizing systems and the inequality otherwise.

Many more constraints on Mueller matrix elements have been recorded. However, we shall not attempt to list or even to discuss them further here. The reason for this is that they may be largely irrelevant when one is making measurements with real optical systems. The measured Mueller matrices are a mixture of pure (nondepolarizing) states, depolarization, and certainly noise (optical and electronic). Is the magnitude of a particular Mueller matrix element due to diattenuation or retardance or is it really noise, or is it a mixture? If it is a mixture, what are the proportions? It is the responsibility of the experimenter to reduce noise sources as much as possible, determine the physical realizability of his Mueller matrices, and if they are not physically realizable, find the closest physically realizable Mueller matrices. A method of finding the closest physically realizable Mueller matrix and a method of decomposing nondepolarizing and depolarizing Mueller matrices are discussed in the remaining sections of this chapter. These are very important and provide useful results; however, only so much can be done to reduce noise intrusion. A study was done [17] to follow error propagation in the process of finding the best estimates, and it was found that the noise was reduced by one-third in nondepolarizing systems and reduced by one-tenth in depolarizing systems in going from the nonphysical matrix to the closest physically realizable matrix. The reduction is significant and worth doing, but no method can completely eliminate measurement noise. We will give examples in Section 9.4.

9.3 EIGENVECTOR AND EIGENVALUE ANALYSIS

Cloude [18,19] has formulated a method to obtain polarization characteristics and answer the question of physical realizability. Any 2×2 matrix J (in particular, a Jones matrix) can be expressed as

$$J = \sum_i k_i \sigma_i \tag{9-8}$$

where the σ_i are the Pauli matrices:

$$\sigma_1 = \begin{pmatrix} 1 & 0 \\ 0 & -1 \end{pmatrix} \qquad \sigma_2 = \begin{pmatrix} 0 & 1 \\ 1 & 0 \end{pmatrix} \qquad \sigma_3 = \begin{pmatrix} 0 & -i \\ i & 0 \end{pmatrix} \tag{9-9}$$

with the addition of the identity matrix:

$$\sigma_0 = \begin{pmatrix} 1 & 0 \\ 0 & 1 \end{pmatrix} \tag{9-10}$$

and the k_i are complex coefficients given by

$$k_i = \frac{1}{2}\mathrm{Tr}(J \cdot \sigma_i) \tag{9-11}$$

The components of this vector can also be written:

$$k_0 = \frac{1}{2}(j_{11} + j_{22}) \tag{9-12}$$

$$k_1 = \frac{1}{2}(j_{11} - j_{22}) \tag{9-13}$$

$$k_2 = \frac{1}{2}(j_{12} + j_{21}) \tag{9-14}$$

$$k_3 = \frac{i}{2}(j_{12} - j_{21}) \tag{9-15}$$

Cloude introduces a 4×4 Hermitian "target coherency matrix" obtained from the tensor product of the k's, i.e.,

$$T_c = k \otimes k^{*T} \tag{9-16}$$

The elements of the Mueller matrix are given in terms of the Jones matrix as

$$m_{ij} = \frac{1}{2}\mathrm{Tr}(J\sigma_i J^\dagger \sigma_j) \tag{9-17}$$

and Cloude shows that this can also be written as

$$m_{ij} = \frac{1}{2}\mathrm{Tr}(T_c \eta_{4i+j}) \tag{9-18}$$

where the η are the 16 Dirac matrices, a set of matrices which form a basis for 4×4 matrices. The Dirac matrices are shown in Table 9-1.

The matrix T_c can be expressed as

$$T_c = m_{ij}\sigma_i \otimes \sigma_j \tag{9-19}$$

Table 9-1 Dirac Matrices

η_0	η_1	η_2	η_3
$\begin{pmatrix} 1 & 0 & 0 & 0 \\ 0 & 1 & 0 & 0 \\ 0 & 0 & 1 & 0 \\ 0 & 0 & 0 & 1 \end{pmatrix}$	$\begin{pmatrix} 0 & 1 & 0 & 0 \\ 1 & 0 & 0 & 0 \\ 0 & 0 & 0 & i \\ 0 & 0 & -1 & 0 \end{pmatrix}$	$\begin{pmatrix} 0 & 0 & 1 & 0 \\ 0 & 0 & 0 & -i \\ 1 & 0 & 0 & 0 \\ 0 & i & 0 & 0 \end{pmatrix}$	$\begin{pmatrix} 0 & 0 & 0 & 1 \\ 0 & 0 & i & 0 \\ 0 & -i & 0 & 0 \\ 1 & 0 & 0 & 0 \end{pmatrix}$
η_4	η_5	η_6	η_7
$\begin{pmatrix} 0 & 1 & 0 & 0 \\ 1 & 0 & 0 & 0 \\ 0 & 0 & 0 & -i \\ 0 & 0 & i & 0 \end{pmatrix}$	$\begin{pmatrix} 1 & 0 & 0 & 0 \\ 0 & 1 & 0 & 0 \\ 0 & 0 & -i & 0 \\ 0 & 0 & 0 & -1 \end{pmatrix}$	$\begin{pmatrix} 0 & 0 & 0 & -i \\ 0 & 0 & 1 & 0 \\ 0 & 1 & 0 & 0 \\ i & 0 & 0 & 0 \end{pmatrix}$	$\begin{pmatrix} 0 & 0 & i & 0 \\ 0 & 0 & 0 & 1 \\ -i & 0 & 0 & 0 \\ 0 & 1 & 0 & 0 \end{pmatrix}$
η_8	η_9	η_{10}	η_{11}
$\begin{pmatrix} 0 & 0 & 1 & 0 \\ 0 & 0 & 0 & i \\ 1 & 0 & 0 & 0 \\ 0 & -i & 0 & 0 \end{pmatrix}$	$\begin{pmatrix} 0 & 0 & 0 & i \\ 0 & 0 & 1 & 0 \\ 0 & 1 & 0 & 0 \\ -i & 0 & 0 & 0 \end{pmatrix}$	$\begin{pmatrix} 1 & 0 & 0 & 0 \\ 0 & -1 & 0 & 0 \\ 0 & 0 & 1 & 0 \\ 0 & 0 & 0 & -1 \end{pmatrix}$	$\begin{pmatrix} 0 & -i & 0 & 0 \\ i & 0 & 0 & 0 \\ 0 & 0 & 0 & 1 \\ 0 & 0 & 1 & 0 \end{pmatrix}$
η_{12}	η_{13}	η_{14}	η_{15}
$\begin{pmatrix} 0 & 0 & 0 & 1 \\ 0 & 0 & -i & 0 \\ 0 & i & 0 & 0 \\ 1 & 0 & 0 & 0 \end{pmatrix}$	$\begin{pmatrix} 0 & 0 & -i & 0 \\ 0 & 0 & 0 & 1 \\ i & 0 & 0 & 0 \\ 0 & 1 & 0 & 0 \end{pmatrix}$	$\begin{pmatrix} 0 & i & 0 & 0 \\ -i & 0 & 0 & 0 \\ 0 & 0 & 0 & 1 \\ 0 & 0 & 1 & 0 \end{pmatrix}$	$\begin{pmatrix} 1 & 0 & 0 & 0 \\ 0 & -1 & 0 & 0 \\ 0 & 0 & -1 & 0 \\ 0 & 0 & 0 & 1 \end{pmatrix}$

where

$$\sigma_i \otimes \sigma_j \tag{9-20}$$

are the Dirac matrices. T_c can be written in the parametric form:

$$\begin{pmatrix} A_0 + A & C - iD & H + iG & I - iJ \\ C + iD & B_0 + B & E + iF & K - iL \\ H - iG & E - iF & B_0 - B & M + iN \\ I + iJ & K + iL & M - iN & A_0 - A \end{pmatrix} \tag{9-21}$$

where A through N are real numbers. If these real numbers are arranged into a 4×4 matrix where the ijth element is the expansion coefficient of the Dirac matrix η_{4i+j} then the matrix:

$$\begin{pmatrix} A_0 + B_0 & C + N & H + L & F + I \\ C - N & A + B & E + J & G + K \\ H - L & E - J & A - B & D + M \\ I - F & K - G & M - D & A_0 - B_0 \end{pmatrix} \tag{9-22}$$

is just the Mueller matrix when T_c is expressed in the Pauli base. The coherency matrix is then obtained from the experimental Mueller matrix by solving for the real elements A through N. When this is done the elements of the coherency matrix are found to be

$$t_{11} = \frac{m_{11} + m_{22} + m_{33} + m_{44}}{2} \tag{9-23}$$

$$t_{12} = \frac{m_{12} + m_{21} - i(m_{34} - m_{43})}{2} \tag{9-24}$$

$$t_{13} = \frac{m_{13} + m_{31} + i(m_{24} - m_{42})}{2} \tag{9-25}$$

$$t_{14} = \frac{m_{14} + m_{41} - i(m_{23} - m_{32})}{2} \tag{9-26}$$

$$t_{21} = \frac{m_{12} + m_{21} + i(m_{34} - m_{43})}{2} \tag{9-27}$$

$$t_{22} = \frac{m_{11} + m_{22} - m_{33} - m_{44}}{2} \tag{9-28}$$

$$t_{23} = \frac{m_{23} + m_{32} + i(m_{14} - m_{41})}{2} \tag{9-29}$$

$$t_{24} = \frac{m_{24} + m_{42} - i(m_{13} - m_{31})}{2} \tag{9-30}$$

$$t_{31} = \frac{m_{13} + m_{31} - i(m_{24} - m_{42})}{2} \tag{9-31}$$

$$t_{32} = \frac{m_{23} + m_{32} - i(m_{14} - m_{41})}{2} \tag{9-32}$$

$$t_{33} = \frac{m_{11} - m_{22} + m_{33} - m_{44}}{2} \tag{9-33}$$

$$t_{34} = \frac{m_{34} + m_{43} + i(m_{12} - m_{21})}{2} \tag{9-34}$$

$$t_{41} = \frac{m_{14} + m_{41} + i(m_{23} - m_{32})}{2} \tag{9-35}$$

$$t_{42} = \frac{m_{24} + m_{42} + i(m_{13} - m_{31})}{2} \tag{9-36}$$

$$t_{43} = \frac{m_{34} + m_{43} - i(m_{12} - m_{21})}{2} \tag{9-37}$$

$$t_{44} = \frac{m_{11} - m_{22} - m_{33} + m_{44}}{2} \tag{9-38}$$

The eigensystem for the coherency matrix T_c can be found and provides the decomposition of T_c into four components i.e.,

$$T_c = \lambda_1 T_{c1} + \lambda_2 T_{c2} + \lambda_3 T_{c3} + \lambda_4 T_{c4} \tag{9-39}$$

where the λ are the eigenvalues of T_c and

$$T_{ci} = k_i \otimes k_i^{*T} \tag{9-40}$$

are the eigenvectors. The eigenvalues of T_c are real since T_c is Hermitian. The eigenvectors are in general complex. Each eigenvalue/eigenvector corresponds to a Jones matrix (and every Jones matrix corresponds to a physically realizable

Table 9-2 Meaning of the C-vector Components

Matrix	Coefficient	Meaning	
σ_0	ρ_0	Amplitude	Absorption
σ_0	ϕ_0	Phase	Phase
σ_1	ρ_1	Amplitude	Linear diattenuation along axes
σ_1	ϕ_1	Phase	Linear retardance along axes
σ_2	ρ_2	Amplitude	Linear diattenuation 45°
σ_2	ϕ_2	Phase	Linear retardance 45°
σ_3	ρ_3	Amplitude	Circular diattenuation
σ_3	ϕ_3	Phase	Circular retardance

polarization element). The Jones matrix corresponding to the dominant eigenvalue is the matrix that describes the dominant polarizing action of the element. Extraction of this Jones matrix may be of interest for some applications: however, here the properties of the sample are most important.

These properties may be found with the realization that the eigenvector corresponding to the dominant eigenvalue is the quantity known as the C-vector [20]. The eigenvector components are the coefficients of the Pauli matrices in the decomposition of the Jones matrix: this is identical to the definition of the C-vector. The components of the C-vector give the information shown in Table 9.2.

Cloude has shown that for an experimental Mueller matrix to be physically realizable, the eigenvalues of the corresponding coherency matrix must be non-negative. The ratio of negative to positive eigenvalues is a quantitative measure of the realizability of the measured matrix. Further, a matrix that is not physically realizable can be "filtered," or made realizable by subtracting the component corresponding to the negative eigenvalue from the coherency matrix. Calculation of a new Mueller matrix then yields one that may include errors and scattering, but one that can be constructed from real polarization components.

9.4 EXAMPLE OF EIGENVECTOR ANALYSIS

In this section, a simple example of the calculations described in Section 9.3 is given. We will also give examples of the calculations to derive the closest physically realizable Mueller matrix from experimentally measured matrices.

The Mueller matrix for a partial linear polarizer with principal intensity transmission coefficients $k_1 = 0.64$ and $k_2 = 0.36$ along the principal axes and having an orientation $\theta = 0$ is given by

$$\begin{bmatrix} 0.50 & 0.14 & 0.0 & 0.0 \\ 0.14 & 0.50 & 0.0 & 0.0 \\ 0.0 & 0.0 & 0.48 & 0.0 \\ 0.0 & 0.0 & 0.0 & 0.48 \end{bmatrix} \tag{9-41}$$

The equivalent Jones matrix is

$$\begin{bmatrix} 0.8 & 0.0 \\ 0.0 & 0.6 \end{bmatrix} \tag{9-42}$$

The Cloude coherency matrix is

$$\begin{bmatrix} 0.98 & 0.14 & 0.0 & 0.0 \\ 0.14 & 0.02 & 0.0 & 0.0 \\ 0.0 & 0.0 & 0.0 & 0.0 \\ 0.0 & 0.0 & 0.0 & 0.0 \end{bmatrix} \tag{9-43}$$

There is only one nonzero eigenvalue of this matrix and it has a value of one. The eigenvector corresponding to this eigenvalue is

$$\begin{bmatrix} 0.9899 \\ 0.1414 \\ 0.000 \\ 0.000 \end{bmatrix} \tag{9-44}$$

where the second element of this vector is the measure of the linear diattenuation. Note that the terms corresponding to diattenuation at 45° and circular diattenuation are zero. Now suppose that the polarizer with the same principal transmission coefficients is rotated 40°. The Mueller matrix is

$$\begin{bmatrix} 0.500000 & 0.024311 & 0.137873 & 0.000000 \\ 0.024360 & 0.480725 & 0.003578 & 0.000000 \\ 0.137900 & 0.003270 & 0.499521 & 0.000000 \\ 0.000000 & 0.000000 & 0.000000 & 0.480000 \end{bmatrix} \tag{9-45}$$

The dominant eigenvalue is approximately one, and the corresponding eigenvector is

$$\begin{bmatrix} 0.9899 \\ 0.0246 \\ 0.1393 \\ 0.0002i \end{bmatrix} \tag{9-46}$$

With the rotation, the original linear polarization has coupled with polarization at 45° and circular polarization, and, in fact, the polarization at 45° is now the largest.

The linear diattenuation can now be calculated from (1) the original Mueller matrix, (2) the Jones matrix as found by Gerrard and Burch, and (3) the Cloude coherency matrix eigenvector. The linear diattenuation is given by

$$\frac{k_1 - k_2}{k_1 + k_2} = \frac{0.64 - 0.36}{0.64 + 0.36} = 0.28 \tag{9-47}$$

Calculation of the linear diattenuation from the Jones matrix derived directly from the Mueller matrix gives

$$\frac{r_1^2 - r_2^2}{r_1^2 + r_2^2} = \frac{0.8^2 - 0.6^2}{0.8^2 + 0.6^2} = 0.28 \tag{9-48}$$

In the method of Cloude, the components of the eigenvector corresponding to the dominant eigenvalue (i.e., the components of the C-vector) are given by

$$k_0 = \frac{(r_1 + r_2)}{2} \tag{9-49}$$

and

$$k_1 = \frac{(r_1 - r_2)}{2} \tag{9-50}$$

so that, solving for r_1 and r_2, and calculating diattenuation, a value of 0.28 is again obtained.

Let us now examine experimental Mueller matrices that have noise and are not likely to be physically realizable, and convert these into the closest possible physically realizable Mueller matrix. We will follow a slightly different prescription (D.M. Hayes, Pers. Commun., 1996) from that given above [19]. First, create the covariance matrix n for the experimental Mueller matrix m from the following equations:

$$n_{11} = m_{11} + m_{22} + m_{12} + m_{21} \tag{9-51}$$

$$n_{12} = n_{21} \tag{9-52}$$

$$n_{13} = n_{31} \tag{9-53}$$

$$n_{14} = n_{41} \tag{9-54}$$

$$n_{21} = m_{13} + m_{23} - i(m_{14} + m_{24}) \tag{9-55}$$

$$n_{22} = m_{11} - m_{22} - m_{12} + m_{21} \tag{9-56}$$

$$n_{23} = n_{32} \tag{9-57}$$

$$n_{24} = n_{42} \tag{9-58}$$

$$n_{31} = m_{31} + m_{32} + i(m_{41} + m_{42}) \tag{9-59}$$

$$n_{32} = m_{33} - m_{44} + i(m_{34} + m_{43}) \tag{9-60}$$

$$n_{33} = m_{11} - m_{22} + m_{12} - m_{21} \tag{9-61}$$

$$n_{34} = n_{43} \tag{9-62}$$

$$n_{41} = m_{33} + m_{44} - i(m_{34} - m_{43}) \tag{9-63}$$

$$n_{42} = m_{31} - m_{32} + i(m_{41} - m_{42}) \tag{9-64}$$

$$n_{43} = m_{13} - m_{23} - i(m_{14} - m_{24}) \tag{9-65}$$

$$n_{44} = m_{11} + m_{22} - m_{12} - m_{21} \tag{9-66}$$

Since this results in a Hermitian matrix, the eigenvalues will be real and the eigenvectors orthogonal. Now find the eigenvalues and eigenvectors of this matrix, and form a diagonal matrix from the eigenvalues, i.e.,

$$\Lambda = \begin{bmatrix} \lambda_1 & 0 & 0 & 0 \\ 0 & \lambda_2 & 0 & 0 \\ 0 & 0 & \lambda_3 & 0 \\ 0 & 0 & 0 & \lambda_4 \end{bmatrix} \tag{9-67}$$

We now set any negative eigenvalues in Λ equal to zero because negative eigenvalues correspond to nonphysical components. Construct a matrix V composed of the eigenvectors of n and perform the similarity transform:

$$N = V\Lambda V^{-1} \tag{9-68}$$

where N is the covariance matrix corresponding to the closest physical Mueller matrix to m. Finally, construct the physical Mueller matrix using the linear transformation:

$$M_{21} = \frac{N_{11} + N_{22} - N_{33} - N_{44}}{2} \tag{9-69}$$

$$M_{12} = M_{21} + N_{33} - N_{22} \tag{9-70}$$

$$M_{22} = N_{11} - N_{22} - M_{12} \tag{9-71}$$

$$M_{11} = 2N_{11} - M_{22} - M_{12} - M_{21} \tag{9-72}$$

$$M_{13} = \mathrm{Re}(N_{21} + N_{43}) \tag{9-73}$$

$$M_{23} = \mathrm{Re}(2N_{21}) - M_{13} \tag{9-74}$$

$$M_{31} = \mathrm{Re}(N_{31} + N_{42}) \tag{9-75}$$

$$M_{32} = \mathrm{Re}(2N_{31}) - M_{31} \tag{9-76}$$

$$M_{33} = \mathrm{Re}(N_{41} + N_{32}) \tag{9-77}$$

$$M_{44} = \mathrm{Re}(2N_{41}) - M_{33} \tag{9-78}$$

$$M_{14} = -\mathrm{Im}(N_{21} + N_{43}) \tag{9-79}$$

$$M_{24} = \mathrm{Im}(2N_{43}) + M_{14} \tag{9-80}$$

$$M_{41} = \mathrm{Im}(N_{31} + N_{42}) \tag{9-81}$$

$$M_{42} = \mathrm{Im}(2N_{31}) - M_{41} \tag{9-82}$$

$$M_{43} = \mathrm{Im}(N_{41} + N_{32}) \tag{9-83}$$

$$M_{34} = \mathrm{Im}(2N_{32}) - M_{43} \tag{9-84}$$

Let us now show numerical examples. The first example is an experimental calibration matrix for a rotating retarder polarimeter. The (normalized) matrix, which should ideally be the identity matrix, is

$$\begin{bmatrix} 0.978 & 0 & 0.003 & 0.005 \\ 0 & 1.000 & -0.007 & 0.006 \\ 0 & 0.007 & 0.999 & -0.007 \\ 0.005 & -0.003 & -0.002 & 0.994 \end{bmatrix} \tag{9-85}$$

The eigenvalues of the corresponding coherency matrix are, written in vector form,

$$\begin{bmatrix} 1.986 & -0.016 & -0.007 & -0.005 \end{bmatrix} \tag{9-86}$$

Three of these eigenvalues are negative so that the three corresponding eigenvalues must be removed (substracted) from the diagonal matrix formed by the set of four eigenvalues. In this case, the filtered matrix is

$$\begin{bmatrix} 0.993 & 0 & 0.002 & 0.005 \\ 0 & 0.993 & 0 & 0 \\ 0.002 & 0 & 0.993 & 0 \\ 0.005 & 0 & 0 & 0.993 \end{bmatrix} \tag{9-87}$$

The eigenvalue ratio, the ratio of the negative eigenvalue to the dominant eigenvalue in decibels, is a measure of the closeness to realizability. For this example the ratio of the largest negative eigenvalue to the dominant eigenvalue is approximately −21 dB. The original matrix was quite close to being physically realizable.

In a second example we have the case of a quartz plate that has its optic axis misaligned from the optical axis, inducing a small birefringence. The measured matrix was

$$\begin{bmatrix} 1.000 & 0.019 & 0.021 & -0.130 \\ -0.024 & -0.731 & -0.726 & 0.005 \\ 0.008 & 0.673 & -0.688 & -0.351 \\ -0.009 & 0.259 & -0.247 & 0.965 \end{bmatrix} \tag{9-88}$$

The eigenvalues of the corresponding coherency matrix are

$$\begin{bmatrix} 2.045 & -0.073 & 0.046 & -0.017 \end{bmatrix} \tag{9-89}$$

and the eigenvalue ratio is approximately −14.5 dB. In this case there are two negative eigenvalues that must be subtracted. The filtered matrix becomes

$$\begin{bmatrix} 0.737 & -0.005 & 0.006 & -0.067 \\ -0.005 & -0.987 & -0.024 & 0.131 \\ 0.006 & -0.024 & -0.989 & -0.304 \\ -0.067 & 0.131 & -0.304 & 0.674 \end{bmatrix} \tag{9-90}$$

9.5 THE LU–CHIPMAN DECOMPOSITION

Given an experimental Mueller matrix, we would like to be able to separate the diattenuation, retardance, and depolarization. A number of researchers had addressed this issue e.g., [21, 22] for nondepolarizing matrices. A general decomposition, a significant and extremely useful development, was only derived with the work of Lu and Chipman. This polar decomposition, which we call the Lu–Chipman decomposition [23, 24], allows a Mueller matrix to be decomposed into the product of the three factors.

Let us first review the nondepolarizing factors of diattenuation and retardance in this context. Diattenuation changes the intensity transmittances of the incident polarization states. The diattenuation is defined as

$$D \equiv \frac{T_{\max} - T_{\min}}{T_{\max} + T_{\min}} \tag{9-91}$$

and takes values from 0 to 1. Eigenpolarizations are polarization states that are transmitted unchanged by an optical element except for a change in phase and intensity. A diattenuator has two eigenpolarizations. For example, a horizontal polarizer has the eigenpolarizations of horizontal polarization and vertical polarization. If the eigenpolarizations are orthogonal, the element is a homogeneous polarization element, and is inhomogeneous otherwise. The axis of diattenuation is along the direction of the eigenpolarization with the larger transmittance. Let this diattenuation axis be along the eigenpolarization described by the Stokes vector:

$$(1 \quad d_1 \quad d_2 \quad d_3)^T = (1, \hat{D}^T)^T \tag{9-92}$$

where

$$\sqrt{d_1^2 + d_2^2 + d_3^2} = |\hat{D}| = 1 \tag{9-93}$$

Let us define a diattenuation vector:

$$\vec{\mathbf{D}} \equiv D\hat{D} = \begin{pmatrix} Dd_1 \\ Dd_2 \\ Dd_3 \end{pmatrix} = \begin{pmatrix} D_H \\ D_{45} \\ D_C \end{pmatrix} \tag{9-94}$$

where D_H is the horizontal diattenuation, D_{45} is the 45° linear diattenuation, and D_C is the circular diattenuation. The linear diattenuation is defined as

$$D_L \equiv \sqrt{D_H^2 + D_{45}^2} \tag{9-95}$$

and the total diattenuation is

$$D = \sqrt{D_H^2 + D_{45}^2 + D_C^2} = \sqrt{D_L^2 + D_C^2} = |\vec{\mathbf{D}}| \tag{9-96}$$

The diattenuation vector provides a complete description of the diattenuation properties of a diattenuator.

The intensity transmittance can be written as the ratio of energies in the exiting to incident Stokes vector:

$$T = \frac{s_0'}{s_0} = \frac{m_{00}s_0 + m_{01}s_1 + m_{02}s_2 + m_{03}s_3}{s_0} \tag{9-97}$$

where there is an intervening element with Mueller matrix M. The first row of the Mueller matrix completely determines the intensity transmittance. Equation (9-97) can be rewritten as

$$T = m_{00} + \frac{\vec{\mathbf{m}} \cdot \vec{\mathbf{s}}}{s_0} \tag{9-98}$$

where the vectors are defined as $\vec{\mathbf{m}} \equiv (m_{01}, m_{02}, m_{03})$ and $\vec{\mathbf{s}} \equiv (s_1, s_2, s_3)$. The maximum and minimum values of the dot product can be taken to be

$$\vec{\mathbf{s}} \cdot \vec{\mathbf{m}} = s_0|\vec{\mathbf{m}}| \tag{9-99}$$

and

$$\vec{\mathbf{s}} \cdot \vec{\mathbf{m}} = -s_0|\vec{\mathbf{m}}| \tag{9-100}$$

so that the maximum and minimum transmittances $T_{\max}$ and $T_{\min}$ are

$$T_{\max} = m_{00} + \sqrt{m_{01}^2 + m_{02}^2 + m_{03}^2} \tag{9-101}$$

$$T_{\min} = m_{00} - \sqrt{m_{01}^2 + m_{02}^2 + m_{03}^2} \tag{9-102}$$

The normalized Stokes vectors associated with $T_{\max}$ and $T_{\min}$ are

$$\hat{S}_{\max} = \begin{pmatrix} 1 \\ m_{01}\Big/\sqrt{m_{01}^2 + m_{02}^2 + m_{03}^2} \\ m_{02}\Big/\sqrt{m_{01}^2 + m_{02}^2 + m_{03}^2} \\ m_{03}\Big/\sqrt{m_{01}^2 + m_{02}^2 + m_{03}^2} \end{pmatrix} \tag{9-103}$$

and

$$\hat{S}_{\min} = \begin{pmatrix} 1 \\ -m_{01}\Big/\sqrt{m_{01}^2 + m_{02}^2 + m_{03}^2} \\ -m_{02}\Big/\sqrt{m_{01}^2 + m_{02}^2 + m_{03}^2} \\ -m_{03}\Big/\sqrt{m_{01}^2 + m_{02}^2 + m_{03}^2} \end{pmatrix} \tag{9-104}$$

The diattenuation of the Mueller matrix is

$$D = \frac{T_{\max} - T_{\min}}{T_{\max} + T_{\min}} = \frac{1}{m_{00}}\sqrt{m_{01}^2 + m_{02}^2 + m_{03}^2} \tag{9-105}$$

and the axis of diattenuation is along the maximum transmittance and thus the direction of $\hat{S}_{\max}$. The axis of diattenuation is along the state $\hat{S}_{\max}$ and the diattenuation vector of the Mueller matrix is then given by

$$\vec{\mathbf{D}} = \begin{pmatrix} D_H \\ D_{45} \\ D_C \end{pmatrix} = \frac{1}{m_{00}}\begin{pmatrix} m_{01} \\ m_{02} \\ m_{03} \end{pmatrix} \tag{9-106}$$

so that the first row of a Mueller matrix gives its diattenuation vector. The expressions for $\hat{S}_{\max}$ and $\hat{S}_{\min}$ can be written as

$$\hat{S}_{\max} = \begin{pmatrix} 1 \\ \hat{D} \end{pmatrix} \tag{9-107}$$

and

$$\hat{S}_{\min} = \begin{pmatrix} 1 \\ -\hat{D} \end{pmatrix} \tag{9-108}$$

Operational definitions for the components of the diattenuation vector are given by

$$\frac{T_H - T_V}{T_H + T_V} = \frac{m_{01}}{m_{00}} = D_H \tag{9-109}$$

$$\frac{T_{45} - T_{135}}{T_{45} + T_{135}} = \frac{m_{02}}{m_{00}} = D_{45} \tag{9-110}$$

$$\frac{T_R - T_L}{T_R + T_L} = \frac{m_{03}}{m_{00}} = D_C \tag{9-111}$$

where T_H is the transmittance for horizontally polarized light, T_V is the transmittance for vertically polarized light, T_{45} is the transmittance for linear 45° polarized light, T_{135} is the transmittance for linear 135° polarized light, T_R is the transmittance for right circularly polarized light, and T_L is the transmittance for left circularly polarized light.

Now consider that we have incident unpolarized light, i.e., only one element of the incident Stokes vector is nonzero. The exiting state is determined completely by the first column of the Mueller matrix. The polarization resulting from changing completely unpolarized light to polarized light is called polarizance. The polarizance is given by

$$P = \frac{1}{m_{00}} \sqrt{m_{10}^2 + m_{20}^2 + m_{30}^2} \tag{9-112}$$

and can take values from 0 to 1. A normalized polarizance vector is given by

$$\vec{\mathbf{P}} \equiv \begin{pmatrix} P_H \\ P_{45} \\ P_R \end{pmatrix} = \frac{1}{m_{00}} \begin{pmatrix} m_{10} \\ m_{20} \\ m_{30} \end{pmatrix} \tag{9-113}$$

The components of the polarizance vector are equal to the horizontal degree of polarization, 45° linear degree of polarization, and circular degree of polarization resulting from incident unpolarized light.

Retarders are phase-changing devices and have constant intensity transmittance for any incident polarization state. Eigenpolarizations are defined for retarders according to the phase changes they produce. The component of light with leading phase has its eigenpolarization along the fast axis (see Chaps. 24 and 26) of the retarder. Let us define a vector along this direction:

$$(1, \quad a_1, \quad a_2, \quad a_3)^T = (1, \quad \hat{R}^T)^T \tag{9-114}$$

where

$$\sqrt{a_1^2 + a_2^2 + a_3^2} = |\hat{R}| = 1 \tag{9-115}$$

The retardance vector and the fast axis are described by

$$\vec{\mathbf{R}} \equiv R\hat{R} = \begin{pmatrix} Ra_1 \\ Ra_2 \\ Ra_3 \end{pmatrix} \equiv \begin{pmatrix} R_H \\ R_{45} \\ R_C \end{pmatrix} \tag{9-116}$$

where the components of $\vec{\mathbf{R}}$ give the horizontal, 45° linear, and circular retardance components. The net linear retardance is

$$R_L = \sqrt{R_H^2 + R_{45}^2} \tag{9-117}$$

and the total retardance is

$$R = \sqrt{R_H^2 + R_{45}^2 + R_C^2} = \sqrt{R_L^2 + R_C^2} = |\vec{\mathbf{R}}| \tag{9-118}$$

Now that we have laid the groundwork for nondepolarizing Mueller matrices, let us consider the decomposition of these matrices. Nondepolarizing Mueller matrices can be written as the product of a retarder and diattenuator, i.e.,

$$M = M_R M_D \tag{9-119}$$

where M_R is the Mueller matrix of a pure retarder and M_D is the Mueller matrix of a pure diattenuator. A normalized Mueller matrix M can be written:

$$M = \begin{pmatrix} 1 & m_{01} & m_{02} & m_{03} \\ m_{10} & m_{11} & m_{12} & m_{13} \\ m_{20} & m_{21} & m_{22} & m_{23} \\ m_{30} & m_{31} & m_{32} & m_{33} \end{pmatrix} = \begin{pmatrix} 1 & \vec{\mathbf{D}}^T \\ \vec{\mathbf{P}} & m \end{pmatrix} \tag{9-120}$$

where the submatrix m is

$$m = \begin{pmatrix} m_{11} & m_{12} & m_{13} \\ m_{21} & m_{22} & m_{23} \\ m_{31} & m_{32} & m_{33} \end{pmatrix} \tag{9-121}$$

and $\vec{\mathbf{D}}$ and $\vec{\mathbf{P}}$ are the diattenuation and polarizance vectors as given in (9.106) and (9.113). The diattenuator M_D is calculated from the first row of M, and M_D^{-1} can then be multiplied by M to obtain the retarder matrix $M_R = MM_D^{-1}$. The diattenuator matrix is given by

$$M_D = \begin{pmatrix} 1 & \vec{\mathbf{D}}^T \\ \vec{\mathbf{D}} & m_D \end{pmatrix} \tag{9-122}$$

where

$$m_D = a\mathbf{I}_3 + b\left(\vec{\mathbf{D}} \cdot \vec{\mathbf{D}}^T\right) \tag{9-123}$$

and where $\mathbf{I}_3$ is the 3×3 identity matrix, and a and b are scalars derived from the norm of the diattenuation vector, i.e.,

$$D = |\vec{\mathbf{D}}| \tag{9-124}$$

$$a = \sqrt{1 - D^2} \tag{9-125}$$

$$b = \frac{1 - \sqrt{1 - D^2}}{D^2} \tag{9-126}$$

Writing the diattenuator matrix out, we have

$$M_D = \begin{pmatrix} 1 & m_{01} & m_{02} & m_{03} \\ m_{01} & a + bm_{01}^2 & bm_{01}m_{02} & bm_{01}m_{03} \\ m_{02} & bm_{02}m_{01} & a + bm_{02}^2 & bm_{02}m_{03} \\ m_{03} & bm_{03}m_{01} & bm_{03}m_{02} & a + bm_{03}^2 \end{pmatrix} \tag{9-127}$$

where

$$a = \sqrt{1 - (m_{01}^2 + m_{02}^2 + m_{03}^2)} \tag{9-128}$$

and

$$b = \frac{1 - \sqrt{1 - (m_{01}^2 + m_{02}^2 + m_{03}^2)}}{(m_{01}^2 + m_{02}^2 + m_{03}^2)} \tag{9-129}$$

M_D^{-1} is then given by

$$M_D^{-1} = \frac{1}{a^2}\begin{pmatrix} 1 & -\vec{D}^T \\ -\vec{D} & \mathbf{I}_3 \end{pmatrix} + \frac{1}{a^2(a+1)}\begin{pmatrix} 0 & \vec{0}^T \\ \vec{0} & (\vec{D} \cdot \vec{D}^T) \end{pmatrix} \tag{9-130}$$

The retarder matrix is

$$M_R = \begin{pmatrix} 1 & \vec{\mathbf{0}}^T \\ \vec{\mathbf{0}} & m_R \end{pmatrix} \tag{9-131}$$

where

$$m_R = \frac{1}{a}\left[m - b(\vec{\mathbf{P}} \cdot \vec{\mathbf{D}}^T)\right] \tag{9-132}$$

The retarder matrix can be written explicitly as

$$M_R = \frac{1}{a}\begin{bmatrix} a & 0 & 0 & 0 \\ 0 & m_{11} - b(m_{10}m_{01}) & m_{12} - b(m_{10}m_{02}) & m_{13} - b(m_{10}m_{03}) \\ 0 & m_{21} - b(m_{20}m_{01}) & m_{22} - b(m_{20}m_{02}) & m_{23} - b(m_{20}m_{03}) \\ 0 & m_{31} - b(m_{30}m_{01}) & m_{32} - b(m_{30}m_{02}) & m_{33} - b(m_{30}m_{03}) \end{bmatrix} \tag{9-133}$$

The total retardance R and the retardance vector can be found from the equations:

$$R = |\vec{\mathbf{R}}| = \cos^{-1}\left(\frac{\mathrm{Tr}(m_R) - 1}{2}\right) \qquad 0 \le R \le \pi \tag{9-134}$$

$$R = |\vec{\mathbf{R}}| = 2\pi - \cos^{-1}\left(\frac{\mathrm{Tr}(m_R) - 1}{2}\right) \qquad \pi \le R \le 2\pi \tag{9-135}$$

$$\vec{\mathbf{R}} = \begin{pmatrix} R_H \\ R_{45} \\ R_C \end{pmatrix} = \begin{pmatrix} (M_R)_{23} - (M_R)_{32} \\ (M_R)_{31} - (M_R)_{13} \\ (M_R)_{12} - (M_R)_{21} \end{pmatrix} \frac{R}{2\sin(R)} \tag{9-136}$$

The total retardance is then given explicitly as

$$R = \cos^{-1}\left(\frac{1}{2a}[m_{11} + m_{22} + m_{33} - b(m_{10}m_{01} + m_{20}m_{02} + m_{30}m_{03}) - a]\right) \tag{9-137}$$

and the retardance vector is given by

$$\vec{\mathbf{R}} = \begin{bmatrix} m_{23} - m_{32} - b(m_{20}m_{03} - m_{30}m_{02}) \\ m_{31} - m_{13} - b(m_{30}m_{01} - m_{10}m_{03}) \\ m_{12} - m_{21} - b(m_{10}m_{02} - m_{20}m_{01}) \end{bmatrix} \times \frac{\cos^{-1}(1/2a[m_{11} + m_{22} + m_{33} - b(m_{10}m_{01} + m_{20}m_{02} + m_{30}m_{03}) - a])}{\sqrt{4a^2 - [m_{11} + m_{22} + m_{33} - b(m_{10}m_{01} + m_{20}m_{02} + m_{30}m_{03}) - a]^2}} \tag{9-138}$$

A pure depolarizer can be represented by the matrix:

$$\begin{bmatrix} 1 & 0 & 0 & 0 \\ 0 & a & 0 & 0 \\ 0 & 0 & b & 0 \\ 0 & 0 & 0 & c \end{bmatrix} \tag{9-139}$$

where $|a|$, $|b|$, $|c| \leq 1$. The principal depolarization factors are $1 - |a|$, $1 - |b|$, and $1 - |c|$, and these are measures of the depolarization of this depolarizer along its principal axes. The parameter Δ given by

$$\Delta \equiv 1 - \frac{|a| + |b| + |c|}{3}, \qquad 0 \leq \Delta \leq 1 \tag{9-140}$$

is the average of the depolarization factors, and this parameter is called the depolarization power of the depolarizer. An expression for a depolarizer can be written as

$$\begin{bmatrix} 1 & \vec{\mathbf{0}}^T \\ \vec{\mathbf{0}} & m_\Delta \end{bmatrix}, \quad m_\Delta^T = m_\Delta \tag{9-141}$$

where m_Δ is a symmetric 3×3 submatrix. The eigenvalues of m_Δ are the principal depolarization factors, and the eigenvectors are the three orthogonal principal axes. This last expression is not the complete description of a depolarizer, because it contains only six degrees of freedom when we require nine. The most general expression for a depolarizer can be written as

$$M_\Delta = \begin{bmatrix} 1 & \vec{\mathbf{0}}^T \\ \vec{\mathbf{P}}_\Delta & m_\Delta \end{bmatrix}, \quad m_\Delta^T = m_\Delta \tag{9-142}$$

where $\vec{\mathbf{P}}_\Delta$ is the polarizance vector, and with this expression we have the required nine degrees of freedom and no diattenuation or retardance. Thus, we see that a depolarizer with a nonzero polarizance may actually have polarizing properties according to our definition here.

Depolarizing Mueller matrices can be written as the product of the three factors of diattenuation, retardance, and depolarization, i.e.,

$$M = M_\Delta M_R M_D \tag{9-143}$$

where M_Δ is the depolarization, and this equation is the generalized polar decomposition for depolarizing Mueller matrices. It is useful for the decomposition of experimental Mueller matrices to allow the depolarizing component to follow the nondepolarizing component. As in the nondepolarizing case, we first find the matrix for the diattenuator. We then define a matrix M' such that

$$M' = MM_D^{-1} = M_\Delta M_R \tag{9-144}$$

This expression can be written out as the product of the 2×2 matrices:

$$\begin{aligned} M_\Delta M_R &= \begin{bmatrix} 1 & \vec{\mathbf{0}}^T \\ \vec{\mathbf{P}}_\Delta & m_\Delta \end{bmatrix} \begin{bmatrix} 1 & \vec{\mathbf{0}}^T \\ \vec{\mathbf{0}} & m_R \end{bmatrix} = \begin{bmatrix} 1 & \vec{\mathbf{0}}^T \\ \vec{\mathbf{P}}_\Delta & m_\Delta m_R \end{bmatrix} \\ &= \begin{bmatrix} 1 & \vec{\mathbf{0}}^T \\ \vec{\mathbf{P}}_\Delta & m' \end{bmatrix} = M' \end{aligned} \tag{9-145}$$

Let λ_1, λ_2, and λ_3 be the eigenvalues of

$$m'(m')^T = m_\Delta m_R (m_\Delta m_R)^T = m_\Delta^2 \tag{9-146}$$

We can obtain the relations:

$$\vec{\mathbf{P}}_\Delta = \frac{\vec{\mathbf{P}} - m\vec{\mathbf{D}}}{1 - D^2} \tag{9-147}$$

and

$$m' = m_\Delta m_R \tag{9-148}$$

from (9-144) and (9-145).

The eigenvalues of m_Δ are then $\sqrt{\lambda_1}$, $\sqrt{\lambda_2}$, and $\sqrt{\lambda_3}$. It should be pointed out that there is an ambiguity in the signs of the eigenvalues [17]. The retarder submatrix m_R is a rotation matrix and has a positive determinant so that the sign of the determinant of m' indicates the sign of the determinant of m_Δ. The assumption that the eigenvalues all have the same sign is reasonable, especially since depolarization in measured systems is usually small and the eigenvalues are close to one. This assumption simplifies the expression for m_Δ. An expression for m_Δ is given by, from the Cayley–Hamilton theorem (a matrix is a root of its characteristic polynomial),

$$m_\Delta = \pm[m'(m')^T + \kappa_2 \mathbf{I}]^{-1}[\kappa_1 m'(m')^T + \kappa_3 \mathbf{I} \tag{9-149}$$

where

$$\kappa_1 = \sqrt{\lambda_1} + \sqrt{\lambda_2} + \sqrt{\lambda_3} \tag{9-150}$$

$$\kappa_2 = \sqrt{\lambda_1 \lambda_2} + \sqrt{\lambda_2 \lambda_3} + \sqrt{\lambda_3 \lambda_1} \tag{9-151}$$

and

$$\kappa_3 = \sqrt{\lambda_1 \lambda_2 \lambda_3} \tag{9-152}$$

The sign in front of the expression on the right-hand side in Eq. (9-149) follows the sign of the determinant of m'. We can now find m_R from the application of m_Δ^{-1} to m', i.e.,

$$m_r = m_\Delta^{-1} m' = \pm[\kappa_1 m'(m')^T + \kappa_3 \mathbf{I}]^{-1}[m'(m')^T m' + \kappa_2 m'] \tag{9-153}$$

The eigenvalues λ_1, λ_2, and λ_3 can be found in terms of the original Mueller matrix elements by solving a cubic equation, but the expressions that result are long and complicated. It is more feasible to find the κ's. We have

$$\kappa_3 = \det(m_\Delta) = \sqrt{\det(m_\Delta^2)} = \sqrt{\det[m'(m')^T]} = \det(m') \tag{9-154}$$

Recall that $M' = M(M_\Delta)^{-1}$ has the form:

$$M' = \begin{bmatrix} 1 & \vec{\mathbf{0}}^T \\ \vec{\mathbf{P}}_\Delta & m' \end{bmatrix} \tag{9-155}$$

so that

$$\kappa_3 = \det(m') = \det(M') = \det(M)\det(M_\Delta^{-1}) = \frac{\det(M)}{\det(M_\Delta)} = \frac{\det(M)}{a^4} \tag{9-156}$$

Let us define a τ_1 and τ_2 such that

$$\tau_1 = \mathrm{Tr}[m_\Delta^2] = \lambda_1 + \lambda_2 + \lambda_3 \tag{9-157}$$

and

$$\tau_2 = \mathrm{Tr}[\kappa_3^2 (m_\Delta^2)^{-1}] = \lambda_1\lambda_2 + \lambda_1\lambda_3 + \lambda_2\lambda_3 \tag{9-158}$$

Then κ_1 satisfies the recursive equation:

$$\kappa_1 = \sqrt{\tau_1 + 2\sqrt{\tau_2 + \kappa_3\kappa_1}} \tag{9-159}$$

This can be approximated by

$$\kappa_1 \approx \sqrt{\tau_1 + 2\sqrt{\tau_2 + 2\kappa_3\sqrt{\tau_1}}} \tag{9-160}$$

Since

$$\kappa_2 = \frac{1}{2}[\kappa_1^2 - \tau_1] \tag{9-161}$$

we can use the approximation for κ_1 to obtain the approximation for κ_2:

$$\kappa_2 \approx \sqrt{\tau_2 + 2\kappa_3\sqrt{\tau_1}} \tag{9-162}$$

Expressions for τ_1 and τ_2 are given in terms of the original Mueller matrix elements and the elements of m_Δ^2:

$$\tau_1 = \frac{1}{a^2}\left[\sum_{i,j=1}^{3} m_{i,j}^2 - \sum_{i=1}^{3} m_{i,0}^2\right] + \frac{1}{a^4}\left[\sum_{i=1}^{3}\left(m_{i,0} - \sum_{j=1}^{3} m_{i,j}m_{0,j}\right)^2\right] \tag{9-163}$$

and

$$\tau_2 = m_{\Delta_{2,2}} m_{\Delta_{3,3}} + m_{\Delta_{1,1}} m_{\Delta_{3,3}} + m_{\Delta_{1,1}} m_{\Delta_{2,2}} - \left(m^2_{\Delta_{2,3}} + m^2_{\Delta_{1,3}} + m^2_{\Delta_{1,2}} \right) \tag{9-164}$$

where the elements of m^2_Δ are

$$m^2_\Delta = \begin{bmatrix} m_{\Delta_{1,1}} & m_{\Delta_{1,2}} & m_{\Delta_{1,3}} \\ m_{\Delta_{2,1}} & m_{\Delta_{2,2}} & m_{\Delta_{2,3}} \\ m_{\Delta_{3,1}} & m_{\Delta_{3,2}} & m_{\Delta_{3,3}} \end{bmatrix} \tag{9-165}$$

where we note that $m_{\Delta_{i,j}} = m_{\Delta_{j,i}}$
and

$$m_{\Delta_{i,j}} = \frac{1}{a^2}\left[\left(\sum_{k=1}^{3} m_{ik} m_{jk}\right) - m_{i0} m_{j0}\right] + \frac{1}{a^4}\left[m_{i0} - \sum_{k=1}^{3} m_{ik} m_{0k}\right]\left[m_{j0} - \sum_{k=1}^{3} m_{jk} m_{0k}\right] \tag{9-166}$$

We can then write:

$$\kappa_3^2 (m^2_\Delta)^{-1} = \begin{bmatrix} m_{\Delta_{2,2}} m_{\Delta_{3,3}} - m^2_{\Delta_{2,3}} & m_{\Delta_{1,3}} m_{\Delta_{2,3}} - m_{\Delta_{1,2}} m_{\Delta_{3,3}} & m_{\Delta_{1,2}} m_{\Delta_{2,3}} - m_{\Delta_{2,2}} m_{\Delta_{2,3}} \\ m_{\Delta_{1,3}} m_{\Delta_{2,3}} - m_{\Delta_{1,2}} m_{\Delta_{3,3}} & m_{\Delta_{1,1}} m_{\Delta_{3,3}} - m^2_{\Delta_{1,3}} & m_{\Delta_{1,2}} m_{\Delta_{1,3}} - m_{\Delta_{1,1}} m_{\Delta_{2,3}} \\ m_{\Delta_{1,2}} m_{\Delta_{2,3}} - m_{\Delta_{2,2}} m_{\Delta_{2,3}} & m_{\Delta_{1,2}} m_{\Delta_{1,3}} - m_{\Delta_{1,1}} m_{\Delta_{2,3}} & m_{\Delta_{1,1}} m_{\Delta_{2,2}} - m^2_{\Delta_{1,2}} \end{bmatrix} \tag{9-167}$$

and the retarder rotation matrix is given by

$$m_R = m_\Delta^{-1} m' = \frac{1}{\kappa_1}\left[\alpha \mathbf{I} - \beta m^2_\Delta + \gamma \kappa_3^2 (m^2_\Delta)^{-1}\right] m' \tag{9-168}$$

If we can find approximations for the depolarizer eigenvalues $\sqrt{\lambda_1}$, $\sqrt{\lambda_2}$, and $\sqrt{\lambda_3}$, then we can write an expression for m_Δ^{-1} as

$$m_\Delta^{-1} = \frac{1}{\kappa_1}\left[\alpha \mathbf{I} - \beta m^2_\Delta + \gamma \kappa_3^2 (m^2_\Delta)^{-1}\right] \tag{9-169}$$

where

$$\alpha = \frac{(\lambda_1 + \lambda_2 + \lambda_3)(\sqrt{\lambda_1} + \sqrt{\lambda_2} + \sqrt{\lambda_3}) - \sqrt{\lambda_1 \lambda_2 \lambda_3}}{(\sqrt{\lambda_1} + \sqrt{\lambda_2})(\sqrt{\lambda_1} + \sqrt{\lambda_3})(\sqrt{\lambda_2} + \sqrt{\lambda_3})} + 1 \tag{9-170}$$

$$\beta = \frac{1}{(\sqrt{\lambda_1} + \sqrt{\lambda_2})(\sqrt{\lambda_1} + \sqrt{\lambda_3})(\sqrt{\lambda_2} + \sqrt{\lambda_3})} \tag{9-171}$$

and

$$\gamma = \frac{(\sqrt{\lambda_1} + \sqrt{\lambda_2} + \sqrt{\lambda_3})}{(\sqrt{\lambda_1 \lambda_2 \lambda_3})(\sqrt{\lambda_1} + \sqrt{\lambda_2})(\sqrt{\lambda_1} + \sqrt{\lambda_3})(\sqrt{\lambda_2} + \sqrt{\lambda_3})} \tag{9-172}$$

9.6 SUMMARY

We have answered the questions posed at the beginning of this chapter. With the material presented here, we now have the tools to determine whether or not a Mueller matrix is physically realizable and we have a method to bring it to the closest physically realizable matrix. We can then separate the matrix into its constituent components of diattenuation, retardance, and depolarization. We must remember, however, that noise, once introduced into the system, is impossible to remove entirely. The experimentalist must take prudent precautions to minimize the influence of errors peculiar to the system at hand.

REFERENCES

1. Jones, R. C., "A new calculus for the treatment of optical systems: I. Description and discussion of the calculus," *J. Opt. Soc. Am.*, *31*, 488–493 (1941).
2. Jones, R. C., "A new calculus for the treatment of optical systems: IV," *J. Opt. Soc. Am.*, *32*, 486–493 (1942).
3. Jones, R. C., "A new calculus for the treatment of optical systems: V. A more general formulation, and description of another calculus," *J. Opt. Soc. Am.*, *37*, 107–110 (1947).
4. Swindell, W., *Polarized Light in Optics*, Dowden, Hutchinson, and Ross, Stroudsberg, PA, 1975.
5. Stokes, G. G., "On the composition and resolution of streams of polarized light from different sources," *Trans. Cambridge Phil. Soc.*, *9*, 399 (1852).
6. Soleillet, P., "Sur les parameters caracterisant la polarization partielle de la lumiere dans les phenomenes de fluorescence," *Ann. Phys.*, *12*, 23 (1929).
7. Perrin, F., "Polarization of light scattered by isotropic opalescent media," *J. Chem. Phys.*, *10*, 415 (1942).
8. Mueller, H., "Memorandum on the polarization optics of the photoelastic shutter," Report No. 2 of the OSRD project OEMsr-576, Nov. 15, 1943.
9. van de Hulst, H. C., *Light Scattering by Small Particles*, Dover, New York, 1981.
10. Anderson, D. G. M. and Barakat, R., "Necessary and sufficient conditions for a Mueller matrix to be derivable from a Jones matrix," *J. Opt. Soc. Am. A*, *11*, 2305–2319 (1994).
11. Brosseau, C., Givens, C. R. and Kostinski, A. B., "Generalized trace condition on the Mueller–Jones polarization matrix," *J. Opt. Soc. Am A*, *10*, 2248–2251 (1993).
12. Kostinski, A. B., Givens, C. R. and Kwiatkowski, J. M., "Constraints on Mueller matrices of polarization optics," *Appl. Opt.*, *32*(9), 1646–1651 (1993).
13. Givens, C. R. and Kostinski, A. B., "A simple necessary and sufficient condition on physically realizable Mueller matrices," *J. Mod. Opt.*, *40*(3), 471–481 (1993).
14. Hovenier, J. W., van de Hulst, H. C. and van der Mee, C. V. M., "Conditions for the elements of the scattering matrix," Astron. Astrophys, *157*, 301–310 (1986).
15. Barakat, R., "Bilinear constraints between elements of the 4×4 Mueller–Jones transfer matrix of polarization theory," *Opt. Comm.*, *38*(3), 159–161 (1981).
16. Fry, E. S. and Kattawar, G. W., "Relationships between elements of the Stokes matrix," *Appl. Opt.*, *20*, 2811–2814 (1981).
17. Hayes, D. M., "Error propagation in decomposition of Mueller matrices," Proc. SPIE, 3121, 112–123 (1997).
18. Cloude, S. R., "Group theory and polarisation algebra," *Optik*, *75*(1), 26–36 (1986).
19. Cloude, S. R., "Conditions for the physical realisability of matrix operators in polarimetry," *Proc. SPIE*, 1166, 177–185 (1989).
20. Chipman, R. A., "Polarization aberrations," PhD thesis, University of Arizona (1987).

20a. Gerrard, A., and Burch, J. M., *Introduction to Matrix Methods in Optics*, Wiley, London, 1975.
22. Gil, J. J. and Bernabeu, E., "Obtainment of the polarizing and retardation parameters of a non-depolarizing optical system from the polar decomposition of its Mueller matrix," *Optik*, *76*, 26–36 (1986).
23. Xing, Z. -F., "On the deterministic and non-deterministic Mueller matrix," *J. Mod. Opt.*, *39*, 461–484 (1992).
24. Lu, S. -Y., "An interpretation of polarization matrices," PhD dissertion, Dept. of Physics, University of Alabama at Huntsville (1995).
25. Lu, S. -Y. and Chipman, R. A., "Interpretation of Mueller matrices based on polar decomposition," *J. Opt. Soc. Am. A*, *13*, 1106–1113 (1996).

10

The Mueller Matrices for Dielectric Plates

10.1 INTRODUCTION

In Chapter 8, Fresnel's equations for reflection and transmission of waves at an air–dielectric interface were cast in the form of Mueller matrices. In this chapter we use these results to derive the Mueller matrices for dielectric plates. The study of dielectric plates is important because all materials of any practical importance are of finite thickness and so at least have upper and lower surfaces. Furthermore, dielectric plates always change the polarization state of a beam that is reflected or transmitted. One of their most important applications is to create linearly polarized light from unpolarized light in the infrared region. While linearly polarized light can be created in the visible and near-infrared regions using calcite polarizers or Polaroid, there are no corresponding materials in the far-infrared region. However, materials such as germanium and silicon, as well as others, do transmit very well in the infrared region. By making thin plates of these materials and then constructing a "pile of plates," it is possible to create light in the infrared that is highly polarized. This arrangement therefore requires that the Mueller matrices for transmission play a more prominent role than the Mueller matrices for reflection.

In order to use the Mueller matrices to characterize a single plate or multiple plates, we must carry out matrix multiplications. The presence of off-diagonal terms of the Mueller matrices create a considerable amount of work. We know, on the other hand, that if we use diagonal matrices the calculations are simplified; the product of diagonalized matrices leads to another diagonal matrix.

10.2 THE DIAGONAL MUELLER MATRIX AND THE *ABCD* POLARIZATION MATRIX

When we apply the Mueller matrices to problems in which there are several polarizing elements, each of which is described by its own Mueller matrix, we soon discover that the appearance of the off-diagonal elements complicates the matrix multiplications. The multiplications would be greatly simplified if we were to use diagonalized forms of the Mueller matrices. In particular, the use of diagonalized matrices enables us to determine more easily the Mueller matrix raised to the mth power, M^m, an important problem when we must determine the transmission of a polarized beam through m dielectric plates.

In this chapter we develop the diagonal Mueller matrices for a polarizer and a retarder. To reduce the amount of calculations, it is simpler to write a single matrix that simultaneously describes the behavior of a polarizer or a retarder or a combination of both. This simplified matrix is called the *ABCD* polarization matrix.

The Mueller matrix for a polarizer is

$$M_P = \frac{1}{2}\begin{pmatrix} p_s^2 + p_p^2 & p_s^2 - p_p^2 & 0 & 0 \\ p_s^2 - p_p^2 & p_s^2 + p_p^2 & 0 & 0 \\ 0 & 0 & 2p_s p_p & 0 \\ 0 & 0 & 0 & 2p_s p_p \end{pmatrix} \tag{10-1}$$

and the Mueller matrix for a phase shifter is

$$M_C = \begin{pmatrix} 1 & 0 & 0 & 0 \\ 0 & 1 & 0 & 0 \\ 0 & 0 & \cos\phi & \sin\phi \\ 0 & 0 & -\sin\phi & \cos\phi \end{pmatrix} \tag{10-2}$$

where p_s and p_p are the absorption coefficients of the polarizer along the s (or x) and p (or y) axes, respectively, and ϕ is the phase shift of the retarder.

The form of (10-1) and (10-2) suggests that the matrices can be represented by a single matrix of the form:

$$\Phi = \begin{pmatrix} A & B & 0 & 0 \\ B & A & 0 & 0 \\ 0 & 0 & C & D \\ 0 & 0 & -D & C \end{pmatrix} \tag{10-3}$$

which we call the *ABCD* polarization matrix. We see that for a polarizer:

$$A = \frac{1}{2}(p_s^2 + p_p^2) \tag{10-4a}$$

$$B = \frac{1}{2}(p_s^2 - p_p^2) \tag{10-4b}$$

$$C = \frac{1}{2}(2p_s p_p) \tag{10-4c}$$

$$D = 0 \tag{10-4d}$$

and for the retarder

$$A = 1 \tag{10-5a}$$

$$B = 0 \tag{10-5b}$$

$$C = \cos\phi \tag{10-5c}$$

$$D = \sin\phi \tag{10-5d}$$

If we multiply (10-1) by (10-2), we see that we still obtain a matrix which can be represented by an $ABCD$ matrix; the matrix describes an absorbing retarder.

The matrix elements $ABCD$ are not all independent; that is, there is a unique relationship between the elements. To find this relationship, we see that (10-3) transforms the Stokes parameters of an incident beam S_i to the Stokes parameters of an emerging beam S_i' so that we have

$$S_0' = AS_0 + BS_1 \tag{10-6a}$$

$$S_1' = BS_0 + AS_1 \tag{10-6b}$$

$$S_2' = CS_2 + DS_3 \tag{10-6c}$$

$$S_3' = -DS_2 + CS_3 \tag{10-6d}$$

We know that for completely polarized light the Stokes parameters of the incident beam are related by

$$S_0^2 = S_1^2 + S_2^2 + S_3^2 \tag{10-7}$$

and, similarly,

$$S_0'^2 = S_1'^2 + S_2'^2 + S_3'^2 \tag{10-8}$$

Substituting (10-6) into (10-8) leads to

$$(A^2 - B^2)(S_0^2 - S_1^2) = (C^2 + D^2)(S_2^2 + S_3^2) \tag{10-9}$$

But, from (10-7),

$$S_0^2 - S_1^2 = S_2^2 + S_3^2 \tag{10-10}$$

Substituting (10-10) into the right side of (10-9) gives

$$(A^2 - B^2 - C^2 - D^2)(S_0^2 - S_1^2) = 0 \tag{10-11}$$

and

$$A^2 = B^2 + C^2 + D^2 \tag{10-12}$$

We see that the elements of (10-4) and (10-5) satisfy (10-12). This is a very useful relation because it serves as a check when measuring the Mueller matrix elements.

The rotation of a polarizing device described by the $ABCD$ matrix is given by the matrix equation:

$$M = M(-2\theta)\Phi M(2\theta) \tag{10-13}$$

which in its expanded form is

$$M = \begin{pmatrix} A & B\cos 2\theta & B\sin 2\theta & 0 \\ B\cos 2\theta & A^2\cos^2 2\theta + C\sin^2 2\theta & (A-C)\sin 2\theta\cos 2\theta & -D\sin 2\theta \\ B\sin 2\theta & (A-C)\sin 2\theta\cos 2\theta & A\sin^2 2\theta + C\cos^2 2\theta & D\cos 2\theta \\ 0 & D\sin 2\theta & -D\cos 2\theta & C \end{pmatrix} \tag{10-14}$$

In carrying out the expansion of (10-13), we used

$$M(2\theta) = \begin{pmatrix} 1 & 0 & 0 & 0 \\ 0 & \cos 2\theta & \sin 2\theta & 0 \\ 0 & -\sin 2\theta & \cos 2\theta & 0 \\ 0 & 0 & 0 & 1 \end{pmatrix} \tag{10-15}$$

We now find the diagonalized form of the $ABCD$ matrix. This can be done using the well-known methods in matrix algebra. We first express (10-3) as an eigenvalue/eigenvector equation, namely,

$$\Phi S = \lambda S \tag{10-16a}$$

or

$$(\Phi - \lambda)S = 0 \tag{10-16b}$$

where λ and S are the eigenvalues and the eigenvectors corresponding to Φ. In order to find the eigenvalues and the eigenvectors, the determinant of (10-3) must be taken; that is,

$$\begin{vmatrix} A-\lambda & B & 0 & 0 \\ B & A-\lambda & 0 & 0 \\ 0 & 0 & C-\lambda & D \\ 0 & 0 & -D & C-\lambda \end{vmatrix} = 0 \tag{10-17}$$

The determinant is easily expanded and leads to an equation called the secular equation:

$$[(A-\lambda)^2 - B^2][(C-\lambda)^2 + D^2] = 0 \tag{10-18}$$

The solution of (10-18) yields the eigenvalues:

$$\lambda_1 = A + B \tag{10-19a}$$

$$\lambda_2 = A - B \tag{10-19b}$$

$$\lambda_3 = C + iD \tag{10-19c}$$

$$\lambda_4 = C - iD \tag{10-19d}$$

Substituting these eigenvalues into (10-17), we easily find that the eigenvector corresponding to each of the respective eigenvalues in (10-19) is

$$S^1 = \frac{1}{\sqrt{2}}\begin{pmatrix} 1 \\ 1 \\ 0 \\ 0 \end{pmatrix} \quad S^2 = \frac{1}{\sqrt{2}}\begin{pmatrix} 1 \\ -1 \\ 0 \\ 0 \end{pmatrix} \quad S^3 = \frac{1}{\sqrt{2}}\begin{pmatrix} 0 \\ 0 \\ 1 \\ i \end{pmatrix} \quad S^4 = \frac{1}{\sqrt{2}}\begin{pmatrix} 0 \\ 0 \\ 1 \\ -i \end{pmatrix} \tag{10-20}$$

The factor $1/\sqrt{2}$ has been introduced to normalize each of the eigenvectors.

We now construct a new matrix K, called the modal matrix, whose columns are formed from each of the respective eigenvectors in (10-20):

$$K = \frac{1}{\sqrt{2}}\begin{pmatrix} 1 & 1 & 0 & 0 \\ 1 & -1 & 0 & 0 \\ 0 & 0 & 1 & 1 \\ 0 & 0 & i & -i \end{pmatrix} \tag{10-21a}$$

The inverse matrix is easily found to be

$$K^{-1} = \frac{1}{\sqrt{2}}\begin{pmatrix} 1 & 1 & 0 & 0 \\ 1 & -1 & 0 & 0 \\ 0 & 0 & 1 & -i \\ 0 & 0 & 1 & i \end{pmatrix} \tag{10-21b}$$

We see that $KK^{-1} = I$, where I is the unit matrix. We now construct a diagonal matrix from each of the eigenvalues in (10-19) and write

$$M_D = \begin{pmatrix} A+B & 0 & 0 & 0 \\ 0 & A-B & 0 & 0 \\ 0 & 0 & C+iD & 0 \\ 0 & 0 & 0 & C-iD \end{pmatrix} \tag{10-22}$$

From (10-4) the diagonal Mueller matrix for a polarizer $M_{D,P}$ is then

$$M_{D,P} = \begin{pmatrix} p_s^2 & 0 & 0 & 0 \\ 0 & p_p^2 & 0 & 0 \\ 0 & 0 & p_s p_p & 0 \\ 0 & 0 & 0 & p_s p_p \end{pmatrix} \tag{10-23}$$

and from (10-5) the diagonal matrix for a retarder is

$$M_{D,C} = \begin{pmatrix} 1 & 0 & 0 & 0 \\ 0 & 1 & 0 & 0 \\ 0 & 0 & e^{i\phi} & 0 \\ 0 & 0 & 0 & e^{-i\phi} \end{pmatrix} \tag{10-24}$$

A remarkable relation now emerges. From (10-21) and (10-22) one readily sees that the following identity is true:

$$\Phi K = K M_D \tag{10-25}$$

Postmultiplying both sides of (10-25) by K^{-1}, we see that

$$\Phi = KM_D K^{-1} \tag{10-26a}$$

or

$$M_D = K^{-1}\Phi K \tag{10-26b}$$

where we have used $KK^{-1} = I$. We now square both sides of (10-26a) and find that

$$\Phi^2 = KM_D^2 K^{-1} \tag{10-27}$$

which shows that Φ^m is obtained from

$$\Phi^m = KM_D^m K^{-1} \tag{10-28}$$

Thus, by finding the eigenvalues and the eigenvectors of Φ and then constructing the diagonal matrix and the modal matrix (and its inverse), the mth power of the $ABCD$ matrix Φ can be found from (10-28). Equation (10-26b) also allows us to determine the diagonalized $ABCD$ matrix Φ.

Equation (10-28) now enables us to find the mth power of the $ABCD$ matrix Φ:

$$\Phi^m = \begin{pmatrix} A & B & 0 & 0 \\ B & A & 0 & 0 \\ 0 & 0 & C & -D \\ 0 & 0 & D & C \end{pmatrix}^m$$
$$= K\begin{pmatrix} (A+B)^m & 0 & 0 & 0 \\ 0 & (A-B)^m & 0 & 0 \\ 0 & 0 & (C+iD)^m & 0 \\ 0 & 0 & 0 & (C-iD)^m \end{pmatrix}K^{-1} \tag{10-29}$$

Carrying out the matrix multiplication using (10-21) then yields

$$\Phi^m = \frac{1}{2}\begin{pmatrix} \left[(A+B)^m + (A-B)^m\right] & \left[(A+B)^m - (A-B)^m\right] & 0 & 0 \\ \left[(A+B)^m - (A-B)^m\right] & \left[(A+B)^m + (A-B)^m\right] & 0 & 0 \\ 0 & 0 & \left[(C+iD)^m + (C-iD)^m\right] & \left[-i(C+iD)^m + i(C-iD)^m\right] \\ 0 & 0 & \left[i(C+iD)^m - i(C-iD)^m\right] & \left[(C+iD)^m + (C-iD)^m\right] \end{pmatrix} \tag{10-30}$$

Using (10-30) we readily find that the mth powers of the Mueller matrix of a polarizer and a retarder are, respectively,

$$M_P^m = \frac{1}{2}\begin{pmatrix} p_s^{2m} + p_p^{2m} & p_s^{2m} - p_p^{2m} & 0 & 0 \\ p_s^{2m} - p_p^{2m} & p_s^{2m} + p_p^{2m} & 0 & 0 \\ 0 & 0 & 2p_s^m p_p^m & 0 \\ 0 & 0 & 0 & 2p_s^m p_p^m \end{pmatrix} \tag{10-31}$$

and

$$M_C^m = \begin{pmatrix} 1 & 0 & 0 & 0 \\ 0 & 1 & 0 & 0 \\ 0 & 0 & \cos m\phi & \sin m\phi \\ 0 & 0 & -\sin m\phi & \cos m\phi \end{pmatrix} \tag{10-32}$$

The diagonalized Mueller matrices will play an essential role in the following section when we determine the Mueller matrices for single and multiple dielectric plates.

Before we conclude this section we discuss another form of the Mueller matrix for a polarizer. We recall that the first two Stokes parameters, S_0 and S_1, are the sum and difference of the orthogonal intensities. The Stokes parameters can then be written as

$$S_0 = I_x + I_y \tag{10-33a}$$

$$S_1 = I_x - I_y \tag{10-33b}$$

$$S_2 = S_2 \tag{10-33c}$$

$$S_3 = S_3 \tag{10-33d}$$

where

$$I_x = E_x E_x^* \qquad I_y = E_y E_y^* \tag{10-33e}$$

We further define

$$I_x = I_0 \tag{10-34a}$$

$$I_y = I_1 \tag{10-34b}$$

$$S_2 = I_2 \tag{10-34c}$$

$$S_3 = I_3 \tag{10-34d}$$

Then, we can relate S to I by

$$\begin{pmatrix} S_0 \\ S_1 \\ S_2 \\ S_3 \end{pmatrix} = \begin{pmatrix} 1 & 1 & 0 & 0 \\ 1 & -1 & 0 & 0 \\ 0 & 0 & 1 & 0 \\ 0 & 0 & 0 & 1 \end{pmatrix} \begin{pmatrix} I_0 \\ I_1 \\ I_2 \\ I_3 \end{pmatrix} \tag{10-35a}$$

or I to S,

$$\begin{pmatrix} I_0 \\ I_1 \\ I_2 \\ I_3 \end{pmatrix} = \frac{1}{2} \begin{pmatrix} 1 & 1 & 0 & 0 \\ 1 & -1 & 0 & 0 \\ 0 & 0 & 2 & 0 \\ 0 & 0 & 0 & 2 \end{pmatrix} \begin{pmatrix} S_0 \\ S_1 \\ S_2 \\ S_3 \end{pmatrix} \tag{10-35b}$$

The column matrix:

$$I = \begin{pmatrix} I_0 \\ I_1 \\ I_2 \\ I_3 \end{pmatrix} \tag{10-36}$$

is called the intensity vector. The intensity vector is very useful because the 4×4 matrix which connects I to I' is diagonalized, thus making the calculations simpler. To show that this is true, we can formally express (10-35a) and (10-35b) as

$$S = K_A I \tag{10-37a}$$

$$I = K_A^{-1} S \tag{10-37b}$$

where K_A and K_A^{-1} are defined by the 4×4 matrices in (10-35), respectively. The Mueller matrix M can be defined in terms of an incident Stokes vector S and an emerging Stokes vector S':

$$S' = M S \tag{10-38}$$

Similarly, we can define the intensity vector relationship:

$$I' = P I \tag{10-39}$$

where P is a 4×4 matrix.

We now show that P is diagonal. We have from (10-37a)

$$S' = K_A I' \tag{10-40}$$

Substituting (10-40) into (10-38) along with (10-37a) gives

$$I' = (K_A^{-1} M K_A) I \tag{10-41}$$

or, from (10-39)

$$P = K_A^{-1} M K_A \tag{10-42}$$

We now show that for a polarizer P is a diagonal matrix. The Mueller matrix for a polarizer in terms of the $ABCD$ matrix elements can be written as

$$M = \begin{pmatrix} A & B & 0 & 0 \\ B & A & 0 & 0 \\ 0 & 0 & C & 0 \\ 0 & 0 & 0 & C \end{pmatrix} \tag{10-43}$$

Substituting (10-43) into (10-42) and using K_A and K_A^{-1} from (10-35), we readily find that

$$P = \begin{pmatrix} A+B & 0 & 0 & 0 \\ 0 & A-B & 0 & 0 \\ 0 & 0 & C & 0 \\ 0 & 0 & 0 & C \end{pmatrix} \tag{10-44}$$

Thus, P is a diagonal polarizing matrix; it is equivalent to the diagonal Mueller matrix for a polarizer. The diagonal form of the Mueller matrix was first used by the Nobel laureate S. Chandrasekhar in his classic papers in radiative transfer in the late 1940s. It is called Chandrasekhar's phase matrix in the literature. In particular, for the Mueller matrix of a polarizer we see that (10-44) becomes

$$P = \begin{pmatrix} p_s^2 & 0 & 0 & 0 \\ 0 & p_p^2 & 0 & 0 \\ 0 & 0 & p_s p_p & 0 \\ 0 & 0 & 0 & p_s p_p \end{pmatrix} \tag{10-45}$$

which is identical to the diagonalized Mueller matrix given by (10-23). In Part II we shall show that the Mueller matrix for scattering by an electron is proportional to

$$M_p = \frac{1}{2}\begin{pmatrix} 1+\cos^2\theta & -\sin^2\theta & 0 & 0 \\ -\sin^2\theta & 1+\cos^2\theta & 0 & 0 \\ 0 & 0 & 2\cos\theta & 0 \\ 0 & 0 & 0 & 2\cos\theta \end{pmatrix} \tag{10-46}$$

where θ is the observation angle in spherical coordinates and is measured from the z axis ($\theta = 0^\circ$). Transforming (10-46) to Chandrasekhar's phase matrix, we find

$$P = \begin{pmatrix} \cos^2\theta & 0 & 0 & 0 \\ 0 & 1 & 0 & 0 \\ 0 & 0 & \cos\theta & 0 \\ 0 & 0 & 0 & \cos\theta \end{pmatrix} \tag{10-47}$$

which is the well-known representation for Chandrasekhar's phase matrix for the scattering of polarized light by an electron.

Not surprisingly, there are other interesting and useful transformations which can be developed. However, this development would take us too far from our original goal, which is to determine the Mueller matrices for single and multiple dielectric plates. We now apply the results in this section to the solution of this problem.

10.3 MUELLER MATRICES FOR SINGLE AND MULTIPLE DIELECTRIC PLATES

In the previous sections, Fresnel's equations for reflection and transmission at an air–dielectric interface were cast into the form of Mueller matrices. In this section we use these results to derive the Mueller matrices for dielectric plates. We first treat the problem of determining the Mueller matrix for a single dielectric plate. The formalism is then easily extended to multiple reflections within a single dielectric plate and then to a pile of m parallel transparent dielectric plates.

For the problem of transmission of a polarized beam through a single dielectric plate, the simplest treatment can be made by assuming a single transmission through the upper surface followed by another transmission through the lower surface. There are, of course, multiple reflections within the dielectric plates, and, strictly speaking, these should be taken into account. While this treatment of multiple internal reflections is straightforward, it turns out to be quite involved. In

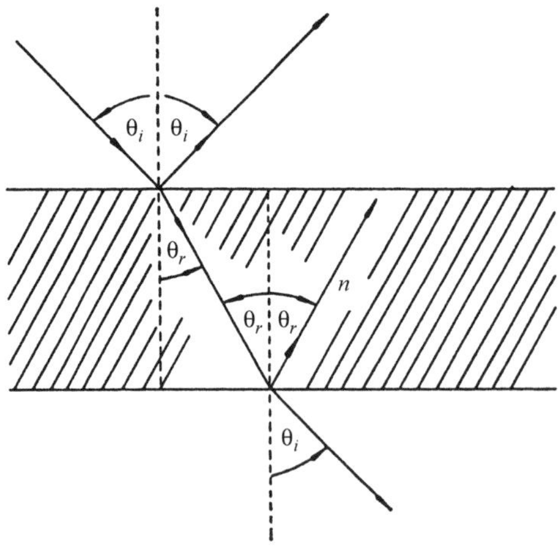

Figure 10-1 Beam propagation through a single dielectric plate.

the treatment presented here, we choose to ignore these effects. The completely correct treatment is given in the papers quoted in the references at the end of this chapter. The difference between the exact results and the approximate results is quite small, and very good results are still obtained by ignoring the multiple internal reflections. Consequently, only the resulting expressions for multiple internal reflections are quoted. We shall also see that the use of the diagonalized Mueller matrices developed in the previous section greatly simplifies the treatment of all of these problems.

In Fig. 10-1 a single dielectric (glass) plate is shown. The incident beam is described by the Stokes vector S. Inspection of the figure shows that the Stokes vector S' of the beam emerging from the lower side of the dielectric plate is related to S by the matrix relation:

$$S' = M_T^2 S \tag{10-48}$$

where M_T is the Mueller matrix for transmission and is given by (8-13) in Section 8.3. We easily see, using (8-13), that M_T^2 is then

$$M_T^2 = \frac{1}{2}\left[\frac{\sin 2\theta_i \sin 2\theta_r}{(\sin\theta_+ \cos\theta_-)^2}\right]^2 \begin{pmatrix} \cos^4\theta_- + 1 & \cos^4\theta_- - 1 & 0 & 0 \\ \cos^4\theta_- - 1 & \cos^4\theta_- + 1 & 0 & 0 \\ 0 & 0 & 2\cos^2\theta_- & 0 \\ 0 & 0 & 0 & 2\cos^2\theta_- \end{pmatrix} \tag{10-49}$$

where θ_i is the angle of incidence, θ_r is the angle of refraction, and $\theta_\pm = \theta_i \pm \theta_r$.

Equation (10-49) is the Mueller matrix (transmission) for a single dielectric plate. We can immediately extend this result to the transmission through m parallel dielectric plates by raising M_T^2 to the mth power, this is, M_T^{2m}. The easiest way to do this is to transform (10-49) to the diagonal form and raise the diagonal matrix to the mth power as described earlier. After this is done we transform back to the

Mueller matrix form. Upon doing this we then find that the Mueller matrix for transmission through m parallel dielectric plates is

$$M_T^{2m} = \frac{1}{2}\left[\frac{\sin 2\theta_i \sin 2\theta_r}{(\sin\theta_+ \cos\theta_-)^2}\right]^{2m} \begin{pmatrix} \cos^{4m}\theta_- + 1 & \cos^{4m}\theta_- - 1 & 0 & 0 \\ \cos^{4m}\theta_- - 1 & \cos^{4m}\theta_- + 1 & 0 & 0 \\ 0 & 0 & 2\cos^{2m}\theta_- & 0 \\ 0 & 0 & 0 & 2\cos^{2m}\theta_- \end{pmatrix} \tag{10-50}$$

Equation (10-50) includes the result for a single dielectric plate by setting $m = 1$. We now consider that the incident beam is unpolarized. Then, the Stokes vector of a beam emerging from m parallel plates is, from (10-50),

$$S' = \frac{1}{2}\left[\frac{\sin 2\theta_i \sin 2\theta_r}{(\sin\theta_+ \cos\theta_-)^2}\right]^{2m} \begin{pmatrix} \cos^{4m}\theta_- + 1 \\ \cos^{4m}\theta_- - 1 \\ 0 \\ 0 \end{pmatrix} \tag{10-51}$$

The degree of polarization P of the emerging beam is then

$$P = \left|\frac{1 - \cos^{4m}\theta_-}{1 + \cos^{4m}\theta_-}\right| \tag{10-52}$$

In Fig. 10-2 a plot of (10-52) is shown for the degree of polarization as a function of the incident angle θ_i. The plot shows that at least six or eight parallel plates are required in order for the degree of polarization to approach unity. At normal incidence the degree of polarization is always zero, regardless of the number of plates.

The use of parallel plates to create linearly polarized light appears very often outside the visible region of the spectrum. In the visible and near-infrared region ($<2\ \mu$m) Polaroid and calcite are available to create linearly polarized light. Above 2 μm, parallel plates made from other materials are an important practical way of creating linearly polarized light. Fortunately, natural materials such as germanium are available and can be used; germanium transmits more than 95% of the incident light up to 20 μm.

According to (10-51) the intensity of the beam emerging from m parallel plates, I_T, is

$$I_T = \frac{1}{2}\left[\frac{\sin 2\theta_i \sin 2\theta_r}{(\sin\theta_+ \cos\theta_-)^2}\right]^{2m} (1 + \cos^{4m}\theta_-) \tag{10-53}$$

Figure 10-3 shows a plot of (10-53) for m dielectric plates.

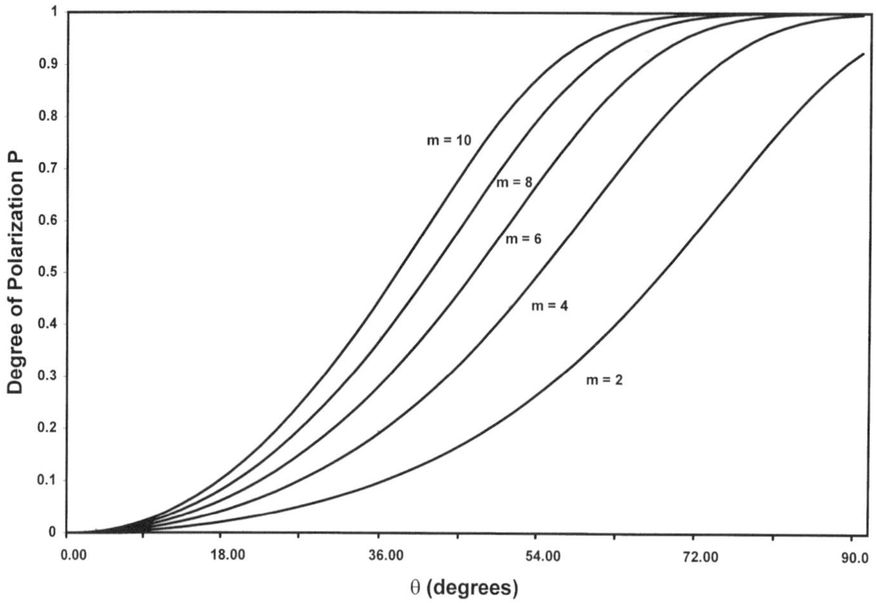

Figure 10-2 Plot of (10-52), the degree of polarization P versus incident angle and the number or parallel plates. The refractive index n is 1.5.

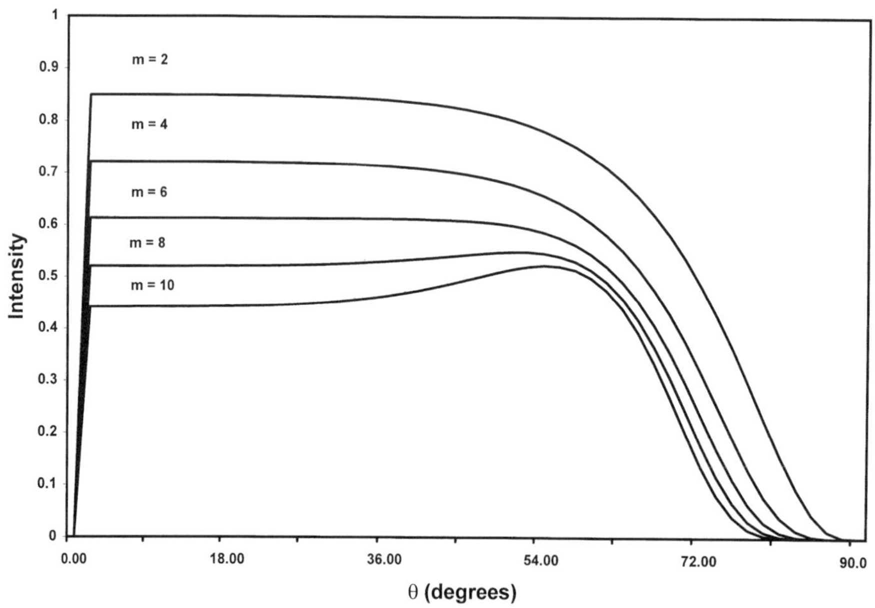

Figure 10-3 The intensity of a beam emerging from m parallel plates as a function of the angle of incidence. The refractive index is 1.5.

At the Brewster angle the Mueller matrix for transmission through m dielectric plates is readily shown from the results given in Chapter 8 and Section 10.2 to be

$$M_{T,B}^{2m} = \frac{1}{2}\begin{pmatrix} \sin^{4m} 2\theta_{i_B} + 1 & \sin^{4m} 2\theta_{i_B} - 1 & 0 & 0 \\ \sin^{4m} 2\theta_{i_B} - 1 & \sin^{4m} 2\theta_{i_B} + 1 & 0 & 0 \\ 0 & 0 & 2\sin^{2m} 2\theta_{i_B} & 0 \\ 0 & 0 & 0 & 2\sin^{2m} 2\theta_{i_B} \end{pmatrix} \tag{10-54}$$

For a single dielectric plate $m = 1$, (10-54) reduces to

$$M_{T,B}^{2} = \frac{1}{2}\begin{pmatrix} \sin^{4} 2\theta_{i_B} + 1 & \sin^{4} 2\theta_{i_B} - 1 & 0 & 0 \\ \sin^{4} 2\theta_{i_B} - 1 & \sin^{4} 2\theta_{i_B} + 1 & 0 & 0 \\ 0 & 0 & 2\sin^{2} 2\theta_{i_B} & 0 \\ 0 & 0 & 0 & 2\sin^{2} 2\theta_{i_B} \end{pmatrix} \tag{10-55}$$

If the incident beam is unpolarized, the Stokes vector for the transmitted beam after passing through m parallel dielectric plates will be

$$S' = \frac{1}{2}\begin{pmatrix} \sin^{4m} 2\theta_{i_B} + 1 \\ \sin^{4m} 2\theta_{i_B} - 1 \\ 0 \\ 0 \end{pmatrix} \tag{10-56}$$

The degree of polarization is then

$$P = \left|\frac{1 - \sin^{4m} 2\theta_{i_B}}{1 + \sin^{4m} 2\theta_{i_B}}\right| \tag{10-57}$$

A plot of (10-57) is shown in Fig. 10-4 for m dielectric plates.

The intensity of the transmitted beam is given by S_0 in (10-56) and is

$$I_T = \frac{1}{2}(1 + \sin^{4m} 2\theta_{i_B}) \tag{10-58}$$

Equation (10-58) has been plotted in Fig. 10-5.

From Figs. 10-4 and 10-5 the following conclusions can be drawn. In Fig. 10-4, there is a significant increase in the degree of polarization up to $m = 6$. Figure 10-5, on the other hand, shows that the intensity decreases and then begins to "level off" for $m = 6$. Thus, these two figures show that after five or six parallel plates there is very little to be gained in using more plates to increase the degree of polarization and still maintain a "constant" intensity. In addition, the cost for making such large assemblies of dielectric plates, the materials, and mechanical alignment becomes considerable.

Dielectric plates can also rotate the orientation of the polarization ellipse. At first this behavior may be surprising, but this is readily shown. Consider the

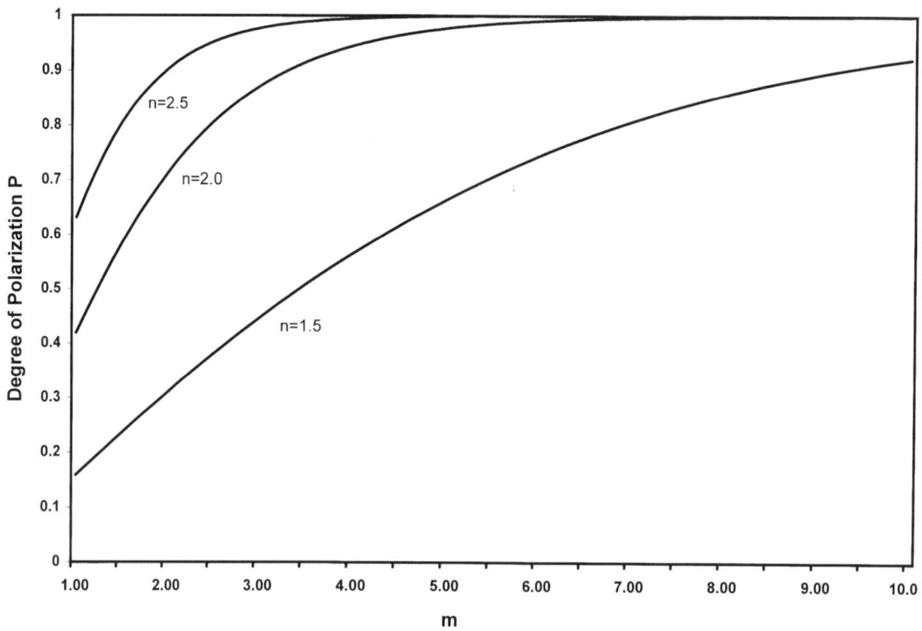

Figure 10-4 Plot of the degree of polarization P versus number of dielectric plates at the Brewster angle for refractive indices of 1.5, 2.0, and 2.5.

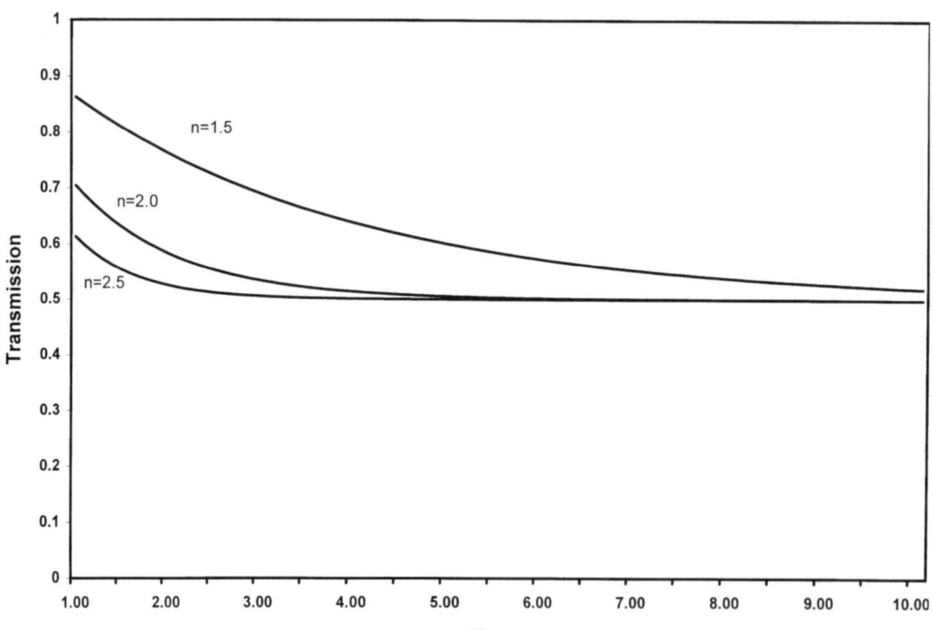

Figure 10-5 Plot of the transmitted intensity of a beam propagating through m parallel plates at the Brewster angle θ_{i_B}. The refractive indices are 1.5, 2.0, and 2.5, respectively.

situation when the incident beam is linear +45° polarized light. The normalized Stokes vector of the beam emerging from m dielectric plates is then, from (10-54),

$$S' = \frac{1}{2}\begin{pmatrix} \sin^{4m} 2\theta_{i_B} + 1 \\ \sin^{4m} 2\theta_{i_B} - 1 \\ 2\sin^{2m} 2\theta_{i_B} \\ 0 \end{pmatrix} \tag{10-59}$$

The emerging light is still linearly polarized. However, the orientation angle ψ is

$$\psi = \frac{1}{2}\tan^{-1}\left(\frac{2\sin^{2m} 2\theta_{i_B}}{\sin^{4m} 2\theta_{i_B} - 1}\right) \tag{10-60}$$

We note that for $m = 0$ (no dielectric plates) the absolute magnitude of the angle of rotation is $\psi = 45°$, as expected. Figure 10-6 illustrates the change in the angle of rotation as the number of parallel plates increases. For five parallel plates the orientation angle rotates from +45° to +24.2°.

Equation (10-57) can also be expressed in terms of the refractive index, n. We recall that (10-57) is

$$P = \left|\frac{1 - \sin^{4m} 2\theta_{i_B}}{1 + \sin^{4m} 2\theta_{i_B}}\right| \tag{10-57}$$

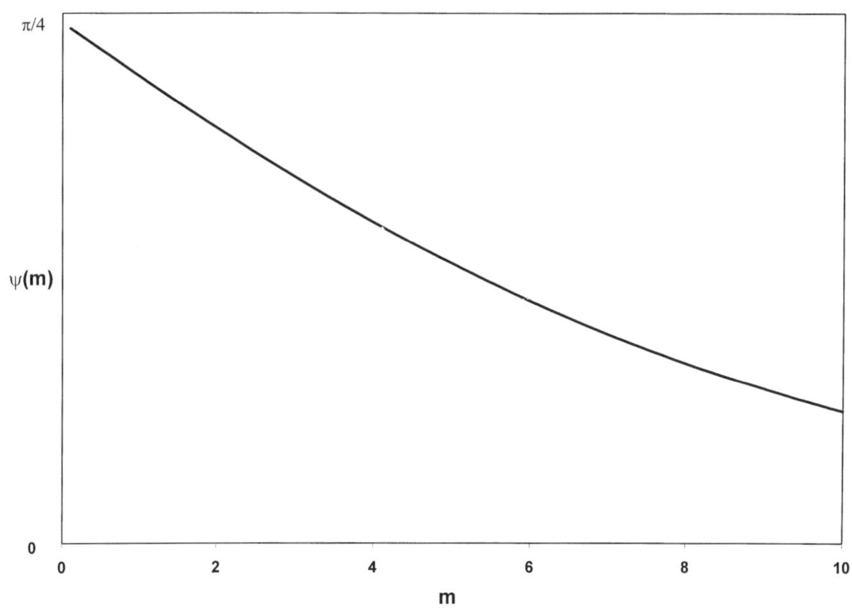

Figure 10-6 Rotation of the polarization ellipse by m parallel dielectric plates according to (10-60). The refractive index is 1.5.

At the Brewster angle we have

$$\tan\theta_{i_B} = n \tag{10-61a}$$

and we see that we can then write

$$\sin\theta_{i_B} = \frac{n}{\sqrt{n^2+1}} \tag{10-61b}$$

and

$$\cos\theta_{i_B} = \frac{1}{\sqrt{n^2+1}} \tag{10-61c}$$

so

$$\sin 2\theta_{i_B} = \frac{2n}{n^2+1} \tag{10-61d}$$

Substituting (10-61d) into (10-59) yields

$$P = \left|\frac{1-[2n/(n^2+1)]^{4m}}{1+[2n/(n^2+1)]^{4m}}\right| \tag{10-62}$$

Equation (10-62) is a much-quoted result in the optical literature and optical handbooks. In Figure 10-7 a plot is made of (10-62) in terms of m and n. Of course,

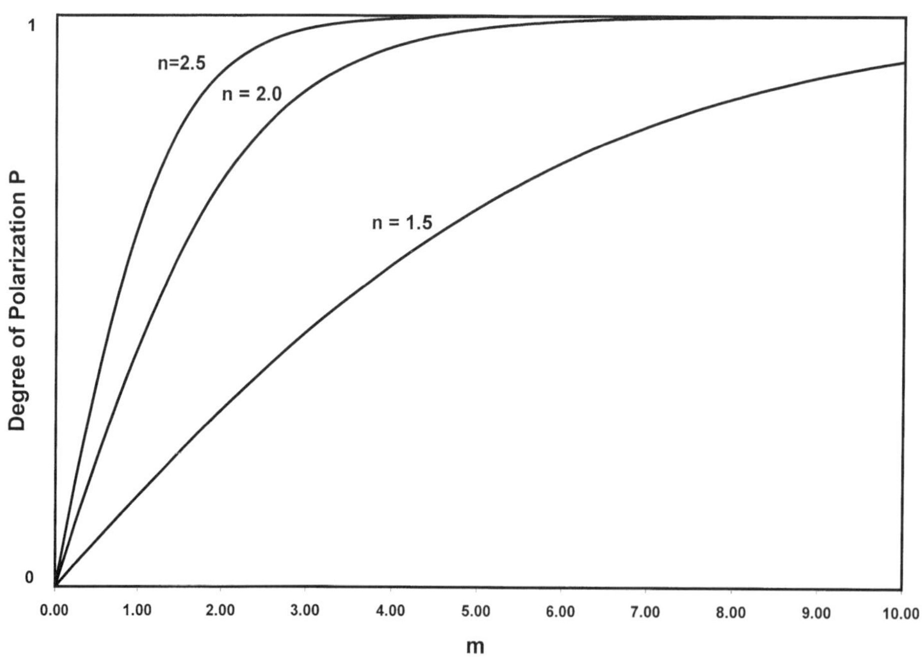

Figure 10-7 Plot of the degree of polarization as a function of the number of parallel plates; multiple reflections are ignored.

as inspection of Fig. 10-7 shows, the curves are identical to those in Fig. 10-4 except in the former figure the abscissa begins with $m = 1$.

In the beginning of this section we pointed out that the Mueller matrix formalism can also be extended to the problem of including multiple reflections within a single dielectric plate as well as the multiple plates. G. G. Stokes (1862) was the first to consider this problem and showed that the inclusion of multiple reflections within the plates led to the following equation for the degree of polarization for m parallel plates at the Brewster angles:

$$P = \left| \frac{m}{m + [2n^2/(n^2 - 1)]^2} \right| \tag{10-63}$$

The derivation of (10-63) along with similar expressions for completely and partially polarized light has been given by Collett (1972), using the Jones matrix formalism (Chapter 11) and the Mueller matrix formalism. In Fig. 10-8, (10-63) has been plotted as a function of m and n, the refractive index.

It is of interest to compare (10-62) and (10-63). In Fig. 10-9 we have plotted these two equations for $n = 1.5$. We see immediately that the degree of polarization is very different. Starting with 0 parallel plates, that is, the unpolarized light source by itself, we see the degree of polarization is zero, as expected. As the number of parallel plates increases, the degree of polarization increases for both (10-62) and (10-63). However, the curves diverge and the magnitudes differ by approximately a factor of two so that for 10 parallel plates the degree of polarization is 0.93 for (10-62) and 0.43 for (10-63). In addition, for (10-63), the

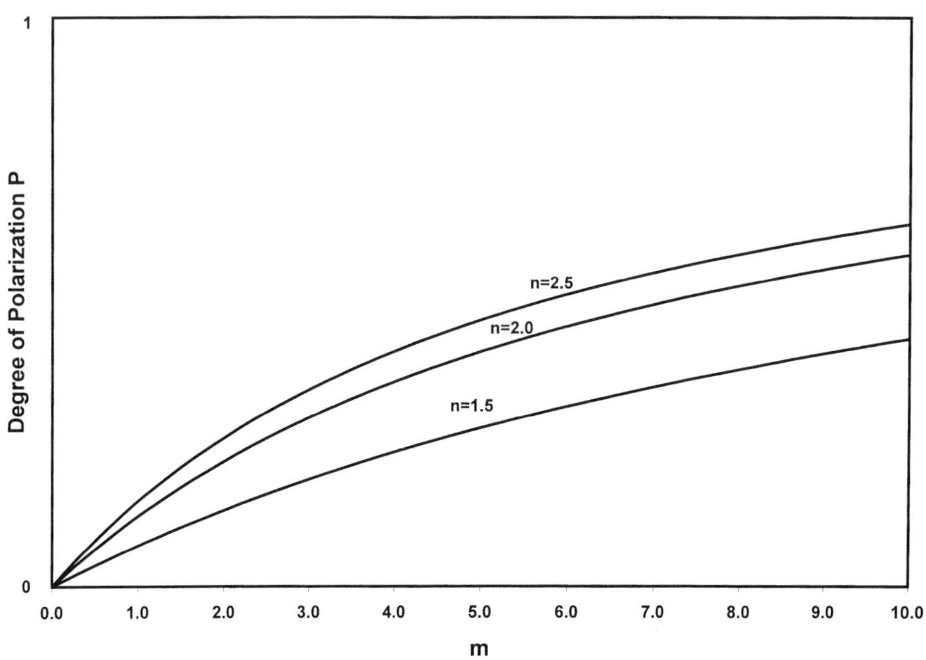

Figure 10-8 Plot of the degree of polarization as a function of the number of parallel plates for the case where multiple reflections are included.

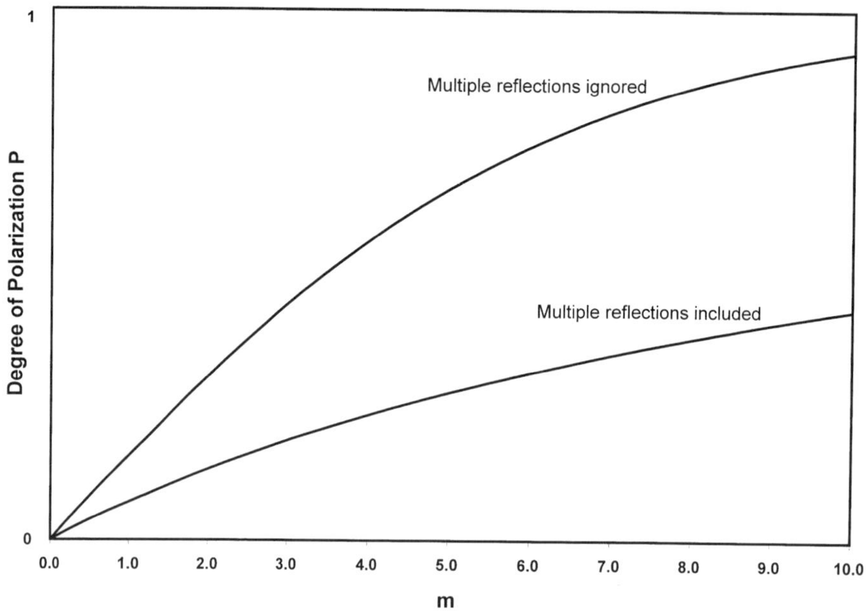

Figure 10-9 Degree of polarization for m parallel plates for $n = 1.5$. The upper curve corresponds to (10-62), and the lower corresponds to (10-63).

lower curve is almost linear with a very shallow slope, and shows that there is very little to be gained by increasing the number of parallel plates in order to increase the degree of polarization.

A final topic that we discuss is the use of a simpler notation for the Mueller matrices for reflection and transmission by representing the matrix elements in terms of the Fresnel reflection and transmission coefficients. These coefficients are defined to be

$$\rho_s = \left(\frac{R_s}{E_s}\right)^2 = \left(\frac{\sin\theta_-}{\sin\theta_+}\right)^2 \tag{10-64a}$$

$$\rho_p = \left(\frac{R_p}{E_p}\right)^2 = \left(\frac{\tan\theta_-}{\tan\theta_+}\right)^2 \tag{10-64b}$$

and

$$\begin{aligned}\tau_s &= \frac{n\cos\theta_r}{\cos\theta_i}\left(\frac{T_s}{E_s}\right)^2 = \frac{\tan\theta_i}{\tan\theta_r}\left(\frac{2\sin\theta_r\cos\theta_i}{\sin\theta_+}\right)^2 \\ &= \frac{\sin 2\theta_i \sin 2\theta_r}{\sin^2\theta_+}\end{aligned} \tag{10-65a}$$

$$\begin{aligned}\tau_p &= \frac{n\cos\theta_r}{\cos\theta_i}\left(\frac{T_p}{E_p}\right)^2 = \frac{\tan\theta_i}{\tan\theta_r}\left(\frac{2\sin\theta_r\cos\theta_i}{\sin\theta_+\cos\theta_-}\right)^2 \\ &= \frac{\sin 2\theta_i \sin 2\theta_r}{\sin^2\theta_+\cos^2\theta_-}\end{aligned} \tag{10-65b}$$

One can readily show that the following relations hold for Fresnel coefficients:

$$\rho_s + \tau_s = 1 \tag{10-66a}$$

and

$$\rho_p + \tau_p = 1 \tag{10-66b}$$

At the Brewster angle, written as θ_{i_B}, Fresnel's reflection and transmission coefficients (10-65) and (10-66) reduce to

$$\rho_{s,B} = \cos^2 2\theta_{i_B} \tag{10-67a}$$

$$\rho_{p,B} = 0 \tag{10-67b}$$

$$\tau_{s,B} = \sin^2 2\theta_{i_B} \tag{10-68a}$$

$$\tau_{p,B} = 1 \tag{10-68b}$$

We see immediately that

$$\rho_{s,B} + \tau_{s,B} = 1 \tag{10-69a}$$

and

$$\rho_{p,B} + \tau_{p,B} = 1 \tag{10-69b}$$

Equations (10-69a) and (10-69b) are, of course, merely special cases of (10-66a) and (10-66b).

With these definitions the Mueller matrices for reflection and transmission can be written, respectively, as

$$M_\rho = \frac{1}{2}\begin{pmatrix} \rho_s + \rho_p & \rho_s - \rho_p & 0 & 0 \\ \rho_s - \rho_p & \rho_s + \rho_p & 0 & 0 \\ 0 & 0 & 2(\rho_s\rho_p)^{1/2} & 0 \\ 0 & 0 & 0 & 2(\rho_s\rho_p)^{1/2} \end{pmatrix} \tag{10-70a}$$

and

$$M_\tau = \frac{1}{2}\begin{pmatrix} \tau_s + \tau_p & \tau_s - \tau_p & 0 & 0 \\ \tau_s - \tau_p & \tau_s + \tau_p & 0 & 0 \\ 0 & 0 & 2(\tau_p\tau_s)^{1/2} & 0 \\ 0 & 0 & 0 & 2(\tau_s\tau_p)^{1/2} \end{pmatrix} \tag{10-70b}$$

The reflection coefficients ρ_s and ρ_p, (10-64a) and (10-64b), are plotted as a function of the incident angle for a range of refractive indices in Figs. 10-10 and 10-11. Similar plots are shown in Figs. 10-12 and 10-13 for τ_s and τ_p, (10-65a) and (10-65b).

In a similar manner the reflection and transmission coefficients at the Brewster angle, (10-67) and (10-68), are plotted as a function of the refractive index n in Figs. 10-14 and 10-15.

The great value of the Fresnel coefficients is that their use leads to simpler forms for the Mueller matrices for reflection and transmission. For example, instead

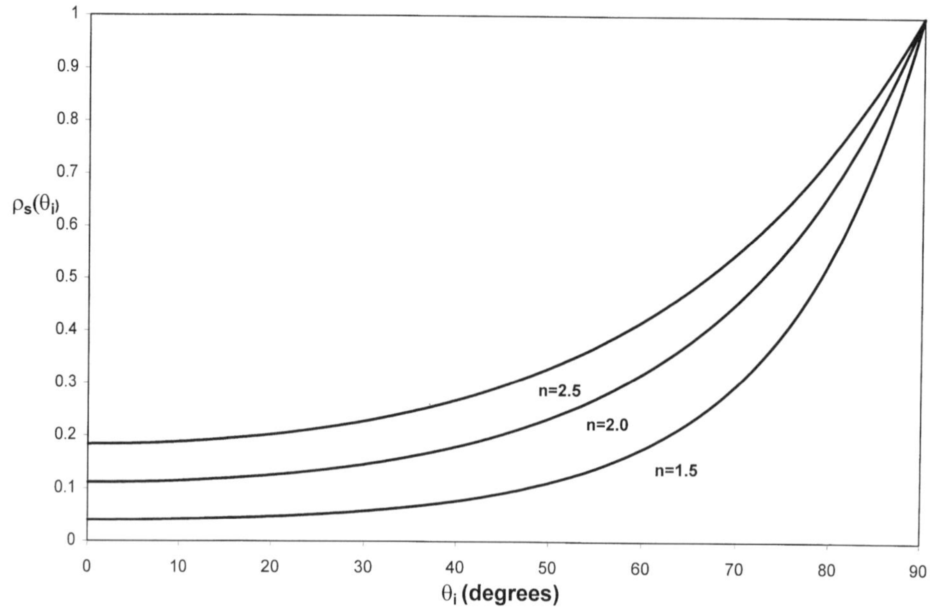

Figure 10-10 Plot of the Fresnel reflection coefficient ρ_s as a function of incidence angle θ_i, (10-64a).

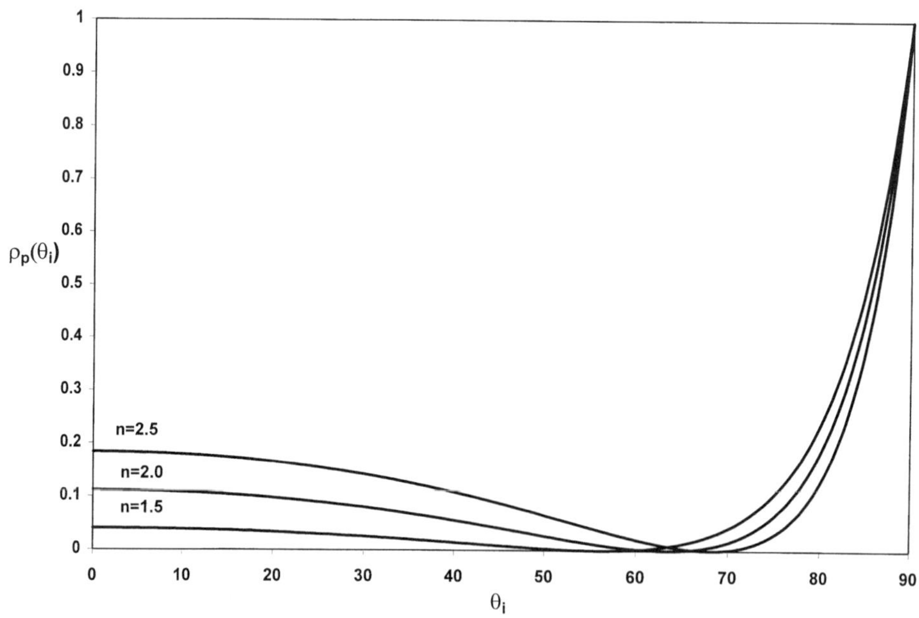

Figure 10-11 Plot of the Fresnel reflection coefficient ρ_p as a function of incidence angle θ_i, (10-64b).

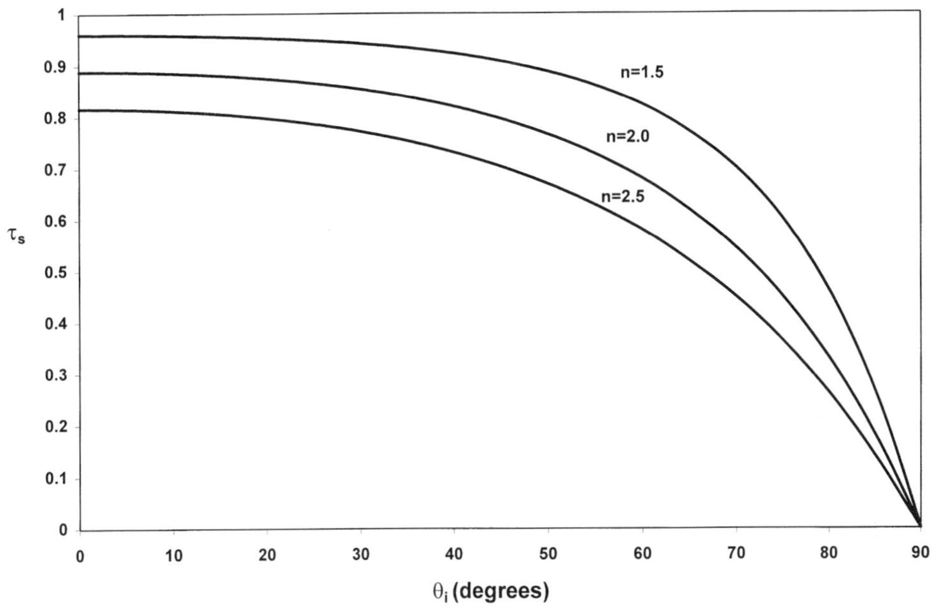

Figure 10-12 Plot of the Fresnel reflection coefficient τ_s as a function of incidence angle θ_i, (10-65a).

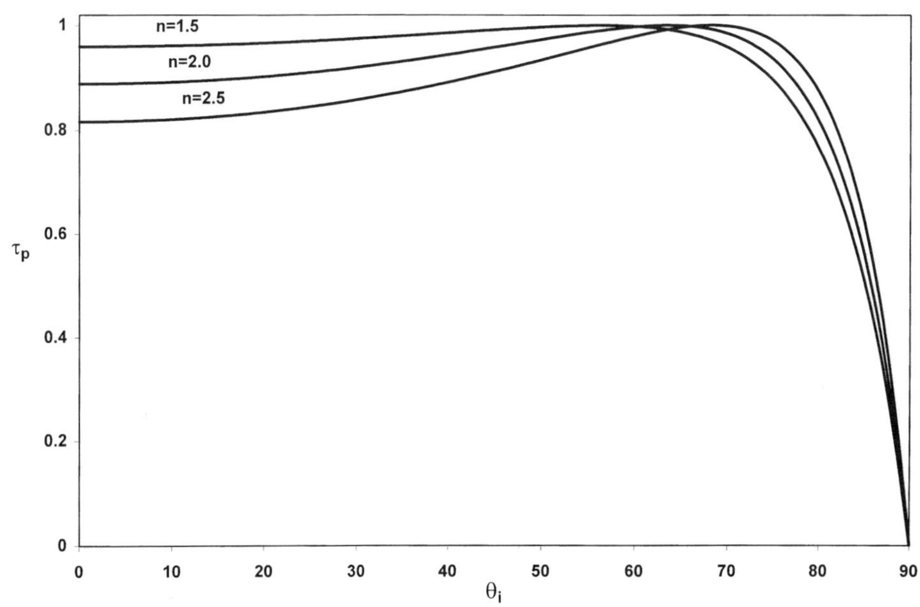

Figure 10-13 Plot of the Fresnel reflection coefficient τ_p as a function of incidence angle θ_i, (10-65b).

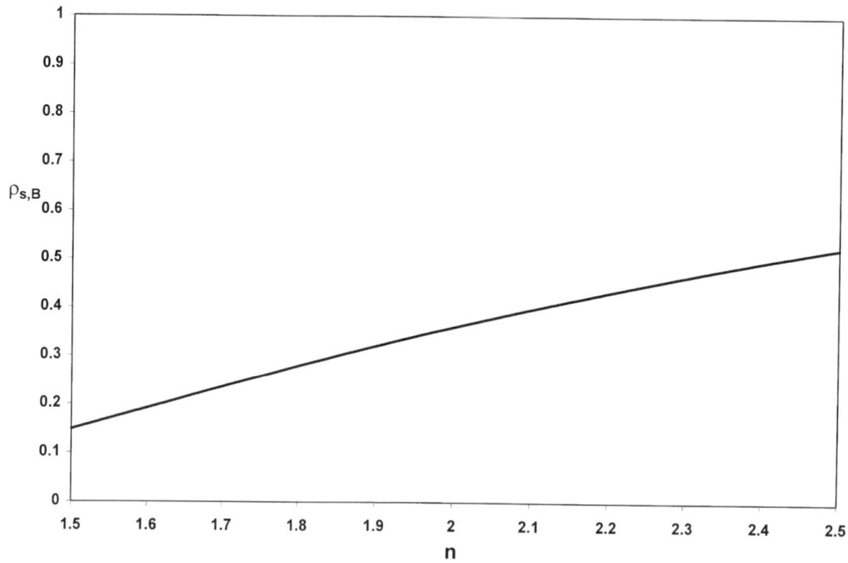

Figure 10-14 Plot of the reflection coefficients at the Brewster angle, (10-67).

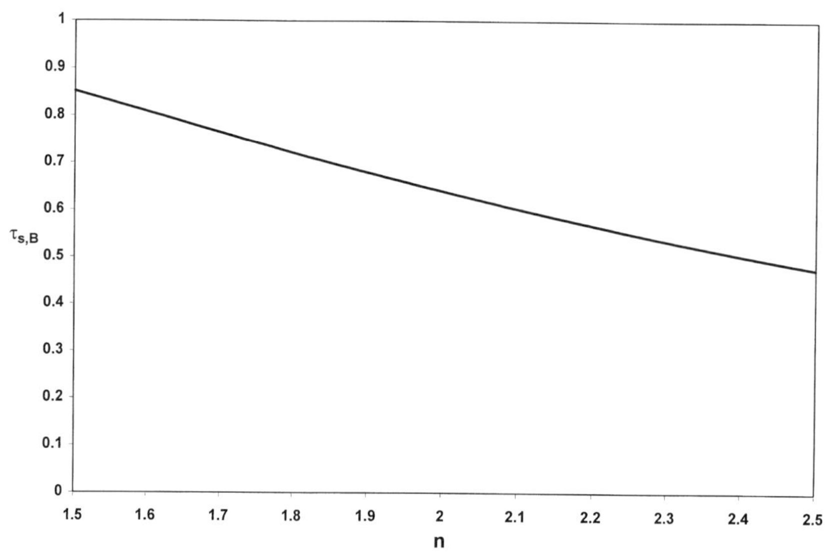

Figure 10-15 Plot of the transmission coefficients at the Brewster angle, (10-68).

of the complicated matrix entries given above, we can write, say, the diagonalized form of the Mueller matrices as

$$M_{\rho,D} = \begin{pmatrix} \rho_s & 0 & 0 & 0 \\ 0 & \rho_p & 0 & 0 \\ 0 & 0 & (\rho_s\rho_p)^{1/2} & 0 \\ 0 & 0 & 0 & (\rho_s\rho_p)^{1/2} \end{pmatrix} \tag{10-71a}$$

and

$$M_{\tau,D} = \begin{pmatrix} \tau_s & 0 & 0 & 0 \\ 0 & \tau_p & 0 & 0 \\ 0 & 0 & (\tau_s\tau_p)^{1/2} & 0 \\ 0 & 0 & 0 & (\tau_s\tau_p)^{1/2} \end{pmatrix} \tag{10-71b}$$

For treating problems at angles other than the Brewster angle it is much simpler to use either (10-71a) or (10-71b) rather than the earlier forms of the Mueller matrices because the matrix elements ρ_s, ρ_p, τ_s, and τ_p are far easier to work with.

In this chapter we have applied the Mueller matrix formalism to the problem of determining the change in the polarization of light by single and multiple dielectric plates. We have treated the problems in the simplest way by ignoring the thickness of the plates and multiple reflections within the plates. Consequently, the results are only approximately correct. Nevertheless, the results are still useful and allow us to predict quite accurately the expected behavior of polarized light and its interaction with dielectric plates. In particular, we have presented a number of formulas, much quoted in the optical literature and handbooks, which describe the degree of polarization for an incident unpolarized beam of light. These formulas describe the number of parallel plates required to obtain any degree of polarization. A fuller discussion of the behavior of multiple plates can be found in the references.

REFERENCES

Papers

1. Stokes, G. G., *Proc. Roy. Soc. (London)*, *11*, 545 (1862).
2. Tuckerman, L. B., *J. Opt. Soc. Am.*, *37*, 818 (1947).
3. Jones, R. Clark, *J. Opt. Soc. Am.*, *31*, 488 (1941).
4. Jones, R. Clark, *J. Opt. Soc. Am.*, *46*, 126 (1956).
5. Collett, E., *Am. J. Phys.*, *39*, 517 (1971).
6. Collett, E., *Appl. Opt.*, *5*, 1184 (1972).
7. McMaster, W. H., *Rev. Mod. Phys.*, *33*, 8 (1961).
8. Schmieder, R. W., *J. Opt. Soc. Am.*, *59*, 297 (1969).

Books

1. Shurcliff, W. A., *Polarized Light*, Harvard University Press, Cambridge, MA, 1962.
2. Born, M. and Wolf, E., *Principles of Optics*, 3rd ed., Pergamon Press, New York, 1965.
3. Chandrasekhar, S., *Radiative Transfer*, Dover, New York, 1960, pp. 24–34.
4. Driscoll, W. G., ed., *Handbook of Optics*, McGraw-Hill, New York, 1978.
5. Sokolnikoff, I. S. and Redheffer, R. M., *Mathematics of Physics and Modern Engineering*, McGraw-Hill, New York, 1966.
6. Menzel, D. H., *Mathematical Physics*, Prentice-Hall, New York, 1953.
7. Hecht, E. and Zajac, A., *Optics*, Addison-Wesley, Reading, MA, 1974.

11

The Jones Matrix Calculus

11.1 INTRODUCTION

We have seen that the Stokes polarization parameters and the Mueller matrix formalism can be used to describe any state of polarization. In particular, if we are dealing with a single beam of polarized light, then the formalism of the Stokes parameters is completely capable of describing any polarization state ranging from completely polarized light to completely unpolarized light. In addition, the formalism of the Stokes parameters can be used to describe the superposition of several polarized beams, provided that there is no amplitude or phase relation between them; that is, the beams are incoherent with respect to each other. This situation arises when optical beams are emitted from several independent sources and are then superposed.

However, there are experiments where several beams must be added and the beams are not independent of each other, e.g., beam superposition in interferometers. There we have a single optical source and the single beam is divided by a beam splitter. Then, at a later stage, the beams are "reunited," that is, superposed. Clearly, there is an amplitude and phase relation between the beams. We see that we must deal with amplitudes and phase and superpose the amplitudes of each of the beams. After the amplitudes of the beam are superposed, the intensity of the combined beams is then found by taking the time average of the square of the total amplitude. If there were no amplitude or phase relations between the beams, then we would arrive at the same result as we obtained for the Stokes parameters. However, if there is a relation between the amplitude and the phase of the optical beams, an interference term will arise.

Of course, as pointed out earlier, the description of the polarizing behavior of the optical field in terms of amplitudes was one of the first great successes of the wave theory of light. The solution of the wave equation in terms of transverse components leads to elliptically polarized light and its degenerate linear and circular forms. On the basis of the amplitude results, many results could be understood (e.g., Young's interference experiment, circularly polarized light). However, even using the amplitude formulation, numerous problems become difficult to treat, such as the propagation of the field through several polarizing components. To facilitate the treatment of complicated polarization problems at the amplitude level, R. Clark Jones, in the early 1940s, developed a matrix calculus for treating these problems, commonly called the Jones matrix calculus. It is most appropriately used when we must superpose amplitudes. The Jones calculus involves complex quantities contained in 2×1 column matrices (the Jones vector) and 2×2 matrices (the Jones matrices). At first sight it would seem that the use of the 2×2 matrices would be simpler than the use of the 4×4 Mueller matrices. Oddly enough, this is not the case. This is due primarily to the fact that even the matrix multiplication of several complex 2×2 matrices can be tedious. Furthermore, even after the complete matrix calculation has been carried out, additional steps are still required. For example, it is often necessary to separate the real and imaginary parts (e.g., E_x and E_y) and superpose the respective amplitudes. This can involve a considerable amount of effort. Another problem is that to find the intensity one must take the complex transpose of the Jones vector and then carry out the matrix multiplication between the complex transpose of the Jones vector and Jones vector itself. All this is done using complex quantities, and the possibility of making a computational error is very real. While the 4×4 Mueller matrix formalism appears to be more complicated, all the entries are real quantities and there are many zero entries, as can be seen by inspecting the Mueller matrix for the polarizer, the retarder and the rotator. This fact greatly simplifies the matrix multiplications, and, of course, the Stokes vector is real.

There are, nevertheless, many instances where the amplitudes must be added (superposed), and so the Jones matrix formalism must be used. There are many problems where either formalism can be used with success. As a general rule, the most appropriate choice of matrix method is to use the Jones calculus for amplitude superposition problems and the Mueller formalism for intensity superposition problems. Experience will usually indicate the best choice to make.

In this chapter we develop the fundamental matrices for the Jones calculus along with its application to a number of problems.

11.2 THE JONES VECTOR

The plane-wave components of the optical field in terms of complex quantities can be written as

$$E_x(z, t) = E_{0x} e^{i(\omega t - kz + \delta_x)} \tag{11-1a}$$

$$E_y(z, t) = E_{0y} e^{i(\omega t - kz + \delta_y)} \tag{11-1b}$$

The propagator $\omega t - kz$ is now suppressed, so (11-1) is then written as

$$E_x = E_{0x} e^{i\delta_x} \tag{11-2a}$$

$$E_y = E_{0y} e^{i\delta_y} \tag{11-2b}$$

Equation (11-2) can be arranged in a 2×1 column matrix $\mathbf{E}$:

$$\mathbf{E} = \begin{pmatrix} E_x \\ E_y \end{pmatrix} = \begin{pmatrix} E_{0x} e^{i\delta_x} \\ E_{0y} e^{i\delta_y} \end{pmatrix} \tag{11-3}$$

called the Jones column matrix or, simply, the Jones vector. The column matrix on the right-hand side of (11-3), incidentally, is the Jones vector for elliptically polarized light.

In the Jones vector (11-3), the maximum amplitudes E_{0x} and E_{0y} are real quantities. The presence of the exponent with imaginary arguments causes E_x and E_y to be complex quantities. Before we proceed to find the Jones vectors for various states of polarized light, we discuss the normalization of the Jones vector; it is customary to express the Jones vector in normalized form. The total intensity I of the optical field is given by

$$I = E_x E_x^* + E_y E_y^* \tag{11-4}$$

Equation (11-4) can be obtained by the following matrix multiplication:

$$I = \begin{pmatrix} E_x^* & E_y^* \end{pmatrix} \begin{pmatrix} E_x \\ E_y \end{pmatrix} \tag{11-5}$$

The row matrix $(E_x^* \; E_y^*)$ is the complex transpose of the Jones vector (column matrix $\mathbf{E}$) and is written $\mathbf{E}^\dagger$; thus,

$$\mathbf{E}^\dagger = \begin{pmatrix} E_x^* & E_y^* \end{pmatrix} \tag{11-6}$$

so

$$I = \mathbf{E}^\dagger \mathbf{E} \tag{11-7}$$

yields (11-4). Carrying out the matrix multiplication of (11-7), using (11-3), yields

$$E_{0x}^2 + E_{0y}^2 = I = E_0^2 \tag{11-8}$$

It is customary to set $E_0^2 = 1$, whereupon we say that the Jones vector is normalized. The normalized condition for (11-5) can then be written as

$$\mathbf{E}^\dagger \mathbf{E} = 1 \tag{11-9}$$

We note that the Jones vector can *only* be used to describe completely polarized light. We now find the Jones vector for the following states of completely polarized light.

1. Linear horizontally polarized light. For this state $E_y = 0$, so (11-3) becomes

$$\mathbf{E} = \begin{pmatrix} E_{0x} e^{i\delta_x} \\ 0 \end{pmatrix} \tag{11-10}$$

From the normalization condition (11-9) we see that $E_{0x}^2 = 1$. Thus, suppressing $e^{i\delta_x}$ because it is unimodular, the normalized Jones vector for linearly horizontally polarized light is written

$$\mathbf{E} = \begin{pmatrix} 1 \\ 0 \end{pmatrix} \tag{11-11}$$

In a similar manner the Jones vectors for the other well-known polarization states are easily found.

2. Linear vertically polarized light. $E_x = 0$, so $E_{0y}^2 = 1$ and

$$\mathbf{E} = \begin{pmatrix} 0 \\ 1 \end{pmatrix} \tag{11-12}$$

3. Linear $+45^\circ$ polarized light. $E_x = E_y$, so $2E_{0x}^2 = 1$ and

$$\mathbf{E} = \frac{1}{\sqrt{2}}\begin{pmatrix} 1 \\ 1 \end{pmatrix} \tag{11-13}$$

4. Linear -45° polarized light. $E_x = -E_y$, so $2E_{0x}^2 = 1$ and

$$\mathbf{E} = \frac{1}{\sqrt{2}}\begin{pmatrix} 1 \\ -1 \end{pmatrix} \tag{11-14}$$

5. Right-hand circularly polarized light. For this case $E_{0x} = E_{0y}$ and $\delta_y - \delta_x = +90^\circ$. Then, $2E_{0x}^2 = 1$ and we have

$$\mathbf{E} = \frac{1}{\sqrt{2}}\begin{pmatrix} 1 \\ +i \end{pmatrix} \tag{11-15}$$

6. Left-hand circularly polarized light. We again have $E_{0x} = E_{0y}$, but $\delta_y - \delta_x = -90^\circ$. The normalization condition gives $2E_{0x}^2 = 1$, and we have

$$\mathbf{E} = \frac{1}{\sqrt{2}}\begin{pmatrix} 1 \\ -i \end{pmatrix} \tag{11-16}$$

Each of the Jones vectors (11-11) through (11-16) satisfies the normalization condition (11-9).

An additional property is the orthogonal or orthonormal property. Two vectors **A** and **B** are said to be orthogonal if **AB** $= 0$ or, in complex notation, $\mathbf{A}^\dagger \mathbf{B} = 0$. If this condition is satisfied, we say that the Jones vectors are orthogonal. For example, for linearly horizontal and vertical polarized light we find that

$$\begin{pmatrix} 1 & 0 \end{pmatrix}^* \begin{pmatrix} 0 \\ 1 \end{pmatrix} = 0 \tag{11-17a}$$

so the states are orthogonal or, since we are using normalized vectors, orthonormal. Similarly, for right and left circularly polarized light:

$$\begin{pmatrix} 1 & +i \end{pmatrix}^* \begin{pmatrix} 1 \\ -i \end{pmatrix} = 0 \tag{11-17b}$$

Thus, the orthonormal condition for two Jones vectors $\mathbf{E}_1$ and $\mathbf{E}_2$ is

$$\mathbf{E}_i^\dagger \mathbf{E}_j = 0 \tag{11-18}$$

We see that the orthonormal condition (11-18) and the normalizing condition (11-9) can be written as a single equation, namely

$$\mathbf{E}_i^\dagger \mathbf{E}_j = \delta_{ij} \qquad i, j = 1, 2 \tag{11-19a}$$

where δ_{ij} is the Kronecker delta and has the property:

$$\delta_{ij} = 1 \qquad i = j \tag{11-19b}$$

$$\delta_{ij} = 0 \qquad i \neq j \tag{11-19c}$$

In a manner analogous to the superposition of incoherent intensities or Stokes vectors, we can superpose coherent amplitudes, that is, Jones vectors. For example, the Jones vector for horizontal polarization is $\mathbf{E}_H$ and that for vertical polarization is $\mathbf{E}_V$, so

$$\mathbf{E}_H = \begin{pmatrix} E_{0x}e^{i\delta_x} \\ 0 \end{pmatrix} \qquad \mathbf{E}_V = \begin{pmatrix} 0 \\ E_{0y}e^{i\delta_y} \end{pmatrix} \tag{11-20}$$

Adding $\mathbf{E}_H$ and $\mathbf{E}_V$ gives

$$\mathbf{E} = \mathbf{E}_H + \mathbf{E}_V = \begin{pmatrix} E_{0x}e^{i\delta_x} \\ E_{0y}e^{i\delta_y} \end{pmatrix} \tag{11-21}$$

which is the Jones vector for elliptically polarized light. Thus, superposing two orthogonal linear polarizations give rise to elliptically polarized light. For example, if $E_{0x} = E_{0y}$ and $\delta_y = \delta_x$, then, from (11-21), we can write

$$\mathbf{E} = E_{0x}e^{i\delta_x}\begin{pmatrix} 1 \\ 1 \end{pmatrix} \tag{11-22}$$

which is the Jones vector for linear +45° polarized light. Equation (11-22) could also be obtained by superposing (11-11) and (11-12):

$$\mathbf{E} = \mathbf{E}_H + \mathbf{E}_V = \begin{pmatrix} 1 \\ 0 \end{pmatrix} + \begin{pmatrix} 0 \\ 1 \end{pmatrix} = \begin{pmatrix} 1 \\ 1 \end{pmatrix} \tag{11-23}$$

which, aside from the normalizing factor, is identical to (11-13).

As another example let us superpose left and right circularly polarized light of equal amplitudes. Then, from (11-15) and (11-16) we have

$$\mathbf{E} = \frac{1}{\sqrt{2}}\begin{pmatrix} 1 \\ -i \end{pmatrix} + \frac{1}{\sqrt{2}}\begin{pmatrix} 1 \\ i \end{pmatrix} = \frac{2}{\sqrt{2}}\begin{pmatrix} 1 \\ 0 \end{pmatrix} \tag{11-24}$$

which, aside from the normalizing factor, is the Jones vector for linear horizontally polarized light (11-11).

As a final Jones vector example, we show that elliptically polarized light can be obtained by superposing two opposite circularly polarized beams of unequal amplitudes. The Jones vectors for two circular polarized beams of unequal amplitudes a and b can be represented by

$$\mathbf{E}_+ = a\begin{pmatrix} 1 \\ +i \end{pmatrix} \qquad \mathbf{E}_- = b\begin{pmatrix} 1 \\ -i \end{pmatrix} \tag{11-25}$$

According to the principle of superposition, the resultant Jones vector for (11-25) is

$$\mathbf{E} = \mathbf{E}_+ + \mathbf{E}_- = \begin{pmatrix} a+b \\ i(a-b) \end{pmatrix} = \begin{pmatrix} E_x \\ E_y \end{pmatrix} \tag{11-26}$$

In component form (11-26) is written as

$$E_x = a + b \tag{11-27a}$$

$$E_y = (a - b)e^{i\pi/2} \tag{11-27b}$$

We now restore the propagator $\omega t - kz$, so (11-27) is then written as

$$E_x = (a + b)e^{i(\omega t - kz)} \tag{11-28a}$$

$$E_y = (a - b)e^{i(\omega t - kz + \pi/2)} \tag{11-28b}$$

Taking the real part of (11-28), we have

$$E_x(z, t) = (a + b)\cos(\omega t - kz) \tag{11-29a}$$

$$E_y(z, t) = (a - b)\cos\left(\omega t - kz + \frac{\pi}{2}\right) \tag{11-29b}$$

$$= (a - b)\sin(\omega t - kz) \tag{11-29c}$$

Equations (11-28a) and (11-28b) are now written as

$$\frac{E_x(z, t)}{a + b} = \cos(\omega t - kz) \tag{11-30a}$$

$$\frac{E_y(z, t)}{a - b} = \sin(\omega t - kz) \tag{11-30b}$$

Squaring and adding (11-30a) and (11-30b) yields

$$\frac{E_x^2(z, t)}{(a + b)^2} + \frac{E_y^2(z, t)}{(a - b)^2} = 1 \tag{11-31}$$

Equation (11-31) is the equation of an ellipse whose major and minor axes lengths are a + b and $a - b$, respectively. Thus, the superposition of two oppositely circularly polarized beams of unequal magnitudes gives rise to a (nonrotated) ellipse with its locus vector moving in a counterclockwise direction.

11.3 JONES MATRICES FOR THE POLARIZER, RETARDER, AND ROTATOR

We now determine the matrix forms for polarizers (diattenuators), retarders (phase shifters), and rotators in the Jones matrix calculus. In order to do this, we assume that the components of a beam emerging from a polarizing element are linearly related to the components of the incident beam. This relation is written as

$$E'_x = j_{xx}E_x + j_{xy}E_y \tag{11-32a}$$

$$E'_y = j_{yx}E_x + j_{yy}E_y \tag{11-32b}$$

where E'_x and E'_y are the components of the emerging beam and E_x and E_y are the components of the incident beam. The quantities $j_{ik}, i, k = x, y$, are the transforming factors (elements). Equation (11-32) can be written in matrix form as

$$\begin{pmatrix} E'_x \\ E'_y \end{pmatrix} = \begin{pmatrix} j_{xx} & j_{xy} \\ j_{yx} & j_{yy} \end{pmatrix} \begin{pmatrix} E_x \\ E_y \end{pmatrix} \tag{11-33a}$$

or

$$\mathbf{E}' = \mathbf{JE} \tag{11-33b}$$

where

$$\mathbf{J} = \begin{pmatrix} j_{xx} & j_{xy} \\ j_{yx} & j_{yy} \end{pmatrix} \tag{11-33c}$$

The 2×2 matrix $\mathbf{J}$ is called the Jones instrument matrix or, simply, the Jones matrix. We now determine the Jones matrices for a polarizer, retarder, and rotator.

A polarizer is characterized by the relations:

$$E'_x = p_x E_x \tag{11-34a}$$

$$E'_y = p_y E_y \qquad 0 \le p_{x,y} \le 1 \tag{11-34b}$$

For complete transmission $p_{x,y} = 1$, and for complete attenuation $p_{x,y} = 0$. In terms of the Jones vector, (11-34) can be written as

$$\begin{pmatrix} E'_x \\ E'_y \end{pmatrix} = \begin{pmatrix} p_x & 0 \\ 0 & p_y \end{pmatrix} \begin{pmatrix} E_x \\ E_y \end{pmatrix} \tag{11-35}$$

so the Jones matrix (11-33c) for a polarizer is

$$\mathbf{J}_p = \begin{pmatrix} p_x & 0 \\ 0 & p_y \end{pmatrix} \qquad 0 \le p_{x,y} \le 1 \tag{11-36}$$

For an ideal linear horizontal polarizer there is complete transmission along the horizontal x axis and complete attenuation along the vertical y axis. This is expressed by $p_x = 1$ and $p_y = 0$, so (11-36) becomes

$$\mathbf{J}_{PH} = \begin{pmatrix} 1 & 0 \\ 0 & 0 \end{pmatrix} \tag{11-37}$$

Similarly, for a linear vertical polarizer, (11-36) becomes

$$\mathbf{J}_{PV} = \begin{pmatrix} 0 & 0 \\ 0 & 1 \end{pmatrix} \tag{11-38}$$

In general, it is useful to know the Jones matrix for a linear polarizer rotated through an angle θ. This is readily found by using the familiar rotation transformation, namely,

$$\mathbf{J}' = \mathbf{J}(-\theta)\mathbf{J}\mathbf{J}(\theta) \tag{11-39a}$$

where $\mathbf{J}(\theta)$ is the rotation matrix:

$$\mathbf{J}(\theta) = \begin{pmatrix} \cos\theta & \sin\theta \\ -\sin\theta & \cos\theta \end{pmatrix} \tag{11-39b}$$

and **J** is given by (11-33c). For a rotated linear polarizer represented by (11-36) and rotated by angle θ we have from (11-39) that

$$\mathbf{J}' = \begin{pmatrix} \cos\theta & -\sin\theta \\ \sin\theta & \cos\theta \end{pmatrix} \begin{pmatrix} p_x & 0 \\ 0 & p_y \end{pmatrix} \begin{pmatrix} \cos\theta & \sin\theta \\ -\sin\theta & \cos\theta \end{pmatrix} \tag{11-40}$$

Carrying out the matrix multiplication in (11-40) we find that the Jones matrix for a rotated polarizer is

$$\mathbf{J}_P(\theta) = \begin{pmatrix} p_x\cos^2\theta + p_y\sin^2\theta & (p_x - p_y)\sin\theta\cos\theta \\ (p_x - p_y)\sin\theta\cos\theta & p_x\sin^2\theta + p_y\cos^2\theta \end{pmatrix} \tag{11-41}$$

For an ideal linear horizontal polarizer we can set $p_x = 1$ and $p_y = 0$ in (11-41), so that the Jones matrix for a rotated linear horizontal polarizer is

$$\mathbf{J}_P(\theta) = \begin{pmatrix} \cos^2\theta & \sin\theta\cos\theta \\ \sin\theta\cos\theta & \sin^2\theta \end{pmatrix} \tag{11-42}$$

The Jones matrix for a linear polarizer rotated through $+45^\circ$ is then seen from (11-42) to be

$$\mathbf{J}_P(45^\circ) = \frac{1}{2}\begin{pmatrix} 1 & 1 \\ 1 & 1 \end{pmatrix} \tag{11-43}$$

If the linear polarizer is not ideal, then the Jones matrix for a polarizer (11-36) at $+45^\circ$ is seen from (11-41) to be

$$\mathbf{J}_P(45^\circ) = \frac{1}{2}\begin{pmatrix} p_x + p_y & p_x - p_y \\ p_x - p_y & p_x + p_y \end{pmatrix} \tag{11-44}$$

We note that for $\theta = 0^\circ$ and 90°, (11-42) gives the Jones matrices for a linear horizontal and vertical polarizer, Eqs. (11-37) and (11-38) respectively.

Equation (11-41) also describes a neutral density (ND) filter. The condition for a ND filter is $p_x = p_y = p$, so (11-41) reduces to

$$\mathbf{J}_{\mathrm{ND}}(\theta) = p\begin{pmatrix} 1 & 0 \\ 0 & 1 \end{pmatrix} \tag{11-45}$$

Thus, $\mathbf{J}_{\mathrm{ND}}(\theta)$ is independent of rotation (θ), and the amplitudes are equally attenuated by an amount p. This is, indeed, the behavior of a ND filter. The presence of the unit (diagonal) matrix in (11-45) confirms that a ND filter does not affect the polarization state of the incident beam.

The next polarizing element of importance is the retarder. The retarder increases the phase by $+\phi/2$ along the fast (x) axis and retards the phase by $-\phi/2$, along the slow (y) axis. This behavior is described by

$$E'_x = e^{+i\phi/2}E_x \tag{11-46a}$$

$$E'_y = e^{-i\phi/2}E_y \tag{11-46b}$$

where E'_x and E'_y are the components of the emerging beam and E_x and E_y are the components of the incident beam. We can immediately express (11-46) in the Jones formalism as

$$\mathbf{J}' = \begin{pmatrix} E'_x \\ E'_y \end{pmatrix} = \begin{pmatrix} e^{+i\phi/2} & 0 \\ 0 & e^{-i\phi/2} \end{pmatrix} \begin{pmatrix} E_x \\ E_y \end{pmatrix} \tag{11-47}$$

The Jones matrix for a retarder (phase shifter) is then

$$\mathbf{J}_R(\phi) = \begin{pmatrix} e^{+i\phi/2} & 0 \\ 0 & e^{-i\phi/2} \end{pmatrix} \tag{11-48}$$

where ϕ is the total phase shift between the field components. The two most common types of phase shifters (retarders) are the quarter-wave retarder and the half-wave retarder. For these devices $\phi = 90^\circ$ and 180°, repectively, and (11-48) becomes

$$\mathbf{J}_R\left(\frac{\lambda}{4}\right) = \begin{pmatrix} e^{i\pi/4} & 0 \\ 0 & e^{-i\pi/4} \end{pmatrix} = e^{i\pi/4} \begin{pmatrix} 1 & 0 \\ 0 & e^{-i\pi/2} \end{pmatrix} = e^{i\pi/4} \begin{pmatrix} 1 & 0 \\ 0 & -i \end{pmatrix} \tag{11-49a}$$

and

$$\mathbf{J}_R\left(\frac{\lambda}{2}\right) = \begin{pmatrix} e^{i\pi/2} & 0 \\ 0 & e^{-i\pi/2} \end{pmatrix} = \begin{pmatrix} i & 0 \\ 0 & -i \end{pmatrix} = i \begin{pmatrix} 1 & 0 \\ 0 & -1 \end{pmatrix} \tag{11-49b}$$

The Jones matrix for a rotated retarder is found from (11-48) and (11-39) to be

$$\mathbf{J}_R(\phi, \theta) = \begin{pmatrix} e^{i\phi/2}\cos^2\theta + e^{-i\phi/2}\sin^2\theta & (e^{i\phi/2} - e^{-i\phi/2})\sin\theta\cos\theta \\ (e^{i\phi/2} - e^{-i\phi/2})\sin\theta\cos\theta & e^{i\phi/2}\sin^2\theta + e^{-i\phi/2}\cos^2\theta \end{pmatrix} \tag{11-50}$$

With the half-angle formulas, (11-50) can also be written in the form:

$$\mathbf{J}_R(\phi, \theta) = \begin{pmatrix} \cos\frac{\phi}{2} + i\sin\frac{\phi}{2}\cos 2\theta & i\sin\frac{\phi}{2}\sin 2\theta \\ i\sin\frac{\phi}{2}\sin 2\theta & \cos\frac{\phi}{2} - i\sin\frac{\phi}{2}\cos 2\theta \end{pmatrix} \tag{11-51}$$

For quarter-wave retarder and a half-wave retarder (11-51) reduces, respectively, to

$$\mathbf{J}_R\left(\frac{\lambda}{4}, \theta\right) = \begin{pmatrix} \frac{1}{\sqrt{2}} + \frac{i}{\sqrt{2}}\cos 2\theta & \frac{i}{\sqrt{2}}\sin 2\theta \\ \frac{i}{\sqrt{2}}\sin 2\theta & \frac{1}{\sqrt{2}} - \frac{i}{\sqrt{2}}\cos 2\theta \end{pmatrix} \tag{11-52}$$

and

$$\mathbf{J}_R\left(\frac{\lambda}{2}, \theta\right) = i \begin{pmatrix} \cos 2\theta & \sin 2\theta \\ \sin 2\theta & -\cos 2\theta \end{pmatrix} \tag{11-53}$$

The factor i in (11-53) is unimodular and can be suppressed. It is common, therefore, to write (11-53) simply as

$$\mathbf{J}_R\left(\frac{\lambda}{2}, \theta\right) = \begin{pmatrix} \cos 2\theta & \sin 2\theta \\ \sin 2\theta & -\cos 2\theta \end{pmatrix} \tag{11-54}$$

Inspecting (11-54) we see that it is very similar to the matrix for rotation, namely,

$$\mathbf{J}(\theta) = \begin{pmatrix} \cos\theta & \sin\theta \\ -\sin\theta & \cos\theta \end{pmatrix} \tag{11-39b}$$

However, (11-54) differs from (11-39b) in two ways. First, in (11-54) we have 2θ rather than θ. Thus, a rotation of a retarder through θ rotates the polarization ellipse through 2θ. Second, a clockwise mechanical rotation θ in (11-54) leads to a counter-clockwise rotation of the polarization ellipse. In order to see this behavior clearly, consider that we have incident linear horizontally polarized light. Its Jones vector is

$$\mathbf{J} = \begin{pmatrix} E_x \\ 0 \end{pmatrix} \tag{11-55}$$

The components of the beam emerging from a true rotator (11-39b) are then

$$E'_x = (\cos\theta)E_x \tag{11-56a}$$

$$E'_y = -(\sin\theta)E_x \tag{11-56b}$$

The angle of rotation α is then

$$\tan\alpha = \frac{E'_y}{E'_x} = \frac{-\sin\theta}{\cos\theta} = \tan(-\theta) \tag{11-57}$$

In a similar manner, multiplying (11-55) by (11-54) leads to

$$E'_x = (\cos 2\theta)E_x \tag{11-58a}$$

$$E'_y = (\sin 2\theta)E_x \tag{11-58b}$$

so we now have

$$\tan\alpha = \frac{E'_y}{E'_x} = \frac{\sin 2\theta}{\cos 2\theta} = \tan 2\theta \tag{11-59}$$

Comparing (11-59) with (11-57), we see that the direction of rotation for a rotated retarder is opposite to the direction of true rotation. Equation (11-59) also shows that the angle of rotation is twice that of a true rotation. Because of this similar but analytically incorrect behavior of a rotated half-wave retarder, (11-54) is called a *pseudorotator*. We note that an alternative form of a half-wave retarder, which is the more common form, is given by factoring out i in (11-49b) or simply setting $\theta = 0°$ in (11-54):

$$\mathbf{J}\left(\frac{\lambda}{2}\right) = \begin{pmatrix} 1 & 0 \\ 0 & -1 \end{pmatrix} \tag{11-60}$$

The final matrix of interest is the Jones matrix for a rotator. The defining equations are

$$E'_x = \cos\beta E_x + \sin\beta E_y \tag{11-61a}$$

$$E'_y = -\sin\beta E_x + \cos\beta E_y \tag{11-61b}$$

where β is the angle of rotation. Equation (11-61) is written in matrix form as

$$\mathbf{J}' = \begin{pmatrix} E'_x \\ E'_y \end{pmatrix} = \begin{pmatrix} \cos\beta & \sin\beta \\ -\sin\beta & \cos\beta \end{pmatrix}\begin{pmatrix} E_x \\ E_y \end{pmatrix} \tag{11-62}$$

so the Jones matrix for a rotator is

$$\mathbf{J}_{\mathrm{ROT}} = \begin{pmatrix} \cos\beta & \sin\beta \\ -\sin\beta & \cos\beta \end{pmatrix} \tag{11-63}$$

It is interesting to see the effect of rotating a true rotator. According to (11-39), the rotation of a rotator, (11-63), is given by

$$\mathbf{J}_{\mathrm{ROT}}(\theta) = \begin{pmatrix} \cos\theta & -\sin\theta \\ \sin\theta & \cos\theta \end{pmatrix} \begin{pmatrix} \cos\beta & \sin\beta \\ -\sin\beta & \cos\beta \end{pmatrix} \begin{pmatrix} \cos\theta & \sin\theta \\ -\sin\theta & \cos\theta \end{pmatrix} \tag{11-64}$$

Carrying out the matrix multiplication in (11-64) yields

$$\mathbf{J}_{\mathrm{ROT}}(\theta) = \begin{pmatrix} \cos\beta & \sin\beta \\ -\sin\beta & \cos\beta \end{pmatrix} = \mathbf{J}_{\mathrm{ROT}} \tag{11-65}$$

Thus, we have the interesting result that the mechanical rotation of a rotator does not affect the rotation of the polarization ellipse. The polarization ellipse can only be rotated by an amount intrinsic to the rotator, which is the rotation angle β. We conclude that the only way to create a rotation of the polarization ellipse mechanically is to use a half-wave retarder placed in a mechanical rotating mount.

11.4 APPLICATIONS OF THE JONES VECTOR AND JONES MATRICES

We now turn our attention to applying the results of Sections 11.2 and 11.3 to several problems of interest. One of the first problems is to determine the Jones vector for a beam emerging from a rotated linear polarizer and its intensity. The Jones vector of the incident beam is

$$\mathbf{E} = \begin{pmatrix} E_x \\ E_y \end{pmatrix} \tag{11-66}$$

The Jones matrix of a rotated (ideal) linear polarizer was shown in (11-42) to be

$$\mathbf{J}_P(\theta) = \begin{pmatrix} \cos^2\theta & \sin\theta\cos\theta \\ \sin\theta\cos\theta & \sin^2\theta \end{pmatrix} \tag{11-42}$$

While it is straightforward to determine the Jones vector and the intensity of the emerging beam, it is of interest to restrict ourselves to the case where the incident beam is linearly horizontally polarized, so

$$\mathbf{E} = \begin{pmatrix} E_x \\ 0 \end{pmatrix} = E_x \begin{pmatrix} 1 \\ 0 \end{pmatrix} \tag{11-67}$$

Multiplying (11-67) by (11-42) yields

$$\mathbf{E}' = \begin{pmatrix} E_x\cos^2\theta \\ E_x\sin\theta\cos\theta \end{pmatrix} \tag{11-68}$$

We now interpret, formally, the state of polarization of (11-68). We can express (11-68) as a Jones vector for elliptically polarized light, namely,

$$\mathbf{E}' = \begin{pmatrix} ae^{i\delta x} \\ be^{i\delta_y} \end{pmatrix} \tag{11-69}$$

where a and b are real. Equating (11-68) and (11-69), we have

$$E'_x = E_x \cos^2\theta = ae^{i\delta_x} \tag{11-70a}$$

$$E'_y = E_x \cos\theta \sin\theta = be^{i\delta_y} \tag{11-70b}$$

Dividing (11-70b) by (11-70a) then gives

$$\frac{E'_y}{E'_x} = \frac{\sin\theta}{\cos\theta} = \left(\frac{b}{a}\right)e^{i\delta} \tag{11-71}$$

where $\delta = \delta_y - \delta_x$. Finally, taking the real and imaginary parts of (11-71) leads to

$$\frac{\sin\theta}{\cos\theta} = \frac{b}{a}\cos\delta \tag{11-72a}$$

$$0 = \frac{b}{a}\sin\delta \qquad b \neq a \tag{11-72b}$$

We conclude immediately from (11-72b) and $\delta = 0°$, so (11-72a) is

$$\frac{b}{a} = \frac{\sin\theta}{\cos\theta} \tag{11-73}$$

The polarization ellipse corresponding to (11-69) is

$$\frac{x^2}{a^2} + \frac{y^2}{b^2} - \frac{2xy\cos\delta}{ab} = \sin^2\delta \tag{11-74}$$

For $\delta = 0°$, (11-74) reduces to

$$y = \frac{b}{a}x = \frac{\sin\theta}{\cos\theta}x \tag{11-75}$$

Thus, the Jones vector (11-68) describes a beam that is linearly polarized with a slope equal to

$$m = \tan\alpha = \tan\theta \tag{11-76}$$

The intensity of the emerging beam is

$$I' = \mathbf{E}^{\dagger}\mathbf{E}$$

$$= (E_x^* \cos^2\theta \quad E_x^* \sin\theta\cos\theta)\begin{pmatrix} E_x \cos^2\theta \\ E_x \sin\theta\cos\theta \end{pmatrix} \tag{11-77a}$$

so

$$I' = I\cos^2\theta \tag{11-77b}$$

where $I = E_x^* E_x$. Equation (11-77b) is Malus' law. It was discovered by E. Malus while observing unpolarized light through a rotating calcite crystal. We recall he discovered that unpolarized light became partially polarized when it was reflected from a plate of glass. He found the form of (11-77b) solely from geometrical considerations.

This problem can be expanded further by allowing the beam emerging from the polarizer (11-68) to be incident on a linear vertical polarizer. The Jones matrix is found by setting $\theta = 90°$ in (11-42):

$$\mathbf{J}_P(90°) = \begin{pmatrix} 0 & 0 \\ 0 & 1 \end{pmatrix} \tag{11-78}$$

The Jones vector of the beam emerging from the second linear polarizer, found by multiplying (11-68) by (11-78) is

$$E'_p = E_x \cos\theta \sin\theta \begin{pmatrix} 0 \\ 1 \end{pmatrix} \tag{11-79}$$

and the intensity is immediately found to be

$$I' = \frac{I}{8}(1 - \cos 4\theta) \tag{11-80}$$

where $I = E_x^* E_x$. Thus, as the second polarizer is rotated, a null intensity is observed at $\theta = 0°$, $90°$, $180°$, and $270°$. Equation (11-80) is, of course, the same as obtained using the Mueller–Stokes calculus.

We now apply the Jones formalism to several other problems of interest. We recall from Section 6.6 that we used the method of Kent and Lawson to determine the Stokes parameters of an incident elliptically polarized beam. We can also treat the problem in the amplitude domain and apply the Kent–Lawson method to determine the phase and orientation of the beam. The incident beam can be written in the form:

$$E = \begin{pmatrix} \cos\alpha \\ \sin\alpha e^{i\delta} \end{pmatrix} \tag{11-81}$$

The beam (11-81) is incident on a retarder of arbitrary phase ϕ oriented at an angle θ. The phase and orientation of the retarder are now adjusted until circularly polarized light is obtained. We recall that this is detected by allowing the circularly polarized beam to be incident on a rotating linear polarizer directly in front of the detector; circular polarization is obtained when a constant intensity is detected. We can write this condition as

$$\mathbf{J}(-\theta)\mathbf{J}_R\mathbf{J}(\theta)\mathbf{E} = \frac{1}{\sqrt{2}} \begin{pmatrix} 1 \\ i \end{pmatrix} \tag{11-82}$$

The column matrix on the right-hand side of (11-82) is the Jones vector for right circularly polarized light; **E** is given by (11-81) and $\mathbf{J}(\theta)$ and $\mathbf{J}_R$ are the Jones matrices for rotation and a retarder, respectively. Again, it is simplest to find **E** in (11-82) by multiplying through $\mathbf{J}(\theta)$, etc. Carrying out this process, we arrive at

$$\mathbf{E} = \begin{pmatrix} \cos\alpha \\ \sin\alpha e^{i\delta} \end{pmatrix} = \frac{1}{\sqrt{2}} \begin{pmatrix} \cos\theta - ie^{i\phi}\sin\theta \\ \sin\theta + ie^{i\phi}\cos\theta \end{pmatrix} \tag{11-83}$$

Equation (11-83) is easily checked because a retarder, even if rotated, does not affect the total intensity. Thus, it is easy to see that taking the complex transpose of (11-83) for each Jones vector and multiplying by its normal Jones vector gives a unit intensity as required.

We now equate the components in each of the column matrices in (11-83) and divide these equations to find

$$\tan \alpha e^{i\delta} = \frac{\sin\theta + ie^{i\phi}\cos\theta}{\cos\theta - ie^{i\phi}\sin\theta} \tag{11-84}$$

Rationalizing the denominator in (11-84), we easily find that

$$\tan \alpha e^{i\delta} = \frac{-\sin\phi\cos 2\theta + i\cos\phi}{1 + \sin 2\theta\sin\phi} \tag{11-85}$$

Equating real and imaginary parts in (11-85) yields

$$\tan\alpha\cos\delta = \frac{-\sin\phi\cos 2\theta}{1 + \sin 2\theta\sin\phi} \tag{11-86a}$$

$$\tan\alpha\sin\delta = \frac{\cos\phi}{1 + \sin 2\theta\sin\phi} \tag{11-86b}$$

Dividing (11-86a) by (11-86b), we obtain

$$\cot\delta = -\tan\phi\cos 2\theta \tag{11-87}$$

Squaring and adding (11-86a) and (11-86b) then leads to

$$\tan\alpha = \frac{\sqrt{1 - \sin 2\theta\sin\phi}}{\sqrt{1 + \sin 2\theta\sin\phi}} \tag{11-88}$$

Equation (11-88) can be rewritten by using the relations:

$$\cos\alpha = \frac{\sqrt{1 + \sin 2\theta\sin\phi}}{\sqrt{2}} \tag{11-89a}$$

$$\sin\alpha = \frac{\sqrt{1 - \sin 2\theta\sin\phi}}{\sqrt{2}} \tag{11-89b}$$

Squaring (11-89a) and (11-89b) and subtracting, we find that

$$\cos 2\alpha = \sin 2\theta\sin\phi \tag{11-90}$$

We now write (11-87) and (11-90) as the pair

$$\cos 2\alpha = \sin 2\theta\sin\phi \tag{11-91a}$$

$$\cot\delta = -\tan\phi\cos 2\theta \tag{11-91b}$$

Equations (11-91a) and (11-91b) are the Kent–Lawson equations which were derived using the Mueller–Stokes formalism in Section 6.6, equations (6-60a) and (6-60b).

This treatment using the Jones formalism illustrates a very important point. At first glance the use of 2×2 rather than 4×4 matrices might lead us to believe that calculations are simpler with the Jones calculus. The example illustrated by the

Kent–Lawson problem shows that this is not necessarily so. We see that even though it is relatively easy to solve for E_x and E_y there is still a considerable amount of algebra to be carried out. Furthermore, because complex quantities are used, the chance of making a calculating error is increased. Consequently, because the Mueller formalism contains only real quantities it is actually easier to use; invariably, the algebra is considerably less. Experience usually indicates which is the preferable formalism to use in order to solve a problem.

These remarks can be illustrated further by considering another problem. Suppose we wish to create elliptically polarized light of arbitrary orientation and phase (α and δ) from, say, linear horizontally polarized light. This can be done by using only a Babinet–Soleil compensator and adjusting its phase and orientation. For the purpose of comparison we address this problem first by using the Mueller formalism and then by using the Jones formalism. The problem is simply stated mathematically by

$$M_R(2\theta)\begin{pmatrix} 1 \\ 1 \\ 0 \\ 0 \end{pmatrix} = \begin{pmatrix} 1 \\ \cos 2\alpha \\ \sin 2\alpha \cos\delta \\ \sin 2\alpha \sin\delta \end{pmatrix} \tag{11-92}$$

where $M_R(2\theta)$ is the Mueller matrix of a rotated retarder (5-57) in Section 5.5. Carrying out the matrix multiplication in (11-92) and equating matrix elements, we have

$$\cos^2 2\theta + \cos\phi \sin^2 2\theta = \cos 2\alpha \tag{11-93a}$$

$$(1 - \cos\phi)\sin 2\theta \cos 2\theta = \sin 2\alpha \cos\delta \tag{11-93b}$$

$$\sin\phi \sin 2\theta = \sin 2\alpha \sin\delta \tag{11-93c}$$

We now solve (11-93) for ϕ and θ. Equation (11-93a) can be rewritten immediately as

$$(1 - \cos\phi)\sin^2 2\theta = 2\sin^2\alpha \tag{11-94}$$

Dividing (11-93b) by (11-94) then gives

$$\cot 2\theta = \cot\alpha \cos\delta \tag{11-95}$$

Next, (11-93c) is divided by (11-93b) to obtain

$$\cot\frac{\phi}{2} = \cos 2\theta \tan\delta \tag{11-96}$$

where we have used the trigonometric half-angle formulas for ϕ. The $\cos 2\theta$ term can be expressed in terms of α and δ. From (11-95) we see that

$$\cos 2\theta = \frac{\cos\alpha\cos\delta}{\sqrt{1 - \cos^2\alpha \sin^2\delta}} \tag{11-97a}$$

$$\sin 2\theta = \frac{\sin\alpha}{\sqrt{1 - \cos^2\alpha \sin^2\delta}} \tag{11-97b}$$

We now substitute (11-97a) into (11-96) and write the result along with (11-95) as the pair:

$$\cot 2\theta = \cot\alpha\cos\delta \tag{11-98a}$$

$$\cot\frac{\phi}{2} = \frac{\cos\alpha\sin\delta}{\sqrt{1-\cos^2\alpha\sin^2\delta}} \tag{11-98b}$$

We can provide two simple numerical checks on (11-98). We know that if we start with linear horizontally polarized light and wish to rotate the linearly polarized light to +45°, this can be done by rotating a half-wave retarder through +22.5°. We can show this formally by writing

$$\begin{pmatrix}1\\0\\1\\0\end{pmatrix} = \begin{pmatrix}1\\ \cos 2\alpha\\ \sin 2\alpha\cos\delta\\ \sin 2\alpha\sin\delta\end{pmatrix} \tag{11-99}$$

We see that linear +45° polarized light corresponds to $2\alpha = 90^\circ$ and $\delta = 0^\circ$ in (11-99). Substituting these conditions into (11-98a) and (11-98b) yields

$$\tan 2\theta = 1 \tag{11-100a}$$

$$\tan\frac{\phi}{2} = \infty \tag{11-100b}$$

from which we immediately find that $\theta = 22.5^\circ$ and $\phi = 180^\circ$ as required.

The other check on (11-98) is to consider the conditions to create right circularly polarized light from linear horizontally polarized light. We know that a quarter-wave retarder rotated through 45° will generate right circularly polarized light. Therefore, we again write

$$\begin{pmatrix}1\\0\\0\\1\end{pmatrix} = \begin{pmatrix}1\\ \cos 2\alpha\\ \sin 2\alpha\cos\delta\\ \sin 2\alpha\sin\delta\end{pmatrix} \tag{11-101}$$

which is satisfied for $2\alpha = 90^\circ$ and $\delta = 90^\circ$. Substituting these conditions into (11-98) gives

$$\tan 2\theta = \infty \tag{11-102a}$$

$$\tan\frac{\phi}{2} = 1 \tag{11-102b}$$

from which we see that we must set the Babinet–Soleil compensator to $\theta = 45^\circ$ and $\phi = 90^\circ$, which is exactly what we would expect.

We now consider the same problem of the rotated Babinet–Soleil compensator using the Jones formalism. The mathematical statement for this problem is written as

$$\mathbf{J}_R(\theta)\begin{pmatrix}1\\0\end{pmatrix} = \begin{pmatrix}\cos\alpha\\ \sin\alpha e^{i\delta}\end{pmatrix} \tag{11-103}$$

where $\mathbf{J}_R(\theta)$ is given by

$$\mathbf{J}_R(\theta) = \begin{pmatrix} \cos\theta & -\sin\theta \\ \sin\theta & \cos\theta \end{pmatrix} \begin{pmatrix} e^{i\phi/2} & 0 \\ 0 & e^{-i\phi/2} \end{pmatrix} \begin{pmatrix} \cos\theta & \sin\theta \\ -\sin\theta & \cos\theta \end{pmatrix} \tag{11-104}$$

Carrying out the matrix multiplication and equating terms, we find

$$e^{i\phi/2}\cos^2\theta + e^{-i\phi/2}\sin^2\theta = \cos\alpha \tag{11-105a}$$

$$(e^{i\phi/2} - e^{-i\phi/2})\sin\theta\cos\theta = \sin\alpha e^{i\delta} \tag{11-105b}$$

We first rewrite (11-105b) as

$$i\sin\frac{\phi}{2}\sin 2\theta = \sin\alpha e^{i\delta} \tag{11-106}$$

Next, we divide (11-105a) by (11-106), group terms, then equate the real and imaginary terms and find

$$\cot 2\theta = \cot\alpha\cos\delta \tag{11-107a}$$

$$\cot\frac{\phi}{2} = \sin 2\theta\cot\alpha\sin\delta \tag{11-107b}$$

From (11-97b) we see that sin 2θ can be replaced, so (11-107) can be written as the pair

$$\cot 2\theta = \cot\alpha\cos\delta \tag{11-108a}$$

$$\cot\frac{\phi}{2} = \frac{\cos\alpha\sin\delta}{\sqrt{1-\cos^2\alpha\sin^2\delta}} \tag{11-108b}$$

Equation (11-108) is identical to the result (11-98) obtained using the Mueller formalism. The reader will see that a considerable amount of increased effort is required to obtain (11-108) using the Jones formalism.

One of the fundamental problems continuously encountered in the field of polarized light is to determine the orientation and ellipticity of an incident (polarized) beam. This can be done by analyzing the beam using a quarter-wave retarder and a linear polarizer, where both elements are capable of being rotated through the angles α and β, respectively. Thus, the Jones matrix for a rotated quarter-wave retarder and a rotated ideal polarizer in sequence using Equations (11-52) and (11-42), is

$$\mathbf{J} = \mathbf{J}_P(\beta)\mathbf{J}_R\left(\frac{\lambda}{4}, \alpha\right) \tag{11-109a}$$

where

$$\mathbf{J}_R\left(\frac{\lambda}{4}, \alpha\right) = \frac{1}{\sqrt{2}}\begin{pmatrix} 1 + i\cos 2\alpha & i\sin 2\alpha \\ i\sin 2\alpha & 1 - i\cos 2\alpha \end{pmatrix} \tag{11-109b}$$

and

$$\mathbf{J}_P(\beta) = \begin{pmatrix} \cos^2\beta & \cos\beta\sin\beta \\ \cos\beta\sin\beta & \sin^2\beta \end{pmatrix} \tag{11-109c}$$

The matrix product (11-109a) is then written out as

$$\mathbf{J} = \frac{1}{\sqrt{2}}\begin{pmatrix} \cos^2\beta & \cos\beta\sin\beta \\ \cos\beta\sin\beta & \sin^2\beta \end{pmatrix}\begin{pmatrix} 1+i\cos 2\alpha & i\sin 2\alpha \\ i\sin 2\alpha & 1-i\cos 2\alpha \end{pmatrix} \tag{11-110}$$

The product in (11-110) is a matrix from which it is clear there is extinction at specific values of α and β. These extinction angles determine the ellipticity and orientation of the incident beam. Rather than giving a general solution of this problem, we consider a specific example.

Suppose we find that extinction occurs at $\alpha = 45^\circ$ and $\beta = 30^\circ$. Then, (11-109b) and (11-109c) become

$$\mathbf{J}_R\left(\frac{\lambda}{4}, 45^\circ\right) = \frac{1}{\sqrt{2}}\begin{pmatrix} 1 & i \\ i & 1 \end{pmatrix} \tag{11-111a}$$

and

$$\mathbf{J}_P(30^\circ) = \frac{1}{4}\begin{pmatrix} 3 & \sqrt{3} \\ \sqrt{3} & 1 \end{pmatrix} \tag{11-111b}$$

Multiplying (11-111a) and (11-111b) according to (11-109a) yields

$$\mathbf{J} = \frac{1}{4\sqrt{2}}\begin{pmatrix} 3+i\sqrt{3} & i3+\sqrt{3} \\ \sqrt{3}+i & i\sqrt{3}+1 \end{pmatrix} \tag{11-112}$$

Equation (11-112) describes the propagation of the incident beam first through the rotated quarter-wave retarder followed by a linear polarizer. The purpose of the rotated quarter-wave retarder is to transform the incident elliptically polarized beam to linearly polarized light. The linear polarizer is then rotated until a null intensity, i.e., extinction, occurs. This, incidentally, is the fundamental basis of ellipsometry. In order to have a null intensity we must have from (11-112)

$$\frac{1}{4\sqrt{2}}\begin{pmatrix} 3+i\sqrt{3} & i3+\sqrt{3} \\ \sqrt{3}+i & i\sqrt{3}+1 \end{pmatrix}\begin{pmatrix} E_x \\ E_y \end{pmatrix} = \begin{pmatrix} 0 \\ 0 \end{pmatrix} \tag{11-113}$$

Writing (11-113) out in component form gives

$$\left(3+i\sqrt{3}\right)E_x + \left(i3+\sqrt{3}\right)E_y = 0 \tag{11-114a}$$

$$\left(\sqrt{3}+i\right)E_x + \left(i\sqrt{3}+1\right)E_y = 0 \tag{11-114b}$$

We see that (11-114a) differs from (11-114b) only by a factor of $\sqrt{3}$, so the equations are identical. We now solve (11-114b) for E_y/E_x and find that

$$\frac{E_y}{E_x} = -\frac{\sqrt{3}}{2} + i\left(\frac{1}{2}\right) \tag{11-115}$$

Now E_y/E_x can be expressed as

$$\frac{E_y}{E_x} = \left(\frac{a}{b}\right)e^{i\delta} \tag{11-116}$$

where a/b is real. Equating the real and imaginary parts in (11-115) and (11-116), we have

$$\frac{a}{b}\cos\delta = \frac{-\sqrt{3}}{2} \tag{11-117a}$$

$$\frac{a}{b}\sin\delta = \frac{1}{2} \tag{11-117b}$$

Squaring (11-117a) and (11-117b) and adding gives

$$\frac{a}{b} = \pm 1 \tag{11-118a}$$

Similarly, dividing (11-117b) by (11-117a) yields

$$\delta = \tan^{-1}\left(\frac{-1}{\sqrt{3}}\right) = -30^\circ \tag{11-118b}$$

Thus, the orthogonal amplitudes of the incident beam are equal, and the phase shift between the orthogonal components is -30°.

The Jones vector of the original beam is then

$$\mathbf{E}' = \begin{pmatrix} E_x \\ E_y \end{pmatrix} = \frac{1}{\sqrt{2}}\begin{pmatrix} a \\ be^{i\delta} \end{pmatrix} = \frac{1}{\sqrt{2}}\begin{pmatrix} 1 \\ \pm e^{i30^\circ} \end{pmatrix} \tag{11-119}$$

where we have introduced a factor of $1/\sqrt{2}$ so that (11-119) is normalized. In terms of the polarization ellipse, (11-119) gives

$$x^2 - \sqrt{3}xy + y^2 = \left(\frac{1}{2\sqrt{2}}\right)^2 \tag{11-120}$$

which is the equation of a rotated ellipse, Equation (11-120) can be rotated to a nonrotated (standard) ellipse by using the well-known equations of analytical geometry. Thus, the left-hand side of (11-120) is of the form:

$$Ax^2 + 2Bxy + Cy^2 \tag{11-121}$$

By using the well-known rotation equations, (11-121) can be transformed to

$$a_1u^2 + 2b_1uv + c_1v^2 \tag{11-122a}$$

where

$$a_1 = A\cos^2\phi + 2B\sin\phi\cos\phi + C\sin^2\phi \tag{11-122b}$$

$$2b_1 = 2B\cos\phi - (A - C)\sin 2\phi \tag{11-122c}$$

$$c_1 = A\sin^2\phi - 2B\sin\phi\cos\phi + C\cos^2\phi \tag{11-122d}$$

The "cross" term $2b_1$ will vanish, and the standard form of the ellipse is obtained for

$$\cot 2\phi = \frac{A - C}{2B} \tag{11-123}$$

From (11-120) we see that $A = C = 1$ and $B = -\sqrt{3}/2$. Thus, (11-123) shows that the angle of rotation ϕ is -45°. Equations (11-122b) and (11-122d) then reduce to

$$a_1 = \frac{2 - \sqrt{3}}{2} \tag{11-124a}$$

$$c_1 = \frac{2 + \sqrt{3}}{2} \tag{11-124b}$$

The ellipticity angle is seen to be

$$\tan \chi = \sqrt{\frac{2 - \sqrt{3}}{2 + \sqrt{3}}} = \sqrt{\frac{1 - \sqrt{3}/2}{1 + \sqrt{3}/2}} \tag{11-125a}$$

Equation (11-125a) can be reduced further by noting that $\cos 30^\circ = \sqrt{3}/2$ and using the half-angle formulas:

$$\tan \chi = \sqrt{\frac{1 - \cos 30^\circ}{1 + \cos 30^\circ}} = \sqrt{\frac{2\sin^2 15^\circ}{2\cos^2 15^\circ}} \tag{11-125b}$$

so $\chi = 15^\circ$. Thus, (11-120) describes an ellipse that is rotated 45° from the x axis. The axial length is $L_2/L_1 = \sqrt{c_1/a_1} = \sqrt{(2+\sqrt{3})/(2-\sqrt{3})} = 3.7321$.

The last problem that we consider is to show that a linear polarizer can be used to measure the major and minor axes of the polarization ellipse in standard form, i.e., the major and minor axes of the ellipse are along the x and y axes, respectively. This is described by setting $\delta = 90^\circ$ in (11-81), so

$$\mathbf{E} = \begin{pmatrix} \cos\alpha \\ i\sin\alpha \end{pmatrix} = \begin{pmatrix} a \\ ib \end{pmatrix} \tag{11-126}$$

The amplitude equations corresponding to (11-126) are

$$E_x = \cos\alpha \cos\omega t \tag{11-127a}$$

$$E_y = \sin\alpha \sin\omega t \tag{11-127b}$$

We can eliminate ωt between (11-127a) and (11-127b), so

$$\frac{E_x^2}{a^2} + \frac{E_y}{b^2} = 1 \tag{11-128}$$

where $a = \cos\alpha$ and $b = \sin\alpha$. Thus, a and b are the lengths of the semimajor and semiminor axes of the polarization ellipse (11-128). We now return to the measurement of a and b.

The Jones matrix of a rotated polarizer is

$$\mathbf{J}_p(\theta) = \begin{pmatrix} \cos^2\theta & \sin\theta\cos\theta \\ \sin\theta\cos\theta & \sin^2\theta \end{pmatrix} \tag{11-42}$$

so the Jones vector of the emerging beam is (multiplying (11-126) by (11-42))

$$\mathbf{E}' = \begin{pmatrix} a\cos^2\theta + ib\sin\theta\cos\theta \\ a\sin\theta\cos\theta + ib\sin^2\theta \end{pmatrix} \tag{11-129}$$

The intensity corresponding to (11-129) is readily seen to be

$$I(\theta) = a^2 \cos^2\theta + b^2 \sin^2\theta \tag{11-130}$$

where the prime on the intensity has been dropped. Setting $\theta = 0^\circ$ and 90°, respectively, (11-130) gives

$$I(0^\circ) = a^2 = \cos^2\alpha \tag{11-131a}$$

$$I(90^\circ) = b^2 = \sin^2\alpha \tag{11-131b}$$

Thus, by measuring the orthogonal intensities, the square of the major and minor axes can be found. It is usually convenient to express (11-131) simply as the ratio:

$$\frac{a}{b} = \sqrt{\frac{I(0^\circ)}{I(90^\circ)}} \tag{11-132}$$

Numerous problems using the Jones and Mueller matrices can be found in the references at the end of this chapter. In particular, Gerrard and Burch treat a number of interesting problems.

11.5 JONES MATRICES FOR HOMOGENEOUS ELLIPTICAL POLARIZERS AND RETARDERS

We now end this chapter with a discussion of a topic of importance. We have described polarizers, retarders, circular polarizers, etc., in terms of the Mueller and Jones matrices. In particular, we have pointed out that a linear polarizer and a circular polarizer derive their names from the fact that, regardless of the polarization state of the incident beam, the polarization state of the emerging beam is always linearly and circularly polarized, respectively. Let us look at this behavior more closely. The Jones matrix of a rotated linear polarizer is given by

$$J_P(\theta) = \begin{pmatrix} \cos^2\theta & \sin\theta\cos\theta \\ \sin\theta\cos\theta & \sin^2\theta \end{pmatrix} \tag{11-42}$$

The incident beam is represented by

$$\mathbf{E} = \begin{pmatrix} E_x \\ E_y \end{pmatrix} \tag{11-66}$$

Multiplying (11-66) by (11-42) yields

$$\mathbf{E}' = (E_x\cos\theta + E_y\sin\theta)\begin{pmatrix} \cos\theta \\ \sin\theta \end{pmatrix} \tag{11-133}$$

which is the Jones matrix for linearly polarized light. We note that if, say, $\theta = 0^\circ$, (11-42) reduces to

$$\mathbf{J}_{PH} = \begin{pmatrix} 1 & 0 \\ 0 & 0 \end{pmatrix} \tag{11-37}$$

We observe that if the incident beam is elliptically polarized, (11-66), then multiplying (11-66) by (11-37) gives

$$\mathbf{E}' = E_x\begin{pmatrix} 1 \\ 0 \end{pmatrix} \tag{11-134}$$

Equation (11-134) shows that we obtain linearly polarized light. This can be written in normalized form as

$$\mathbf{E}' = \begin{pmatrix} 1 \\ 0 \end{pmatrix} \tag{11-135}$$

If we now try to transmit the orthogonal state, namely, linear vertically polarized light:

$$\mathbf{E} = \begin{pmatrix} 0 \\ 1 \end{pmatrix} \tag{11-136}$$

we find from (11-37) and (11-136) that

$$\mathbf{E}' = \begin{pmatrix} 0 \\ 0 \end{pmatrix} \tag{11-137}$$

so there is no emerging beam. This behavior of the polarizer (11-37) can be summarized by writing

$$\mathbf{E}' = \begin{pmatrix} 1 & 0 \\ 0 & 0 \end{pmatrix}\begin{pmatrix} 1 \\ 0 \end{pmatrix} = 1\begin{pmatrix} 1 \\ 0 \end{pmatrix} \tag{11-138a}$$

$$\mathbf{E}' = \begin{pmatrix} 1 & 0 \\ 0 & 0 \end{pmatrix}\begin{pmatrix} 0 \\ 1 \end{pmatrix} = 0\begin{pmatrix} 0 \\ 1 \end{pmatrix} \tag{11-138b}$$

Written in this way we see that the problem of transmission by a polarizer can be thought of in terms of an eigenvector/eigenvalue problem. Thus, we see that the eigenvectors of the 2×2 Jones matrix (11-37) are

$$\mathbf{E}_1 = \begin{pmatrix} 1 \\ 0 \end{pmatrix} \qquad \mathbf{E}_2 = \begin{pmatrix} 0 \\ 1 \end{pmatrix} \tag{11-139}$$

and the corresponding eigenvalues are 1 and 0. A linear polarizer has the property that it transmits one of its eigenvectors perfectly and rejects the orthogonal eigenvector completely.

Let us now consider the same problem using a circular polarizer. We have seen that a circular polarizer can be constructed by using a linear polarizer set at $+45^\circ$ followed by a quarter-wave retarder. The Jones matrix is

$$\mathbf{J} = \frac{1}{2}\begin{pmatrix} 1 & 1 \\ i & i \end{pmatrix} \tag{11-140}$$

We now multiply (11-66) by (11-140) and find that

$$\mathbf{E}' = \frac{E_x + E_y}{2}\begin{pmatrix} 1 \\ i \end{pmatrix} = \frac{1}{\sqrt{2}}\begin{pmatrix} 1 \\ i \end{pmatrix} \tag{11-141}$$

in its normalized form. Thus, again, regardless of the polarization state of the incident beam, the emerging beam is always right circularly polarized. In the case of a linear polarizer the transmission of the orthogonal polarization state was completely blocked by the linear polarizer. Let us now see what happens when we try to transmit the orthogonal polarization state, namely, left circularly polarized light through the circular polarizer (11-140). The Jones vector of the orthogonal state (which is left circularly polarized) is

$$\mathbf{E} = \frac{1}{\sqrt{2}}\begin{pmatrix} 1 \\ -i \end{pmatrix} \tag{11-16}$$

Multiplying (11-16) by (11-140), we find that the Jones vector of the emerging beam is

$$\mathbf{E}' = \frac{1-i}{2\sqrt{2}}\begin{pmatrix} 1 \\ i \end{pmatrix} \tag{11-142}$$

The emerging beam is right circularly polarized. The circular polarizer (11-140) does not block the left circularly polarized beam! Equation (11-142), therefore, is not an eigenvector of (11-140). The reason for this seemingly anomalous behavior, which is unlike the linear polarizer, is that the circular polarizer is constructed from a linear $+45^\circ$ polarizer and a quarter-wave retarder. That is, it is not a homogeneous polarizing element. The eigenvectors of (11-140) are easily shown actually to be

$$\mathbf{E}_1 = \frac{1}{\sqrt{2}}\begin{pmatrix} 1 \\ -1 \end{pmatrix} \qquad \mathbf{E}_2 = \frac{1}{\sqrt{2}}\begin{pmatrix} 1 \\ i \end{pmatrix} \tag{11-143}$$

which are linear -45° and right circularly polarized light, respectively; the corresponding eigenvalues are 0 and $1+i$. Consequently, (11-140) does not describe a true "circular" polarizer. We would expect that a true circular polarizer would behave in a manner identical to that of the linear polarizer. Namely, only one state of polarized light always emerges and this corresponds to one of the two eigenvectors. Furthermore, the other eigenvector is orthogonal to the transmitted eigenvector, but it is completely blocked by the polarizing element, that is, the eigenvalues are 1 and 0. A polarizing element that exhibits these two properties simultaneously is called homogeneous. We now wish to construct the homogeneous polarizing elements not only for circularly polarized light but also for the more general state, elliptically polarized light.

The key to solving this problem is to recall our earlier work on raising the matrix to the *m*th power. There we saw that the Mueller matrix could be diagonalized and that it was possible to represent the Mueller matrix in terms of its eigenvalues, eigenvectors, and another matrix, which we called the modal matrix. Let us now consider this problem again, now using the Jones vector.

Let us represent the Jones vector of a beam by

$$\mathbf{E}_1 = \begin{pmatrix} p \\ q \end{pmatrix} \tag{11-144a}$$

Then, the orthogonal state is given by

$$\mathbf{E}_2 = \begin{pmatrix} -q^* \\ p^* \end{pmatrix} \tag{11-144b}$$

The reader can easily prove that Eqs. (11-144) are orthogonal by applying the orthogonality condition:

$$\mathbf{E}_1^\dagger \mathbf{E}_2 = \mathbf{E}_2^\dagger \mathbf{E}_1 = 0 \tag{11-145}$$

where † represents complex transpose. We also know that the corresponding eigenvalues are λ_1 and λ_2. Earlier, we saw that we could construct a new matrix K, which we called the modal matrix, from the eigenvectors and written

$$K = \begin{pmatrix} p & -q^* \\ q & p^* \end{pmatrix} \tag{11-146a}$$

The inverse modal matrix K^{-1} is easily found to be

$$K^{-1} = \frac{1}{pp^* + qq^*}\begin{pmatrix} p^* & q^* \\ -q & p \end{pmatrix} \tag{11-146b}$$

It is easily shown that $K\,K^{-1} = K^{-1}K = I$ if we normalize $pp^* + qq^*$ to 1.

Now, we saw earlier that there is a unique relationship between a matrix Φ ($\equiv \mathbf{J}$) to its eigenvalues and eigenvectors expressed by

$$\Phi K = K\Lambda \tag{11-147}$$

where Λ is the diagonal eigenvalue matrix:

$$\Lambda = \begin{pmatrix} \lambda_1 & 0 \\ 0 & \lambda_2 \end{pmatrix} \tag{11-148}$$

We now solve (11-147) for Φ to obtain

$$\Phi = K\Lambda K^{-1} \tag{11-149}$$

Equation (11-149) is a rather remarkable result because it shows that a matrix Φ can be constructed completely from its eigenvectors and eigenvalues. We can write (11-149) (Φ is replaced by $\mathbf{J}$) so that

$$\mathbf{J} = \frac{1}{pp^* + qq^*}\begin{pmatrix} p & -q^* \\ q & p \end{pmatrix}\begin{pmatrix} \lambda_1 & 0 \\ 0 & \lambda_2 \end{pmatrix}\begin{pmatrix} p^* & q^* \\ -q & p \end{pmatrix} \tag{11-150}$$

Carrying out the multiplication yields

$$\mathbf{J} = \frac{1}{pp^* + qq^*}\begin{pmatrix} \lambda_1 pp^* + \lambda_2 qq^* & (\lambda_1 - \lambda_2)pq^* \\ (\lambda_1 - \lambda_2)qp^* & \lambda_1 qq^* + \lambda_2 pp^* \end{pmatrix} \tag{11-151}$$

To check (11-150) and (11-151), let us consider linearly polarized light. We know its eigenvectors are

$$\mathbf{E}_1 = \begin{pmatrix} 0 \\ 1 \end{pmatrix} \qquad \mathbf{E}_2 = \begin{pmatrix} 0 \\ 1 \end{pmatrix} \tag{11-139}$$

and its eigenvalues are 1 and 0. The modal matrix K is then

$$K = \begin{pmatrix} 1 & 0 \\ 0 & 1 \end{pmatrix} \tag{11-152a}$$

The inverse modal matrix K^{-1} is easily found to be

$$K^{-1} = \begin{pmatrix} 1 & 0 \\ 0 & 1 \end{pmatrix} \tag{11-152b}$$

From (11-150) we then can write

$$\mathbf{J} = \begin{pmatrix} 1 & 0 \\ 0 & 1 \end{pmatrix}\begin{pmatrix} 1 & 0 \\ 0 & 0 \end{pmatrix}\begin{pmatrix} 1 & 0 \\ 0 & 1 \end{pmatrix} = \begin{pmatrix} 1 & 0 \\ 0 & 0 \end{pmatrix} \tag{11-153}$$

which is identical to the Jones matrix of a linear horizontal polarizer; it is a homogeneous polarizing element.

Let us now construct a homogeneous right circular polarizer. The orthogonal eigenvectors are

$$\mathbf{E}_1 = \frac{1}{\sqrt{2}}\begin{pmatrix} 1 \\ i \end{pmatrix} \qquad \mathbf{E}_2 = \frac{1}{\sqrt{2}}\begin{pmatrix} i \\ 1 \end{pmatrix} \tag{11-154}$$

Thus, from (11-150) and (11-154) the Jones matrix for a right circular homogeneous polarizer will be ($p = 1$, $q = i$)

$$\mathbf{J} = \frac{1}{2}\begin{pmatrix} 1 & i \\ i & 1 \end{pmatrix}\begin{pmatrix} 1 & 0 \\ 0 & 0 \end{pmatrix}\begin{pmatrix} 1 & -i \\ -i & 1 \end{pmatrix} = \frac{1}{2}\begin{pmatrix} 1 & -i \\ i & 1 \end{pmatrix} \tag{11-155}$$

We can check to see if (11-155) is the Jones matrix for a homogeneous right circular polarizer. First, we consider an elliptically polarized beam represented by

$$\mathbf{E} = \begin{pmatrix} E_x \\ E_y \end{pmatrix} \tag{11-156}$$

We multiply (11-156) by (11-155), and we find that

$$\mathbf{E} = \frac{E_x - iE_y}{2}\begin{pmatrix} 1 \\ i \end{pmatrix} \tag{11-157}$$

so only right circularly polarized light emerges, as required. Next, we take the product of (11-155) and the eigenvector for right circularly polarized light and the eigenvector for left circularly polarized light, respectively:

$$\mathbf{E}' = \frac{1}{2}\begin{pmatrix} 1 & -i \\ i & 1 \end{pmatrix}\frac{1}{\sqrt{2}}\begin{pmatrix} 1 \\ i \end{pmatrix} = (1)\frac{1}{\sqrt{2}}\begin{pmatrix} 1 \\ i \end{pmatrix} \tag{11-158a}$$

$$\mathbf{E}' = \frac{1}{2}\begin{pmatrix} 1 & -i \\ i & 1 \end{pmatrix}\frac{1}{\sqrt{2}}\begin{pmatrix} 1 \\ -i \end{pmatrix} = (0)\frac{1}{\sqrt{2}}\begin{pmatrix} 1 \\ -i \end{pmatrix} \tag{11-158b}$$

which is exactly what we require for a homogeneous right circular polarizer.

We can now turn our attention to constructing a homogeneous elliptical polarizer. For convenience, we describe this by the Jones vector:

$$E_1 = \begin{pmatrix} p \\ q \end{pmatrix} \tag{11-159a}$$

and describe its orthogonal vector (eigenvector) by

$$E_2 = \begin{pmatrix} -q^* \\ p^* \end{pmatrix} \tag{11-159b}$$

From (11-151) we then have immediately, by setting $\lambda_1 = 1$ and $\lambda_2 = 0$,

$$\mathbf{J} = \begin{pmatrix} pp^* & pq^* \\ qp^* & qq^* \end{pmatrix} \tag{11-160}$$

There are two other ways to represent an elliptical polarizer. The first is to write the incident Jones vector in the form:

$$\mathbf{E}_1 = \begin{pmatrix} a_x e^{i\phi_x} \\ a_y e^{i\phi_y} \end{pmatrix} \tag{11-161}$$

The orthogonal state and eigenvalues are constructed as shown earlier in this section. Then, we easily see from (11-150) and (11-161) that the Jones matrix for an elliptical polarizer is

$$\mathbf{J} = \begin{pmatrix} a_x^2 & a_x a_y e^{-i\phi} \\ a_x a_y e^{+i\phi} & a_y^2 \end{pmatrix} \qquad \phi = \phi_y - \phi_x \tag{11-162}$$

The other representation of an elliptical polarizer can be obtained by using the Jones vector:

$$\mathbf{E}_1 = \begin{pmatrix} \cos\alpha \\ \sin\alpha e^{i\delta} \end{pmatrix} \tag{11-163}$$

as the eigenvector. Again, the orthogonal state and eigenvalues are constructed as shown earlier. Then, from (11-150) we see that the Jones matrix for the elliptical polarizer is

$$\mathbf{J} = \begin{pmatrix} \cos\alpha & -\sin\alpha e^{-i\delta} \\ \sin\alpha e^{i\delta} & \cos\alpha \end{pmatrix} \begin{pmatrix} 1 & 0 \\ 0 & 0 \end{pmatrix} \begin{pmatrix} \cos\alpha & \sin\alpha e^{-i\delta} \\ -\sin\alpha e^{i\delta} & \cos\alpha \end{pmatrix} \tag{11-164a}$$

or

$$\mathbf{J} = \begin{pmatrix} \cos^2\alpha & \sin\alpha\cos\alpha e^{-i\delta} \\ \sin\alpha\cos\alpha e^{+i\delta} & \sin^2\alpha \end{pmatrix} \tag{11-164b}$$

The form expressed by (11-164b) enables us to determine the Jones matrix for any type of elliptical polarizer including, for example, a linear polarizer and a circular polarizer. For a linear horizontal polarizer $\alpha = 0^\circ$ and (11-164b) reduces to

$$\mathbf{J} = \begin{pmatrix} 1 & 0 \\ 0 & 0 \end{pmatrix} \tag{11-165}$$

which is, indeed, the Jones matrix for a linear horizontal polarizer given earlier by (11-37). Similarly, for $\alpha = 45^\circ$ and $\delta = 90^\circ$, which are the conditions for right circularly polarized light, we see that (11-164b) reduces to

$$\mathbf{J} = \frac{1}{2}\begin{pmatrix} 1 & -i \\ i & 1 \end{pmatrix} \tag{11-166}$$

which is the Jones matrix for a homogeneous right circular polarizer, in agreement with (11-155).

While we have considered only ideal polarizers, it is simple to extend this analysis to the general case where the polarizer is described by

$$\mathbf{J} = \begin{pmatrix} p_x & 0 \\ 0 & p_y \end{pmatrix} \tag{11-167}$$

where p_x, p_y range between 0 and 1.
For an ideal linear horizontal polarizer $p_x = 1$ and $p_y = 0$. The terms used to describe nonideal behavior of polarizers are diattenuation or dichroism. We shall use the preferable diattenuation. For an explanation of the term dichroism and the origin of its usage, see Shurcliff. Equation (11-167) describes a diattenuator. We see immediately that the eigenvalues of a diattenuator are p_x and p_y. Therefore, the Jones matrix for a nonideal (diattenuating) elliptical polarizer is

$$\mathbf{J} = \begin{pmatrix} \cos\alpha & -\sin\alpha e^{-i\delta} \\ \sin\alpha e^{i\delta} & \cos\alpha \end{pmatrix} \begin{pmatrix} p_x & 0 \\ 0 & p_y \end{pmatrix} \begin{pmatrix} \cos\alpha & \sin\alpha e^{-i\delta} \\ -\sin\alpha e^{i\delta} & \cos\alpha \end{pmatrix} \tag{11-168a}$$

or

$$\mathbf{J} = \begin{pmatrix} p_x \cos^2\alpha + p_y \sin^2\alpha & (p_x - p_y)\sin\alpha\cos\alpha e^{-i\delta} \\ (p_x - p_y)\sin\alpha\cos\alpha e^{+i\delta} & p_x \sin^2\alpha + p_y \cos^2\alpha \end{pmatrix} \tag{11-168b}$$

Equation (11-168b) enables us to describe any type of elliptical polarizer and is the most useful of all representations of homogeneous elliptical polarizers.

There is, of course, the other importance type of polarizing element, which is the retarder. We now treat the problem of representing homogeneous linear, circular, and elliptical retarders. We begin this discussion by recalling that the Jones matrix for a retarder was given by

$$\mathbf{J} = \begin{pmatrix} e^{+i\phi/2} & 0 \\ 0 & e^{-i\phi/2} \end{pmatrix} \tag{11-169}$$

We now determine the eigenvectors and the eigenvalues of (11-169). We do this by forming the familiar eigenvector/eigenvalue equation:

$$\begin{pmatrix} e^{+i\phi/2} - \lambda & 0 \\ 0 & e^{-i\phi/2} - \lambda \end{pmatrix} \begin{pmatrix} p \\ q \end{pmatrix} = 0 \tag{11-170}$$

The eigenvalues are

$$\lambda_1 = e^{+i\phi/2} \qquad \lambda_2 = e^{-i\phi/2} \tag{11-171a}$$

and the corresponding eigenvectors are

$$\mathbf{E}_1 = \begin{pmatrix} 1 \\ 0 \end{pmatrix} \qquad \mathbf{E}_2 = \begin{pmatrix} 0 \\ 1 \end{pmatrix} \tag{11-171b}$$

which are the Jones vectors for linear horizontally and linear vertically polarized light, respectively. Thus, the respective eigenvector/eigenvalue equations are

$$\begin{pmatrix} e^{+i\phi/2} & 0 \\ 0 & e^{-i\phi/2} \end{pmatrix}\begin{pmatrix} 1 \\ 0 \end{pmatrix} = e^{+i\phi/2}\begin{pmatrix} 1 \\ 0 \end{pmatrix} \tag{11-172a}$$

and

$$\begin{pmatrix} e^{+i\phi/2} & 0 \\ 0 & e^{-i\phi/2} \end{pmatrix}\begin{pmatrix} 0 \\ 1 \end{pmatrix} = e^{-i\phi/2}\begin{pmatrix} 0 \\ 1 \end{pmatrix} \tag{11-172b}$$

Because the eigenvectors of (11-169) are orthogonal states of linear polarized light, the retarder is called a linear retarder. We can now immediately find the Jones matrix for an elliptical retarder. For an elliptical retarder we must obtain the same eigenvalues given by (11-171a). If we use the Jones vector for elliptically polarized light given by

$$\mathbf{E}_1 = \begin{pmatrix} \cos\alpha \\ \sin\alpha e^{i\delta} \end{pmatrix} \tag{11-173}$$

then the Jones matrix for an elliptical retarder must be

$$\mathbf{J} = \begin{pmatrix} \cos\alpha & -\sin\alpha e^{-i\delta} \\ \sin\alpha e^{i\delta} & \cos\alpha \end{pmatrix}\begin{pmatrix} e^{+i\phi/2} & 0 \\ 0 & e^{-i\phi/2} \end{pmatrix}\begin{pmatrix} \cos\alpha & \sin\alpha e^{-i\delta} \\ -\sin\alpha e^{i\delta} & \cos\alpha \end{pmatrix} \tag{11-174a}$$

or

$$\mathbf{J} = \begin{pmatrix} e^{i\phi/2}\cos^2\alpha + e^{-i\phi/2}\sin^2\alpha & (e^{i\phi/2} - e^{-i\phi/2})\sin\alpha\cos\alpha e^{-i\delta} \\ (e^{i\phi/2} - e^{-i\phi/2})\sin\alpha\cos\alpha e^{+i\delta} & e^{i\phi/2}\sin^2\alpha + e^{-i\phi/2}\cos^2\alpha \end{pmatrix} \tag{11-174b}$$

Equation (11-174b) can be checked immediately by observing that for $\alpha = 0^\circ$ (linear horizontally polarized light) it reduces to

$$\mathbf{J} = \begin{pmatrix} e^{+i\phi/2} & 0 \\ 0 & e^{-i\phi/2} \end{pmatrix} \tag{11-175}$$

which is the Jones matrix for a linear retarder (11-169), as we expect.

We can use (11-174b) to find, say, the Jones matrix for a homogeneous right circular retarder. We do this by using the familiar conditions of $\alpha = 45^\circ$ and $\delta = 90^\circ$. Substituting these values in (11-174b) then gives

$$\mathbf{J} = \begin{pmatrix} \cos\dfrac{\phi}{2} & \sin\dfrac{\phi}{2} \\ -\sin\dfrac{\phi}{2} & \cos\dfrac{\phi}{2} \end{pmatrix} \tag{11-176}$$

which is, indeed, the Jones matrix for a homogeneous right circular retarder.

These results can be summarized by writing the Jones matrices for a homogeneous elliptical polarizer and the Jones matrix for a homogeneous elliptical retarder as the pair:

$$\mathbf{J} = \begin{pmatrix} p_x\cos^2\alpha + p_y\sin^2\alpha & (p_x - p_y)\sin\alpha\cos\alpha e^{-i\delta} \\ (p_x - p_y)\sin\alpha\cos\alpha e^{+i\delta} & p_x\sin^2\alpha + p_y\cos^2\alpha \end{pmatrix} \tag{11-168b}$$

$$\mathbf{J} = \begin{pmatrix} e^{i\phi/2}\cos^2\alpha + e^{-i\phi/2}\sin^2\alpha & (e^{i\phi/2} - e^{-i\phi/2})\sin\alpha\cos\alpha e^{-i\delta} \\ (e^{i\phi/2} - e^{-i\phi/2})\sin\alpha\cos\alpha e^{+i\delta} & e^{i\phi/2}\sin^2\alpha + e^{-i\phi/2}\cos^2\alpha \end{pmatrix} \tag{11-174b}$$

Shurcliff and, more recently, Kliger et al., have tabulated the Jones matrices and the Mueller matrices for elliptical polarizers and retarders as well as their degenerate forms. All of their forms can, of course, be obtained from (11-168b) and (11-174b).

REFERENCES

Papers

1. Jones, R. Clark, *J. Opt. Soc. Am.*, *31*, 488 (1941).
2. Jones, R. Clark and Hurwitz, H., *J. Opt. Soc. Am.*, *31*, 493 (1941).
3. Jones, R. Clark, *J. Opt. Soc. Am.*, *31*, 500 (1941).
4. Jones, R. Clark, *J. Opt. Soc. Am.*, *32*, 486 (1942).
5. Jones, R. Clark, *J. Opt. Soc. Am.*, *37*, 107 (1947).
6. Jones, R. Clark, *J. Opt. Soc. Am.*, *37*, 110 (1947).
7. Jones, R. Clark, *J. Opt. Soc. Am.*, *38*, 671 (1948).
8. Jones, R. Clark, *J. Opt. Soc. Am.*, *46*, 126 (1956).
9. Boerner, W. M., Yan, Wei-Ling, and Xi, An-Qing, *Proc. SPIE*, *1317*, 16 (1990).
10. Chipman, R. A., *Proc. SPIE*, *891*, 10 (1988).
11. Challener, W. A. and Rinehart, T. A., *Appl. Opt.*, *26*, 3974 (1987).

Books

1. Hecht, E. and Zajac, A., *Optics*, Addison-Wesley, Reading, MA., 1974.
2. Azzam, R. M. A. and Bashara, N. M., *Ellipsometry and Polarized Light*, North-Holland, Amsterdam, 1977.
3. Simmons, J. W. and Guttman, M. J., *States, Waves and Photons*, Addison-Wesley, Reading, MA, 1970.
4. Shurcliff, W. A., *Polarized Light*, Harvard University Press, Cambridge, MA, 1962.
5. Gerrad, A. and Burch, J. M., *Introduction to Matrix Methods in Optics*, Wiley, London, 1975.
6. O'Neil, E. L., *Introduction to Statistical Optics*, Addison-Wesley, Reading, MA, 1963.
7. Shurcliff, W. A. and Ballard, S. S., *Polarized Light*, Van Nostrand, New York, 1962.
8. Clarke, D. and Grainger, J. F., *Polarized Light and Optical Measurement*, Pergamon Press, Oxford, 1971.
9. Kliger, D. S., Lewis, J. W. and Randall, C. E., *Polarized Light in Optics and Spectroscopy*, Academic Press, New York, 1990.

12

The Poincaré Sphere

12.1 INTRODUCTION

In the previous chapters we have seen that the Mueller matrix formalism and the Jones matrix formalism enable us to treat many complex problems involving polarized light. The use of matrices, however, only slowly made its way into physics and optics. In fact, before the advent of quantum mechanics in 1925 matrix algebra was rarely used. It is clear that matrix algebra greatly simplifies the treatment of many difficult problems. In the optics of polarized light even the simplest problem of determining the change in polarization state of a beam propagating through several polarizing elements becomes surprisingly difficult to do without matrices. Before the advent of matrices only direct and very tedious algebraic methods were available. Consequently, other methods were sought to simplify these difficult calculations.

The need for simpler ways to carry out difficult calculations began in antiquity. Around 150 BC the Greek astronomer Hipparchus was living in Alexandria, Egypt, and working at the famous library of Alexandria. There, he compiled a catalog of stars and also plotted the positions of these stars in terms of latitude and longitude (in astronomy, longitude and latitude are called right ascension and declination) on a large globe which we call the celestial sphere. In practice, transporting a large globe for use at different locations is cumbersome. Therefore, he devised a method for projecting a three-dimensional sphere on to a two-dimensional plane. This type of projection is called a stereographic projection. It is still one of the most widely used projections and is particularly popular in astronomy. It has many interesting properties, foremost of which is that the longitudes and latitudes (right ascension and declination) continue to intersect each other at right angles on the plane as they do on the sphere. It appears that the stereographic projection was forgotten for many centuries and then rediscovered during the European Renaissance when the ancient writings of classical Greece and Rome were rediscovered. With the advent of the global exploration of the world by the European navigators and explorers there was a need for accurate charts, particularly charts that were mathematically correct. This need led not only to the rediscovery and use of the stereographic projection but also to the invention of new types of projections, e.g., the famous Mercator projection.

Henri Poincaré, a famous nineteenth-century French mathematician and physicist, discovered around 1890 that the polarization ellipse could be represented on a complex plane. Further, he discovered that this plane could be projected on to a sphere in exactly the same manner as the stereographic projection. In effect, he reversed the problem of classical antiquity, which was to project a sphere on to a plane. The sphere that Poincaré devised is extremely useful for dealing with polarized light problems and, appropriately, it is called the Poincaré sphere.

In 1892, Poincaré introduced his sphere in his text *Traité de Lumierè*. Before the advent of matrices and digital computers it was extremely difficult to carry out calculations involving polarized light. As we have seen, as soon as we go beyond the polarization ellipse, e.g., the interaction of light with a retarder, the calculations become difficult. Poincaré showed that the use of his sphere enabled many of these difficulties to be overcome. In fact, Poincaré's sphere not only simplifies many calculations but also provides remarkable insight into the manner in which polarized light behaves in its interaction with polarizing elements.

While the Poincaré sphere became reasonably well known in the optical literature in the first half of the twentieth century, it was rarely used in the treatment of polarized light problems. This was probably due to the considerable mathematical effort required to understand its properties. In fact, its use outside of France appears to have been virtually nonexistent until the 1930s. Ironically, the appreciation of its usefulness only came *after* the appearance of the Jones and Mueller matrix formalisms. The importance of the Poincaré sphere was finally established in the optical literature in the long review article by Ramachandran and Ramaseshan on crystal optics in 1961.

The Poincaré sphere is still much discussed in the literature of polarized light. In larger part this is due to the fact that it is really surprising how simple it is to use once it is understood. In fact, despite its introduction nearly a century ago, new properties and applications of the Poincaré sphere are still being published and appearing in the optical literature. The two most interesting properties of the Poincaré sphere are that any point on the sphere corresponds to the three Stokes parameters S_1, S_2, and S_3 for elliptically polarized light, and the magnitude of the interaction of a polarized beam with an optical polarizing element corresponds to a rotation of the sphere; the final point describes the new set of Stokes parameters. In view of the continued application of the Poincaré sphere we present a detailed discussion of it. This is followed by simple applications of the sphere to describing the interaction of polarized light with a polarizer, retarder, and rotator. More complicated and involved applications of the Poincaré sphere are listed in the references.

12.2 THEORY OF THE POINCARÉ SPHERE

Consider a Cartesian coordinate system with axes x, y, z and let the direction of propagation of a monochromatic elliptically polarized beam of light be in the

z direction. The equations of propagation are described by

$$E_x(z, t) = E_x \exp i(\omega t - kz) \tag{12-1a}$$

$$E_y(z, t) = E_y \exp i(\omega t - kz) \tag{12-1b}$$

where E_x and E_y are the complex amplitudes:

$$E_x = E_{0x} \exp(i\delta_x) \tag{12-2a}$$

$$E_y = E_{0y} \exp(i\delta_y) \tag{12-2b}$$

and E_{0x} and E_{0y} are real quantities. We divide (12-2b) by (12-2a) and write

$$\frac{E_y}{E_x} = \frac{E_{0y}}{E_{0x}} e^{i\delta} \tag{12-3a}$$

$$= \frac{E_{0y}}{E_{0x}} \cos\delta + i\left(\frac{E_{0y}}{E_{0x}}\right) \sin\delta$$

$$= u + iv \tag{12-3b}$$

where $\delta = \delta_y - \delta_x$ and u and v are orthogonal axes in the complex plane. On eliminating the propagator in (12-1) and (12-2), we obtain the familiar equation of the polarization ellipse:

$$\frac{E_x^2}{E_{0x}^2} + \frac{E_y^2}{E_{0y}^2} - 2\frac{E_x E_y}{E_{0x} E_{0y}} \cos\delta = \sin^2\delta \tag{3-7a}$$

We have shown in Section 3.2 that the maximum values of E_x and E_y are E_{0x} and E_{0y}, respectively. Equation (3-7a) describes an ellipse inscribed in a rectangle of sides $2E_{0x}$ and $2E_{0y}$.

This is shown in Fig. 12-1.

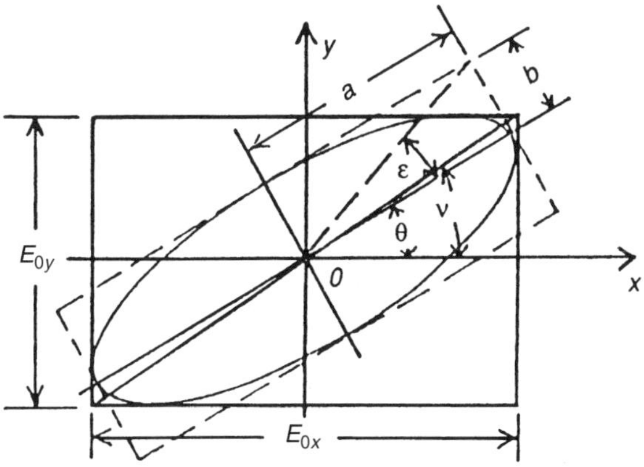

Figure 12-1 Parameters of the polarization ellipse having amplitude components E_{0x} and E_{0y} along x and y axes, respectively. The angle ν is related to E_{0x} and E_{0y} by $\tan\nu = E_{0y}/E_{0x}$. The major and minor axes of the ellipse are $2a$ and $2b$, and the ellipticity is $e = b/a = \tan\varepsilon$; the azimuth angle θ is with respect to the x axis. (From Jerrard.)

In general, we recall, the axes of the ellipse are not necessarily along the x and y axes but are rotated, say, along x' and y'. Thus, we can write the oscillation along x' and y' as

$$x' = a\cos\phi \tag{12-4a}$$

$$y' = b\sin\phi \tag{12-4b}$$

where $\phi = \omega t - kz$. The ellipticity e, which is the ratio of the minor axis to the major axis, is $e = b/a$. The orientation of the ellipse is given by the azimuth angle $\theta(0 \leq \theta \leq 180^\circ)$; this is the angle between the major axis and the positive x axis. From Fig. 12-1 the angles ε and ν are defined by the equations:

$$\tan\varepsilon = \frac{b}{a} \qquad (0 \leq \varepsilon \leq 90^\circ) \tag{12-5a}$$

$$\tan\nu = \frac{E_{0y}}{E_{0x}} \qquad (0 \leq \nu \leq 90^\circ) \tag{12-5b}$$

The sense of the ellipse or the direction of rotation of the light vector depends on δ; it is designated right or left according to whether $\sin\delta$ is negative or positive. The sense will be indicated by the sign of the ratio of the principal axes. Thus, $\tan\varepsilon = +b/a$ or $-b/a$ refers to left (counterclockwise) or right (clockwise) rotation, respectively.

By using the methods presented earlier (see Section 3.4), we see that the following relations exist with respect to the parameters of the polarization ellipse, namely,

$$E_{0x}^2 + E_{0y}^2 = a^2 + b^2 \tag{12-6a}$$

$$E_{0x}^2 - E_{0y}^2 = (a^2 - b^2)\cos 2\theta \tag{12-6b}$$

$$E_{0x}E_{0y}\sin\delta = \pm ab \tag{12-6c}$$

$$2E_{0x}E_{0y}\cos\delta = (a^2 - b^2)\sin 2\theta \tag{12-6d}$$

By adding and subtracting (12-6a) and (12-6b), we can relate E_{0x} and E_{0y} to a, b, and θ. Thus, we find that

$$E_{0x}^2 = a^2\cos^2\theta + b^2\sin^2\theta \tag{12-7a}$$

$$E_{0y}^2 = a^2\sin^2\theta + b^2\cos^2\theta \tag{12-7b}$$

We see that when the polarization ellipse is not rotated, so $\theta = 0^\circ$, (12-7a) and (12-7b) become

$$E_{0x} = \pm a \qquad E_{0y} = \pm b \tag{12-8}$$

which is to be expected, as Fig. 12-1 shows. The ellipticity is then seen to be

$$e = \frac{b}{a} = \frac{E_{0y}}{E_{0x}} \tag{12-9}$$

when $\theta = 0^\circ$.

We can now obtain some interesting relations between the foregoing parameters. The first one can be obtained by dividing (12-6d) by (12-6b). Then

$$\frac{\sin 2\theta}{\cos 2\theta} = \tan 2\theta = \frac{E_{0x}E_{0y}}{E_{0x}^2 - E_{0y}^2}\cos\delta \tag{12-10}$$

Substituting (12-5b) into (12-10) then yields

$$\tan 2\theta = \left(\frac{2\tan\nu}{1-\tan^2\nu}\right)\cos\delta \tag{12-11}$$

The factor in parentheses is equal to tan 2ν. We then have

$$\tan 2\theta = \tan 2\nu \cos\delta \tag{12-12}$$

The next important relationship is obtained by dividing (12-6c) by (12-6a), whence

$$\frac{\pm ab}{a^2+b^2} = \frac{E_{0x}E_{0y}}{E_{0x}^2 + E_{0y}^2}\sin\delta \tag{12-13}$$

Using both (12-5a) and (12-5b), we find that (12-13) becomes

$$\pm\sin 2\varepsilon = \sin 2\nu \sin\delta \tag{12-14}$$

Another important relation is obtained by dividing (12-6b) by (12-6a). Then

$$\frac{E_{0x}^2 - E_{0y}^2}{E_{0x}^2 + E_{0y}^2} = \frac{a^2-b^2}{a^2+b^2}\cos 2\theta \tag{12-15}$$

Again, substituting (12-5a) and (12-5b) into (12-15), we find that

$$\cos 2\nu = \cos 2\varepsilon \cos 2\theta \tag{12-16}$$

Equation (12-16) can be used to obtain still another relation. We divide (12-6d) by (12-6a) to obtain

$$\frac{2E_{0x}E_{0y}\cos\delta}{E_{0x}^2 + E_{0y}^2} = \frac{a^2-b^2}{a^2+b^2}\sin 2\theta \tag{12-17}$$

Next, using (12-5a) and (12-5b), we find that (12-17) can be written as

$$\sin 2\nu \cos\delta = \cos 2\varepsilon \sin 2\theta \tag{12-18}$$

Equation (12-18) can be solved for cos2ε by multiplying through by sin2θ so that

$$\sin 2\theta \sin 2\nu \cos\delta = \cos 2\varepsilon \sin^2 2\theta = \cos 2\varepsilon - \cos 2\varepsilon \cos^2 2\theta \tag{12-19}$$

or

$$\cos 2\varepsilon = (\cos 2\varepsilon \cos 2\theta)\cos 2\theta + \sin 2\theta \sin 2\nu \cos\delta \tag{12-20}$$

We see that the term in parentheses is identical to (12-16), so (12-20) can finally be written as

$$\cos 2\varepsilon = \cos 2\nu \cos 2\theta + \sin 2\theta \sin 2\nu \cos\delta \tag{12-21}$$

Equation (12-21) represents the law of cosines for sides from spherical trigonometry. Consequently, it represents our first hint or suggestion that the foregoing results can be related to a sphere. We shall not discuss (12-21) at this time, but defer its discussion until we have developed some further relations.

Equation (12-21) can be used to find a final relation of importance. We divide (12-14) by (12-21):

$$\pm\tan 2\varepsilon = \frac{\sin 2\nu \sin\delta}{\cos 2\nu\cos\theta + \sin 2\nu\sin 2\theta\cos\delta} \tag{12-22}$$

Dividing the numerator and the denominator of (12-22) by $\sin 2\nu\cos\delta$ yields

$$\pm\tan 2\varepsilon = \frac{\tan\delta}{\sin 2\theta + (\cos 2\nu\cos 2\theta)/(\sin 2\nu\cos\delta)} \tag{12-23}$$

We now observe that (12-12) can be written as

$$\cos 2\nu\tan 2\theta = \sin 2\nu\cos\delta \tag{12-24}$$

so

$$\cos\delta = \frac{\cos 2\nu\tan 2\theta}{\sin 2\nu} \tag{12-25}$$

Substituting (12-25) into the second term in the denominator of (12-23) yields the final relation:

$$\pm\tan 2\varepsilon = \sin 2\theta\tan\delta \tag{12-26}$$

For convenience we now collect relations (12-12), (12-14), (12-16), (12-21), and (12-26) and write them as a set of relations:

$$\tan 2\theta = \tan 2\nu\cos\delta \tag{12-27a}$$

$$\pm\sin 2\varepsilon = \sin 2\nu\sin\delta \tag{12-27b}$$

$$\cos 2\nu = \cos 2\varepsilon\cos 2\theta \tag{12-27c}$$

$$\cos 2\varepsilon = \cos 2\nu\cos 2\theta + \sin 2\theta\sin 2\nu\cos\delta \tag{12-27d}$$

$$\pm\tan 2\varepsilon = \sin 2\theta\tan\delta \tag{12-27e}$$

The equations in (12-27) have very familiar forms. Indeed, they are well-known relations, which appear in spherical trigonometry.

Figure 12-2 shows a spherical triangle formed by three great circle arcs, $\overline{AB}$, $\overline{BC}$, and $\overline{CA}$ on a sphere. At the end of this section the relations for a spherical triangle are derived by using vector analysis. There it is shown that 10 relations exist for a so-called right spherical triangle. For an oblique spherical triangle there exists, analogous to plane triangles, the law of sines and the law of cosines. With respect to the law of cosines, however, there is a law of cosines for the angles (uppercase letters)

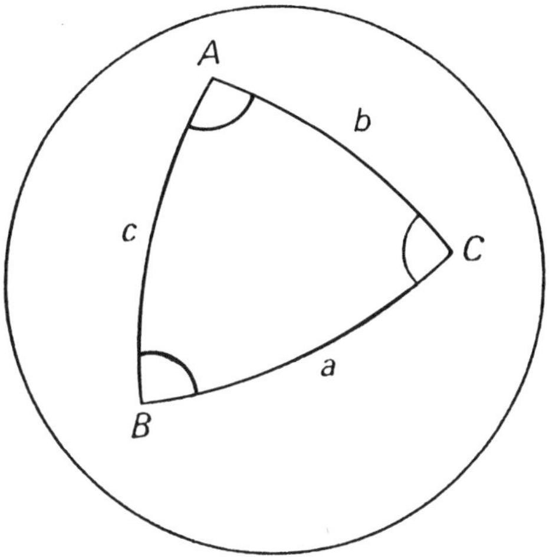

Figure 12-2 Spherical triangle on a sphere. The vertex angles are designated by A, B, C. The side opposite to each angle is represented by a, b, and c, respectively.

and a law of cosines for sides (lower case letters). Of particular interest are the following relations derived from Fig. 12-2.

$$\cos c = \cos a \cos b \tag{12-28a}$$

$$\sin a = \sin c \sin A \tag{12-28b}$$

$$\tan b = \tan c \cos A \tag{12-28c}$$

$$\cos a = \cos b \cos c + \sin b \sin c \cos A \tag{12-28d}$$

$$\tan a = \sin b \tan A \tag{12-28e}$$

If we now compare (12-28a) with (12-27a), etc., we see that the equations can be made completely compatible by constructing the right spherical triangle in Fig. 12-3. If, for example, we equate the spherical triangles in Figs. 12-2 and 12-3, we have

$$a = 2\varepsilon \quad b = 2\theta \quad \delta = A \tag{12-29}$$

Substituting (12-29) into, say, (12-28a) gives

$$\cos 2\nu = \cos 2\varepsilon \cos 2\theta \tag{12-30}$$

which corresponds to (12-27c). In a similar manner by substituting (12-29) into the remaining equations in (12-28), we obtain (12-27). Thus, we arrive at the very interesting result that the polarization ellipse on a plane can be transformed to a spherical triangle on a sphere. We shall return to these equations after we have discussed some further transformation properties of the rotated polarization ellipse in the complex plane.

The ratio E_y/E_x in (12-3) defines the shape and orientation of the elliptical vibration given by (3-7a). This vibration may be represented by a point m on a plane in which the abscissa and ordinate are u and v, respectively. The diagram in the complex plane is shown in Fig. 12-4.

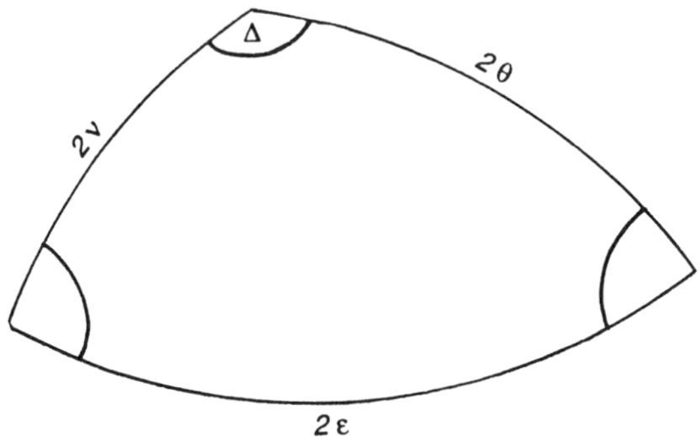

Figure 12-3 Right spherical triangle for the parameters of the polarization ellipse.

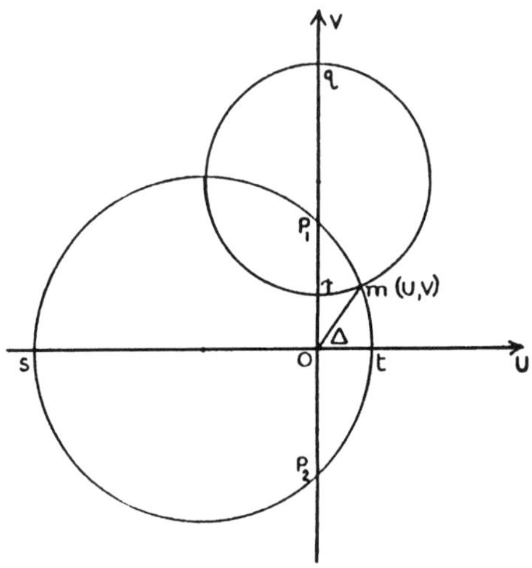

Figure 12-4 Representation of elliptically polarized light by a point m on a plane; δ is the plane difference between the components of the ellipse (From Jerrard.)

From (12-3b) we have

$$u = \frac{E_{0y}}{E_{0x}} \cos \delta \tag{12-31a}$$

$$v = \frac{E_{0y}}{E_{0x}} \sin \delta \tag{12-31b}$$

The point $m(u, v)$ is described by the radius Om and the angle δ. The angle δ is found from (12-31) to be

$$\tan \delta = \frac{v}{u} \tag{12-32a}$$

or

$$\delta = \tan^{-1} \frac{v}{u} \tag{12-32b}$$

Squaring (12-31a) and (12-31b) and adding yields

$$u^2 + v^2 = \left(\frac{E_{0y}}{E_{0x}}\right)^2 = \rho^2 \tag{12-33}$$

which is the square of the distance from the origin to m. We see that we can also write (12-33) as

$$u^2 + v^2 = (u + iv)(u - iv) = \left(\frac{E_y}{E_x}\right)\left(\frac{E_y}{E_x}\right)^* = \rho\rho^* = \frac{E_{0y}^2}{E_{0x}^2} \tag{12-34a}$$

so

$$u + iv = \frac{E_y}{E_x} = \rho \tag{12-34b}$$

Thus, the radius vecor Om and the angle mOu represent the ratio E_y/E_x and the phase difference δ, respectively. It is postulated that the polarization is left- or right-handed according to whether δ is between 0 and π or π and 2π.

We now show that (12-34a) can be expressed either in terms of the rotation angle θ or the ellipticity angle ε. To do this we have from (12-33) that

$$u^2 + v^2 = \left(\frac{E_{0y}}{E_{0x}}\right)^2 = \rho^2 \tag{12-35a}$$

We also have, from (12-5b)

$$\frac{E_{0y}}{E_{0x}} = \tan \nu \tag{12-35b}$$

Squaring (12-35b) gives

$$\frac{E_{0y}^2}{E_{0x}^2} = \tan^2 \nu \tag{12-35c}$$

Now,

$$\tan 2\nu = \frac{2 \tan \nu}{1 - \tan^2 \nu} \tag{12-35d}$$

so

$$\tan^2 \nu = 1 - \frac{2 \tan \nu}{\tan 2\nu} \tag{12-35e}$$

But, from (12-27a) we have

$$\tan 2\theta = \tan 2\nu \cos \delta \tag{12-35f}$$

Substituting (12-35f) into (12-35e) gives

$$\tan^2 \nu = 1 - \frac{2 \tan \nu}{\tan 2\theta} \cos \delta \tag{12-35g}$$

Equating (12-35g) to (12-35c) and (12-35a) we have

$$\begin{aligned} u^2 + v^2 &= 1 - 2(\tan \nu \cos \delta / \tan 2\theta) \\ &= 1 - 2 \cot 2\theta (\tan \nu \cos \delta) \end{aligned} \tag{12-35h}$$

Finally, substituting (12-35b) into (12-35h) and using (12-31a), we find that

$$u^2 + v^2 + 2u \cot 2\theta - 1 = 0 \tag{12-36}$$

Thus, we have expressed u and v in terms of the rotation angle θ of the polarization ellipse. It is also possible to find a similar relation to (12-36) in terms of the ellipticity angle ε rather than θ. To show this we again use (12-35a), (12-35b), and (12-35d) to form

$$u^2 + v^2 = 1 - \frac{2 \tan \nu}{\sin 2\nu} \cos 2\nu \tag{12-37a}$$

Substituting (12-27a) and (12-27b) into (12-37a) then gives

$$u^2 + v^2 = 1 \mp 2v \csc 2\varepsilon \cos 2\nu \tag{12-37b}$$

After replacing $\cos 2\nu$ with its half-angle equivalent and choosing the upper sign, we are led to

$$u^2 + v^2 - 2v \csc 2\varepsilon + 1 = 0 \tag{12-38}$$

Thus, we can describe (12-35a),

$$u^2 + v^2 = \left(\frac{E_{0y}}{E_{0x}}\right)^2 = \rho^2 \tag{12-35a}$$

in terms of either θ or ε, respectively, by

$$u^2 + v^2 + 2u \cot 2\theta - 1 = 0 \tag{12-39a}$$

$$u^2 + v^2 - 2v \csc 2\varepsilon + 1 = 0 \tag{12-39b}$$

At this point it is useful to remember that the two most important parameters describing the polarization ellipse are the rotation angle θ and the ellipticity angle ε, as shown in Fig. 12-1. Equations (12-39a) and (12-39b) describe the polarization ellipse in terms of each of the parameters.

Equations (12-39a) and (12-39b) are recognized as the equations of a circle. They can be rewritten in standard forms as

$$(u + \cot 2\theta)^2 + v^2 = (\csc 2\theta)^2 \tag{12-40a}$$

$$u^2 + (v - \csc 2\varepsilon)^2 = (\cot 2\varepsilon)^2 \tag{12-40b}$$

Equation (12-40a) describes, for a constant value of θ, a family of circles each of radius $\csc 2\theta$ with centers at the point $(-\cot 2\theta, 0)$. Similarly (12-40b) describes, for a constant value of ε, a family of circles each of radius $\cot 2\theta$ and centers at the point $(0, \csc 2\theta)$. The circles in the two systems are orthogonal to each other. To show this we recall that if we have a function described by a differential equation of the form

$$M(x, y)dx + N(x, y)dy = 0 \tag{12-41a}$$

then the differential equation for the orthogonal trajectory is given by

$$N(x, y)dx - M(x, y)dy = 0 \tag{12-41b}$$

We therefore consider (12-39a) and show that (12-39b) describes the orthogonal trajectory. We first differentiate (12-39a)

$$u\,du + v\,dv + \cot 2\theta\,du = 0 \tag{12-42}$$

We eliminate the constant parameter cot2θ from (12-42) by writing (12-39a) as

$$\cot 2\theta = \frac{1 - u^2 - v^2}{2u} \tag{12-43}$$

Substituting (12-43) into (12-42) and grouping terms, we find that

$$(1 + u^2 - v^2)\,du + 2uv\,dv = 0 \tag{12-44}$$

According to (12-41a) and (12-41b), the trajectory orthogonal to (12-44) must, therefore, be

$$2uv\,du - (1 + u^2 - v^2)\,dv = 0 \tag{12-45}$$

We now show that (12-39b) reduces to (12-45). We differentiate (12-39b) to obtain

$$u\,du + v\,dv - \csc 2\varepsilon\,dv = 0 \tag{12-46a}$$

Again, we eliminate the constant parameter $\csc 2\varepsilon$ by solving for $\csc 2\varepsilon$ in (12-39b):

$$\csc 2\varepsilon = \frac{1 + u^2 + v^2}{2v} \tag{12-46b}$$

We now substitute (12-46b) into (12-46a), group terms, and find that

$$2uv\,du - (1 + u^2 - v^2)\,dv = 0 \tag{12-47}$$

Comparing (12-47) with (12-45) we see that the equations are identical so the trajectories are indeed orthogonal to each other. In Fig. 12-5 we have plotted the family

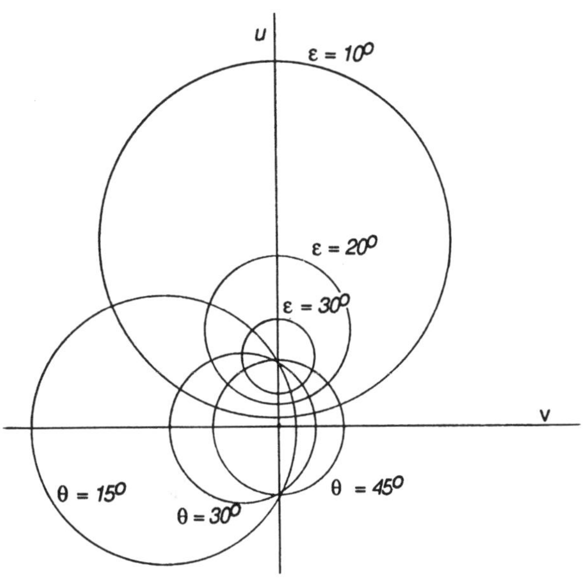

Figure 12-5 Orthogonal circles of the polarization ellipse in the uv plane.

of circles for $\theta = 15^\circ$ to 45° and for $\varepsilon = 10^\circ$ to 30°. We note that the circles intersect at m and that at this intersection each circle has the same value of ρ and δ.

Each of the circles, (12-40a) and (12-40b), has an interesting property which we now consider. If $v = 0$, for example, then (12-40a) reduces to

$$(u + \cot 2\theta)^2 = (\csc 2\theta)^2 \tag{12-48a}$$

Solving for u, we find that

$$u = -\cot\theta \quad \text{or} \quad u = \tan\theta \tag{12-48b}$$

Referring to Fig. 12-4, these points occur at s and t and correspond to linearly polarized light in azimuth $\cot^{-1}u$ and $\tan^{-1}u$, respectively; we also note from (12-3a) and (12-3b) that because $v = 0$ we have $\delta = 0$, so $u = E_{0y}/E_{0x}$. Similarly, if we set $u = 0$ in (12-40a), we find that

$$v = \pm 1 \tag{12-48c}$$

Again, referring to (12-3b), (12-48c) corresponds to $E_{0y}/E_{0x} = 1$ and $\delta = \pm\pi/2$, that is, right- and left-circularly polarized light, respectively. These points are plotted as P_1 and P_2 in Fig. 12-4. Thus, the circle describes linearly polarized light along the u axis, circularly polarized light along the v axis, and elliptically polarized light everywhere else in the uv plane.

From these results we can now project the point m in the complex uv plane on to a sphere, the Poincaré sphere. This is described in the following section.

12.2.1 Note on the Derivation of Law of Cosines and Law of Sines in Spherical Trigonometry

In this section we have used a number of formulas that originate from spherical trigonometry. The two most important formulas are the law of cosines and the law of sines for spherical triangles and the formulas derived by setting one of the angles to 90° (a right angle). We derive these formulas by recalling the following vector identities:

$$\mathbf{A} \times (\mathbf{B} \times \mathbf{C}) = (\mathbf{A} \cdot \mathbf{C})\mathbf{B} - (\mathbf{A} \cdot \mathbf{B})\mathbf{C} \tag{12-N1a}$$

$$(\mathbf{A} \times \mathbf{B}) \times \mathbf{C} = (\mathbf{A} \cdot \mathbf{C})\mathbf{B} - (\mathbf{B} \cdot \mathbf{C})\mathbf{A} \tag{12-N1b}$$

$$(\mathbf{A} \times \mathbf{B}) \times (\mathbf{C} \times \mathbf{D}) = [\mathbf{A} \cdot (\mathbf{C} \times \mathbf{D})]\mathbf{B} - [\mathbf{B} \cdot (\mathbf{C} \times \mathbf{D})]\mathbf{A} \tag{12-N1c}$$

$$(\mathbf{A} \times \mathbf{B}) \cdot (\mathbf{C} \times \mathbf{D}) = (\mathbf{A} \cdot \mathbf{C})(\mathbf{B} \cdot \mathbf{D}) - (\mathbf{A} \cdot \mathbf{D})(\mathbf{B} \cdot \mathbf{C}) \tag{12-N1d}$$

The terms in brackets in (12-N1c) are sometimes written as

$$[\mathbf{A} \cdot (\mathbf{C} \times \mathbf{D})] = [\mathbf{A}, \mathbf{C}, \mathbf{D}] \tag{12-N1e}$$

$$[\mathbf{B} \cdot (\mathbf{C} \times \mathbf{D})] = [\mathbf{B}, \mathbf{C}, \mathbf{D}] \tag{12-N1f}$$

A spherical triangle is a three-sided figure drawn on the surface of a sphere as shown in Fig. 12-N1. The sides of a spherical triangle are required to be arcs of great circles. We recall that a great circle is obtained by intersecting the sphere with a plane passing through its center. Two great circles always intersect at two distinct points, and their angle of intersection is defined to be the angle between their corresponding planes. This is equivalent to defining the angle to be equal to the plane angle between two lines tangent to the corresponding great circles at a point of intersection.

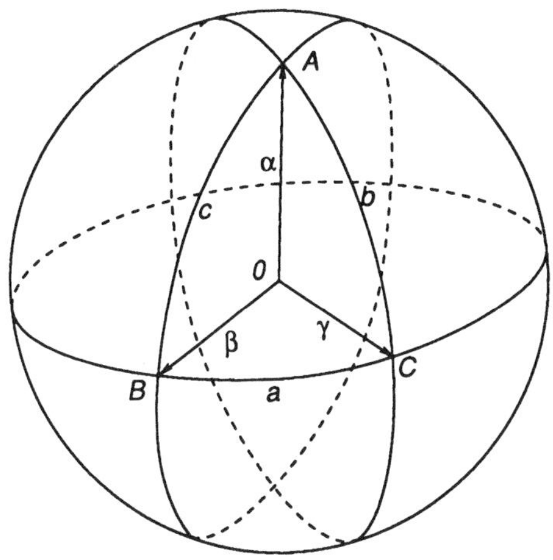

Figure 12-N1 Fundamental angles and arcs on a sphere.

The magnitude of a side of a spherical triangle may be measured in two ways. Either we can take its arc length, or we can take the angle it subtends at the center of the sphere. These two methods give the same numerical result if the radius of the sphere is unity. We shall adopt the second of the two methods. In other words, if A, B, and C are the vertices of a spherical triangle with opposite sides a, b, and c, respectively, the numerical value of, say, a will be taken to be the plane angle BOC, where O is the center of the sphere in Fig. 12-N1.

In the following derivations we assume that the sphere has a radius $R = 1$ and the center of the sphere is at the origin. The unit vectors extending from the center to A, B, and C are α, β, and γ, respectively; the vertices are labeled in such a way that α, β, and γ are positively oriented.

We now refer to Fig. 12-N2. We introduce another set of unit vectors α', β', and γ' extending from the origin and defined so that

$$\alpha \times \beta = \gamma = \sin c\gamma' \tag{12-N2a}$$

$$\beta \times \gamma = \alpha = \sin a\alpha' \tag{12-N2b}$$

$$\gamma \times \alpha = \beta = \sin b\beta' \tag{12-N2c}$$

In Fig. 12-N2 only α' is shown. However, in Fig. 12-N3 all three unit vectors are shown. The unit vectors α', β', and γ' determine a spherical triangle $A'B'C'$ called the polar triangle of ABC; this is shown in Fig. 12-N4. We now let the sides of the polar triangle be a', b', and c', respectively. We see that B' is a pole corresponding to the great circle joining A and C. Also, C' is a pole corresponding to the great circle AB. If these great circles are extended to intersect the side $B'C'$, we see that this side is composed of two overlapping segments $B'E$ and DC' each of magnitude of 90°. Their common overlap has a magnitude A, so we see that

$$a' + A = \pi \tag{12-N3a}$$

$$b' + B = \pi \tag{12-N3b}$$

$$c' + C = \pi \tag{12-N3c}$$

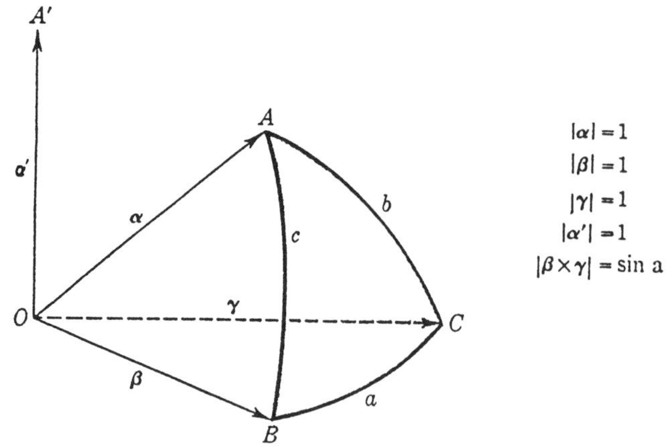

Figure 12-N2 The construction of a spherical triangle on the surface of a sphere.

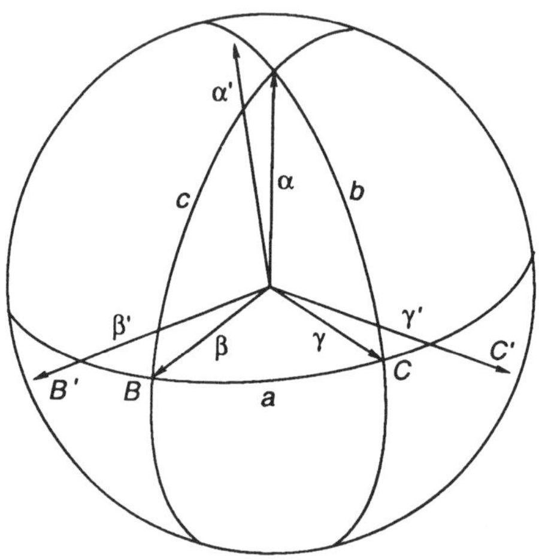

Figure 12-N3 Unit vectors within a unit sphere.

Equation (12-N3) is useful for relating the angles of a spherical triangle to the sides of the corresponding polar triangle. We now derive the law of cosines and law of sines for spherical trigonometry.

In the identity (12-N1d):

$$(\mathbf{A}\times\mathbf{B})\cdot(\mathbf{C}\times\mathbf{D}) = (\mathbf{A}\cdot\mathbf{C})(\mathbf{B}\cdot\mathbf{D}) - (\mathbf{A}\cdot\mathbf{D})(\mathbf{B}\cdot\mathbf{C}) \tag{12-N1d}$$

we substitute α for **A**, β for **B**, α for **C**, γ for **D**. Since α is a unit vector, we see that (12-N1d) becomes

$$(\alpha\times\beta)\cdot(\alpha\times\gamma) = \beta\cdot\gamma - (\alpha\cdot\gamma)(\beta\cdot\alpha) \tag{12-N4}$$

In Fig. 12-N2 we have $\beta\cdot\gamma = \cos a$, $\alpha\cdot\beta = \cos c$, and $\alpha\cdot\gamma = \cos b$. Hence, the right-hand side of (12-N4) becomes

$$\cos a - \cos b\cos c \tag{12-N5}$$

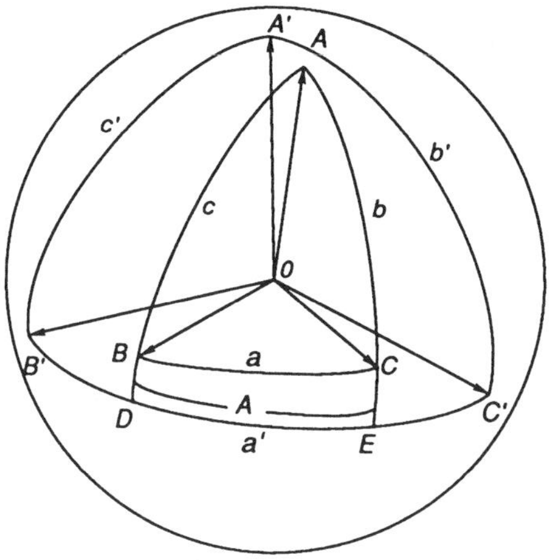

Figure 12-N4 The polar triangle on a sphere.

From (12-N2) we see that the left-hand side of (12-N4) becomes

$$(\sin c\gamma') \cdot (-\sin b\beta') = -\sin c \sin b(\gamma' \cdot \beta') \tag{12-N6}$$

Now, just as $\gamma \cdot \beta$ is equal to $\cos a$, we see from the polar triangle in Figs. 12-N3 and 12-N4 that $\gamma' \cdot \beta' = \cos a'$. From (12-N3a) $\cos a'$ is $\cos(\pi - A)$, which equals $-\cos A$. Thus, the left-hand side of (12-N4) equals

$$\sin c \sin b \sin A \tag{12-N7}$$

Equating the two sides we obtain the law of cosines:

$$\cos a = \cos b \cos c + \sin b \sin c \cos A \tag{12-N8a}$$

We can, of course, imagine that Fig. 12-N2 is rotated so that the roles previously played by a, b, and c, respectively, are now replaced by b, c, and a, so we can write

$$\cos b = \cos c \cos a + \sin c \sin a \cos B \tag{12-N8b}$$

$$\cos c = \cos a \cos b + \sin a \sin b \cos C \tag{12-N8c}$$

Three other versions of the cosine law are obtained by applying the law of cosines to the polar triangle by merely changing a to a', b to b', etc., according to (12-N3), namely,

$$\cos A = -\cos B \cos C + \sin B \sin C \cos a \tag{12-N9a}$$

$$\cos B = -\cos C \cos A + \sin C \sin A \cos b \tag{12-N9b}$$

$$\cos C = -\cos A \cos B + \sin A \sin B \cos c \tag{12-N9c}$$

We now turn to the law of sines. Here, we make use of the identity:

$$(\mathbf{A} \times \mathbf{B}) \times (\mathbf{C} \times \mathbf{D}) = [\mathbf{A} \cdot (\mathbf{C} \times \mathbf{D})]\mathbf{B} - [\mathbf{B} \cdot (\mathbf{C} \times \mathbf{D})]\mathbf{A} \tag{12-N1c}$$

Replacing **A** by α, **B** by β, **C** by α, and **D** by γ, (12-N1c) becomes

$$(\alpha \times \beta) \times (\alpha \times \gamma) = [\alpha \cdot (\alpha \times \gamma)]\beta - [\beta \cdot (\alpha \times \gamma)]\alpha \tag{12-N10}$$

From the relations given by (12-N2) the left-hand side of (12-N10) becomes

$$\begin{aligned}(\sin c\gamma') \times (-\sin b\beta') &= -\sin b \sin c(\gamma' \times \beta') \\ &= -\sin b \sin c(-\sin a'\alpha) \\ &= (\sin b \sin c \sin A)\alpha \end{aligned} \tag{12-N11}$$

In this manner we obtain

$$(\sin b \sin c \sin A)\alpha = [\alpha, \beta, \gamma]\alpha \tag{12-N12a}$$

$$(\sin c \sin a \sin B)\beta = [\beta, \gamma, \alpha]\beta \tag{12-N12b}$$

$$(\sin a \sin b \sin C)\gamma = [\gamma, \alpha, \beta]\gamma \tag{12-N12c}$$

We see from either (12-N1e) or (12-N1f) that $[\alpha, \beta, \gamma] = [\beta, \gamma, \alpha] = [\gamma, \alpha, \beta]$ and, hence, the left-hand sides of (12-N12) are all equal. Thus, for example, we can write

$$\sin b \sin c \sin A = \sin c \sin a \sin B \tag{12-N13}$$

which yields

$$\frac{\sin b}{\sin B} = \frac{\sin a}{\sin A} \tag{12-N14}$$

Similarly, we obtain from (12-N12) that

$$\frac{\sin a}{\sin A} = \frac{\sin c}{\sin C} \tag{12-N15}$$

so that we can finally write the law of sines:

$$\frac{\sin a}{\sin A} = \frac{\sin b}{\sin B} = \frac{\sin c}{\sin C} \tag{12-N16}$$

From the law of cosines and the law of sines we can derive the equations for a right spherical triangle. To show this let us first summarize the previous results by writing

The law of cosines for sides a, b, and c:

$$\cos a = \cos b \cos c + \sin b \sin c \cos A \tag{12-N17a}$$

$$\cos b = \cos c \cos a + \sin c \sin a \cos B \tag{12-N17b}$$

$$\cos c = \cos a \cos b + \sin a \sin b \cos C \tag{12-N17c}$$

The law of cosines for angles A, B, and C:

$$\cos A = -\cos B \cos C + \sin B \sin C \cos a \tag{12-N18a}$$

$$\cos B = -\cos C \cos A + \sin C \sin A \cos b \tag{12-N18b}$$

$$\cos C = -\cos A \cos B + \sin A \sin B \cos c \tag{12-N18c}$$

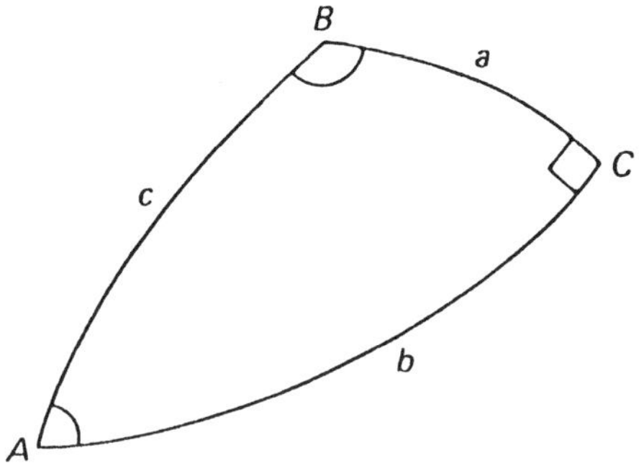

Figure 12-N5 Arc length and angle relations for a right spherical triangle.

The law of sines:

$$\frac{\sin a}{\sin A} = \frac{\sin b}{\sin B} = \frac{\sin c}{\sin C} \tag{12-N19}$$

We can use (12-N17), (12-N18) and (12-N19) to obtain the equations for a right spherical triangle. In order to derive these equations, we assume that the angle C is the right angle. The spherical right triangle is shown in Fig. 12-N5.

In (12-N17c) we set $C = 90°$, and we see that

$$\cos c = \cos a \cos b \tag{12-N20a}$$

Similarly, from the law of sines (12-N19) we find that

$$\sin a = \sin c \sin A \tag{12-N20b}$$

$$\sin b = \sin c \sin B \tag{12-N20c}$$

From the law of cosines for angles (12-N18) we then have

$$\cos A = \cos a \sin B \tag{12-N20d}$$

$$\cos B = \cos b \sin A \tag{12-N20e}$$

$$\cos A \cos B = \sin A \sin B \cos c \tag{12-N20f}$$

We note that (12-N20f) can also be derived by multiplying (12-N20d) by (12-N20e) and using (12-N20a).

Next, we divide (12-N20b) by $\cos a$ so that

$$\begin{aligned}\sin a/\cos a = \tan a &= \sin c \sin A/\cos a \\ &= \tan c[\cos b \sin A] \\ &= \tan c \cos B\end{aligned} \tag{12-N20g}$$

where we have used (12-N20a) and (12-N20e). We see that we have found six relations. Further analysis shows that there are four more relations for a right spherical triangle, so there are ten relations altogether. We therefore find that for a right spherical triangle we have the following relations:

$$\cos c = \cos a \cos b \tag{12-N21a}$$

$$\sin a = \sin c \sin A \tag{12-N21b}$$

$$\sin b = \sin c \sin B \tag{12-N21c}$$

$$\tan a = \sin b \tan A \tag{12-N21d}$$

$$\tan b = \sin a \tan B \tag{12-N21e}$$

$$\tan b = \tan c \cos A \tag{12-N21f}$$

$$\tan a = \tan c \cos B \tag{12-N21g}$$

$$\cos A = \cos a \sin B \tag{12-N21h}$$

$$\cos B = \cos b \sin A \tag{12-N21i}$$

$$\cos c = \cot A \cot B \tag{12-N21j}$$

These relations are important because they are constantly appearing in the study of polarized light.

12.3 PROJECTION OF THE COMPLEX PLANE ONTO A SPHERE

We now consider the projection of the point m in the complex plane on to the surface of a sphere. This projection is shown in Fig. 6. Specifically, the point m in the uv plane is projected as point M on the sphere. A sphere of unit diameter (the radius r is equal to 1/2) is constructed such that point O is tangential to the uv plane, and points in the plane on the u axis project on to the surface of the sphere by joining them to O'. The line OO' is the diameter, and the points p_1 and p_2 project on to the poles P_1 and P_2 of the sphere. Then, by the principles of sterographic projection the family of circles given by (12-39a) and (12-39b) project into meridians of longitude and parallels of latitude, respectively. The point O' (see Fig. 12-6) is called the antipode of a sphere. If all the projected lines come from this point, the projection is called stereographic. The vector Om projects into the arc $\overline{OM}$ of length 2ν, and the spherical angle $\overline{MOO'}$ is δ. Thus, any point M on the sphere will, as does m on the plane, represent the state of polarization of light.

In order to find the relationship between the coordinates of M and the parameters of the light, i.e., the ellipticity, azimuth, sense, and phase difference, it is necessary to determine the coordinates in terms of θ and ε. We must, therefore, express (transform) the coordinates of m on the uv plane to M on the sphere. If the center of the sphere (Fig. 12-6) is taken as the origin of the coordinate system, then the coordinates of m and O', referenced to this origin are found as follows: the coordinates of m, in terms of x, y, z, are seen from the figure to be

$$x = -\frac{1}{2} \tag{12-49a}$$

$$y = \tan \nu \cos \delta \tag{12-49b}$$

$$z = \tan \nu \sin \delta \tag{12-49c}$$

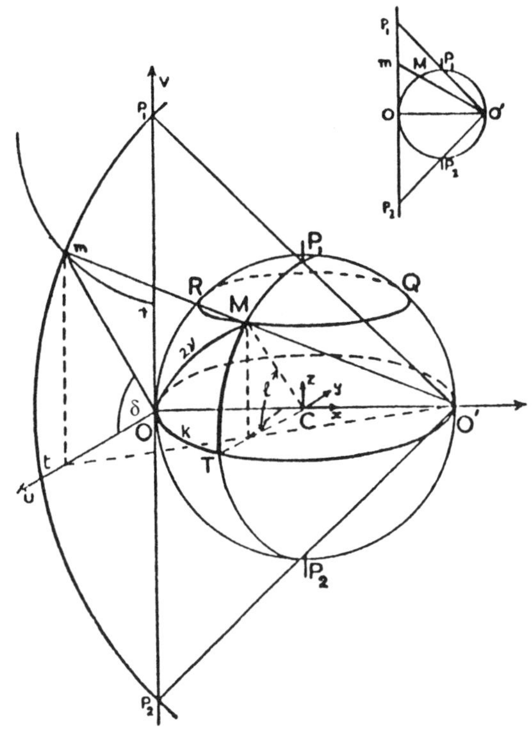

Figure 12-6 Stereographic projection of the complex plane on a sphere. Elliptically polarized light is represented by the points m and M on the plane and the sphere, respectively. The vector Om projects into the arc $\overline{OM}$ of length 2ν; the angle δ projects into the spherical angle $\overline{MOT}$. The latitude and longitude of M are l and k, respectively. (From Jerrard.)

The coordinates of the point O' are

$$x' = \frac{1}{2} \tag{12-50a}$$

$$y' = 0 \tag{12-50b}$$

$$z' = 0 \tag{12-50c}$$

The point m is projected along the straight line mMO' on to M. That is, we must determine the coordinates of the straight line mMO' and the point M on the sphere. The equation of the sphere is

$$x^2 + y^2 + z^2 = \left(\frac{1}{2}\right)^2 \tag{12-51}$$

We must now find the equation of the straight line mMO'. In order to do this, we must digress for a moment and determine the general equation of a straight line in three-dimensional space. This is most easily done using vector analysis.

Consider Fig. 12-7. A straight line is drawn through the point $\mathbf{R}_0$ and parallel to a constant vector $\mathbf{A}$. If the point $\mathbf{R}$ is also on the line, then the vector $\mathbf{R} - \mathbf{R}_0$ is

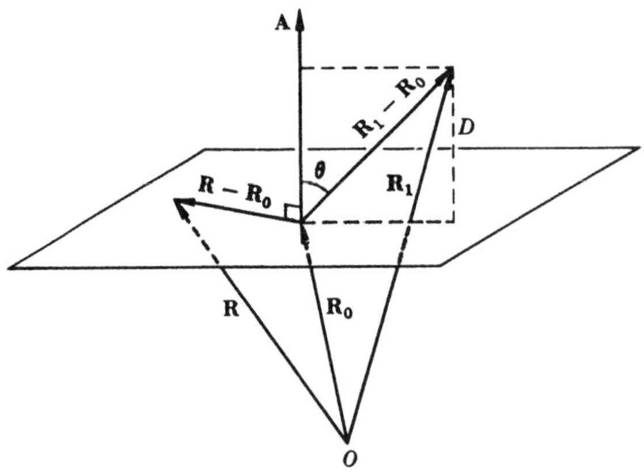

Figure 12-7 Vector equation of a straight line in three-dimensional space.

parallel to **A**. This is expressed by

$$(\mathbf{R} - \mathbf{R}_0) \times \mathbf{A} = 0 \tag{12-52}$$

which is the equation of a straight line. The fact that $\mathbf{R} - \mathbf{R}_0$ is parallel to **A** may also be expressed by the vector equation:

$$\mathbf{R} - \mathbf{R}_0 = \mathbf{A}t \tag{12-53}$$

where t is a scalar. Thus, the equation of a straight line in parametric form is

$$\mathbf{R} = \mathbf{R}_0 + \mathbf{A}t \qquad -\infty < t < \infty \tag{12-54}$$

We can deduce the Cartesian form of (12-54) by setting

$$\mathbf{R} = x\mathbf{i} + y\mathbf{j} + z\mathbf{k} \tag{12-55a}$$

$$\mathbf{R}_0 = x_0\mathbf{i} + y_0\mathbf{j} + z_0\mathbf{k} \tag{12-55b}$$

$$\mathbf{A} = a\mathbf{i} + b\mathbf{j} + c\mathbf{k} \tag{12-55c}$$

where **i**, **j**, **k** are the Cartesian unit vectors. Thus, we have

$$x = x_0 + at \tag{12-56a}$$

$$y = y_0 + bt \tag{12-56b}$$

$$z = z_0 + ct \tag{12-56c}$$

Eliminating t in (12-56), we find

$$\frac{x - x_0}{a} = \frac{y - y_0}{b} = \frac{z - z_0}{c} \tag{12-57}$$

We now return to our original problem. We have

$$\mathbf{A} = (1)\mathbf{i} - (\tan\nu\cos\delta)\mathbf{j} - (\tan\nu\sin\delta)\mathbf{k} \tag{12-58}$$

Similarly, $\mathbf{R}_0$ is

$$\mathbf{R}_0 = -\frac{1}{2}\mathbf{i} + (\tan\nu\cos\delta)\mathbf{j} + (\tan\nu\sin\delta)\mathbf{k} \tag{12-59}$$

Thus, from (12-55), (12-58) and (12-59) we find from (12-57) the relation:

$$\frac{x+1/2}{1} = \frac{y - \tan\nu\cos\delta}{-\tan\nu\cos\delta} = \frac{z - \tan\nu\sin\delta}{-\tan\nu\sin\delta} \tag{12-60}$$

which is the equation of the line $O'm$.

The coordinates (x, y, z) of M, the point of intersection of $O'm$ and the sphere, are obtained by solving (12-51) and (12-60) simultaneously. To do this, let us first solve for x. We first write, from (12-60),

$$z = \frac{1}{2}(\tan\nu\sin\delta)(1-2x) \tag{12-61a}$$

$$y = \frac{1}{2}(\tan\nu\cos\delta)(1-2x) \tag{12-61b}$$

We now substitute (12-61a) and (12-61b) into (12-51):

$$x^2 + y^2 + z^2 = \left(\frac{1}{2}\right)^2 \tag{12-51}$$

and find that

$$4(\tan^2\nu + 1)x^2 - (4\tan^2\nu)x + (\tan^2\nu - 1) = 0 \tag{12-62}$$

The solution of this quadratic equation is

$$x = \frac{1}{2}\cos 2\nu, \ \frac{1}{2} \tag{12-63a}$$

In a similar manner the solutions for y and z are found to be

$$y = \frac{1}{2}\sin 2\nu\cos\delta, 0 \tag{12-63b}$$

$$z = \frac{1}{2}\sin 2\nu\sin\delta, 0 \tag{12-63c}$$

The first set of x, y, z coordinates in (12-63) refers to the intersection of the straight line at M on the surface of the sphere. Thus, the coordinates of M are

$$M(x, y, z) = \left(\frac{1}{2}\cos 2\nu, \frac{1}{2}\sin 2\nu\cos\delta, \frac{1}{2}\sin 2\nu\sin\delta\right) \tag{12-64a}$$

The second set of coordinates in (12-63) describes the intersection of the line at the origin O', that is, the antipode of the sphere:

$$O'(x, y, z) = \left(\frac{1}{2}, 0, 0\right) \tag{12-64b}$$

We note that for $\nu = 0$ that (12-64a) reduces to (12-64b). Using (12-27a), (12-27b), and (12-27c), we can express the coordinates for M as

$$x = \frac{1}{2}\cos 2\nu = \frac{1}{2}\cos 2\varepsilon\cos 2\theta \tag{12-65a}$$

$$y = \frac{1}{2}\sin 2\nu\cos\delta = \frac{1}{2}\cos 2\varepsilon\sin 2\theta \tag{12-65b}$$

$$z = \frac{1}{2}\sin 2\nu\sin\delta = \frac{1}{2}\sin 2\varepsilon \tag{12-65c}$$

Equations (12-65a), (12-65b), and (12-65c) have a familiar appearance. We recall that the orthogonal field components E_x and E_y (12-2a) and (12-2b) are

$$E_x = E_{0x} \exp(i\delta_x) \tag{12-2a}$$

$$E_y = E_{0y} \exp(i\delta_y) \tag{12-2b}$$

where the propagator has been suppressed. The Stokes parameters for (12-2) are then defined in the usual way:

$$S_0 = E_x E_x^* + E_y E_y^* \tag{12-66a}$$

$$S_1 = E_x E_x^* - E_y E_y^* \tag{12-66b}$$

$$S_2 = E_x E_y^* + E_y E_x^* \tag{12-66c}$$

$$S_3 = i(E_x E_y^* - E_y E_x^*) \tag{12-66d}$$

Substituting (12-2) into (12-66) gives

$$S_0 = E_{0x}^2 + E_{0y}^2 \tag{12-67a}$$

$$S_1 = E_{0x}^2 - E_{0y}^2 \tag{12-67b}$$

$$S_2 = 2E_{0x}E_{0y} \cos\delta \tag{12-67c}$$

$$S_3 = 2E_{0x}E_{0y} \sin\delta \tag{12-67d}$$

where we have written $\delta = \delta_y - \delta_x$. From (12-7b) in Section 12.2 we have

$$\tan\nu = \frac{E_{0y}}{E_{0x}} \qquad (0 \leq \nu \leq 90^\circ) \tag{12-5b}$$

We now set

$$S_0 = A^2 + B^2 = C^2 \tag{12-68}$$

where $A = E_{0x}$ and $B = E_{0y}$ and construct the right triangle in Fig. 12-8. We see immediately that (12-67) can be rewritten in the form:

$$S_0 = C^2 \tag{12-69a}$$

$$S_1 = C^2 \cos 2\nu \tag{12-69b}$$

$$S_2 = C^2 \sin 2\nu \cos\delta \tag{12-69c}$$

$$S_3 = C^2 \sin 2\nu \sin\delta \tag{12-69d}$$

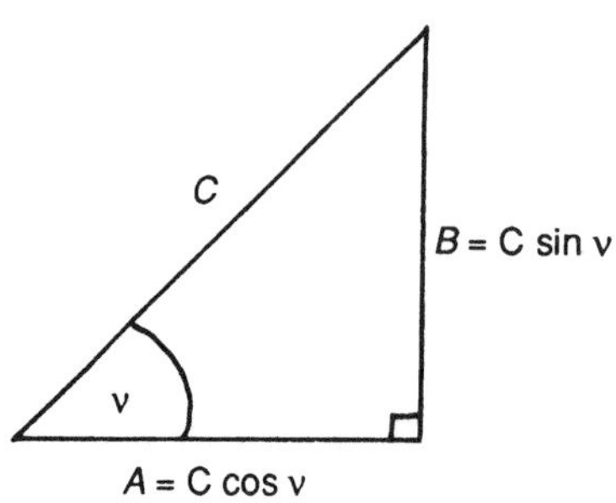

Figure 12-8 Construction of a right triangle.

Finally, we set $C^2 = 1/2$ in (12-69), so we have

$$S_0 = \frac{1}{2} \tag{12-70a}$$

$$S_1 = \frac{1}{2}\cos 2\nu \tag{12-70b}$$

$$S_2 = \frac{1}{2}\sin 2\nu \cos\delta \tag{12-70c}$$

$$S_3 = \frac{1}{2}\sin 2\nu \sin\delta \tag{12-70d}$$

We now compare (12-70) with the coordinates of M in (12-65), and we see that the equations for S_1, S_2, and S_3 and x, y, and z are identical. Thus, the coordinates of the point M on the Poincaré sphere correspond exactly to the Stokes parameters S_1, S_2, and S_3 of the optical beam and S_0 corresponds to the radius of the sphere.

On the Poincaré sphere we see that for a unit intensity we can write the Stokes parameters as [see (12-65)]

$$S_0 = 1 \tag{12-71a}$$

$$S_1 = \cos 2\varepsilon \cos 2\theta \tag{12-71b}$$

$$S_2 = \cos 2\varepsilon \sin 2\theta \tag{12-71c}$$

$$S_3 = \sin 2\varepsilon \tag{12-71d}$$

where of course, ε and θ are the ellipticity and azimuth (rotation) of the polarized beam.

In Fig. 12-9 we have drawn the Poincaré sphere in terms of the Stokes parameters given in (12-71). The point M on the surface of the Poincaré sphere is described in terms of its latitude (2ε), where $-\pi/2 \leq 2\varepsilon \leq \pi/2$, and its longitude ($2\theta$),

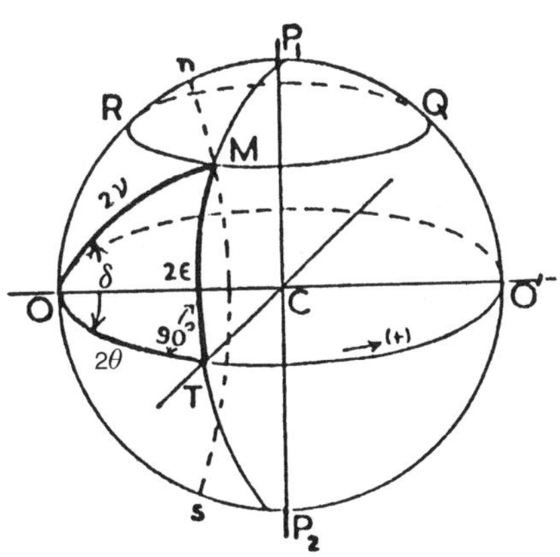

Figure 12-9 The Poincaré sphere showing the representation of an elliptically polarized vibration by a point M. From the spherical triangle OMT, the parameters of the vibration can be found. Points on the equator OO' represent linearly polarized light. The sense of rotation of the ellipse is left and right in the upper and lower hemispheres, respectively. The poles P_1 and P_2 represent left and right circularly polarized light, respectively. (From Jerrard.)

where $-\pi \leq 2\theta \leq \pi$. We see immediately that for $2\varepsilon = 0$, which corresponds to the equator on the Poincaré sphere, (12-71) reduces to

$$S_0 = 1 \tag{12-72a}$$

$$S_1 = \cos 2\theta \tag{12-72b}$$

$$S_2 = \sin 2\theta \tag{12-72c}$$

$$S_3 = 0 \tag{12-72d}$$

Equation (12-72) is the Stokes parameters for linearly polarized light oriented at an angle θ. For $2\theta = 0$, (12-72) reduces to linear horizontally polarized light, for $2\theta = \pi/2$ we find linear $+45^\circ$ light, and for $2\theta = \pi$ linear vertically polarized light. Thus, as we move counterclockwise on the equator, we pass through different states of linearly polarized light.

If we now set $2\theta = 0$ so we move along the prime meridian (longitude), then (12-71) reduces to

$$S_0 = 1 \tag{12-73a}$$

$$S_1 = \cos 2\varepsilon \tag{12-73b}$$

$$S_2 = 0 \tag{12-73c}$$

$$S_3 = \sin 2\varepsilon \tag{12-73d}$$

Equation (12-73) is recognized as elliptically polarized light for the polarization ellipse in its standard form. We see that if we start from the equator ($2\varepsilon = 0$) and move up in latitude then at the pole we have $2\varepsilon = \pi/2$, and (12-73) reduces to right circularly polarized light. Similarly, moving down from the equator at the lower pole $2\varepsilon = -\pi/2$, and we have left circularly polarized light.

We can now summarize the major properties of the Poincaré sphere:

1. The coordinates of a point M on the Poincaré sphere are represented by latitude angle 2ε and longitude angle 2θ. A polarization state is described by $P(2\varepsilon, 2\theta)$.
2. The latitude $2\varepsilon = 0^\circ$ corresponds to the equator and for this angle the Stokes vector, (12-71), is seen to reduce to the Stokes vector for linearly polarized light, (12-72). Thus, linearly polarized light is always restricted to the equator. The angles $2\theta = 0^\circ$, 90°, 180°, and 270° correspond to the linear polarization states linear horizontal, linear $+45^\circ$, linear vertical, and linear -45°, respectively.
3. The longitude $2\theta = 0^\circ$ corresponds to the prime meridian and for this angle the Stokes vector, (12-71), is seen to reduce to the Stokes vector for elliptically polarized light for an nonrotated polarization ellipse, (12-73). According to (12-73) we see that for $2\varepsilon = 0^\circ$ we have linear horizontally polarized light and as we move up along the prime meridian we pass from right elliptically polarized light to right circularly polarized light at $2\varepsilon = 90^\circ$ (the north pole). Similarly, moving down the meridian from the equator we pass from left elliptically polarized light to left circularly polarized light at $2\varepsilon = -90^\circ$ at the south pole.

4. The points along a given parallel represent ellipses of the same form (ellipticities) but different orientations (azimuths).
5. The points on a given meridian represent vibrations of the same orientation (azimuth) whose eccentricity varies from 0 on the equator to ± 1 at the north and south poles, respectively.

The real power of the Poincaré sphere is that it enables us to determine the state of polarization of an optical beam after it has propagated through a polarizing element or several polarizing elements without carrying out the calculations. In the following section we apply the sphere to the problem of propagation of a polarized beam through (1) a polarizer, (2) a retarder, (3) a rotator, and (4) an elliptical polarizer consisting of a linear polarizer and a retarder.

12.4 APPLICATIONS OF THE POINCARÉ SPHERE

In Section 12.1 we pointed out that the Poincaré sphere was introduced by Poincaré in order to treat the problem of determining the polarization state of an optical beam after it had propagated through a number of polarizing elements. Simply put, given the Stokes parameters (vector) of the input beam, the problem is to determine the Stokes parameters of the output beam after it has propagated through a polarizing element or several polarizing elements. In this section we apply the Poincaré sphere to the problem of describing the effects of polarizing elements on an incident polarized beam. In order to understand this behavior, we first consider the problem using the Mueller matrix formalism, and then discuss the results in terms of the Poincaré sphere.

The Mueller matrix for an ideal linear polarizer rotated through an angle β is

$$M_P = \frac{1}{2}\begin{pmatrix} 1 & \cos 2\beta & \sin 2\beta & 0 \\ \cos 2\beta & \cos^2 2\beta & \sin 2\beta \cos 2\beta & 0 \\ \sin 2\beta & \sin 2\beta \cos 2\beta & \sin^2 2\beta & 0 \\ 0 & 0 & 0 & 0 \end{pmatrix} \tag{12-74}$$

In the previous section we saw that the Stokes vector of a beam of unit intensity and written in terms of its ellipticity ε and its azimuth (orientation) θ is given by

$$S = \begin{pmatrix} 1 \\ \cos 2\varepsilon \cos 2\theta \\ \cos 2\varepsilon \sin 2\theta \\ \sin 2\varepsilon \end{pmatrix} \tag{12-75}$$

The incident beam is now represented by (12-75) and is plotted as a point P on the Poincaré sphere, specifically $P(2\varepsilon, 2\theta)$. The polarized beam now propagates through the rotated polarizer, and the Stokes vector of the emerging beam is found by multiplying (12-75) by (12-74) to obtain the result:

$$S' = \frac{1}{2}[1 + \cos 2\varepsilon \cos 2(\beta - \theta)]\begin{pmatrix} 1 \\ \cos 2\beta \\ \sin 2\beta \\ 0 \end{pmatrix} \tag{12-76}$$

The Stokes vector of the emerging beam aside from the intensity factor in (12-76) can also be described in terms of ellipticity and orientation:

$$S' = \begin{pmatrix} 1 \\ \cos 2\varepsilon' \cos 2\theta' \\ \cos 2\varepsilon' \sin 2\theta' \\ \sin 2\varepsilon' \end{pmatrix} \tag{12-77}$$

The polarization state is described only by the parameters within the column matrix and (12-76) shows that, regardless of the polarization state of the incident beam, the polarization state of the emerging beam is a function only of β, the orientation angle of the linear polarizer. From (12-77) we see that $2\varepsilon' = 0$, so the ellipticity is zero and the point P', that is, $P(2\varepsilon', 2\theta')$, is always on the equator. Thus, regardless of the polarization state of the incident beam and its position on the Poincare spheré the final point P' is always located on the equator at the position 2β, that is $P'(0, 2\beta)$. We also see from (12-77) and (12-76) that $\tan 2\theta' = \tan 2\theta$, that is, $\theta' = \beta$, so the final longitude is β.

It is also possible to use the Poincaré sphere to obtain the intensity factor in (12-76). For the Poincaré sphere, 2ε corresponds to the parallels and 2θ corresponds to the longitudes. Within the factor in (12-76) we see that we have

$$\cos 2\varepsilon \cos(2\beta - 2\theta) \tag{12-78}$$

Now (12-27c) is

$$\cos 2\nu = \cos 2\varepsilon \cos 2\theta \tag{12-79}$$

Thus, (12-78) is obtained by constructing a right spherical triangle on the Poincaré sphere. In order to determine the magnitude of the arc on a great circle (2ν), we need only measure the length of the angle 2ε on the meridian (longitude) followed by measuring the length $(2\beta - 2\theta)$ on the equator (latitude). The length (2ν) of the arc of the great circle is then measured from the initial point of the meridian to the final point along the equator. This factor is then added to 1 and the final result is divided by 2, as required by (12-76). The Poincaré sphere, therefore, can also be used to determine the final intensity as well as the change in the polarization state.

The next case of interest is a retarder. The Mueller matrix is given by

$$M_R = \begin{pmatrix} 1 & 0 & 0 & 0 \\ 0 & 1 & 0 & 0 \\ 0 & 0 & \cos\phi & -\sin\phi \\ 0 & 0 & \sin\phi & \cos\phi \end{pmatrix} \tag{12-80}$$

In order to determine the point P' on the Poincaré sphere, we consider first the case where the incident beam is linearly polarized. For linearly polarized light with its azimuth plane at an angle α the Stokes vector is

$$S = \begin{pmatrix} 1 \\ \cos 2\alpha \\ \sin 2\alpha \\ 0 \end{pmatrix} \tag{12-81}$$

In terms of 2ε and 2θ (latitude and longitude), (12-81) can be expressed in terms of (12-75) as

$$S = \begin{pmatrix} 1 \\ \cos 2\varepsilon \cos 2\theta \\ \cos 2\varepsilon \sin 2\theta \\ \sin 2\varepsilon \end{pmatrix} \tag{12-75}$$

Equating the terms in (12-75) and (12-81):

$$\cos 2\varepsilon \cos 2\theta = \cos 2\alpha \tag{12-82a}$$

$$\cos 2\varepsilon \sin 2\theta = \sin 2\alpha \tag{12-82b}$$

$$\sin 2\varepsilon = 0 \tag{12-82c}$$

We immediately find from (12-82) that $2\varepsilon = 0$ and $2\theta = 2\alpha$. The latter result allows us to express (12-75) as

$$S = \begin{pmatrix} 1 \\ \cos 2\theta \\ \sin 2\theta \\ 0 \end{pmatrix} \tag{12-83}$$

The Stokes vector S' of the emerging beam is found by multiplying (12-83) by (12-80), so

$$S' = \begin{pmatrix} 1 \\ \cos 2\theta \\ \cos \phi \sin 2\theta \\ \sin \phi \sin 2\theta \end{pmatrix} \tag{12-84}$$

The corresponding Stokes vector in terms of $2\varepsilon'$ and $2\theta'$ is

$$S' = \begin{pmatrix} 1 \\ \cos 2\varepsilon' \cos 2\theta' \\ \cos 2\varepsilon' \sin 2\theta' \\ \sin 2\varepsilon' \end{pmatrix} \tag{12-85}$$

Equating terms in (12-84) and (12-85) gives

$$\sin 2\varepsilon' = \sin 2\theta \sin \phi \tag{12-86a}$$

$$\tan 2\theta' = \tan 2\theta \cos \phi \tag{12-86b}$$

Equations (12-86a) and (12-86b) can be expressed in terms of the right spherical triangle shown in Fig. 12-10. The figure is constructed using the equations for a right spherical triangle given at the end of Section 12.2 (compare Figure 12-10 to Figure 12-N5). Figure 12-10 shows how the retarder moves the initial point $P(2\varepsilon, 2\theta)$ to $P'(2\varepsilon', 2\theta')$ on the Poincaré sphere. To carry out the operations equivalent to the right spherical triangle, the following steps are performed:

1. Determine the initial point $P(2\varepsilon = 0, 2\theta)$ on the equator and label it A.
2. Draw an angle at A from the equator of magnitude ϕ, the phase shift of the retarder.

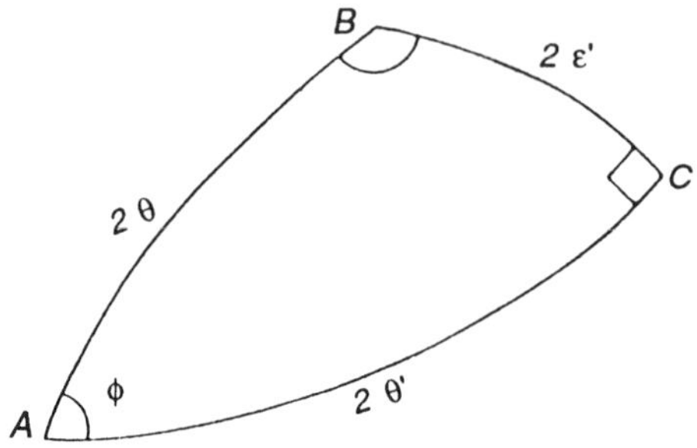

Figure 12-10 Right spherical triangle for a retarder.

3. Measure the arc length 2θ along the equator from A. Then draw this arc length from A to B. The end of this arc corresponds to the point P' $(2\varepsilon', 2\theta')$.
4. The meridian $2\varepsilon'$ is drawn down to the equator; this arc length corresponds to the ellipticity angle $2\varepsilon'$. The intersection of the meridian with the equator is the orientation angle $2\theta'$. Three cases are of special interest: linear horizontally polarized light, linear $+45^\circ$ polarized light, and linear vertically polarized light. We discuss each of these cases and their interaction of a retarder as they are described on the Poincaré sphere.
   a. Linear horizontally polarized light. For this case $2\alpha = 2\theta = 0^\circ$. We see from (12-86a) and (12-86b) that $2\varepsilon'$ and $2\theta'$ are zero. Thus, the linear horizontally polarized light is unaffected by the retarder and P is identical to P'.
   b. Linear $+45^\circ$ light. Here, $2\alpha = 2\theta = \pi/2$, and from (12-86a) and (12-86b) we have

$$\sin 2\varepsilon' = \sin \phi \tag{12-87a}$$

$$\tan 2\theta' = \infty \tag{12-87b}$$

   Thus, the arc length (the longitude or the meridian) is $2\varepsilon' = \phi$ and $2\theta' = \pi/2$. We see that as ϕ increases, $2\varepsilon'$ increases, so that when $2\varepsilon' = \pi/2$, which corresponds to right circularly polarized light, the arc length $2\varepsilon'$ extends from the equator to the pole.
   c. Linear vertically polarized light. For this final case $2\alpha = 2\theta = \pi$. We see from (12-86a) that $2\varepsilon' = 0$, that is, P' is on the equator. However, tan $2\theta = -\infty$, so $2\theta' = -\pi$. Thus, P' is on the equator but diametrically opposite to P on the Poincaré sphere.

The Stokes vector confirms this behavior for these three cases, since we have from (12-84) that

$$S' = \begin{pmatrix} 1 \\ \cos 2\theta \\ \cos\phi \sin 2\theta \\ \sin\phi \sin 2\theta \end{pmatrix} \tag{12-84}$$

which reduces to linear horizontally polarized, linear $+45^\circ$, and linear vertically polarized light for $2\theta = 0, \pi/2$, and π respectively.

We now consider the case where the incident light is elliptically polarized. In order to understand the behavior of the elliptically polarized light and the effect of a retarder on its polarization state in terms of the Poincaré sphere, we write the Stokes vector for the incident beam as

$$S = \begin{pmatrix} S_0 \\ S_1 \\ S_2 \\ S_3 \end{pmatrix} \tag{12-88}$$

Multiplying (12-88) by (12-80) then gives

$$S' = \begin{pmatrix} S_0 \\ S_1 \\ S_2 \cos\phi - S_3 \sin\phi \\ S_2 \sin\phi - S_3 \cos\phi \end{pmatrix} \tag{12-89}$$

Now, the third and fourth elements (S_2' and S_3') describe rotation through the angle ϕ. To see this behavior more clearly, let us consider (12-89) for a quarter-wave retarder ($\phi = \pi/2$) and a half-wave retarder ($\phi = \pi$). For these cases (12-89) reduces, respectively, to

$$S' = \begin{pmatrix} S_0 \\ S_1 \\ -S_3 \\ S_2 \end{pmatrix} \tag{12-90a}$$

and

$$S' = \begin{pmatrix} S_0 \\ S_1 \\ -S_2 \\ -S_3 \end{pmatrix} \tag{12-90b}$$

Let us now consider the Poincaré sphere in which we show the axes labeled as $S_1, S_2,$ and S_3. We see that, according to (12-90a) and (12-90b), S_1 remains invariant, but $S_2 \to -S_3 \to -S_2$ and $S_3 \to S_2 \to -S_3$. As can be seen from Fig. 12-11 for the

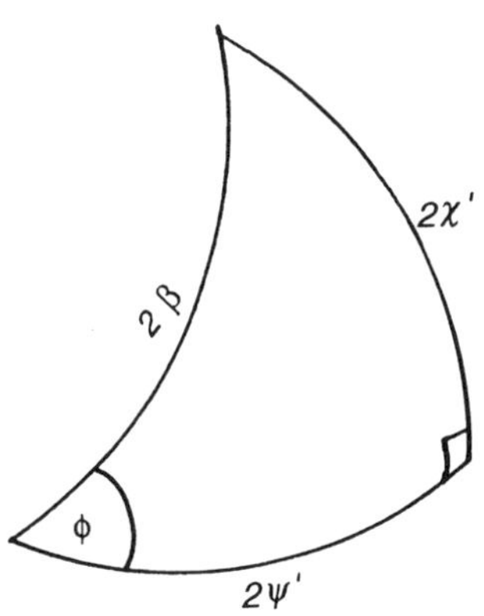

Figure 12-11 Right spherical triangle for a linear polarizer–retarder combination.

Poincaré sphere, this corresponds to rotating the sphere around the S_1 axis sequentially through $\pi/2$ and then again through another $\pi/2$ for a total rotation of π. Thus, the effect of the retarder can be expressed merely by rotating the Poincaré sphere around the S_1 axis; the magnitude of the rotation is equal to the phase shift ϕ. It is this remarkably simple property of the Poincaré sphere which has led to its great use and interest.

In terms of the equation for $2\varepsilon'$ and $2\theta'$, we can obtain these values by determining the Stokes vector S' of the emerging beam, namely, by multiplying (12-75) by (12-80). The result is easily seen to be

$$S' = \begin{pmatrix} 1 \\ \cos 2\varepsilon \cos 2\theta \\ \cos\phi \cos 2\varepsilon \sin 2\theta - \sin\phi \sin 2\varepsilon \\ \sin\phi \cos 2\varepsilon \sin 2\theta + \cos\phi \sin 2\varepsilon \end{pmatrix} \tag{12-91}$$

We immediately find by equating the elements of (12-91) to (12-85) that

$$\tan 2\theta' = \frac{\cos\phi \cos 2\varepsilon \sin 2\theta - \sin\phi \sin 2\varepsilon}{\cos 2\varepsilon \cos 2\theta} \tag{12-92a}$$

$$\sin 2\varepsilon' = \sin\phi \cos 2\varepsilon \sin 2\theta + \cos\phi \sin 2\varepsilon \tag{12-92b}$$

In the Stokes vector (12-91) the element S_1 is recognized as the relation for a right spherical triangle, namely, (12-N21a). The elements S_2 and S_3 are the relations for an oblique spherical triangle if the angle C shown in Fig. 12-N5 is an oblique angle. We can use the Poincaré sphere to obtain the orientation angle θ' and the ellipticity angle χ'. For example, if we set $a = 90^\circ - 2\varepsilon$, $b = \phi$, $c = 90^\circ - 2\varepsilon'$, and $C = 90^\circ - 2\theta$ in (12-N17c) we obtain the ellipticity angle (12-92b) of the emerging beam; a similar set of angles leads to the orientation angle (12-92a).

We now turn to the problem of describing the interaction of an elliptically polarized beam with a rotator using the Poincaré sphere. The Mueller matrix for a rotator is

$$M_{\text{rot}} = \begin{pmatrix} 1 & 0 & 0 & 0 \\ 0 & \cos 2\beta & \sin 2\beta & 0 \\ 0 & -\sin 2\beta & \cos 2\beta & 0 \\ 0 & 0 & 0 & 1 \end{pmatrix} \tag{12-93}$$

where β is the angle of rotation. The Stokes vector of the emerging beam is found by multiplying (12-88) by (12-93):

$$S' = \begin{pmatrix} 1 \\ S_1 \cos 2\beta + S_2 \sin 2\beta \\ -S_1 \sin 2\beta + S_2 \cos 2\beta \\ S_3 \end{pmatrix} \tag{12-94}$$

or in terms of ε and θ, (12-75) by (12-93),

$$S' = \begin{pmatrix} 1 \\ \cos 2\varepsilon \cos 2(\theta - \beta) \\ \cos 2\varepsilon \sin 2(\theta - \beta) \\ \sin 2\varepsilon \end{pmatrix} \tag{12-95}$$

We see that from (12-94) we have a rotation around the S_3 axis, i.e., starting from $2\beta = 0°$ and moving to $270°$ in increments of $90°$, $S_1 \rightarrow S_2 \rightarrow -S_1 \rightarrow -S_2$ and, similarly, $S_2 \rightarrow -S_1 \rightarrow -S_2 \rightarrow S_1$. Thus, rotating the Poincaré sphere around the S_3 axis by β transforms $P(2\varepsilon, 2\theta)$ to $P'(2\varepsilon, 2(\theta - \beta))$; the ellipticity angle ε remains unchanged and only the orientation of the polarization ellipse is changed.

To summarize, the rotation around the S_1 axis describes the change in phase, i.e., propagation through a birefringent medium, and the rotation around the S_3 axis describes the change in azimuth, i.e., propagation through an optically active medium.

The final problem we consider is the propagation of a polarized beam through a linear polarizer oriented at an angle θ to the x axis followed by a retarder with its fast axis along the x axis. For the linear polarizer we have from (12-76) and (12-77) that

$$S' = \frac{1}{2}[1 + \cos 2\varepsilon \cos 2(\beta - \theta)] \begin{pmatrix} 1 \\ \cos 2\beta \\ \sin 2\beta \\ 0 \end{pmatrix} \tag{12-76}$$

$$S' = \begin{pmatrix} 1 \\ \cos 2\varepsilon' \cos 2\theta' \\ \cos 2\varepsilon' \sin 2\theta' \\ \sin 2\varepsilon' \end{pmatrix} \tag{12-77}$$

We, of course, immediately see that $2\varepsilon'$ is zero and $2\theta' = 2\beta$. The point $P(2\varepsilon, 2\theta)$, the incident beam, is moved along the equator through an angle β to the point $P'(2\varepsilon', 2\theta')$. Equivalently, we need only rotate the Poincaré sphere around its polar axis. Next, the beam propagates through the retarder. Using (12-76) and the Mueller matrix for a retarder (12-80), we see that the Stokes vector is

$$S' = \begin{pmatrix} 1 \\ \cos 2\beta \\ \sin 2\beta \cos \phi \\ \sin 2\beta \sin \phi \end{pmatrix} \tag{12-96}$$

We now equate the elements in (12-96) with (12-77) to obtain

$$\cos 2\varepsilon' \cos 2\theta' = \cos 2\beta \tag{12-97a}$$

$$\cos 2\varepsilon' \sin 2\theta' = \sin 2\beta \cos \phi \tag{12-97b}$$

$$\sin 2\varepsilon' = \sin 2\beta \sin \phi \tag{12-97c}$$

Alternatively, we can equate the elements to the Stokes vector representation of elliptically polarized light using χ' and ψ' notation for (12-77):

$$S' = \begin{pmatrix} 1 \\ \cos 2\chi' \cos 2\psi' \\ \cos 2\chi' \sin 2\psi' \\ \sin 2\chi' \end{pmatrix} \tag{12-98}$$

Equating elements of (12-96) with (12-98) yields

$$\cos 2\chi' \cos 2\psi' = \cos 2\beta \tag{12-99a}$$

$$\cos 2\chi' \sin 2\psi' = \sin 2\beta \cos \phi \tag{12-99b}$$

$$\sin 2\chi' = \sin 2\beta \sin \phi \tag{12-99c}$$

We now divide (12-99c) by (12-99b) to obtain

$$\tan 2\chi' = \sin 2\psi' \tan \phi \tag{12-100}$$

Similarly, we divide (12-99b) by (12-99a) and find that

$$\tan 2\psi' = \tan 2\beta \cos \phi \tag{12-101}$$

We now collect (12-99a), (12-100) and (12-101) and write

$$\cos 2\chi' \cos 2\psi' = \cos 2\beta \tag{12-102a}$$

$$\tan 2\chi' = \sin 2\psi \tan \phi \tag{12-102b}$$

$$\tan 2\psi' = \tan 2\beta \cos \phi \tag{12-102c}$$

Not surprisingly, (12-102a), (12-102b), and (12-102c) correspond to (12-N21a), (12-N21d), and (12-N21f), respectively. These equations are satisfied by the right spherical triangle in Fig. 12-11. The arc 2β and the angle ϕ determine the magnitudes $2\chi'$ and $2\psi'$. We see that all that is required to determine these latter two angles is to rotate the sphere through an angle ϕ around the S_1 axis and then to measure the arc length 2β. We note that the magnitude of the angle ϕ is then confirmed by the intersection of the arcs 2β and $2\psi'$.

A number of further applications of the Poincaré sphere have been given in the optical literature. A very good introduction to some of the simplest aspects of the Poincaré sphere and certainly one of the clearest descriptions is found in Shurcliff. An excellent and very detailed description, as well as a number of applications, has been given by Jerrard; much of the material presented in this chapter is based on Jerrard's excellent paper. Further applications have been considered by Ramaseshan and Ramachandran, who have also described the Poincaré sphere and its application in a very long and extensive review article entitled "Crystal Optics" in the *Handbuch der Physik*. This is not an easy article to read, however, and requires much time and study to digest fully. Finally, E. A. West et al. have given an excellent discussion of the application of the Poincaré sphere to the design of a polarimeter to measure solar vector magnetic fields.

Remarkably, even though the Poincaré sphere was introduced a century ago, papers on the subject continue to appear. A recent paper of interest on a planar graphic representation of the state of polarization, a planar Poincaré chart, is given by Tedjojuwono et al. Finally, a very good review has been recently published by Boerner et al. on polarized light and includes other projections analogous to the stereographic projection (the mercator, the azimuthal, etc.).

REFERENCES

Papers

1. Jerrard, H. G., *J. Opt. Soc, Am.*, *44*, 634 (1954).
2. Ramachandran, G. N. and Rameseshan, S., *J. Opt. Soc. Am.*, *42*, 49 (1952).
3. Jerrard, H. G., *J. Opt. Soc. Am.*, *38*, 35 (1948).
4. West, E. A., Reichman, E. J., Hagyard, M. J., and Gary, G. A., *Opt. Eng.*, *29*, 131 (1989).
5. Tedjojuwono, K. K., Hunter, W. W. Jr. and Ocheltree, S. L., *Appl. Opt.*, *28*, 2614 (1989).
6. Boerner, W. M., Yan, Wei-Ling and Xi, An-Qing, *Proc. SPIE*, *1317*, 16 (1990).

Books

1. Poincaré H., *Théorie Mathematique de la Lumieré*, Gauthiers-Villars, Paris, 1892, Vol. 2, Chap. 12.
2. Shurcliff, W. A., *Polarized Light*, Harvard University Press, Cambridge, MA, 1962.
3. Ramachandran, G. N. and Ramaseshan, S., "*Crystal Optics*," *Encyclopedia of Physics*, Vol. XXV/1 (S. Flügge, ed.), Springer-Verlag, Berlin, 1961.
4. Azzam, R. M. A. and Bashara, N. M., *Ellipsometry and Polarized Light*, North-Holland, Amsterdam, 1977.
5. Simmons, J. W. and Guttman, M. J., *States, Waves and Photons*, Addison-Wesley, Reading, MA, 1970.

13

The Interference Laws of Fresnel and Arago

13.1 INTRODUCTION

In this last chapter of the first part, we now turn to the topic that led Stokes to introduce his polarization parameters, namely, the mathematical formulation of unpolarized light and its application to the interference laws of Fresnel and Arago. In this section the events that led up to Stokes' investigation are described. We briefly review these events.

The investigation by Stokes that led to his paper in 1852 began with the experiments performed by Fresnel and Arago in 1817. At the beginning of these experiments both Fresnel and Arago held the view that light vibrations were longitudinal. However, one of the results of these experiments, namely, that two rays that are polarized at right angles could in no way give rise to interference, greatly puzzled Fresnel. Such a result was impossible to understand on the basis of light vibrations that are longitudinal. Young heard of the experiments from Arago and suggested that the results could be completely understood if the light vibrations were transverse. Fresnel immediately recognized that this condition would indeed make the experiments intelligible. Indeed, as J. Strong has correctly pointed out, only after these experiments had been performed was the transverse nature of light as well as the properties of linearly, circularly, and elliptically polarized light fully understood.

The results of the Fresnel–Arago experiments have been succinctly stated as the interference laws of Fresnel and Arago. These laws, of which there are four, can be summarized as follows:

1. Two waves linearly polarized in the same plane can interfere.
2. Two waves linearly polarized with perpendicular polarizations cannot interfere.
3. Two waves linearly polarized with perpendicular polarizations, if derived from perpendicular components of unpolarized light and subsequently brought into the same plane, cannot interfere.

4. Two waves linearly polarized with perpendicular polarizations, if derived from the same linearly polarized wave and subsequently brought into the same plane, can interfere.

The fact that orthogonally polarized rays cannot be made to interfere can be taken as a proof that light vibrations are transverse. This leads to a complete understanding of laws 1 and 2. The same confidence in understanding cannot be made with respect to laws 3 and 4, however. For these laws involve unpolarized light, a quantity that Fresnel and Arago were unable to understand completely or to characterize mathematically. As a consequence, they never attempted a mathematical formulation of these laws and merely presented them as experimental facts.

Having established the basic properties of unpolarized, as well as partially polarized light, along with their mathematical formulation, Stokes then took up the question of the interference laws of Fresnel and Arago. The remarkable fact now emerges that Stokes made no attempt to formulate these laws. Rather, he analyzed a related experiment that Stokes states is due to Sir John Herschel. This experiment is briefly discussed at the end of this chapter.

The analysis of the interference laws is easily carried to completion by means of the Mueller matrix formalism. The lack of a matrix formalism does not preclude a complete analysis of the experiments, but the use of matrices does make the calculations far simpler to perform. We shall first discuss the mathematical statements of unpolarized light. With these statements we then analyze the experiments through the use of matrices, and we present the final results in the form of the Stokes vectors.

The apparatus that was used by Fresnel and Arago is similar to that devised by Young to demonstrate the phenomenon of interference arising from two slits. In their experiments, however, polarizers are appropriately placed in front of the light source and behind the slits in order to obtain various interference effects. Another polarizer is placed behind the observation screen in two of the experiments in order to bring the fields into the same plane of polarization. The optical configuration will be described for each experiment as we go along.

13.2 MATHEMATICAL STATEMENTS FOR UNPOLARIZED LIGHT

In most optics texts very little attention is paid to the subject of unpolarized light. This subject was the source of numerous investigations during the nineteenth century and first half of the twentieth century. One of the major reasons for this interest was that until the invention of the laser practically every known optical source emitted only unpolarized light. Ironically, when the subject of unpolarized light was finally "understood" in the late 1940s and 1950s, a new optical source, the laser, was invented and it was completely polarized! While there is a natural tendency to think of lasers as the optical source of choice, the fact is that unpolarized light sources continue to be widely used in optical laboratories. This observation is supported by looking into any commercial optics catalog. One quickly discovers that manufacturers continue to develop and build many types of optical sources, including black-body sources, deuterium lamps, halogen lamps, mercury lamps, tungsten lamps, etc., all of which are substantially unpolarized. Consequently, the subject of unpolarized light is still of major importance not only for understanding the Fresnel–Arago laws but because of the existence and use of these optical sources. Hence, we should keep in mind that the subject of unpolarized light is far from being only of academic interest.

In all of the experiments of Fresnel and Arago an unpolarized source of light is used. The mathematical statements that characterize unpolarized light will now be developed, and these expressions will then be used in the analysis of the Fresnel–Arago experiments and the formulation of their laws.

The Stokes parameters of a beam of light, as first shown by Stokes, can be determined experimentally by allowing a beam of light to propagate through a retarder of retardance ϕ and then through a polarizer with its transmission axis at an angle θ from the x axis. The observed intensity $I(\theta, \phi)$ of the beam is found to be

$$I(\theta, \phi) = \frac{1}{2}[S_0 + S_1 \cos 2\theta + S_2 \sin 2\theta \cos \phi - S_3 \sin 2\theta \sin \phi] \tag{13-1}$$

where S_0, S_1, S_2, and S_3 are the Stokes parameters of the incident beam. In order to use (13-1) to characterize unpolarized light, Stokes invoked the experimental fact that the observed intensity of unpolarized light is unaffected by the presence of the retarder and the orientation of the polarizer. In other words, $I(\theta, \phi)$ must be independent of θ and ϕ. This condition can only be satisfied if

$$S_1 = S_2 = S_3 = 0, \quad S_0 \neq 0 \tag{13-2a}$$

so

$$I(\theta, \phi) = S_0/2 \tag{13-2b}$$

The Stokes parameters for a time-varying field with orthogonal components $E_x(t)$ and $E_y(t)$ in a linear basis are defined to be

$$S_0 = \langle E_x(t)E_x^*(t)\rangle + \langle E_y(t)E_y^*(t)\rangle \tag{13-3a}$$

$$S_1 = \langle E_x(t)E_x^*(t)\rangle - \langle E_y(t)E_y^*(t)\rangle \tag{13-3b}$$

$$S_2 = \langle E_x(t)E_y^*(t)\rangle + \langle E_y(t)E_x^*(t)\rangle \tag{13-3c}$$

$$S_3 = i\langle E_x(t)E_y^*(t)\rangle - i\langle E_y(t)E_x^*(t)\rangle \tag{13-3d}$$

where $\langle \cdots \rangle$ means a time average and an asterisk signifies the complex conjugate. The Stokes parameters for an unpolarized beam (13-2) can be expressed in terms of the definition of (13-3) so we have

$$\langle E_x(t)E_x^*(t)\rangle + \langle E_y(t)E_y^*(t)\rangle = S_0 \tag{13-4a}$$

$$\langle E_x(t)E_x^*(t)\rangle - \langle E_y(t)E_y^*(t)\rangle = 0 \tag{13-4b}$$

$$\langle E_x(t)E_y^*(t)\rangle + \langle E_y(t)E_x^*(t)\rangle = 0 \tag{13-4c}$$

$$i\langle E_x(t)E_y^*(t)\rangle - i\langle E_y(t)E_x^*(t)\rangle = 0 \tag{13-4d}$$

From (13-4a) and (13-4b) we see that

$$\langle E_x(t)E_x^*(t)\rangle = \langle E_y(t)E_y^*(t)\rangle = \frac{1}{2}S_0 \tag{13-5}$$

Thus, we conclude from (13-5) that the time-averaged orthogonal quadratic field components are equal, and so for unpolarized light we tentatively set

$$E_x(t) = E_y(t) = A(t) \tag{13-6}$$

This expression indeed satisfies (13-4a) and (13-4b). However, from (13-4c) and (13-4d) we have

$$\langle E_x(t)E_y^*(t)\rangle = \langle E_y(t)E_x^*(t)\rangle = 0 \tag{13-7a}$$

and this cannot be satisfied by (13-6). Therefore, we must set

$$E_x(t) = A_x(t) \tag{13-7b}$$

$$E_y(t) = A_y(t) \tag{13-7c}$$

in order to satisfy (13-4a) through (13-4d). We see that unpolarized light can be represented by

$$\langle A_x(t)A_x^*(t)\rangle = \langle A_y(t)A_y^*(t)\rangle = \langle A(t)A^*(t)\rangle \tag{13-8a}$$

and

$$\langle A_x(t)A_y^*(t)\rangle = \langle A_y(t)A_x^*(t)\rangle = 0 \tag{13-8b}$$

Equations (13-8) are the classical mathematical statements for unpolarized light. The condition (13-8b) is a statement that the orthogonal components of unpolarized light have no permanent phase relation. In the language of statistical analysis, (13-8b) states that the orthogonal field components of unpolarized light are uncorrelated. We can express (13-8a) and (13-8b) as a single statement by writing

$$\langle A_i(t)A_j^*(t)\rangle = \langle A(t)A^*(t)\rangle \cdot \delta_{ij} \qquad i,j = x, y \tag{13-9a}$$

where δ_{ij} is the Kronecker delta defined by

$$\delta_{ij} = 1 \qquad \text{if } i = j \tag{13-9b}$$

$$\delta_{ij} = 0 \qquad \text{if } i \neq j \tag{13-9c}$$

13.3 YOUNG'S INTERFERENCE EXPERIMENT WITH UNPOLARIZED LIGHT

Before we treat the Fresnel–Arago experiments, we consider Young's interference experiment with an unpolarized light source using the results of the previous section. In many treatments of Young's interference experiments, a discussion of the nature of the light source is avoided. In fact, nearly all descriptions of the experiment in many textbooks begin with the fields at each of the slits and then proceed to show that interference occurs because of the differences in path lengths between the slits and the screen. It is fortuitous, however, that *regardless* of the nature of the light source and its state of polarization, interference will always be observed. It was fortunate for the science of optics that the phenomenon of interference could be described without having to understand the nature of the optical source. Had optical physicists been forced to attack the problem of the polarization of sources before proceeding, the difficulties might have been insurmountable and, possibly, greatly impeded further progress. Fortunately, this did not occur. Nevertheless, the problem of characterizing the polarization of light remained a problem well into the twentieth century as a reading of the papers in the references at the end of this chapter show.

Many beginning students of physical optics sometimes believe that Young's experiment must be performed with light that is specially prepared; i.e., initially the light source is unpolarized and then is transformed to linear polarized light before it arrives at the slits. The fact is, however, that interference phenomena can be observed with unpolarized light. This can be easily shown with the mathematical statements derived in the previous section.

In Young's experiment an unpolarized light source is symmetrically placed between the slits A and B as shown in Fig. 13-1. The Stokes vector of the unpolarized light can again be decomposed in the following manner:

$$S_\sigma = \langle AA^* \rangle \begin{pmatrix} 1 \\ 0 \\ 0 \\ 0 \end{pmatrix} = \frac{1}{2}\langle A_x A_x^* \rangle \begin{pmatrix} 1 \\ 1 \\ 0 \\ 0 \end{pmatrix} + \frac{1}{2}\langle A_y A_y^* \rangle \begin{pmatrix} 1 \\ -1 \\ 0 \\ 0 \end{pmatrix} \tag{13-10}$$

The Stokes vector at slit A is

$$S_A = \frac{1}{2}\langle A_x A_x^* \rangle \begin{pmatrix} 1 \\ 1 \\ 0 \\ 0 \end{pmatrix}_A + \frac{1}{2}\langle A_y A_y^* \rangle \begin{pmatrix} 1 \\ -1 \\ 0 \\ 0 \end{pmatrix}_A \tag{13-11a}$$

and at slit B is

$$S_B = \frac{1}{2}\langle A_x A_x^* \rangle \begin{pmatrix} 1 \\ 1 \\ 0 \\ 0 \end{pmatrix}_B + \frac{1}{2}\langle A_y A_y^* \rangle \begin{pmatrix} 1 \\ -1 \\ 0 \\ 0 \end{pmatrix}_B \tag{13-11b}$$

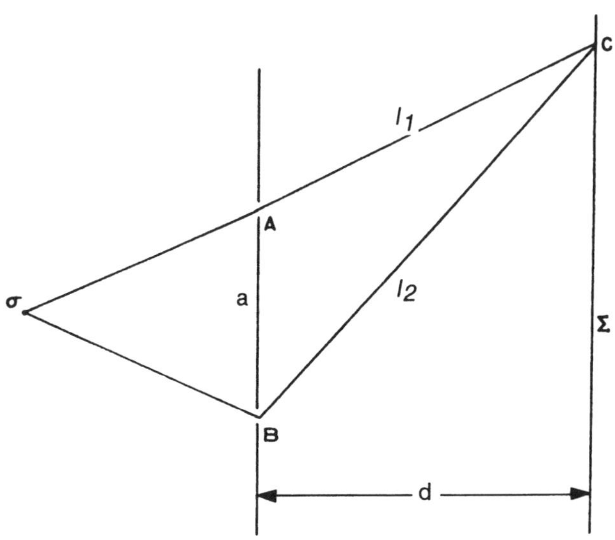

Figure 13-1 Young's interference experiment with unpolarized light.

where the subscripts A and B remind us that these are the Stokes vectors of the field at the respective slits.

The fields which satisfy the Stokes vector S_A are

$$E_{xA}(t) = \frac{A_x(t)}{\sqrt{2}} \qquad E_{yA}(t) = \frac{A_y(t)}{\sqrt{2}} \tag{13-12a}$$

and S_B

$$E_{xB}(t) = \frac{A_x(t)}{\sqrt{2}} \qquad E_{yB}(t) = \frac{A_y(t)}{\sqrt{2}} \tag{13-12b}$$

The field components at point C on the screen arising from the field propagating from slit A is

$$E_{xA}(t) = \frac{A_x(t)}{\sqrt{2}} \exp(i\phi_A) \tag{13-13a}$$

$$E_{yA}(t) = \frac{A_y(t)}{\sqrt{2}} \exp(i\phi_A) \tag{13-13b}$$

and, similarly, that due to slit B

$$E_{xB}(t) = \frac{A_x(t)}{\sqrt{2}} \exp(i\phi_B) \tag{13-14a}$$

$$E_{yB}(t) = \frac{A_y(t)}{\sqrt{2}} \exp(i\phi_B) \tag{13-14b}$$

The total field in the x and y directions is

$$E_x(t) = E_{xA}(t) + E_{xB}(t) = \frac{A_x(t)}{\sqrt{2}} [\exp(i\phi_A) + \exp(i\phi_B)] \tag{13-15a}$$

and

$$E_y(t) = E_{yA}(t) + E_{yB}(t) = \frac{A_y(t)}{\sqrt{2}} [\exp(i\phi_A) + \exp(i\phi_B)] \tag{13-15b}$$

or

$$E_x(t) = \frac{A_x(t)}{\sqrt{2}} (1 + e^{i\phi}) \tag{13-16a}$$

$$E_y(t) = \frac{A_y(t)}{\sqrt{2}} (1 + e^{i\phi}) \tag{13-16b}$$

where $\phi = \phi_B - \phi_A$ and the constant factor $\exp(i\phi_A)$ has been dropped. Equation (13-16) describes the field components at a point C on the observing screen. It is interesting to note that it is not necessary at this point to know the relation between the slit separation and the distance between the slits and the observing screen. Later, this relation will have to be known to obtain a *quantitative* description of the interference phenomenon. We shall see shortly that interference is predicted with the information presented above.

The Stokes vector for (13-16) is now formed in accordance with (13-3) and applying the conditions for unpolarized light (13-8) or (13-9). We then find that the Stokes vector for the field at C is

$$S = \langle AA^* \rangle (1 + \cos\phi) \begin{pmatrix} 1 \\ 0 \\ 0 \\ 0 \end{pmatrix} \tag{13-17}$$

Thus, we see from (13-17) that light observed on the screen is still unpolarized. Furthermore, the intensity is

$$I = \langle AA^* \rangle (1 + \cos\phi) \tag{13-18}$$

Equation (13-18) is the familiar statement for describing interference. According to (13-18), the interference pattern on the screen will consist of bright and dark (null intensity) lines.

In order to use (13-18) for a quantitative measurement, the specific relation between the slit separation and the distance from the slits to the screen must be known. This is described by $\phi = \phi_B - \phi_A = k\Delta l$, where $k = 2\pi/\lambda$ and Δl is the path difference between the fields propagating from A and B to C. The phase shift can be expressed in terms of the parameters shown in Fig. 13-1.

$$l_2^2 = d^2 + \left(y + \frac{a}{2}\right)^2 \tag{13-19a}$$

$$l_1^2 = d^2 + \left(y - \frac{a}{2}\right)^2 \tag{13-19b}$$

Subtracting (13-19b) from (13-19a) yields

$$l_2^2 - l_1^2 = 2ay \tag{13-20}$$

We can assume that a is small, $d \gg a$, and c is not far from the origin so that

$$l_2 + l_1 \cong 2d \tag{13-21}$$

so (13-20) becomes

$$\Delta l = l_2 - l_1 = \frac{ay}{d} \tag{13-22}$$

The phase shift ϕ is then

$$\phi = \phi_B - \phi_A = k\Delta l = \frac{2\pi ay}{\lambda d} \tag{13-23}$$

where $k = 2\pi/\lambda$ is the wavenumber and λ is the wavelength of the optical field. The maximum intensities are, of course, observed when $\cos\phi = 1$, $\phi = 2\pi m$ so that

$$y = \left(\frac{\lambda d}{a}\right) m \quad m = 0, 1, 2, \ldots \tag{13-24a}$$

and the minimum (null) intensities are observed when $\cos\phi = -1$, $\phi = (2m+1)\pi$ so that

$$y = \left(\frac{\lambda d}{a}\right)m \qquad m = \frac{1}{2}, \frac{3}{2}, \frac{5}{2}, \ldots \tag{13-24b}$$

One can easily show that, regardless of the state of polarization of the incident beam, interference will be observed. Historically, this was first done by Young and then by Fresnel and Arago, using unpolarized light.

We now consider the mathematical formulation of the Fresnel–Arago interference laws.

13.4 THE FIRST EXPERIMENT: FIRST AND SECOND INTERFERENCE LAWS

We consider a source of unpolarized light σ symmetrically placed between slits A and B as shown in Fig. 13-2. A linear polarizer P_σ with its transmission axis parallel to the x axis is placed in front of the light source. A pair of similar polarizers P_A and P_B are also placed behind slits A and B, respectively. The transmission axes of these polarizers P_A and P_B are at angles α and β with respect to the x axis, respectively. We wish to determine the intensity and polarization of the light on the screen Σ.

The Stokes vector for unpolarized light of intensity AA^* can be represented by

$$S_\sigma = \langle A(t)A^*(t)\rangle \begin{pmatrix} 1 \\ 0 \\ 0 \\ 0 \end{pmatrix} \tag{13-25}$$

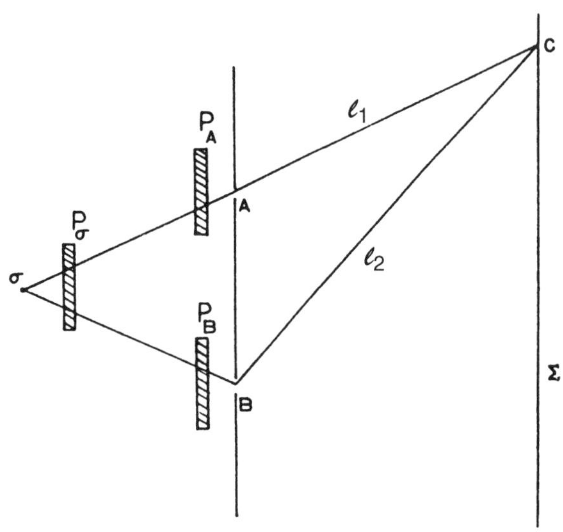

Figure 13.2 The first experiment. The transmission axis of the P_σ is parallel to the x axis. The transmission axes of P_A and P_B are at angles α and β from the x axis.

Equation (13-25) can be decomposed into two orthogonally linearly polarized beams. We then write

$$S_\sigma = \langle AA^* \rangle \begin{pmatrix} 1 \\ 0 \\ 0 \\ 0 \end{pmatrix} = \frac{1}{2}\langle A_x A_x^* \rangle \begin{pmatrix} 1 \\ 1 \\ 0 \\ 0 \end{pmatrix} + \frac{1}{2}\langle A_y A_y^* \rangle \begin{pmatrix} 1 \\ -1 \\ 0 \\ 0 \end{pmatrix} \tag{13-26}$$

where we have used (13-8a).
The Mueller matrix for P_σ is

$$M_\sigma = \frac{1}{2}\begin{pmatrix} 1 & 1 & 0 & 0 \\ 1 & 1 & 0 & 0 \\ 0 & 0 & 0 & 0 \\ 0 & 0 & 0 & 0 \end{pmatrix} \tag{13-27}$$

The output beam from P_σ is obtained from the multiplication of (13-27) and (13-26):

$$S_{P_\sigma} = \frac{1}{2}\langle A_x A_x^* \rangle \begin{pmatrix} 1 \\ 1 \\ 0 \\ 0 \end{pmatrix} \tag{13-28}$$

Thus, the polarizer P_σ transmits the horizontal and rejects the vertical component of the unpolarized light, (13-26). The light is now linearly horizontally polarized.

The matrix of a polarizer, M_P, with its transmission axis at an angle θ from the x axis, is determined from

$$M_P(2\theta) = M(-2\theta) M_P M(2\theta) \tag{13-29}$$

where $M_P(2\theta)$ is the matrix of the rotated polarizer and $M(2\theta)$ is the rotation matrix:

$$M(2\theta) = \begin{pmatrix} 1 & 0 & 0 & 0 \\ 0 & \cos 2\theta & \sin 2\theta & 0 \\ 0 & -\sin 2\theta & \cos 2\theta & 0 \\ 0 & 0 & 0 & 1 \end{pmatrix} \tag{13-30}$$

The Mueller matrix for P_A is then found by setting $\theta = \alpha$ in (13-30) and then substituting (13-27) into (13-29). The result is

$$M_{P_A} = \frac{1}{2}\begin{pmatrix} 1 & \cos 2\alpha & \sin 2\alpha & 0 \\ \cos 2\alpha & \cos^2 2\alpha & \cos 2\alpha \sin 2\alpha & 0 \\ \sin 2\alpha & \cos 2\alpha \sin 2\alpha & \sin^2 2\alpha & 0 \\ 0 & 0 & 0 & 0 \end{pmatrix} \tag{13-31}$$

A similar result holds for M_{P_B} with α replaced by β. The Stokes vector S_A that emerges from P_A is obtained by the multiplication of (13-28) by (13-31):

$$S_A = \frac{1}{2}\langle A_x A_x^* \rangle \cos^2\alpha \begin{pmatrix} 1 \\ \cos 2\alpha \\ \sin 2\alpha \\ 0 \end{pmatrix} \tag{13-32a}$$

In a similar manner the Stokes vector S_B is found to be

$$S_B = \frac{1}{2}\langle A_x A_x^* \rangle \cos^2\beta \begin{pmatrix} 1 \\ \cos 2\beta \\ \sin 2\beta \\ 0 \end{pmatrix} \tag{13-32b}$$

Inspection of (13-32a) and (13-32b) shows that both beams are linearly polarized at slits A and B.

In order to describe interference phenomena at the screen Σ, we must now determine the fields at slits A and B in the following manner. From the definition of the Stokes vector given by (13-3) and the Stokes vector that we have just found at slit A, Eq. (13-32a), we can write

$$\langle E_x(t)E_x^*(t)\rangle_A + \langle E_y(t)E_y^*(t)\rangle_A = \frac{1}{2}\langle A_x(t)A_x^*(t)\rangle \cos^2\alpha \tag{13-33a}$$

$$\langle E_x(t)E_x^*(t)\rangle_A - \langle E_y(t)E_y^*(t)\rangle_A = \frac{1}{2}\langle A_x(t)A_x^*(t)\rangle \cos^2\alpha \cos 2\alpha \tag{13-33b}$$

$$\langle E_x(t)E_y^*(t)\rangle_A + \langle E_y(t)E_x^*(t)\rangle_A = \frac{1}{2}\langle A_x(t)A_x^*(t)\rangle \cos^2\alpha \sin 2\alpha \tag{13-33c}$$

$$i\langle E_x(t)E_y^*(t)\rangle_A - i\langle E_y(t)E_x^*(t)\rangle_A = 0 \tag{13-33d}$$

where the subscript A on the angle brackets reminds us that we are at slit A. We now solve these equations and find that

$$\langle E_x(t)E_x^*(t)\rangle_A = \frac{1}{2}\langle A_x(t)A_x^*(t)\rangle \cos^4\alpha \tag{13-34a}$$

$$\langle E_y(t)E_y^*(t)\rangle_A = \frac{1}{2}\langle A_x(t)A_x^*(t)\rangle \cos^2\alpha \sin^2\alpha \tag{13-34b}$$

$$\langle E_x(t)E_y^*(t)\rangle_A = \langle E_y(t)E_x^*(t)\rangle_A = \frac{1}{4}\langle A_x(t)A_x^*(t)\rangle \cos^2\alpha \sin 2\alpha \tag{13-34c}$$

We see that the following fields will then satisfy (13-34):

$$E_{xA}(t) = \frac{A_x(t)}{\sqrt{2}}\cos^2\alpha \tag{13-35a}$$

$$E_{yA}(t) = \frac{A_x(t)}{\sqrt{2}}\cos\alpha \sin\alpha \tag{13-35b}$$

where $A_x(t)$ is the time-varying amplitude. The quantity $A_x(t)$ is assumed to vary slowly in time. In view of the fact that the Stokes vector at slit B is identical in form with that at slit A, the field at slit B, following (13-35), will be

$$E_{xB}(t) = \frac{A_x(t)}{\sqrt{2}} \cos^2 \beta \tag{13-36a}$$

$$E_{yB}(t) = \frac{A_x(t)}{\sqrt{2}} \cos \beta \sin \beta \tag{13-36b}$$

The propagation of the beams along the paths $\overline{AC}$ and $\overline{BC}$ as shown in Fig. 13-2 increases the phase of the fields by an amount $\phi_A = kl_1$ and $\phi_B = kl_2$, respectively, where $k = 2\pi/\lambda$ and λ is the wavelength. Thus, at point C on the screen Σ, the s and p field components will be, by the principle of superposition,

$$E_x(t) = E_{xA}(t) \exp(i\phi_A) + E_{xB}(t) \exp(i\phi_B) \tag{13-37a}$$

$$E_y(t) = E_{yA}(t) \exp(i\phi_A) + E_{yB}(t) \exp(i\phi_B) \tag{13-37b}$$

or

$$E_x(t) = \exp(i\phi_A)[E_{xA}(t) + e^{i\phi} E_{xB}(t)] \tag{13-38a}$$

$$E_y(t) = \exp(i\phi_A)[E_{yA}(t) + e^{i\phi} E_{yB}(t)] \tag{13-38b}$$

where $\phi = \phi_B - \phi_A = k(l_2 - l_1)$. The factor $\exp(i\phi_A)$ will disappear when the Stokes parameters are formed, and so it can be dropped. We now substitute (13-35) and (13-36) into (13-38), and we find that

$$E_x(t) = \frac{A_x(t)}{\sqrt{2}} (\cos^2 \alpha + e^{i\phi} \cos^2 \beta) \tag{13-39a}$$

$$E_y(t) = \frac{A_x(t)}{\sqrt{2}} (\cos \alpha \sin \alpha + e^{i\phi} \cos \beta \sin \beta) \tag{13-39b}$$

The Stokes parameters for $E_x(t)$ and $E_y(t)$ are now formed in the same manner as in (13-3). The Stokes vector observed on the screen will then be

$$S = \frac{1}{2} \langle AA^* \rangle \begin{pmatrix} \cos^2 \alpha + \cos^2 \beta + 2\cos(\alpha - \beta) \cos \alpha \cos \beta \cos \phi \\ \cos^2 \alpha \cos 2\alpha + \cos^2 \beta \cos 2\beta + 2\cos(\alpha + \beta) \cos \alpha \cos \beta \cos \phi \\ \cos^2 \alpha \sin 2\alpha + \cos^2 \beta \sin 2\beta + (\cos^2 \alpha \sin 2\beta + \cos^2 \beta \sin 2\alpha) \cos \phi \\ (\cos^2 \alpha \sin 2\beta - \cos^2 \beta \sin 2\alpha) \sin \phi \end{pmatrix} \tag{13-40}$$

We now examine the Stokes vector, (13-40), for some special cases.

Case I. The transmission axes of the polarizers P_A and P_B are parallel. For this condition $\alpha = \beta$ and (13-40) reduces to

$$S = \langle AA^* \rangle (1 + \cos \phi) \cos^2 \alpha \begin{pmatrix} 1 \\ \cos 2\alpha \\ \sin 2\alpha \\ 0 \end{pmatrix} \tag{13-41}$$

The factor $1 + \cos \phi$ tells us that we will always have perfect interference. Furthermore, the beam intensity is proportional to $(1 + \cos \phi) \cos^2 \alpha$ and the light is always linearly polarized. Thus, (13-41) is the mathematical statement of the first interference law of Fresnel and and Arago.

There are two further subcases of interest.

Case I(a). The axes of the polarizers P_A and P_B are parallel to the axis of the polarizer P_σ. Then $\alpha = 0$, and (13-41) reduces to

$$S = \langle AA^* \rangle (1 + \cos\phi) \begin{pmatrix} 1 \\ 1 \\ 0 \\ 0 \end{pmatrix} \tag{13-42a}$$

The beam is linearly horizontally polarized, and the intensity is at a maximum.

Case I(b). The axes of the polarizers P_A and P_B are perpendicular to the axis of the polarizer P_σ. Then $\alpha = \pi/2$, and (13-41) reduces to

$$S = \langle AA^* \rangle (1 + \cos\phi)(0) \begin{pmatrix} 1 \\ 1 \\ 0 \\ 0 \end{pmatrix} \tag{13-42b}$$

Thus the observed intensity of the beam will be zero at all points on the observation screen.

Case II. The transmission axes of P_A and P_B are perpendicular to each other. For this condition $\beta = \alpha + \pi/2$ and (13-40) reduces to

$$S = \frac{1}{2} \langle AA^* \rangle \begin{pmatrix} 1 \\ \cos^2 2\alpha + \sin^2 2\alpha \cos\phi \\ \sin 2\alpha \cos 2\alpha (1 + \cos\phi) \\ -\sin 2\alpha \sin\phi \end{pmatrix} \tag{13-43}$$

We now see that the interference term $1 + \cos\phi$ is missing in S_0, the intensity. Equation (13-43) is the mathematical statement of the second law of Fresnel and Arago, i.e., we do not have interference in this case. In general, the light is elliptically polarized as the presence of S_3 in (13-43) shows. Again there are some interesting subcases of (13-43).

Case II(a). The axis of P_A is parallel to the axis of P_σ. For this condition $\alpha = 0$, and (13-43) reduces to

$$S = \frac{1}{2} \langle AA^* \rangle \begin{pmatrix} 1 \\ 1 \\ 0 \\ 0 \end{pmatrix} \tag{13-44}$$

There is no interference, and the intensity and polarization of the observed beam and the polarized light from the source are identical.

Case II(b). The axis of P_A is perpendicular to the axis of P_σ. In this case the axis of P_B is parallel to P_σ. Then (13-43) reduces again to (13-44).

Case II(c). The transmission axes of P_A and P_B are at $+\pi/4$ and $-\pi/4$ from the transmission axis of P_σ. For this last case (13-43) reduces to

$$S = \frac{1}{2}\langle AA^*\rangle \begin{pmatrix} 1 \\ \cos\phi \\ 0 \\ -\sin\phi \end{pmatrix} \tag{13-45}$$

Again, there will be no interference, but the light is elliptically polarized. The Stokes vector degenerates into circularly or linearly polarized light for $\phi = (m \pm 1/2)\pi$ and $\pm m\pi$, respectively, where $m = 0, \ \pm 1, \ \pm 2, \ldots.$

13.5 THE SECOND EXPERIMENT: THIRD INTERFERENCE LAW

In order to determine the mathematical statement that corresponds to the third law, we consider the following experiment represented by Fig. 13-3. The polarizer P is placed, with its transmission axis at an angle θ from the x axis, behind the screen Σ to enable the fields that orginate at A and B to be brought into the same plane of polarization.

Here, σ is again an unpolarized light source and the transmission axes of P_A and P_B are placed parallel and perpendicular, respectively, to the x axis. The

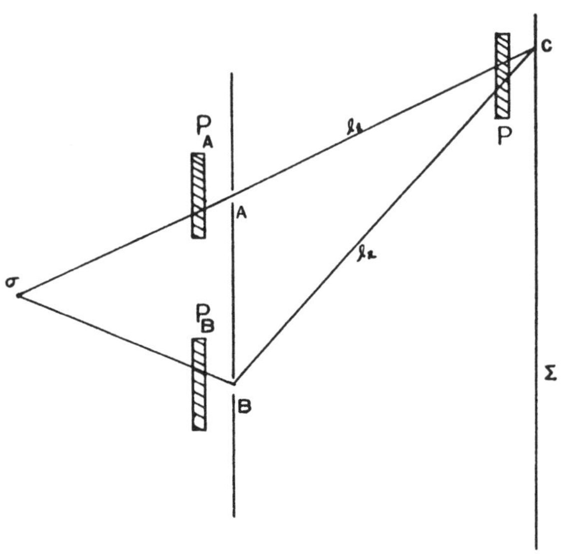

Figure 13.3 The transmission axes of P_A and P_B are along the x and y axes, respectively. The transmission axis of P is at an angle θ from the x axis.

matrices for P_A and P_B are then [we set $\alpha = 0$ and then $\pi/2$ in (13-31)]

$$M_{P_A}(\alpha = 0) = \frac{1}{2}\begin{pmatrix} 1 & 1 & 0 & 0 \\ 1 & 1 & 0 & 0 \\ 0 & 0 & 0 & 0 \\ 0 & 0 & 0 & 0 \end{pmatrix} \tag{13-46}$$

$$M_{P_B}\left(\alpha = \frac{\pi}{2}\right) = \frac{1}{2}\begin{pmatrix} 1 & -1 & 0 & 0 \\ -1 & 1 & 0 & 0 \\ 0 & 0 & 0 & 0 \\ 0 & 0 & 0 & 0 \end{pmatrix} \tag{13-47}$$

The Stokes vector at slit A is then found by multiplication of the Stokes vector for unpolarized light, (13-10), by (13-46). The result is

$$S_A = \frac{1}{2}\langle A_x A_x^* \rangle \begin{pmatrix} 1 \\ 1 \\ 0 \\ 0 \end{pmatrix} \tag{13-48}$$

Similarly, the Stokes vector at slit B is obtained by multiplication of the Stokes vector for unpolarized light by (13-47):

$$S_B = \frac{1}{2}\langle A_y A_y^* \rangle \begin{pmatrix} 1 \\ -1 \\ 0 \\ 0 \end{pmatrix} \tag{13-49}$$

Thus, the beams are linearly and orthogonally polarized; they are derived from the perpendicular components of the unpolarized light. The fields which satisfy (13-48) and (13-49) are, respectively,

$$E_{xA}(t) = \frac{A_x(t)}{\sqrt{2}} \qquad E_{yA}(t) = 0 \tag{13-50a}$$

$$E_{xB}(t) = 0 \qquad E_{yB}(t) = \frac{A_y(t)}{\sqrt{2}} \tag{13-50b}$$

These fields now propagate to the screen Σ, where they are intercepted by the polarizer P. At the polarizer the fields are

$$E_x(t) = \frac{A_x(t)}{\sqrt{2}} \exp(i\phi_A) \tag{13-51a}$$

$$E_y(t) = \frac{A_y(t)}{\sqrt{2}} \exp(i\phi_B) \tag{13-51b}$$

or

$$E_x(t) = \frac{A_x(t)}{\sqrt{2}} \tag{13-52a}$$

$$E_y(t) = \frac{A_y(t)}{\sqrt{2}} \exp(i\phi) \tag{13-52b}$$

where again $\phi = \phi_B - \phi_A$ and we have dropped the factor $\exp(i\phi_A)$. The transmission axis of the polarizer P is at an angle θ with respect to the x axis. Since we are now dealing with fields, we can conveniently use the Jones calculus to find the field on the screen after the beam has passed through the polarizer P.

The Jones matrix of the rotated polarizer, $J(\theta)$, is

$$J(\theta) = \begin{pmatrix} \cos\theta & -\sin\theta \\ \sin\theta & \cos\theta \end{pmatrix} \begin{pmatrix} 1 & 0 \\ 0 & 0 \end{pmatrix} \begin{pmatrix} \cos\theta & \sin\theta \\ -\sin\theta & \cos\theta \end{pmatrix}$$

$$= \begin{pmatrix} \cos^2\theta & \sin\theta\cos\theta \\ \sin\theta\cos\theta & \sin^2\theta \end{pmatrix} \tag{13-53}$$

The field which is now at the screen can be obtained if we write (13-52) as a column matrix. Multiplication of the vector composed of Eqs. (13-52) by (13-53) then gives the field at the screen as

$$E_x(t) = \frac{A_x(t)}{\sqrt{2}}\cos^2\theta + \frac{A_y(t)}{\sqrt{2}}e^{i\phi}\cos\theta\sin\theta \tag{13-54a}$$

$$E_y(t) = \frac{A_x(t)}{\sqrt{2}}\cos\theta\sin\theta + \frac{A_y(t)}{\sqrt{2}}e^{i\phi}\sin^2\theta \tag{13-54b}$$

We now form the Stokes vector and apply the conditions for unpolarized light given by (13-3) and (13-8) and find that

$$S = \frac{1}{2}\langle AA^*\rangle \begin{pmatrix} 1 \\ \cos 2\theta \\ \sin 2\theta \\ 0 \end{pmatrix} \tag{13-55}$$

Thus, we see that under no circumstances can there be interference. Equation (13-55) is the mathematical statement of the third interference law of Fresnel and Arago. In general, the light is linearly polarized. In particular, for $\theta = 0$ the light is linearly horizontally polarized, and for $\theta = \pi/2$ it is linearly vertically polarized, as expected.

13.6 THE THIRD EXPERIMENT: FOURTH INTERFERENCE LAW

In this final experiment the arrangement of the polarizers is identical to the previous experiment except that a linear polarizer P_σ, with its transmission axis at $+\pi/4$ from the x axis, is placed in front of the unpolarized light source (see Fig. 13-4). In this case we take the axes of the unpolarized light source to be at an angle of $+\pi/4$ from the horizontal x axis.

The Stokes vector of the unpolarized light for this new direction is related to the old direction by the transformation:

$$S'_\sigma = M(2\theta) \cdot S_\sigma \tag{13-56}$$

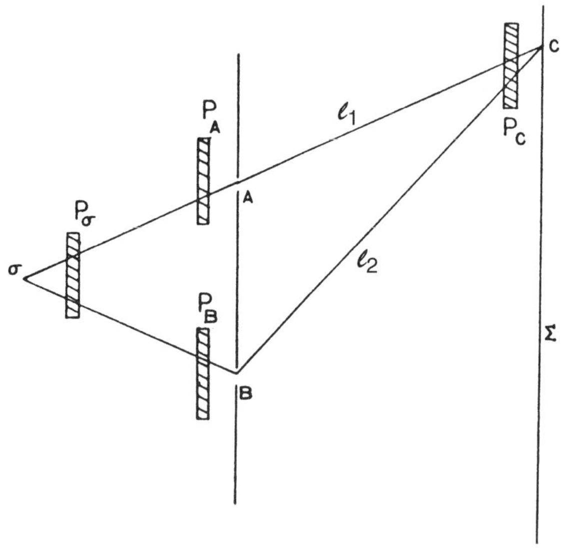

Figure 13-4 The third experiment. The transmission axis of P_σ is at $+\pi/4$ from the x axis. The directions of the axes of P_A, P_B, and P_C are identical to the second experiment.

where $M(2\theta)$ is the rotation matrix. We now multiply out (13-56) (and set $\theta = \pi/4$), and we find, after decomposing the unpolarized light in the familiar manner, that

$$S'_\sigma = \frac{1}{2}\langle A_x A_x^* \rangle \begin{pmatrix} 1 \\ 0 \\ -1 \\ 0 \end{pmatrix} + \frac{1}{2}\langle A_y A_y^* \rangle \begin{pmatrix} 1 \\ 0 \\ 1 \\ 0 \end{pmatrix} \tag{13-57}$$

Another way to arrive at (13-57) is to use the fact that unpolarized light is equivalent to two *independent* beams of light of equal intensities and polarized in orthogonal directions. Then we could simply take the statement for unpolarized light, (13-10) directly and resolve it into (13-57) without the introduction of (13-56). Either way we obtain (13-57).

The Mueller matrix for the polarizer P_σ [with α set to $\pi/4$ in (13-31)] is

$$M_{y\sigma} = \frac{1}{2}\begin{pmatrix} 1 & 0 & 1 & 0 \\ 0 & 0 & 0 & 0 \\ 1 & 0 & 1 & 0 \\ 0 & 0 & 0 & 0 \end{pmatrix} \tag{13-58}$$

We now multiply (13-57) and (13-58), and the beam that emerges from P_σ is

$$S_{y\sigma'} = \frac{1}{2}\langle A_y A_y^* \rangle \begin{pmatrix} 1 \\ 0 \\ 1 \\ 0 \end{pmatrix} \tag{13-59}$$

Thus, the light is linearly polarized ($+\pi/4$ preference) and derived from a single orthogonal component of the unpolarized light. The beam (13-59) now passes through P_A and P_B, and the Stokes vectors at the slits are

$$S_A = \frac{1}{2}\langle A_y A_y^* \rangle \begin{pmatrix} 1 \\ 1 \\ 0 \\ 0 \end{pmatrix} \tag{13-60a}$$

$$S_B = \frac{1}{2}\langle A_y A_y^* \rangle \begin{pmatrix} 1 \\ -1 \\ 0 \\ 0 \end{pmatrix} \tag{13-60b}$$

Thus, both beams are orthogonally linearly polarized but are derived from the same component of the unpolarized light.

The fields at slits A and B which satisfy (13-60a) and (13-60b) are then

$$E_{xA}(t) = \frac{A_y(t)}{\sqrt{2}} \qquad E_{yA}(t) = 0 \tag{13-61a}$$

$$E_{xB}(t) = 0 \qquad E_{yB}(t) = \frac{A_y(t)}{\sqrt{2}} \tag{13-61b}$$

The fields at the polarizer P_C will then be

$$E_x(t) = \frac{A_y(t)}{\sqrt{2}} \tag{13-62a}$$

$$E_y(t) = \frac{A_y(t)}{\sqrt{2}} e^{i\phi} \tag{13-62b}$$

After the field passes through the polarizer P_C the components become

$$E_x(t) = \frac{A_y(t)}{\sqrt{2}} \cos^2\theta + \frac{A_y(t)}{\sqrt{2}} e^{i\phi} \cos\theta \sin\theta \tag{13-63a}$$

$$E_y(t) = \frac{A_y(t)}{\sqrt{2}} \cos\theta \sin\theta + \frac{A_y(t)}{\sqrt{2}} e^{i\phi} \sin^2\theta \tag{13-63b}$$

The Stokes vector observed on the screen is then, from (13-63),

$$S = \frac{1}{2}\langle AA^* \rangle (1 + \sin 2\theta \cos\phi) \begin{pmatrix} 1 \\ \cos 2\theta \\ \sin 2\theta \\ 0 \end{pmatrix} \tag{13-64}$$

An inspection of this Stokes vector shows that interference can be observed. Equation (13-64) is the mathematical statement of the fourth and last of the interference laws of Fresnel and Arago. There are again some interesting subcases.

Case III(a). The axis of P_C is parallel to the axis of P_A and orthogonal to the axis of P_B. Then $\theta = 0$ and (13-64) reduces to

$$S = \frac{1}{2}\langle AA^* \rangle \begin{pmatrix} 1 \\ 1 \\ 0 \\ 0 \end{pmatrix} \tag{13-65a}$$

The light is linearly horizontally polarized, and there is no interference (the beam from P_B is not contributing to the field).

Case III(b). The axis of P_C is $+\pi/4$ from the axis of P_A. In this case $\theta = \pi/4$, and so (13-64) reduces to

$$S = \frac{1}{2}\langle AA^*\rangle(1+\cos\phi)\begin{pmatrix}1\\0\\1\\0\end{pmatrix} \tag{13-65b}$$

The light shows maximum interference and is linearly polarized ($+\pi/4$ preference).

Case III(c). The axis of P_C is perpendicular to the axis of P_A so $\theta = \pi/2$. Then (13-64) becomes

$$S = \frac{1}{2}\langle AA^*\rangle\begin{pmatrix}1\\-1\\0\\0\end{pmatrix} \tag{13-65c}$$

The light is linearly vertically polarized, an again there is no interference (now the beam from P_A is not contributing).

At this point we can summarize the Fresnel–Arago laws. However, we defer this in order to consider one more interesting related problem.

13.7 THE HERSCHEL–STOKES EXPERIMENT

In Section 13.1 we pointed out that Stokes did not formulate the Fresnel–Arago interference laws, but treated a related experiment suggested by Sir John Herschel. This experiment is represented in Fig. 13-5. In this experiment an unpolarized source of light, σ, is again used. The transmission axis of polarizer P_B is fixed in the direction of the x axis, while the polarizer P_A is rotated through an angle α. The Stokes vector on the screen Σ is to be determined.

The Stokes vector at the slit B is, following the methods developed earlier,

$$S_B = \frac{1}{2}\langle A_xA_x^*\rangle\begin{pmatrix}1\\1\\0\\0\end{pmatrix} \tag{13-66a}$$

while the Stokes vector at slit A is

$$S_A = \frac{1}{2}\langle A_xA_x^*\rangle\cos^2\alpha\begin{pmatrix}1\\\cos 2\alpha\\\sin 2\alpha\\0\end{pmatrix} + \frac{1}{2}\langle A_yA_y^*\rangle\sin^2\alpha\begin{pmatrix}1\\\cos 2\alpha\\\sin 2\alpha\\0\end{pmatrix} \tag{13-66b}$$

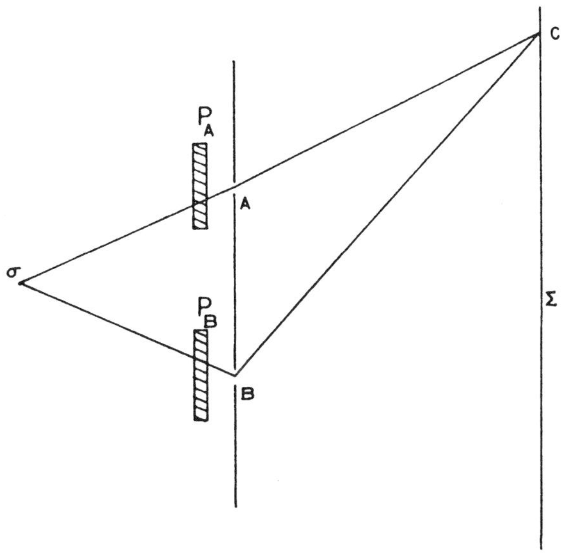

Figure 13-5 The Herschel–Stokes experiment. The transmission axis of P_B is fixed along the x axis, and the transmission axis of P_A is rotated through an angle α from the x axis.

The fields at B and A that satisfy (13-66a) and (13-66b) are

$$E_{xB}(t) = \frac{A_x(t)}{\sqrt{2}} \qquad E_{yB}(t) = 0 \tag{13-67a}$$

$$E'_{xA}(t) = \frac{A_x(t)}{\sqrt{2}}\cos^2\alpha \qquad E'_{yA}(t) = \frac{A_x(t)}{\sqrt{2}}\cos\alpha\sin\alpha \tag{13-67b}$$

$$E''_{xA}(t) = \frac{A_y(t)}{\sqrt{2}}\cos\alpha\sin\alpha \qquad E''_{yA}(t) = \frac{A_y(t)}{\sqrt{2}}\sin^2\alpha \tag{13-67c}$$

The primed and double primed fields correspond to the first and second Stokes vector in (13-66b) and arise because the Stokes vectors in (13-66b) are *independent*. The fields at the screen Σ are then

$$E_x(t) = \frac{A_x(t)}{\sqrt{2}}(e^{i\phi} + \cos^2\alpha) + \frac{A_y(t)}{\sqrt{2}}\sin\alpha\cos\alpha \tag{13-68a}$$

$$E_y(t) = \frac{A_x(t)}{\sqrt{2}}\cos\alpha\sin\alpha + \frac{A_y(t)}{\sqrt{2}}\sin^2\alpha \tag{13-68b}$$

We now form the Stokes vector in the usual way and apply the condition for unpolarized light and find that

$$S = \langle AA^*\rangle \begin{pmatrix} 1 + \cos^2\alpha\cos\phi \\ \cos^2\alpha(1+\cos\phi) \\ \sin\alpha\cos\alpha(1+\cos\phi) \\ -\sin\alpha\cos\alpha\sin\phi \end{pmatrix} \tag{13-69}$$

Stokes actually obtained only S_0 and took ϕ to be equal to zero or π. We examine (13-69) at some special values of α.

Case IV(a). The transmission axis of the polarizer P_A is parallel to the transmission axis of P_B, so $\alpha = 0$. Then (13-69) reduces to

$$S = \langle AA^* \rangle (1 + \cos\phi) \begin{pmatrix} 1 \\ 1 \\ 0 \\ 0 \end{pmatrix} \tag{13-70}$$

We have perfect interference, and the light is linearly horizontally polarized.

Case IV(b). The transmission axis of P_A is perpendicular to the transmission axis of P_B, so $\alpha = \pi/2$. Then (13-69) reduces to

$$S = \langle AA^* \rangle \begin{pmatrix} 1 \\ 0 \\ 0 \\ 0 \end{pmatrix} \tag{13-71}$$

There is no interference, and the light is unpolarized.

This problem shows why it was selected by Stokes. Within the confines of a single problem he was able to show that one could obtain complete interference along with completely polarized light, (13-70), and, conversely, no interference and completely unpolarized light, (13-71). It was this "peculiar" behavior of polarized light which was a source of great confusion. Stokes, however, by his investigation was able to show that with his parameters all these questions could be answered, and, equally important, this could be done within the structure of the wave theory of light.

13.8 SUMMARY OF THE FRESNEL–ARAGO INTERFERENCE LAWS

In view of the rather lengthy analysis required to obtain the mathematical statements for the Fresnel–Arago interference laws, it is worthwhile to summarize these results. We restate each of the laws and the corresponding Stokes vector.

13.8.1 The First Interference Law

Two waves, linearly polarized in the same plane, can interfere.

$$S = \langle AA^* \rangle (1 + \cos\phi) \cos^2\alpha \begin{pmatrix} 1 \\ \cos 2\alpha \\ \sin 2\alpha \\ 0 \end{pmatrix} \tag{13-41}$$

The angle α refers to the condition when the transmission axes of the two polarizers behind the slits are parallel. We see that the light is always linearly polarized and there will always be interference.

13.8.2 The Second Interference Law

Two waves, linearly polarized with perpendicular polarizations, cannot interfere.

$$S = \frac{1}{2}\langle AA^* \rangle \begin{pmatrix} 1 \\ \cos^2 2\alpha + \sin^2 2\alpha \cos\phi \\ \sin 2\alpha \cos 2\alpha (1 + \cos\phi) \\ -\sin 2\alpha \sin\phi \end{pmatrix} \tag{13-43}$$

The interference term $1 + \cos\phi$ is missing in S_0, the intensity. Equation (13-43) shows that the light is always elliptically polarized, but there is never any interference.

13.8.3 The Third Interference Law

Two waves, linearly polarized with perpendicular polarizations, if derived from perpendicular components of unpolarized light and subsequently brought into the same plane, cannot interfere.

$$S = \frac{1}{2}\langle AA^* \rangle \begin{pmatrix} 1 \\ \cos 2\theta \\ \sin 2\theta \\ 0 \end{pmatrix} \tag{13-55}$$

Equation (13-55) shows that interference is never seen under these conditions.

13.8.4 The Fourth Interference Law

Two waves, linearly polarized with perpendicular polarizations, if derived from the same linearly polarized wave and subsequently brought into the same plane, can interfere.

$$S = \frac{1}{2}\langle AA^* \rangle (1 + \sin 2\theta \cos\phi) \begin{pmatrix} 1 \\ \cos 2\theta \\ \sin 2\theta \\ 0 \end{pmatrix} \tag{13-64}$$

Only if $\theta = 0^\circ$ or 90° does the interference term in (13-64) vanish; otherwise interference will always be observed; the Stokes vector is always linearly polarized.

This concludes our discussion of the fundamental properties of polarized light. At this point the reader can certainly see that a great deal of knowledge can be obtained about the properties and behavior of polarized light without having to resort to the equations of the electromagnetic field. However, this is as far as we can go. Ultimately, we must deal with the source of the polarized radiation fields. In order to do this, we must now turn to the theory of the electromagnetic field, i.e., Maxwell's equations and the source of polarized light. We shall see that the Stokes parameters and Mueller formalism play a major and very interesting role.

REFERENCES

Papers

1. Stokes, G. G., *Trans. Camb. Phil. Soc.*, *9*, 399 (1852). Reprinted in *Mathematical and Physical Papers*, Cambridge University Press, London, 1901, Vol. 3, p. 233.
2. Soleillet, P., *Ann. Phys.*, *12* (10) 23 (1929).
3. Langsdorf, A. and DuBridge, L., *J. Opt. Soc. Am.*, *24*, 1 (1934).
4. Birge, R. T., *J. Opt. Soc. Am.*, *25*, 179 (1935).
5. Perrin, F. *J. Chem. Phys.*, *10*, 415 (1942).
6. Hurwitz, H. *J. Opt. Soc. Am.*, *35*, 525 (1945).
7. Parke, N. G., III, *Statistical Optics*. II: *Mueller Phenomenological Algebra*, RLE TR-119, Research Laboratory of Elect. at M.I.T. (1949).
8. Wolf, E., *Nuovo Cimento*, *12*, 884 (1954).
9. Hannau, R., *Am. J. Phys.*, *31*, 303 (1962).
10. Collett, E., *Am. J. Phys.*, *39*, 1483 (1971).

Books

1. Fresnel, A. J., *L'Oeuvres Complètes*, Publiées par Henri de Senarmont, Emile Verdet et Leonor Fresnel, Paris, 1866, Vol I.
2. Whittaker, E., *A History of the Theories of Aether and Electricity*, Philosophical Society of New York, 1951, Vol I.
3. Wood, R. W., *Physical Optics*, 3rd ed., Optical Society of America, Washington, DC, 1988.
4. Born, M., and Wolf, E., *Principles of Optics*, 3rd ed., Pergamon Press, New York, 1965.
5. Ditchburn, R. W., *Light*, 2nd ed., Blackie, London, 1963.
6. Strong, J., *Concepts of Classical Optics*, Freeman, San Francisco, 1959.
7. Jenkins, F. S., and White, H. E., *Fundamental of Optics*, McGraw-Hill, New York, 1957.
8. Shurcliff, W. A., *Polarized light*, Harvard University Press, Cambridge, MA, 1962.
9. Hecht, E. and Zajac, A., *Optics*, Addison-Wesley, Reading, MA, 1974.

14

Introduction to the Classical and Quantum Theory of Radiation by Accelerating Charges

In Part I, Chapters 1–13, we dealt with the polarization of the optical field and the phenomenological interaction of polarized light with optical components, namely, polarizers, retarders, and rotators. All this was accomplished with only the classical theory of light. By the mid-nineteenth century Fresnel's theory of light was a complete triumph. The final acceptance of the wave theory took place when Stokes showed that the Fresnel–Arago interference laws could also be explained and understood on the basis of classical optics. Most importantly, Stokes showed that unpolarized light and partially polarized light were completely compatible with the wave theory of light. Thus, polarized light played an essential role in the acceptance of this theory. We shall now see how polarized light was again to play a crucial role in the acceptance of an entirely new theory of the optical field, namely, Maxwell's theory of the electrodynamic field.

In spite of all of the successes of Frensel's theory there was an important problem that classical optics could not treat. We saw earlier that the classical optical field was described by the wave equation. This equation, however, says nothing about the source of the optical field. In 1865 James Clerk Maxwell introduced a totally new and unexpected theory of light. Maxwell's new theory was difficult to understand because it arose not from the description of optical phenomena but from a remarkable synthesis of the laws of the electromagnetic field. This theory was summarized by expressing all of the known behavior of the electromagnetic field in the form of four differential equations. In these equations a source term existed in the form of a current $\mathbf{j}(\mathbf{r}, t)$ along with a new term postulated by Maxwell, namely, the displacement current $\partial \mathbf{D}(\mathbf{r}, t)/\partial t$.

After Maxwell had formulated his equations, he proceeded to solve them. He was completely surprised at his results. First, when either the magnetic or electric field was eliminated between the equations, he discovered that in free space the electromagnetic field was described by the wave equation of classical optics. The next result surprised him even more. It appeared that the electromagnetic field propagated at the same speed as light. This led him to speculate that, perhaps, the optical field and the electromagnetic field were actually manifestations of the same disturbance, being different only in their frequency (wavelength).

Maxwell died in 1879. Nearly 10 years later Heinrich Hertz (1888) carried out a set of very sophisticated and brilliant experiments and confirmed Maxwell's theory. In spite of Hertz's verification, however, Maxwell's theory was not immediately adopted by the optics community. There were several reasons for this. One reason was due to the simple fact that Hertz confirmed Maxwell's theory not at optical wavelengths but at millimeter wavelengths. For the optical community this was not enough. In order for them to accept Maxwell's theory, it would have to be proved at optical wavelengths. Another reason for the slow acceptance of Maxwell's theory was that for 30 years after the publication of Maxwell's theory in 1865 nothing had been found which could clearly differentiate between the classical wave theory and Maxwell's theory. Nothing had appeared in optics which was not known or understood using Fresnel's theory; no one yet understood exactly what fluorescence or the photoelectric effect was. There was, however, one very slim difference between the two theories. Maxwell's theory, in contrast to Fresnel's theory, showed that in free space only transverse waves existed. It was this very slim difference that sustained the "Maxwellians" for several decades. A third important reason why Maxwell's theory was not readily embraced by the optics community was that a considerable effort had to be expended to study electromagnetism–a nonoptical subject–in order to understand fundamental optical phenomena. Furthermore, as students to this day know, a fair degree of mathematical training is required to understand and manipulate Maxwell's equations (this was especially true before the advent of vector analysis). It was, therefore, very understandable why the optics community was reluctant to abandon a theory that explained everything in a far simpler way and accounted for all the known facts.

In 1896, less than a decade after Hertz's experiments, two events took place which overthrew Fresnel's elastic theory of light and led to the complete acceptance of Maxwell's theory. The first was the discovery by J. J. Thomson of the electron, the long-sought source of the optical field, and the second was the splitting of unpolarized spectral lines which became polarized when an electron was placed in a magnetic field (the Lorentz–Zeeman effect). In this part we shall see how polarized light played a crucial role in the acceptance of Maxwell's theory. We shall use the Stokes parameters to describe the radiation by accelerating electrons and see how the Stokes parameters and the Stokes vector take on a surprising new role in all of this. In the final chapter of this part we shall show that the Stokes vector can be used to describe both classical and quantum radiating systems, thereby providing a single description of radiation phenomena.

REFERENCES

Books

1. Maxwell, J. C., *A Treatise on Electricity and Magnetism*, 3rd ed., Clarendon Press, Oxford, 1892.
2. Hertz, H., *Electric Waves*, Macmillan, London, 1893.
3. Lorentz, H. A., *The Theory of the Electron*, Dover, New York, 1952.

4. Born, M. and Wolf, E., *Principles of Optics*, 3rd ed., Pergamon Press, New York, 1965.
5. Whittaker, E., *A History of the Theories of Aether and Electricity*, Philosophical Society, New York, 1951, Vol. I.
6. Sommerfeld, A., *Lectures on Theoretical Physics,* Vols. I–V, Academic Press, New York, 1952.
7. Stratton, J. A., *Electrodynamic Theory*, McGraw-Hill, New York, 1941.
8. Feuer, L. S., *Einstein and the Generations of Science*, Basic Books, New York, 1974.
9. Jammer, M., *The Philosophy of Quantum Mechanics*, Wiley, New York, 1974.
10. Schwinger, J., *Einstein's Legacy*, Scientific American Books, Inc., New York, 1986.

15

Maxwell's Equations for the Electromagnetic Field

Maxwell's equations describe the basic laws of the electromagnetic field. Over the 40 years preceding Maxwell's enunciation of his equations (1865) the four fundamental laws describing the electromagnetic field had been discovered. These are known as Ampère's law, Faraday's law, Coulomb's law, and the magnetic continuity law. These four laws were cast by Maxwell, and further refined by his successors, into four differential equations:

$$\nabla \times \mathbf{H} = \mathbf{j} + \frac{\partial \mathbf{D}}{\partial t} \tag{15-1a}$$

$$\nabla \times \mathbf{E} = -\frac{\partial \mathbf{B}}{\partial t} \tag{15-1b}$$

$$\nabla \cdot \mathbf{D} = \rho \tag{15-1c}$$

$$\nabla \cdot \mathbf{B} = 0 \tag{15-1d}$$

These are Maxwell's famous equations for fields and sources in macroscopic media: $\mathbf{E}$ and $\mathbf{H}$ are the instantaneous electric and magnetic fields, $\mathbf{D}$ and $\mathbf{B}$ are the displacement vector and the magnetic induction vector, and $\mathbf{j}$ and ρ are the current and the charge density, respectively. We note that (15-1a) without the term $\partial \mathbf{D}/\partial t$ is Ampère's law; the second term in (15-1a) was added by Maxwell and is called the displacement current. A very thorough and elegant discussion of Maxwell's equations is given in the text *Classical Electrodynamics* by J. D. Jackson, and the reader will find the required background to Maxwell's equations there.

When Maxwell first arrived at his equations, the term $(\partial \mathbf{D}/\partial t)$ was not present. He added this term because he observed that (15-1a) did not satisfy the continuity equation. To see that the addition of this term leads to the continuity equation, we take the divergence $\nabla\cdot$, of both sides of (15-1a).

$$\nabla \cdot [\nabla \times \mathbf{H}] = (\nabla \cdot \mathbf{j}) + \frac{\partial}{\partial t}(\nabla \cdot \mathbf{D}) \tag{15-2a}$$

The divergence of curl is zero, so the left-hand side is zero and we have

$$(\nabla \cdot \mathbf{j}) + \frac{\partial}{\partial t}(\nabla \cdot \mathbf{D}) = 0 \tag{15-2b}$$

Next, we substitute (15-1c) into (15-2b) and find that

$$\left[\nabla \cdot \mathbf{j} + \frac{\partial \rho}{\partial t}\right] = 0 \tag{15-3a}$$

or

$$\nabla \cdot \mathbf{j} + \frac{\partial \rho}{\partial t} = 0 \tag{15-3b}$$

which is the continuity equation. Equation (15-3b) states that the divergence of the current $(\nabla \cdot \mathbf{j})$ is equal to the time rate of change of the creation of charge $(-\partial\rho/\partial t)$. What Maxwell saw, as Jackson has pointed out, was that the continuity equation could be converted into a vanishing divergence by using Coulomb's law, (15-1c). Thus, (15-1c) could only be satisfied if

$$\nabla \cdot \mathbf{j} + \frac{\partial \rho}{\partial t} = \nabla \cdot \left(\mathbf{j} + \frac{\partial \mathbf{D}}{\partial t}\right) = 0 \tag{15-4}$$

Maxwell replaced $\mathbf{j}$ in Ampère's law by its generalization, and arrived at a new type of current for the electromagnetic field, namely,

$$\mathbf{j} \rightarrow \mathbf{j} + \frac{\partial \mathbf{D}}{\partial t} \tag{15-5}$$

for time-dependent fields. The additional term $\partial\mathbf{D}/\partial t$ in (15-5) is called the displacement current.

Maxwell's equations form the basis for describing all electromagnetic phenomena. When combined with the Lorentz force equation (which shall be discussed shortly) and Newton's second law of motion, these equations provide a complete description of the classical dynamics of interacting charged particles and electromagnetic fields. For macroscopic media the dynamical response of the aggregates of atoms is summarized in the constitutive relations that connect $\mathbf{D}$ and $\mathbf{j}$ with $\mathbf{E}$, and $\mathbf{H}$ with $\mathbf{B}$; that is, $\mathbf{D} = \varepsilon\mathbf{E}$, $\mathbf{j} = \sigma\mathbf{E}$, and $\mathbf{B} = \mu\mathbf{H}$, respectively, for an isotropic, permeable, conducting dielectric.

We can now solve Maxwell's equations. The result is remarkable and was the primary reason for Maxwell's belief in the validity of his equations. In order to do this, we first use the constitutive relations:

$$\mathbf{D} = \varepsilon\mathbf{E} \tag{15-6a}$$

$$\mathbf{B} = \mu\mathbf{H} \tag{15-6b}$$

Equations (15-6a) and (15-6b) are substituted into (15-1a) and (15-1b), respectively, to obtain

$$\nabla \times \mathbf{H} = \mathbf{j} + \varepsilon \frac{\partial \mathbf{E}}{\partial t} \tag{15-7a}$$

$$\nabla \times \mathbf{E} = -\mu \frac{\partial \mathbf{H}}{\partial t} \tag{15-7b}$$

Next, we take the curl ($\nabla \times$) of both sides of (15-7b):

$$\nabla \times (\nabla \times \mathbf{E}) = -\mu \frac{\partial}{\partial t}(\nabla \times \mathbf{H}) \tag{15-8}$$

We can eliminate $\nabla \times \mathbf{H}$ in (15-8) by using (15-7a), and find that

$$\nabla \times (\nabla \times \mathbf{E}) = \mu \frac{-\partial}{\partial t}\left(\mathbf{j} + \varepsilon \frac{\partial \mathbf{E}}{\partial t}\right)$$

so

$$\nabla \times (\nabla \times \mathbf{E}) = -\mu \frac{\partial \mathbf{j}}{\partial t} - \mu\varepsilon \frac{\partial^2 \mathbf{E}}{\partial t^2} \tag{15-9}$$

The left-hand side is known from vector analysis to reduce to

$$\nabla \times \nabla \times \mathbf{E} = \nabla(\nabla \cdot \mathbf{E}) - \nabla^2 \mathbf{E} \tag{15-10}$$

Equation (15-9) then reduces to

$$\nabla(\nabla \cdot \mathbf{E}) - \nabla^2 \mathbf{E} = -\mu \frac{\partial \mathbf{j}}{\partial t} - \mu\varepsilon \frac{\partial^2 \mathbf{E}}{\partial t^2} \tag{15-11}$$

Finally, if there are no free charges then $\rho = 0$ and (15-1c) becomes

$$\nabla \cdot \mathbf{D} = \varepsilon \nabla \cdot \mathbf{E} = 0$$

or

$$\nabla \cdot \mathbf{E} = 0 \tag{15-12}$$

Thus, (15-11) can be written as

$$\nabla^2 \mathbf{E} - \mu\varepsilon \frac{\partial^2 \mathbf{E}}{\partial t^2} = -\mu \frac{\partial \mathbf{j}}{\partial t} \tag{15-13}$$

Inspection of (15-13) quickly reveals the following. If there are no currents, then $\mathbf{j} = 0$ and (15-13) becomes

$$\nabla^2 \mathbf{E} = \mu\varepsilon \frac{\partial^2 \mathbf{E}}{\partial t^2} \tag{15-14}$$

which is the wave equation of classical optics. Thus, the electric field $\mathbf{E}$ propagates exactly according to the classical wave equation. Furthermore, if we write (15-14) as

$$\nabla^2 \mathbf{E} = \frac{1}{1/\mu\varepsilon} \frac{\partial^2 \mathbf{E}}{\partial t^2} \tag{15-15}$$

then we have

$$\nabla^2 \mathbf{E} = \frac{1}{v^2}\frac{\partial^2 \mathbf{E}}{\partial t^2} \tag{15-16}$$

where $v^2 = c^2$. The propagation of the electromagnetic field is not only governed by the wave equation but propagates at the speed of light. It was this result that led Maxwell to the belief that the electromagnetic field and the optical field were one and the same.

Maxwell's equations showed that the wave equation for optics, if his theory was correct, was no longer a hypothesis but rested on firm experimental and theoretical foundations.

The association of the electromagnetic field with light was only a speculation on Maxwell's part. In fact, there was only a single bit of evidence for its support, initially. We saw that in a vacuum we have

$$\nabla \cdot \mathbf{E} = 0 \tag{15-12}$$

Now it is easy to show that the solution of Maxwell's equation gives rise to an electric field whose form is

$$\mathbf{E} = \mathbf{E}_0 e^{i(\mathbf{k}\cdot\mathbf{r}-\omega t)} \tag{15-17a}$$

where

$$\mathbf{E} = E_x \mathbf{u}_x + E_y \mathbf{u}_y + E_z \mathbf{u}_z \tag{15-17b}$$

$$\mathbf{E}_0 = E_{0x} \mathbf{u}_x + E_{0y} \mathbf{u}_y + E_{0z} \mathbf{u}_z \tag{15-17c}$$

$$\mathbf{k} = k_x \mathbf{u}_x + k_y \mathbf{u}_y + k_z \mathbf{u}_z \tag{15-17d}$$

$$\mathbf{r} = x \mathbf{u}_x + y \mathbf{u}_y + z \mathbf{u}_z \tag{15-17e}$$

$$\mathbf{k}\cdot\mathbf{r} = k_x x + k_y y + k_z z \tag{15-17f}$$

Substituting (15-17a) into (15-12) quickly leads to the relation:

$$\mathbf{k}\cdot\mathbf{E} = 0 \tag{15-18}$$

where we have used the remaining equations in (15-17) to obtain (15-18). The wave vector is **k** and is in the direction of propagation of the field, **E**. Equation (15-18) is the condition for orthogonality between **k** and **E**. Thus, if the direction of propagation is taken along the z axis, we can only have field components along the x and y axes; that is, the field in free space is transverse. This is exactly what is observed in the Fresnel–Arago interference equations. Thus, in Maxwell's theory this result is an immediate consequence of his equations, whereas in Fresnel's theory it is a defect. This fact was the only known difference between Maxwell's theory and Fresnel's theory when Maxwell's theory appeared in 1865. For most of the scientific community and, especially, the optics community this was not a sufficient reason to overthrow the highly successful Fresnel theory. Much more evidence would be needed to do this.

Maxwell's equations differ from the classical wave equation in another very important respect, however. Returning to (15-13), Maxwell's equations lead to

$$\nabla^2 \mathbf{E} - \mu\varepsilon \frac{\partial^2 \mathbf{E}}{\partial t^2} = -\mu \frac{\partial \mathbf{j}}{\partial t} \tag{15-13}$$

The right-hand term in (15-13) is something very new. It describes the *source* of the electromagnetic field *or the optical field.* Maxwell's theory now describes not only the propagation of the field but also enables one to say something about the source of these fields, something which no one had been able to say with certainty before Maxwell. According to (15-13) the field **E** arises from a term $\partial \mathbf{j}/\partial t$. More specifically the field arises not from **j**, the current per se, but from the *time rate of change of the current*. Now this can be interpreted, as follows, by noting that the current can be written as

$$\mathbf{j} = e\mathbf{v} \tag{15-19}$$

where e is the charge and $\mathbf{v}$ is the velocity of the charge. Substituting (15-19) into (15-13), we have

$$\nabla^2 \mathbf{E} - \mu\varepsilon \frac{\partial^2 \mathbf{E}}{\partial t^2} = \mu e \frac{\partial \mathbf{v}}{\partial t} = \mu e \dot{\mathbf{v}} \tag{15-20}$$

The term $\partial \mathbf{v}/\partial t$ is obviously an acceleration. Thus, the field arises from *accelerating* charges. In 1865 no one knew of the existence of actual charges, let alone accelerating charges, and certainly no one knew how to generate or control accelerating charges. In other words, the term $(\mu e)\partial \mathbf{v}/\partial t$ in 1865 was "superfluous," and so we are left just with the classical wave equation in optics:

$$\nabla^2 \mathbf{E} - \mu\varepsilon \frac{\partial^2 \mathbf{E}}{\partial t^2} = 0 \tag{15-21}$$

Thus, we arrive at the same result from Maxwell's equations after a considerable amount of effort, as we do by introducing (15-21) as an hypothesis or deriving it from mechanics. This difference is especially sharp when we recall that it takes only a page to obtain the identical result from classical mechanics! Aside from the existence of the transverse waves and the source term in (15-13), there was very little motivation to replace the highly successful Fresnel theory with Maxwell's theory. The only difference between the two theories was that in Fresnel's theory the wave equation was the starting point, whereas Maxwell's theory led up to it.

Gradually, however, the nature of the source term began to become clearer. These investigations, e.g., Lorentz's theory of the electron, led physicists to search for the source of the optical field. Thus, (15-13) became a fundamental equation of interest. Because it plays such an important role in the discussion of the optical field, (15-13) is also known as the radiation equation, a name that will soon be justified. In general, (15-13) has the form of the inhomogeneous wave equation.

The solution of the radiation equation can be obtained by a technique called Green's function method. This is a very elegant and powerful method for solving differential equations, in general. However, it is quite involved and requires a considerable amount of mathematical background. Consequently, in order not to detract from our discussions on polarized light, we refer the reader to its solution

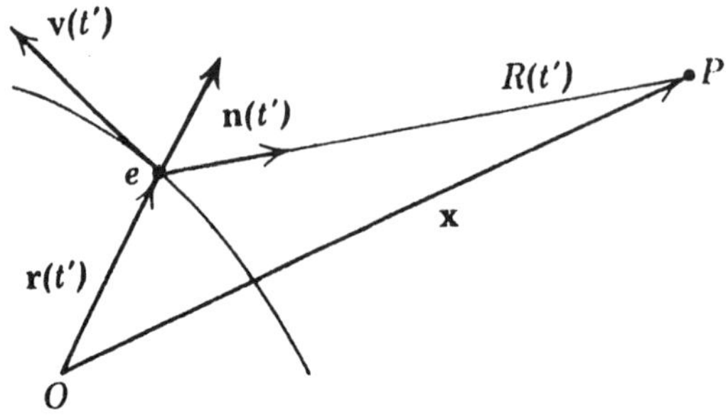

Figure 15-1 Radiating field coordinates arising from an accelerating charge; P is the observation point (From Jackson).

by Jackson (*Classical Electrodynamics*). Here, we merely state the result. Using Green's function method, the solution of the radiation equation in the form given by (15-20) is found to be

$$\mathbf{E}(\mathbf{r}, t) = \frac{e}{4\pi\varepsilon_0 c^2}\left[\frac{n}{\kappa^3 R} \times \{(n - \mathbf{v}) \times \dot{\mathbf{v}}\}\right] \tag{15-22a}$$

where

$$\kappa = 1 - \mathbf{n} \cdot \mathbf{v} \tag{15-22b}$$

and $\mathbf{n} = \mathbf{R}/R$ is a unit vector directed from the position of the charge to the observation. The geometry of the moving charge is shown in Fig. 15-1.

In the following chapter we determine the field components of the radiated field for (15-22) in terms of the accelerating charges.

REFERENCES

1. Jackson, J. D., *Classical Electrodynamics*, Wiley, New York, 1962.
2. Sommerfeld, A., *Lectures on Theoretical Physics*, Vols. I–V, Academic Press, New York, 1952.
3. Born, M. and Wolf, E., *Principles of Optics*, 3rd ed., Pergamon Press, New York, 1965.

16

The Classical Radiation Field

16.1 FIELD COMPONENTS OF THE RADIATION FIELD

Equation (15-22a) is valid for any acceleration of the electron. However, it is convenient to describe (15-22a) in two different regimes, namely, for nonrelativistic speeds ($\mathbf{v}/c \ll 1$) and for relativistic speeds ($\mathbf{v}/c \simeq 1$). The field emitted by an accelerating charge observed in a reference frame where the velocity is much less than the speed of light, that is, the nonrelativistic regime, is seen from (15-22a) to reduce to

$$\mathbf{E}(\mathbf{X}, t) = \left(\frac{e}{4\pi\varepsilon_0 c^2 R}\right)[\mathbf{n} \times (\mathbf{n} \times \dot{\mathbf{v}})] \tag{16-1}$$

where $\mathbf{E}(\mathbf{X}, t)$ is the field vector of the radiated field measured from the origin, e is the charge, c is the speed of light, R is the distance from the charge to the observer, $\mathbf{n} = \mathbf{R}/R$ is the unit vector directed from the position of the charge to the observation point, and $\dot{\mathbf{v}}$ is the acceleration (vector) of the charge. The relation between the vectors $\mathbf{X}$ and $\mathbf{n}$ is shown in Fig. 16-1.

To apply (16-1), we consider the (radiated) electric field $\mathbf{E}$ in spherical coordinates. Since the field is transverse, we can write

$$\mathbf{E} = E_\theta \mathbf{u}_\theta + E_\phi \mathbf{u}_\phi \tag{16-2}$$

where $\mathbf{u}_\theta$ and $\mathbf{u}_\phi$ are unit vectors in the θ and ϕ directions, respectively. Because we are relatively far from the source, we can take $\mathbf{n}$ to be directed from the origin and write $\mathbf{n} = \mathbf{u}_r$, where $\mathbf{u}_r$ is the radial unit vector directed from the origin. The triple vector product in (16-1) can then be expanded and written as

$$\mathbf{u}_r \times (\mathbf{u}_r \times \dot{\mathbf{v}}) = \mathbf{u}_r(\mathbf{u}_r \cdot \dot{\mathbf{v}}) - \dot{\mathbf{v}} \tag{16-3}$$

For many problems of interest it is preferable to express the acceleration of the charge $\dot{\mathbf{v}}$ in Cartesian coordinates, thus

$$\dot{\mathbf{v}} = \ddot{x}\mathbf{u}_x + \ddot{y}\mathbf{u}_y + \ddot{z}\mathbf{u}_z \tag{16-4}$$

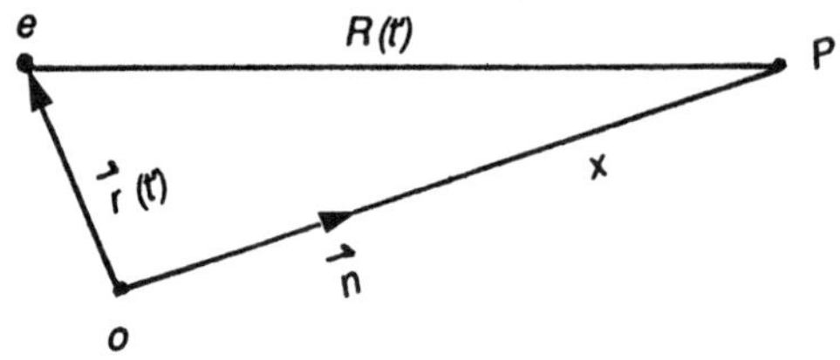

Figure 16-1 Vector relation for a moving charge and the radiation field.

where the double dot refers to twofold differentiation with respect to time. The unit vectors $\mathbf{u}$ in spherical and Cartesian coordinates are shown later to be related by

$$\mathbf{u}_r = \sin\theta\cos\phi\mathbf{u}_x + \sin\theta\sin\phi\mathbf{u}_y + \cos\theta\mathbf{u}_z \tag{16-5a}$$

$$\mathbf{u}_\theta = \cos\theta\cos\phi\mathbf{u}_x + \cos\theta\sin\phi\mathbf{u}_y - \sin\theta\mathbf{u}_z \tag{16-5b}$$

$$\mathbf{u}_\phi = -\sin\phi\mathbf{u}_x + \cos\phi\mathbf{u}_y \tag{16-5c}$$

or

$$\mathbf{u}_x = \sin\theta\cos\phi\mathbf{u}_r + \cos\theta\cos\phi\mathbf{u}_\theta - \sin\phi\mathbf{u}_\phi \tag{16-6a}$$

$$\mathbf{u}_y = \sin\theta\sin\phi\mathbf{u}_r + \cos\theta\sin\phi\mathbf{u}_\theta + \cos\phi\mathbf{u}_\phi \tag{16-6b}$$

$$\mathbf{u}_z = \cos\theta\mathbf{u}_r - \sin\theta\mathbf{u}_\theta \tag{16-6c}$$

Using (16-5) and (16-6), we readily find that (16-3) expands to

$$\begin{aligned}\mathbf{u}_r(\mathbf{u}_r\cdot\dot{\mathbf{v}}) - \dot{\mathbf{v}} = &-\mathbf{u}_\theta(\ddot{x}\cos\theta\cos\phi + \ddot{y}\cos\theta\sin\phi - \ddot{z}\sin\theta)\\ &+ \mathbf{u}_\phi(-\ddot{x}\sin\phi + \ddot{y}\cos\phi)\end{aligned} \tag{16-7}$$

We see that $\mathbf{u}_r$ is not present in (16-7), so the field components are indeed transverse to the direction of the propagation $\mathbf{u}_r$.

An immediate simplification in (16-7) can be made by noting that we shall only be interested in problems that are symmetric in ϕ. Thus, we can conveniently take $\phi = 0$. Then, from (16-1), (16-2), and (16-7) the transverse field components of the radiation field are found to be

$$E_\theta = \frac{e}{4\pi\varepsilon_0 c^2 R}[\ddot{x}\cos\theta - \ddot{z}\sin\theta] \tag{16-8}$$

$$E_\phi = \frac{e}{4\pi\varepsilon_0 c^2 R}[\ddot{y}] \tag{16-9}$$

Equations (16-8) and (16-9) are the desired relations between the transverse radiation field components, E_θ and E_ϕ, and the accelerating charge described by $\ddot{x}$, $\ddot{y}$, and $\ddot{z}$. We note that E_θ, E_ϕ, and θ refer to the observer's coordinate system, and $\ddot{x}$, $\ddot{y}$, and $\ddot{z}$ refer to the charge's coordinate system.

Because we are interested in field quantities that are actually measured, namely, the Stokes parameters, in spherical coordinates the Stokes parameters are defined by

$$S_0 = E_\phi E_\phi^* + E_\theta E_\theta^* \tag{16-10a}$$

$$S_1 = E_\phi E_\phi^* - E_\theta E_\theta^* \tag{16-10b}$$

$$S_2 = E_\phi E_\theta^* + E_\theta E_\phi^* \tag{16-10c}$$

$$S_3 = i(E_\phi E_\theta^* - E_\theta E_\phi^*) \tag{16-10d}$$

where $i = \sqrt{-1}$. While it is certainly possible to substitute (16-8) and (16-9) directly into (16-10) and find an expression for the Stokes parameters in terms of the acceleration, it is simpler to break the problem into two parts. Namely, we first determine the acceleration and the field components and then form the Stokes parameters according to (16-10).

16.2 RELATION BETWEEN THE UNIT VECTOR IN SPHERICAL COORDINATES AND CARTESIAN COORDINATES

We derive the relation between the vector in a spherical coordinate system and a Cartesian coordinate system.

The rectangular coordinates x, y, z are expressed in terms of spherical coordinates r, θ, ϕ by the equations:

$$x = x(r, \theta, \phi) \qquad y = y(r, \theta, \phi) \qquad z = z(r, \theta, \phi) \tag{16-11}$$

Conversely, these equations can be expressed so that r, θ, ϕ can be written in terms of x, y, z. Then, any point with coordinates (x, y, z) has corresponding coordinates (r, θ, ϕ). We assume that the correspondence is unique. If a particle moves from a point P in such a way that θ and ϕ are held constant and only r varies, a curve in space is generated. We speak of this curve as the r curve. Similarly, two other coordinate curves, the θ curves and the ϕ curves, are determined at each point as shown in Fig. 16-2. If only one coordinate is held constant, we determine successively three surfaces passing through a point in space, these surfaces intersecting in the coordinate curves. It is generally convenient to choose the new coordinates in such a way that the coordinate curves are mutually perpendicular to each other at each point in space. Such coordinates are called orthogonal curvilinear coordinates.

Let $\mathbf{r}$ represent the position vector of a point P in space. Then

$$\mathbf{r} = x\mathbf{i} + y\mathbf{j} + z\mathbf{k} \tag{16-12}$$

From Fig. 16-2 we see that a vector $\mathbf{v}_r$ tangent to the r curve at P is given by

$$\mathbf{v} = \frac{\partial \mathbf{r}}{\partial r} = \left(\frac{\partial \mathbf{r}}{\partial s_r}\right) \cdot \left(\frac{ds_r}{dr}\right) \tag{16-13}$$

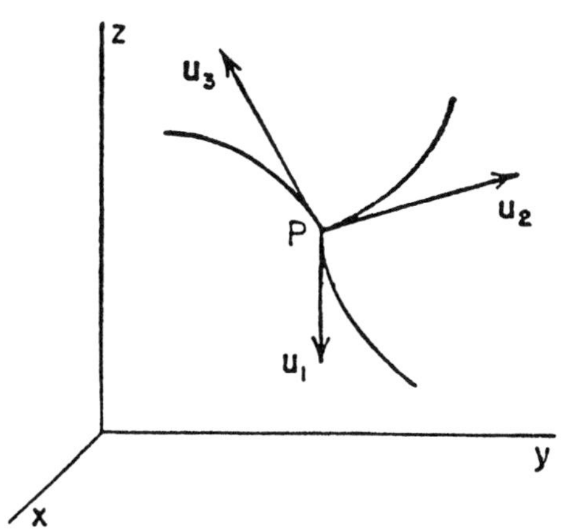

Figure 16-2 Determination of the r, θ, and ϕ curves in space.

where s_r is the arc length along the r curve. Since $\partial\mathbf{r}/\partial s_r$ is a unit vector (this ratio is the vector chord length $\Delta\mathbf{r}$, to the arc length Δs_r such that in the limit as $\Delta s_r \to 0$ the ratio is 1), we can write (16-13) as

$$\mathbf{v}_r = h_r \mathbf{u}_r \tag{16-14}$$

where $\mathbf{u}_r$ is the unit vector tangent to the r curves in the direction of increasing arc length. From (16-14) we see then that $h_r = ds_r/dr$ is the length of $\mathbf{v}_r$.

Considering now the other coordinates, we write

$$\mathbf{v}_r = h_r \mathbf{u}_r \qquad \mathbf{v}_\theta = h_\theta \mathbf{u}_\theta \qquad \mathbf{v}_\phi = h_\phi \mathbf{u}_\phi \tag{16-15}$$

so (16-14) can be simply written as

$$\mathbf{v}_k = h_k \mathbf{u}_k \qquad k = r, \theta, \phi \tag{16-16}$$

where $\mathbf{u}_k (k = r, \theta, \phi)$ is the unit vector tangent to the u_k curve. Furthermore, we see from (16-13) that

$$h_r = \frac{ds_r}{dr} = \left|\frac{\partial\mathbf{r}}{\partial r}\right| \tag{16-17a}$$

$$h_\theta = \frac{ds_\theta}{d\theta} = \left|\frac{\partial\mathbf{r}}{\partial \theta}\right| \tag{16-17b}$$

$$h_\phi = \frac{ds_\phi}{d\phi} = \left|\frac{\partial\mathbf{r}}{\partial \phi}\right| \tag{16-17c}$$

Equation (16-17) can be written in differential form as

$$ds_r = h_r dr \qquad ds_\theta = h_\theta d\theta \qquad ds_\phi = h_\phi d\phi \tag{16-18}$$

We thus see that h_r, h_θ, h_ϕ are scale factors, giving the ratios of differential distances to the differentials of the coordinate parameters. The calculations of $\mathbf{v}_k$ from (16-15) leads to the determination of the scale factors from $h_k = |\mathbf{v}_k|$ and the unit vector from $\mathbf{u}_k = \mathbf{v}_k/h_k$.

We now apply these results to determining the unit vectors for a spherical coordinate system. In Fig. 16-3 we show a spherical coordinate system with unit vectors $\mathbf{u}_r$, $\mathbf{u}_\theta$, and $\mathbf{u}_\phi$. The angles θ and ϕ are called the polar and azimuthal angles,

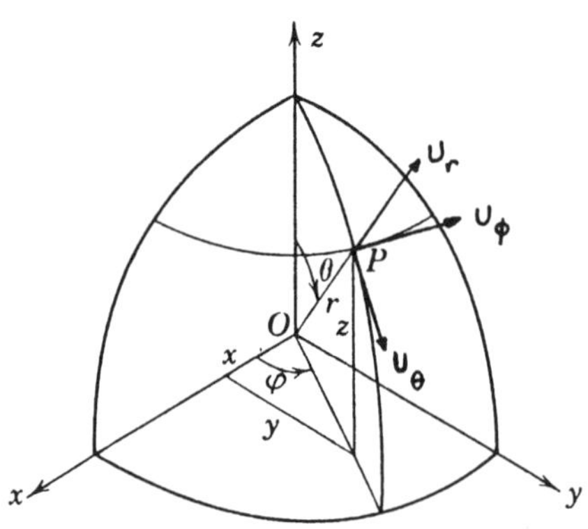

Figure 16-3 Unit vectors for a spherical coordinate system.

respectively. We see from the figure that x, y, and z can be expressed in terms of r, θ and ϕ by

$$x = r\sin\theta\cos\phi \qquad y = r\sin\theta\sin\phi \qquad z = r\cos\theta \tag{16-19}$$

Substituting (16-19) into (16-12) the position vector $\mathbf{r}$ becomes

$$\mathbf{r} = (r\sin\theta\cos\phi)\mathbf{i} + (r\sin\theta\sin\phi)\mathbf{j} + (r\cos\theta)\mathbf{k} \tag{16-20}$$

From (16-13) we find that

$$\mathbf{v}_r = \frac{\partial\mathbf{r}}{\partial r} = \sin\theta\cos\phi\mathbf{i} + \sin\theta\sin\phi\mathbf{j} + \cos\theta\mathbf{k} \tag{16-21a}$$

$$\mathbf{v}_\theta = \frac{\partial\mathbf{r}}{\partial\theta} = r\cos\theta\cos\phi\mathbf{i} + r\cos\theta\sin\phi\mathbf{j} - r\sin\theta\mathbf{k} \tag{16-21b}$$

$$\mathbf{v}_\phi = \frac{\partial\mathbf{r}}{\partial\phi} = -r\sin\theta\sin\phi\mathbf{i} + r\sin\theta\cos\phi\mathbf{j} \tag{16-21c}$$

The scale factors are, from (16-17),

$$h_r = \left|\frac{\partial\mathbf{r}}{\partial r}\right| = 1 \tag{16-22a}$$

$$h_\theta = \left|\frac{\partial\mathbf{r}}{\partial\theta}\right| = r \tag{16-22b}$$

$$h_\phi = \left|\frac{\partial\mathbf{r}}{\partial\phi}\right| = r\sin\theta \tag{16-22c}$$

Finally, from (16-21) and (16-22) the unit vectors are

$$\mathbf{u}_r = \frac{\mathbf{v}_r}{h_r} = \sin\theta\cos\phi\mathbf{i} + \sin\theta\sin\phi\mathbf{j} + \cos\theta\mathbf{k} \tag{16-23a}$$

$$\mathbf{u}_\theta = \frac{\mathbf{v}_\theta}{h_\theta} = \cos\theta\cos\phi\mathbf{i} + \cos\theta\sin\phi\mathbf{j} - \sin\theta\mathbf{k} \tag{16-23b}$$

$$\mathbf{u}_\phi = \frac{\mathbf{v}_\phi}{h_\phi} = -\sin\phi\mathbf{i} + \cos\phi\mathbf{j} \tag{16-23c}$$

which corresponds to the result given by (16-6) (it is customary to express $\mathbf{u}_x$, $\mathbf{u}_y$, $\mathbf{u}_z$ as $\mathbf{i}$, $\mathbf{j}$, $\mathbf{k}$).

We can easily check the direction of the unit vectors shown in Fig. 16-3 by considering (16-23) at, say, $\theta = 0^\circ$ and $\phi = 90^\circ$. For this condition (16-23) reduces to

$$\mathbf{u}_r = \mathbf{k} \tag{16-24a}$$

$$\mathbf{u}_\theta = \mathbf{j} \tag{16-24b}$$

$$\mathbf{u}_\phi = -\mathbf{i} \tag{16-24c}$$

which is exactly what we would expect according to Fig. 16-3.

An excellent discussion of the fundamentals of vector analysis can be found in the text of Hilderbrand given in the references at the end of this chapter. The material presented here was adapted from his Chapter 6.

16.3 RELATION BETWEEN THE POYNTING VECTOR AND THE STOKES PARAMETERS

Before we proceed to use the Stokes parameters to describe the field radiated by accelerating charges, it is useful to see how the Stokes parameters are related to the Poynting vector and Larmor's radiation formula in classical electrodynamics.

In Chapter 13, in the discussion of Young's interference experiment the fact was pointed out that two ideas were borrowed from mechanics. The first was the wave equation. Its solution alone, however, was found to be insufficient to arrive at a mathematical description of the observed interference fringes. In order to describe these fringes, another concept was borrowed from mechanics, namely, energy. Describing the optical field in terms of energy or, as it is called in optics, intensity, did lead to results in complete agreement with the observed fringes with respect to their intensity and spacing. However, the wave equation and the intensity formulation were accepted as hypotheses. In particular, it was not at all clear why the quadratic averaging of the amplitudes of the optical field led to the correct results. In short, neither aspect of the optical field had a theoretical basis.

With the introduction of Maxwell's equations, which were a mathematical formulation of the fundamental laws of the electromagnetic field, it was possible to show that these two hypotheses were a direct consequence of his theory. The first success was provided by Maxwell himself, who showed that the wave equation of optics arose directly from his field equations. In addition, he was surprised that his wave equation showed that the waves were propagating with the speed of light. The other hypothesis, namely, the intensity formed by taking time averages of the quadratic field components was also shown around 1885 by Poynting to be a direct consequence of Maxwell's equations. We now show this by returning to Maxwell's equations in free space [see Eqs.(15-1)],

$$\nabla \times \mathbf{E} = -\mu \frac{\partial \mathbf{H}}{\partial t} \tag{16-25a}$$

$$\nabla \times \mathbf{H} = \varepsilon \frac{\partial \mathbf{E}}{\partial t} \tag{16-25b}$$

$$\nabla \cdot \mathbf{E} = 0 \tag{16-25c}$$

$$\nabla \cdot \mathbf{B} = 0 \tag{16-25d}$$

and where we have also used the constitutive equations, (15-6). First, we take the scalar product of (16-25a) and $\mathbf{H}$ so that we have

$$\mathbf{H} \cdot \nabla \times \mathbf{E} = -\mu \mathbf{H} \cdot \frac{\partial \mathbf{H}}{\partial t} \tag{16-26a}$$

Next, we take the scalar product of (16-25b) and $\mathbf{E}$ so that we have

$$\mathbf{E} \cdot \nabla \times \mathbf{H} = \varepsilon \mathbf{E} \cdot \frac{\partial \mathbf{E}}{\partial t} \tag{16-26b}$$

We now subtract (16-26b) from (16-26a):

$$\mathbf{H} \cdot \nabla \times \mathbf{E} - \mathbf{E} \cdot \nabla \times \mathbf{H} = -\mu \mathbf{H} \cdot \frac{\partial \mathbf{H}}{\partial t} - \varepsilon \mathbf{E} \cdot \frac{\partial \mathbf{E}}{\partial t} \tag{16-27}$$

The left-hand side of (16-27) is recognized as the identity:

$$\nabla \cdot (\mathbf{E} \times \mathbf{H}) = \mathbf{H} \cdot (\nabla \times \mathbf{E}) - \mathbf{E} \cdot (\nabla \times \mathbf{H}) \tag{16-28}$$

The terms on the right-hand side of (16-27) can be written as

$$\mathbf{H} \cdot \frac{\partial \mathbf{H}}{\partial t} = \frac{1}{2} \frac{\partial}{\partial t} (\mathbf{H} \cdot \mathbf{H}) \tag{16-29a}$$

and

$$\mathbf{E} \cdot \frac{\partial \mathbf{E}}{\partial t} = \frac{1}{2} \frac{\partial}{\partial t} (\mathbf{E} \cdot \mathbf{E}) \tag{16-29b}$$

Then, using (16-28) and (16-29), (16-27) can be written as

$$\nabla \cdot (\mathbf{E} \times \mathbf{H}) + \frac{\partial}{\partial t} \left[\frac{\mu(\mathbf{H} \cdot \mathbf{H}) + \varepsilon(\mathbf{E} \cdot \mathbf{E})}{2} \right] = 0 \tag{16-30}$$

Inspection of (16-30) shows that it is identical in form to the continuity equation for current and charge:

$$\nabla \cdot \mathbf{j} + \frac{\partial \rho}{\partial t} = 0 \tag{16-31}$$

In (16-31) $\mathbf{j}$ is a current, that is, a flow of charge. Thus, we write the corresponding term for current in (16-30) as

$$\mathbf{S} = (\mathbf{E} \times \mathbf{H}) \tag{16-32}$$

The vector $\mathbf{S}$ is known as Poynting's vector and represents, as we shall show, the flow of energy.

The second term in (16-30) is interpreted as the time derivative of the sum of the electrostatic and magnetic energy densities. The assumption is now made that this sum represents the total electromagnetic energy even for time–varying fields, so the energy density w is

$$w = \frac{\mu H^2 + \varepsilon E^2}{2} \tag{16-33a}$$

where

$$\mathbf{H} \cdot \mathbf{H} = H^2 \tag{16-33b}$$

$$\mathbf{E} \cdot \mathbf{E} = E^2 \tag{16-33c}$$

Thus, (16-30) can be written as

$$\nabla \cdot \mathbf{S} + \frac{\partial w}{\partial t} = 0 \tag{16-34}$$

The meaning of $\mathbf{S}$ is now clear. It is the flow of energy, analogous to the flow of charge $\mathbf{j}$ (the current). Furthermore, if we write (16-34) as

$$\nabla \cdot \mathbf{S} = -\frac{\partial w}{\partial t} \tag{16-35}$$

then the physical meaning of (16-35) (and (16-34)) is that the decrease in the time rate of change of electromagnetic energy within a volume is equal to the flow of energy out of the volume. Thus, (16-34) is a conservation statement for energy.

We now consider the Poynting vector further:

$$\mathbf{S} = (\mathbf{E} \times \mathbf{H}) \tag{16-32}$$

In free space the solution of Maxwell's equations yields plane-wave solutions:

$$\mathbf{E}(\mathbf{r}, t) = \mathbf{E}_0 e^{i(\mathbf{k}\cdot\mathbf{r}-\omega t)} \tag{16-36a}$$

$$\mathbf{H}(\mathbf{r}, t) = \mathbf{H}_0 e^{i(\mathbf{k}\cdot\mathbf{r}-\omega t)} \tag{16-36b}$$

We can use (16-25a) to relate **E** to **H**:

$$\nabla \times \mathbf{E} = -\mu \frac{\partial \mathrm{H}}{\partial t} \tag{16-25a}$$

Thus, for the left-hand side of (16-25a) we have, using (16-36a),

$$\begin{aligned} \nabla \times \mathbf{E} &= \nabla \times [\mathbf{E}_0 e^{i(\mathbf{k}\cdot\mathbf{r}-\omega t)}] \\ &= i\mathbf{k} \times \mathbf{E} \end{aligned} \tag{16-37a}$$

where we have used the vector identity

$$\nabla \times (\phi \mathbf{a}) = \nabla\phi \times \mathbf{a} + \phi \nabla \times \mathbf{a} \tag{16-38}$$

Similarly, for the right-hand side we have

$$-\mu \frac{\partial \mathbf{H}}{\partial t} = i\omega \mathbf{H} \tag{16-39}$$

Thus (16-25a) becomes

$$\mathbf{n} \times \mathbf{E} = \frac{\mathbf{H}}{c\varepsilon_0} \tag{16-40a}$$

where

$$\mathbf{n} = \frac{\mathbf{k}}{k} \tag{16-40b}$$

since $k = \omega/c$. The vector **n** is the direction of propagation of **S**. Equation (16-40a) shows that **n**, **E**, and **H** are perpendicular to one another. Thus, if **n** is in the direction of propagation, then **E** and **H** are perpendicular to **n**, that is, in the transverse plane. We now substitute (16-40a) into (16-32) and we have

$$\mathbf{S} = c\varepsilon_0[\mathbf{E} \times (\mathbf{n} \times \mathbf{E})] \tag{16-41}$$

From the vector identity:

$$\mathbf{a} \times (\mathbf{b} \times \mathbf{c}) = (\mathbf{a} \cdot \mathbf{c})\mathbf{b} - (\mathbf{a} \cdot \mathbf{b})\mathbf{c} \tag{16-42}$$

we see that (16-41) reduces to

$$\mathbf{S} = c\varepsilon_0(\mathbf{E} \cdot \mathbf{E})\mathbf{n} \tag{16-43}$$

In Cartesian coordinates the quadratic term in (16-43) is written out as

$$\mathbf{E} \cdot \mathbf{E} = E_x E_x + E_y E_y \tag{16-44}$$

Thus, Maxwell's theory leads to quadratic terms, which we associate with the flow of energy.

For more than 20 years after Maxwell's enunciation of his theory in 1865, physicists constantly sought to arrive at other well-known results from his theory, e.g., Snell's law of refraction, or Fresnel's equations for reflection and transmission at an interface. Not only were these fundamental formulas found but their derivations led to new insights into the nature of the optical field. Nevertheless, this did not give rise to the acceptance of this theory. An experiment would have to be undertaken which only Maxwell's theory could explain. Only then would his theory be accepted.

If we express **E** and **H** in complex terms, then the time-averaged flux of energy is given by the real part of the complex Poynting vector, so

$$\langle S \rangle = \frac{1}{2}(\mathbf{E} \times \mathbf{H}^*) \tag{16-45}$$

From (16-40) we have

$$\mathbf{n} \times \mathbf{E}^* = \mathbf{H}^* \tag{16-46}$$

and substituting (16-46) into (16-45) leads immediately to

$$\langle S \rangle = \frac{1}{2} c\varepsilon_0 (\mathbf{E} \cdot \mathbf{E}^*)\mathbf{n} \tag{16-47}$$

Thus, Maxwell's theory justifies the use of writing the intensity I as

$$I = E_x E_x^* + E_y E_y^* \tag{16-48}$$

for the time-averaged intensity of the optical field.

In spherical coordinates the field is written as

$$\mathbf{E} = E_\theta \mathbf{u}_\theta + E_\phi \mathbf{u}_\phi \tag{16-49}$$

so the Poynting vector (16-47) becomes

$$\langle S \rangle = \frac{c\varepsilon_0}{2}(E_\phi E_\phi^* + E_\theta E_\theta^*)\mathbf{n} \tag{16-50}$$

The quantity within parentheses is the total intensity of the radiation field, i.e., the Stokes parameter S_0. Thus, the Poynting vector is directly proportional to the first Stokes parameter.

Another quantity of interest is the power radiated per unit solid angle, written as

$$\frac{dP}{d\Omega} = \frac{c\varepsilon_0}{2}(\mathbf{E} \cdot \mathbf{E}^*)R^2 \tag{16-51}$$

We saw that the field radiated by accelerating charges is given by

$$\mathbf{E} = \frac{e}{4\pi\varepsilon_0 c^2 R}[\mathbf{n} \times (\mathbf{n} \times \dot{\mathbf{v}})] \tag{16-1}$$

Expanding (16-1) by the vector triple product:

$$\mathbf{E} = \frac{e}{4\pi\varepsilon_0 c^2 R}[\mathbf{n}(\mathbf{n}\cdot\dot{\mathbf{v}}) - \dot{\mathbf{v}}] \tag{16-52}$$

We denote

$$\mathbf{n}\cdot\dot{\mathbf{v}} = |\mathbf{n}||\dot{\mathbf{v}}|\cos\Theta \tag{16-53}$$

where Θ is the angle between $\mathbf{n}$ and $\dot{\mathbf{v}}$ and $|\cdots|$ denotes that the absolute magnitude is to be taken. Using (16-52) and (16-53), we then find (16-51) becomes

$$\frac{dP}{d\Omega} = e^2|\dot{\mathbf{v}}|\sin^2\Theta \tag{16-54}$$

We saw that the field radiated by accelerating charges is given by

$$E_\theta = \frac{e}{4\pi\varepsilon_0 c^2 R}(\ddot{x}\cos\theta\cos\phi + \ddot{y}\cos\theta\sin\phi - \ddot{z}\sin\theta) \tag{16-55a}$$

$$E_\phi = \frac{e}{4\pi\varepsilon_0 c^2 R}(-\ddot{x}\sin\phi + \ddot{y}\cos\phi) \tag{16-55b}$$

The total radiated power over the sphere is given by integrating (16-51) over the solid angle:

$$P = \frac{c\varepsilon_0}{2}\int_0^{2\pi}\int_0^{\pi}(E_\phi E_\phi^* + E_\theta E_\theta^*)R^2\sin\theta\, d\theta\, d\phi \tag{16-56}$$

We easily find that

$$\int_0^{2\pi}\int_0^{\pi}(E_\phi E_\phi^*)R^2\sin\theta\, d\theta\, d\phi = \frac{4\pi e^2}{16\pi^2\varepsilon_0^2 c^4}(|\ddot{x}|^2 + |\ddot{y}|^2) \tag{16-57a}$$

and

$$\int_0^{2\pi}\int_0^{\pi}(E_\theta E_\theta^*)R^2\sin\theta\, d\theta\, d\phi = \frac{4\pi e^2}{3(16\pi^2\varepsilon_0^2 c^4)}(|\ddot{x}|^2 + |\ddot{y}|^2 + 4|\ddot{z}|^2) \tag{16-57b}$$

where $|\ddot{\ }|^2 \equiv (\ddot{\ })(\ddot{\ })^*$. Thus, adding (16-57a) and (16-57b) yields

$$\int_0^{2\pi}\int_0^{\pi}(E_\phi E_\phi^* + E_\theta E_\theta^*)R^2\sin\theta\, d\theta\, d\phi = \frac{4}{3}\frac{e^2}{4\pi\varepsilon_0 c^4}(|\ddot{\mathbf{r}}|^2) \tag{16-58a}$$

where

$$\ddot{\mathbf{r}} = \ddot{x}\mathbf{u}_x + \ddot{y}\mathbf{u}_y + \ddot{z}\mathbf{u}_z \tag{16-58b}$$

Substituting (16-58a) into (16-56) yields the power radiated by an accelerating charge:

$$P = \frac{2}{3}\frac{e^2}{4\pi\varepsilon_0 c^3}|\ddot{\mathbf{r}}|^2 \tag{16-59}$$

Equation (16-59) was first derived by J. J. Larmor in 1900 and, consequently, is known as Larmor's radiation formula.

The material presented in this chapter shows how Maxwell's equations led to the Poynting vector and then to the relation for the power radiated by the acceleration of an electron, that is, Larmor's radiation formula. We now apply these results to obtain the polarization of the radiation emitted by accelerating electrons. Finally, very detailed discussions of Maxwell's equations and the radiation by accelerating electrons are given in the texts by Jackson and Stratton.

REFERENCES

Books

1. Jackson, J. D., *Classical Electrodynamics*, Wilcy, New York, 1962.
2. Sommerfeld, A., *Lectures on Theoretical Physics*, Vols. I–V, Academic Press, New York, 1952.
3. Hildebrand, F. B., *Advanced Calculus for Engineers*, Prentice-Hall, Englewood Cliffs, NJ, 1949.
4. Stratton, J. A., *Electromagnetic Theory*, McGraw-Hill, New York, 1941.
5. Schott, G. A., *Electromagnetic Radiation*, Cambridge University Press, Cambridge, UK, 1912.
6. Jeans, J. H., *Mathematical Theory of Electricity and Magnetism*, 5th ed., Cambridge University Press, 1948.

17

Radiation Emitted by Accelerating Charges

17.1 STOKES VECTOR FOR A LINEARLY OSCILLATING CHARGE

We have shown how Maxwell's equation gave rise to the equations of the radiation field and the power emitted by an accelerating electron. We now discuss the polarization of the radiation emitted by specific electron configurations, e.g., bound charges and charges moving in circular and elliptical paths.

At the beginning of the nineteenth century the nature of electric charges was not fully understood. In 1895 the electron (charge) was discovered by J. J. Thompson. Thus, the long-sought source of the optical field was finally found. A year after Thompson's discovery, P. Zeeman performed a remarkable experiment by placing radiating atoms in a constant magnetic field. He thereupon discovered that the original single spectral line was split into two, or even three, spectral lines.

Shortly thereafter, H. Lorentz heard of Zeeman's results. Using Maxwell's theory and his electron theory, Lorentz then treated this problem. Lorentz's calculations predicted that the spectral lines should not only be split but also completely polarized. On Lorentz's suggestions Zeeman then performed further measurements and completely confirmed the predictions in all respects. It was only after the work of Zeeman and Lorentz that Maxwell's theory was accepted and Fresnel's theory of light replaced.

Not surprisingly, the importance of this work was immediately recognized, and Zeeman and Lorentz received the Nobel Prize in physics in 1902. We should emphasize that the polarization predictions of the spectral lines played a key part in understanding these experiments. This prediction, more than any other factor, was one of the major reasons for the acceptance of Maxwell's theory into optics.

In this chapter we build up to the experiment of Zeeman and the theory of Lorentz. We do this by first applying the Stokes parameters to a number of classical radiation problems. These are the radiation emitted by (1) a charge oscillating along an axis, (2) an ensemble of randomly oriented oscillating charges, (3) a charge moving in a circle, (4) a charge moving in an ellipse, and (5) a charge moving in a magnetic field. In the following chapter we then consider the problem of a randomly oriented oscillating charge moving in a constant magnetic field—the Lorentz–Zeeman effect.

We consider a bound charge oscillating along the z axis as shown in Fig. 17-1. The motion of the charge is described by

$$\frac{d^2z}{dt^2} + \omega_0^2 z = 0 \tag{17-1}$$

The solution of (17-1) is

$$z(t) = z(0)\cos(\omega_0 t + \alpha) \tag{17-2}$$

where $z(0)$ is the maximum amplitude and α is an arbitrary phase constant. Because we shall be using the complex form of the Stokes parameters, we write (17-2) as

$$z(t) = z(0)e^{i(\omega_0 t + \alpha)} \tag{17-3}$$

where it is understood that by taking the real part of (17-3), we recover (17-2); that is,

$$\text{Re}[z(t)] = z(0)\cos(\omega_0 t + \alpha) \tag{17-4}$$

The radiation field equations are given by (16-8) and (16-9) in Section 16.1:

$$E_\theta = \frac{e}{4\pi\varepsilon_0 c^2 R}[\ddot{x}\cos\theta - \ddot{z}\sin\theta] \tag{16-8}$$

$$E_\phi = \frac{e}{4\pi\varepsilon_0 c^2 R}[\ddot{y}] \tag{16-9}$$

Recall that these equations refer to the observation being made in the xz plane, that is, at $\phi = 0$. The angle θ is the polar angle in the observer's reference frame.

Performing the differentiation of (17-3) to obtain $\ddot{z}$, we have

$$\ddot{z} = -\omega_0^2 z(0)e^{i(\omega_0 t + \alpha)} \tag{17-5}$$

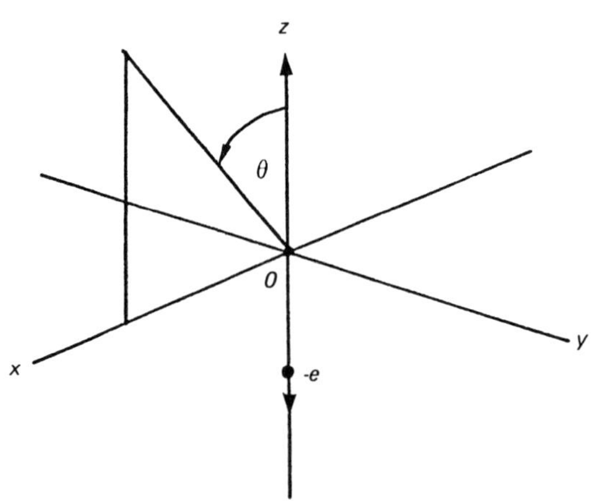

Figure 17-1 Motion of a linear oscillating charge.

Substituting (17-5) into (16-8) yields

$$E_\theta = \frac{e}{4\pi\varepsilon_0 c^2 R}[\omega_0^2 z(0)\sin\theta e^{i(\omega_0 t+\alpha)}] \tag{17-6a}$$

$$E_\phi = 0 \tag{17-6b}$$

The Stokes parameters are defined in a spherical coordinate system to be

$$S_0 = E_\phi E_\phi^* + E_\theta E_\theta^* \tag{16-10a}$$

$$S_1 = E_\phi E_\phi^* - E_\theta E_\theta^* \tag{16-10b}$$

$$S_2 = E_\phi E_\theta^* + E_\theta E_\phi^* \tag{16-10c}$$

$$S_3 = i(E_\phi E_\theta^* - E_\theta E_\phi^*) \tag{16-10d}$$

Substituting (17-6a) and (17-6b) into (16-10) yields

$$S_0 = \left(\frac{ez(0)}{4\pi\varepsilon_0 c^2 R}\right)^2 \omega_0^4 \sin^2\theta \tag{17-7a}$$

$$S_1 = -\left(\frac{ez(0)}{4\pi\varepsilon_0 c^2 R}\right)^2 \omega_0^4 \sin^2\theta \tag{17-7b}$$

$$S_2 = 0 \tag{17-7c}$$

$$S_3 = 0 \tag{17-7d}$$

We now arrange (17-7) in the form of the Stokes vector:

$$S = \left(\frac{ez(0)}{4\pi\varepsilon_0 c^2 R}\right)^2 \sin^2\theta\omega_0^4 \begin{pmatrix} 1 \\ -1 \\ 0 \\ 0 \end{pmatrix} \tag{17-8}$$

Equation (17-8) shows that the observed radiation is always linearly vertically polarized light at a frequency ω_0, the fundamental frequency of oscillation of the bound charge. Furthermore, when we observe the radiation parallel to the z axis ($\theta = 0^\circ$), the intensity is zero. Observing the radiation perpendicular to the z axis ($\theta = 90^\circ$), we note that the intensity is a maximum. This behavior is shown in Fig. 17-2. In order to plot the intensity behavior as a function of θ, we set

$$I(\theta) = \sin^2\theta \tag{17-9a}$$

In terms of $x(\theta)$ and $z(\theta)$ we then have

$$x(\theta) = I(\theta)\sin\theta = \sin^2\theta\sin\theta \tag{17-9b}$$

$$z(\theta) = I(\theta)\cos\theta = \sin^2\theta\cos\theta \tag{17-9c}$$

The term $ez(0)$ in (17-8) is recognized as a dipole moment. A characteristic of dipole radiation is the presence of the $\sin^2\theta$ term shown in (17-8). Hence, (17-8)

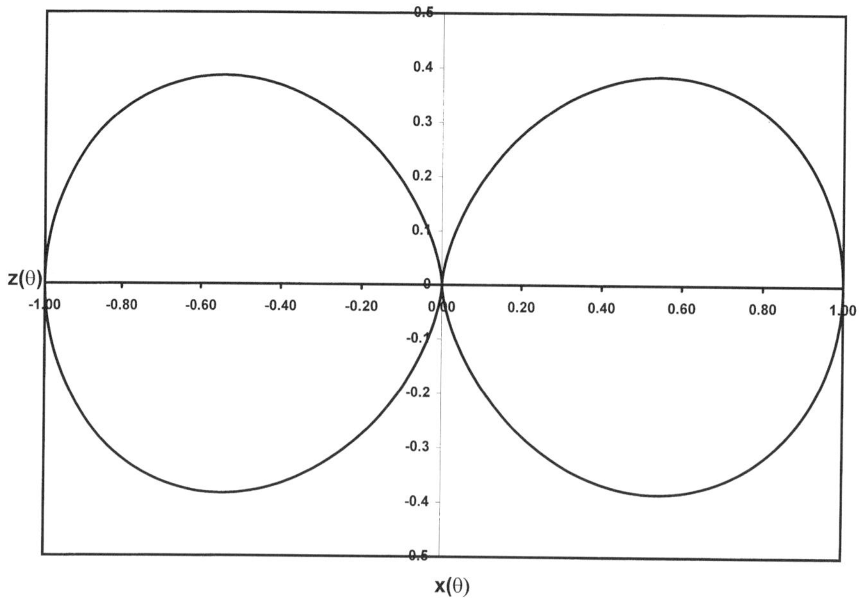

Figure 17-2 Plot of the intensity behavior of a dipole radiation field.

describes the Stokes vector of a dipole radiation field. This type of field is very important because it appears in many types of radiation problems in physics and engineering. Finally, we note that a linearly oscillating charge gives rise to linearly polarized light. Thus, the state of polarization is a manifestation of the fundamental motion of the electron. This observation will be confirmed for other types of radiating systems.

17.2 STOKES VECTOR FOR AN ENSEMBLE OF RANDOMLY ORIENTED OSCILLATING CHARGES

In the previous section, we considered the radiation field emitted by a charge or electron oscillating with an angular frequency ω_0 about an origin. Toward the end of the nineteenth century a model was proposed for the atom in which an oscillating electron was *bound* to a positively charged atom. The electron was believed to be negative (from work with "free" electrons in gases and chemical experiments). The assumption was made that the electron was attracted to the positively charged atom, and the force on the electron was described by Hooke's law, namely, $-k\mathbf{r}$. This model was used by H. Lorentz to solve a number of longstanding problems, e.g., the relation between the refractive index and the wavelength, the so-called dispersion relation.

The motion of the electron was described by the simple force equation:

$$m\ddot{\mathbf{r}} = -k\mathbf{r} \tag{17-10a}$$

or

$$\ddot{\mathbf{r}} + \omega_0^2\mathbf{r} = 0 \tag{17-10b}$$

where m is the mass of the electron, k is the restoring force constant, and the angular frequency is $\omega_0^2 = k/m$. We saw in Part I that the nature of unpolarized light was not well understood throughout most of the nineteenth century. We shall now show that this simple model for the motion of the electron within the atom leads to the correct Stokes vector for unpolarized light.

The treatment of this problem can be considered to be among the first successful applications of Maxwell's equations in optics. This simple atomic model provides a physical basis for the source term in Maxwell's equations. The model leads to the appearance of unpolarized light, a quantity that was a complete mystery up to the time of the electron. Thus, an ensemble of oscillating charges bound to a positive nucleus and randomly oriented gives rise to unpolarized light.

We now determine the Stokes vector of an ensemble of randomly oriented, bound, charged oscillators moving through the origin. This problem is treated by first considering the field emitted by a single charge oriented at the polar angle α and the azimuthal angle β in the reference frame of the charge. An ensemble average is then taken by integrating the radiated field over the solid angle $\sin\alpha\, d\alpha\, d\beta$. The diagram describing the motion of a single charge is given in Fig. 17-3.

The equations of motion of the charged particle can be written immediately from Fig. 17-3 and are

$$x(t) = A \sin\alpha \sin\beta e^{i\omega_0 t} \tag{17-11a}$$

$$y(t) = A \sin\alpha \sin\beta e^{i\omega_0 t} \tag{17-11b}$$

$$z(t) = A \cos\alpha e^{i\omega_0 t} \tag{17-11c}$$

where ω_0 is the angular frequency of natural oscillation. Differentiating (17-11) twice with respect to time gives

$$\ddot{x}(t) = -\omega_0^2 A \sin\alpha \cos\beta e^{i\omega_0 t} \tag{17-12a}$$

$$\ddot{y}(t) = -\omega_0^2 A \sin\alpha \sin\beta e^{i\omega_0 t} \tag{17-12b}$$

$$\ddot{z}(t) = -\omega_0^2 A \cos\alpha e^{i\omega_0 t} \tag{17-12c}$$

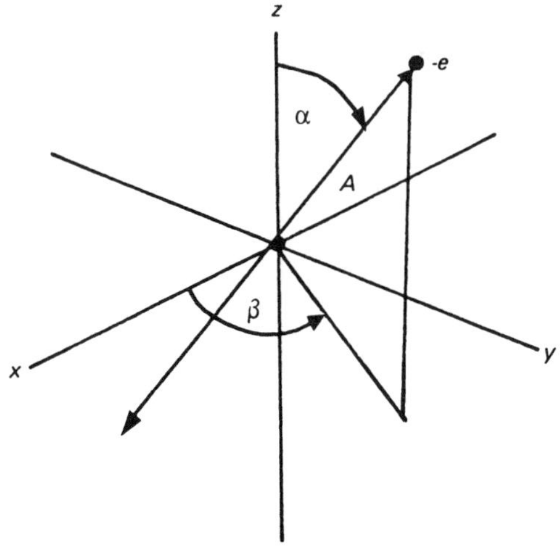

Figure 17-3 Instantaneous motion of an ensemble of oscillating charges.

Substituting (17-12) into the radiation field equations, we find that

$$E_\theta = -\frac{eA\omega_0^2 e^{i\omega_0 t}}{4\pi\varepsilon_0 c^2 R}(\sin\alpha\cos\beta\cos\theta - \cos\alpha\sin\theta) \tag{17-13a}$$

$$E_\phi = -\frac{eA\omega_0^2 e^{i\omega_0 t}}{4\pi\varepsilon_0 c^2 R}(\sin\alpha\sin\beta) \tag{17-13b}$$

where θ is the observer's viewing angle measured from the z axis.

Recall that the Stokes parameters are defined by

$$S_0 = E_\phi E_\phi^* + E_\theta E_\theta^* \tag{16-10a}$$

$$S_1 = E_\phi E_\phi^* - E_\theta E_\theta^* \tag{16-10b}$$

$$S_2 = E_\phi E_\theta^* + E_\theta E_\phi^* \tag{16-10c}$$

$$S_3 = i(E_\phi E_\theta^* - E_\theta E_\phi^*) \tag{16-10d}$$

Substituting (17-13) in (16-10), we then find that the Stokes parameters are

$$S_0 = C[\sin^2\alpha\sin^2\beta + \sin^2\alpha\cos^2\beta\cos^2\theta \\ - 2\sin\alpha\cos\alpha\cos\beta\cos\theta\sin\theta + \cos^2\alpha\sin^2\theta] \tag{17-14a}$$

$$S_1 = C[\sin^2\alpha\sin^2\beta - \sin^2\alpha\cos^2\beta\cos^2\theta \\ + 2\sin\alpha\cos\alpha\cos\beta\cos\theta\sin\theta - \cos^2\alpha\sin^2\theta] \tag{17-14b}$$

$$S_2 = C[2(\sin^2\alpha\sin\beta\cos\beta\cos\theta - \cos\alpha\sin\alpha\sin\beta\sin\theta)] \tag{17-14c}$$

$$S_3 = 0 \tag{17-14d}$$

where

$$C = \left(\frac{eA}{4\pi\varepsilon_0 c^2 R}\right)^2 \omega_0^4 \tag{17-14e}$$

The fact that S_3 is zero in (17-14d) shows that the emitted radiation is always linearly polarized, as we would expect from a model in which the electron only undergoes linear motion.

In order to describe an ensemble of randomly oriented charges we integrate (17-14) over the solid angle $\sin\alpha\, d\alpha\, d\beta$:

$$\langle\cdots\rangle = \int_0^{2\pi}\int_0^{\pi}(\cdots)\sin\alpha\, d\alpha\, d\beta \tag{17-15}$$

where $\langle\cdots\rangle$ is the ensemble average and $(\cdots)$ represents (17-14a), etc. Carrying out the integration of (17-14) by using (17-15) and forming the Stokes vector, we find that

$$S = \frac{8\pi}{3}\left(\frac{eA}{4\pi\varepsilon_0 c^2 R}\right)^2 \omega_0^4 \begin{pmatrix} 1 \\ 0 \\ 0 \\ 0 \end{pmatrix} \tag{17-16}$$

which is the Stokes vector for unpolarized light. This is exactly what is observed from natural light sources. Note that the polarization state is always independent of the observer's viewing angle θ; the observed light always appears to be unpolarized.

Thus, this simple model explains the appearance of unpolarized light from optical sources. Unpolarized light can only arise from an ensemble of randomly oriented accelerating charges, which can be the case for bound electrons. Electrons moving at a constant velocity, even if the motion is random, cannot give rise to unpolarized light.

This simple atomic model received further support when it was used by Lorentz to explain the Lorentz–Zeeman effect, namely, the radiation field emitted by a *bound* electron moving in a constant magnetic field. We emphasize that the motion of a *free* accelerating electron gives rise to a different result, as we shall see.

17.2.1 Note on Use of Hooke's Law for a Simple Atomic System

At first glance the use of Hooke's law to describe the motion of a negative electron bound to a positive charge (nucleus) within an atom may appear to be quite arbitrary. The use of Hooke's law is based, however, on the following simple atomic model.

The force of attraction between two opposite but equal charges e separated by a distance r is given by

$$F = \frac{(+e)(-e)}{4\pi\varepsilon_0 r^2}\mathbf{u}_r \tag{17-17}$$

where $\mathbf{u}_r$ is a unit radius vector. The positive charge is located at the origin of a spherical coordinate system.

We now assume that the positive charge is distributed over a sphere of volume V and radius r, so the charge density ρ is

$$\rho = \frac{+e}{V} = \frac{+e}{4\pi r^3/3} \tag{17-18}$$

or

$$+e = \frac{4\pi\rho r^3}{3} \tag{17-19}$$

Substituting (17-19) into (17-17) gives

$$\mathbf{F} = -k\mathbf{r} \tag{17-20}$$

where $\mathbf{r} = r\mathbf{u}_r$, and $k = e\rho/3\varepsilon_0$. Equation (17-20) is Hooke's law. Thus, on the basis of this very simple atomic model the motion of the electron is expected to undergo simple harmonic motion.

17.3 STOKES VECTOR FOR A CHARGE ROTATING IN A CIRCLE

We now continue with our application of the Stokes parameters to describe radiation problems. In this section we turn our attention to the determination of the field

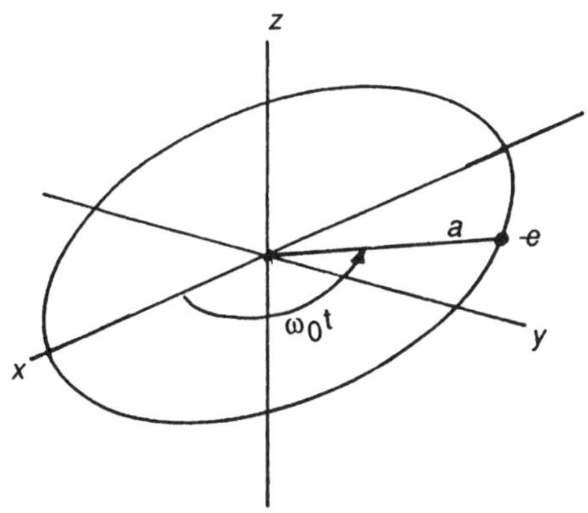

Figure 17-4 Motion of a charge moving counterclockwise in a circle of radius a in the xy plane with an angular frequency ω_0.

radiated by a charge moving in a circle. This is shown in Fig. 17-4. The coordinates of the charge are

$$x(t) = a\cos\omega_0 t \tag{17-21a}$$

$$y(t) = a\sin\omega_0 t \tag{17-21b}$$

$$z(t) = 0 \tag{17-21c}$$

That (17-21) describe counterclockwise motion is easily checked by first setting $\theta = \omega_0 t$. Then, as t increases, θ increases. Choosing $\theta = 0$, $\pi/2$, π and $3\pi/2$, the reader will easily see that plotting the position of the charge describes a counterclockwise motion as it moves in a circle of radius a.

To use the complex form of the Stokes parameters, the coordinates (17-21) must also be expressed in complex form. We have (Euler's relation)

$$e^{i\omega_0 t} = \cos\omega_0 t + i\sin\omega_0 t \tag{17-22}$$

The real part of (17-22) is cos $\omega_0 t$. We can also express sin $\omega_0 t$ in terms of the real part of (17-22), Re{ }, by multiplying (17-22) by $-i$. Then, we see that

$$\mathrm{Re}\{e^{i\omega_0 t}\} = \cos\omega_0 t \tag{17-23a}$$

$$\mathrm{Re}\{-ie^{i\omega_0 t}\} = \sin\omega_0 t \tag{17-23b}$$

Thus, in complex notation (17-21a) and (17-21b) become

$$x(t) = ae^{i\omega_0 t} \tag{17-24a}$$

$$y(t) = -iae^{i\omega_0 t} \tag{17-24b}$$

and the acceleration is then

$$\ddot{x}(t) = -a\omega_0^2 e^{i\omega_0 t} \tag{17-25a}$$

$$\ddot{y}(t) = +ia\omega_0^2 e^{i\omega_0 t} \tag{17-25b}$$

Substituting (17-25a) and (17-25b) into the radiation field equations (16-8) and (16-9) we find that

$$E_\theta = \frac{e}{4\pi\varepsilon_0 c^2 R}\left[-a\omega_0^2 \cos\theta e^{i\omega_0 t}\right] \tag{17-26a}$$

$$E_\phi = \frac{e}{4\pi\varepsilon_0 c^2 R}\left[ia\omega_0^2 e^{i\omega_0 t}\right] \tag{17-26b}$$

Again, we express (17-26a) and (17-26b) in terms of the Stokes parameters and form the Stokes vector. The result is

$$S = \left(\frac{ea}{4\pi\varepsilon_0 c^2 R}\right)^2 \omega_0^4 \begin{pmatrix} 1+\cos^2\theta \\ 1-\cos^2\theta \\ 0 \\ 2\cos\theta \end{pmatrix} \tag{17-27}$$

Equation (17-27) is the Stokes vector for elliptically polarized light. Thus, we see that the radiation is elliptically polarized and is characterized by a frequency ω_0, the frequency of rotation of the electron. Furthermore, we see that we have the factor ea in (17-27), the familiar expression for the dipole moment. We observe that (17-27) shows that the orientation angle ψ of the polarization ellipse is always zero. Similarly, the ellipticity angle χ is

$$\chi = \frac{1}{2}\sin^{-1}\left(\frac{S_3}{S_0}\right) \tag{17-28}$$

so from (17-27) we have

$$\chi = \frac{1}{2}\sin^{-1}\left(\frac{2\cos\theta}{1+\cos^2\theta}\right) \tag{17-29}$$

The ellipticity angle is a function of the observation angle θ. We see that for $\theta = 0°$, that is, we view the rotating electron along the z axis, (17-29) becomes $\chi = 45°$ and we observe right circularly polarized light. The Stokes vector (17-27) reduces to

$$S = 2\left(\frac{ea}{4\pi\varepsilon_0 c^2 R}\right)^2 \omega_0^4 \begin{pmatrix} 1 \\ 0 \\ 0 \\ 1 \end{pmatrix} \tag{17-30}$$

If we now view the rotating electron perpendicular to the z axis, that is, $\theta = 90°$, we find that $\chi = 0°$ and we observe linearly horizontally polarized light. The corresponding Stokes vector is

$$S = \left(\frac{ea}{4\pi\varepsilon_0 c^2 R}\right)^2 \omega_0^4 \begin{pmatrix} 1 \\ 1 \\ 0 \\ 0 \end{pmatrix} \tag{17-31}$$

These results agree with our earlier observation that the polarization of the emitted radiation is a manifestation of the motion of the charge. Thus, if we look along the z axis we would see an electron moving counterclockwise in a circle, so we observe right circularly polarized light. If we look perpendicular to the z axis, the electron appears to behave as a linear oscillator and we observe linearly horizontally polarized light, in agreement with our earlier conclusion. The linear polarization is to be expected, because if we view the motion of the charge as it moves in a circle at $\theta = 90^\circ$ it appears to move from left to right and then from right to left, identical to the behavior of a linear oscillator described in Section 17.1. Finally, for $\theta = 180^\circ$ we see that (17-29) becomes $\chi = -45^\circ$, so we observe left circularly polarized light.

Also observe that (17-27) satisfies the equality:

$$S_0^2 = S_1^2 + S_2^2 + S_3^2 \tag{17-32}$$

The equals sign shows that the emitted radiation is always completely polarized. Furthermore, the degree of polarization is independent of the observation angle θ.

17.4 STOKES VECTOR FOR A CHARGE MOVING IN AN ELLIPSE

It is of interest to consider the case where an electron moves in an elliptical orbit. The equations of motion are

$$x(t) = a\cos\omega_0 t \tag{17-33a}$$

$$y(t) = b\sin\omega_0 t \tag{17-33b}$$

where a and b are the semimajor and semiminor axes lengths, respectively. In complex notation (17-33) becomes

$$x(t) = ae^{i\omega_0 t} \tag{17-34a}$$

$$y(t) = -ibe^{i\omega_0 t} \tag{17-34b}$$

The acceleration is then

$$\ddot{x}(t) = -a\omega_0^2 e^{i\omega_0 t} \tag{17-35a}$$

$$\ddot{y}(t) = ib\omega_0^2 e^{i\omega_0 t} \tag{17-35b}$$

Again using the radiation field equations (16-8) and (16-9), the radiated fields are found to be

$$E_\theta = \left(\frac{e\omega_0^2}{4\pi\varepsilon_0 c^2 R}\right) e^{i\omega_0 t}[-a\cos\theta] \tag{17-36a}$$

$$E_\phi = \left(\frac{e\omega_0^2}{4\pi\varepsilon_0 c^2 R}\right) e^{i\omega_0 t}[ib] \tag{17-36b}$$

We now form the Stokes vector for (17-36) and find that

$$S = \left(\frac{e}{4\pi\varepsilon_0 c^2 R}\right)^2 \omega_0^4 \begin{pmatrix} b^2 + a^2\cos^2\theta \\ b^2 - a^2\cos^2\theta \\ 0 \\ 2ab\cos\theta \end{pmatrix} \tag{17-37}$$

Equation (17-37) is the Stokes vector for elliptically polarized light. We see immediately that if $a = b$ then (17-37) reduces to the Stokes vector for an electron moving in a circle. The orientation angle ψ of the polarization ellipse is seen from (17-37) to be 0°. The ellipticity angle χ is

$$\chi = \frac{1}{2}\sin^{-1}\left(\frac{2ab\cos\theta}{b^2 + a^2\cos\theta}\right) \tag{17-38}$$

The radiation is always elliptically polarized with one exception; the exception will be discussed in a moment. We see that for $\theta = 0^\circ$, (17-37) reduces to

$$S = \left(\frac{e}{4\pi\varepsilon_0 c^2 R}\right)^2 \omega_0^4 \begin{pmatrix} b^2 + a^2 \\ b^2 - a^2 \\ 0 \\ 2ab \end{pmatrix} \tag{17-39}$$

which is the Stokes vector for elliptically polarized light. The other case of interest is to observe the radiation perpendicular to the z axis, that is, $\theta = 90^\circ$. For this angle (17-37) reduces to

$$S = \left(\frac{e}{4\pi\varepsilon_0 c^2 R}\right)^2 \omega_0^4 b^2 \begin{pmatrix} 1 \\ 1 \\ 0 \\ 0 \end{pmatrix} \tag{17-40}$$

which is the Stokes vector for linear horizontally polarized light. Again, this is perfectly understandable, because at this angle the moving charge appears to be oscillating in a straight line as it moves in its elliptical path.

The Stokes vectors derived here will reappear when we discuss the Lorentz–Zeeman effect.

REFERENCES

Books

1. Jackson, J. D., *Classical Electrodynamics*, Wiley, New York, 1962.
2. Lorentz, H. A., *Theory of Electrons*, Dover reprint, New York, 1952.
3. Sommerfeld, A., *Lectures on Theoretical Physics*, Vols. I–V, Academic Press, New York, 1952.
4. Jeans, J. H., *Mathematical Theory of Electricity and Magnetism*, 5th ed., Cambridge Unversity Press, Cambridge, UK, 1948.
5. Wood, R. W., *Physical Optics*, 3rd ed., Optical Society of America, Washington, DC, 1988.

18

The Radiation of an Accelerating Charge in the Electromagnetic Field

18.1 MOTION OF A CHARGE IN AN ELECTROMAGNETIC FIELD

In previous chapters the Stokes vectors were determined for charges moving in a linear, circular, or elliptical path. At first sight the examples chosen appear to have been made on the basis of simplicity. However, the examples were chosen because charged particles actually move in these paths in an electromagnetic field; that is, the examples are based on physical reality. In this section we show from Lorentz's force equation that in an electromagnetic field charged particles follow linear and circular paths. In the following section we determine the Stokes vectors corresponding to these physical configurations.

The reason for treating the motion of a charge in this chapter as well as in the previous chapter is that the material is necessary to understand and describe the Lorentz–Zeeman effect. Another reason for discussing the motion of charged particles in the electromagnetic field is that it has many important applications. Many physical devices of importance to science, technology, and medicine are based on our understanding of the fundamental motion of charged particles. In particle physics these include the cyclotron, betatron, and synchrotron, and in microwave physics the magnetron and traveling-wave tubes. While these devices, per se, will not be discussed here, the mathematical analysis presented is the basis for describing all of them. Our primary interest is to describe the motion of charges as they apply to atomic and molecular systems and to determine the intensity and polarization of the emitted radiation.

In this chapter we treat the motion of a charged particle in three specific configurations of the electromagnetic field: (1) the acceleration of a charge in an electric field, (2) the acceleration of a charge in a magnetic field, and (3) the acceleration of a charge in perpendicular electric and magnetic fields. In particular, the motion of a charged particle in perpendicular electric and magnetic fields is extremely interesting not only from the standpoint of its practical importance but because the paths taken by the charged particle are quite beautiful and remarkable.

In an electromagnetic field the motion of a charged particle is governed by the Lorentz force equation:

$$\mathbf{F} = q[\mathbf{E} + (\mathbf{v} \times \mathbf{B})] \tag{18-1}$$

where q is the magnitude of the charge, $\mathbf{E}$ is the applied electric field, $\mathbf{B}$ is the applied magnetic field, and $\mathbf{v}$ is the velocity of the charge. The background to the Lorentz force equation can be found in the texts given in the references. The text by G. P. Harnwell on electricity and magnetism is especially clear and illuminating. Quite understandably, because of the importance of the phenomenon of the radiation of accelerating charges in the design and fabrication of instruments and devices, many articles and textbooks are devoted to the subject. Several are listed in the references.

18.1.1 Motion of an Electron in a Constant Electric Field

The first and simplest example of the motion of an electron in an electromagnetic field is for a charge moving in a constant electric field. The field is directed along the z axis and is of strength E_0. The vector representation for the general electric field $\mathbf{E}$ is

$$\mathbf{E} = E_x\mathbf{u}_x + E_y\mathbf{u}_y + E_z\mathbf{u}_z \tag{18-2}$$

Since the electric field is directed only in the z direction, $\mathbf{E}_x = E_y = 0$, so

$$\mathbf{E} = E_z\mathbf{u}_z = E_0\mathbf{u}_z \tag{18-3}$$

For simplicity the motion of the electron is restricted to the xz plane and is initially moving with a velocity v_0 at an angle α from the z axis. This is shown in Fig. 18-1.

Because there is no magnetic field, the Lorentz force equation (18-1) reduces to

$$m\ddot{\mathbf{r}} = -e\mathbf{E} \tag{18-4}$$

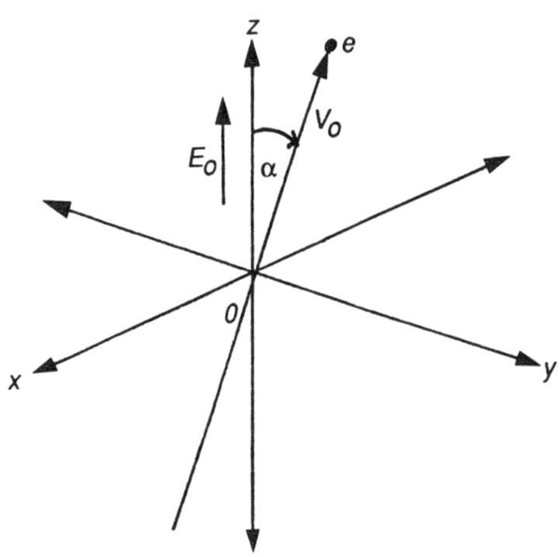

Figure 18-1 Motion of an electron in the xz plane in a constant electric field directed along the z axis.

where m is the mass of the electron. In component form (18-4) is

$$m\ddot{x} = 0 \tag{18-5a}$$

$$m\ddot{y} = 0 \tag{18-5b}$$

$$m\ddot{z} = -eE_z = -eE_0 \tag{18-5c}$$

At the initial time $t = 0$ the electron is assumed to be at the origin of the coordinate system, so

$$x(0) = z(0) = 0 \tag{18-6}$$

Similarly, the velocity at the initial time is assumed to be

$$\dot{x}(0) = v_x = v_0 \sin\alpha \tag{18-7a}$$

$$\dot{z}(0) = v_z = v_0 \cos\alpha \tag{18-7b}$$

There is no force in the y direction, so (18-5b) can be ignored. We integrate (18-5a) and (18-5c) and find

$$\dot{x}(t) = C_1 \tag{18-8a}$$

$$\dot{z}(t) = -\frac{-eE_0 t}{m} + C_2 \tag{18-8b}$$

where C_1 and C_2 are constants of integration. From the initial conditions, C_1 and C_2 are easily found, and the specific solution of (18-8) is

$$\dot{x}(t) = v_0 \sin\alpha \tag{18-9a}$$

$$\dot{z}(t) = \frac{-eE_0 t}{m} + v_0 \cos\alpha \tag{18-9b}$$

Integrating (18-9) once more yields

$$x(t) = v_0 t \sin\alpha \tag{18-10a}$$

$$z(t) = \frac{-eE_0 t^2}{2m} + v_0 t \cos\alpha \tag{18-10b}$$

where the constants of integration are found from (18-6) to be zero. We can eliminate t between (18-10a) and (18-10b) to obtain

$$z(t) = -\left(\frac{eE_0}{2mv_0^2 \sin^2\alpha}\right)x^2 + (\cot\alpha)x \tag{18-11}$$

which is the equation of a parabola. The path is shown in Fig. 18-2.

Inspecting (18-11) we see that if $\alpha = 0$ then $z(t) = \infty$. That is, the electron moves in a straight line starting from the origin 0 along the z axis and "intercepts" the z axis at infinity (∞). However, if α is not zero, then we can determine the

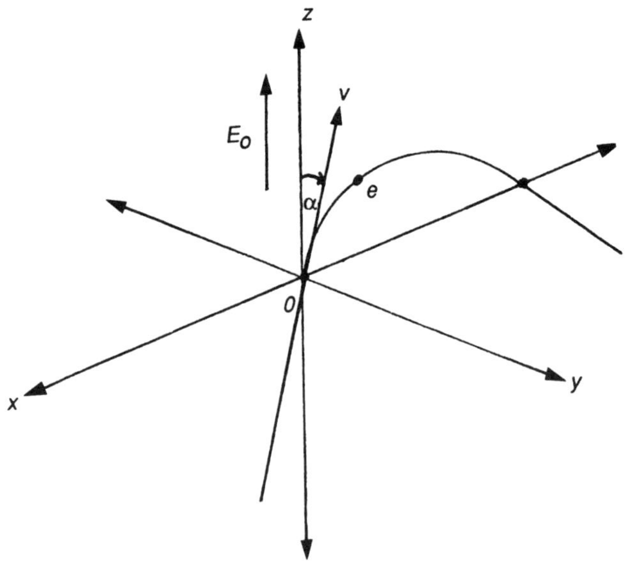

Figure 18-2 Parabolic path of an electron in a constant electric field.

positions $x(t)$ where the electron intercepts the z axis by setting $z(t) = 0$ in (18-11). On doing this the intercepts are found to occur at

$$x(t) = 0 \tag{18-12a}$$

$$x(t) = \frac{mv_0^2}{eE_0} \sin 2\alpha \tag{18-12b}$$

The first value corresponds to our initial condition $x(0) = z(0) = 0$. Equation (18-12b) shows that the maximum value of x is attained by setting $\alpha = 45^\circ$, so

$$x_{\max} = \frac{mv_0^2}{eE_0} \tag{18-13}$$

This result is not at all surprising, since (18-11) is identical in form to the equation for describing a projectile moving in a constant gravitational field. Finally, the maximum value of z is found from (18-11) to be

$$z(t) = \frac{1}{2}\left(\frac{mv_0^2}{eE}\right) \sin 2\alpha \tag{18-14a}$$

or

$$z_{\max} = \frac{1}{2} x_{\max} \tag{18-14b}$$

where we have used (18-12b).

18.1.2 Motion of a Charged Particle in a Constant Magnetic Field

We now consider the motion of an electron moving in a constant magnetic field. The coordinate configuration is shown in Fig. 18-3. In the figure **B** is the magnetic field directed in the positive z direction. The Lorentz force equation (18-1) then reduces to, where the charge on an electron is $q = -e$,

$$\mathbf{F} = -e(\mathbf{v} \times \mathbf{B}) \tag{18-15}$$

Equation (18-15) can be expressed as a differential equation:

$$m\ddot{\mathbf{r}} = -e(\mathbf{v} \times \mathbf{B}) \tag{18-16}$$

where m and $\ddot{\mathbf{r}}$ are the mass and acceleration vector of the charged particle, respectively. In component form (18-16) is

$$m\ddot{x} = -e(\mathbf{v} \times \mathbf{B})_x \tag{18-17a}$$

$$m\ddot{y} = -e(\mathbf{v} \times \mathbf{B})_y \tag{18-17b}$$

$$m\ddot{z} = -e(\mathbf{v} \times \mathbf{B})_z \tag{18-17c}$$

where the subscript on ($\mathbf{v} \times \mathbf{B}$) refers to the appropriate component to be taken. The vector product $\mathbf{v} \times \mathbf{B}$ can be expressed as a determinant

$$\mathbf{v} \times \mathbf{B} = \begin{vmatrix} \mathbf{u}_x & \mathbf{u}_y & \mathbf{u}_z \\ \dot{x} & \dot{y} & \dot{z} \\ \mathbf{B}_x & \mathbf{B}_y & \mathbf{B}_z \end{vmatrix} \tag{18-18}$$

where $\mathbf{u}_x$, $\mathbf{u}_y$, and $\mathbf{u}_z$ are the unit vectors pointing in the positive x, y, and z directions, respectively and the velocities have been expressed as $\dot{x}$, $\dot{y}$, and $\dot{z}$. The constant magnetic field is directed only along z, so $\mathbf{B}_z = \mathbf{B}$ and $\mathbf{B}_x = \mathbf{B}_y = 0$. Then, (18-18) and (18-17) reduce to

$$m\ddot{x} = -e(\dot{y}\mathbf{B}) \tag{18-19a}$$

$$m\ddot{y} = -e(-\dot{x}\mathbf{B}) \tag{18-19b}$$

$$m\ddot{z} = 0 \tag{18-19c}$$

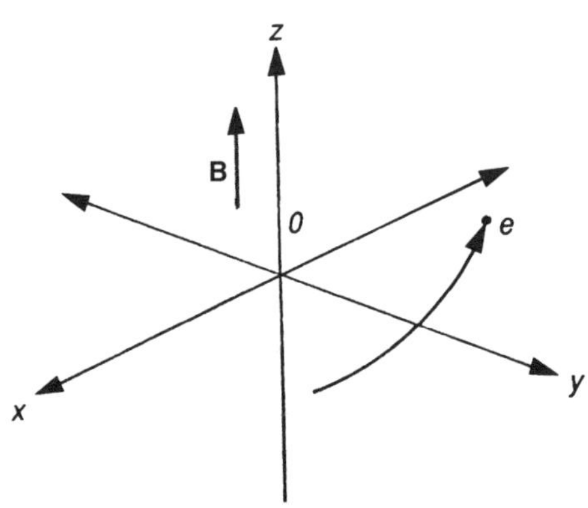

Figure 18-3 Motion of an electron in a constant magnetic field.

Equation (18-19c) is of no interest because the motion along z is not influenced by the magnetic field. The equations of motion are then

$$\ddot{x} = \frac{-e\mathbf{B}}{m}\dot{y} \tag{18-20a}$$

$$\ddot{y} = \frac{e\mathbf{B}}{m}\dot{x} \tag{18-20b}$$

Equation (18-20a) and (18-20b) can be written as a single equation by introducing the complex variable $\zeta(t)$:

$$\zeta(t) = x(t) + iy(t) \tag{18-21}$$

Differentiating (18-21) with respect to time, we have

$$\dot{\zeta} = \dot{x} + i\dot{y} \tag{18-22a}$$

$$\ddot{\zeta} = \ddot{x} + i\ddot{y} \tag{18-22b}$$

Multiplying (18-20b) by i and adding this result to (18-20a) and using (18-22a) leads to

$$\ddot{\zeta} - \frac{ie\mathbf{B}}{m}\dot{\zeta} = 0 \tag{18-23}$$

The solution of (18-23) is readily found by assuming a solution of the form:

$$\zeta(t) = e^{\omega t} \tag{18-24}$$

Substituting (18-24) into (18-23) we find that

$$\omega(\omega - i\omega_c) = 0 \tag{18-25}$$

where $\omega_c = e\mathbf{B}/m$ is the frequency of rotation, known as the cyclotron frequency.

Equation (18-25) is called the auxiliary or characteristic equation of (18-23), and from (18-25) the roots are $\omega = 0$, $i\omega_c$. The general solution of (18-23) can be written immediately as

$$\zeta(t) = c_1 + c_2 e^{i\omega_c t} \tag{18-26}$$

where c_1 and c_2 are constants of integration. To provide a specific solution for (18-23), we assume that, initially, the charge is at the origin and moving along the x axis with a velocity v_0. Thus, we have

$$x(0) = 0 \qquad y(0) = 0 \tag{18-27a}$$

$$\dot{x}(0) = v_0 \qquad \dot{y}(0) = 0 \tag{18-27b}$$

which can be expressed in terms of (18-21) and (18-22a) as

$$\zeta(0) = x(0) + iy(0) = 0 \tag{18-28a}$$

$$\dot{\zeta}(0) = \dot{x}(0) + i\dot{y}(0) = v_0 \tag{18-28b}$$

This leads immediately to

$$c_1 = -c_2 \tag{18-29a}$$

$$c_2 = \frac{iv_0}{\omega_c} \tag{18-29b}$$

so the specific solution of (18-26) is

$$\zeta(t) = -\frac{iv_0}{\omega_c}(1 - e^{i\omega_c t}) \tag{18-30}$$

Taking the real and imaginary part of (18-30) then yields

$$x(t) = \frac{v_0}{\omega_c}\sin\omega_c t \tag{18-31a}$$

$$y(t) = -\frac{v_0}{\omega_c}(1 - \cos\omega_c t) \tag{18-31b}$$

or

$$x(t) = \frac{v_0}{\omega_c}\sin\omega_c t \tag{18-32a}$$

$$y + \frac{v_0}{\omega_c} = \frac{v_0}{\omega_c}\cos\omega_c t \tag{18-32b}$$

Squaring and adding (18-32a) and (18-32b) give

$$x^2 + \left(y + \frac{v_0}{\omega_c}\right)^2 = \left(\frac{v_0}{\omega_c}\right)^2 \tag{18-33}$$

which is an equation of a circle with radius v_0/ω_c and center at $x = 0$ and $y = -v_0/\omega_c$.

Equations (18-32) and (18-33) show that in a constant magnetic field a charged particle does indeed move in a circle. Also, (18-32) describes a charged particle moving in a clockwise direction as viewed along the positive axis toward the origin. Equation (18-33) is of great historical and scientific interest, because it is the basis of one of the first methods and instruments used to measure the ratio e/m, namely, the mass spectrometer. To see how this measurement is made, we note that since the electron moves in a circle, (18-33) can be solved for the condition where it crosses the y axis, which is $x = 0$. We see from (18-33) that this occurs at

$$y = 0 \tag{18-34a}$$

$$y = -\frac{2v_0}{\omega_c} \tag{18-34b}$$

We note that (18-34b) is twice the radius ρ ($\rho = v_0/\omega_c$). This is to be expected because the charged particle moves in a circle. Since $\omega_c = e\mathbf{B}/m$, we can solve (18-34b) for e/m to find that

$$\frac{e}{m} = -\left(\frac{2v_0}{\mathbf{B}y}\right) \tag{18-35}$$

The initial velocity v_0 is known from equating the kinetic energy of the electron with the voltage applied to the charged particle as it enters the chamber of the mass spectrometer. The magnitude of y where the charged particle is intercepted ($x = 0$) is measured. Finally, the strength of the magnetic field $\mathbf{B}$ is measured with a magnetic flux meter. Consequently, all the quantities on the right side of (18-35) are known, so the ratio e/m can then be found. The value of this ratio found in this manner agrees with those of other methods.

18.1.3 Motion of an Electron in a Crossed Electric and Magnetic Field

The final configuration of interest is to determine the path of an electron which moves in a constant magnetic field directed along the z axis and in a constant electric field directed along the y axis, a so-called crossed, or perpendicular, electric and magnetic field. This configuration is shown in Fig. 18-4.

For this case Lorentz's force equation (18-1) reduces to

$$m\ddot{x} = -e(\dot{y}\mathbf{B}) \tag{18-36a}$$

$$m\ddot{y} = -eE + e(\dot{x}\mathbf{B}) \tag{18-36b}$$

$$m\ddot{z} = 0 \tag{18-36c}$$

From (18-21) and (18-22), (18-36) can be written as a single equation:

$$\ddot{\zeta} - i\omega_c\dot{\zeta} = -\frac{ieE}{m} \tag{18-37}$$

where $\omega_c = e\mathrm{B}/m$. Equation (18-37) is easily solved by noting that if we multiply by $e^{-i\omega_c t}$ then (18-37) can be rewritten as

$$\frac{d}{dt}(e^{-i\omega_c t}\dot{\zeta}) = \left(\frac{-ieE}{m}\right)e^{-i\omega_c t} \tag{18-38}$$

Straightforward integration of (18-38) yields

$$\zeta = \left(\frac{eE}{m\omega_c}\right)t - \left(\frac{ic_1}{\omega_c}\right)e^{i\omega_c t} + c_2 \tag{18-39}$$

where c_1 and c_2 are constants of integration. We choose the initial conditions to be

$$x(0) = 0 \qquad y(0) = 0 \tag{18-40a}$$

$$\dot{x}(0) = v_0 \qquad \dot{y}(0) = 0 \tag{18-40b}$$

The specific solution of (18-39) is

$$\zeta = a\phi + ib(1 - \cos\phi) + b\sin\phi \tag{18-41a}$$

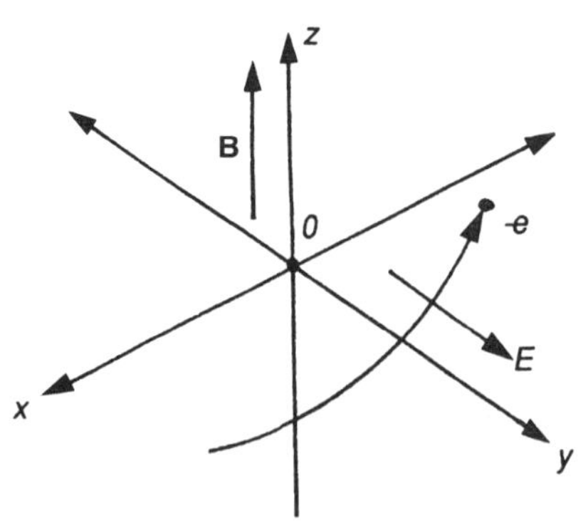

Figure 18-4 Motion of an electron in a crossed electric and magnetic field.

where

$$\phi = \omega_c t \tag{18-41b}$$

$$a = \frac{eE}{m\omega_c^2} \tag{18-41c}$$

$$b = \frac{v_0 - eE/m\omega_c}{\omega_c} \tag{18-41d}$$

Equating the real and imaginary parts of (18-41a) and (18-21), we then find that

$$x(\phi) = a\phi + b\sin\phi \tag{18-42a}$$

$$y(\phi) = b(1 - \cos\phi) \tag{18-42b}$$

Equation (18-42) is well known from analytical geometry and describes a general cycloid or trochoid. Specifically, the trochoidal path is a prolate cycloid, cycloid, or curtate cycloid, depending on whether $a < b$, $a = b$, or $a > b$, respectively. We can easily understand the meaning of this result. First, we note that if the applied electric field E were not present then (18-42) would reduce to the equation of a circle of radius b, so the electron moves along a circular path. However, an electric field in the y direction forces the electron to move in the same direction continuously as the electron moves in the circular path. Consequently, the path is stretched, so the circle becomes a general cycloid or trochoid. This "stretching" factor is represented by the term a_ϕ in (18-42a). We note that (18-40) shows $\phi = 0$ corresponds to the origin. Thus, ϕ is measured from the origin and increases in a clockwise motion.

We can easily find the maximum and minimum values of $x(\phi)$ and $y(\phi)$ over a single cycle of ϕ. The maximum and minimum values of $y(\phi)$ are simply 0 and $2b$ and occur at $\phi = 0$ and π, respectively. For $x(\phi)$ the situation is more complicated. From (18-42a) the angles where the minimum and maximum values of $x(\phi)$ occur are

$$\phi = \tan^{-1}\left(\frac{\pm\sqrt{b^2 - a^2}}{a}\right) \tag{18-43}$$

The negative sign refers to the minimum value of $x(\phi)$, and the positive sign refers to the maximum value of $x(\phi)$. The corresponding maximum and minimum values of $x(\phi)$ are then found to be

$$x(a, b) = a\tan^{-1}\left(\frac{\pm\sqrt{b^2 - a^2}}{a}\right) \pm \sqrt{b^2 - a^2} \tag{18-44}$$

In particular, if we set $b = 1$ in (18-43) and (18-44) we have

$$\phi = \tan^{-1}\left(\frac{\pm\sqrt{1 - a^2}}{a}\right) \tag{18-45}$$

$$x(a) = a \cdot \tan^{-1}\left(\frac{\pm\sqrt{1 - a^2}}{a}\right) \pm \sqrt{1 - a^2} \tag{18-46}$$

Equation (18-46) shows that $x(a)$ is imaginary for $a > 1$; that is, a maximum and a minimum do not exist. This behavior is confirmed in Fig. 18-13 and 18-14 for $a = 1.25$ and $a = 1.5$.

Equation (18-45) ranges from $a = 0$ to 1; for $a = 0$ (no applied electric field) $\phi = \pi/2$ and $3\pi/2$ (or $-\pi/2$), respectively. This is exactly what we would expect for a circular path. Following the conventional notation the path of the electron moves counterclockwise, so $\pi/2$ is the angle at the maximum point and $3\pi/2$ $(-\pi/2)$ corresponds to the angle at the minimum point. Figure 18-5 shows the change in $\phi(a)$ as the electric field (a) increases. The upper curve corresponds to the positive sign of the argument in (18-45), and the lower curve corresponds to the negative sign, respectively. We see that at $a = 1$ the maximum and minimum values converge. The point of convergence corresponds to a cycloid. This behavior is confirmed by the curve for $x(a)$ in the figure for $a = 1$, as we shall soon see.

The maximum and minimum points of the (prolate) cycloid are given by (18-46). We see immediately that for $a = 0$ we have $x(0) = \pm 1$. This, of course, applies to a circle. For $0 < a < 1$ we have a prolate cycloid. For a cycloid $a = 1$, and (18-46) gives $x(1) = 0$ and π; that is, the maximum and minimum points coincide. This behavior is also confirmed for the plot of $x(a)$ versus a at the value where $a = 1$. In Fig. 18-6 we have plotted the change in the maximum and minimum values of $x(a)$ as a increases from 0 to 1. The upper curve corresponds to the positive sign in (18-46), and the lower curve corresponds to the negative sign.

It is of interest to determine the points on the x axis where the electron path intersects or is tangent to the x axis. This is found by setting $y = 0$ in (18-42b). We see that this is satisfied by $\phi = 0$ or $\phi = 2\pi$. Setting $b = 1$ in (18-42a), the points of intersection on the x axis are given by $x = 0$ and $x = 2\pi a$; the point $x = 0$ and $y = 0$,

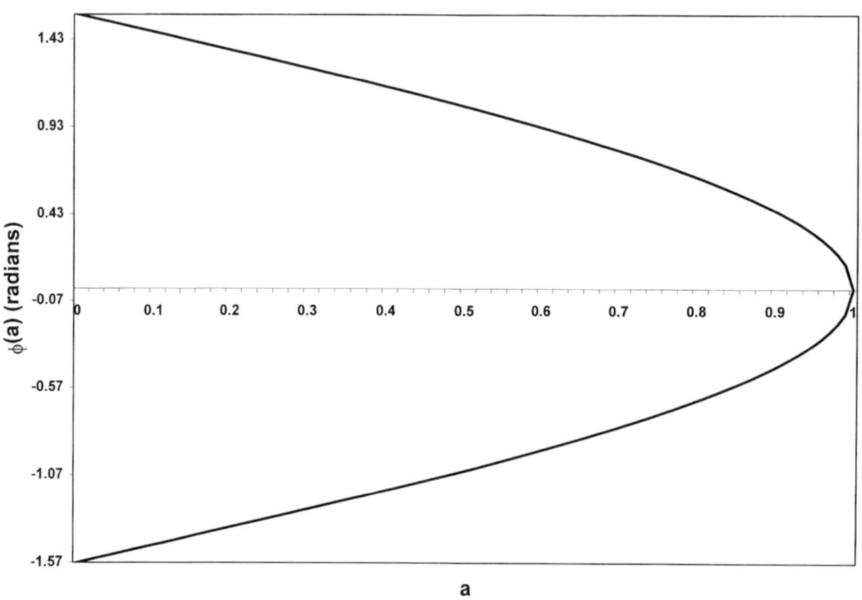

Figure 18-5 Plot of the angle $\phi(a)$, Eq. (18-45), for the maximum and minimum points as the electric field (a) increases.

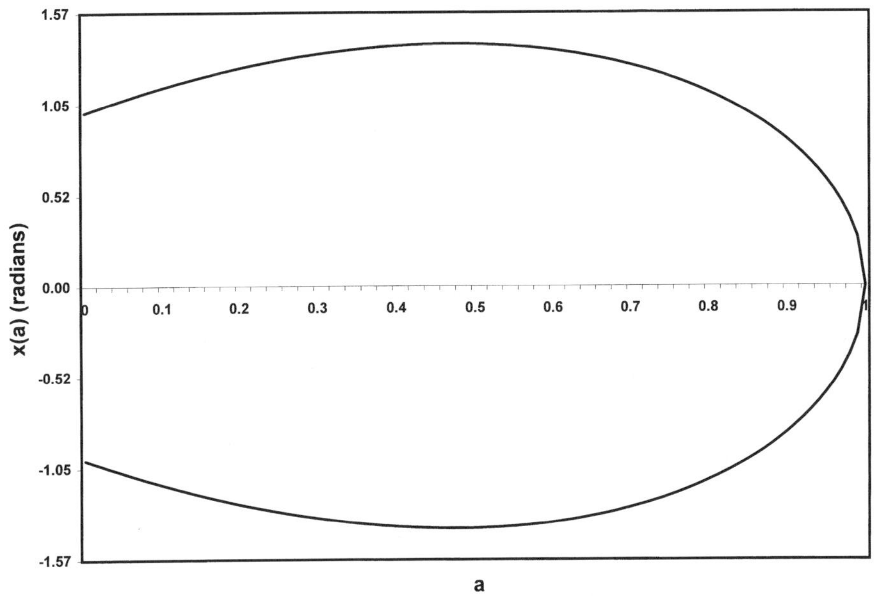

Figure 18-6 Plot of the maximum and minimum values of $x(\phi)$ written as $x(a)$, Eq. (18-46) as the electric field (a) increases from 0 to 1.

we recall, is the position of the electron at the initial time $t = 0$. Thus, setting $b = 1$ in (18-42a), the initial and final positions of the electron for $a = 0$ are at $x(i) = 0$ and $x(f) = 0$, which is the case for a circle. For the other extreme, obtained by setting $a = 1$, the initial and final intersections are 0 and 2π, respectively. Thus, as the magnitude of the electric field increases, the final point of intersection on the x axis increases. In addition, as a increases, the prolate cycloid advances so that for $a = 0$ (a circle) the midpoint of the path is at $x = 0$ and for $a = 1$ the midpoint is at $x = \pi$.

We now plot the evolution of the trochoid as the electric field $E(a)$ increases. The equations used are, from (18-42) with $b = 1$,

$$x(\phi) = a\phi + \sin\phi \tag{18-47a}$$

$$y(\phi) = 1 - \cos\phi \tag{18-47b}$$

It is of interest to plot (18-47a) from $\phi = 0$ to 2π for $a = 0$, 0.25, 0.50, 0.75, and 1.0. Figure 18-7 is a plot of the evolution of $x(\phi)$ from a pure sinusoid for $a = 0$ to a cycloid for $a = 1$.

The most significant feature of Fig. 18-7 is that the maxima shift to the right as a increases. This behavior continues until $a = 1$, whereupon the maximum point virtually disappears. Similarly, the minima shift to the left, so that at $a = 1$ the minimum point virtually disappears. This behavior is later confirmed for $a = 1$, a cycloid.

The paths of the electrons are specifically shown in Figs. 18-8 to 18-15. The curves are plotted over a single cycle of ϕ (0 to 2π). For these values (18-45) shows

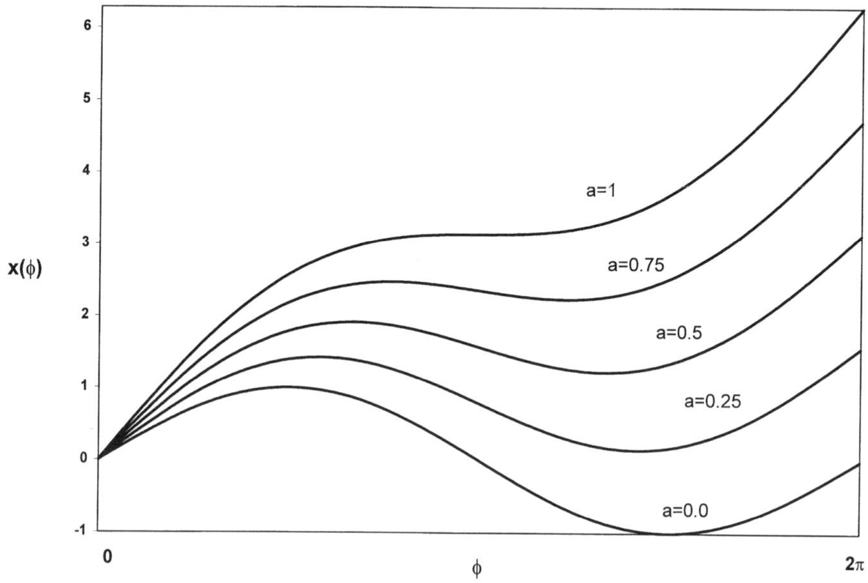

Figure 18-7 Plot of $x(\phi)$, Eq. (18-47a), for $a = 0$ to 1.

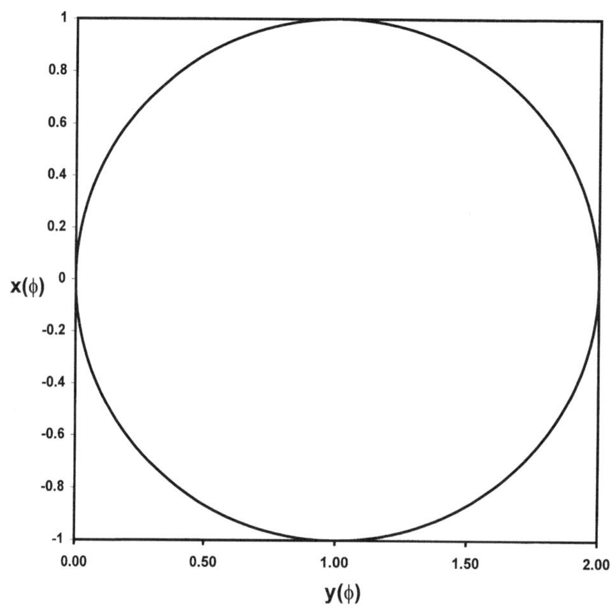

Figure 18-8 The trochoidal path of an electron, $a = 0$ (a circle).

that the path intersects the x axis at 0 and $2\pi a$, respectively. We select a to be 0, 0.25, 0.5, ..., 1.5. The corresponding intersections of the path on the x axis are then (0, 0), (0, $\pi/2$), (0, π), ..., (0, 3π). With these values of a, Figs. 18-8 to 18-15 show the evolutionary change in the path. Figure 18-15 shows the path of the electron as it moves over four cycles.

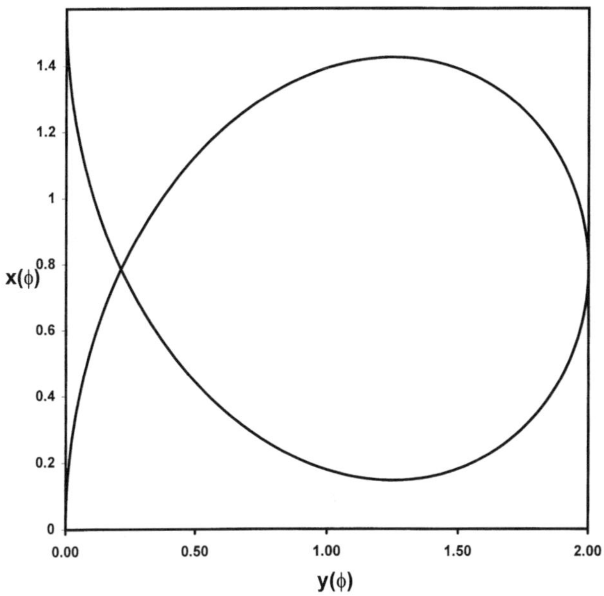

Figure 18-9 The trochoidal path of an electron $a = 0.25$.

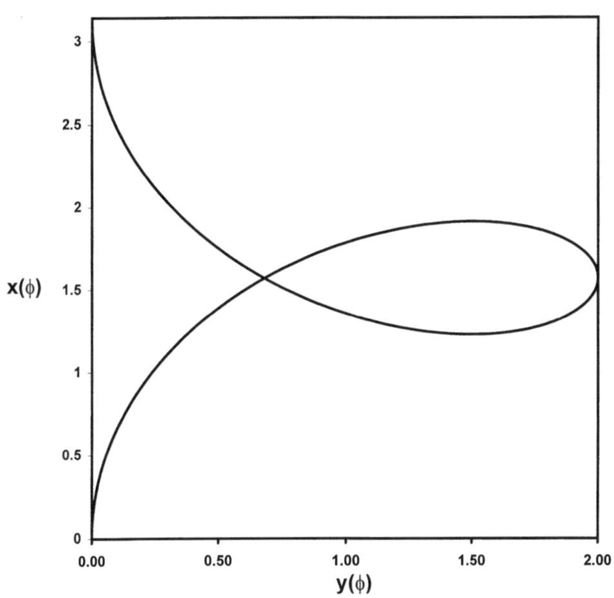

Figure 18-10 The trochoidal path of an electron, $a = 0.5$.

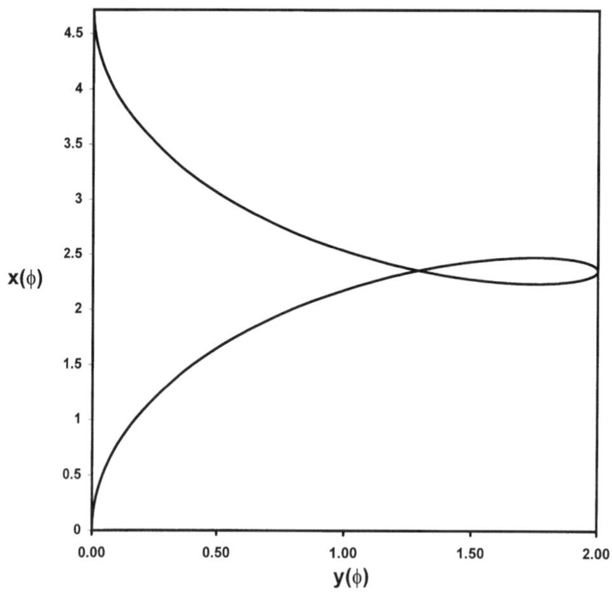

Figure 18-11 The trochoidal path of an electron, $a = 0.75$.

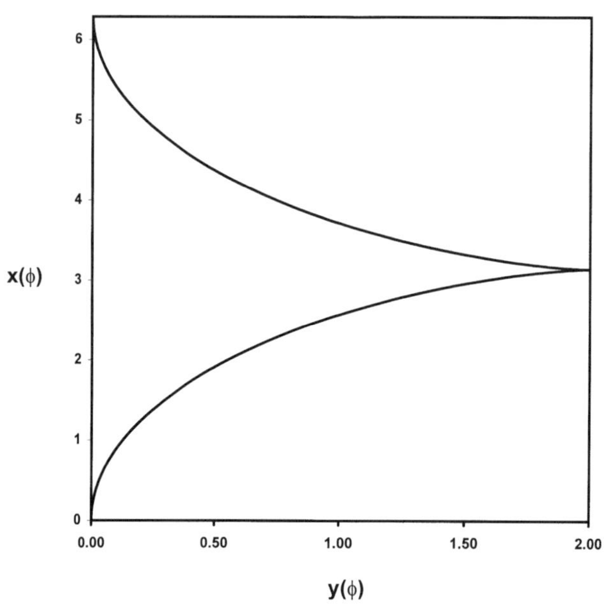

Figure 18-12 The trochoidal path of an electron, $a = 1.0$.

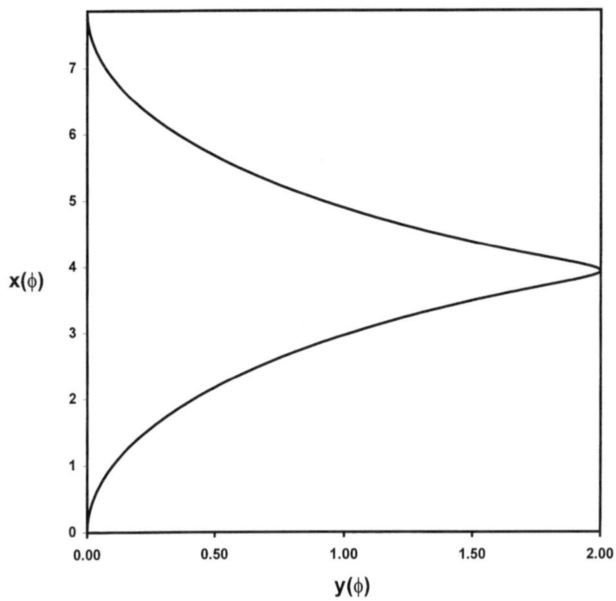

Figure 18-13 The trochoidal path of an electron, $a = 1.25$.

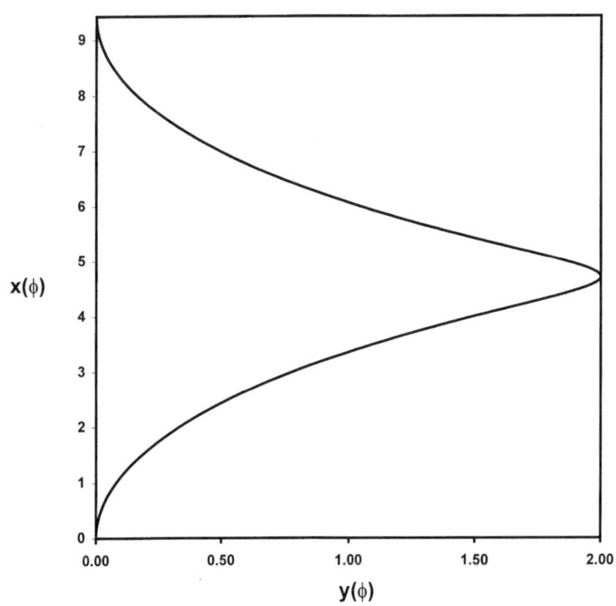

Figure 18-14 The trochoidal path of an electron, $a = 1.5$.

18.2 STOKES VECTORS FOR RADIATION EMITTED BY ACCELERATING CHARGES

We now determine the Stokes vectors for the radiation emitted by the accelerating charges undergoing the motions described in the previous section, namely, (1) the

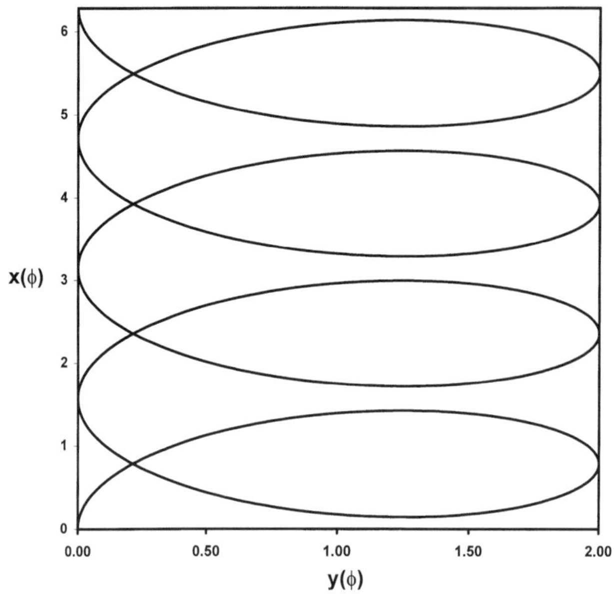

Figure 18-15 The trochoidal path of an electron over four cycles, $a = 0.25$.

motion of an electron in a constant electric field, (2) the motion of an electron in a constant magnetic field, and (3) the motion of the electron in a crossed electric and magnetic field.

The components of the radiation field in spherical coordinates were shown in Chapter 16 to be

$$E_\theta = \frac{e}{4\pi\varepsilon_0 c^2 R}[\ddot{x}\cos\theta - \ddot{z}\sin\theta] \tag{16-8}$$

$$E_\phi = \frac{e}{4\pi\varepsilon_0 c^2 R}[\ddot{y}] \tag{16-9}$$

These equations refer to the observation being made in the xz plane; that is, at $\phi = 0$. The angle θ is the polar angle in the observer's reference frame.

Recall that the Stokes parameters of the radiation field are defined by

$$S_0 = E_\phi E_\phi^* + E_\theta E_\theta^* \tag{16-10a}$$

$$S_1 = E_\phi E_\phi^* - E_\theta E_\theta^* \tag{16-10b}$$

$$S_2 = E_\phi E_\theta^* + E_\theta E_\phi^* \tag{16-10c}$$

$$S_3 = i(E_\phi E_\theta^* - E_\theta E_\phi^*) \tag{16-10d}$$

In the following problems we represent the emitted radiation and its polarization in the form of Stokes vectors.

18.2.1 Stokes Vector for a Charge Moving in an Electric Field

The path of the charge moving in a constant electric field in the xz plane was found to be

$$x(t) = v_0 t \sin\alpha \tag{18-10a}$$

$$z(t) = \frac{-eE_0 t^2}{2m} + v_0 t \cos\alpha \tag{18-10b}$$

We see that the accelerations of the charge in the x and z directions are then

$$\ddot{x}(t) = 0 \tag{18-48a}$$

$$\ddot{z}(t) = -\frac{eE_0}{m} \tag{18-48b}$$

Substituting (18-48) into (16-8) and (16-9) yields

$$E_\theta = \frac{e^2 E_0}{m4\pi\varepsilon_0 c^2 R}\sin\theta \tag{18-49a}$$

$$E_\phi = 0 \tag{18-49b}$$

and we immediately find from (18-49) that the Stokes vector is

$$S = \left(\frac{e^2 E_0}{m4\pi\varepsilon_0 c^2 R}\right)^2 \sin^2\theta \begin{pmatrix} 1 \\ -1 \\ 0 \\ 0 \end{pmatrix} \tag{18-50}$$

Equation (18-50) shows that the emitted radiation is linearly vertically polarized. It also shows the accelerating electron emits the familiar dipole radiation pattern described by $\sin^2\theta$, so the intensity observed along the z axis is zero ($\theta = 0$) and is a maximum when viewed along the x axis ($\theta = \pi/2$).

Before we finish the discussion of (18-50) there is another point of interest that should be noted. We observe that in (18-50) there is a factor of $e^2/4\pi\varepsilon_0 mc^2$. We now ask the question, what, if any, is the meaning of this quantity? The answer can be obtained by recalling that the electric field **E** "outside" of an electron is given by

$$\mathbf{E} = \frac{e}{4\pi\varepsilon_0 r^2}\mathbf{u}_r \tag{18-51}$$

where r is the distance from the center of the electron and $\mathbf{u}_r$ is the unit radius vector. We now imagine the electron has a radius a and compute the work that must be done to move another (positive) charge of the same magnitude from the surface of this electron to infinity. The total work, or energy, required to do this is

$$W = -e\int_a^\infty \mathbf{E}\cdot d\mathbf{r} \tag{18-52}$$

where $d\mathbf{r}$ is $dr\mathbf{u}_r$. Substituting (18-51) into (18-52) gives

$$W = \frac{e^2}{4\pi\varepsilon_0}\int_a^\infty \frac{dr}{r^2} = \frac{e^2}{4\pi\varepsilon_0 a} \tag{18-53}$$

We now equate (18-53) to the rest mass of the electron mc^2 and find that

$$a = \frac{e^2}{4\pi\varepsilon_0 mc^2} \tag{18-54}$$

Thus, the factor $e^2/4\pi\varepsilon_0 mc^2$ is the classical radius of the electron. The value of a is readily calculated from the values $e = 1.60 \times 10^{-19}$ C, $m = 9.11 \times 10^{-31}$ kg, and $c = 2.997 \times 10^8$ m/sec, which yields

$$a = 2.82 \times 10^{-15}\text{m} \tag{18-55}$$

We see that the radius of the electron is extremely small. The factor $e^2/4\pi\varepsilon_0 mc^2$ appears repeatedly in radiation problems. Later, it will appear again when we consider the problem where radiation is incident on an electron and is then re-emitted, that is, the scattering of radiation by an electron.

18.2.2 Stokes Vector for a Charge Accelerating in a Constant Magnetic Field

In the previous section we saw that the path described by an electron moving in a constant magnetic field is given by the equations:

$$x(t) = \frac{v_0}{\omega_c}\sin\omega_c t \tag{18-31a}$$

$$y(t) = -\frac{v_0}{\omega_c}(1 - \cos\omega_c t) \tag{18-31b}$$

where v_0 is the initial velocity and $\omega_c = eB/m$ is the cyclotron frequency. Using the exponential representation:

$$\text{Re}\{e^{i\omega_c t}\} = \cos\omega_c t \tag{18-56a}$$

$$\text{Re}\{-ie^{i\omega_c t}\} = \sin\omega_c t \tag{18-56b}$$

we can then write

$$x = \alpha_c(-ie^{i\omega_c t}) \tag{18-57a}$$

$$y + \alpha_c = \alpha_c(e^{i\omega_c t}) \tag{18-57b}$$

where

$$\alpha_c = \frac{v_0}{\omega_c} \tag{18-57c}$$

The accelerations $\ddot{x}(t)$ and $\ddot{y}(t)$ are then

$$\ddot{x}(t) = i\alpha_c\omega_c^2 e^{i\omega_c t} \tag{18-58a}$$

$$\ddot{y}(t) = -\alpha_c\omega_c^2 e^{i\omega_c t} \tag{18-58b}$$

and the radiation field components become

$$E_\theta = \frac{-ie\alpha_c\omega_c^2}{4\pi\varepsilon_0 c^2 R}\cos\theta e^{i\omega_c t} \tag{18-59a}$$

$$E_\phi = \frac{e\alpha_c\omega_c^2}{4\pi\varepsilon_0 c^2 R} e^{i\omega_c t} \tag{18-59b}$$

From the definition of the Stokes parameters in (16-10) the Stokes vector is

$$S = \left(\frac{e\alpha_c}{4\pi\varepsilon_0 c^2 R}\right)^2 \omega_c^4 \begin{pmatrix} 1+\cos^2\theta \\ 1-\cos^2\theta \\ 0 \\ 2\cos\theta \end{pmatrix} \tag{18-60}$$

which is the Stokes vector for elliptically polarized light radiating at the same frequency as the cyclotron frequency ω_c. Thus, the Stokes vector found earlier for a charge moving in a circle is based on physical reality. We see that (18-60) reduces to right circularly polarized light, linearly horizontally polarized light, and left circularly polarized light for $\theta = 0$, $\pi/2$, and π, respectively.

18.2.3 Stokes Vector for a Charge Moving in a Crossed Electric and Magnetic Field

The path of the electron was seen to be a trochoid described by

$$x(\phi) = a\phi + b\sin\phi \tag{18-42a}$$

$$y(\phi) = b(1-\cos\phi) \tag{18-42b}$$

where

$$\phi = \omega_c t \tag{18-41b}$$

$$a = \frac{eE}{m\omega_c^2} \tag{18-41c}$$

$$b = \frac{v_0 - eE/m\omega_c}{\omega_c} \tag{18-41d}$$

Differentiating (18-42a) and (18-42b) twice with respect to time and using (18-56) then gives

$$\ddot{x}(t) = ib\omega_c^2 e^{i\omega_c t} \tag{18-61a}$$

$$\ddot{y}(t) = b\omega_c^2 e^{i\omega_c t} \tag{18-61b}$$

and we immediately find that the Stokes vector is

$$S = b^2\omega_c^4 \begin{pmatrix} 1+\cos^2\theta \\ 1-\cos^2\theta \\ 0 \\ 2\cos\theta \end{pmatrix} \tag{18-62}$$

which, again, is the Stokes vector for elliptically polarized light.

With this material behind us we now turn our attention to the Lorentz–Zeeman effect and see how the role of polarized light led to the acceptance of Maxwell's electrodynamical theory in optics.

REFERENCES

Books

1. Jackson, J. D., *Classical Electrodynamics*, John Wiley, New York, 1962.
2. Sommerfeld, A., *Lectures on Theoretical Physics*, Vols. I–V, Academic Press, New York, 1952.
3. Harnwell, G. P., *Principles of Electricity and Electromagnetism*, McGraw-Hill, New York, 1949.
4. Humphries, S., Jr., *Charged Particle Beams*, John Wiley, New York, 1990.
5. Hutter, R. C. E. and Harrison, S. W., *Beam and Wave Electronics in Microwave Tubes*, D. Van Nostrand Princeton, 1960.
6. Panofsky, W. K. H. and Phillips, M., *Classical Electricity and Magnetism*, Addison-Wesley, Reading, MA, 1955.
7. Goldstein, H., *Classical Mechanics*, Addison-Wesley Reading, MA, 1950.
8. Corben, H. C. and Stehle, P., *Classical Mechanics*, John Wiley, New York, 1957.

19

The Classical Zeeman Effect

19.1 HISTORICAL INTRODUCTION

In 1846, Michael Faraday discovered that by placing a block of heavy lead glass between the poles of an electromagnet and passing a linearly polarized beam through the block in the direction of the lines of force, the plane of polarization of the linearly polarized beam was rotated by the magnetic medium; this is called the Faraday effect. Thus, he established that there was a link between electromagnetism and light. It was this discovery that stimulated J. C. Maxwell, a great admirer of Faraday, to begin to think of the relation between the electromagnetic field and the optical field.

Faraday was very skillful at inverting questions in physics. In 1819, H. Oersted discovered that a current gives rise to a magnetic field. Faraday then asked the inverse question of how can a magnetic field give rise to a current? After many years of experimentation Faraday discovered that a *changing* magnetic field rather than a steady magnetic field generates a current (Faraday's law). In the Faraday effect, Faraday had shown that a magnetic medium affects the polarization of light as it propagates through the medium. Faraday now asked the question, how, if at all, does the magnetic field affect the source of light itself? To answer this question, he placed a sodium flame between the poles of a large electromagnet and observed the D lines of the sodium radiation when the magnetic field was "on" and when it was "off." After many attempts, by 1862 he was still unable to convince himself that any change resulted in the appearance of the lines, a circumstance which we now know was due to the insufficient resolving power of his spectroscope.

In 1896, P. Zeeman, using a more powerful magnet and an improved spectroscope, repeated Faraday's experiment. This time there was success. He established that the D lines were *broadened* when a constant magnetic field was applied. H. Lorentz heard of Zeeman's discovery and quickly developed a theory to explain the phenomenon.

The fact has been pointed out that, even with the success of Hertz's experiments in 1888, Maxwell's theory was still not accepted by the optics community, because Hertz had carried out his experiments not at optical frequencies but at microwave frequencies; he developed a source which operated at microwaves. For Maxwell's theory to be accepted by the optical community, it would be necessary to prove the theory at optical frequencies (wavelengths); that is, an optical source which could be characterized in terms of a current would have to be created. There was nothing in Fresnel's wave theory which enabled this to be done. Lorentz recognized that at long last an optical source could be created which could be understood in terms of the simple electron theory (sodium has only a single electron in its outer shell). Therefore, he used the simple model of the (sodium) atom in which an electron was bound to the nucleus and its motion governed by Hooke's law. With this model he then discovered that Zeeman's line broadening should actually consist of two or even three spectral lines. Furthermore, using Maxwell's theory he was able to predict that the lines would be linearly, circularly, or elliptically polarized in a completely predictable manner. Lorentz communicated his theoretical conclusions to Zeeman, who investigated the edges of his broadened lines and confirmed Lorentz's predictions in all respects.

Lorentz's spectacular predictions with respect to the splitting, intensity, and polarization of the spectral lines led to the complete acceptance of Maxwell's theory. Especially impressive were the polarization predictions, because they were very complicated. It was virtually impossible without Maxwell's theory and the electron theory even remotely to understand the polarization behavior of the spectral lines. Thus, polarization played a critical role in the acceptance of Maxwell's theory. In 1902, Zeeman and Lorentz shared the Nobel Prize in physics for their work. The prize was given not just for their discovery of and understanding of the Zeeman effect but, even more importantly, for the verification of Maxwell's theory at optical wavelengths. It is important to recognize that Lorentz's contribution was of critical importance. Zeeman discovered that the D lines of the sodium were *broadened, not split*. Because Lorentz predicted that the spectral lines would be split, further experiments were conducted and the splitting was observed. Soon after Zeeman's discovery, however, it was discovered that additional spectral lines appeared. In fact, just as quickly as Lorentz's theory was accepted, it was discovered that it was inadequate to explain the appearance of the numerous spectral lines. The explanation would only come with the advent of quantum mechanics in 1925.

The Zeeman effect and the Faraday effect belong to a class of optical phenomena that are called magneto-optical effects. In this chapter we analyze the Zeeman effect in terms of the Stokes vector. We shall see that the Stokes vector takes on a new and very interesting interpretation. In Chapter 20 we describe the Faraday effect along with other related phenomena in terms of the Mueller matrices.

19.2 MOTION OF A BOUND CHARGE IN A CONSTANT MAGNETIC FIELD

To describe the Zeeman effect and determine the Stokes vector of the emitted radiation, it is necessary to analyze the motion of a bound electron in a constant magnetic field, that is, determine $x(t)$, $y(t)$, $z(t)$ of the electron and then the corresponding accelerations. The model proposed by Lorentz to describe the Zeeman effect was a charge bound to the nucleus of an atom and oscillating with an amplitude A through the origin. The motion is shown in Fig. 19-1; χ is the polar angle and ψ is the

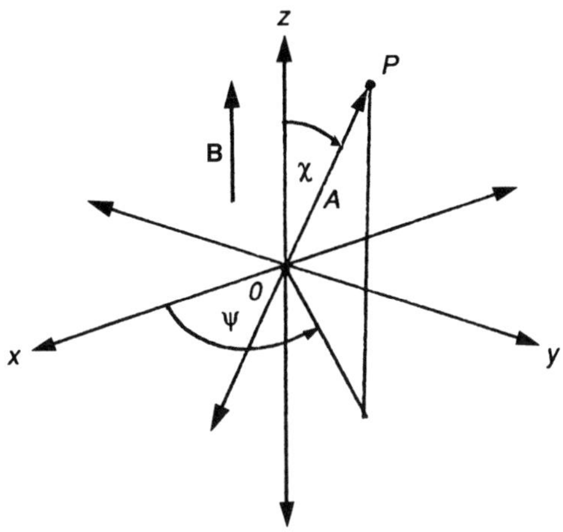

Figure 19-1 Motion of bound charge in a constant magnetic field; χ is the polar angle and ψ is the azimuthal angle. In particular, the angle ψ describes the projection of OP on to the xy plane.

azimuthal angle. In particular, the angle ψ describes the projection of OP on to the xy plane. The significance of emphasizing this will appear shortly.

The equation of motion of the bound electron in the magnetic field is governed by the Lorentz force equation:

$$m\ddot{\mathbf{r}} + k\mathbf{r} = -e[\mathbf{v} \times \mathbf{B}] \tag{19-1}$$

where m is the mass of the electron, $k\mathbf{r}$ is the restoring force (Hooke's law), $\mathbf{v}$ is the velocity of the electron, and $\mathbf{B}$ is the strength of the applied magnetic field. In component form (19-1) can be written

$$m\ddot{x} + kx = -e[\mathbf{v} \times \mathbf{B}]_x \tag{19-2a}$$

$$m\ddot{y} + ky = -e[\mathbf{v} \times \mathbf{B}]_y \tag{19-2b}$$

$$m\ddot{z} + kz = -e[\mathbf{v} \times \mathbf{B}]_z \tag{19-2c}$$

We saw in the previous chapter that for a constant magnetic field directed along the positive z axis ($\mathbf{B} = \mathbf{B}u_z$), (19-2) becomes

$$m\ddot{x} + kx = -e[\dot{y}\mathbf{B}] \tag{19-3a}$$

$$m\ddot{y} + ky = -e[-\dot{x}\mathbf{B}] \tag{19-3b}$$

$$m\ddot{z} + kz = 0 \tag{19-3c}$$

Equation (19-3) can be rewritten further as

$$\ddot{x} + \omega_0^2 x = -\left(\frac{e\mathbf{B}}{m}\right)\dot{y} \tag{19-4a}$$

$$\ddot{y} + \omega_0^2 y = -\left(\frac{e\mathbf{B}}{m}\right)\dot{x} \tag{19-4b}$$

$$\ddot{z} + \omega_0^2 z = 0 \tag{19-4c}$$

where $\omega_0 = \sqrt{k/m}$ is the natural frequency of the charge oscillating along the line OP.

Equation (19-4c) can be solved immediately. We assume a solution of the form $z(t) = e^{\omega t}$. Then, the auxiliary equation for (19-4c) is

$$\omega^2 + \omega_0^2 = 0 \tag{19-5a}$$

so

$$\omega = \pm i\omega_0 \tag{19-5b}$$

The general solution of (19-4c) is then

$$z(t) = c_1 e^{i\omega_0 t} + c_2 e^{-i\omega_0 t} \tag{19-6}$$

To find a specific solution of (19-6), the constants c_1 and c_2 must be found from the initial conditions on $z(0)$ and $\dot{z}(0)$. From Fig. 19-1 we see that when the charge is at P we have

$$z(0) = A \cos \chi \tag{19-7a}$$

$$\dot{z}(0) = 0 \tag{19-7b}$$

Using (19-7) we find the solution of (19-6) to be

$$z(t) = A \cos \chi \cos \omega_0 t \tag{19-8}$$

Next, we solve (19-4a) and (19-4b). We again introduce the complex variable:

$$\zeta = x + iy \tag{19-9}$$

In the same manner as in the previous chapters (19-4a) and (19-4b) can be written as a single equation:

$$\ddot{\zeta} + \left(\frac{-ie\mathbf{B}}{m}\right)\dot{\zeta} + \omega_0^2 \zeta = 0 \tag{19-10}$$

Again, assuming a solution of the form $z(t) = e^{\omega t}$, the solution of the auxiliary equation is

$$\omega = i\left(\frac{e\mathbf{B}}{2m}\right) \pm i\left(\omega_0^2 - \left(\frac{e\mathbf{B}}{2m}\right)^2\right)^{1/2} \tag{19-11}$$

The term $(e\mathbf{B}/2m)^2$ in (19-11) is orders of magnitude smaller than ω_0^2, so (19-11) can be written as

$$\omega_\pm = i(\omega_L \pm \omega_0) \tag{19-12a}$$

where

$$\omega_L = \frac{e\mathbf{B}}{2m} \tag{19-12b}$$

The frequency ω_L is known from the Larmor precession frequency; the reason for the term *precession* will soon become clear. The solution of (19-10) is then

$$z(t) = c_1 e^{i\omega_+ t} + c_2 e^{i\omega_- t} \tag{19-13}$$

where ω_+ is given by (19-12a).

To obtain a specific solution of (19-13), we must again use the initial conditions. From Fig. 19-1 we see that

$$x(0) = A\sin\chi\cos\psi \tag{19-14a}$$

$$y(0) = A\sin\chi\sin\psi \tag{19-14b}$$

so

$$\zeta(0) = x(0) + iy(0) = A\sin\chi\exp(i\psi) \tag{19-14c}$$

$$\dot{\zeta}(0) = 0 \tag{19-14d}$$

After a little algebraic manipulation we find that the conditions (19-14c) and (19-14d) lead to the following specific relations for $x(t)$ and $y(t)$:

$$x(t) = \frac{A\sin\chi}{\omega_0}[\omega_0\cos(\psi + \omega_L t)\cos\omega_0 t + \omega_L\sin(\psi + \omega_L t)\sin\omega_0 t] \tag{19-15a}$$

$$y(t) = \frac{A\sin\chi}{\omega_0}[\omega_0\sin(\psi + \omega_L t)\cos\omega_0 t - \omega_L\cos(\psi + \omega_L t)\sin\omega_0 t] \tag{19-15b}$$

Because the Larmor frequency is much smaller than the fundamental oscillation frequency of the bound electron, $\omega_L \ll \omega_0$, the second term in (19-15a) and (19-15b) can be dropped. The equations of motion for $x(t)$, $y(t)$, and $z(t)$ are then simply

$$x(t) = A\sin\chi\cos(\psi + \omega_L t)\cos\omega_0 t \tag{19-16a}$$

$$y(t) = A\sin\chi\sin(\psi + \omega_L t)\cos\omega_0 t \tag{19-16b}$$

$$z(t) = A\cos\chi\cos\omega_0 t \tag{19-16c}$$

In (19-16) we have also included $z(t)$ from (19-8) as (19-16c). We see that $\omega_L t$, the angle of precession, is coupled only with ψ and is completely independent of χ. To show this precessional behavior we deliberately chose to show ψ in Fig. 19-1. The angle ψ is completely arbitrary and is symmetric around the z axis. We could have chosen its value immediately to be zero. However, to demonstrate clearly that $\omega_L t$ is restricted to the xy plane, we chose to include ψ in the formulation. We therefore see from (19-16) that, as time increases, the factor ψ increases by $\omega_L t$. Thus, while the bound charge is oscillating to and fro along the radius OP there is a simultaneous counterclockwise rotation in the xy plane. This motion is called precession, and we see $\omega_L t$ is the angle of precession. The precession caused by the presence of the magnetic field is very often called the Larmor precession, after J. Larmor, who, around 1900, first pointed out this behavior of an electron in a magnetic field.

The angle ψ is arbitrary, so we can conveniently set $\psi = 0$ in (19-16). The equations then become

$$x(t) = A\sin\chi\cos\omega_L t\cos\omega_0 t \tag{19-17a}$$

$$y(t) = A\sin\chi\sin\omega_L t\cos\omega_0 t \tag{19-17b}$$

$$z(t) = A\cos\chi\cos\omega_0 t \tag{19-17c}$$

We note immediately that (19-17) satisfies the equation:

$$r^2(t) = x^2(t) + y^2(t) + z^2(t) \tag{19-18a}$$

$$= A^2\cos^2\omega_0 t \tag{19-18b}$$

This result is completely expected because the radial motion is due only to the natural oscillation of the electron. The magnetic field has no effect on this radial motion, and, indeed, we see that there is no contribution.

Equations (19-17) are the fundamental equations which describe the path of the bound electron. From them the accelerations can then be obtained as is done in the following section. However, we consider (19-17a) and (19-17b) further. If we plot these equations, we can "follow" the precessional motion of the bound electron as it oscillates along OP. Equations (19-17a) and (19-17b) give rise to a remarkably beautiful pattern called a *petal* plot. Physically, we have the electron oscillating very rapidly along the radius OP while the magnetic field forces the electron to move relatively slowly counterclockwise in the xy plane. Normally, $\omega_L \ll \omega_0$ and $\omega_L \simeq \omega_0/10^7$. Thus, the electron oscillates about 10 million times through the origin during one precessional revolution. Clearly, this is a practical impossibility to illustrate or plot. However, if we artificially take ω_L to be close to ω_0, we can demonstrate the precessional behavior and still lose none of our physical insight. To show this behavior we first arbitrarily set the factor $A \sin\chi$ to unity. Then, using the well-known trigonometric sum and difference formulas, (19-17a) and (19-17b) can be written as

$$x(t) = \frac{1}{2}[\cos(\omega_0 + \omega_L)t + \cos(\omega_0 - \omega_L)t] \tag{19-19a}$$

$$y(t) = \frac{1}{2}[\sin(\omega_0 + \omega_L)t - \sin(\omega_0 - \omega_L)t] \tag{19-19b}$$

We now set

$$\theta_0 = \omega_0 t \qquad \text{and} \qquad \theta_L = \omega_L t \tag{19-20}$$

so (19-19) becomes

$$x(\theta_0) = \frac{1}{2}[\cos(\theta_0 + \theta_L) + \cos(\theta_0 - \theta_L)] \tag{19-21a}$$

$$y(\theta_0) = \frac{1}{2}[\sin(\theta_0 + \theta_L) - \sin(\theta_0 - \theta_L)] \tag{19-21b}$$

To plot the precessional motion, we set $\theta_L = \theta_0/p$, where p can take on any integer value. Equation (19-21) then can be written as

$$x(\theta_0) = \frac{1}{2}\left[\cos\left(\frac{p+1}{p}\right)\theta_0 + \cos\left(\frac{p-1}{p}\right)\theta_0\right] \tag{19-22a}$$

$$y(\theta_0) = \frac{1}{2}\left[\sin\left(\frac{p+1}{p}\right)\theta_0 - \sin\left(\frac{p-1}{p}\right)\theta_0\right] \tag{19-22b}$$

where we have dropped the subscript L. As a first example of (19-22) we set $\omega_L = \omega_0/5$, so $\theta_L = 0.2\theta_0$. In Fig. 19-2, (19-22) has been plotted over 360° for $\theta_L = 0.2\theta_0$ (in which time the electron makes $5 \times 360 = 1800$ radial oscillations, which is equivalent to θ taking on values from 0 to 1800°. The figure shows that the electron describes five petals over a single precessional cycle. The actual path and direction taken by the electron can be followed by starting, say, at the origin, facing the three

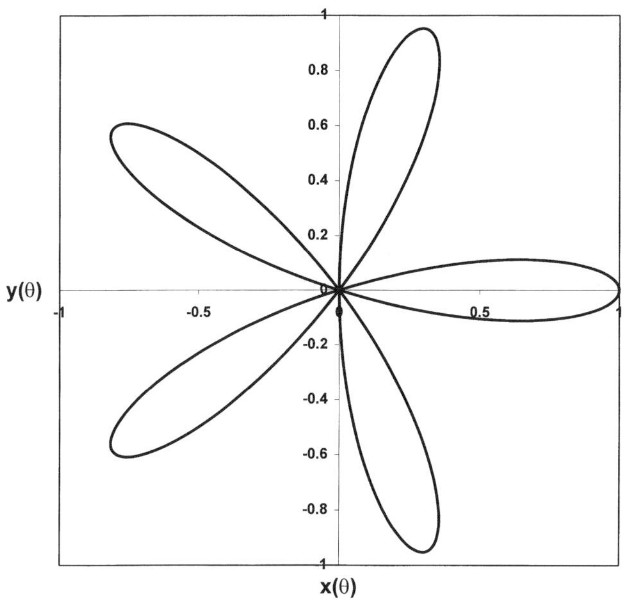

Figure 19-2 Petal diagram for a precessing electron; $\omega_L = \omega_0/5$, $\theta_L = \theta_0/5$.

o'clock position and following the arrows while keeping the "surface" of the petal to the left of the electron as it traverses the path.

One can readily consider other values of ω_L. In Fig. 19-3 through Fig. 19-6 other petal diagrams are shown for four additional values of ω_L, namely, ω_0, $\omega_0/2$, $\omega_0/4$, and $\omega_0/8$, respectively. The result shows a proportional increase in the number of petals and reveals a very beautiful pattern for the precessional motion of the bound electron.

Equations (19-21) (or (19-19)) can be transformed in an interesting manner by a rotational transformation. The equations are

$$x' = x\cos\theta + y\sin\theta \tag{19-23a}$$

$$y' = -x\sin\theta + y\cos\theta \tag{19-23b}$$

where θ is the angle of rotation. We now substitute (19-21) into (19-23), group terms, and find that

$$x' = (1/2)[\cos(\theta_0 + \theta') + \cos(\theta_0 - \theta')] \tag{19-24a}$$

$$y' = (1/2)[\sin(\theta_0 + \theta') - \sin(\theta_0 - \theta')] \tag{19-24b}$$

where

$$\theta' = \theta_L - \theta \tag{19-24c}$$

Inspecting (19-24) we see that the equations are identical in form with (19-21); that is, under a rotation of coordinates x and y are invariant. In a (weak) magnetic field (19-24) shows that the equations of motion with respect to axes rotating with an angular velocity ω_L are the same as those in a nonrotating system when **B** is zero. This is known as Larmor's theorem. The result expressed by (19-24) allows us to

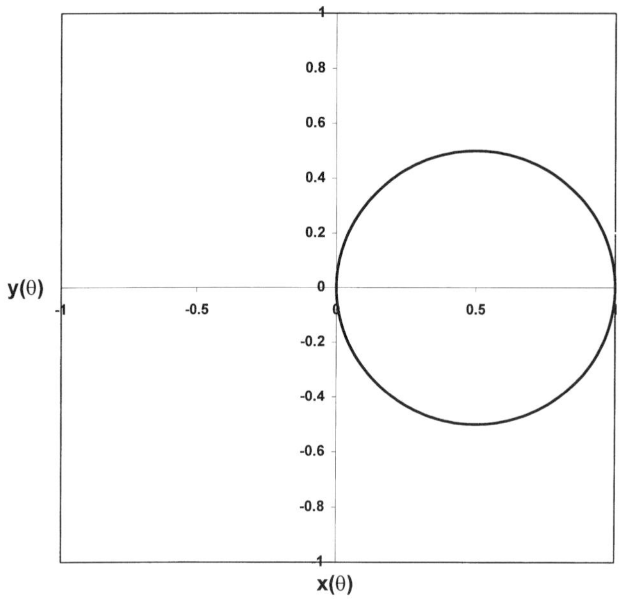

Figure 19-3 Petal diagram for a precessing electron; $\omega_L = \omega_0$, $\theta_L = \theta_0$.

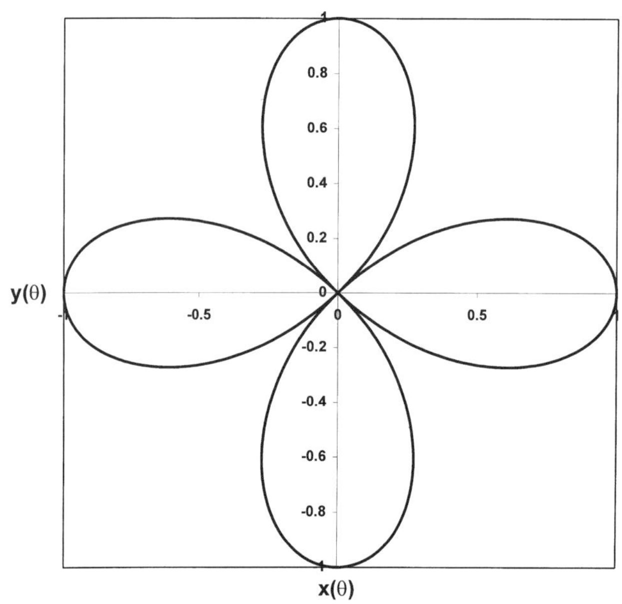

Figure 19-4 Petal diagram for a precessing electron; $\omega_L = \omega_0/2$, $\theta_L = \theta_0/2$.

describe x' and y' in a very simple way. If we set $\theta = \theta_L - \theta_0$ then $\theta' = \theta_0$ and (19-24a) and (19-24b) reduce, respectively, to

$$x' = (1/2)[1 + \cos 2\theta_0] \tag{19-25a}$$

$$y' = (1/2)\sin 2\theta_0 \tag{19-25b}$$

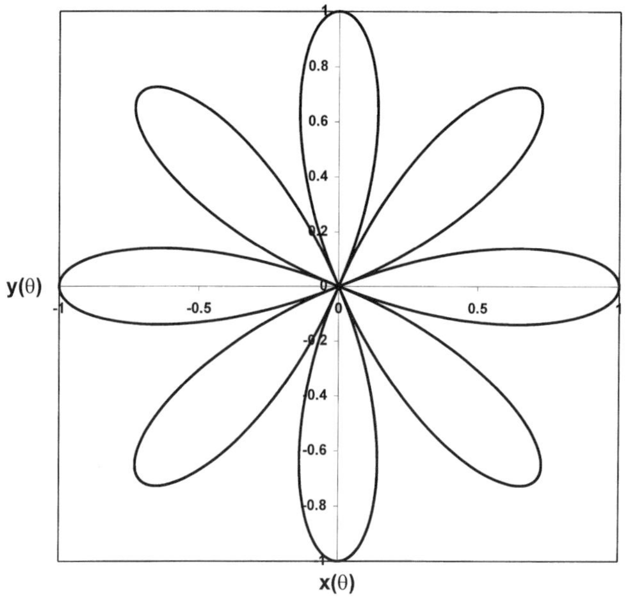

Figure 19-5 Petal diagram for a precessing electron; $\omega_L = \omega_0/4$, $\theta_L = \theta_0/4$.

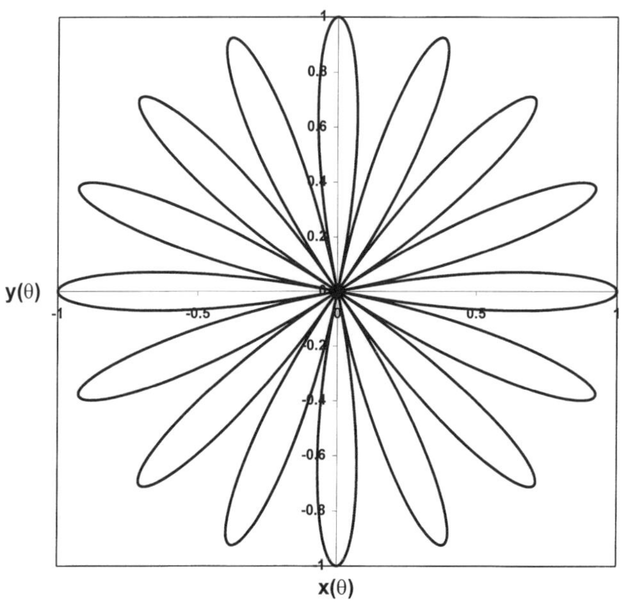

Figure 19-6 Petal diagram for a precessing electron; $\omega_L = \omega_0/8$, $\theta_L = \theta_0/8$.

Thus, in the primed coordinate system only θ_0, the natural oscillation angle, appears. The angle θ_L can be eliminated and we find that

$$(x' - 1/2)^2 + y'^2 = (1/2)^2 \tag{19-26}$$

which is a circle of unit diameter with intercepts on the x' axis at 0 and 1.

A final observation can be made. The petal diagrams for precession based on (19-21) and shown in the figures appear to be remarkably similar to the rose diagrams which arise in analytical geometry, described by the equation:

$$\rho = \cos k\theta \qquad k = 1, 2, \ldots, N \tag{19-27}$$

where there are $2N$ petals if N is even and N petals if N is odd. We can express (19-27) in terms of x and y from the relations:

$$x = \rho \cos\theta \tag{19-28a}$$

$$y = \rho \sin\theta \tag{19-28b}$$

so

$$x = \cos k\theta \cos\theta = (1/2)[\cos(k+1)\theta + \cos(k-1)\theta] \tag{19-29a}$$

$$y = \cos k\theta \sin\theta = (1/2)[\sin(k+1)\theta - \sin(k-1)\theta] \tag{19-29b}$$

where we have used the sum and difference formulas for the cosine and sine functions.

We can show that the precession equations (19-21a) and (19-21b) reduce to either (19-27) or (19-29) by writing them as

$$x = (1/2)[\cos p + \cos q] \tag{19-30a}$$

$$y = (1/2)[\sin p - \sin q] \tag{19-30b}$$

where

$$p = \theta_0 + \theta_L \tag{19-30c}$$

$$q = \theta_0 - \theta_L \tag{19-30d}$$

Equation (19-30) can be transformed to polar coordinates by squaring and adding (19-30a) and (19-30b)

$$\rho^2 = x^2 + y^2 = (1/2)[1 + \cos(p+q)] \tag{19-31}$$

We now set θ_0 to

$$\theta_0 = k\theta_L = k\theta \qquad k = 1, 2, \ldots, N \tag{19-32a}$$

so

$$p = \theta_0 + \theta_L = (k+1)\theta \qquad k = 1, 2, \ldots, N \tag{19-32b}$$

$$q = \theta_0 - \theta_L = (k-1)\theta \qquad k = 1, 2, \ldots, N \tag{19-32c}$$

Thus,

$$p + q = 2k\theta \tag{19-33}$$

Substituting (19-32) into (19-30) and (19-33) into (19-31) then yields

$$x = (1/2)[\cos(k+1)\theta + \cos(k-1)\theta] \tag{19-34a}$$

$$y = (1/2)[\sin(k+1)\theta - \sin(k-1)\theta] \tag{19-34b}$$

and substituting (19-33) into (19-31) yields,

$$\rho^2 = (1/2)[1 + \cos 2k\theta] = \cos^2 k\theta \tag{19-35}$$

or

$$\rho = \cos k\theta \qquad k = 1, 2, \ldots, N \tag{19-36}$$

We see that (19-36) (or, equivalently, (19-34)) is the well-known rose equation of analytical geometry. Thus, the rose equation describes the phenomenon of the precession of a bound electron in a magnetic field, an interesting fact that does not appear to be pointed out in courses in analytical geometry.

19.3 STOKES VECTOR FOR THE ZEEMAN EFFECT

We now determine the Stokes vector for the Zeeman effect. We repeat Eqs. (19-17), which describe the path of the oscillating electron bound to an atom.

$$x(t) = A \sin\chi \cos\omega_L t \cos\omega_0 t \tag{19-17a}$$

$$y(t) = A \sin\chi \sin\omega_L t \cos\omega_0 t \tag{19-17b}$$

$$z(t) = A \cos\chi \cos\omega_0 t \tag{19-17c}$$

where

$$\omega_L = \frac{e\mathbf{B}}{2m} \tag{19-12b}$$

Equations (19-17) can be represented in complex form by first rewriting them by using the trigonometric identities for sums and differences:

$$x(t) = \frac{A}{2}\sin\chi(\cos\omega_+ t + \cos\omega_- t) \tag{19-37a}$$

$$y(t) = \frac{A}{2}\sin\chi(\sin\omega_+ t - \sin\omega_- t) \tag{19-37b}$$

$$z(t) = A \cos\chi \cos\omega_0 t \tag{19-37c}$$

where

$$\omega_\pm = \omega_0 \pm \omega_L \tag{19-37d}$$

Using the familiar rule of writing (19-37) in complex notation, we have

$$x(t) = \frac{A}{2}\sin\chi[\exp(i\omega_+ t) + \exp(i\omega_- t)] \tag{19-38a}$$

$$y(t) = -i\left(\frac{A}{2}\right)\sin\chi[\exp(i\omega_+ t) - \exp(i\omega_- t)] \tag{19-38b}$$

$$z(t) = A \cos\chi \exp(i\omega_0 t) \tag{19-38c}$$

Twofold differentiation of (19-38) with respect to time yields

$$\ddot{x}(t) = -\frac{A}{2}\sin\chi[\omega_+^2 \exp(i\omega_+ t) + \omega_-^2 \exp(i\omega_- t)] \tag{19-39a}$$

$$\ddot{y}(t) = i\left(\frac{A}{2}\right)\sin\chi[\omega_+^2 \exp(i\omega_+ t) - \omega_-^2 \exp(i\omega_- t)] \tag{19-39b}$$

$$\ddot{z}(t) = -(A\cos\chi)\omega_0^2 \exp(i\omega_0 t) \tag{19-39c}$$

The radiation field equations are

$$E_\theta = \frac{e}{4\pi\varepsilon_0 c^2 R}[\ddot{x}(t)\cos\theta - \ddot{z}(t)\sin\theta] \tag{19-40a}$$

$$E_\phi = \frac{e}{4\pi\varepsilon_0 c^2 R}[\ddot{y}(t)] \tag{19-40b}$$

Substituting (19-39) into (19-40) yields

$$E_\theta = \frac{eA}{8\pi\varepsilon_0 c^2 R}[\sin\chi\cos\theta\{\omega_+^2 \exp(i\omega_+ t) + \omega_-^2 \exp(i\omega_- t)\} + 2\omega_0^2 \cos\chi\sin\theta\exp(i\omega_0 t)] \tag{19-41a}$$

and

$$E_\phi = \frac{ieA\sin\chi}{8\pi\varepsilon_0 c^2 R}\{\omega_+^2 \exp(i\omega_+ t) - \omega_-^2 \exp(i\omega_- t)\} \tag{19-41b}$$

The Stokes parameters are defined in spherical coordinates to be

$$S_0 = E_\phi E_\phi^* + E_\theta E_\theta^* \tag{16-10a}$$

$$S_1 = E_\phi E_\phi^* - E_\theta E_\theta^* \tag{16-10b}$$

$$S_2 = E_\phi E_\theta^* + E_\theta E_\phi^* \tag{16-10c}$$

$$S_3 = i(E_\phi E_\theta^* - E_\theta E_\phi^*) \tag{16-10d}$$

We now form the quadratic field products of (19-41) according to (16-10), drop all cross-product terms, and average χ over a sphere of unit radius. Finally, we group terms and find that the Stokes vector for the classical Zeeman effect is

$$S = \left(\frac{eA}{8\pi\varepsilon_0 c^2 R}\right)^2 \begin{pmatrix} \frac{2}{3}(\omega_+^4 + \omega_-^4)(1+\cos^2\theta) + \frac{4}{3}\omega_0^4 \sin^2\theta \\ -\frac{2}{3}(\omega_+^4 + \omega_-^4)\sin^2\theta + \frac{4}{3}\omega_0^4\sin^2\theta \\ 0 \\ \frac{4}{3}(\omega_+^4 - \omega_-^4)\cos\theta \end{pmatrix} \tag{19-42}$$

The form of (19-42) suggests that we can decompose the column matrix according to frequency. This implies that the converse of the principle of incoherent superposition is valid; namely, (19-42) can be decomposed according to a principle that we call the principle of spectral incoherent decomposition. Therefore, (19-42) is decomposed into column matrices in terms of ω_-, ω_0, and ω_+. We now do this and find that

$$S = \frac{2}{3}\left(\frac{eA}{8\pi\varepsilon_0 c^2 R}\right)^2 \left(\omega_-^4 \begin{pmatrix} 1+\cos^2\theta \\ -\sin^2\theta \\ 0 \\ -2\cos\theta \end{pmatrix} + \omega_0^4 \begin{pmatrix} 2\sin^2\theta \\ 2\sin^2\theta \\ 0 \\ 0 \end{pmatrix} + \omega_+^4 \begin{pmatrix} 1+\cos^2\theta \\ -\sin^2\theta \\ 0 \\ 2\cos\theta \end{pmatrix} \right) \tag{19-43}$$

The meaning of (19-43) is now immediately evident. According to (19-43), we will observe three spectral lines at frequency ω_-, ω_0, and ω_+, respectively. This is exactly what is observed in a spectroscope. Furthermore, we see that the Stokes vectors associated with ω_- and ω_+ correspond to elliptically polarized light with their polarization ellipses oriented at 90° and of opposite ellipticity. Similarly, the Stokes vector associated with the ω_0 spectral line is always linearly horizontally polarized.

In Fig. 19-7 we represent the spectral lines corresponding to (19-43) as they would be observed in a spectroscope.

Thus, by describing the Zeeman effect in terms of the Stokes vector, we have obtained a mathematical formulation that corresponds exactly to the observed spectrum, that is, each of the column matrices in (19-43) corresponds to a spectral line. Furthermore, the column matrix (Stokes vector) contains all of the information which can be measured, namely, the frequency (wavelength), intensity, and polarization. In this way we have extended the usefulness of the Stokes vector.

Originally, the Stokes parameters were introduced to obtain a formulation of the optical field whereby the polarization could be measured in terms of the intensity, a measurable quantity. The Stokes vector was then constructed and introduced to facilitate the mathematical analyses of polarized light via the Mueller matrix formalism. The Stokes vector now takes on another meaning. It can be used to represent

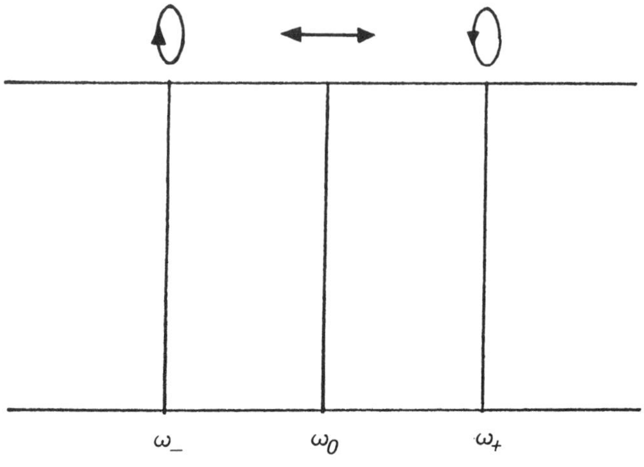

Figure 19-7 The Zeeman effect observed in a spectroscope.

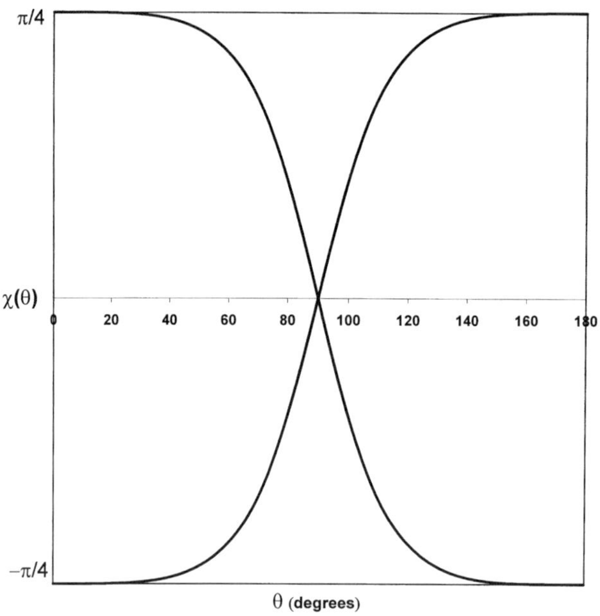

Figure 19-8 Plot of the ellipticity angle $\chi(\theta)$ versus the viewing angle θ of the spectral lines associated with the ω_- and ω_+ frequencies in (19-43).

the observed spectral lines. In a sense we have finally reached a goal enunciated first by W. Heisenberg (1925) in his formulation of quantum mechanics and, later, for optics by E. Wolf (1954)—the description of atomic and optical phenomena in terms of observables.

We see from (19-43) that the ellipticity angle is a function of the observation angle θ. In Fig. 19-8 a plot is made of the ellipticity angle versus θ. We observe that from $\theta = 0^\circ$ (viewing down along the magnetic field) to $\theta = 180^\circ$ (viewing up along the magnetic field) there is a reversal in the ellipticity.

Equation (19-43) reduces to special forms when the radiation is observed parallel to the magnetic field ($\theta = 0^\circ$) and perpendicular to the magnetic field ($\theta = 90^\circ$). For $\theta = 0^\circ$ we see from (19-43) that the Stokes vector associated with the ω_0 column matrix vanishes, and only the Stokes vectors associated with ω_- and ω_+ remain. We then have

$$S = \frac{4}{3}\left(\frac{eA}{8\pi\varepsilon_0 c^2 R}\right)^2 \left(\omega_-^4 \begin{pmatrix} 1 \\ 0 \\ 0 \\ -1 \end{pmatrix} + \omega_+{}^4 \begin{pmatrix} 1 \\ 0 \\ 0 \\ 1 \end{pmatrix} \right) \tag{19-44}$$

Thus, we observe two radiating components (spectral lines) at ω_- and ω_+, which are left and right circularly polarized, respectively. Also, the intensities are equal; the magnitudes of the frequencies $\omega_\pm^4$ are practically equal. The observation of only two spectral lines parallel to the magnetic field is sometimes called the *longitudinal* Zeeman effect. Figure 19-9 corresponds to (19-44) as viewed in a spectroscope.

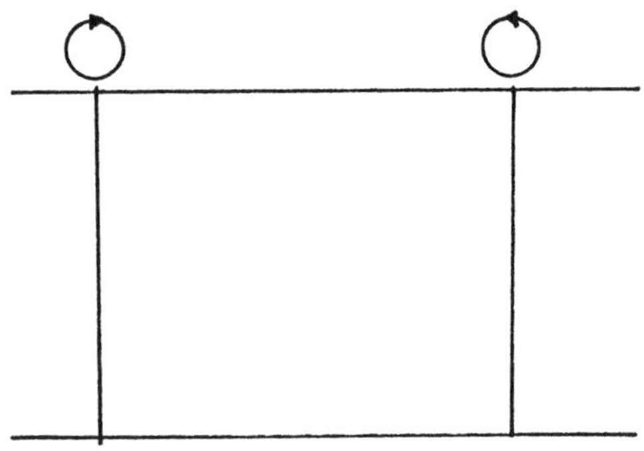

Figure 19-9 The longitudinal Zeeman effect. The spectral lines observed in a spectroscope for the Zeeman effect parallel to the magnetic field ($\theta = 0°$).

Next, we consider the case where the radiation is observed perpendicular to the magnetic field ($\theta = 90°$). Equation (19-43) now reduces to

$$S = \frac{2}{3}\left(\frac{eA}{8\pi\varepsilon_0 c^2 R}\right)^2 \left[\omega_-^4 \begin{pmatrix} 1 \\ -1 \\ 0 \\ 0 \end{pmatrix} + 2\omega_0^4 \begin{pmatrix} 1 \\ 1 \\ 0 \\ 0 \end{pmatrix} + \omega_+^4 \begin{pmatrix} 1 \\ -1 \\ 0 \\ 0 \end{pmatrix}\right] \tag{19-45}$$

Three components (spectral lines) are observed at ω_-, ω_0, and ω_+, respectively. The spectral lines observed at ω_- and ω_+ are linearly vertically polarized, and the spectral line at ω_0 is linearly horizontally polarized. Furthermore, we see that the intensity of the center spectral line (ω_0) is twice that of ω_- and ω_+. The observation of the Zeeman effect perpendicular to the magnetic field is sometimes called the *transverse* Zeeman effect or the Zeeman triplet. The appearance of the spectra corresponding to (19-45) is shown in Fig. 19-10.

Finally, it is of interest to determine the form of the Stokes vector (19-43) when the applied magnetic field is removed. We set $\mathbf{B} = 0$, and we have $\omega_- = \omega_+ = \omega_0$. Adding the elements of each row of matrices gives

$$S = \frac{8}{3}\left(\frac{eA}{8\pi\varepsilon_0 c^2 R}\right)^2 \omega_0^4 \begin{pmatrix} 1 \\ 0 \\ 0 \\ 0 \end{pmatrix} \tag{19-46}$$

which is the Stokes vector for unpolarized light. Thus, we observe a single spectral line radiating at the frequency ω_0, the natural frequency of oscillation of the bound atom. This is exactly what we would expect for an electron oscillating randomly about the nucleus of an atom. In a spectroscope we would, therefore, observe Fig. 19-11.

In the following chapter we extend the observable formulation to describing the intensity and polarization of the radiation emitted by relativistically moving

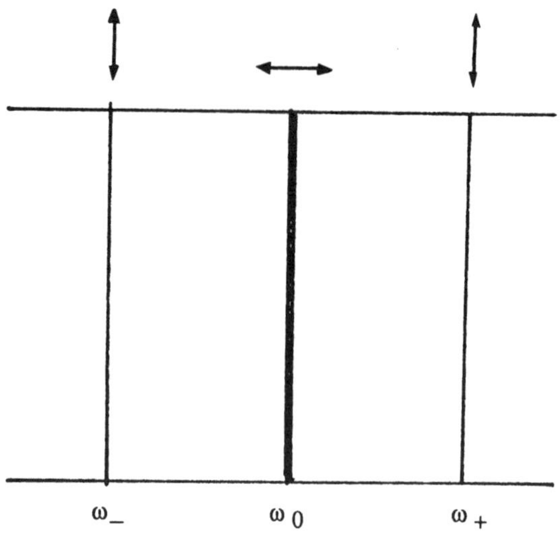

Figure 19-10 The transverse Zeeman effect. The spectral lines observed in a spectroscope for the Zeeman effect perpendicular to the magnetic field ($\theta = 90^\circ$).

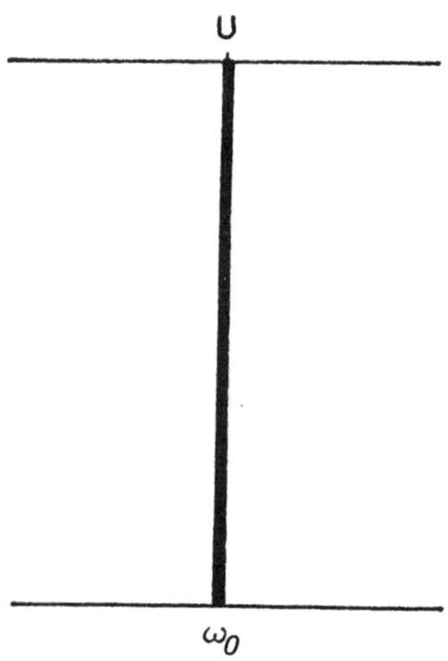

Figure 19-11 The Zeeman effect with the magnetic field removed. A single unpolarized spectral line is observed radiating at a frequency ω_0.

electrons. In Chapter 22 we use the Stokes vectors to describe the emission of radiation by quantized atomic systems.

REFERENCES

Papers

1. Heisenberg, W., *Z. Phys.*, *33*, 879 (1925).
2. Wolf, E., *Nuovo Cimento*, *12*, 884 (1954).

3. Collett, E., *Am. J. Phys.*, *36*, 713 (1968).
4. McMaster, W. H., *Am. J. Phys.*, *22*, 351 (1954).

Books

1. Jackson, J. D., *Classical Electrodynamics*, Wiley, New York, 1962.
2. Sommerfeld, A., *Lectures on Theoretical Physics*, Vols. I–V, Academic Press, New York, 1952.
3. Born, M. and Wolf, E., *Principles of Optics*, 3rd ed., Pergamon Press, New York, 1965.
4. Wood, R. W., *Physical Optics*, 3rd ed., Optical Society of America, Washington, DC, 1988.
5. Strong, J., *Concepts of Classical Optics*, Freeman, San Francisco, 1959.
6. Jenkins, F. S. and White H. E., *Fundamentals of Optics*, McGraw-Hill, New York, 1957.
7. Stone, J. M., *Radiation and Optics*, McGraw-Hill, New York, 1963.

20

Further Applications of the Classical Radiation Theory

20.1 RELATIVISTIC RADIATION AND THE STOKES VECTOR FOR A LINEAR OSCILLATOR

In previous chapters we have considered the emission of radiation by nonrelativistic moving particles. In particular, we determined the Stokes parameters for particles moving in linear or curvilinear paths. Here and in Section 20.2 and 20.3 we reconsider these problems in the relativistic regime. It is customary to describe the velocity of the charge relative to the speed of light by $\beta = v/c$.

For a linearly oscillating charge we saw that the emitted radiation was linearly polarized and its intensity dependence varied as $\sin^2\theta$. This result was derived for the nonrelativistic regime ($\beta \ll 1$). We now consider the same problem, using the relativistic form of the radiation field. Before we can do this, however, we must first show that for the relativistic regime ($\beta \sim 1$) the radiation field continues to consist only of transverse components, E_θ and E_ϕ, and the radial or longitudinal electric component E_r is zero. If this is true, then we can continue to use the same definition of the Stokes parameters for a spherical radiation field.

The relativistic radiated field has been shown by Jackson to be

$$E(x, t) = \frac{e}{4\pi\varepsilon_0 c^2}\left[\frac{\mathbf{n}}{\kappa^3 R} \times \{(\mathbf{n} - \beta) \times (\dot{\beta})\}\right]_{\text{ret}} \tag{20-1a}$$

where

$$\kappa = 1 - \mathbf{n} \cdot \beta \tag{20-1b}$$

The brackets $[\cdots]_{\text{ret}}$ means that the field is to be evaluated at an earlier or retarded time, $t' = t - R(t')/c$ where R/c is just the time of propagation of the disturbance from one point to the other. Furthermore, $c\beta$ is the instantaneous velocity of the particle, $c\dot{\beta}$ is the instantaneous acceleration, and $\mathbf{n} = \mathbf{R}/R$. The quantity $\kappa \to 1$ for nonrelativistic motion. For relativistic motion the fields depend on the velocity as

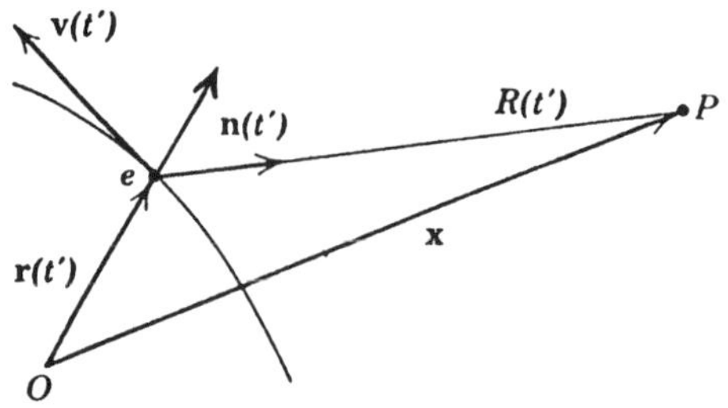

Figure 20-1 Coordinate relations for an accelerating electron. P is the observation point and O is the origin. (From Jackson.)

well as the acceleration. Consequently, as we shall soon clearly see the angular distribution is more complicated.

In Fig. 20-1 we show the relations among the coordinates given in (20-1a) We recall that the Poynting vector **S** is given by

$$\mathbf{S} = \frac{1}{2} c\varepsilon_0 |\mathbf{E}|^2 \mathbf{n} \tag{20-2}$$

Thus, we can write, using (20-1a),

$$[\mathbf{S} \cdot \mathbf{n}] = \frac{e^2}{32\pi^2 \varepsilon_0 c^3} \left[\frac{1}{\kappa^6 R^2} \left| \mathbf{n} \times [(\mathbf{n} - \beta) \times \dot{\beta}] \right|^2 \right]_{\text{ret}} \tag{20-3}$$

There are two types of relativistic effects present. The first is the effect of the specific spatial relationship between β and $\dot{\beta}$, which determines the detailed angular distribution. The other is a general relativistic effect arising from the transformation from the rest frame of the particle to the observer's frame and manifesting itself by the presence of the factor κ in the denominator of (20-3). For ultrarelativistic particles the latter effect dominates the whole angular distribution.

In (20-3), $\mathbf{S} \cdot \mathbf{n}$ is the energy per unit area per unit time detected at an observation point at time t due to radiation emitted by the charge at time $t' = t - \mathbf{R}(t')/c$. To calculate the energy radiated during a finite period of acceleration, say from $t' = T_1$ *to* $t' = T_2$, we write

$$W = \int_{t=T_1+R(T_1)/c}^{t=T_2+R(T_2)/c} [\mathbf{S} \cdot \mathbf{n}]_{\text{ret}}\, dt = \int_{t'=T_1}^{t'=T_2} (\mathbf{S} \cdot \mathbf{n}) \frac{dt}{dt'} dt' \tag{20-4}$$

The quantity $(\mathbf{S} \cdot \mathbf{n})(dt/dt')$ is the power radiated per unit area in terms of the charge's own time. The terms t' and t are related by

$$t' = t - \frac{R(t')}{c} \tag{20-5}$$

Furthermore, as Jackson has also shown,

$$\kappa = 1 + \frac{1}{c} \frac{dR(t')}{dt'} \tag{20-6}$$

Differentiating (20-5) yields

$$\frac{dt}{dt'} = \kappa \tag{20-7}$$

The power radiated per unit solid angle is

$$\frac{dP(t')}{d\Omega} = R^2(\mathbf{S} \cdot \mathbf{n})\frac{dt}{dt'} = \kappa R^2 \mathbf{S} \cdot \mathbf{n} \tag{20-8}$$

These results show that we will obtain a set of Stokes parameters consistent with (20-8) by defining the Stokes parameters as

$$S_0 = \frac{1}{2} c\varepsilon_0 \kappa R^2 \left[E_\phi E_\phi^* + E_\theta E_\theta^*\right] \tag{20-9a}$$

$$S_1 = \frac{1}{2} c\varepsilon_0 \kappa R^2 \left[E_\phi E_\phi^* - E_\theta E_\theta^*\right] \tag{20-9b}$$

$$S_2 = \frac{1}{2} c\varepsilon_0 \kappa R^2 \left[E_\phi E_\theta^* + E_\theta E_\phi^*\right] \tag{20-9c}$$

$$S_3 = \frac{1}{2} c\varepsilon_0 \kappa R^2 \left[i(E_\phi E_\theta^* - E_\theta E_\phi^*)\right] \tag{20-9d}$$

where the electric field $\mathbf{E}(\mathbf{x}, t)$ is calculated from (20-1a).

Before we proceed to apply these results to various problems of interest, we must demonstrate that the definition of the Stokes parameters (20-9) is valid for relativistic motion. That is, the field is transverse and there is no longitudinal component ($E_r = 0$). We thus write (20-1a) as

$$\mathbf{E}(\mathbf{x}, t) = \frac{e}{4\pi\varepsilon_0 c^2 R}\left[\frac{[\mathbf{n} \times (\mathbf{n} \times \dot{\beta})] - [\mathbf{n} \times (\beta \times \dot{\beta})]}{\kappa^3}\right]_{\text{ret}} \tag{20-10}$$

Because the unit vector $\mathbf{n}$ is practically in the same direction as u_r, (20-10) is rewritten as

$$\mathbf{E}(\mathbf{r}, t) = \frac{e}{4\pi\varepsilon_0 c^2 \kappa^3 R}\{[\mathbf{u}_r \times (\mathbf{u}_r \times \dot{\beta})] - [\mathbf{u}_r \times (\beta \times \dot{\beta})]\} \tag{20-11}$$

The triple vector product relation can be expressed as

$$\mathbf{a} \times (\mathbf{b} \times \mathbf{c}) = \mathbf{b}(\mathbf{a} \cdot \mathbf{c}) - \mathbf{c}(\mathbf{a} \cdot \mathbf{b}) \tag{20-12}$$

so (20-11) can be rewritten as

$$\mathbf{E}(\mathbf{r}, t) = \frac{e}{4\pi\varepsilon_0 c^2 \kappa^3 R}\left[\mathbf{u}_r(\mathbf{u}_r \cdot \dot{\beta}) - \dot{\beta}(\mathbf{u}_r \cdot \mathbf{u}_r) - \dot{\beta}(\mathbf{u}_r \cdot \dot{\beta}) + \dot{\beta}(\mathbf{u}_r \cdot \beta)\right] \tag{20-13}$$

In spherical coordinates the field $\mathbf{E}(r, t)$ is

$$\mathbf{E}(\mathbf{r}, t) = E_r \mathbf{u}_r + E_\theta \mathbf{u}_\theta + E_\phi \mathbf{u}_\phi \tag{20-14}$$

Taking the dot product of both sides of (20-13) with $\mathbf{u}_r$ and using (20-14), we see that

$$E_r = (\mathbf{u}_r \cdot \dot{\beta}) - (\mathbf{u}_r \cdot \dot{\beta}) - (\mathbf{u}_r \cdot \beta)(\mathbf{u}_r \cdot \dot{\beta}) + (\mathbf{u}_r \cdot \dot{\beta})(\mathbf{u}_r \cdot \beta) = 0 \tag{20-15}$$

so the longitudinal (radial) component is zero. Thus, the radiated field is *always* transverse in both the nonrelativistic and relativistic regimes. Hence, the Stokes parameters definition for spherical coordinates continues to be valid.

The components E_θ and E_ϕ are readily found for the relativistic regime. We have

$$\beta = \frac{\dot{x}\mathbf{i} + \dot{y}\mathbf{j} + \dot{z}\mathbf{k}}{c} \tag{20-16a}$$

$$\dot{\beta} = \frac{\ddot{x}\mathbf{i} + \ddot{y}\mathbf{j} + \ddot{z}\mathbf{k}}{c} \tag{20-16b}$$

The Cartesian unit vectors in (20-16a) and (20-16b) can be replaced with the unit vectors in spherical coordinates, namely,

$$\mathbf{i} = \sin\theta\mathbf{u}_r + \cos\theta\mathbf{u}_\theta \tag{20-17a}$$

$$\mathbf{j} = \mathbf{u}_\phi \tag{20-17b}$$

$$\mathbf{k} = \cos\theta\mathbf{u}_r - \sin\theta\mathbf{u}_\theta \tag{20-17c}$$

In (20-17) the azimuthal angle has been set to zero because we assume that we always have symmetry around the z axis. Substituting (20-17) into (20-16) yields

$$c\beta = (\dot{x}\sin\theta + \dot{z}\cos\theta)\mathbf{u}_r + \dot{y}\mathbf{u}_\phi + (\dot{x}\cos\theta - \dot{z}\sin\theta)\mathbf{u}_\theta \tag{20-18a}$$

$$c\dot{\beta} = (\ddot{x}\sin\theta + \ddot{z}\cos\theta)\mathbf{u}_r + \ddot{y}\mathbf{u}_\phi + (\ddot{x}\cos\theta - \ddot{z}\sin\theta)\mathbf{u}_\theta \tag{20-18b}$$

The transverse components E_θ and E_ϕ are then

$$E_\theta = \frac{-e}{4\pi\varepsilon_0 c^2\kappa^3 R}\left[(\ddot{x}\cos\theta - \ddot{z}\sin\theta) - \frac{\ddot{x}\dot{z} - \dot{x}\ddot{z}}{c}\right] \tag{20-19a}$$

$$E_\phi = \frac{-e}{4\pi\varepsilon_0 c^2\kappa^3 R}\left[\ddot{y} - \frac{(\ddot{y}\dot{x} - \dot{y}\ddot{x})\sin\theta + (\ddot{y}\dot{z} + \dot{y}\ddot{z})\cos\theta}{c}\right] \tag{20-19b}$$

We see that the factors divided by c in (20-19a) and (20-19b) are the relativistic contributions. For β and $\dot{\beta} \ll 1$, (20-19) reduces to the nonrelativistic forms used in previous chapters.

We now apply these results to determining the radiation and the polarization emitted by charges undergoing linear and circular motion. In the following sections we treat synchrotron radiation and the motion of a charge moving in a dielectric medium (Čerenkov radiation). In the final section we deal with the scattering of radiation by electric charges.

For a linear charge that is accelerating along the z axis, β and $\dot{\beta}$ are parallel, so

$$\beta \times \dot{\beta} = 0 \tag{20-20}$$

Equation (20-1a) then reduces to

$$\mathbf{E}(\mathbf{x}, t) = \frac{e}{4\pi\varepsilon_0 c^2\kappa^3 R}\left[\mathbf{n} \times (\mathbf{n} \times \dot{\beta})\right] \tag{20-21a}$$

or

$$\mathbf{E}(\mathbf{r}, t) = \frac{e}{4\pi\varepsilon_0 c^2 R}\left[\frac{\mathbf{u}_r \times (\mathbf{u}_r \times \dot{v})}{\kappa^3}\right] \tag{20-21b}$$

According to (20-11) and (20-14), the field components of (20-21b) are

$$E_\theta = \frac{e}{4\pi\varepsilon_0 c^2 \kappa^3 R}[\ddot{x}\cos\theta - \ddot{z}\sin\theta] \tag{20-22a}$$

$$E_\phi = \frac{e}{4\pi\varepsilon_0 c^2 \kappa^3 R}\ddot{y} \tag{20-22b}$$

From the definition of the Stokes parameters (20-9) we then find the Stokes vector for the relativistic accelerating charge [from (20-22) and (20-1b)] is

$$S = \frac{e^2\ddot{z}^2}{32\pi^2\varepsilon_0 c^3}\left[\frac{\sin^2\theta}{(1-\beta\cos\theta)^5}\right]\begin{pmatrix}1\\1\\0\\0\end{pmatrix} \tag{20-23}$$

where $\mathbf{n}\cdot\beta = \beta\cos\theta$. We see immediately that the radiation is linearly horizontally polarized as in the nonrelativistic case.

The intensity of the radiation field is seen from (20-23) to be

$$I(\theta, \beta) = I_0\left(\frac{\sin^2\theta}{(1-\beta\cos\theta)^5}\right) \tag{20-24}$$

where $I_0 = e^2\ddot{z}^2/32\pi^2\varepsilon_0 c^3$. For the nonrelativistic case $\beta \to 0$, and (20-24) reduces to

$$I(\theta) = I_0 \sin^2\theta \tag{20-25}$$

which is the well-known dipole radiation distribution. In (20-25), the nonrelativistic result, the minimum intensity $I(\theta)$ is at $\theta = 0°$, where $I(0°) = 0$, and the maximum intensity is at $\theta = 90°$, where $I(90°) = I_0$.

Equation (20-24), on the other hand, shows that the maximum intensity shifts toward the z axis as β increases. To determine the positions of the maximum and minimum of (20-24), we differentiate (20-24) with respect to θ, set the result equal to zero, and find that

$$\sin\theta = 0 \tag{20-26a}$$

$$3\beta\cos^2\theta + 2\cos\theta - 5\beta = 0 \tag{20-26b}$$

The solution of the quadratic equation (20-26b) is

$$\cos\theta = \frac{\sqrt{15\beta^2+1}-1}{3\beta} \tag{20-27}$$

where we have taken the positive root because of the requirement that $|\cos\theta| \leq 1$. For small values of β, (20-27) reduces to

$$\cos\theta \simeq \frac{5\beta}{2} \tag{20-28}$$

so that for $\beta = 0$ the angle θ is 90° as before . For extreme relativistic motion $\beta \simeq 1$, and (20-28) then reduces to

$$\cos\theta \simeq 1 \tag{20-29}$$

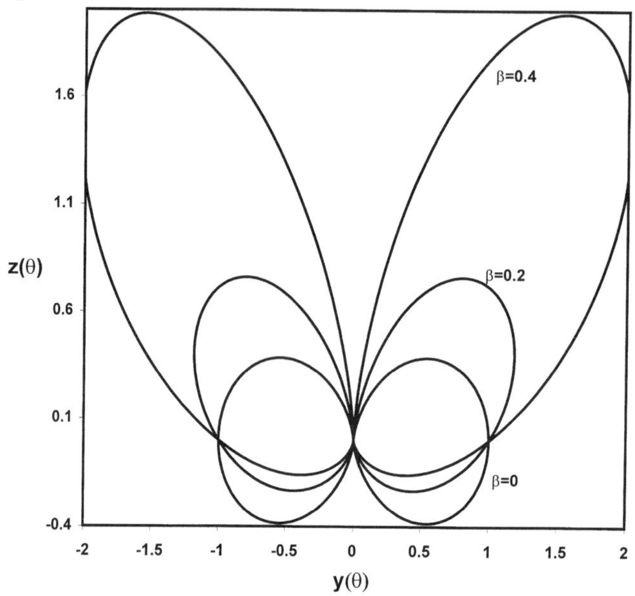

Figure 20-2 Intensity distribution of a relativistic moving charge for $\beta = 0,\ 0.2,$ and 0.4.

so $\theta \simeq 0^\circ$. We see that the maximum intensity has moved from $\theta = 90^\circ (\beta \ll 1)$ to $\theta = 0^\circ$ $(\beta \simeq 1)$, that is, the direction of the maximum intensity moves toward the charge moving along the z axis.

In Fig. 20-2 the intensity contours for various values of β are plotted. The contours clearly show the shift of the maximum intensity toward the z axis for increasing β. In the figure the charge is moving up the z axis from the origin, and the horizontal axis corresponds to the y direction. To make the plot we equated $I(\theta)$ with ρ, so

$$y(\theta) = \rho \sin\theta \tag{20-30a}$$

$$z(\theta) = \rho \cos\theta \tag{20-30b}$$

where

$$\rho = I(\theta) = \frac{\sin^2\theta}{(1 - \beta\cos\theta)^5} \tag{20-30c}$$

We see that, as β increases, the familiar $\sin^2\theta$ distribution becomes lobelike, a characteristic behavior of relativistically moving charges.

The formulation we have derived is readily extended to an oscillating charge. The motion of a charge undergoing linear oscillation is described by

$$z(t) = z_0 e^{i\omega_0 t} \tag{20-31}$$

In vector form (20-31) can be written as

$$z(t) = z(t)\mathbf{u}_z = z_0 e^{i\omega_0 t}\mathbf{u}_z \tag{20-32}$$

Furthermore, using β notation, $\beta = \dot{z}/c$, we can express the velocity and acceleration in vector form as

$$\beta = \frac{\dot{z}}{c}\mathbf{u}_z \qquad \dot{\beta} = \frac{\ddot{z}}{c}\mathbf{u}_z \tag{20-33}$$

We see immediately that

$$\beta \times \dot{\beta} = 0 \tag{23-34}$$

which is identical to (20-20). Hence, we have the same equations for an oscillating charge as for a unidirectional relativistic moving charge. We easily find that the corresponding Stokes vector is

$$S = \frac{1}{2c\varepsilon_0}\left(\frac{ez_0}{4\pi c}\right)^2 \left(\frac{\sin^2\theta}{(1-\beta\cos\theta)^5}\right)\omega_0^4 \begin{pmatrix} 1 \\ 1 \\ 0 \\ 0 \end{pmatrix} \tag{20-35}$$

Thus, the radiation appears at the same frequency as the frequency of oscillation. With respect to the intensity distribution we now have radiation also appearing below the $z = 0$ axis because the charge is oscillating above and below the xy plane. Thus, the intensity pattern is identical to the unidirectional case but is now symmetrical with respect to the xy plane. In Fig. 20-3 we show a plot of the intensity contour for $\beta = 0.4$.

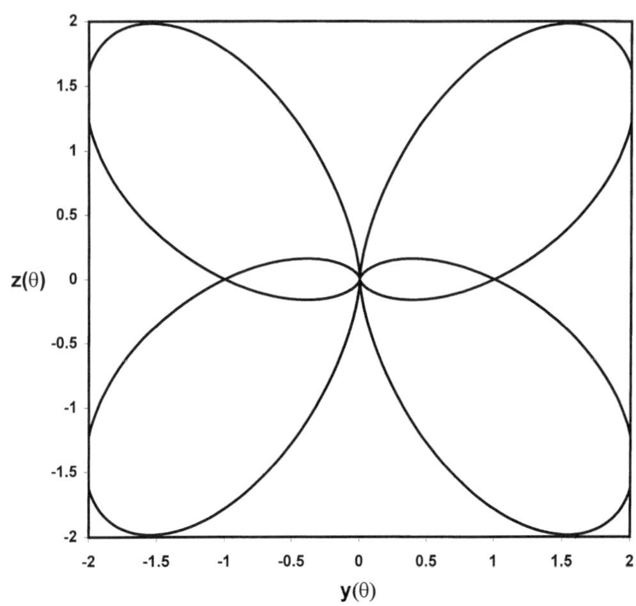

Figure 20-3 Intensity contours for a relativistic oscillating charge ($\beta = 0.4$).

20.2 RELATIVISTIC MOTION OF A CHARGE MOVING IN A CIRCLE: SYNCHROTRON RADIATION

In the previous section we dealt with the relativistic motion of charges moving in a straight line and with the intensity and polarization of the emitted radiation. This type of radiation is emitted by electrons accelerated in devices known as linear accelerators. We have determined the radiation emitted by nonrelativistic charges moving in circular paths as well. In particular, we saw that a charge moves in a circular path when a constant magnetic field is applied to a region in which the free charge is moving.

In this section we now consider the radiation emitted by relativistically moving charges in a constant magnetic field. The radiation emitted from highly relativistic charges is known as synchrotron radiation, after its discovery in the operation of the synchrotron. A charge moving in a circle of radius a is shown in Fig. 20-4.

The coordinates of the electron are

$$x(t) = a\cos\omega t \qquad y(t) = a\sin\omega t \tag{20-36}$$

Using the familiar complex notation, we can express (20-36) as

$$x(t) = ae^{i\omega t} \qquad y(t) = -iae^{i\omega t} \tag{20-37a}$$

$$\dot{x}(t) = i\omega ae^{i\omega t} \qquad \dot{y}(t) = a\omega e^{i\omega t} \tag{20-37b}$$

$$\ddot{x}(t) = -a\omega^2 e^{i\omega t} \qquad \ddot{y}(t) = ia\omega^2 e^{i\omega t} \tag{20-37c}$$

For the nonrelativistic case we saw that ω, the cyclotron frequency, was given by

$$\omega = \frac{eB}{m} \tag{20-38}$$

where e is the magnitude of the charge, B is the strength of the applied magnetic field, m is the mass of the charge, and c is the speed of light in free space. We can obtain

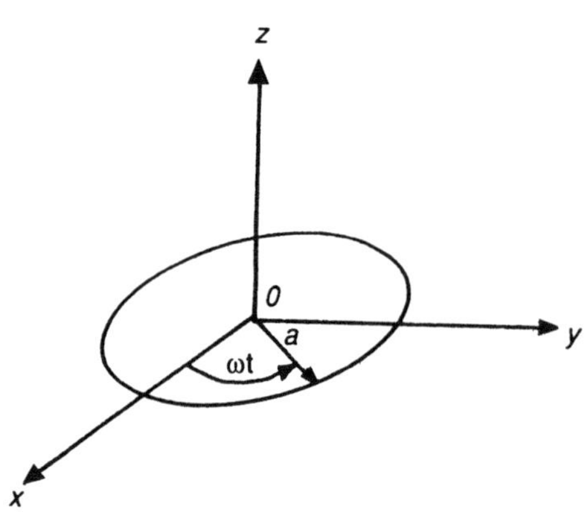

Figure 20-4 Motion of a relativistic charge moving in a circle of radius a in the xy plane with an angular frequency ω.

the corresponding form for ω for relativistic motion by merely replacing m in (20-38), the rest mass, with the relativistic mass m by

$$m \rightarrow \frac{m}{(1-\beta^2)^{1/2}} \tag{20-39}$$

Thus, (20-38) becomes

$$\omega = \frac{eB}{m}(1-\beta^2)^{1/2} \tag{20-40}$$

The frequency ω in (20-40) is now called the synchrotron frequency.

To find the Stokes vector of the emitted radiatin, we recall from Section 20-1 that the relativistic field components are

$$E_\theta = \frac{-e}{4\pi\varepsilon_0 c^2 \kappa^3 R}\left[(\ddot{x}\cos\theta - \ddot{z}\sin\theta) - \frac{\ddot{x}\dot{z} - \dot{x}\ddot{z}}{c}\right] \tag{20-41a}$$

$$E_\phi = \frac{-e}{4\pi\varepsilon_0 c^2 \kappa^3 R}\left[\ddot{y} - \frac{(\ddot{y}\dot{x} - \dot{y}\ddot{x})\sin\theta + (\ddot{y}\dot{x} + \dot{y}\ddot{x})\cos\theta}{c}\right] \tag{20-41b}$$

Because there is no motion in the z direction, (20-41) reduces to

$$E_\theta = \frac{-e}{4\pi\varepsilon_0 c^2 \kappa^3 R}[\ddot{x}\cos\theta] \tag{20-42a}$$

$$E_\phi = \frac{-e}{4\pi\varepsilon_0 c^2 \kappa^3 R}\left[\ddot{y} - \frac{(\ddot{y}\dot{x} - \dot{y}\ddot{x})\sin\theta}{c}\right] \tag{20-42b}$$

Substituting (20-37b) and (20-37c) into (20-42), we find

$$E_\theta = \frac{-e}{4\pi\varepsilon_0 c^2 \kappa^3 R}[a\omega^2\cos\theta] \tag{20-43a}$$

$$E_\phi = \frac{-e}{4\pi\varepsilon_0 c^2 \kappa^3 R}\left[ia\omega^2 - \frac{a^2\omega^3}{c}\sin\theta\right] \tag{20-43b}$$

where we have suppressed the exponential time factor $e^{i\omega t}$. From the definition of the Stokes parameters given in Section 20.1, we then find that the Stokes vector for synchrotron radiation is

$$S = \frac{e^2\beta^4\omega^4}{a^2(1-\beta\cos\theta)^5}\begin{pmatrix} 1+\cos^2\theta+\beta^2\sin^2\theta \\ -(1-\beta^2)\sin^2\theta \\ 2\beta\sin\theta \\ -2\cos\theta \end{pmatrix} \tag{20-44}$$

where we emphasize that θ is the observer's angle measured from the z axis. Equation (20-44) shows that for synchrotron radiation the radiation is, in general, elliptically polarized. The Stokes vector (20-44) is easily shown to be correct because the matrix elements satisfy the equality:

$$S_0^2 = S_1^2 + S_2^2 + S_3^2 \tag{20-45}$$

We saw earlier when dealing with the motion of a charge moving in a circle for the nonrelativistic case that the Stokes vector reduces to simpler (degenerate) forms.

A similar situation arises with relativistically moving charges. Thus, when we observe the radiation at $\theta = 0^\circ$, the Stokes vector (20-44) reduces to

$$S = \frac{2e^2\beta^4\omega^4}{a^2(1-\beta)^5}\begin{pmatrix} 1 \\ 0 \\ 0 \\ -1 \end{pmatrix} \tag{20-46}$$

which is the Stokes vector for left circularly polarized light. Similarly, for $\theta = \pi/2$ the Stokes vector is

$$S = \frac{e^2a^2\omega^4}{c^4}\begin{pmatrix} 1+\beta^2 \\ -(1-\beta^2) \\ 2\beta \\ 0 \end{pmatrix} \tag{20-47}$$

At this observation angle the radiation is linearly polarized. Finally, at $\theta = \pi$, we see that the radiation is right circularly polarized.

For $\beta \ll 1$, the nonrelativistic regime, (20-44) reduces to

$$S = \frac{e^2a^2\omega^4}{c^4}\begin{pmatrix} 1+\cos^2\theta \\ -\sin^2\theta \\ 0 \\ -2\cos\theta \end{pmatrix} \tag{20-48}$$

where $\omega = eB/m = \omega_c$ is the cyclotron frequency. This is the Stokes vector we found in Section 17.3 for a charge rotating in the xy plane. We now examine the intensity, orientation angle, and ellipticity of the polarization ellipse for synchrotron radiation (20-44).

The intensity of the radiation field, $I(\theta)$, can be written from (20-44) as

$$I(\theta) = \frac{e^2\omega_c^4}{a^2}\left[\frac{(1-\beta^2)^2(1+\cos^2\theta+\beta^2\sin^2\theta}{(1-\beta\cos\theta)^5}\right] \tag{20-49}$$

where we have set $\omega = \omega_c(1-\beta^2)^{1/2}$. The presence of the factor $(1-\beta\cos\theta)^5$ in the denominator of (20-49) shows that a lobelike structure will again emerge; this behavior will be shown shortly when a plot is made of (20-49).

The orientation angle ψ and the ellipticity angle χ are

$$\psi = \frac{1}{2}\tan^{-1}\left[\frac{-2\beta}{(1-\beta^2)\sin\theta}\right] \tag{20-50}$$

and

$$\chi = \frac{1}{2}\sin^{-1}\left[\frac{-2\cos\theta}{1+\cos^2\theta+\beta^2\sin^2\theta}\right] \tag{20-51}$$

In Fig. 20-5 a plot of the intensity, (20-49) has been made as a function of the observation angle θ for $\beta = 0,\ 0.1,$ and 0.2. We see that for $\beta = 0$ (i.e., the nonrelativistic radiation pattern) the intensity contour follows a bubblelike distribution. However, as β increases, the bubblelike contour becomes lobelike. This behavior is further emphasized in Figs. 20-6 and 20-7. Figure 20-6 shows (20-49) for

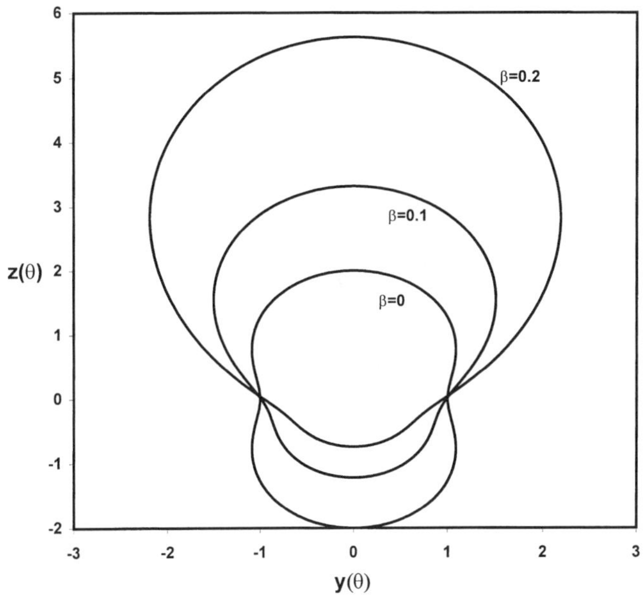

Figure 20-5 Relativistic intensity contours for $\beta = 0.0,\ 0.1,$ and $0.2.$

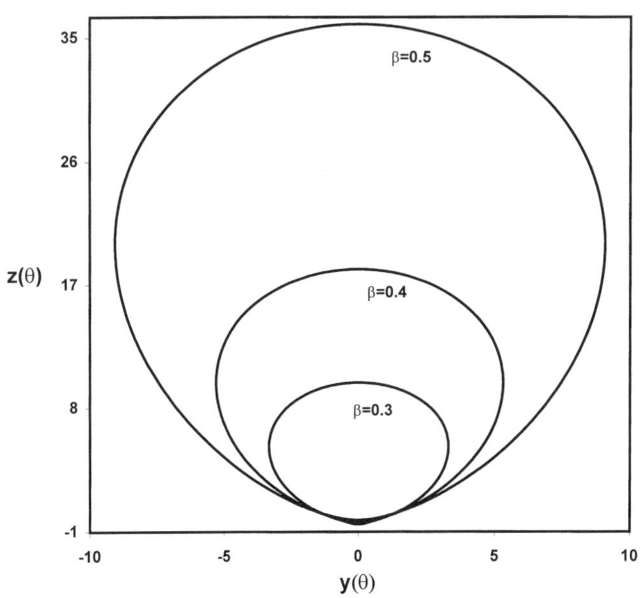

Figure 20-6 Relativistic intensity contours for $\beta = 0.3,\ 0.4,$ and $0.5.$

$\beta = 0.3,\ 0.4,$ and 0.5. Similarly, Figure 20-7 shows (20-49) for $\beta = 0.6,\ 0.7,$ and 0.8. For β, say, equal to 0.99 we have $I(0^\circ)/I(90^\circ) = 2 \times 10^9$, which is an extraordinarily narrow beam.

In Fig. 20-8 we have plotted the logarithm of the intensity $I(\theta)$ from $\theta = 0^\circ$ to $180^\circ$for $\beta = 0$ to 0.9 in steps of 0.3.

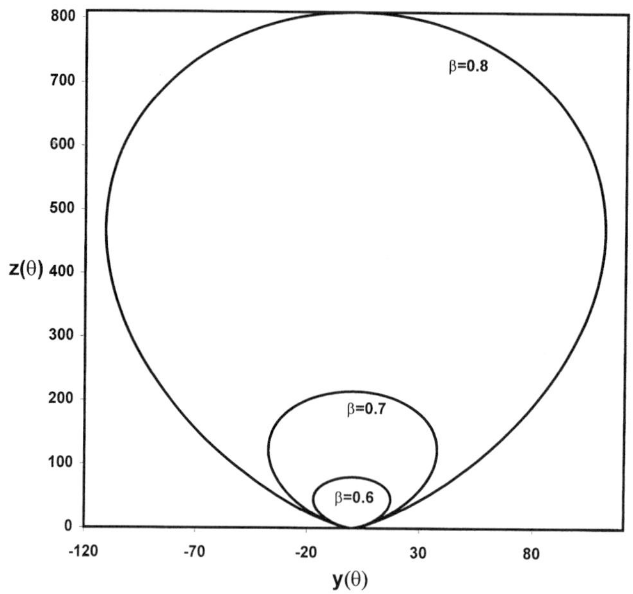

Figure 20-7 Relativistic intensity contours for $\beta = 0.6$, 0.7, and 0.8.

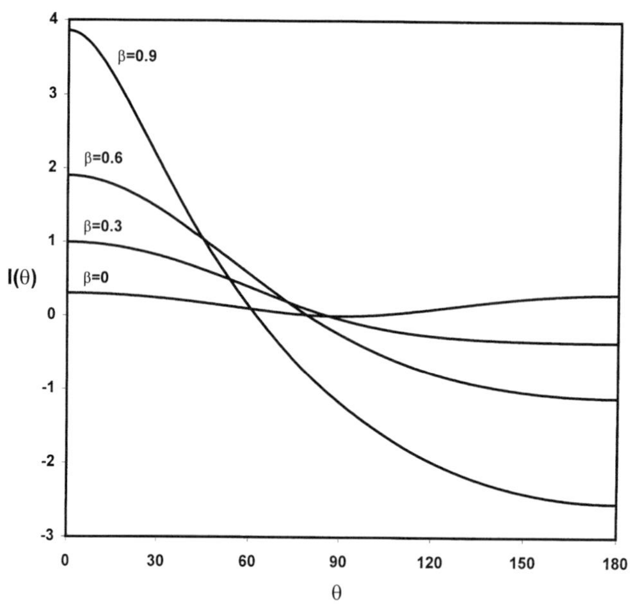

Figure 20-8 Logarithmic plot of the intensity for $\beta = 0.0$ through 0.9.

In order to plot the orientation angle ψ, (20-50), as a function of θ, we note that, for $\beta = 0.0$ and 1.0, $\psi = 0$ and $-\pi/4$, respectively. In Fig. 20-9 we plot ψ as a function of θ where the contours correspond to $\beta = 0.0, 0.1, \ldots, 1.0$ for decreasing ψ.

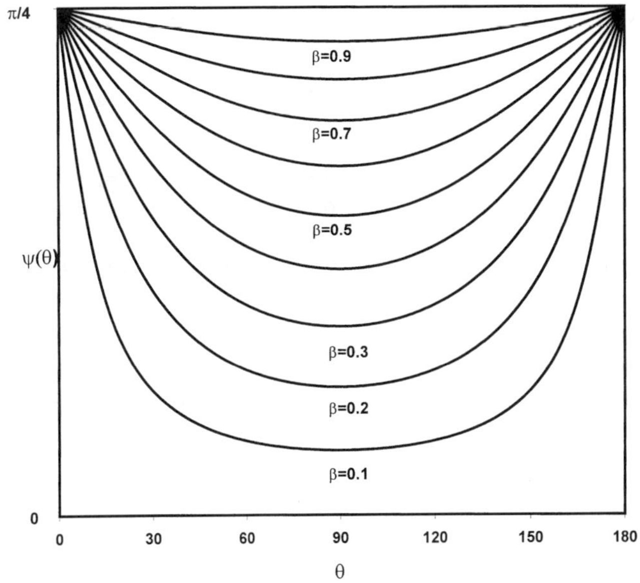

Figure 20-9 Orientation angle ψ for synchrotron radiation.

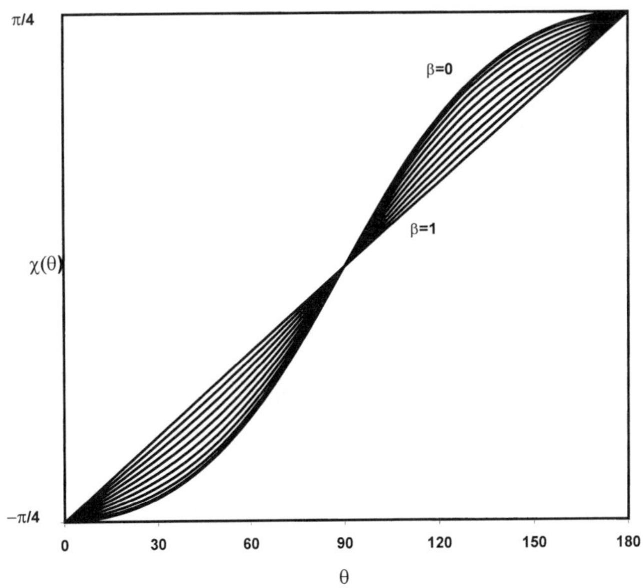

Figure 20-10 Ellipticity angle χ for synchrotron radiation.

In Fig. 20-10 the ellipticity angle χ, (20-51), is plotted for $\beta = 0$ to 1.0 over a range of $\theta = 0^\circ$ to 180°. For the extreme relativistic case (20-51) becomes

$$\chi = -\frac{1}{2}\sin^{-1}(\cos\theta) \tag{20-52}$$

It is straightforward to show that (20-52) can be rewritten in the form of an equation for a straight line, namely,

$$\chi = \frac{\theta}{2} - 45^\circ \tag{20-53}$$

and this behavior is confirmed in Fig. 20-10. We see that Fig. 20-10 shows that the ellipticity varies from $\chi = -45^\circ$ (a circle) at $\theta = 0^\circ$ and $\chi = 45^\circ$ (a counterclockwise circle) at $\theta = 180^\circ$.

Finally, it is of interest to compare the Stokes vector for $\beta = 0$ and for $\beta = 1$. The Stokes vectors are, respectively,

$$S = K\begin{pmatrix} 1+\cos^2\theta \\ \sin^2\theta \\ 0 \\ 2\cos\theta \end{pmatrix} \qquad \beta = 0 \tag{20-54a}$$

and

$$S' = K'\begin{pmatrix} 1 \\ \sin^2\theta \\ \sin\theta\cos\theta \\ \cos\theta \end{pmatrix} \qquad \beta = 1 \tag{20-54b}$$

where K and K' are constants [see (20-44)]. For $\theta = 0^\circ$ and 90°, S and S' become, respectively,

$$S = 2K\begin{pmatrix} 1 \\ 0 \\ 0 \\ 1 \end{pmatrix} \qquad S' = K'\begin{pmatrix} 1 \\ 0 \\ 0 \\ 1 \end{pmatrix} \qquad \theta = 0^\circ \tag{20-55a}$$

$$S = K\begin{pmatrix} 1 \\ 1 \\ 0 \\ 0 \end{pmatrix} \qquad S' = K'\begin{pmatrix} 1 \\ 1 \\ 0 \\ 0 \end{pmatrix} \qquad \theta = 90^\circ \tag{20-55b}$$

Thus, in the extreme cases of $\beta = 0$ and $\beta = 1$, the Stokes vectors—that is, the polarization states—are identical! However, *between* these two extremes the polarization states are very different.

Synchrotron radiation was first observed in the operation of synchrotrons. However, many astronomical objects emit synchrotron radiation, and it has been associated with sunspots, the crab nebula in the constellation of Taurus, and radiation from Jupiter. Numerous papers and discussions of synchrotron radiations have appeared in the literature, and further information can be found in the references.

20.3 ČERENKOV EFFECT

A charged particle in uniform motion and traveling in a straight line in free space does not radiate. However, if the particle is moving with a constant velocity through a material medium, it can radiate if its velocity is greater than the phase velocity of light in the medium. Such radiation is called Čerenkov radiation, after its discoverer, P. A. Čerenkov (1937). According to the great German physicist A. Sommerfeld, the problem of the emission of radiation by charged particles moving in an optical medium characterized by a refractive index n was studied as early as the beginning of the last century.

The emission of Čerenkov radiation is a cooperative phenomenon involving a large number of atoms of the medium whose electrons are accelerated by the fields of the passing particle and so emit radiation. Because of the collective aspects of the process, it is convenient to use the macroscopic concept of a dielectric constant ε rather than the detailed properties of individual atoms.

In this section our primary concern is to determine the polarization of Čerenkov radiation. The mathematical background as well as additional information on the Čerenkov effect can be found in Jackson's text on classical electrodynamics. Here, we shall determine the radiated field $\mathbf{E}(\mathbf{x}, t)$ for the Čerenkov effect, whereupon we then find the Stokes parameters (vector).

A qualitative explanation of the Čerenkov effect can be obtained by considering the fields of the fast particle in the dielectric medium as a function of time. The medium is characterized by a refractive index n, so the phase velocity of the light is c/n, where c is the speed of light in a vacuum. The particle velocity is denoted by v. In order to understand the Čerenkov effect it is not necessary to include the refractive index in the analysis, however. Therefore, we set $n = 1$, initially. At the end of the analysis we shall see the significance of n.

If we have a charged particle that is stationary but capable of emitting spherical waves, then after the passage of time t the waves are described by

$$x^2 + y^2 + z^2 = r^2(t) = (ct)^2 \tag{20-56}$$

If the charge is moving along the positive x axis with a velocity v, then the coordinate x is replaced by $x - vt$, so (20-56) becomes

$$(x - vt)^2 + y^2 + z^2 = (ct)^2 \tag{20-57}$$

We can consider the form of (20-57) in the xy plane by setting $z = 0$, so the two-dimensional representation of the spherical wave is

$$(x - vt)^2 + y^2 = (ct)^2 \tag{20-58}$$

We now write $\beta = v/c$. Furthermore, for convenience we set $c = 1$, so (20-58) becomes

$$(x - \beta t)^2 + y^2 = t^2 \tag{20-59}$$

The intercepts of the spherical wave on the x axis are found by setting $y = 0$ in (20-59). Then

$$x_{\pm} = (\beta \pm 1)t \tag{20-60}$$

The intercept of the leading edge of the spherical wave front is then

$$x_{+} = (\beta + 1)t \tag{20-61a}$$

and, similarly, the intercept of the trailing edge of the spherical wave front is

$$x_{-} = (\beta - 1)t \tag{20-61b}$$

The maximum and minimum values of the spherical wave along the y axis are found from the condition $dy/dx = 0$. From Eq. (20-59) we can then show that the maximum and minimum values of y occur at

$$x = \beta t \tag{20-62a}$$

This result is to be expected for a wave source propagating with a velocity $v = \beta$. The corresponding maximum and minimum values of y are then found from (20-59) to be

$$y_{\pm} = \pm t \tag{20-62b}$$

Since the radius of the spherical wave front is $r(t) = ct = (1)t$, this, too, is to be expected. We see that at $t = 0$ both x and $y = 0$ correspond to the particle's position $(x - \beta t)$ at the origin. The phase velocity v_p of the spherical wave is determined from $r(t) = ct$ and $v_p = dr(t)/dt = c(= 1)$.

Solving (20-59) for $y(t)$ we have

$$y(t) = \pm\sqrt{t^2 - (x - \beta t)^2} \tag{20-63}$$

It is of interest to plot (20-63) for $\beta = 0$, 0.5, and 1.0. We see from (20-61b) that for $\beta = 1$ we have $x_- = 0$; that is, the trailing edges of the spherical wave fronts coincide. In Figs. 20-11 to 20-13. we have made plots of (20-63) for $\beta = 0$, 0.5, and 1.0. However, to describe the expansion of the spherical wave with the passage of time as the particle moves, the coordinates of the x axis have been reversed. That is, the largest circle corresponds to 4 sec and appears first, followed by decreasing circles for 3, 2, and 1 sec. For completeness we have included a plot for $\beta = 0$. The plot for $\beta = 1$, Fig. 20-13, confirms that when the particle is moving with the speed of light the trailing edges, which are shown as the leading edges in the plot, coincide.

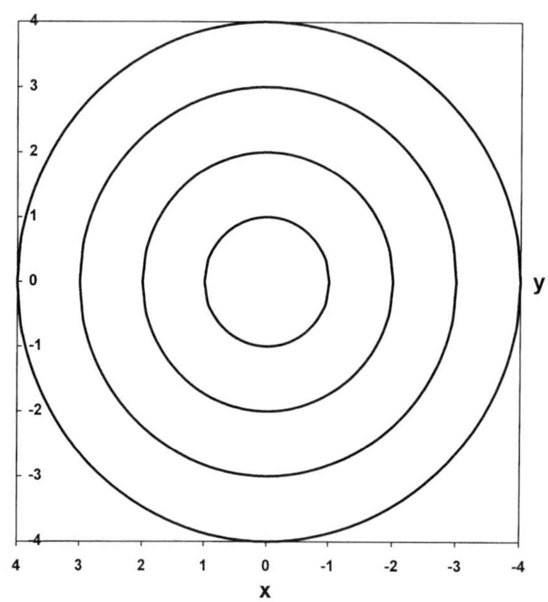

Figure 20-11 Propagation of a spherical wave for a staionary particle ($\beta = 0$).

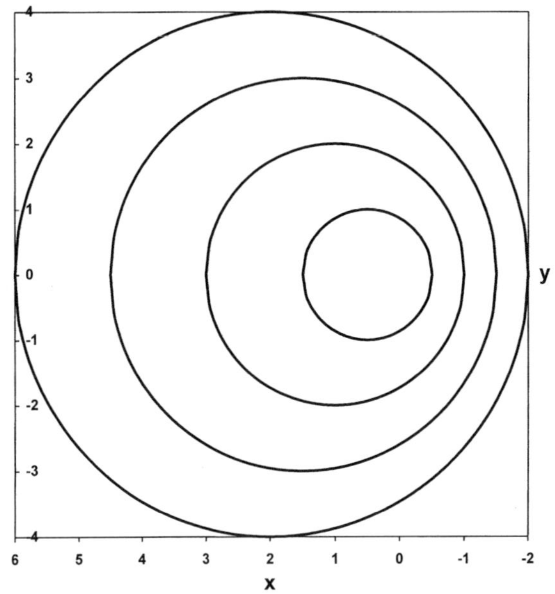

Figure 20-12 Propagation of a spherical wave for a particle moving with a velocity $\beta = 0.5$.

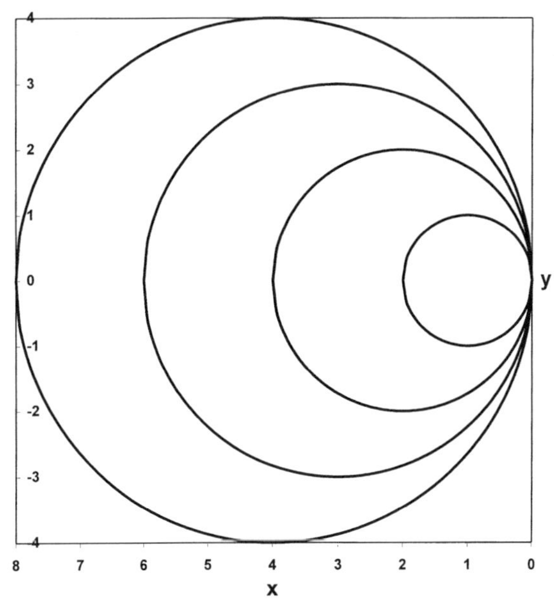

Figure 20-13 Propagation of a spherical wave for a particle moving with a velocity $\beta = 1.0$.

Figure 20-13 is especially interesting because it shows that the wave fronts only coincide for $\beta = 1$. The question now arises, what happens when $\beta > 1$? To answer this question we return to (20-63). We observe that $y(t)$ is *imaginary* if

$$x_+ > (\beta + 1)t \tag{20-64a}$$

and

$$x_{-} > (\beta - 1)t \tag{20-64b}$$

If we now choose, say, $\beta = 1.5$, then (20-64a) and (20-64b) become

$$x_{+} > 2.5t \tag{20-65a}$$

and

$$x_{-} > 0.5t \tag{20-65b}$$

Equation (20-65b) is especially interesting. We see from the condition (20-61b) that for $\beta = 0$, x is always less than 0 and for $\beta = 1$ it is exactly 0. However, (20-65b) now shows that if the speed of the particle *exceeds* the speed of light then there is a reversal of sign. However, so long as x_{-} is *less* than $0.5t$, $y(t)$ is real, so the wave can propagate! In Figs. 20-14 and 20-15 we show this behavior for $\beta = 1.5$ and $\beta = 2.5$.

If we now observe Figs. 20-11 to 20-13 we see that the spherical wave fronts do not interfere for $0 \leq \beta \leq 1$. Furthermore, we observe from Figs. 20-14 and 20-15 that if we extend a straight line from the origin through the tangents of the spherical wave fronts, then a new wave front appears, which is linear. This behavior is exactly what is observed when a boat moves quickly through water. It should be clearly understood that for $\beta < 1$ or $\beta > 1$ spherical waves are always generated. However, for $\beta < 1$ the waves cannot interfere, whereas for $\beta > 1$ the waves can interfere. Furthermore, this reinforcement of waves for $\beta > 1$ appears suddenly as soon as this condition appears. Hence, we experience a "shock," and so the straight line or tangent line is called a shock wave. The appearance of this shock wave does *not*

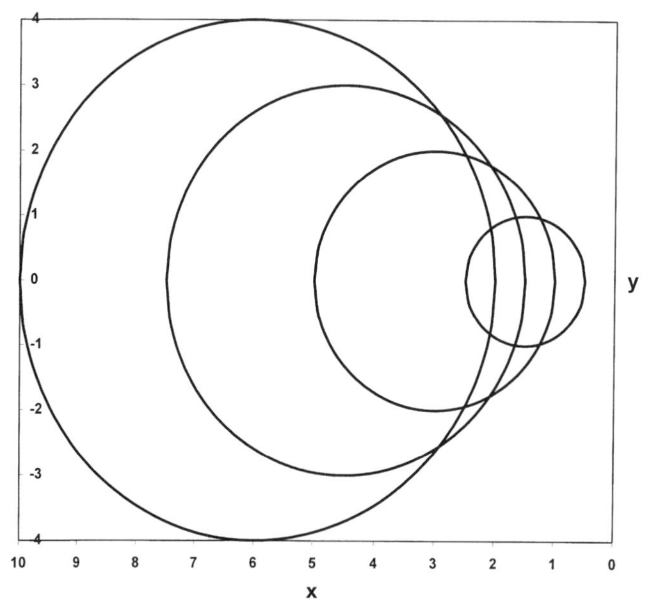

Figure 20-14 Propagation of a spherical wave for a particle moving with a velocity $\beta = 1.5$.

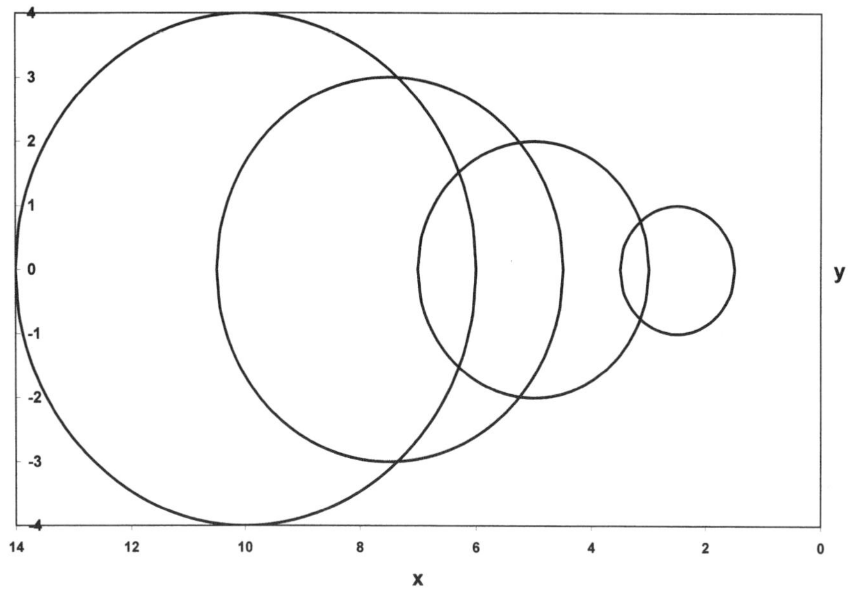

Figure 20-15 Propagation of a spherical wave for a particle moving with a velocity $\beta = 2.5$.

occur because there is a sudden change in the medium (the medium is unaffected), but because the waves, which were previously noninterfering ($\beta \leq 1$), now interfere ($\beta > 1$). In Fig. 20-16 we have drawn the straight line from the origin through the tangents of the spheres.

The tangents line in Fig. 20-16 is called a *wake*. The normal to the wake makes an angle θ_c, which is called the critical angle. From the figure we see that it can be expressed as

$$\cos\theta_c = \frac{c}{v} \tag{20-66}$$

In free space a particle cannot propagate equal to or faster than the speed of light. However, in an optical medium the phase velocity of the light is less than c and is given by c/n. Thus, if a particle moves with a speed greater than c/n it will generate an interference phenomenon exactly in the same manner as we have been describing. This behavior was first observed by Čerenkov, and, consequently, in optics the phenomenon is called the Čerenkov effect and the emitted radiation, Čerenkov radiation. Furthermore, the critical angle θ_c is now called the Čerenkov angle; the shock wave is in the direction given by θ_c.

The Čerenkov radiation is characterized by a cone. Its most important application is to measure the velocity of fast particles; that is, θ_c is measured by moving a detector such that the maximum intensity is observed. At this condition θ_c is determined, and v can then be immediately found.

With this background we now determine the intensity and polarization of the Čerenkov radiation. Our analysis draws heavily on Jackson's treatment of the Čerenkov effect and classical radiation in general.

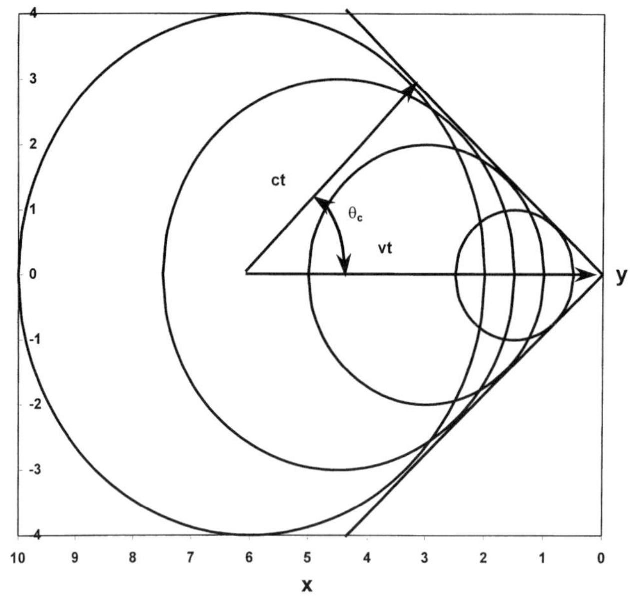

Figure 20-16 Construction of the tangent line for $\beta = 1.5$.

We restate the first two equations of the chapter. The electric field emitted by an accelerating charge is given by

$$\mathbf{E}(\mathbf{x}, t) = \frac{e}{4\pi\varepsilon_0 c^2}\left[\frac{\mathbf{n}}{\kappa^3 R} \times \{(\mathbf{n} - \beta) \times \dot{\beta}\}\right]_{\text{ret}} \tag{20-1a}$$

where $[\cdots]_{\text{ret}}$ means that the quantity in the brackets is to be evaluated at the retarded time, $t' = t - R(t')/c$. The quantity κ is given by

$$\kappa = 1 - \mathbf{n} \cdot \beta \tag{20-1b}$$

where $c\beta$ is the instantaneous velocity of the particle, $c\dot{\beta}$ is the instantaneous acceleration, and $\mathbf{n} = \mathbf{R}/R$. The quantity $\kappa \to 1$ for nonrelativistic motion. See Fig. 20-1 for the relations among the coordinates.

The instantaneous energy flux is given by the Poynting vector:

$$\mathbf{S} = [\mathbf{E} \times \mathbf{H}] \tag{16-32}$$

or

$$\mathbf{S} = \frac{1}{2} c\varepsilon_0 |\mathbf{E}|^2 \mathbf{n} \tag{20-2}$$

The power radiated per unit solid angle is then

$$\frac{dP}{d\Omega} = \frac{1}{2} c\varepsilon_0 |R\mathbf{E}|^2 \tag{20-67}$$

The total energy radiated per unit solid angle is the time integral of (20-67), namely,

$$\frac{dW}{d\Omega} = \frac{1}{2} c\varepsilon_0 R^2 \int_{-\infty}^{\infty} |\mathbf{E}(\mathbf{x}, \mathbf{t})|^2 dt \tag{29-68}$$

Equation (20-68) describes the radiation of energy in the time domain. A similar expression can be obtained in the temporal frequency domain (Parseval's theorem), and (20-68) can be expressed as

$$\frac{dW}{d\Omega} = \frac{1}{2} c\varepsilon_0 R^2 \int_{-\infty}^{\infty} |\mathbf{E}(\mathbf{x}, \omega)|^2 d\omega \tag{29-69}$$

We now introduce the Fourier transform pair:

$$\mathbf{E}(\mathbf{x}, \omega) = \frac{1}{\sqrt{2\pi}} \int_{-\infty}^{\infty} \mathbf{E}(\mathbf{x}, t) e^{i\omega t} dt \tag{20-70a}$$

$$\mathbf{E}(\mathbf{x}, t) = \frac{1}{\sqrt{2\pi}} \int_{-\infty}^{\infty} \mathbf{E}(\mathbf{x}, \omega) e^{-i\omega t} d\omega \tag{20-70b}$$

By substituting (20-1a), the electric field of an accelerated charge, into (20-70a), we obtain a general epression for the energy radiated per unit solid angle per unit frequency interval in terms of an integral over the trajectory of the particle. Thus, we find that

$$\mathbf{E}(\mathbf{x}, \omega) = \frac{e}{4\pi\varepsilon_0 \sqrt{2\pi} c^3} \int_{-\infty}^{\infty} \left[\frac{\mathbf{n}}{\kappa^3 R} \times \{(\mathbf{n} - \beta) \times \dot{\beta}\} \right]_{\text{ret}} e^{i\omega t} dt \tag{20-71}$$

We now change the variable of integration from t and t' by using the relation between the retarded time t' and the observer's time t, namely,

$$t' + \frac{R(t')}{c} = t \tag{20-72}$$

Using (20-72) in (20-71), we then find that

$$\mathbf{E}(\mathbf{x}, \omega) = \frac{e}{4\pi\varepsilon_0 \sqrt{2\pi} c^3} \int_{-\infty}^{\infty} \left[\frac{\mathbf{n}}{\kappa^2 R} \times \{(\mathbf{n} - \beta) \times \dot{\beta}\} \right] e^{i\omega(t' + R(t')/c)} dt' \tag{20-73}$$

In obtaining (20-73) we have used the relation from (20-72) that $dt = \kappa dt'$. We also observe that transforming to t' in (20-73) requires that the "ret" be dropped because the integral is no longer being evaluated at the 'ret" time. Since the observation point is assumed to be far away from the region where the acceleration occurs, the unit vector **n** is sensibly constant in time. Furthermore, referring to Fig. 20-1 the distance $R(t')$ can be approximated as

$$R(t') \simeq x - \mathbf{n} \cdot \mathbf{r} \tag{20-74}$$

Substituting this relation into (20-73), we then have

$$\mathbf{E}(\mathbf{x}, \omega) = \frac{e}{4\pi\varepsilon_0 \sqrt{2\pi} c^3} \int_{-\infty}^{\infty} \left[\frac{\mathbf{n}}{\kappa^2 R} \times \{(\mathbf{n} - \beta) \times \dot{\beta}\} \right] e^{i\omega(t - \mathbf{n}\cdot\mathbf{r}(t)/c)} dt \tag{20-75}$$

where x is the distance from the origin O to the observation point P and $r(t')$ is the position of the particle relative to O as shown in Fig. 20-17.

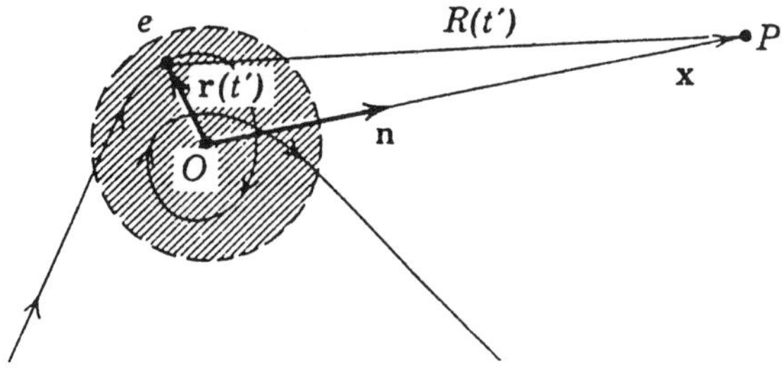

Figure 20-17 Coordinate relations for a moving charge. (From Jackson.)

In the integral (20-75) we have neglected the unimodular phase factor and dropped the primes on the time variable for the sake of brevity.

The integral in (20-75) can be simplified further. One can show that the factor within the integrand of (20-75) can be rewritten as

$$\mathbf{n} \times \frac{(\mathbf{n} - \beta) \times \dot{\beta}}{\kappa^2} = \frac{d}{dt}\left\{\frac{\mathbf{n} \times (\mathbf{n} \times \beta)}{\kappa}\right\} \tag{20-76}$$

Thus, the integrand in (20-75) can be replaced with the right-hand side of (20-76), and we have

$$\mathbf{E}(\mathbf{x}, \omega) = \frac{e}{4\pi\varepsilon_0\sqrt{2\pi}c^3 R}\int_{-\infty}^{\infty} \frac{d}{dt}\left\{\frac{\mathbf{n} \times (\mathbf{n} \times \beta)}{\kappa}\right\} e^{i\omega(t - \mathbf{n}\cdot\mathbf{r}(t)/c)}\, dt \tag{20-77}$$

We note that $d/dt = (dt'/dt)\,(d/dt) = \kappa(d/dt')$, and we recall that we have dropped the prime on the final dt in (20-77). Thus, (20-77) becomes

$$\mathbf{E}(\mathbf{x}, \omega) = \frac{e}{4\pi\varepsilon_0\sqrt{2\pi}c^3 R}\int_{-\infty}^{\infty} d\{\mathbf{n} \times (\mathbf{n} \times \beta)\} e^{i\omega(t - \mathbf{n}\cdot\mathbf{r}(t)/c)} \tag{20-78}$$

Equation (20-78) can now be integrated by parts to obtain

$$\mathbf{E}(\mathbf{x}, \omega) = \frac{e\omega}{4\pi\varepsilon_0\sqrt{2\pi}c^4 R}\int_{-\infty}^{\infty} [\mathbf{n} \times (\mathbf{n} \times \mathbf{v})] e^{i\omega(t - \mathbf{n}\cdot\mathbf{r}(t)/c)}\, dt \tag{20-79}$$

For a nonpermeable medium the correct fields and energy radiated for a particle moving in free space with a velocity $v > c$ require that at the end of the calculation we make the replacement:

$$c \to \frac{c}{\sqrt{\varepsilon}} \qquad e \to \frac{e}{\sqrt{\varepsilon}} \tag{20-80}$$

Thus, (20-79) becomes

$$\mathbf{E}(\mathbf{x}, \omega) = \frac{e\omega\varepsilon^{1/2}}{4\pi\varepsilon_0\sqrt{2\pi}c^4 R}\int_{-\infty}^{\infty} [\mathbf{n} \times (\mathbf{n} \times \mathbf{v})] e^{i\omega(t - \mathbf{n}\cdot\mathbf{r}(t)\varepsilon^{1/2}/c)}\, dt \tag{20-81}$$

To describe the Čerenkov effect, we have a charged particle moving in a straight line whose motion is described by

$$\mathbf{r}(t) = vt \tag{20-82}$$

Since the velocity is constant, the triple vector product in (20-81) can be factored out and we have

$$\mathbf{E}(\mathbf{x}, \omega) = \frac{e\omega\varepsilon^{1/2}}{4\pi\varepsilon_0\sqrt{2\pi}c^4 R}[\mathbf{n} \times (\mathbf{n} \times \mathbf{v})] \int_{-\infty}^{\infty} e^{i\omega t(1-\mathbf{n}\cdot\mathbf{v}\varepsilon^{1/2}/c)} dt \tag{20-83}$$

The integral is a Dirac delta function. We finally find that we have, using (20-83),

$$\mathbf{E}(\mathbf{x}, \omega) = \frac{e\varepsilon^{1/2}}{2\varepsilon_0\sqrt{2\pi}c^4 R}[\mathbf{n} \times (\mathbf{n} \times \mathbf{v})]\delta\left(1 - \varepsilon^{1/2}\frac{v}{c}\cos\theta\right) \tag{20-84}$$

where θ is measured relative to the velocity $\mathbf{v}$. The delta function only leads to a nonzero result when its argument is zero; that is

$$\cos\theta_c = \frac{1}{\beta\varepsilon^{1/2}} \tag{20-85}$$

which is the condition we found earlier for the emission of radiation at θ_c, the critical or Čerenkov angle. Thus, the delta function guarantees that the radiation is emitted only at the Čerenkov angle.

The significance of the delta function in (20-85) is that the field, i.e., the total energy radiated per unit frequency interval, is infinite. This infinity occurs because the particle has been moving through the medium forever. To obtain a meaningful result, we assume that the particle passes through a slab of dielectric in a time interval $2T$. Then, the infinite integral in (20-83) is replaced by

$$\frac{\omega}{2\pi}\int_{-T}^{T} e^{i\omega t(1-\mathbf{n}\cdot\mathbf{v}\varepsilon^{1/2}/c)} dt = \frac{\omega T}{\pi}\frac{\sin[\omega T(1-\varepsilon^{1/2}\beta\cos\theta)]}{\omega T(1-\varepsilon^{1/2}\beta\cos\theta)} \tag{20-86}$$

For the moment we shall represent the right-hand side of (20-86) by $f(\omega, T)$ and write (20-84) as

$$\mathbf{E}(\mathbf{x}, \omega) = \frac{e\varepsilon^{1/2}}{2\varepsilon_0\sqrt{2\pi}c^4 R}[\mathbf{n} \times (\mathbf{n} \times \mathbf{v})]f(\omega, T) \tag{20-87}$$

To find the Stokes parameters for (20-87) we must expand the triple vector product. The vector $\mathbf{n}$ can be set to $\mathbf{u}_r$, the unit vector in the radial direction. Then

$$\mathbf{u}_r \times (\mathbf{u}_r \times \mathbf{v}) = \mathbf{u}_r(\mathbf{v}_r \cdot \mathbf{u}) - \mathbf{v} \tag{20-88}$$

As before, we express the velocity in Cartesian coordinates:

$$\mathbf{v} = \dot{x}\mathbf{i} + \dot{y}\mathbf{j} + \dot{z}\mathbf{k} \tag{20-89}$$

where $\mathbf{i}$, $\mathbf{j}$, and $\mathbf{k}$, are unit vectors in the x, y, and z directions, respectively.

We now express $\mathbf{i}$, $\mathbf{j}$, and $\mathbf{k}$ in spherical coordinates. We assume that we have symmetry around the z axis, so we can arbitrarily take $\phi = 0°$. Then, the unit vectors in Cartesian coordinates are related to the unit vectors in spherical coordinates

$\mathbf{u}_r$, $\mathbf{u}_\theta$, and $\mathbf{u}_\phi$, by

$$\mathbf{i} = \sin\theta\mathbf{u}_r + \cos\theta\mathbf{u}_\theta \tag{20-90a}$$

$$\mathbf{j} = \mathbf{u}_\phi \tag{20-90b}$$

$$\mathbf{k} = \cos\theta\mathbf{u}_r - \sin\theta\mathbf{u}_\theta \tag{20-90c}$$

Then (20-89) becomes

$$\mathbf{v} = \mathbf{u}_r(\dot{x}\sin\theta + \dot{z}\cos\theta) + \mathbf{u}_\theta(\dot{x}\cos\theta - \dot{z}\sin\theta) + \dot{y}\mathbf{u}_\phi \tag{20-91}$$

Substituting (20-91) into the right-hand side of (20-88) yields

$$\mathbf{u}_r \times (\mathbf{u}_r \times \mathbf{v}) = -\mathbf{u}_\theta(\dot{x}\cos\theta - \dot{z}\sin\theta) - \dot{y}\mathbf{u}_\phi \tag{20-92}$$

which shows that the field is transverse to the direction of propagation $\mathbf{u}_r$. We now replace the triple vector product in (20-87) by (20-92), so

$$\mathbf{E}(\mathbf{x}, \omega) = \frac{e\varepsilon^{1/2}}{2\varepsilon_0\sqrt{2\pi c^4}R} f(\omega, T)[-\mathbf{u}_\theta(\dot{x}\cos\theta - \dot{z}\sin\theta) - \dot{y}\mathbf{u}_\phi] \tag{20-93}$$

Finally, the vector $\mathbf{E}(\mathbf{x}, \omega)$ can be expressed in terms of its spherical coordinates:

$$\mathbf{E}(\mathbf{x}, \omega) = E_r\mathbf{u}_r + E_\theta\mathbf{u}_\theta + E_\phi\mathbf{u}_\phi \tag{20-94}$$

Equating the right-hand sides of Eqs. (20-93) and (20-94), we find

$$E_\theta = \frac{-e\varepsilon^{1/2}}{2\varepsilon_0\sqrt{2\pi c^4}R} f(\omega, T)(\dot{x}\cos\theta - \dot{z}\sin\theta) \tag{20-95a}$$

$$E_\phi = \frac{-e\varepsilon^{1/2}}{2\varepsilon_0\sqrt{2\pi c^4}R} f(\omega, T)[\dot{y}] \tag{20-95b}$$

Let us now assume that the charge is moving along the z axis with a velocity $\dot{z} = c\beta$, so $\dot{x} = \dot{y} = 0$. Then, (20-95) reduces to

$$E_\theta = \frac{e\varepsilon^{1/2}\beta}{2\varepsilon_0\sqrt{2\pi c^3}R} f(\omega, T)\sin\theta \tag{20-96a}$$

$$E_\phi = 0 \tag{20-96b}$$

The Stokes polarization parameters are defined by

$$S_0 = E_\phi E_\phi^* + E_\theta E_\theta^* \tag{16-10a}$$

$$S_1 = E_\phi E_\phi^* - E_\theta E_\theta^* \tag{16-10b}$$

$$S_2 = E_\phi E_\theta^* + E_\theta E_\phi^* \tag{16-10c}$$

$$S_3 = i(E_\phi E_\theta^* - E_\theta E_\phi^*) \tag{16-10d}$$

Substituting (20-96) into (16-10) and forming the Stokes vector, we find that

$$S = \frac{e^2 \varepsilon \beta^2}{8\pi\varepsilon_0^2 c^6} f^2(\omega, T) \sin^2\theta \begin{pmatrix} 1 \\ -1 \\ 0 \\ 0 \end{pmatrix} \tag{20-97}$$

The polarization of the radiation emitted by the Čerenkov effect is linearly vertically polarized.

Finally, we can integrate S in (20-97) over the solid angle $d\Omega$. On doing this we find that the Stokes vector for the Čerenkov effect is

$$S = \frac{e^2 \tan^2\theta_c f^2(\omega, T)}{2\varepsilon_0^2 c^6} \begin{pmatrix} 1 \\ -1 \\ 0 \\ 0 \end{pmatrix} \tag{20-98}$$

Thus, we see that the radiation emitted by the Čerenkov effect is linearly vertically polarized. Further information on the Čerenkov effect can be found in the texts by Jackson and Sommerfeld, as well as in the references listed in Jackson's text.

20.4 THOMSON AND RAYLEIGH SCATTERING

Maxwell's original purpose for developing his theory of the electromagnetic field was to encompass all the known phenomena of electromagnetism into a fundamental set of equations. It came as a surprise to Maxwell (and his contemporaries!) that his differential equations led to waves propagating with the speed of light. After the work of Hertz and Lorentz and Zeeman the only conclusion that could be drawn was that Maxwell's theory was a *unifying* theory between the electromagnetic field and the optical field. Furthermore, the phenomena were one and the same in both disciplines, the major difference being the wavelength (or frequency). Electromagnetism phenomena were associated with "low" frequencies, and optical phenomena were associated with "high" frequencies.

Maxwell's theory when coupled with Lorentz's theory of the electron led not only to the correct description of the seemingly complex Lorentz–Zeeman effect but also to a very good understanding of the phenomenon of dispersion. Lorentz's electron theory was able to provide a description of dispersion, which led to a complete understanding of Cauchy's simple empirical relation between the refractive index and the wavelength. This result was another triumph for Maxwell's theory.

But there was still another application for Maxwell's theory, which was totally unexpected. This was in the area of a phenomenon known as *scattering*. It is not clear at all the Maxwell's theory can be applied to this phenomenon, but it can and does lead to results in complete agreement with experiments. The phenomenon of scattering is described within Maxwell's theory as follows. An incident field consisting of transverse components impinges on a free electron. The electron will be accelerated and so emits radiation; that is, it *reradiates* the incident radiation. If the electron is "bound" to a nucleus so that it is oscillating about the nucleus with a fundamental frequency, then it, too, is found to scatter or reradiate the incident radiation. The reradiation or scattering takes place in an extremely short time (nanoseconds or less). Remarkably, one discovers that the scattered radiation manifests two distinct characteristics. The first is that there is a change in the polarization state between the incident and scattered radiation in which the degree of polarization varies with the observer's viewing angle. This behavior is very different from the Lorentz–Zeeman effect. There we saw that the polarization state changes as the observing angle varies, but the degree of polarization remains the same, and, in fact, is unity. The other notable difference is that the incident radiation field propagates along one axis and, ideally, can only be observed along this axis. The scattered radiation, on the other hand, is observed to exist not only along the axis but away from the axis as well. Characteristically, the maximum intensity of the scattered radiation is observed along the axis of the incident radiation and the minimum intensity perpendicular to the direction of the propagation of the incident beam. However, unlike the behavior of dipole radiation the intensity does not go to zero anywhere in the observed scattered radiation field. Maxwell's theory along with Lorentz's electron theory completely account for this behavior. We now treat the problem of scattering and present the results in terms of the Stokes parameters. The scattering behavior is represented by the Mueller matrix.

We first determine the Stokes parameters for the scattering of electromagnetic waves by a so-called free electron located at the origin of a Cartesian coordinate system. This is illustrated in Fig. 20-18. The incident field is represented by $\mathbf{E}(z, t)$ and propagates in the z direction. The motion of a free electron is then described by

$$m\ddot{\mathbf{r}} = -e\mathbf{E} \tag{20-99}$$

or, in component form,

$$\ddot{x} = \frac{-e}{m} E_x(t) \tag{20-100a}$$

$$\ddot{y} = \frac{-e}{m} E_y(t) \tag{20-100b}$$

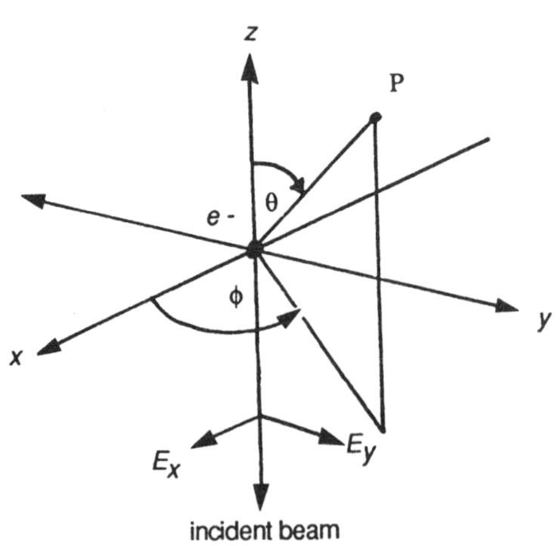

Figure 20-18 Scattering of incident radiation by a free electron.

where m is the mass of the electron, e is the charge, and $E_x(t)$ and $E_y(t)$ are the transverse components of the incident field. The incident field components can be written as

$$E_x(t) = E_{0x}e^{i(\omega t+\delta_x)} \tag{20-101a}$$

$$E_y(t) = E_{0y}e^{i(\omega t+\delta_y)} \tag{20-101b}$$

Equation (20-100) can be written from (20-101) as

$$\ddot{x} = \frac{-e}{m}E_{0x}e^{i\delta_x}e^{i\omega t} \tag{20-102a}$$

$$\ddot{y} = \frac{-e}{m}E_{0y}e^{i\delta_y}e^{i\omega t} \tag{20-102b}$$

The accelerations are now known, so we can substitute these results directly into the Eqs. (16-8) and (16-9) for the radiated field in spherical coordinates. Thus,

$$E_\theta = \frac{-e^2}{4\pi\varepsilon_0 mc^2 R}[E_{0x}e^{i\delta_x}e^{i\omega t}]\cos\theta \tag{20-103a}$$

and

$$E_\phi = \frac{-e^2}{4\pi\varepsilon_0 mc^2 R}[E_{0y}e^{i\delta_y}e^{i\omega t}] \tag{20-103b}$$

The Stokes vector S' corresponding to (20-103) is readily found. In terms of the Poynting vector, a factor of $c/4\pi$ should be included in the definition to obtain complete consistency. However, no essential information is lost by not including this factor, so the Stokes vector S' is given in the usual form:

$$S' = \frac{1}{2}\left(\frac{e^2}{4\pi\varepsilon_0 mc^2 R}\right)^2 \begin{pmatrix} S_0(1+\cos^2\theta) + S_1\sin^2\theta \\ S_0\sin^2\theta + S_1(1+\cos^2\theta) \\ 2S_2\cos\theta \\ 2S_3\cos\theta \end{pmatrix} \tag{20-104}$$

where S_0, etc., are the Stokes parameters for the incident plane wave (20-101).

Equation (20-104) can be readily written in terms of the Stokes vector S of the incident field and the Stokes vector S' of the scattered field, whereupon we find that the Mueller matrix for the scattering process is

$$M = \frac{1}{2}\left(\frac{e^2}{4\pi\varepsilon_0 mc^2 R}\right)^2 \begin{pmatrix} 1+\cos^2\theta & \sin^2\theta & 0 & 0 \\ \sin^2\theta & 1+\cos^2\theta & 0 & 0 \\ 0 & 0 & 2\cos\theta & 0 \\ 0 & 0 & 0 & 2\cos\theta \end{pmatrix} \tag{20-105}$$

We see that (20-105) corresponds to the Mueller matrix of a polarizer. This type of scattering by a free charge is known as Thomson scattering and is applicable to the scattering of x-rays by electrons and gamma rays by protons. Note the term $e^2/4\pi\varepsilon_0 mc^2$, which, as we saw earlier, is the classical electron radius r_0. We observe that the radius enters (20-105) as a *squared* quantity. Thus, the scattered intensity is proportional to the *area* of the electron. We, therefore, look upon the reradiation of the incident radiation as scattering by the area presented by the electron.

Two other facts can be observed. The first is that, according to (20-104), the scattered intensity is

$$I(\theta) = \frac{1}{2}[S_0(1 + \cos^2\theta) + S_1 \sin^2\theta] \tag{20-106}$$

where, for convenience, we have set the factor containing the physical constants to unity. We see immediately that the magnitude of the scattered radiation depends on the contribution of the linear polarization (S_1) of the incident beam. To plot (20-106) we use the normalized Stokes parameters and set S_0 to unity. We see that the two extremes for (20-106) are for linearly polarized light ($S_1 = -1$ and $S_1 = 1$) and midway is unpolarized light ($S_1 = 0$). The corresponding intensities are

$$I(\theta) = \frac{1}{2}[1 + \cos 2\theta] \qquad (S_1 = -1) \tag{20-107a}$$

$$I(\theta) = \frac{1}{2}[1 + \cos^2\theta] \qquad (S_1 = 0) \tag{20-107b}$$

$$I(\theta) = \frac{1}{2}[2] \qquad (S_1 = 1) \tag{20-107c}$$

We see that there is a significant change in the intensity over this range of polarization. In Fig. 20-19 we have plotted (20-106) by setting $S_0 = 1$ and varying

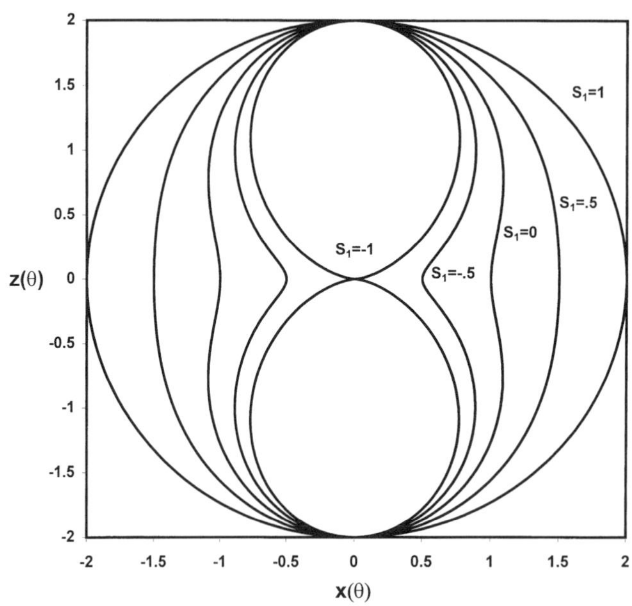

Figure 20-19 Intensity contours for scattering by a free electron for incident linearly polarized light from linear vertically polarized light (innermost contour) to linear horizontally polarized light (outermost contour) in steps of 0.5.

$S_1 = -1$ to 1 in steps of 0.5 over a range of $\theta = 0°$ to $360°$. The inner lobe corresponds to $S_1 = -1$, and the outer lobe to $S_1 = 1$. We note that for $S_1 = 0$ we obtain a "peanutlike" lobe.

The other fact of interest is that we can express the scattering in terms of the scattering cross-section. This is defined by

$$\frac{d\sigma}{d\Omega} = \frac{\text{energy radiated/unit time/unit solid angle}}{\text{incident energy/unit area/unit time}} \tag{20-108}$$

From (20-108) and (20-104) we see that the ratio of the scattered to incident Stokes parameters is the differential cross-section:

$$\frac{d\sigma}{d\Omega} = \frac{1}{2}\left(\frac{e^2}{4\pi\varepsilon_0 mc^2}\right)^2 \frac{S_0[1+\cos^2\theta] + S_1 \sin^2\theta}{S_0} \tag{20-109}$$

For the case of incident unpolarized light (20-108) reduces to

$$\frac{d\sigma}{d\Omega} = \frac{1}{2}\left(\frac{e^2}{4\pi\varepsilon_0 mc^2}\right)^2 (1+\cos^2\theta) \tag{20-110}$$

Equation (20-108) is known as Thomson's formula for the scattering by free charges. The total cross-section is defined to be

$$\sigma_T = \int_{4\pi} \frac{d\sigma}{d\Omega}\, d\Omega \tag{20-111}$$

Integrating (20-110) over the solid angle according to (20-111), the total cross-section for the free electron is

$$\sigma_T = \frac{8\pi}{3}\left(\frac{e^2}{4\pi\varepsilon_0 mc^2}\right)^2 \tag{20-112}$$

The Thomson cross section in equal to $0.665 \times 10^{-28}\,\text{m}^2$ for electrons. The unit of length, $e^2/4\pi\varepsilon_0 mc^2 = 2.82 \times 10^{-15}\,\text{m}$, is the classical electron radius, because a classical distribution of charge totalling the electronic charge must have a radius of this order if its electrostatic self-energy is equal to the electron mass. Finally, we note that the classical Thomson scattering is valid only at low frequencies. The quantum effects become important when the frequency ω becomes comparable to $mc^2/\hbar$, i.e., when the photon energy $\hbar\omega$ is comparable with, or larger than, the particle's rest energy mc^2.

Another quantity of interest is the degree of polarization. According to (20-104) this depends on both the polarization of the incident radiation and the observer's viewing angle. For example, from (20-104) we see that if we have linearly horizontally polarized light, $S_0 = S_1$ and $S_2 = S_3 = 0$, the scattered radiation is also linearly horizontally polarized and the degree of polarization is unity. However, if the incident radiation is unpolarized light, the Stokes vector is $\{I_0, 0, 0, 0\}$, and the Stokes vector of the scattered radiation is

$$S = \frac{1}{2}\left(\frac{e^2}{4\pi\varepsilon_0 mc^2 R}\right)^2 I_0 \begin{pmatrix} 1+\cos\theta \\ \sin^2\theta \\ 0 \\ 0 \end{pmatrix} \tag{20-113}$$

Equation (20-113) shows that the scattered radiation is, in general, partially polarized and the degree of polarization is

$$P = \left|\frac{\sin^2\theta}{1+\cos^2\theta}\right| \tag{20-114}$$

We see that for $\theta = 0^\circ$ (so-called on-axis scattering) the degree of polarization P is zero, whereas for $\theta = 90^\circ$ (off-axis scattering) the degree of polarization is unity. This behavior in the degree of polarization is characteristic of all types of scattering. In Fig. 20-20 we have plotted (20-114) as a function of the angle of scattering.

We now consider the scattering from a bound charge. The equation of motion is

$$m\ddot{\mathbf{r}} + k\mathbf{r} = -e\mathbf{E} \tag{20-115}$$

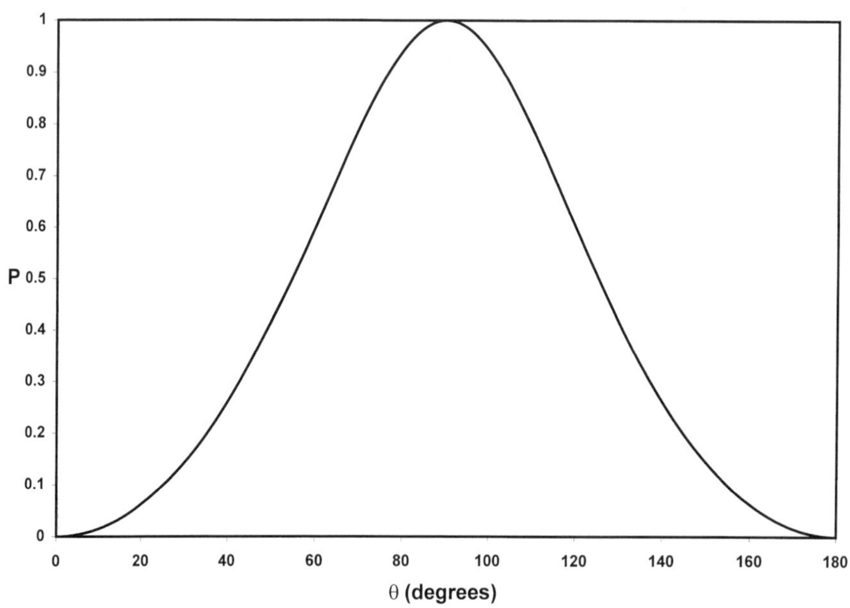

Figure 20-20 The degree of polarization P for scattering of unpolarized light by a free electron.

or, in component form,

$$\ddot{x} + \omega_0^2 x = \frac{-e}{m} E_x \tag{20-116a}$$

$$\ddot{y} + \omega_0^2 y = \frac{-e}{m} E_y \tag{20-116b}$$

$$\ddot{z} + \omega_0^2 z = 0 \tag{20-116c}$$

where $\omega_0 = (k/m)^{1/2}$ and the incident field is again propagating along the z axis and consists of the transverse components $E_x(t)$ and $E_y(t)$. We first consider the solution of (20-116a). In order to solve this equation we know that the solution is

$$x(t) = x_c(t) + x_p(t) \tag{20-117}$$

where $x_c(t)$ is the complementary solution and $x_p(t)$ is the particular solution. Using the notation:

$$D \equiv \frac{d}{dt} \tag{20-118}$$

we can write (20-116a) as

$$(D^2 + \omega_0^2)x(t) = R(t) \tag{20-119a}$$

where

$$R(t) = \left(\frac{-e}{m}\right) E_x(t) = \left(\frac{-e}{m}\right) E_{0x} e^{i\delta_x} e^{i\omega t} \tag{20-119b}$$

and ω is the frequency of the incident light. By using the well-known methods of differential equations for solving nonhomogeneous equations, we obtain the general solution:

$$x(t) = c_1 e^{i\omega_0 t} + c_2 e^{-i\omega_0 t} + c_3 e^{i\omega t} \tag{20-120}$$

where c_1, c_2 and c_3 are arbitrary constants. By substituting (20-120) into (20-116a) we readily find that c_3 is

$$c_3 = \frac{e}{m(\omega^2 - \omega_0^2)} E_{0x} e^{i\delta_x} \tag{20-121}$$

so the solution of (20-116a) is

$$x(t) = c_1 e^{i\omega_0 t} + c_2 e^{-i\omega_0 t} + \frac{e}{m(\omega^2 - \omega_0^2)} E_{0x} e^{i\delta_x} e^{i\omega t} \tag{20-122}$$

The first two terms in (20-122) describe the natural oscillation of the bound electron and are not involved in the scattering process. The last term in (20-122) is the term that arises from the interaction of the incident field **E** with the bound electron and describes the scattering process. Hence, the scattering term is

$$x(t) = \frac{e}{m(\omega^2 - \omega_0^2)} E_{0x} e^{i\delta_x} e^{i\omega t} \tag{20-123a}$$

Similarly, for (20-116b), we have

$$y(t) = \frac{e}{m(\omega^2 - \omega_0^2)} E_{0y} e^{i\delta_y} e^{i\omega t} \tag{20-123b}$$

The x and y accelerations of the bound electron are then found from (20-123a) and (20-123b):

$$\ddot{x}(t) = \frac{-e\omega^2}{m(\omega^2 - \omega_0^2)} E_x(t) \tag{20-124a}$$

$$\ddot{y}(t) = \frac{-e\omega^2}{m(\omega^2 - \omega_0^2)} E_y(t) \tag{20-124b}$$

where

$$E_x(t) = E_{0x} e^{i\omega t + i\delta_x} \tag{20-124c}$$

$$E_y(t) = E_{0y} e^{i\omega t + i\delta_y} \tag{20-124d}$$

The radiation field components, that is, scattered field components, are

$$E_\theta = (-e/4\pi\varepsilon_0 c^2 R)[\ddot{x}\cos\theta] \tag{20-125a}$$

$$E_\phi = (-e/4\pi\varepsilon_0 c^2 R)[\ddot{y}] \tag{20-125b}$$

Substituting (20-124a) and (20-124b) into (20-125a) and (20-125b), respectively, and forming the Stokes parameters, we find the Stokes vector of the scattered radiation:

$$S = \frac{1}{2}\left[\frac{e^2}{m4\pi\varepsilon_0 c^2 R(\omega^2 - \omega_0^2)}\right]^2 \omega^4 \begin{pmatrix} S_0(1+\cos^2\theta) + S_1\sin^2\theta \\ S_0\sin^2\theta + S_1(1+\cos^2\theta) \\ 2S_2\cos\theta \\ 2S_3\cos\theta \end{pmatrix} \tag{20-126}$$

The result is very similar to the one we obtained for scattering by a free electron. In fact, if we set $\omega_0 = 0$ in (20-126) (the free-electron condition), we obtain the same Stokes vector given by (20-104). However, for a bound electron we have an important difference. While the polarization behavior is identical, we see that the scattered intensity is now proportional to ω^4 or $(2\pi c/\lambda)^4$, that is, to the inverse fourth power of the wavelength. This shows that as the wavelength of light decreases, e.g., from the red region to the blue region of the spectrum, the intensity of the scattered light increases. This accounts for the "blue" sky; the sky is blue because of the scattering by bound electrons. This behavior was first explained by Lord Rayleigh in the latter part of the nineteenth century. The scattering process associated with ω^4 (or $1/\lambda^4$) is called, consequently, Rayleigh scattering.

Scattering phenomena play an important role not only in optics but, especially, in nuclear physics. The ideas developed here are readily extended to particle scattering, and the interested reader can find further discussions of other aspects of scattering in the references.

REFERENCES

Books

1. Jackson, J. D., *Classical Electrodynamics*, John Wiley, New York, 1962.
2. Sommerfeld, A., *Lectures on Theoretical Physics*, Vols. I–V, Academic Press, New York, 1952.
3. Strong, J., *Concepts of Classical Optics*, Freeman, San Francisco, 1959.
4. Stone, J. M., *Radiation and Optics*, McGraw-Hill, New York, 1963.
5. Blatt, J. M. and V. F. Wcisskopf, *Theoretical Nuclear Physics*, John Wiley, New York, 1952.
6. Heitler, W., *Quantum Theory of Radiation*, 3rd ed., Oxford University Press, Oxford, UK, 1954.

21

The Stokes Parameters and Mueller Matrices for Optical Activity and Faraday Rotation

21.1 INTRODUCTION

In 1811, Arago discovered that the "plane of polarization" of linearly polarized light was rotated when a beam of light propagated through quartz in a direction parallel to its optic axis. This property of quartz is called optical activity. Shortly afterwards, in 1815, Biot discovered, quite by accident, that many liquids and solutions are also optically active. Among these are sugars, albumens, and fruit acids, to name a few. In particular, the rotation of the plane of polarization as the beam travels through a sugar solution can be used to measure its concentration. The measurement of the rotation in sugar solutions is a widely used method and is called *saccharimetry*. Furthermore, polarization measuring instruments used to measure the rotation are called *saccharimeters*.

The rotation of the optical field occurs because optical activity is a manifestation of an unsymmetric isotropic medium; that is, the molecules lack not only a center of symmetry but also a plane of symmetry as well. Molecules of this type are called enantiomorphic since they cannot be brought into coincidence with their mirror image. Because this rotation takes place naturally, the rotation associated with optically active media is called *natural rotation*.

In this chapter we shall only discuss the optical activity associated with liquids and solutions and the phenomenon of Faraday rotation in transparent media and plasmas. In Chapter 24 we shall discuss optical activity in crystals.

Biot discovered that the rotation was proportional to the concentration and path length. Specifically, for an optically active liquid or for a solution of an optically active substance such as sugar in an inactive solvent, the specific rotation or rotary power γ is defined as the rotation produced by a 10-cm column of liquid containing 1 g of active substance per cubic centimeter (cc) of solution. For a solution containing

m g/cc the rotation for a path length l is given by

$$\theta = \frac{\gamma m l}{10} \tag{21-1a}$$

or, in terms of the rotary power γ,

$$\gamma = \frac{10\theta}{ml} \tag{21-1b}$$

The product of the specific rotation and the molecular weight of the active substance is known as the *molecular rotation.*

In 1845, after many unsuccessful attempts, Faraday discovered that the plane of polarization was also rotated when a beam of light propogates through a medium subjected to a strong magnetic field. Still later, Kerr discovered that very strong electric fields rotate the plane of polarization. These effects are called either *magneto-optical* or *electro-optical.* The magneto-optical effect discovered by Faraday took place when lead glass was subjected to a relatively strong magnetic field; this effect has since become known as the Faraday effect. It was through this discovery that a connection between electromagnetism and light was first made.

The Faraday effect occurs when an optical field propagates through a transparent medium along the direction of the magnetic field. This phenomenon is strongly reminiscent of the rotation that occurs in an optically active uniaxial crystal when the propagation is along its optical axis; we shall defer the discussion of propagation in crystals until Chapter 24.

The magnitude of the rotation angle θ for the Faraday effect is given by

$$\theta = VHl \tag{21-2}$$

where H is the magnetic intensity, l is the path length in the medium, and V is a constant called Verdet's constant, a "constant" that depends weakly on frequency and temperature. In (21-2) H can be replaced by B, the magnetic field strength. If B is in gauss, l in centimeters, and θ in minutes of arc ($'$), then Verdet's constant measured with yellow sodium light is typically about 10^{-5} for gases under standard conditions and about 10^{-2} for transparent liquids and solids. Verdet's constant becomes much larger for ferromagnetic solids or colloidal suspensions of ferromagnetic particles.

The theory of the Faraday effect can be easily worked out for a gas by using the Lorentz theory of the bound electron. This analysis is described very nicely in the text by Stone. However, our interest here is to derive the Mueller matrices that explicitly describe the rotation of the polarization ellipse for optically active liquids and the Faraday effect. Therefore, we derive the Mueller matrices using Maxwell's equations along with the necessary additions from Lorentz's theory.

In addition to the Faraday effect observed in the manner described above, namely, rotation of the polarization ellipse in a transparent medium, we can easily extend the analysis to Faraday rotation in a plasma (a mixture of charged particles).

There is an important difference between natural rotation and Faraday rotation (magneto-optical rotation), however. In the Faraday effect the medium is levorotatory for propagation in the direction of the magnetic field and dextrarotatory for propagation in the opposite direction. If at the end of the path l the light ray is reflected back along the same path, then the natural rotation is canceled while the magnetic rotation is doubled. The magnetic rotation effect is because, for the return path, as we shall see, not only are k_- and k_+ interchanged but i and $-i$ are also interchanged. The result is that the vector direction of a positive rotation is opposite to the direction of the magnetic field. Because of this, Faraday was able to multiply his very minute rotation effect by repeated back-and-forth reflections. In this way he was then able to observe his effect in spite of the relatively weak magnetic field that was used.

21.2 OPTICAL ACTIVITY

In optically active media there are no free charges or currents. Furthermore, the permeability of the medium is, for all practical purposes, unity, so $\mathbf{B} = \mathbf{H}$. Maxwell's equations then become

$$\nabla \times \mathbf{E} = -\frac{\partial \mathbf{H}}{\partial t} \tag{21-3a}$$

$$\nabla \times \mathbf{H} = \frac{\partial \mathbf{D}}{\partial t} \tag{21-3b}$$

$$\nabla \cdot \mathbf{D} = 0 \tag{21-3c}$$

$$\nabla \cdot \mathbf{B} = 0 \tag{21-3d}$$

Eliminating $\mathbf{H}$ between (21-3a) and (21-3b) leads to

$$\nabla \times (\nabla \times \mathbf{E}) = -\frac{\partial}{\partial t}\left(\frac{\partial \mathbf{D}}{\partial t}\right) \tag{21-4a}$$

or

$$\nabla(\nabla \cdot \mathbf{E}) - \nabla^2 \mathbf{E} = \omega^2 \mathbf{D} \tag{21-4b}$$

where we have assumed a sinusoidal time dependence for the fields.

In an optically active medium the relation between $\mathbf{D}$ and $\mathbf{E}$ is

$$\mathbf{D} = \varepsilon \mathbf{E} \tag{21-5}$$

where ε is a tensor whose form is

$$\varepsilon = \begin{pmatrix} \varepsilon_x & -i\alpha_z & i\alpha_y \\ i\alpha_z & \varepsilon_y & -i\alpha_x \\ -i\alpha_y & i\alpha_x & \varepsilon_z \end{pmatrix} \tag{21-6}$$

The parameters ε_x, ε_y, and ε_z correspond to real (on-axis) components of the refractive index and α_x, α_y, and α_z correspond to imaginary (off-axis) components of the refractive index. For isotropic media the diagonal elements are equal, so we have

$$\varepsilon_x = \varepsilon_y = \varepsilon_z = n^2 \tag{21-7}$$

where n is the refractive index. The vector quantity α can be expressed as

$$\alpha = \left(\frac{b}{\lambda}\right)\mathbf{s} \tag{21-8}$$

where b is a constant (actually a pseudoscalar) of the medium, λ is the wavelength, and $\mathbf{s}$ is a unit vector in the direction of propagation equal to $\mathbf{k}/k$. We thus can write (21-5) as

$$\mathbf{D} = n^2\mathbf{E} + \frac{i\beta}{k}(\mathbf{k} \times \mathbf{E}) \tag{21-9}$$

where $\beta = b/\lambda$.

Now from (21-3c) we see that

$$\nabla \cdot \mathbf{D} = i\mathbf{k} \cdot \mathbf{D} = 0 \tag{21-10}$$

Taking the scalar product of $\mathbf{k}$ with $\mathbf{D}$ in (21-9), we then see that

$$\mathbf{k} \cdot \mathbf{D} = n^2\mathbf{k} \cdot \mathbf{E} = 0 \tag{21-11}$$

Thus, the displacement vector and the electric vector are perpendicular to the propagation vector $\mathbf{k}$. This fact is quite important since the formation of the Stokes parameters requires that the direction of energy flow (along $\mathbf{k}$) and the direction of the fields be perpendicular.

With these results (21-4) now becomes (replacing $\mathbf{k}/k$ by $\mathbf{s}$)

$$\nabla^2\mathbf{E} = -\frac{\omega^2}{c^2}(n^2\mathbf{E} + i\beta\mathbf{s} \times \mathbf{E}) \tag{21-12}$$

From the symmetry of this equation we see that we can take the direction of propagation to be along any arbitrary axis. We assume that this is the z axis, so (21-12) then reduces to

$$\frac{\partial^2 E_x}{\partial z^2} = -\frac{\omega^2 n^2}{c^2}E_x + \frac{i\omega^2\beta}{c^2}E_y \tag{21-13a}$$

$$\frac{\partial^2 E_y}{\partial z^2} = -\frac{\omega^2 n^2}{c^2}E_y + \frac{i\omega^2\beta}{c^2}E_x \tag{21-13b}$$

The equation for E_z is trivial and need not be considered further.

We now assume that we have plane waves of the form:

$$E_x = E_{0x}e^{i\delta_x - ik_z z} \tag{21-14a}$$

$$E_y = E_{0y}e^{i\delta_y - ik_z z} \tag{21-14b}$$

and substitute (21-14) into (21-13), whereupon we find that

$$\left(k_z^2 - \frac{\omega^2 n^2}{c^2}\right)E_x + \frac{i\omega^2\beta}{c^2}E_y = 0 \tag{21-15a}$$

$$\frac{i\omega^2\beta}{c^2}E_x + \left(k_z^2 - \frac{\omega^2 n^2}{c^2}\right)E_y = 0 \tag{21-15b}$$

This pair of equations can have a nontrivial solution only if their determinant vanishes:

$$\begin{vmatrix} k_z^2 - \dfrac{\omega^2 n^2}{c^2} & \dfrac{i\omega^2\beta}{c^2} \\ \dfrac{i\omega^2\beta}{c^2} & k_z^2 - \dfrac{\omega^2 n^2}{c^2} \end{vmatrix} = 0 \tag{21-16}$$

so the solution of (21-16) is

$$k_z^2 = k_0^2(n^2 \pm i\beta) \tag{21-17}$$

where $k_0^2 = \omega^2/c^2$. Because we are interested in the propagation along the positive z axis, we take only the positive root of (21-17), so

$$k_z' = k_0(n^2 - i\beta)^{1/2} \tag{21-18a}$$

$$k_z'' = k_0(n^2 + i\beta)^{1/2} \tag{21-18b}$$

Substituting (21-18a) into (21-15a), we find that

$$E_y' = +E_x' \tag{21-19a}$$

while substitution of (21-18b) into (21-15a) yields

$$E_y'' = -E_x'' \tag{21-19b}$$

For the single primed wave field we can write

$$\mathbf{E}' = E_x'\mathbf{i} + E_y'\mathbf{j} = (E_{0x}'e^{i\delta_x'}\mathbf{i} + E_{0y}'e^{i\delta_y'}\mathbf{j})e^{-ik_zz'} \tag{21-20}$$

Now from (21-19a) we see that

$$E_{0x}' = E_{0y}' \tag{21-21a}$$

and

$$\delta_x' = \delta_y' + \frac{\pi}{2} \tag{21-21b}$$

Hence, we can write (21-20) as

$$\mathbf{E}' = (E_{0x}'e^{i\delta_x'}\mathbf{i} + iE_{0x}'e^{i\delta_x'}\mathbf{j})e^{-ik_zz'} \tag{21-22a}$$

In a similar manner the double-primed wave field is found to be

$$\mathbf{E}'' = (E_{0x}''e^{i\delta_x''}\mathbf{i} - iE_{0x}''e^{i\delta_x''}\mathbf{j})e^{-ik_zz''} \tag{21-22b}$$

To simplify notation let $E_{0x}' = E_{01}$, $\delta_x' = \delta_1$, $E_{0x}'' = E_{02}$, and $\delta_x'' = \delta_2$. Then, the fields are

$$\mathbf{E}_1 = (E_{01}e^{i\delta_1}\mathbf{i} + iE_{01}e^{i\delta_1}\mathbf{j})e^{ik_1z} \tag{21-23a}$$

$$\mathbf{E}_2 = (E_{02}e^{i\delta_2}\mathbf{i} - iE_{02}e^{i\delta_2}\mathbf{j})e^{ik_2z} \tag{21-23b}$$

where $k_1 = k_z'$ and $k_2 = k_z''$. We now add the x and y components of (21-23) and obtain

$$E_x = E_{01}e^{i(\delta_1+k_1z)} + E_{02}e^{i(\delta_2+k_2z)} \tag{21-24a}$$

$$E_y = +i(E_{01}e^{i(\delta_1+k_1z)} - E_{02}e^{i(\delta_2+k_2z)}) \tag{21-24b}$$

The Stokes parameters at any point z in the medium are defined to be

$$S_0(z) = E_x(z)E_x^*(z) + E_y(z)E_y^*(z) \tag{21-25a}$$

$$S_1(z) = E_x(z)E_x^*(z) - E_y(z)E_y^*(z) \tag{21-25b}$$

$$S_2(z) = E_x(z)E_y^*(z) + E_y(z)E_x^*(z) \tag{21-25c}$$

$$S_3(z) = i(E_x(z)E_y^*(z) - E_y(z)E_x^*(z)) \tag{21-25d}$$

Straightforward substitution of (21-24) into (21-25) leads to

$$S_0(z) = 2(E_{01}^2 + E_{02}^2) \tag{21-26a}$$
$$S_1(z) = 4E_{01}E_{02}\cos(\delta + kz) \tag{21-26b}$$
$$S_2(z) = 4E_{01}E_{02}\sin(\delta + kz) \tag{21-26c}$$
$$S_3(z) = 2(E_{01}^2 - E_{02}^2) \tag{21-26d}$$

where $\delta = \delta_2 - \delta_1$ and $k = k_2 - k_1$. We can find the incident Stokes parameters by considering the Stokes parameters at $z = 0$. We then find the parameters are

$$S_0(0) = 2(E_{01}^2 + E_{02}^2) \tag{21-27a}$$
$$S_1(0) = 4E_{01}E_{02}\cos\delta \tag{21-27b}$$
$$S_2(0) = 4E_{01}E_{02}\sin\delta \tag{21-27c}$$
$$S_3(0) = 2(E_{01}^2 - E_{02}^2) \tag{21-27d}$$

We now expand (21-26), using the familiar trignometric identities and find that

$$S_0(z) = 2(E_{01}^2 + E_{02}^2) \tag{21-28a}$$
$$S_1(z) = (4E_{01}E_{02}\cos\delta)\cos kz - (4E_{01}E_{02}\sin\delta)\sin kz \tag{21-28b}$$
$$S_2(z) = (4E_{01}E_{02}\sin\delta)\cos kz + (4E_{01}E_{02}\cos\delta)\sin kz \tag{21-28c}$$
$$S_3(z) = 2(E_{01}^2 - E_{02}^2) \tag{21-28d}$$

which can now be written in terms of the incident Stokes parameters, as given by (21-27), as

$$S_0(z) = S_0(0) \tag{21-29a}$$
$$S_1(z) = S_1(0)\cos kz - S_2(0)\sin kz \tag{21-29b}$$
$$S_2(z) = S_1(0)\sin kz + S_2(0)\cos kz \tag{21-29c}$$
$$S_3(z) = S_3(0) \tag{21-29d}$$

or, in matrix form,

$$\begin{pmatrix} S_0(z) \\ S_1(z) \\ S_2(z) \\ S_3(z) \end{pmatrix} = \begin{pmatrix} 1 & 0 & 0 & 0 \\ 0 & \cos kz & -\sin kz & 0 \\ 0 & \sin kz & \cos kz & 0 \\ 0 & 0 & 0 & 1 \end{pmatrix} \begin{pmatrix} S_0(0) \\ S_1(0) \\ S_2(0) \\ S_3(0) \end{pmatrix} \tag{21-30}$$

Thus, the optically active medium is characterized by a Mueller matrix whose form, corresponds to a rotator. The expression for k in (21-30) can be rewritten with the aid of (21-18) as

$$k = k_2 - k_1 = k_z'' - k_z' = k_0(n^2 - \beta)^{1/2} - k_0(n^2 + \beta)^{1/2} \tag{21-31}$$

Since $\beta \ll n^2$ (21-31) can be approximated as

$$k \simeq \frac{k_0\beta}{n} \tag{21-32}$$

The degree of polarization at any point in the medium is defined to be

$$P(z) = \frac{(S_1^2(z) + S_2^2(z) + S_3^2(z))^{1/2}}{S_0(z)} \tag{21-33}$$

On substituting (21-29) into (21-33) we find that

$$P(z) = \frac{(S_1^2(0) + S_2^2(0) + S_3^2(0))^{1/2}}{S_0(0)} = P(0) \tag{21-34}$$

that is, the degree of polarization does not change as the optical beam propogates through the medium.

The ellipticity of the optical beam is given by

$$\sin 2\chi(z) = \frac{S_3(z)}{(S_1^2(z) + S_2^2(z) + S_3^2(z))^{1/2}} \tag{21-35}$$

Substituting (21-29) into (21-35) then shows that the ellipticity is

$$\sin 2\chi(z) = \frac{S_3(0)}{(S_1^2(0) + S_2^2(0) + S_3^2(0))^{1/2}} = \sin 2\chi(0) \tag{21-36}$$

so the ellipticity is unaffected by the medium.

Finally, the orientation angle of the polarization ellipse is given by

$$\tan 2\psi(z) = \frac{S_2(z)}{S_1(z)} \tag{21-37a}$$

$$= \frac{S_1(0)\sin kz + S_2(0)\cos kz}{S_1(0)\cos kz - S_2(0)\sin kz} \tag{21-37b}$$

When the incident beam is linearly vertically or horizontally polarized, the respective Stokes vectors are

$$(1, -1, 0, 0) \qquad \text{and} \qquad (1, 1, 0, 0) \tag{21-38}$$

so $S_1(0) = \pm 1$, $S_2(0) = 0$, and (21-37b) reduces to

$$\tan 2\psi(z) = \pm \tan kz \tag{21-39a}$$

whence

$$\psi(z) = \pm \frac{1}{2} kz = \pm \left(\frac{k_0 \beta}{2n} \right) z = \pm \left(\frac{\pi \beta}{\lambda n} \right) z \tag{21-39b}$$

Thus, the orientation angle $\psi(z)$ is proportional to the distance traveled by the beam through the optically active medium and inversely proportional to wavelength, in agreement with the experimental observation. We can now simply equate (21-39b) with (21-1a) and relate the measured quantities of the medium to each other. As a result we see that Maxwell's equations completely account for the behavior of the optical activity.

Before, we conclude this section one question should still be answered. In section 21.1 we pointed out that for *natural rotation* the polarization of the beam is unaffected by the optically active medium when it is reflected back through the

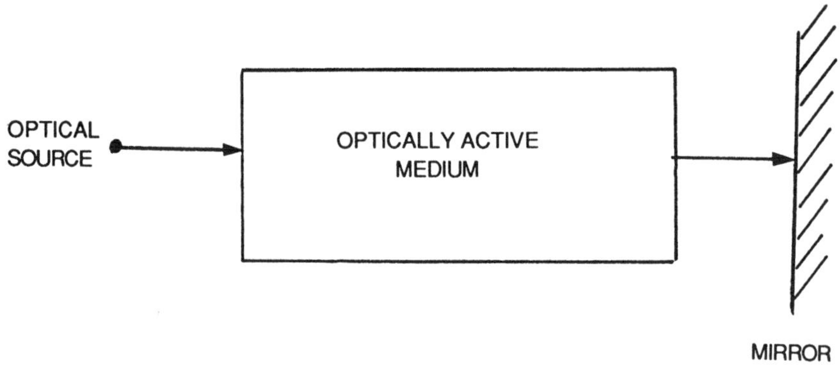

Figure 21-1 Reflection of a polarized beam propagating through an optically active medium.

medium. To study this problem, we consider Fig. 21-1. The Mueller matrix of the optically active medium is, from (21-30)

$$M(kz) = \begin{pmatrix} 1 & 0 & 0 & 0 \\ 0 & \cos kz & -\sin kz & 0 \\ 0 & \sin kz & \cos kz & 0 \\ 0 & 0 & 0 & 1 \end{pmatrix} \tag{21-40}$$

Now for a reflected beam we must replace z by $-z$ and k by $-k$. We thus obtain (21-40). From a physical point of view we must obtain the same Mueller matrix regardless of the direction of propagation of the beam. Otherwise, we would have a preferential direction! The Mueller matrix for a perfect reflector is

$$M_R = \begin{pmatrix} 1 & 0 & 0 & 0 \\ 0 & 1 & 0 & 0 \\ 0 & 0 & -1 & 0 \\ 0 & 0 & 0 & -1 \end{pmatrix} \tag{21-41}$$

Thus, from Fig. 21-1 the Mueller matrix for propagation through the medium, reflection, and propagation back through the medium, is

$$M = M(kz)M_R M(kz) \tag{21-42a}$$

$$= \begin{pmatrix} 1 & 0 & 0 & 0 \\ 0 & \cos kz & -\sin kz & 0 \\ 0 & \sin kz & \cos kz & 0 \\ 0 & 0 & 0 & 1 \end{pmatrix} \begin{pmatrix} 1 & 0 & 0 & 0 \\ 0 & 1 & 0 & 0 \\ 0 & 0 & -1 & 0 \\ 0 & 0 & 0 & -1 \end{pmatrix} \begin{pmatrix} 1 & 0 & 0 & 0 \\ 0 & \cos kz & -\sin kz & 0 \\ 0 & \sin kz & \cos kz & 0 \\ 0 & 0 & 0 & 1 \end{pmatrix} \tag{21-42b}$$

Carrying out the matrix multiplication in (21-42b), we obtain

$$M = \begin{pmatrix} 1 & 0 & 0 & 0 \\ 0 & 1 & 0 & 0 \\ 0 & 0 & -1 & 0 \\ 0 & 0 & 0 & -1 \end{pmatrix} \tag{21-43}$$

Thus, (21-43) shows that the forward and backward propagation, as well as polarization of the beam, are completely unaffected by the presence of the optically active medium.

21.3 FARADAY ROTATION IN A TRANSPARENT MEDIUM

Natural rotation of the plane of polarization was first observed in quartz by Arago in 1811. With the development of electromagnetism, physicists began to investigate the effects of the magnetic field on materials and, in particular, the possible relationship between electromagnetism and light. In 1845, Michael Faraday discovered that when a linearly polarized wave is propagating in a dielectric medium parallel to a static magnetic field the plane of polarization rotates. This phenomenon is known as the Faraday effect. The behavior is similar to that taking place in optically active media. However, there is an important difference. If, at the end of a path l the radiation is reflected backwards, then the rotation in optically active media is opposite to the original direction and cancels out; this was shown at the end of the previous section. For the magnetic case, however, the angle of rotation is doubled. This behavior along with some other important observations, will be shown at the end of this section.

In the present problem we take the direction of the magnetic field to be along the z axis. In addition, the plane waves are propagating along the z axis, and the directions of the electric (optical) vibrations are along the x and y axes. In such a medium (transparent, isotropic, and nonconducting) the displacement current vector is

$$\mathbf{D} = \varepsilon_0 \mathbf{E} + \mathbf{P} \tag{21-44}$$

where $\mathbf{P}$ is the polarization vector (this vector refers to the electric polarizibility of the material) and is related to the position vector $\mathbf{r}$ of the electron by

$$\mathbf{P} = -Ne\mathbf{r} \tag{21-45}$$

Maxwell's equation (21-3) then become

$$\nabla \times \mathbf{E} = -i\omega \mathbf{H} \tag{21-46a}$$

$$\nabla \times \mathbf{H} = i\omega(\varepsilon_0 \mathbf{E} + \mathbf{P}) \tag{21-46b}$$

Eliminating $\mathbf{H}$ between (21-46a) and (21-46b), we find that

$$\nabla^2 \mathbf{E} + \omega^2 \varepsilon_0 \mathbf{E} = -\omega^2 \mathbf{P} \tag{21-47}$$

or, in component form,

$$\nabla^2 \mathbf{E}_x + \omega^2 \varepsilon_0 \mathbf{E}_x = -\omega^2 \mathbf{P}_x \tag{21-48a}$$

$$\nabla^2 \mathbf{E}_y + \omega^2 \varepsilon_0 \mathbf{E}_y = -\omega^2 \mathbf{P}_y \tag{21-48b}$$

The position of the electron can readily be found from the Lorentz force equation to be

$$\zeta_\pm = \left(\frac{e}{m} E_\pm\right)\left(\omega^2 - \omega_0^2 \mp \frac{eH\omega}{m}\right) \tag{21-49a}$$

where

$$\zeta_{\pm} = x \pm iy \tag{21-49b}$$

$$E_{\pm} = E_x \pm iE_y \tag{21-49c}$$

The polarization vector is then expressed as

$$P_{\pm} = Ne\zeta_{\pm} = P_x \pm iP_y \tag{21-50}$$

Solving for P_x and P_y, we find that

$$P_x = AE_x + iBE_y \tag{21-51a}$$

$$P_y = AE_y + iBE_x \tag{21-51b}$$

where

$$A = \frac{Ne^2}{m}(\omega^2 - \omega_0^2)\left[(\omega^2 - \omega_0^2)^2 - \left(\frac{eH\omega}{m}\right)^2\right]^{-1} \tag{21-52a}$$

$$B = \frac{Ne^3 H\omega}{m}\left[(\omega^2 - \omega_0^2)^2 - \left(\frac{eH\omega}{m}\right)^2\right]^{-1} \tag{21-52b}$$

With P_x and P_y now known, (21-48a) and (21-48b) become

$$\frac{\partial^2 E_x}{\partial z^2} + \omega^2\varepsilon_0 E_x + (\omega^2 A)E_x + i(\omega^2 B)E_y = 0 \tag{21-53a}$$

$$\frac{\partial^2 E_y}{\partial z^2} + \omega^2\varepsilon_0 E_y + (\omega^2 A)E_y + i(\omega^2 B)E_x = 0 \tag{21-53b}$$

Since we are assuming that there is propagation only along the z axis, we can rewrite (21-53) as

$$(-k^2 + \omega^2\varepsilon_0 + \omega^2 A)E_x + i(\omega^2 B)E_y = 0 \tag{21-54a}$$

$$(-k^2 + \omega^2\varepsilon_0 + \omega^2 A)E_y + i(\omega^2 B)E_x = 0 \tag{21-54b}$$

If we now compare (21-53) with (21-13), we see that the forms of the equations are identical. Hence, we can proceed directly with the writing of the Mueller rotation matrix and the remaining relations. In addition, we find the wavenumber for the propagating waves to be

$$k^{\prime,\prime\prime} = \frac{\omega}{c}\left[1 - \frac{Ne^2/m}{(\omega^2 - \omega_0^2) \pm eH\omega/m}\right]^{1/2} \tag{21-55}$$

where the single and double primes correspond to the (+) and (−) solutions in (21-55), respectively. The orientation angle for linearly polarized radiation is then determined from (21-37) to be

$$\psi = \frac{1}{2}(k'' - k')z \tag{21-56}$$

Since the second term under the square root in (21-55) is small compared with unity, we easily find that

$$k'' - k' \simeq \frac{2Ne^3\omega^2}{m^2}\frac{H}{\omega^2 - \omega_0^2} \tag{21-57}$$

so the orientation angle of the radiation is

$$\psi \simeq \frac{Ne^3\omega^2 Hz}{m^2(\omega^2 - \omega_0^2)} = VHz \tag{21-58}$$

where Verdet's constant V is

$$V = \frac{Ne^3\omega^2}{m^2(\omega^2 - \omega_0^2)} \tag{21-59}$$

We thus see that the Mueller matrix for the Faraday effect is

$$M(z) = \begin{pmatrix} 1 & 0 & 0 & 0 \\ 0 & \cos VHz & -\sin VHz & 0 \\ 0 & \sin VHz & \cos VHz & 0 \\ 0 & 0 & 0 & 1 \end{pmatrix} \tag{21-60}$$

Thus, the rotation (21-58), is proportional to the path length, in agreement with the experimental observation.

Before concluding, let us again consider the problem where the beam propagates through the magneto-optical medium and is reflected back toward the optical source. For convenience, we replace VHz with θ and we write (21-60) as

$$M(z) = \begin{pmatrix} 1 & 0 & 0 & 0 \\ 0 & \cos\theta & -\sin\theta & 0 \\ 0 & \sin\theta & \cos\theta & 0 \\ 0 & 0 & 0 & 1 \end{pmatrix} \tag{21-61}$$

Now for a reflected beam we must replace z by $-z$. However, VH is unaffected. Unlike natural rotation, in the Faraday effect we have superposed an asymmetry in the problem with the unidirectional magnetic field. Thus, θ transforms to $-\theta$, and the Mueller matrix $M(z)$ for the beam propagating back to the source becomes

$$M(-z) = \begin{pmatrix} 1 & 0 & 0 & 0 \\ 0 & \cos\theta & \sin\theta & 0 \\ 0 & -\sin\theta & \cos\theta & 0 \\ 0 & 0 & 0 & 1 \end{pmatrix} \tag{21-62}$$

The Mueller matrix for a reflector (mirror) is

$$M_R = \begin{pmatrix} 1 & 0 & 0 & 0 \\ 0 & 1 & 0 & 0 \\ 0 & 0 & -1 & 0 \\ 0 & 0 & 0 & -1 \end{pmatrix} \tag{21-63}$$

Thus, the Mueller matrix for the propagation through the medium, reflection, and propagation back through the medium, is

$$M = M(-z)M_R M(z) \tag{21-64a}$$

$$= \begin{pmatrix} 1 & 0 & 0 & 0 \\ 0 & \cos\theta & \sin\theta & 0 \\ 0 & -\sin\theta & \cos\theta & 0 \\ 0 & 0 & 0 & 1 \end{pmatrix} \begin{pmatrix} 1 & 0 & 0 & 0 \\ 0 & 1 & 0 & 0 \\ 0 & 0 & -1 & 0 \\ 0 & 0 & 0 & -1 \end{pmatrix} \begin{pmatrix} 1 & 0 & 0 & 0 \\ 0 & \cos\theta & -\sin\theta & 0 \\ 0 & \sin\theta & \cos\theta & 0 \\ 0 & 0 & 0 & 1 \end{pmatrix} \tag{21-64b}$$

Carrying out the matrix multiplication in (21-64b), we find that

$$M = \begin{pmatrix} 1 & 0 & 0 & 0 \\ 0 & \cos 2\theta & -\sin 2\theta & 0 \\ 0 & -\sin 2\theta & -\cos 2\theta & 0 \\ 0 & 0 & 0 & -1 \end{pmatrix} \tag{21-65}$$

Since θ is arbitrary, we can replace θ by $-\theta$ in (21-65) without changing its meaning, and we then have

$$M = \begin{pmatrix} 1 & 0 & 0 & 0 \\ 0 & \cos 2\theta & \sin 2\theta & 0 \\ 0 & \sin 2\theta & -\cos 2\theta & 0 \\ 0 & 0 & 0 & -1 \end{pmatrix} \tag{21-66}$$

Equation (21-66) is recognized as the Mueller matrix for a *pseudorotator*. That is, the rotation as well as the ellipticity are opposite to the true behavior of a rotator. Thus, unlike natural rotation the angle of rotation is *doubled* upon reflection, so that for n reflections, 2θ in (21-66) is replaced by $2n\theta$ and a relatively large rotation angle can then be measured.

21.4 FARADAY ROTATION IN A PLASMA

While we have used Maxwell's equations to describe the propagation and polarization of light in optical media, the fact is that Maxwell's equations are universally applicable. In this section we briefly wish to show that the phenomenon of Faraday rotation appears when waves propagate in plasmas. Plasmas are gaseous matter consisting of charged particles. They appear not only in the laboratory but throughout the universe.

In a plasma the fields are again described by Maxwell's equations, which we write here as

$$\nabla \times \mathbf{E} = -i\omega\mathbf{H} \tag{21-67a}$$

$$\nabla \times \mathbf{H} = i\omega\varepsilon \cdot \mathbf{E} \tag{21-67b}$$

where ε is the plasma dielectric tensor. For a plasma having a static magnetic field along the z axis we have $\mathbf{H} = H\mathbf{k}$, and the tensor ε is then found to be (see the book by Bekefi)

$$\varepsilon = \begin{pmatrix} \varepsilon_{xx} & \varepsilon_{xy} & 0 \\ -\varepsilon_{xy} & \varepsilon_{xx} & 0 \\ 0 & 0 & \varepsilon_{zz} \end{pmatrix} \tag{21-68}$$

where

$$\varepsilon_{xx} = \varepsilon_{yy} = 1 - \frac{\omega_p^2}{\omega^2 - \omega_g^2} \tag{21-69a}$$

$$\varepsilon_{xy} = -\varepsilon_{yx} = \frac{-i\omega_g\omega_p^2}{\omega(\omega^2 - \omega_g^2)} \tag{21-69b}$$

$$\varepsilon_{zz} = 1 - \frac{\omega_p^2}{\omega^2} \tag{21-69c}$$

and

$$\omega_p^2 = \frac{e^2}{\varepsilon_0 m} = \text{plasma frequency} \tag{21-69d}$$

$$\omega_g = \frac{eH}{m} = \text{electron gyrofrequency} \tag{21-69e}$$

Eliminating $\mathbf{H}$ between (21-67a) and (21-67b) gives

$$\nabla(\nabla \cdot \mathbf{E}) - \nabla^2\mathbf{E} = \omega^2\varepsilon\mathbf{E} \tag{21-70}$$

We now consider the wave to be propagating along the z axis, i.e., in the direction of the static magnetic field. For this case it is not difficult to show that $\nabla \cdot \mathbf{E} = 0$. Equation (21-70) then reduces to

$$\frac{\partial^2 E_x}{\partial z^2} = -\omega^2[\varepsilon_{xx}E_x + \varepsilon_{xy}E_y] \tag{21-71a}$$

$$\frac{\partial^2 E_y}{\partial z^2} = -\omega^2[\varepsilon_{xx}E_y - \varepsilon_{xy}E_x] \tag{21-71b}$$

These equations are identical to (21-54), and again we obtain the same results as in that section. In the present problem we now find that the wavenumbers for individual waves are

$$k' = \frac{\omega}{c}(\varepsilon_{xx} + i\varepsilon_{xy})^{1/2} \tag{21-72a}$$

$$k'' = \frac{\omega}{c}(\varepsilon_{xx} + i\varepsilon_{xy})^{1/2} \tag{21-72b}$$

so

$$k' - k'' \simeq \frac{\omega_g\omega_p^2}{c(\omega^2 - \omega_g^2)} \tag{21-73}$$

and the angle of rotation is

$$\psi = \left(\frac{k' - k''}{2}\right) z = \frac{\omega_g \omega_p^2 z}{2c(\omega^2 - \omega_g^2)} \tag{21-74}$$

Thus, we see that for a plasma we obtain the rotation Mueller matrix:

$$M = \begin{pmatrix} 1 & 0 & 0 & 0 \\ 0 & \cos\psi & \sin\psi & 0 \\ 0 & -\sin\psi & \cos\psi & 0 \\ 0 & 0 & 0 & 1 \end{pmatrix} \tag{21-75}$$

The subject of optical activity and magneto-optical phenomena is vast. Many of the details of particular aspects as well as general treatments of the subject can be found in the references.

REFERENCES

Paper

1. Condon, E. U., *Rev. Mod. Phys.*, *9*, 432 (1937).

Books

1. Born, M., *Optik*, Springer Verlag, Berlin, 1933.
2. Sommerfeld, A., *Lectures on Theoritical Physics*, Vols I-V, Academic Press, New York, 1952.
3. Stone, J. M., *Radiation and Optics*, McGraw-Hill, New York, 1963.
4. Papas, C. H., *Theory of Electromagnetic Wave Propagation*, McGraw-Hill, New York, 1965.
5. Bekefi, G., *Radiation Processes in Plasmas*, Wiley, New York, 1966.
6. Wood, R. W., *Physical Optics*, 3rd ed., Optical Society of America, Washington, DC, 1988.
7. Jenkins, F. S., and White, H. E., *Fundamentals of Optics*, McGraw-Hill, New York, 1957.

22

The Stokes Parameters for Quantum Systems

22.1 INTRODUCTION

In previous chapters we saw that classical radiating systems could be represented in terms of the Stokes parameters and the Stokes vector. In addition, we saw that the representation of spectral lines in terms of the Stokes vector enabled us to arrive at a formulation of spectral lines which corresponds exactly to spectroscopic observations, namely, the frequency, intensity, and polarization. Specifically, when this formulation was applied to describing the motion of a bound electron moving in a constant magnetic field, there was a complete agreement between the Maxwell–Lorentz theory and Zeeman's experimental observations. Thus, by the end of the nineteenth century the combination of Maxwell's theory of radiation (Maxwell's equations) and the Lorentz theory of the electron appeared to be completely triumphant. The triumph was short-lived, however.

The simple fact was that while the electrodynamic theory explained the appearance of spectral lines in terms of frequency, intensity, and polarization there was still a very serious problem. Spectroscopic observations actually showed that even for the simplest element, ionized hydrogen gas, there was a *multiplicity* of spectral lines. Furthermore, as the elements increased in atomic number the number of spectral lines for each element greatly increased. For example, the spectrum of iron showed hundreds of lines whose intensities and frequencies appeared to be totally irregular. In spite of the best efforts of nineteenth-century theoreticians, no theory was ever devised within classical concepts, e.g., nonlinear oscillators, which could account for the number and position of the spectral lines.

Nevertheless, the fact that the Lorentz–Zeeman effect was completely explained by the electrodynamic theory clearly showed that in many ways the theory was on the right track. One must not forget that Lorentz's theory not only predicted the polarizations and the frequencies of the spectral lines, but even showed that the intensity of the central line in the "three line linear spectrum ($\theta = 90°$)"

would be twice as bright as the outer lines. It was this quasi-success that was so puzzling for such a long time.

Intense efforts were carried on for the first 25 years of the twentieth century on this problem of the multiplicity of spectral lines. The first real breakthrough was by Niels Bohr in a paper published in 1913. Using Planck's quantum ideas (1900) and the Rutherford model of an atom (1911) in which an electron rotated around a nucleus, Bohr was able to predict with great accuracy the spectrum of ionized hydrogen gas. A shortcoming of this model, however, was that even though the electron rotated in a circular orbit it did not appear to radiate, in violation of classical electrodynamics; we saw earlier that a charged particle moving in a circular orbit radiates. According to Bohr's model the "atomic system" radiated only when the electron dropped to a lower orbit; the phenomenon of absorption corresponded to the electron moving to a higher orbit. In spite of the difficulty with the Bohr model of hydrogen, it worked successfully. It was natural to try to treat the next element, the two-electron helium atom, in the same way. The attempt was unsuccessful.

Finally, in 1925, Werner Heisenberg published a new theory of the atom, which has since come to be known as quantum mechanics. This theory was a radical departure from classical physics. In this theory Heisenberg avoided all attempts to introduce those quantities that are not subject to experimental observation, e.g., the motion of an electron moving in an orbit. In its simplest form he constructed a theory in which only observables appeared. In the case of spectral lines this was, of course, the frequency, intensity, and polarization. This approach was considered even then to be extremely novel. By now, however, physicists had long forgotten that a similar approach had been taken nearly 75 years earlier by Stokes. The reader will recall that to describe unpolarized light Stokes had abandoned a model based on amplitudes (nonobservables) and succeeded by using an intensity formulation (observables). Heisenberg applied his new theory to determining the energy levels of the harmonic oscillator and was delighted when he arrived at the formula $E_n = \hbar\omega(n + 1/2)$. The significance of this result was that for the first time the factor of 1/2 arose directly out of the theory and not as a factor to be added to obtain the right result. Heisenberg noted at the end of his paper, however, that his formulation "might" be difficult to apply even to the "simplest" of problems such as the hydrogen atom because of the very formidable mathematical complexities.

At the same time that Heisenberg was working, an entirely different approach was being taken by another physicist, Erwin Schrödinger. Using an idea put forth in a thesis by Louis de Broglie, he developed a new equation to describe quantum systems. This new equation was a partial differential equation, which has since come to be known as Schrödinger's wave equation. On applying his equation to a number of outstanding problems, such as the harmonic oscillator, he also arrived at the same result for the energy as Heisenberg. Remarkably, Schrödinger's formulation of quantum mechanics was totally different from Heisenberg's. His formulation, unlike Heisenberg's, used the pictorial representation of electrons moving in orbits in a wavelike motion, an idea proposed by de Broglie.

The question then arose, how could two seemingly different theories arrive at the same results? The answer was provided by Schrödinger. He discovered that Heisenberg's quantum mechanics, which was now being called quantum matrix mechanics, and his wave mechanics were mathematically identical. In a very remarkable result Schrödinger showed that Heisenberg's matrix elements could be obtained by simply integrating the absolute magnitude squared of his wave equation solution multiplied by the variable over the volume of space. This result is extremely important for our present problem because it provides the mechanism for calculating the variables $\ddot{x}$, $\ddot{y}$ and $\ddot{z}$ in our radiation equation.

We saw that the radiation equations for E_θ and E_ϕ were proportional to the acceleration components $\ddot{x}$ $\ddot{y}$, and $\ddot{z}$. To obtain the corresponding equations for quantum mechanical radiating systems, we must calculate these quantities using the rules of quantum mechanics. In Section 22.4 we transform the radiation equations so that they also describe the radiation emitted by quantum systems. In Section 22.5 we determine the Stokes vectors for several quantized systems. We therefore see that we can describe both classical and quantum radiating systems by using the Stokes vector.

Before we carry this out, however, we describe some relationships between classical and quantum radiation fields.

22.2 RELATION BETWEEN STOKES POLARIZATION PARAMETERS AND QUANTUM MECHANICAL DENSITY MATRIX

In quantum mechanics the treatment of partially polarized light and the polarization of the radiation emitted by quantum mechanical systems appears to be very different from the classical methods. In classical optics the radiation field is described in terms of the polarization ellipse and amplitudes. On the other hand, in quantum optics the radiation field is described in terms of density matrices. Furthermore, the polarization of the radiation emitted by quantum systems is described in terms of intensities and selection rules rather than the familiar amplitude and phase relations of the optical field. Let us examine the descriptions of polarization in classical and quantum mechanical terms. We start with a historical review and then present the mathematics for the quantum mechanical treatment.

It is a remarkable fact that after the appearance of Stokes' paper (1852) and his introduction of his parameters, they were practically forgotten for nearly a century! It appears that only in France was the significance of his work fully appreciated. After the publication of Stokes' paper, E. Verdet expounded upon them (1862). It appears that the Stokes parameters were thereafter known to French students of opitcs, e.g., Henri Poincaré (ca. 1890) and Paul Soleillet (1927). The Stokes parameters did not reappear in any publication in the English-speaking world until 1942, in a paper by Francis Perrin. (Perrin was the son of the Nobel laureate Jean Perrin. Both father and son fled to the United States after the fall of France in June 1940. Jean Perrin was a scientist of international standing, and he also appears to have been a very active voice against fascism in prewar France. Had both father and son remained in France, they would have very probably been killed during the occupation.)

Perrin's 1942 paper is very important because he (1) reintroduced the Stokes parameters to the English-speaking world, (2) presented the relation between the Stokes parameters for a beam that underwent rotation or was phase shifted, (3) showed the connection between the Stokes parameters and the wave statistics of John von Neumann, and (4) derived conditions on the Mueller matrix elements for scattering (the Mueller matrix had not been named at that date). Perrin also stated that Soleillet (1927) had pointed out that only a linear relation could exist between the Stokes parameter for an incident beam (S_i) and the transmitted (or scattered) beam (S_i'). According to Perrin the argument for a linear relation was a direct consequence of the superposition of the Stokes parameters for n independent beams; only a linear relation would satisfy this requirement. This is discussed further in this section. The impact of his paper did not appear for several years, because of its publication during the Second World War. As a result, even by 1945 the Stokes parameters were still not generally known.

The question of the relation between the classical and quantum representation of the radiation field only appears to have arisen after the "rediscovery" of Stokes' 1852 paper and the Stokes parameters by the Nobel laureate Subrahmanyan Chandrasekhar in 1947, while writing his fundamental papers on radiative transfer. Chandrasekhar's astrophysical research was well known, and consequently, his papers were immediately read by the scientific community.

Shortly after the appearance of Chandrasekhar's radiative transfer papers, U. Fano (1949) showed that the Stokes parameters are a very suitable analytical tool for treating problems of polarization in both classical optics and quantum mechanics. He appears to have been the first to give a quantum mechanical description of the electromagnetic field in terms of the Stokes parameters; he also used the formalism of the Stokes parameters to determine the Mueller matrix for Compton scattering. Fano also noted that the reason for the successful application of the Stokes parameters to the quantum theoretical treatment of electromagnetic radiation problems is that they are the observable quantities of phenomenological optics.

The appearance of the Stokes parameters of classical optics in quantum physics appears to have come as a surprise at the time. The reason for their appearance was pointed out by Falkoff and MacDonald (1951) shortly after the publication of Fano's paper. In classical and quantum optics the representations of completely (i.e., elliptically) polarized light are identical (this was also first pointed out by Perrin) and can be written as

$$\psi = c_1\psi_1 + c_2\psi_2 \tag{22-1}$$

However, the classical and quantum interpretations of this equation are quite different. In classical optics ψ_1 and ψ_2 represent perpendicular unit vectors, and the resultant polarization vector ψ for a beam is characterized by the complex amplitudes c_1 and c_2. The absolute magnitude squared of these coefficients then yields the intensities $|c_1|^2$ and $|c_2|^2$ that one would measure through an analyzer in the direction of ψ_1 and ψ_2. In the quantum interpretation ψ_1 and ψ_2 represent orthogonal polarization states for a *photon*, but now $|c_1|^2$ and $|c_2|^2$ yield the relative probabilities for a single photon to pass through an analyzer which admits only quanta in the states ψ_1 and ψ_2, respectively.

In both interpretations the polarization of the beam (photon) is completely determined by the complex amplitudes c_1 and c_2. In terms of these quantities one can define a 2×2 matrix with elements:

$$\rho_{ij} = c_i c_j^* \qquad i, j = 1, 2 \tag{22-2}$$

In quantum mechanics an arbitrary wave equation can be expanded into any desired complete set of orthonormal eigenfunctions; that is,

$$\psi = \sum_i c_i \psi_i \tag{22-3}$$

Then

$$|\psi|^2 = \psi\psi^* = \sum_{ij} c_i c_j^* \psi_i \psi_j^* \tag{22-4}$$

From the expansion coefficients we can form a matrix ρ by the rule:

$$\rho_{ij} = c_i c_j^* \qquad i, j = 1, 2 \tag{22-5}$$

According to (22-1), we can then express (22-5) in a 2×2 matrix:

$$\rho = \begin{pmatrix} \rho_{11} & \rho_{12} \\ \rho_{21} & \rho_{22} \end{pmatrix} \tag{22-6}$$

The matrix ρ is known as the density matrix and has a number of interesting properties; it is usually associated with von Neumann (1927). First, we note that $\rho_{ii} = c_i c_i^*$ gives the probability of finding the system in the state characterized by the eigenfunction ψ_i. If we consider the ψ function as being normalized, then

$$\int \psi\psi^* d\tau = \sum_{ij} c_i c_j^* \int \psi_i \psi_j \, d\tau = \sum_i c_i c_i^* = \rho_{11} + \rho_{22} = 1 \tag{22-7}$$

Thus, the sum of the diagonal matrix elements is 1. The process of summing these elements is known as taking the *trace* of the matrix and is written as $\mathrm{Tr}(\cdots)$, so we have

$$\mathrm{Tr}(\rho) = 1 \tag{22-8}$$

If we measure some variable F in the system described by ψ, the result is given by

$$\begin{aligned} \langle F \rangle &= \int \psi F \psi^* \, d\tau = \sum_{ij} \int c_i \psi_i F c_j^* \psi_j^* \, d\tau \\ &= \sum_{ij} c_i c_j^* F_{ij} \end{aligned} \tag{22-9a}$$

where the matrix F_{ij} is defined by the formula:

$$F_{ij} = \int \psi_i F \psi_j^* \, d\tau \tag{22-9b}$$

However,

$$\sum_i F_{ij} \rho_{ij} = (F\rho)_{ii} \tag{22-10}$$

Therefore,

$$\langle F \rangle = \sum_i (F\rho)_{ii} \tag{22-11a}$$

or

$$\langle F \rangle = \mathrm{Tr}(F\rho) \tag{22-11b}$$

Thus, the expectation value of F, $\langle F \rangle$, is determined by taking the trace of the matrix product of F and ρ.

In classical statistical mechanics the density function $\rho(\mathbf{p}, \mathbf{q})$ in phase space, where $\mathbf{p}$ and $\mathbf{q}$ are the momentum and the position, respectively, is normalized by the condition:

$$\int \rho(\mathbf{p}, \mathbf{q}) \, d\mathbf{p} \, d\mathbf{q} = 1 \tag{22-12a}$$

and the average value of a variable is given by

$$\langle F \rangle = \int F\rho(\mathbf{p}, \mathbf{q}) \, d\mathbf{p} \, d\mathbf{q} \tag{22-12b}$$

We see immediately that a similar role is played by the density matrix in quantum mechanics by comparing (22-7) and (22-11b) with (22-12a) and (22-12b).

The polarization of electromagnetic radiation can be described by the vibration of the electric vector. For a complete description the field may be represented by two independent beams of orthogonal polarizations. That is, the electric vector can be represented by

$$\mathbf{E} = c_1\mathbf{e}_1 + c_2\mathbf{e}_2 \tag{22-13}$$

where $\mathbf{e}_1$ and $\mathbf{e}_2$ are two orthogonal unit vectors and c_1 and c_2, which are in general complex, describe the amplitude and phase of the two vibrations. From the two expansion coefficients in (22-13) we can form a 2×2 density matrix. Furthermore, from the viewpoint of quantum mechanics the equation analogous to (22-13) is given by (22-1), which is rewritten here:

$$\psi = c_1\psi_1 + c_2\psi_2 \tag{22-1}$$

We now consider the representation of an optical beam in terms of its density matrix. An optical beam can be represented by

$$\mathbf{E} = E_1\mathbf{e}_1 + E_2\mathbf{e}_2 \tag{22-14a}$$

where

$$E_1 = a_1 \cos(\omega t + \delta_1) \tag{22-14b}$$

$$E_2 = a_2 \cos(\omega t + \delta_2) \tag{22-14c}$$

In complex notation, (22-14) is written as

$$E_1 = a_1 \exp i(\omega t + \delta_1) \tag{22-15a}$$

$$E_2 = a_2 \exp i(\omega t + \delta_2) \tag{22-15b}$$

We now write

$$a_1 = \cos\theta \tag{22-16a}$$

$$a_2 = \sin\theta \tag{22-16b}$$

$$\delta = \delta_2 - \delta_1 \tag{22-16c}$$

Equation (22-14) can then be expressed as

$$\mathbf{E} = \cos\theta e^{-i\delta}\mathbf{e}_1 + \sin\theta\mathbf{e}_2 \tag{22-17}$$

so we have

$$c_1 = \cos\theta e^{-i\delta} \tag{22-18a}$$

$$c_2 = \sin\theta \tag{22-18b}$$

The density matrix is now explicitly written out as

$$\rho = \begin{pmatrix} \rho_{11} & \rho_{12} \\ \rho_{21} & \rho_{22} \end{pmatrix} = \begin{pmatrix} c_1c_1^* & c_1c_2^* \\ c_2c_1^* & c_2c_2^* \end{pmatrix} = \begin{pmatrix} \cos^2\theta & \cos\theta\sin e^{-i\delta} \\ \sin\theta\cos\theta e^{i\delta} & \sin^2\theta \end{pmatrix} \tag{22-19}$$

Complete polarization can be described by writing (22-1) in terms of a single eigenfunction for each of the two orthogonal states. Thus, we write

$$\psi = c_1\psi_1 \tag{22-20a}$$

or

$$\psi = c_2\psi_2 \tag{22-20b}$$

where ψ_i refers to a state of pure polarization. The corresponding density matrices are then, respectively,

$$\rho_1 = \begin{pmatrix} c_1c_1^* & 0 \\ 0 & 0 \end{pmatrix} = \begin{pmatrix} 1 & 0 \\ 0 & 0 \end{pmatrix} \tag{22-21a}$$

and

$$\rho_2 = \begin{pmatrix} 0 & 0 \\ 0 & c_2c_2^* \end{pmatrix} = \begin{pmatrix} 0 & 0 \\ 0 & 1 \end{pmatrix} \tag{22-21b}$$

where we have set $c_1c_1^*$ and $c_2c_2^*$ equal to 1 to represent a beam of unit intensity.

We can use (22-21a) and (22-21b) to obtain the density matrix for unpolarized light. Since an unpolarized beam may be considered to be the incoherent superposition of two polarized beams with equal intensity, if we add (22-21a) and (22-21b) the density matrix is

$$\rho_U = \frac{1}{2}\begin{pmatrix} 1 & 0 \\ 0 & 1 \end{pmatrix} \tag{22-22}$$

The factor 1/2 has been introduced because the normalization condition requires that the trace of the density matrix be unity. Equation (22-22) can also be obtained from (22-19) by averaging the angles θ and δ over π and 2π, respectively.

In general, a beam will have an arbitrary degree of polarization, and we can characterize such a beam by the incoherent superposition of an unpolarized beam and a totally polarized beam. From (22-19) the polarized contribution is described by

$$\rho_P = \begin{pmatrix} c_1c_1^* & c_1c_2^* \\ c_2c_1^* & c_2c_2^* \end{pmatrix} \tag{22-23}$$

The density matrix for a beam with arbitrary polarization can then be written in the form:

$$\rho = U\begin{pmatrix} 1 & 0 \\ 0 & 1 \end{pmatrix} + P = \begin{pmatrix} c_1c_1^* & c_1c_2^* \\ c_2c_1^* & c_2c_2^* \end{pmatrix} \tag{22-24}$$

where U and P are the factors to be determined. In particular, P is the degree of polarization; it is a real quantity and its range is $0 \leq P \leq 1$. We now note the following three cases:

1. If $0 < P < 1$, then the beam is partially polarized.
2. If $P = 0$, then the beam is unpolarized.
3. If $P = 1$, then the beam is totally polarized.

For $P = 0$, we know that

$$\rho_U = \frac{1}{2}\begin{pmatrix} 1 & 0 \\ 0 & 1 \end{pmatrix} \tag{22-22}$$

Thus, $U = 1/2$ and $P = 0$. For $P = 1$, the density matrix is given by (22-23), so $U = 0$ when $P = 1$. We can now easily determine the explicit relation between U and P by writing

$$U = aP + b \tag{22-25}$$

From the condition on U and P just given we find that $b = 1/2$ and $a = -b$ so the explicit form of (22-25) is

$$U = -\frac{1}{2}P + \frac{1}{2} \tag{22-26}$$

Thus (22-24) becomes

$$\rho = \frac{1}{2}(1 - P)\begin{pmatrix} 1 & 0 \\ 0 & 1 \end{pmatrix} + P\begin{pmatrix} c_1c_1^* & c_1c_2^* \\ c_2c_1^* & c_2c_2^* \end{pmatrix} \tag{22-27}$$

Equation (22-27) is the density matrix for a beam of arbitrary polarization.

By the proper choice of pure states of polarization ψ_i, the part of the density matrix representing total polarization can be written in one of the forms given by (22-20). Therefore, we may write the general density matrix as

$$\rho = \frac{1}{2}(1 - P)\begin{pmatrix} 1 & 0 \\ 0 & 1 \end{pmatrix} + P\begin{pmatrix} 1 & 0 \\ 0 & 0 \end{pmatrix} \tag{22-28}$$

or

$$\rho = \frac{1}{2}\begin{pmatrix} 1 + P & 0 \\ 0 & 1 - P \end{pmatrix} \tag{22-29}$$

Hence, any intensity measurement made in relation to these pure states will yield the eigenvalues:

$$I_+ = \frac{1}{2}(1 + P) \tag{22-30a}$$

$$I_- = \frac{1}{2}(1 - P) \tag{22-30b}$$

Classical optics requires that to determine experimentally the state of polarization of an optical beam four measurements must be made. The optical field in classical optics is described by

$$\mathbf{E} = E_1\mathbf{e}_1 + E_2\mathbf{e}_2 \tag{22-14a}$$

where

$$E_1 = a_1 \exp i(\omega t + \delta_1) \tag{22-15a}$$

$$E_2 = a_2 \exp i(\omega t + \delta_2) \tag{22-15b}$$

In quantum optics the optical field is described by

$$\psi = c_1\psi_1 + c_2\psi_2 \tag{22-1}$$

Comparing c_1 an c_2 in (22-1) with E_1 and E_2 in (22-15) suggests that we set

$$c_1 = a_1 \exp i(\omega t + \delta_1) \tag{22-31a}$$

$$c_2 = a_2 \exp i(\omega t + \delta_2) \tag{22-31b}$$

We now define the Stokes polarization parameters for a beam to be

$$S_0 = c_1c_1^* + c_2c_2^* \tag{22-32a}$$

$$S_1 = c_1c_1^* - c_2c_2^* \tag{22-32b}$$

$$S_2 = c_1c_2^* + c_2c_1^* \tag{22-32c}$$

$$S_3 = i(c_1c_2^* - c_2c_1^*) \tag{22-32d}$$

We now substitute (22-31) into (22-32) and find that

$$S_0 = a_1^2 + a_2^2 \tag{22-33a}$$

$$S_1 = a_1^2 - a_2^2 \tag{22-33b}$$

$$S_2 = 2a_1a_2 \cos\delta \tag{22-33c}$$

$$S_3 = 2a_1a_2 \sin\delta \tag{22-33d}$$

We see that (22-33) are exactly the classical Stokes parameters (with a_1 and a_2 replacing, e.g., E_{0x} and E_{0y} as previously used in this text). Expressing (22-32) in terms of the density matrix elements, $\rho_{11} = c_1c_1^*$ etc., the Stokes parameters are linearly related to the density matrix elements by

$$S_0 = \rho_{11} + \rho_{22} \tag{22-34a}$$

$$S_1 = \rho_{11} - \rho_{22} \tag{22-34b}$$

$$S_2 = \rho_{12} + \rho_{21} \tag{22-34c}$$

$$S_3 = i(\rho_{12} - \rho_{21}) \tag{22-34d}$$

Thus, the Stokes parameters are linear combinations of the elements of the 2 × 2 density matrix.

It will be convenient to express (22-34a) by the symbol I for the intensity and the remaining parameters of the beam by P_1, P_2 and P_3, so

$$I = \rho_{11} + \rho_{22} \tag{22-35a}$$

$$P_1 = \rho_{11} - \rho_{22} \tag{22-35b}$$

$$P_2 = \rho_{12} + \rho_{21} \tag{22-35c}$$

$$P_3 = i(\rho_{12} - \rho_{21}) \tag{22-35d}$$

In terms of the density matrix (22-19) we can then write

$$\rho = \begin{pmatrix} \rho_{11} & \rho_{12} \\ \rho_{21} & \rho_{22} \end{pmatrix} = \frac{1}{2}\begin{pmatrix} 1 + P_1 & P_2 - iP_3 \\ P_2 + iP_3 & 1 - P_1 \end{pmatrix} \tag{22-36}$$

where we have set $I = 1$. From the point of view of measurement both the classical and quantum theories yield the same results. However, the interpretations, as pointed out above, are completely different.

We also recall that the Stokes parameters satisfy the condition:

$$I^2 \geq P_1^2 + P_2^2 + P_3^2 \tag{22-37}$$

Substituting (22-35) into (22-37), we find that

$$\det(\rho) = \rho_{11}\rho_{22} - \rho_{12}\rho_{21} \geq 0 \tag{22-38}$$

where "det" stands for the determinant. Similarly, the degree of polarization P is given by

$$P = \frac{\sqrt{(\rho_{11} - \rho_{22})^2 + 4\rho_{12}\rho_{21}}}{\rho_{11} + \rho_{22}} \tag{22-39}$$

There is one further point that we wish to make. The wave function ψ can be expanded in a complete set of orthonormal eigenfunctions. For electromagnetic radiation (optical field) this consists only of the terms:

$$\psi = c_1\psi_1 + c_2\psi_2 \tag{22-1}$$

The wave functions describing pure states may be chosen in the form:

$$\psi_1 = \begin{pmatrix} 1 \\ 0 \end{pmatrix} \quad \text{and} \quad \psi_2 = \begin{pmatrix} 0 \\ 1 \end{pmatrix} \tag{22-40}$$

Substituting (22-40) into (22-1), we have

$$\psi = \begin{pmatrix} c_1 \\ c_2 \end{pmatrix} \tag{22-41}$$

Using this wave function leads to the following expressions for the *expectation values* (see 22-9a) of the unit matrix and the Pauli spin matrices:

$$I = \langle 1 \rangle = (c_1^* \; c_2^*) \begin{pmatrix} 1 & 0 \\ 0 & 1 \end{pmatrix} \begin{pmatrix} c_1 \\ c_2 \end{pmatrix} = c_1 c_1^* + c_2 c_2^* \tag{22-42a}$$

$$P_1 = \langle \sigma_z \rangle = (c_1^* \; c_2^*) \begin{pmatrix} 1 & 0 \\ 0 & -1 \end{pmatrix} \begin{pmatrix} c_1 \\ c_2 \end{pmatrix} = c_1 c_1^* - c_2 c_2^* \tag{22-42b}$$

$$P_2 = \langle \sigma_x \rangle = (c_1^* \; c_2^*) \begin{pmatrix} 0 & 1 \\ 1 & 0 \end{pmatrix} \begin{pmatrix} c_1 \\ c_2 \end{pmatrix} = c_1 c_2^* + c_1 c_2^* \tag{22-42c}$$

$$P_3 = \langle \sigma_y \rangle (c_1^* c_2^*) \begin{pmatrix} 0 & -i \\ i & 0 \end{pmatrix} \begin{pmatrix} c_1 \\ c_2 \end{pmatrix} = i(c_1 c_2^* - c_2 c_1^*) \tag{22-42d}$$

We see that the terms on the right hand side of (22-42) are exactly the Stokes polarization parameters. The Pauli spin matrices are usually associated with particles of spin 1/2, e.g., the electron. However, for both the electromagnetic radiation field and for particles of spin 1/2 the wave function can be expanded in a complete set of orthonormal eigenfunctions consisting of only two terms (22-1). Thus, the quantum mechanical expectation values correspond exactly to observables.

Further information on the quantum mechanical density matrices and the application of the Stokes parameters to quantum problems, e.g., Compton scattering, can be found in the numerous papers cited in the references.

22.3 NOTE ON PERRIN'S INTRODUCTION OF STOKES PARAMETERS, DENSITY MATRIX, AND LINEARITY OF THE MUELLER MATRIX ELEMENTS

It is worthwhile to discuss Perrin's observations further. It is rather remarkable that he discussed the Stokes polarization parameters and their relationship to the Poincaré sphere without any introduction or background. While they appear to have been known by French optical physicists, the only English-speaking references to them are in the papers of Lord Rayleigh and a textbook by Walker. Walker's textbook is remarkably well written, but does not appear to have had a wide circulation. It was in this book, incidentally, that Chandrasekhar found the Stokes polarization parameters and recognized that they could be used to incorporate the phenomenon of polarization in the (intensity) radiative transfer equations.

As is often the case, because Perrin's paper was one of the first papers on the Stokes parameters, his presentation serves as a very good introduction to the subject. Furthermore, he briefly described their relation to the quantum mechanical density matrix.

For completely polarized monochromatic light the optical vibrations may be represented along the two rectangular axes as

$$E_1 = a_1 \cos(\omega t + \delta_1) \tag{22-14b}$$

$$E_2 = a_2 \cos(\omega t + \delta_2) \tag{22-14c}$$

where a_1 and a_2 are the maximum amplitudes and δ_1 and δ_2 are the phases. The phase difference between these components is

$$\delta = \delta_2 - \delta_1 \tag{22-16c}$$

and the total intensity of the vibration is

$$I = a_1^2 + a_2^2 \tag{22-43}$$

In nature, light is not strictly monochromatic. Furthermore, as we have seen, because of the rapid vibrations of the optical field only mean values can be measured. To analyze polarized light, we must use analyzers, that is, polarizers (with transmission factors k_1 and k_2 along the axes) and phase shifters (with phase shifts of η_1 and η_2 along the fast and slow axes respectively). These analyzers then yield the mean intensity of a vibration E_a obtained as a linear combination, with given changes in phase, of the two components E_1 and E_2 of the initial vibration as

$$E_a = k_1 a_1 \cos(\omega t + \delta_1 + \eta_1) + k_2 a_2 \cos(\omega t + \delta_2 + \eta_2) \tag{22-44}$$

We note that this form is identical to the quantum mechanical form given by (22-1). The mean intensity of (22-44) is then

$$\begin{aligned} I_a = \frac{1}{2}\Big[&(k_1^2 + k_2^2)(\langle a_1^2\rangle + \langle a_2^2\rangle). + (k_1^2 - k_2^2)(\langle a_1^2\rangle - \langle a_2^2\rangle) \\ &+ 2k_1k_2\cos(\eta_1 - \eta_2)(\langle 2a_1a_2\cos\delta\rangle) \\ &+ 2k_1k_2\sin(\eta_1 - \eta_2)(\langle 2a_1a_2\sin\delta\rangle)\Big] \end{aligned} \tag{22-45}$$

We can write the terms within parentheses as

$$S_0 = \langle a_1^2\rangle + \langle a_2^2\rangle \tag{22-46a}$$

$$S_1 = \langle a_1^2\rangle - \langle a_2^2\rangle \tag{22-46b}$$

$$S_2 = \langle 2a_1a_2\cos\delta\rangle \tag{22-46c}$$

$$S_3 = \langle 2a_1a_2\sin\delta\rangle \tag{22-46d}$$

where $\langle\cdots\rangle$ refers to the mean or average value, and S_0, S_1, S_2 and S_3 are the four Stokes parameters of the optical beam. Equation (22-45) can then be rewritten as

$$\begin{aligned} I_a = \frac{1}{2}\Big[&(k_1^2 + k_2^2)S_0 + (k_1^2 - k_2^2)S_1 + 2k_1k_2\cos(\eta_1 - \eta_2)S_2. \\ &+ 2k_1k_2\sin(\eta_1 - \eta_2)S_3\Big] \end{aligned} \tag{22-47}$$

As we have seen, by choosing different combinations of a_1 and a_2 and η_1 and η_2 we can determine S_0, S_1, S_2, and S_3. Equation (22-47) is essentially the equation first derived by Stokes.

The method used by Stokes to characterize a state of polarization may be generalized and connected with the wave statistics of von Neumann. Consider a system of n harmonic oscillations of the same frequency subjected to small random perturbations. This may be represented by the complex expression:

$$E_k = P_k \exp(i\omega t) \tag{22-48a}$$

where

$$P_k = p_k \exp(i\phi_k) \tag{22-48b}$$

and the modulus p_k and the argument ϕ_k vary slowly over time in comparison with the period of oscillation but quickly with respect to the period of measurement. Suppose we can measure the mean intensity of an oscillation E linearly dependent on these oscillations:

$$E = \sum_k C_k E_k = \sum_k C_k P_k \exp(i\omega t) \tag{22-49a}$$

where

$$C_k = c_k \exp(i\eta_k) \tag{22-49b}$$

The mean intensity corresponding to (22-49a) is then

$$\langle EE^* \rangle = \sum_{kl} C_k C_l \langle P_k P_l^* \rangle \tag{22-50}$$

The mean intensity depends on the particular oscillations involving only the von Neumann matrix elements (the density matrix):

$$\rho_{kl} = \langle P_k P_l^* \rangle \tag{22-51}$$

The knowledge of these matrix elements determines all that we can know about the oscillations by such measurements. Since this matrix is Hermitian, we can set

$$\rho_{kk} = \mu_k \qquad \rho_{kl} = \gamma_{kl} + i\sigma_{kl} \qquad (k \neq 1) \tag{22-52}$$

where μ_k, $\gamma_{kl} = \gamma_{lk}$, and $\sigma_{kl} = -\sigma_{kl}$ are real quantities. The diagonal terms μ_k are the mean intensities of the oscillations.

$$\mu_k = \langle p_k^2 \rangle \tag{22-53a}$$

and the other terms give the correlations between the oscillations:

$$\gamma_{kl} = \langle p_k p_l \cos(\phi_k - \phi_l) \rangle \tag{22-53b}$$

$$\sigma_{kl} = \langle p_k p_l \sin(\phi_k - \phi_l) \rangle \tag{22-53c}$$

While Perrin did not explicitly show the relation of the Stokes parameters to the density matrix, it is clear, as we have shown, that only an additional step is required to do this.

Perrin made additional observations on the correlation functions for nonharmonic systems. Before we conclude, however, there is one additional remark that we wish to investigate. Perrin noted that Soleillet first pointed out that, when a beam of light passes through some optical arrangement, or, more generally, produces a secondary beam of light, the intensity and the state of polarization of the emergent beam are functions of those of the incident beam. If two independent incident beams are superposed, the new emergent beam will be, if the process is linear, the superposition without interference of the two emergent beams corresponding to the separate incident beams. Consequently, in such a linear process, from the additivity properties of the Stokes parameters, the parameters S_0', S_1', S_2', S_3' which define the polarization of the emergent beam, must be *homogenous linear functions* of the parameters S_0, S_1, S_2, S_3 corresponding to the incident beam; the 16 coefficients of these linear functions will completely characterize the corresponding optical phenomenon.

Perrin offers this statement without proof. We can easily show that from Stokes' law of additivity of independent beams that the relationship between S_0' and S_0 etc., must be linear.

Let us assume a functional relation between S_0', S_1' etc., such that

$$S_0' = f(S_0, S_1, S_2, S_3) \tag{22-54a}$$

$$S_1' = f(S_0, S_1, S_2, S_3) \tag{22-54b}$$

$$S_2' = f(S_0, S_1, S_2, S_3) \tag{22-54c}$$

$$S_3' = f(S_0, S_1, S_2, S_3) \tag{22-54d}$$

To determine the explicit form of this functional relationship, consider only $I = S_0'$ (22-54). Furthermore, assume that I' is simply related to $I = S_0$ only by

$$I' = f(I) \tag{22-55}$$

For two *independent* incident beams with intensities I_1 and I_2 the corresponding emergent beams $I_1'$and I_2' are functionally related by

$$I_1' = f(I_1) \tag{22-56a}$$

$$I_2' = f(I_2) \tag{22-56b}$$

Both equations must have the same functional form. From Stokes' law of additivity we can then write

$$I_1' + I_2' = I = f(I_1) + f(I_2) \tag{22-57}$$

Adding $I_1'$and I_2' the total intensity I must also be a function of $I_1 + I_2$ by Stokes' law of additivity. Thus, we have from (22-57)

$$f(I_1) + f(I_2) = f(I_1 + I_2) \tag{22-58}$$

Equation (22-58) is a functional equation. The equation can be solved for $f(I)$ by expanding $f(I_1)$, $f(I_2)$, and $f(I_1 + I_2)$ in a series so that

$$f(I_1) = a_0 + a_1 I_1 + a_2 I_1^2 + \cdots \tag{22-59a}$$

$$f(I_2) = a_0 + a_1 I_2 + a_2 I_2^2 + \cdots \tag{22-59b}$$

$$f(I_1 + I_2) = a_0 + a_1(I_1 + I_2) + a_2(I_1 + I_2)^2 + \cdots \tag{22-59c}$$

so

$$\begin{aligned} f(I_1) + f(I_2) &= 2a_0 + a_1(I_1 + I_2) + a_2(I_1^2 + I_2^2) + \cdots \\ &= a_0 + a_1(I_1 + I_2) + a_2(I_1 + I_2)^2 + \cdots \end{aligned} \tag{22-60}$$

The left- and right-hand sides of (22-60) are only consistent with Stokes' law of additivity for the linear terms, that is $a_0 = 0$, $a_1 \neq 0$, $a_2 = 0$, etc., so the solution of (22-58) is

$$f(I_1) = a_1 I_1 \tag{22-61a}$$

$$f(I_2) = a_1 I_2 \tag{22-61b}$$

$$f(I_1 + I_2) = a_1(I_1 + I_2) \tag{22-61c}$$

Thus, $f(I)$ is *linearly* related to I; $f(I)$ must be *linear* if Stokes' law of additivity is to apply simultaneously to I_1 and I_2 and I'_1 and I'_2. We can therefore relate S'_0 to S_0, S_1, S_2 and S_3 by a linear relation of the form:

$$S'_0 = f(S_0, S_1, S_2, S_3) = a_1 S_0 + b_1 S_1 + c_1 S_2 + d_1 S_3 \tag{22-62}$$

and similar relations (equations) for S'_1, S'_2, and S'_3. Thus, the Stokes vectors are related by 16 coefficients a_{ik}.

As examples of this linear relationship, Perrin noted that, for a light beam rotated through an angle ψ around its direction of propagation, for instance by passing through a crystal plate with simple rotatory power, we have

$$S'_0 = S_0 \tag{22-63a}$$

$$S'_1 = \cos(2\psi)S_1 - \sin(2\psi)S_2 \tag{22-63b}$$

$$S'_2 = \sin(2\psi)S_1 + \cos(2\psi)S_2 \tag{22-63c}$$

$$S'_3 = S_3 \tag{22-63d}$$

Similarly, when there is a difference in phase ϕ introduced between the components of the vibration along the axes, for instance by birefringent crystals with axes parallel to the reference axes, then

$$S'_0 = S_0 \tag{22-64a}$$

$$S'_1 = S_1 \tag{22-64b}$$

$$S'_2 = \cos(\phi)S_2 - \sin(\phi)S_3 \tag{22-64c}$$

$$S'_3 = \sin(\phi)S_2 + \cos(\phi)S_3 \tag{22-64d}$$

In the remainder of this paper Perrin then determined the number of nonzero (independent) coefficients a_{ik} for different media. These included (1) symmetrical media (8), (2) the scattering of light by an asymmetrical isotropic medium (10), (3) forward axial scattering (5), (4) forward axial scattering for a symmetric medium (3), (5) backward scattering by an asymmetrical medium (4), and (6) scattering by identical spherical particles without mirror symmetry (5).

Perrin's paper is actually quite remarkable because so many of the topics that he discussed have become the basis of much research. Even to this day there is much to learn from it.

22.4 RADIATION EQUATIONS FOR QUANTUM MECHANICAL SYSTEMS

We now turn to the problem of determining the polarization of radiation emitted by atomic and molecular systems. We assume that the reader has been exposed to the rudimentary ideas and methods of quantum mechanics particularly Schrödinger's wave equation and Heisenberg's matrix mechanics.

Experimental evidence of atomic and molecular systems has shown that a dynamical system in an excited state may *spontaneously* go to a state of lower energy, the transition being accompanied by the emission of energy in the form of radiation. In quantum mechanics the interaction of matter and radiation is allowed from the beginning, so that we start with a dynamical system:

$$\text{atom} + \text{radiation} \tag{22-65}$$

Every energy value of the system described by (22-65) can be interpreted as a possible energy of the atom alone plus a possible energy of the radiation alone plus a small interaction energy, so that it is still possible to speak of the energy levels of the atom itself. If we start with a system (22-65) at $t = 0$ in a state that can be described roughly as

$$\text{atom in an excited state } n + \text{no radiation} \tag{22-66}$$

we find at a subsequent time t the system may have gone over into a state described by

$$\text{atom in an excited state } m + \text{radiation} \tag{22-67}$$

which has the same total energy as the initial state (22-66), although the energy of the atom itself is now smaller. Whether or not the transition (22-66) → (22-67) will actually occur, or the precise instant at which it takes place, if it does take place, cannot be inferred from the information that at $t = 0$ the system is certainly in the state given by (22-66). In other words, an excited atom may "jump" spontaneously into a state of lower energy and in the process emit radiation.

To obtain the radiation equations suitable for describing quantum systems, two facts must be established. The first is the Bohr frequency condition, which states that a spontaneous transition of a dynamical system from an energy state of energy E_n to an energy state of lower energy E_m is accompanied by the emission of radiation of spectroscopic frequency $\omega_{n\to m}$ given by the formula:

$$\omega_{n\to m} = \frac{1}{\hbar(E_n - E_m)} \tag{22-68}$$

where $\hbar$ is Planck's constant divided by 2π.

The other fact is that the transition probability $A_{n\to m}$ for a spontaneous quantum jump of a one-dimensional dynamical system from an energy state n to an energy state m of lower energy is, to a high degree of approximation, given by the formula:

$$A_{n\to m} = \frac{e^2}{3\pi\varepsilon_0 c^3 h}\omega_{n\to m}^3 \left| \int \psi_n^* x \psi_m \, dx \right|^2 \tag{22-69}$$

where e is the electric charge and c is the speed of light. The transition probability $A_{n\to m}$ for a spontaneous quantum jump from the nth to the mth energy state is seen to be proportional to the square of the absolute magnitude of the *expectation value* of the variable x. That is, the quantity within the absolute magnitude signs is $\langle x \rangle$. Equation (22-69) shows that to determine $\langle x \rangle$ we must also know the eigenfunction ψ of the atomic system. The expectation value of x is then found by carrying out the required integration.

The importance of this brief discussion of the Bohr frequency condition and the transition probability is that these two facts allow us to proceed from the classical radiation equations to the radiation equations for describing the radiation emitted by quantum systems.

According to classical electrodynamics the radiation field components (spherical coordinates) emitted by an accelerating charge are given by

$$E_\theta = \frac{e}{4\pi\varepsilon_0 c^2 R}[\ddot{x}\cos\theta - \ddot{z}\sin\theta] \tag{16-8}$$

$$E_\phi = \frac{e}{4\pi\varepsilon_0 c^2 R}[\ddot{y}] \tag{16-9}$$

Quantum theory recognized early that these equations were essentially correct. They could also be used to describe the radiation emitted by atomic systems; however, new rules were needed to calculate $\ddot{x}$, $\ddot{y}$, and $\ddot{z}$. Thus, we retain the classical radiation equation (16-8) and (16-9), but we replace $\ddot{x}$, $\ddot{y}$, and $\ddot{z}$ by their quantum mechanical equivalents.

To derive the appropriate form of (16-8) and (16-9) suitable for quantum mechanical systems, we use Bohr's correspondence principle along with the frequency condition given by (22-68). Bohr's correspondence principle states that "in the limit of large quantum numbers quantum mechanics reduces to classical physics". We recall that the energy emitted by an oscillator of moment $\mathbf{p} = e\mathbf{r}$ is

$$I = \frac{1}{6\pi\varepsilon_0 c^3}|\ddot{\mathbf{p}}|^2 \tag{22-70}$$

Each quantum state n has two neighboring states, one above and one below, which for large quantum numbers differ by the same amount of energy $\hbar\omega_{nm}$. Hence, if we replace $\mathbf{p}$ by the matrix element $\mathbf{p}_{nm}$, we must at the same time multiply (22-70) by 2 so that the radiation emitted per unit time is

$$I = \frac{1}{3\pi\varepsilon_0 c^3}|\mathbf{p}_{nm}|^2 = \frac{e^2}{3\pi\varepsilon_0 c^3}\omega_{nm}^4|\mathbf{r}_{nm}|^2 \tag{22-71}$$

We see that the transition probability is simply the intensity of radiation emitted per unit time. Thus, dividing (22-71) by ω_{nm} gives the transition probability stated in (22-69). The quantity $\mathbf{r}_{nm}$ can now be calculated according to the rules of wave mechanics, namely,

$$\mathbf{r}_{nm} = \int_V \Psi_n(\mathbf{r}, t)\mathbf{r}\Psi_m^*(\mathbf{r}, t)d\mathbf{r} \tag{22-72}$$

where $\mathbf{r}$ stands for the radius vector from the nucleus to the field point, $\Psi_m(\mathbf{r},t)$ and $\Psi_n(\mathbf{r},t)$ are the Schrödinger wave functions for the mth and nth states of the quantum system, the asterisk denotes the complex conjugate, $d\mathbf{r}$ is the differential volume element, and V is the volume of integration.

In quantum mechanics $\mathbf{r}_{nm}$ is calculated from (22-72). We now assume that by a twofold differentiation of (22-72) with respect to time we can transform the classical $\mathbf{r}$ to the quantum mechanical form $\mathbf{r}_{nm}$. Thus, according to Bohr's correspondence principle, $\ddot{x}$ is transformed to $\ddot{x}_{nm}$ etc. i.e.,

$$\ddot{x} \Longrightarrow \ddot{x}_{nm} \tag{22-73a}$$

$$\ddot{y} \Longrightarrow \ddot{y}_{nm} \tag{22-73b}$$

$$\ddot{z} \Longrightarrow \ddot{z}_{nm} \tag{22-73c}$$

We now write (22-72) in component form:

$$x_{nm} = \int_V \Psi_n(\mathbf{r}, t) x \Psi^*_m(\mathbf{r}, t)\, d\mathbf{r} \tag{22-74a}$$

$$y_{nm} = \int_V \Psi_n(\mathbf{r}, t) y \Psi^*_m(\mathbf{r}, t)\, d\mathbf{r} \tag{22-74b}$$

$$z_{nm} = \int_V \Psi_n(\mathbf{r}, t) z \Psi^*_m(\mathbf{r}, t)\, d\mathbf{r} \tag{22-74c}$$

The wave functions $\Psi_m(\mathbf{r}, t)$ and $\Psi_n(\mathbf{r}, t)$ can be written as

$$\Psi_m(\mathbf{r}, t) = \Psi_m(\mathbf{r}) e^{i\omega_m t} \tag{22-75a}$$

$$\Psi_n(\mathbf{r}, t) = \Psi_n(\mathbf{r}) e^{i\omega_n t} \tag{22-75b}$$

where $\omega_{mn} = 2\pi f_{mn}$. Substituting (22-75) into (22-74) and then differentiating the result twice with respect to time yields

$$\ddot{x}_{nm} = -(\omega_n - \omega_m)^2 e^{i(\omega_n - \omega_m)t} \int_V \Psi_n(\mathbf{r}) x \Psi^*_m(\mathbf{r})\, d\mathbf{r} \tag{22-76a}$$

$$\ddot{y}_{nm} = -(\omega_n - \omega_m)^2 e^{i(\omega_n - \omega_m)t} \int_V \Psi_n(\mathbf{r}) y \Psi^*_m(\mathbf{r})\, d\mathbf{r} \tag{22-76b}$$

$$\ddot{z}_{nm} = -(\omega_n - \omega_m)^2 e^{i(\omega_n - \omega_m)t} \int_V \Psi_n(\mathbf{r}) z \Psi^*_m(\mathbf{r})\, d\mathbf{r} \tag{22-76c}$$

Now, it is easily proved that the integrals in (22-76) vanish for all states of an atom if $n = m$, so the derivative of the dipole moment vanishes and, accordingly, the emitted radiation also; that is, a stationary state does not radiate. This explains the fact, unintelligible from the standpoint of Bohr's theory, that an electron revolving around the nucleus, which according to the classical laws ought to emit radiation of the same frequency as the revolution, can continue to revolve in its orbit without radiating.

Returning now to the classical radiation equations (16-8) and (16-9), we see that the corresponding equations are, using (22-73)

$$E_\theta = \frac{e}{4\pi\varepsilon_o c^2 R}[\ddot{x}_{nm}\cos\theta - \ddot{z}_{nm}\sin\theta] \tag{22-77a}$$

$$E_\phi = \frac{e}{4\pi\varepsilon_o c^2 R}[\ddot{y}_{nm}] \tag{22-77b}$$

where $\ddot{x}_{nm}$, $\ddot{y}_{nm}$ and $\ddot{z}_{nm}$ are calculated according to (22-76a), (22-76b), and (22-76c), respectively.

The Schrödinger wave function $\Psi(\mathbf{r})$ is found by solving Schrödinger's time independent wave equation:

$$\nabla^2\Psi(\mathbf{r}) + \frac{2m}{\hbar^2}(E - V)\Psi(\mathbf{r}) = 0 \tag{22-78}$$

where ∇^2 is the Laplacian operator; in Cartesian coordinates it is

$$\nabla^2 \equiv \frac{\partial^2}{\partial x^2} + \frac{\partial^2}{\partial y^2} + \frac{\partial^2}{\partial z^2} \tag{22-79}$$

The quantities E and V are the total energy and potential energy, respectively, m is the mass of the particle, and $\hbar = h/2\pi$ is Planck's constant divided by 2π.

Not surprisingly, Schrödingers's equation (22-78) is extremely difficult to solve. Fortunately, several simple problems can be solved exactly, and these can be used to demonstrate the manner in which the quantum radiation equations, (22-77a) and (22-77b), and the Stokes parameters can be used. We now consider these problems.

22.5 STOKES VECTORS FOR QUANTUM MECHANICAL SYSTEMS

In this section we determine the Stokes vectors for several quantum systems of interest. The problems we select are chosen because the mathematics is relatively simple. Nevertheless, the examples presented are sufficiently detailed so that they clearly illustrate the difference between the classical and quantum representations. This is especially true with respect to the so-called selection rules as well as the representation of emission and absorption spectra. The examples presented are (1) a particle in an infinite potential well, (2) a one-dimensional harmonic oscillator, and (3) a rigid rotator restricted to rotating in the xy plate. We make no attempt to develop the solutions to these problems, but merely present the wave function and then determine the expectation values of the coordinates. The details of these problems are quite complicated, and the reader is referred to any of the numerous texts on quantum mechanics given in the references.

22.5.1 Particle in an Infinite Potential Well

The simplest quantum system is that of the motion of a particle in an infinite potential well of width extending from 0 to L. We assume the motion is along the z axis, so Schrödinger's equation for the system is

$$\frac{-\hbar^2}{2m}\frac{d^2\psi(z)}{dz^2} = E\psi(z) \tag{22-80}$$

and vanishes outside of the region. The normalized eigenfunctions are

$$\psi_n(z) = \left(\frac{2}{L}\right)^{1/2} \sin\left(\frac{n\pi z}{L}\right) \qquad 0 \le z \le L \tag{22-81}$$

and the corresponding energy is

$$E_n = \left(\frac{\pi^2\hbar^2}{2mL^2}\right)n^2 \qquad n = 1, 2, 3, \ldots \tag{22-82}$$

Since the motion is only along the z axis, we need only evaluate z_{nm}. Thus,

$$z_{nm} = \int_0^L \psi_n^*(z) z \psi_m(z)\, dz \tag{22-83a}$$

$$= \frac{2}{L}\int_0^L \sin\left(\frac{n\pi z}{L}\right) z \sin\left(\frac{m\pi z}{L}\right) dz \tag{22-83b}$$

Straightforward evaluation of this integral yields

$$z_{nm} = \frac{8Lnm}{\pi^2(n^2 - m^2)^2} \qquad (n + m \text{ odd}) \tag{22-84a}$$

$$= \frac{L}{2} \qquad (n = m) \tag{22-84b}$$

$$= 0 \qquad (\text{otherwise}) \tag{22-84c}$$

Equations (22-84b) and (22-84c) are of no interest because ω_{nm} describes a nonradiating condition and the field components are zero for $z_{nm} = 0$. Equation (22-84) is known as the *selection rule* for a quantum transition. Emission and absorption of radiation only take place in *discrete* amounts. The result is that there will be an infinite number of discrete spectral lines in the observed spectrum.

The field amplitudes are

$$E_\theta = \frac{2eL}{\pi^3 \varepsilon_0 c^2}\omega_{nm}^2 \left[\frac{nm}{(n^2 - m^2)^2}\right] \sin\theta \tag{22-85a}$$

$$E_\phi = 0 \tag{22-85b}$$

where we have set R to unity. We now form the Stokes parameters and then the Stokes vector in the usual way and obtain

$$S = \left(\frac{2eL}{\pi^3 \varepsilon_0 c^2}\right)^2 \sin^2\theta \left[\omega_{nm}^4 \left(\frac{nm}{(n^2 - m^2)^2}\right)^2\right] \begin{pmatrix} 1 \\ 1 \\ 0 \\ 0 \end{pmatrix} \tag{22-86}$$

This is the Stokes vector for linearly horizontally polarized light. We also have the familiar dipole radiation angular factor $\sin^2\theta$. We can observe either absorption or emission spectra, depending on whether we have a transition from a lower energy level to an upper energy level or from an upper to a lower level, respectively. For the absorption case the spectrum that would be observed is obtained by considering all possible combinations of n and m subject to the condition that $n + m$ is odd. Thus, for example, for a maximum number of five we have

$$S = \left(\frac{2eL}{\pi^3 \varepsilon_0 c^2}\right) \sin^2\theta \left\{ \omega_{12}^4 \left(\frac{2^2}{3^4}\right) \begin{pmatrix} 1 \\ 1 \\ 0 \\ 0 \end{pmatrix}, \omega_{14}^4 \left(\frac{4^2}{15^4}\right) \begin{pmatrix} 1 \\ 1 \\ 0 \\ 0 \end{pmatrix}, \omega_{23}^4 \left(\frac{6^2}{5^4}\right) \begin{pmatrix} 1 \\ 1 \\ 0 \\ 0 \end{pmatrix} \right\} \tag{22-87}$$

Similarly, for the emission spectrum we would observe

$$S = \left(\frac{2eL}{\pi^3 \varepsilon_0 c^2}\right)^2 \sin^2\theta \left\{ \omega_{21}^4 \left(\frac{2^2}{3^4}\right) \begin{pmatrix} 1 \\ 1 \\ 0 \\ 0 \end{pmatrix}, \omega_{41}^4 \left(\frac{4^2}{15^4}\right) \begin{pmatrix} 1 \\ 1 \\ 0 \\ 0 \end{pmatrix}, \omega_{32}^4 \left(\frac{6^2}{5^4}\right) \begin{pmatrix} 1 \\ 1 \\ 0 \\ 0 \end{pmatrix} \right\} \tag{22-88}$$

The intensity of the emission lines are in the ratio:

$$\omega_{21}^4 \left(\frac{2^2}{3^4}\right) : \omega_{41}^4 \left(\frac{4^2}{15^4}\right) : \omega_{32}^4 \left(\frac{6^2}{5^4}\right) \tag{22-89}$$

Using the Bohr frequency condition and (22-82), we can write ω_{nm} as

$$\omega_{nm} = \frac{E_n - E_m}{\hbar} = \frac{\pi^2 \hbar}{2mL^2}(n^2 - m^2) \tag{22-90}$$

Thus, the ratio of the intensities of the emission lines are $2^2 : 4^2 : 6^2$ or $1:4:9$, showing that the transition $3 \to 2$ is the most intense.

22.5.2 One-Dimensional Harmonic Oscillator

The potential $V(z)$ of a one-dimensional harmonic oscillator is $V(z) = z^2/2$. Schrödinger's equation then becomes

$$\frac{-\hbar^2}{2m}\frac{d^2\psi(z)}{dz^2} + \frac{m\omega^2 z^2}{2}\psi(z) = E\psi(z) \tag{22-91}$$

The normalized solutions are

$$\psi_n(z) = \frac{2^{-n/2}}{(n!)^{1/2}}\left(\frac{m\omega}{\pi\hbar}\right)^{1/2} \exp\left(\frac{-m\omega z^2}{2\hbar}\right) H_n\left[\left(\frac{2m}{\hbar}\right)^{1/2} z\right] \qquad n = 0, 1, 2 \tag{22-92}$$

where $H_n(u)$ are the Hermite polynomials. The corresponding energy levels are

$$E_n = \left(n + \frac{1}{2}\right)\hbar\omega \tag{22-93}$$

where $\omega^2 = k/m$. The expectation value of z is readily found to be

$$z_{nm} = \left(\frac{\hbar}{m\omega}\right)^{1/2} \left[\frac{n+1}{2}\right]^{1/2} \qquad n \to n+1 \qquad \text{absorption} \tag{22-94a}$$

$$= \left(\frac{\hbar}{m\omega}\right)^{1/2} \left[\frac{n}{2}\right]^{1/2} \qquad n \to n-1 \qquad \text{emission} \tag{22-94b}$$

$$= 0 \qquad \text{otherwise} \tag{22-94c}$$

The field components for the emitted and absorbed fields are then

$$E_\theta = \frac{-e}{4\pi\varepsilon_0 c^2}\left(\frac{\hbar}{m\omega}\right)^{1/2} \sin\theta\left[\omega_{n,n+1}^2\left(\frac{n+1}{2}\right)^{1/2}\right] \quad \text{(22-95a)}$$

$$E_\phi = 0 \quad \text{(22-95b)}$$

and

$$E_\theta = \frac{-e}{4\pi\varepsilon_0 c^2}\left(\frac{\hbar}{m\omega}\right)^{1/2} \sin\theta\left[\omega_{n,n-1}^2\left(\frac{n}{2}\right)^{1/2}\right] \quad \text{(22-96a)}$$

$$E_\phi = 0 \quad \text{(22-96b)}$$

The Stokes vector for the absorption and emission spectra are then

$$S = \left(\frac{e^2\hbar}{16\pi^2\varepsilon_0^2 c^4 m\omega}\right)\sin^2\theta\left[\omega_{n,n+1}^4\frac{n+1}{2}\right]\begin{pmatrix}1\\1\\0\\0\end{pmatrix} \quad \text{(22-97)}$$

$$S = \left(\frac{e^2\hbar}{16\pi^2\varepsilon_0^2 c^4 m\omega}\right)\sin^2\theta\left[\omega_{n,n-1}^4\frac{n}{2}\right]\begin{pmatrix}1\\1\\0\\0\end{pmatrix} \quad \text{(22-98)}$$

Equations (22-97) and (22-98) show that for both absorption and emission spectra the radiation is linearly horizontally polarized, and, again, we have the familiar $\sin^2\theta$ angular dependence of dipole radiation. To obtain the observed spectral lines we take $n = 0, 1, 2, 3, \ldots$ for the absorption spectrum and $n = 1, 2, 3\ldots$ for the emission spectrum. We then obtain a series of spectral lines similar to (22-89) and (22-90). With respect to the intensities of the spectral lines for, say $n = 5$, the ratio of intensities is 1 : 2 : 3 : 4 : 5 : 6, showing that the strongest transition is $6 \rightarrow 5$ for emission and $5 \rightarrow 6$ for absorption.

22.5.3 Rigid Rotator

The ideal diatomic molecule is represented by a rigid rotator; that is, a molecule can be represented by two atoms with masses m_1 and m_2 rigidly connected so that the distance between them is a constant R. If there are no forces acting on the rotator, the potential may be set to zero and the variable r, the radial distance, to unity. Schrödinger's equation for this case is then

$$(\sin\theta)^{-1}\frac{\partial}{\partial\theta}\left(\sin\theta\frac{\partial\psi}{\partial\theta}\right) + (\sin^2\theta)^{-1}\frac{\partial^2\psi}{\partial\phi^2} + \left(\frac{2IE}{\hbar^2}\right)\psi = 0 \quad \text{(22-99)}$$

where I is the moment of inertia, given by

$$I = m_1 r_1^2 + m_2 r_2^2 \quad \text{(22-100)}$$

The solution of Schrödinger's equation (22-99) is then

$$\psi_{l,m} = Y_{l,\pm m}(\theta,\phi) = \Theta_{l,\pm m}(\theta)\Phi_{\pm m}(\phi) \quad \text{(22-101)}$$

where $l \geq |m|$. The energy levels are given by

$$E = \left(\frac{\hbar^2}{2I}\right) l(l+1) \qquad l = 0, 1, 2, 3, \ldots \tag{22-102}$$

A very important and illustrative example is the case where the motion of the rotator is restricted to the xy plane. For this case the polar angle $\theta = \pi/2$ and (22-99) reduces to

$$\frac{d^2\psi}{d\phi^2} = -\left(\frac{2IE}{\hbar^2}\right)\psi \tag{22-103}$$

with the solutions:

$$\psi = \Phi_{\pm m}(\phi) = (2\pi)^{-1/2}\exp(\pm im\phi) \qquad m = 1, 2, 3, \ldots \tag{22-104}$$

Equation (22-104) can also be obtained from (22-101) by evaluating the associated Legendre polynomial at $\theta = \pi/2$. The energy levels for (22-103) are found to be

$$E = \left(\frac{\hbar}{2I}\right) m^2 \qquad m = 1, 2, 3, \ldots \tag{22-105}$$

We now calculate the Stokes vector corresponding to (22-103). Since we are assuming that ϕ is measured positively in the xy plane, the z component vanishes. Thus, we need only calculate x_{nm} and y_{nm}. The coordinates x and y are related to ϕ by

$$x = a\cos\phi \tag{22-106a}$$

$$y = a\sin\phi \tag{22-106b}$$

where a is the radius of the rigid rotator (molecule). We now calculate the expectation values:

$$\begin{aligned} x_{nm} &= \int_0^{2\pi} \psi_n^* x \psi_m \, d\phi \\ &= \frac{a}{2\pi}\int_0^{2\pi} \exp(-in\phi)\cos\phi\exp(im\phi)\, d\phi \\ &= \frac{a}{4\pi}\int_0^{2\pi} \exp[-i(n-m-1)\phi]\, d\phi \\ &\quad + \frac{a}{4\pi}\int_0^{2\pi} \exp[-i(n-m+1)\phi]\, d\phi \end{aligned} \tag{22-107}$$

The first integral vanishes except for $m = n-1$, while the second integral vanishes except for $m = n+1$; we then have the selection rule that $\Delta m = \pm 1$. Evaluation of the integrals in (22-107) then gives

$$x_{m,m\pm 1} = +\frac{a}{2} \tag{22-108}$$

In a similar manner we find that

$$y_{m,m\pm 1} = \pm\frac{a}{2i} \tag{22-109}$$

Thus, the amplitudes for the absorbed and emitted fields are

$$E_\theta = \left(-\frac{ea}{8\pi\varepsilon_0 c^2}\right)\omega^2_{n,m+1}\cos\theta \tag{22-110a}$$

$$E_\phi = \left(-\frac{ea}{8i\pi\varepsilon_0 c^2}\right)\omega^2_{m,m+1} \tag{22-110b}$$

and

$$E_\theta = \left(-\frac{ea}{8\pi\varepsilon_0 c^2}\right)\omega^2_{m,m-1}\cos\theta \tag{22-111a}$$

$$E_\phi = \left(\frac{ea}{8i\pi\varepsilon_0 c^2}\right)\omega^2_{m,m-1} \tag{22-111b}$$

respectively. The Stokes vectors using (22-106) and (22-107) are then readily found to be

$$S = \left(\frac{ea}{8\pi\varepsilon_0 c^2}\right)^2 \omega^4_{m,m+1}\begin{pmatrix} 1+\cos^2\theta \\ -\sin^2\theta \\ 0 \\ -2\cos\theta \end{pmatrix} \tag{22-112}$$

and

$$S = \left(\frac{ea}{8\pi\varepsilon_0 c^2}\right)^2 \omega^4_{m,m-1}\begin{pmatrix} 1+\cos^2\theta \\ -\sin^2\theta \\ 0 \\ 2\cos\theta \end{pmatrix} \tag{22-113}$$

In general, we see that for both the absorption and emission spectra the spectral lines are *elliptically* polarized and of opposite ellipticity. As usual, if the radiation is observed parallel to the z axis ($\theta = 0^\circ$), then (22-112) and (22-113) reduce to

$$S = 2\left(\frac{ea}{8\pi\varepsilon_0 c^2}\right)^2 \omega^4_{m,m+1}\begin{pmatrix} 1 \\ 0 \\ 0 \\ -1 \end{pmatrix} \tag{22-114}$$

and

$$S = 2\left(\frac{ea}{8\pi\varepsilon_0 c^2}\right)^2 \omega^4_{m,m-1}\begin{pmatrix} 1 \\ 0 \\ 0 \\ 1 \end{pmatrix} \tag{22-115}$$

which are the Stokes vectors for left and right circularly polarized light. For $\theta = 90^\circ$, (22-111) and (22-112) reduce to

$$S = \left(\frac{ea}{8\pi\varepsilon_0 c^2}\right)^2 \omega^4_{m,m+1}\begin{pmatrix} 1 \\ -1 \\ 0 \\ 0 \end{pmatrix} \tag{22-116}$$

and

$$S = \left(\frac{ea}{8\pi\varepsilon_0 c^2}\right)^2 \omega_{m,\,m-1}^4 \begin{pmatrix} 1 \\ -1 \\ 0 \\ 0 \end{pmatrix} \tag{22-117}$$

which are the Stokes vectors for linearly vertically polarized light.

Inspection of (22-116) and (22-117) shows that the Stokes vectors, aside from the frequency $\omega_{m,\,m\pm1}$, are identical to the classical result. Thus, the quantum behavior expressed by Planck's constant is nowhere to be seen in the spectrum! This result is very different from the result for the linear harmonic oscillator where Planck's constant $\hbar$ appears in the intensity. It was this peculiar behavior of the spectra that made their interpretation so difficult for a long time. That is, for some problems (the linear oscillator) the quantum behavior appeared in the spectral intensity, and for other problems (the rigid rotator) it did not. The reason for the disappearance of Planck's constant could usually be traced to the fact that it actually appeared in both the denominator and numerator of many problems and simply canceled out. In all cases, using Bohr's correspondence principle, in the limit of large quantum numbers $\hbar$ always canceled out of the final result.

We now see that the Stokes vector can be used to represent both classical and quantum radiation phenomena. Before we conclude, a final word must be said about the influence of the selection rules on the polarization state. The reader is sometimes led to believe that the selection rule *itself* is the cause for the appearance of either linear, circular, or elliptical polarization. This is not quite correct. We recall that the field equations emitted by an accelerating charge are

$$E_\theta = \frac{e}{4\pi\varepsilon_0 c^2 R}[\ddot{x}\cos\theta - \ddot{z}\sin\theta] \tag{16-8}$$

$$E_\phi = \frac{e}{4\pi\varepsilon_0 c^2 R}[\ddot{y}] \tag{16-9}$$

We have seen that we can replace $\ddot{x}$, $\ddot{y}$, and $\ddot{z}$ by their quantum mechanical equivalents:

$$\ddot{x} \to -\omega_{nm}^2 x_{nm} \tag{22-118a}$$

$$\ddot{y} \to -\omega_{nm}^2 y_{nm} \tag{22-118b}$$

$$\ddot{z} \to -\omega_{nm}^2 z_{nm} \tag{22-118c}$$

so that (16-8) and (16-9) become

$$E_\theta = -\left(\frac{e}{4\pi\varepsilon_0 c^2 R}\right)\omega_{nm}^2[x_{nm}\cos\theta - z_{nm}\sin\theta] \tag{22-119a}$$

$$E_\phi = -\left(\frac{e}{4\pi\varepsilon_0 c^2 R}\right)\omega_{nm}^2 y_{nm} \tag{22-119b}$$

If only a *single* Cartesian variable remains in (22-119), then we have linearly polarized light. If two variables appear, e.g., x_{nm} and y_{nm}, then we obtain elliptically or circularly polarized light. However, if the selection rule is such that either x_{nm} or y_{nm}

were to vanish, then we would obtain linearly polarized light regardless of the presence of the angular factor. In other words, in classical physics the angular factor dominates the state of polarization emitted by the radiation. However, in quantum mechanics the fact that either x_{nm} or y_{nm} can vanish and thus give rise to linearly polarized light shows that the role of the selection rule is equally significant in the polarization of the emitted or absorbed radiation.

Numerous other problems can easily be treated with the methods discussed here, such as the rigid rotator in three dimensions and the Zeeman effect [22]. We refer the reader to the numerous texts on quantum mechanics for further examples and applications.

REFERENCES

Papers

1. Stokes, G. G. *Trans. Camb. Phil. Soc.*, *9*, 399 (1852), Reprinted in *Mathematical and Physical Papers*, Cambridge University Press, London, (1901), Vol. 3, p.233.
2. Stokes, G. G. *Proc. Roy. Soc.*, *6*, 195 (1852). Reprinted in *Mathematical and Physical Papers*, Cambridge University Press, London, 1901, Vol. 3, p.259.
3. Einstein, A., *Ann. Phys.*, *17*, 132 (1905).
4. Heisenberg, W., *Z. Phys.*, *33*, 879 (1925).
5. Landau, L. D., *Z. Phys.*, *45*, 430 (1972).
6. von Neumann, J., *Göttinger Nachr.*, *24* (1) 273 (1927).
7. Soleillet, P., *Ann. Phys.* (10) *23* (1929).
8. Perrin, F., *J. Chem. Phys. 10*, 415 (1942).
9. Chandrashekar, S., *Astrophys. J.*, *104*, 110 (1946).
10. Chandrashekar, S., *Astrophys. J.*, *105*, 424 (1947).
11. Jones, R. Clark. *J. Opt. Soc. Am.*, *37*, 107 (1947).
12. Parke, N. G., III, "Statistical Optics: II: Mueller Phenomenological Algebra," Research Laboratory of Electronics, M. I. T., (June 15 1949).
13. Fano, U., *J. Opt. Soc. Am.*, *39*, 859 (1949).
14. Fano, U. *J. Opt. Soc., Am.*, *41*, 58 (1951).
15. Fano, U. *Phys. Rev.*, *93*, 121 (1954).
16. Falkoff, D. L. and MacDonald, J. E., *J. Opt. Soc. Am.*, *41*, 861 (1951).
17. Wolf, E., *Nuovo Cimento*, *12*, 884 (1954).
18. McMaster, W. H. *Am. J. Phys. 22*, 351 (1954).
19. Walker, M. J., *Am. J. Phys.*, *22*, 170 (1954).
20. McMaster, W. H., *Rev. Mod. Phys.*, *33*, 8 (1961).
21. Collett, E., *Am. J. Phys.*, *36*, 713 (1968).
22. Collett, E., *Am. J. Phys.*, *38*, 563 (1970).
23. Schmieder, R. W., *J. Opt. Soc. Am.*, *59*, 297 (1969).
24. Whitney, C., *J. Opt. Soc. Am.*, *61*, 1207 (1971).

Books

1. Born, M. and Wolf, E., *Principles of Optics*, 3rd ed., Pergamon Press, New York, 1965.
2. Chandrashekar, S., *Radiative Transfer*, Dover, New York, 1960, pp. 24–34.
3. Strong, J., *Concepts of Classical Optics*, Freeman, San Francisco, 1959.
4. Stone, J. M., *Radiation and Optics*, McGraw-Hill, New York, 1963.
5. Shurcliff, W.A., *Polarized Light*, Harvard University Press, Cambridge, MA, 1962.
6. Hecht, E. and Zajac, A., *Optics*, Addison-Wesley, Reading, MA, 1974.

7. Simmons, J. W., and Guttman, M. J. *States, Waves and Photons.*, Addison-Wesley, Reading, MA, 1970.
8. O'Neil, E. L., *Introduction to Statistical Optics*, Addison-Wesley, Reading, MA, 1963.
9. Schiff, L. I., *Quantum Mechanics*, 2nd ed., McGraw-Hill, New York 1955.
10. Rojansky, V. B., *Introductory Quantum Mechanics*, Prentice-Hall, Englewood Cliffs, NJ, 1938.
11. Ruark, A. E. and Urey, H. C., *Atoms, Molecules and Quanta*, Dover, New York, 1964.
12. French, A. P., *Principles of Modern Physics*, Wiley, New York, 1958.
13. Dirac, P. A. M. *Quantum Mechanics*, 3rd ed., Clarendon Press, Oxford, 1947.
14. Pauling, L. and Wilson, E. B., *Introduction to Quantum Mechanics*, McGraw-Hill, New York 1935.
15. Rayleigh, Lord, *Scientific Papers*, 6 Vols. Cambridge University Press, London, 1899–1920.
16. Walker, J., *The Analytical Theory of Light*, Cambridge University Press, Cambridge, UK, 1904.

23

Introduction

Polarized light and its applications appear in many branches of science and engineering. These include astrophysics (synchrotron radiation, solar physics, atmospheric scattering), chemistry (saccharimetry, optical activity, fluorescence polarization), microscopy (the polarizing microsope), and, of course, optics [polarization by reflection from glass (dielectrics) and metals, liquid crystals, thin films, electro-optics, etc.] It is not practical to deal with all these different applications of polarized light in this single textbook. Therefore, in this final part the discussion is restricted to several applications which are of special importance.

We begin with Chapter 24 "Crystal Optics." The polarization of light was first discovered by Bartholinus while investigating the transmission of unpolarized light through a crystal of Iceland spar (calcite). It is a remarkable fact that in spite of all the research on materials over the last 300 years very few natural or synthetic materials have been found which can be used to create and analyze polarized light. The crystals having the widest applications in the visible region of the spectrum are calcite, quartz, mica, and tourmaline. The optics of crystals is quite complicated. Fortunately, calcite and quartz are uniaxial crystals and relatively easy to understand in terms of their polarizing behavior.

In Chapter 24 we discuss a very important application of polarized light, namely, the phenomenon of electro-optical crystals. Many crystals become anisotropic when subjected to an electric field or a magnetic field or both; the associated effects are called the electro-optical and magneto-optical effects, respectively. Of the two phenomena, in crystals the electro-optical effect is the more important, so we consider only this effect in detail.

The polarization of light is changed when light is reflected from dielectric materials. The change in polarization also occurs when light is reflected (and transmitted) by metals and semiconductors. In Chapter 25 we discuss the optics of metals. In particular, we show that the optical constants of the metal can be determined by analyzing the polarization of the reflected light.

Chapter 26 is a summary of some of the most common polarization optical elements that are used in the practice of optics. One of these is Polaroid or sheet polarizer. For many years a synthetic material was sought which could create polarized light. This was finally accomplished with the invention of Polaroid by Edwin Land. Polaroid is a dichroic polarizer that creates polarized light by the differential absorption of an incident beam of light. For many applications Polaroid is a useful substitute for calcite polarizers, which are very expensive. Because Polaroid is so widely used, its parameters and their measurement are presented and discussed in Section 26.2 with other types of polarizers.

Chapter 27 describes modern techniques of measurement of the Stokes vector. Stokes polarimetry is employed when the polarization properties of light are desired. Mueller matrix polarimetry, discussed in Chapter 28, is used when the polarization properties of a sample are needed. This measurement technique implies that one has control over the incident light by means of a polarization state generator, and reflected or transmitted light from the sample is analyzed by a polarization state analyzer.

In Chapter 29 we discuss one of the most important and elegant applications of polarized light, ellipsometry. The objective of ellipsometry is to measure the thickness and real and imaginary refractive indices of thin films. We introduce the fundamental equation of ellipsometry and solve it by using the Stokes parameters and the Mueller matrices.

24

Crystal Optics

24.1 INTRODUCTION

Crystals are among nature's most beautiful and fascinating objects. Even the slightest examination of crystals shows remarkable forms, symmetries, and colors. Some also have the property of being almost immutable, and appear to last forever. It is this property of chemical and physical stability that has allowed them to become so valuable.

Many types of crystals have been known since time immemorial, e.g., diamonds, sapphires, topaz, emeralds, etc. Not surprisingly, therefore, they have been the subject of much study and investigation for centuries. One type of crystal, calcite, was probably known for a very long time before Bartholinus discovered in the late seventeenth century that it was birefringent. Bartholinus apparently obtained the calcite crystal from Iceland (Iceland spar); the specimens he obtained were extremely free of striations and defects. His discovery of double refraction (birefringence) and its properties was a source of wonder to him. According to his own accounts, it gave him endless hours of pleasure—as a crystal he far preferred it to diamond! It was Huygens, however, nearly 30 years later, who explained the phenomenon of double refraction.

In this chapter we describe the fundamental behavior of the optical field propagating in crystals; this behavior can be correctly described by assuming that crystals are anisotropic. Most materials are anisotropic. This anisotropy results from the structure of the material, and our knowledge of the nature of that structure can help us to understand the optical properties.

The interaction of light with matter is a process that is dependent on the geometrical relationships between light and matter. By its very nature, light is asymmetrical. Considering light as a wave, it is a transverse oscillation in which the oscillating quantity, the electric field vector, is oriented in a particular direction in space perpendicular to the propagation direction. Light that crosses the boundary between two materials, isotropic or not, at any angle other than normal to the boundary, will produce an anisotropic result. The Fresnel equations illustrate this, as we saw in Chapter 8. Once light has crossed a boundary separating materials, it experiences the bulk properties of the material through which it is currently traversing, and we are concerned with the effects of those bulk properties on the light.

The study of anisotropy in materials is important to understanding the results of the interaction of light with matter. For example, the principle of operation of many solid state and liquid crystal spatial light modulators is based on polarization modulation. Modulation is accomplished by altering the refractive index of the modulator material, usually with an electric or magnetic field. Crystalline materials are an especially important class of modulator materials because of their use in electro-optics and in ruggedized or space-worthy systems, and also because of the potential for putting optical systems on integrated circuit chips.

We will briefly review the electromagnetics necessary to the understanding of anisotropic materials, and show the source and form of the electro-optic tensor. We will discuss crystalline materials and their properties, and introduce the concept of the index ellipsoid. We will show how the application of electric and magnetic fields alters the properties of materials and give examples. Liquid crystals will be discussed as well.

A brief summary of electro-optic modulation modes using anisotropic materials concludes the chapter.

24.2 REVIEW OF CONCEPTS FROM ELECTROMAGNETISM

Recall from electromagnetics [1–3] that the electric displacement vector $\bar{\mathbf{D}}$ is given by (MKS units)

$$\bar{\mathbf{D}} = \varepsilon\bar{\mathbf{E}} \tag{24-1}$$

where ε is the permittivity and $\varepsilon = \varepsilon_0(1+\chi)$, where ε_0 is the permittivity of free space, χ is the electric susceptibility, $(1+\chi)$ is the dielectric constant, and $\mathbf{n} = (1+\chi)^{1/2}$ is the index of refraction. The electric displacement is also given by

$$\bar{\mathbf{D}} = \varepsilon_0\bar{\mathbf{E}} + \bar{\mathbf{P}} \tag{24-2}$$

but

$$\bar{\mathbf{D}} = \varepsilon_0(1+\chi)\bar{\mathbf{E}} = \varepsilon_0\bar{\mathbf{E}} + \varepsilon_0\chi\bar{\mathbf{E}} \tag{24-3}$$

so $\bar{\mathbf{P}}$, the polarization (also called the electric polarization or polarization density), is $\bar{\mathbf{P}} = \varepsilon_0\boldsymbol{\chi}\bar{\mathbf{E}}$.

The polarization arises because of the interaction of the electric field with bound charges. The electric field can produce a polarization by inducing a dipole moment, i.e., separating charges in a material, or by orienting molecules that possess a permanent dipole moment.

For an isotropic, linear medium:

$$\bar{\mathbf{P}} = \varepsilon_0\boldsymbol{\chi}\bar{\mathbf{E}} \tag{24-4}$$

and χ is a scalar, but note that in

$$\mathbf{D} = \varepsilon_0\bar{\mathbf{E}} + \bar{\mathbf{P}} \tag{24-5}$$

the vectors do not have to be in the same direction, and in fact in anisotropic media, $\bar{\mathbf{E}}$ and $\bar{\mathbf{P}}$ are not in the same direction (and so $\bar{\mathbf{D}}$ and $\bar{\mathbf{E}}$ are not in the same direction). Note that χ does not have to be a scalar nor is $\bar{\mathbf{P}}$ necessarily linearly related to $\bar{\mathbf{E}}$. If the medium is linear but anisotropic:

$$\mathbf{P_i} = \sum_j \varepsilon_o \chi_{ij} \mathbf{E_j} \tag{24-6}$$

where χ_{ij} is the susceptibility tensor, i.e.,

$$\begin{pmatrix} \mathbf{P}_1 \\ \mathbf{P}_2 \\ \mathbf{P}_3 \end{pmatrix} = \varepsilon_o \begin{pmatrix} \chi_{11} & \chi_{12} & \chi_{13} \\ \chi_{21} & \chi_{22} & \chi_{23} \\ \chi_{31} & \chi_{32} & \chi_{33} \end{pmatrix} \begin{pmatrix} \mathbf{E}_1 \\ \mathbf{E}_2 \\ \mathbf{E}_3 \end{pmatrix} \tag{24-7}$$

and

$$\begin{aligned} \begin{pmatrix} \mathbf{D}_1 \\ \mathbf{D}_2 \\ \mathbf{D}_3 \end{pmatrix} &= \varepsilon_o \begin{pmatrix} 1 & 0 & 0 \\ 0 & 1 & 0 \\ 0 & 0 & 1 \end{pmatrix} \begin{pmatrix} \mathbf{E}_1 \\ \mathbf{E}_2 \\ \mathbf{E}_3 \end{pmatrix} + \varepsilon_o \begin{pmatrix} \chi_{11} & \chi_{12} & \chi_{13} \\ \chi_{21} & \chi_{22} & \chi_{23} \\ \chi_{31} & \chi_{32} & \chi_{33} \end{pmatrix} \begin{pmatrix} \mathbf{E}_1 \\ \mathbf{E}_2 \\ \mathbf{E}_3 \end{pmatrix} \\ &= \varepsilon_o \begin{pmatrix} 1+\chi_{11} & \chi_{12} & \chi_{13} \\ \chi_{21} & 1+\chi_{22} & \chi_{23} \\ \chi_{31} & \chi_{32} & 1+\chi_{33} \end{pmatrix} \begin{pmatrix} \mathbf{E}_1 \\ \mathbf{E}_2 \\ \mathbf{E}_3 \end{pmatrix} \end{aligned} \tag{24-8}$$

where the vector indices 1,2,3 represent the three Cartesian directions. This can be written

$$\mathbf{D_i} = \varepsilon_{ij} \mathbf{E_j} \tag{24-9}$$

where

$$\varepsilon_{ij} = \varepsilon_o (1 + \chi_{ij}) \tag{24-10}$$

is variously called the dielectric tensor, permittivity tensor, or dielectric permittivity tensor. Equations (24-9) and (24-10) use the Einstein summation convention, i.e., whenever repeated indices occur, it is understood that the expression is to be summed over the repeated indices. This notation will be used throughout this chapter.

The dielectric tensor is symmetric and real (assuming that the medium is homogeneous and nonabsorbing) so that

$$\varepsilon_{ij} = \varepsilon_{ji} \tag{24-11}$$

and there are at most six independent elements.

Note that for an isotropic medium with nonlinearity (which occurs with higher field strengths):

$$\mathbf{P} = \varepsilon_o \left(\chi \mathbf{E} + \chi_2 \mathbf{E}^2 + \chi_3 \mathbf{E}^3 + \cdots \right) \tag{24-12}$$

where χ_2, χ_3, etc., are the nonlinear terms.

Returning to the discussion of a linear, homogeneous, anisotropic medium, the susceptibility tensor:

$$\begin{pmatrix} \chi_{11} & \chi_{12} & \chi_{13} \\ \chi_{21} & \chi_{22} & \chi_{23} \\ \chi_{31} & \chi_{32} & \chi_{33} \end{pmatrix} = \begin{pmatrix} \chi_{11} & \chi_{12} & \chi_{13} \\ \chi_{12} & \chi_{22} & \chi_{23} \\ \chi_{13} & \chi_{23} & \chi_{33} \end{pmatrix} \tag{24-13}$$

is symmetric so that we can always find a set of coordinate axes (i.e., we can always rotate to an orientation) such that the off-diagonal terms are zero and the tensor is diagonalized thus

$$\begin{pmatrix} \chi'_{11} & 0 & 0 \\ 0 & \chi'_{22} & 0 \\ 0 & 0 & \chi'_{33} \end{pmatrix} \tag{24-14}$$

The coordinate axes for which this is true are called the principal axes, and these χ' are the principal susceptibilities. The principal dielectric constants are given by

$$\begin{pmatrix} 1 & 0 & 0 \\ 0 & 1 & 0 \\ 0 & 0 & 1 \end{pmatrix} + \begin{pmatrix} \chi_{11} & 0 & 0 \\ 0 & \chi_{22} & 0 \\ 0 & 0 & \chi_{33} \end{pmatrix} = \begin{pmatrix} 1+\chi_{11} & 0 & 0 \\ 0 & 1+\chi_{22} & 0 \\ 0 & 0 & 1+\chi_{33} \end{pmatrix}$$
$$= \begin{pmatrix} \mathbf{n}_1^2 & 0 & 0 \\ 0 & \mathbf{n}_2^2 & 0 \\ 0 & 0 & \mathbf{n}_3^2 \end{pmatrix} \tag{24-15}$$

where $\mathbf{n}_1$, $\mathbf{n}_2$, and $\mathbf{n}_3$ are the principal indices of refraction.

24.3 CRYSTALLINE MATERIALS AND THEIR PROPERTIES

As we have seen above, the relationship between the displacement and the field is

$$\mathbf{D}_\mathbf{i} = \varepsilon_{\mathbf{ij}}\mathbf{E}_\mathbf{j} \tag{24-16}$$

where $\varepsilon_{\mathbf{ij}}$ is the dielectric tensor. The impermeability tensor $\eta_{\mathbf{ij}}$ is defined as

$$\eta_{\mathbf{ij}} = \varepsilon_\mathbf{o}(\varepsilon^{-1})_{\mathbf{ij}} \tag{24-17}$$

where ε^{-1} is the inverse of the dielectric tensor. The principal indices of refraction, $\mathbf{n}_1$, $\mathbf{n}_2$, and $\mathbf{n}_3$ are related to the principal values of the impermeability tensor and the principal values of the permittivity tensor by

$$\frac{1}{\mathbf{n}_1^2} = \eta_{\mathbf{ii}} = \frac{\varepsilon_\mathbf{o}}{\varepsilon_{\mathbf{ii}}} \qquad \frac{1}{\mathbf{n}_2^2} = \eta_{\mathbf{jj}} = \frac{\varepsilon_\mathbf{o}}{\varepsilon_{\mathbf{jj}}} \qquad \frac{1}{\mathbf{n}_3^2} = \eta_{\mathbf{kk}} = \frac{\varepsilon_\mathbf{o}}{\varepsilon_{\mathbf{kk}}} \tag{24-18}$$

The properties of the crystal change in response to the force from an externally applied electric field. In particular, the impermeability tensor is a function of the field. The electro-optic coefficients are defined by the expression for the expansion, in terms of the field, of the change in the impermeability tensor from zero field value, i.e.,

$$\eta_{\mathbf{ij}}(\mathbf{E}) - \eta_{\mathbf{ij}}(0) \equiv \Delta\eta_{\mathbf{ij}} = \mathbf{r}_{\mathbf{ijk}}\mathbf{E}_\mathbf{k} + \mathbf{s}_{\mathbf{ijkl}}\mathbf{E}_\mathbf{k}\mathbf{E}_\mathbf{l} + \mathrm{O}(\mathbf{E}^n), \qquad n = 3, 4, \ldots \tag{24-19}$$

where η_{ij} is a function of the applied field E, $\mathbf{r}_{ijk}$ are the linear, or Pockels, electro-optic tensor coefficients, and the $\mathbf{s}_{ijkl}$ are the quadratic, or Kerr, electro-optic tensor coefficients. Terms higher than quadratic are typically small and are neglected.

Note that the values of the indices and the electro-optic tensor coefficients are dependent on the frequency of light passing through the material. Any given indices are specified at a particular frequency (or wavelength). Also note that the external applied fields may be static or alternating fields, and the values of the tensor coefficients are weakly dependent on the frequency of the applied fields. Generally, low- and/or high-frequency values of the tensor coefficients are given in tables. Low frequencies are those below the fundamental frequencies of the acoustic resonances of the sample, and high frequencies are those above. Operation of an electro-optic modulator subject to low (high) frequencies is sometimes described as being unclamped (clamped).

The linear electro-optic tensor is of third rank with 3^3 elements and the quadratic electro-optic tensor is of fourth rank with 3^4 elements; however, symmetry reduces the number of independent elements. If the medium is lossless and optically inactive:

ε_{ij} is a symmetric tensor, i.e., $\varepsilon_{ij} = \varepsilon_{ji}$,

η_{ij} is a symmetric tensor, i.e., $\eta_{ij} = \eta_{ji}$,

$\mathbf{r}_{ijk}$ has symmetry where coefficients with permuted first and second indices are equal, i.e., $\mathbf{r}_{ijk} = \mathbf{r}_{jik}$,

$\mathbf{s}_{ijkl}$ has symmetry where coefficients with permuted first and second indices are equal and coefficients with permuted third and fourth coefficients are equal, i.e., $\mathbf{s}_{ijkl} = \mathbf{s}_{jikl}$ and $\mathbf{s}_{ijkl} = \mathbf{s}_{ijlk}$.

Symmetry reduces the number of linear coefficients from 27 to 18, and reduces the number of quadratic coefficients from 81 to 36. The linear electro-optic coefficients are assigned two indices so that they are $\mathbf{r}_{lk}$ where **l** runs from 1 to 6 and **k** runs from 1 to 3. The quadratic coefficients are assigned two indices so that they become $\mathbf{s}_{ij}$ where **i** runs from 1 to 6 and **j** runs from 1 to 6. For a given crystal symmetry class, the form of the electro-optic tensor is known.

24.4 CRYSTALS

Crystals are characterized by their lattice type and symmetry. There are 14 lattice types. As an example of three of these, a crystal having a cubic structure can be simple cubic, face-centered cubic, or body-centered cubic.

There are 32 point groups corresponding to 32 different symmetries. For example, a cubic lattice has five types of symmetry. The symmetry is labeled with point group notation, and crystals are classified in this way. A complete discussion of crystals, lattice types, and point groups is outside the scope of the present work, and will not be given here; there are many excellent references [4–9]. Table 24-1 gives a summary of the lattice types and point groups, and shows how these relate to optical symmetry and the form of the dielectric tensor.

In order to understand the notation and terminology of Table 24-1, some additional information is required which we now introduce. As we have seen in the previous sections, there are three principal indices of refraction. There are three types of materials; those for which the three principal indices are equal, those where two principal indices are equal, and the third is different, and those

Table 24-1 Crystal Types, Point Groups, and the Dielectric Tensors

Symmetry	Crystal System	Point Group	Dielectric Tensor
Isotropic	Cubic	$\bar{4}3\mathbf{m}$ 432 $\mathbf{m}3$ 23 $\mathbf{m}3\mathbf{m}$	$\varepsilon = \varepsilon_0 \begin{pmatrix} \mathbf{n}^2 & 0 & 0 \\ 0 & \mathbf{n}^2 & 0 \\ 0 & 0 & \mathbf{n}^2 \end{pmatrix}$
Uniaxial	Tetragonal	4 $\bar{4}$ $4/\mathbf{m}$ 422 $4\mathbf{mm}$ $\bar{4}2\mathbf{m}$ $4/\mathbf{mmm}$	
	Hexagonal	6 $\bar{6}$ $6/\mathbf{m}$ 622 $6\mathbf{mm}$ $\bar{6}\mathbf{m}2$ $6/\mathbf{mmm}$	$\varepsilon = \varepsilon_0 \begin{pmatrix} \mathbf{n}_o^2 & 0 & 0 \\ 0 & \mathbf{n}_o^2 & 0 \\ 0 & 0 & \mathbf{n}_e^2 \end{pmatrix}$
	Trigonal	3 $\bar{3}$ 32 $3\mathbf{m}$ $\bar{3}\mathbf{m}$	
Biaxial	Triclinic	1 $\bar{1}$	
	Monoclinic	2 $\mathbf{m}$ $2/\mathbf{m}$	$\varepsilon = \varepsilon_0 \begin{pmatrix} \mathbf{n}_1^2 & 0 & 0 \\ 0 & \mathbf{n}_2^2 & 0 \\ 0 & 0 & \mathbf{n}_3^2 \end{pmatrix}$
	Orthorhombic	222 $2\mathbf{mm}$ $\mathbf{mmm}$	

Source: Ref. 11.

where all three principal indices are different. We will discuss these three cases in more detail in the next section. The indices for the case where there are only two distinct values are named the ordinary index ($\mathbf{n}_o$) and the extraordinary index ($\mathbf{n}_e$). These labels are applied for historical reasons [10]. Erasmus Bartholinus, a Danish mathematician, in 1669 discovered double refraction in calcite. If the calcite crystal, split along its natural cleavage planes, is placed on a typewritten sheet of paper, two images of the letters will be observed. If the crystal is then rotated about an axis perpendicular to the page, one of the two images of the letters will rotate about the other. Bartholinus named the light rays from the letters that do not rotate the ordinary rays, and the rays from the rotating letters he named the extraordinary rays, hence the indices that produce these rays are named likewise. This explains the notation in the dielectric tensor for tetragonal, hexagonal, and trigonal crystals.

Let us consider such crystals in more detail. There is a plane in the material in which a single index would be measured in any direction. Light that is propagating in the direction normal to this plane with equal indices experiences the same refractive index for any polarization (orientation of the **E** vector). The direction for which this occurs is called the optic axis. Crystals having one optic axis are called uniaxial crystals. Materials with three principal indices have two directions in which the **E** vector experiences a single refractive index. These materials have two optic axes and are called biaxial crystals. This will be more fully explained in Section 24.4.1. Materials that have more than one principal index of refraction are called birefringent materials and are said to exhibit double refraction.

Crystals are composed of periodic arrays of atoms. The lattice of a crystal is a set of points in space. Sets of atoms that are identical in composition, arrangement, and orientation are attached to each lattice point. By translating the basic structure attached to the lattice point, we can fill space with the crystal. Define vectors **a**, **b**, and **c** which form three adjacent edges of a parallelepiped which spans the basic atomic structure. This parallelepiped is called a unit cell. We call the axes that lie along these vectors the crystal axes.

We would like to be able to describe a particular plane in a crystal, since crystals may be cut at any angle. The Miller indices are quantities that describe the orientation of planes in a crystal. The Miller indices are defined as follows: (1) locate the intercepts of the plane on the crystal axes—these will be multiples of lattice point spacing; (2) take the reciprocals of the intercepts and form the three smallest integers having the same ratio. For example, suppose we have a cubic crystal so that the crystal axes are the orthogonal Cartesian axes. Suppose further that the plane we want to describe intercepts the axes at the points 4, 3, and 2. The reciprocals of these intercepts are 1/4, 1/3, and 1/2. The Miller indices are then (3,4,6). This example serves to illustrate how the Miller indices are found, but it is more usual to encounter simpler crystal cuts. The same cubic crystal, if cut so that the intercepts are 1, ∞, ∞ (defining a plane parallel to the yz plane in the usual Cartesian coordinates) has Miller indices (1,0,0). Likewise, if the intercepts are 1, 1, ∞ (diagonal to two of the axes), the Miller indices are (1,1,0), and if the intercepts are 1, 1, 1 (diagonal to all three axes), the Miller indices are (1,1,1).

Two important electro-optic crystal types have the point group symbols $\bar{4}3\mathbf{m}$ (this is a cubic crystal, e.g., CdTe and GaAs) and $\bar{4}2\mathbf{m}$ (this is a tetragonal crystal, e.g., $AgGaS_2$). The linear and quadratic electro-optic tensors for these two crystal types, as well as all the other linear and quadratic electro-optic coefficient tensors for all crystal symmetry classes, are given in Tables 24-2 and 24-3. Note from these tables that the linear electro-optic effect vanishes for crystals that retain symmetry under inversion, i.e., centrosymmetric crystals, whereas the quadratic electro-optic effect never vanishes. For further discussion of this point, see Yariv and Yeh, [11].

24.4.1 The Index Ellipsoid

Light propagating in anisotropic materials experiences a refractive index and a phase velocity that depends on the propagation direction, polarization state, and

Table 24-2 Linear Electro-optic Tensors

System	Point group	Tensor
Centrosymmetric	$\bar{1}$, $\mathbf{2/m}$, $\mathbf{mmm}$, $\mathbf{4/m}$, $\mathbf{4/mmm}$, $\bar{3}$, $\bar{3}\mathbf{m}$, $\mathbf{6/m}$, $\mathbf{6/mmm}$, $\mathbf{m3}$, $\mathbf{m3m}$	$\begin{pmatrix} 0 & 0 & 0 \\ 0 & 0 & 0 \\ 0 & 0 & 0 \\ 0 & 0 & 0 \\ 0 & 0 & 0 \\ 0 & 0 & 0 \end{pmatrix}$
Triclinic	1	$\begin{pmatrix} \mathbf{r}_{11} & \mathbf{r}_{12} & \mathbf{r}_{13} \\ \mathbf{r}_{21} & \mathbf{r}_{22} & \mathbf{r}_{23} \\ \mathbf{r}_{31} & \mathbf{r}_{32} & \mathbf{r}_{33} \\ \mathbf{r}_{41} & \mathbf{r}_{42} & \mathbf{r}_{43} \\ \mathbf{r}_{51} & \mathbf{r}_{52} & \mathbf{r}_{53} \\ \mathbf{r}_{61} & \mathbf{r}_{62} & \mathbf{r}_{63} \end{pmatrix}$
Monoclinic	$2\ (2 \parallel \mathbf{x}_2)$	$\begin{pmatrix} 0 & \mathbf{r}_{12} & 0 \\ 0 & \mathbf{r}_{22} & 0 \\ 0 & \mathbf{r}_{32} & 0 \\ \mathbf{r}_{41} & 0 & \mathbf{r}_{43} \\ 0 & \mathbf{r}_{52} & 0 \\ \mathbf{r}_{61} & 0 & \mathbf{r}_{63} \end{pmatrix}$
	$2\ (2 \parallel \mathbf{x}_3)$	$\begin{pmatrix} 0 & 0 & \mathbf{r}_{13} \\ 0 & 0 & \mathbf{r}_{23} \\ 0 & 0 & \mathbf{r}_{33} \\ \mathbf{r}_{41} & \mathbf{r}_{42} & 0 \\ \mathbf{r}_{51} & \mathbf{r}_{52} & 0 \\ 0 & 0 & \mathbf{r}_{63} \end{pmatrix}$
	$\mathbf{m}\ (\mathbf{m} \perp \mathbf{x}_2)$	$\begin{pmatrix} \mathbf{r}_{11} & 0 & \mathbf{r}_{13} \\ \mathbf{r}_{21} & 0 & \mathbf{r}_{23} \\ \mathbf{r}_{31} & 0 & \mathbf{r}_{33} \\ 0 & \mathbf{r}_{42} & 0 \\ \mathbf{r}_{51} & 0 & \mathbf{r}_{53} \\ 0 & \mathbf{r}_{62} & 0 \end{pmatrix}$
	$\mathbf{m}\ (\mathbf{m} \perp \mathbf{x}_3)$	$\begin{pmatrix} \mathbf{r}_{11} & \mathbf{r}_{12} & 0 \\ \mathbf{r}_{21} & \mathbf{r}_{22} & 0 \\ \mathbf{r}_{31} & \mathbf{r}_{32} & 0 \\ 0 & 0 & \mathbf{r}_{43} \\ 0 & 0 & \mathbf{r}_{53} \\ \mathbf{r}_{61} & \mathbf{r}_{62} & 0 \end{pmatrix}$
Orthorhombic	222	$\begin{pmatrix} 0 & 0 & 0 \\ 0 & 0 & 0 \\ 0 & 0 & 0 \\ \mathbf{r}_{41} & 0 & 0 \\ 0 & \mathbf{r}_{52} & 0 \\ 0 & 0 & \mathbf{r}_{63} \end{pmatrix}$

(contd.)

Table 24-2 Continued

	2mm	$\begin{pmatrix} 0 & 0 & \mathbf{r}_{13} \\ 0 & 0 & \mathbf{r}_{23} \\ 0 & 0 & \mathbf{r}_{33} \\ 0 & \mathbf{r}_{42} & 0 \\ \mathbf{r}_{51} & 0 & 0 \\ 0 & 0 & 0 \end{pmatrix}$
Tetragonal	4	$\begin{pmatrix} 0 & 0 & \mathbf{r}_{13} \\ 0 & 0 & \mathbf{r}_{13} \\ 0 & 0 & \mathbf{r}_{33} \\ \mathbf{r}_{41} & \mathbf{r}_{51} & 0 \\ \mathbf{r}_{51} & -\mathbf{r}_{41} & 0 \\ 0 & 0 & 0 \end{pmatrix}$
	$\bar{4}$	$\begin{pmatrix} 0 & 0 & \mathbf{r}_{13} \\ 0 & 0 & -\mathbf{r}_{13} \\ 0 & 0 & 0 \\ \mathbf{r}_{41} & -\mathbf{r}_{51} & 0 \\ \mathbf{r}_{51} & \mathbf{r}_{41} & 0 \\ 0 & 0 & \mathbf{r}_{63} \end{pmatrix}$
	422	$\begin{pmatrix} 0 & 0 & 0 \\ 0 & 0 & 0 \\ 0 & 0 & 0 \\ \mathbf{r}_{41} & 0 & 0 \\ 0 & -\mathbf{r}_{41} & 0 \\ 0 & 0 & 0 \end{pmatrix}$
	4mm	$\begin{pmatrix} 0 & 0 & \mathbf{r}_{13} \\ 0 & 0 & \mathbf{r}_{13} \\ 0 & 0 & \mathbf{r}_{33} \\ 0 & \mathbf{r}_{51} & 0 \\ \mathbf{r}_{51} & 0 & 0 \\ 0 & 0 & 0 \end{pmatrix}$
	$\bar{4}2\mathbf{m}(2\parallel\mathbf{x}_1)$	$\begin{pmatrix} 0 & 0 & 0 \\ 0 & 0 & 0 \\ 0 & 0 & 0 \\ \mathbf{r}_{41} & 0 & 0 \\ 0 & \mathbf{r}_{41} & 0 \\ 0 & 0 & \mathbf{r}_{63} \end{pmatrix}$
Trigonal	3	$\begin{pmatrix} \mathbf{r}_{11} & -\mathbf{r}_{22} & \mathbf{r}_{13} \\ -\mathbf{r}_{11} & \mathbf{r}_{22} & \mathbf{r}_{13} \\ 0 & 0 & \mathbf{r}_{33} \\ \mathbf{r}_{41} & \mathbf{r}_{51} & 0 \\ \mathbf{r}_{51} & -\mathbf{r}_{41} & 0 \\ -\mathbf{r}_{22} & -\mathbf{r}_{11} & 0 \end{pmatrix}$
	32	$\begin{pmatrix} \mathbf{r}_{11} & 0 & 0 \\ -\mathbf{r}_{11} & 0 & 0 \\ 0 & 0 & 0 \\ \mathbf{r}_{41} & 0 & 0 \\ 0 & -\mathbf{r}_{41} & 0 \\ 0 & -\mathbf{r}_{11} & 0 \end{pmatrix}$

(contd.)

Table 24-2 Continued

	3m ($\mathbf{m}\perp\mathbf{x}_1$)	$\begin{pmatrix} 0 & -\mathbf{r}_{22} & \mathbf{r}_{13} \\ 0 & \mathbf{r}_{22} & \mathbf{r}_{13} \\ 0 & 0 & \mathbf{r}_{33} \\ 0 & \mathbf{r}_{51} & 0 \\ \mathbf{r}_{51} & 0 & 0 \\ -\mathbf{r}_{22} & 0 & 0 \end{pmatrix}$
	3**m**($\mathbf{m}\perp\mathbf{x}_2$)	$\begin{pmatrix} \mathbf{r}_{11} & 0 & \mathbf{r}_{13} \\ -\mathbf{r}_{11} & 0 & \mathbf{r}_{13} \\ 0 & 0 & \mathbf{r}_{33} \\ 0 & \mathbf{r}_{51} & 0 \\ \mathbf{r}_{51} & 0 & 0 \\ 0 & -\mathbf{r}_{11} & 0 \end{pmatrix}$
Hexagonal	6	$\begin{pmatrix} 0 & 0 & \mathbf{r}_{13} \\ 0 & 0 & \mathbf{r}_{13} \\ 0 & 0 & \mathbf{r}_{33} \\ \mathbf{r}_{41} & \mathbf{r}_{51} & 0 \\ \mathbf{r}_{51} & -\mathbf{r}_{41} & 0 \\ 0 & 0 & 0 \end{pmatrix}$
	6**mm**	$\begin{pmatrix} 0 & 0 & \mathbf{r}_{13} \\ 0 & 0 & \mathbf{r}_{13} \\ 0 & 0 & \mathbf{r}_{33} \\ 0 & \mathbf{r}_{51} & 0 \\ \mathbf{r}_{51} & 0 & 0 \\ 0 & 0 & 0 \end{pmatrix}$
	622	$\begin{pmatrix} 0 & 0 & 0 \\ 0 & 0 & 0 \\ 0 & 0 & 0 \\ \mathbf{r}_{41} & 0 & 0 \\ 0 & -\mathbf{r}_{41} & 0 \\ 0 & 0 & 0 \end{pmatrix}$
	$\bar{6}$	$\begin{pmatrix} \mathbf{r}_{11} & -\mathbf{r}_{22} & 0 \\ -\mathbf{r}_{11} & \mathbf{r}_{22} & 0 \\ 0 & 0 & 0 \\ 0 & 0 & 0 \\ 0 & 0 & 0 \\ -\mathbf{r}_{22} & -\mathbf{r}_{11} & 0 \end{pmatrix}$
	$\bar{6}$**m**2 ($\mathbf{m}\perp\mathbf{x}_1$)	$\begin{pmatrix} 0 & -\mathbf{r}_{22} & 0 \\ 0 & \mathbf{r}_{22} & 0 \\ 0 & 0 & 0 \\ 0 & 0 & 0 \\ 0 & 0 & 0 \\ -\mathbf{r}_{22} & 0 & 0 \end{pmatrix}$
	$\bar{6}$**m**2 ($\mathbf{m}\perp\mathbf{x}_2$)	$\begin{pmatrix} \mathbf{r}_{11} & 0 & 0 \\ -\mathbf{r}_{11} & 0 & 0 \\ 0 & 0 & 0 \\ 0 & 0 & 0 \\ 0 & 0 & 0 \\ 0 & -\mathbf{r}_{11} & 0 \end{pmatrix}$

(contd.)

Table 24-2 Continued

Cubic	$\bar{4}3\mathbf{m}$ 23	$\begin{pmatrix} 0 & 0 & 0 \\ 0 & 0 & 0 \\ 0 & 0 & 0 \\ \mathbf{r}_{41} & 0 & 0 \\ 0 & \mathbf{r}_{41} & 0 \\ 0 & 0 & \mathbf{r}_{41} \end{pmatrix}$
	432	$\begin{pmatrix} 0 & 0 & 0 \\ 0 & 0 & 0 \\ 0 & 0 & 0 \\ 0 & 0 & 0 \\ 0 & 0 & 0 \\ 0 & 0 & 0 \end{pmatrix}$

Source: Ref. 11

wavelength. The refractive index for propagation (for monochromatic light of some specified frequency) in an arbitrary direction (in Cartesian coordinates):

$$\bar{\mathbf{a}} = \mathbf{x}\hat{\mathbf{i}} + \mathbf{y}\hat{\mathbf{j}} + \mathbf{z}\hat{\mathbf{k}} \tag{24-20}$$

can be obtained from the index ellipsoid, a useful and lucid construct for visualization and determination of the index. (Note that we now shift from indexing the Cartesian directions with numbers to using x, y, and z.) In the principal coordinate system the index ellipsoid is given by

$$\frac{\mathbf{x}^2}{\mathbf{n}_x^2} + \frac{\mathbf{y}^2}{\mathbf{n}_y^2} + \frac{\mathbf{z}^2}{\mathbf{n}_z^2} = 1 \tag{24-21}$$

in the absence of an applied electric field. The lengths of the semimajor and semiminor axes of the ellipse formed by the intersection of this index ellipsoid and a plane normal to the propagation direction and passing through the center of the ellipsoid are the two principal indices of refraction for that propagation direction. Where there are three distinct principal indices, the crystal is defined as biaxial, and the above equation holds. If two of the three indices of the index ellipsoid are equal, the crystal is defined to be uniaxial and the equation for the index ellipsoid is

$$\frac{\mathbf{x}^2}{\mathbf{n}_o^2} + \frac{\mathbf{y}^2}{\mathbf{n}_o^2} + \frac{\mathbf{z}^2}{\mathbf{n}_e^2} = 1 \tag{24-22}$$

Uniaxial materials are said to be uniaxial positive when $\mathbf{n}_o < \mathbf{n}_e$ and uniaxial negative when $\mathbf{n}_o > \mathbf{n}_e$. When there is a single index for any direction in space, the crystal is isotropic and the equation for the ellipsoid becomes that for a sphere:

$$\frac{\mathbf{x}^2}{\mathbf{n}^2} + \frac{\mathbf{y}^2}{\mathbf{n}^2} + \frac{\mathbf{z}^2}{\mathbf{n}^2} = 1 \tag{24-23}$$

The index ellipsoids for isotropic, uniaxial, and biaxial crystals are illustrated in Fig. 24-1.

Examples of isotropic materials are CdTe, NaCl, diamond, and GaAs. Examples of uniaxial positive materials are quartz and ZnS. Materials that are uniaxial negative include calcite, $LiNbO_3$, $BaTiO_3$, and KDP (KH_2PO_4). Examples of biaxial materials are gypsum and mica.

Table 24-3 Quadratic Electro-optic Tensors

System	Class	Tensor
Triclinic	1 $\bar{1}$	$\begin{pmatrix} s_{11} & s_{12} & s_{13} & s_{14} & s_{15} & s_{16} \\ s_{21} & s_{22} & s_{23} & s_{24} & s_{25} & s_{26} \\ s_{31} & s_{32} & s_{33} & s_{34} & s_{35} & s_{36} \\ s_{41} & s_{42} & s_{43} & s_{44} & s_{45} & s_{46} \\ s_{51} & s_{52} & s_{53} & s_{54} & s_{55} & s_{56} \\ s_{61} & s_{62} & s_{63} & s_{64} & s_{65} & s_{66} \end{pmatrix}$
Monoclinic	2 **m** 2/**m**	$\begin{pmatrix} s_{11} & s_{12} & s_{13} & 0 & s_{15} & 0 \\ s_{21} & s_{22} & s_{23} & 0 & s_{25} & 0 \\ s_{31} & s_{32} & s_{33} & 0 & s_{35} & 0 \\ 0 & 0 & 0 & s_{44} & 0 & s_{46} \\ s_{51} & s_{52} & s_{53} & 0 & s_{55} & 0 \\ 0 & 0 & 0 & s_{64} & 0 & s_{66} \end{pmatrix}$
Orthorhombic	2**mm** 222 **mmm**	$\begin{pmatrix} s_{11} & s_{12} & s_{13} & 0 & 0 & 0 \\ s_{21} & s_{22} & s_{23} & 0 & 0 & 0 \\ s_{31} & s_{32} & s_{33} & 0 & 0 & 0 \\ 0 & 0 & 0 & s_{44} & 0 & 0 \\ 0 & 0 & 0 & 0 & s_{55} & 0 \\ 0 & 0 & 0 & 0 & 0 & s_{66} \end{pmatrix}$
Tetragonal	4 $\bar{4}$ 4/**m**	$\begin{pmatrix} s_{11} & s_{12} & s_{13} & 0 & 0 & s_{16} \\ s_{12} & s_{11} & s_{13} & 0 & 0 & -s_{16} \\ s_{31} & s_{31} & s_{33} & 0 & 0 & 0 \\ 0 & 0 & 0 & s_{44} & s_{45} & 0 \\ 0 & 0 & 0 & -s_{45} & s_{44} & 0 \\ s_{61} & -s_{61} & 0 & 0 & 0 & s_{66} \end{pmatrix}$
	422 4**mm** $\bar{4}$2**m** 4/**mm**	$\begin{pmatrix} s_{11} & s_{12} & s_{13} & 0 & 0 & 0 \\ s_{12} & s_{11} & s_{13} & 0 & 0 & 0 \\ s_{31} & s_{31} & s_{33} & 0 & 0 & 0 \\ 0 & 0 & 0 & s_{44} & 0 & 0 \\ 0 & 0 & 0 & 0 & s_{44} & 0 \\ 0 & 0 & 0 & 0 & 0 & s_{66} \end{pmatrix}$
Trigonal	3 $\bar{3}$	$\begin{pmatrix} s_{11} & s_{12} & s_{13} & s_{14} & s_{15} & -s_{61} \\ s_{12} & s_{11} & s_{13} & -s_{14} & -s_{15} & s_{61} \\ s_{31} & s_{31} & s_{33} & 0 & 0 & 0 \\ s_{41} & -s_{41} & 0 & s_{44} & s_{45} & -s_{51} \\ s_{51} & -s_{51} & 0 & -s_{45} & s_{44} & s_{41} \\ s_{61} & -s_{61} & 0 & -s_{15} & s_{14} & \frac{1}{2}(s_{11}-s_{12}) \end{pmatrix}$
	32 3**m** $\bar{3}$**m**	$\begin{pmatrix} s_{11} & s_{12} & s_{13} & s_{14} & 0 & 0 \\ s_{12} & s_{11} & s_{13} & -s_{14} & 0 & 0 \\ s_{13} & s_{13} & s_{33} & 0 & 0 & 0 \\ s_{41} & -s_{41} & 0 & s_{44} & 0 & 0 \\ 0 & 0 & 0 & 0 & s_{44} & s_{41} \\ 0 & 0 & 0 & 0 & s_{14} & \frac{1}{2}(s_{11}-s_{12}) \end{pmatrix}$
Hexagonal	6 $\bar{6}$ 6/**m**	$\begin{pmatrix} s_{11} & s_{12} & s_{13} & 0 & 0 & -s_{61} \\ s_{12} & s_{11} & s_{13} & 0 & 0 & s_{61} \\ s_{31} & s_{31} & s_{33} & 0 & 0 & 0 \\ 0 & 0 & 0 & s_{44} & s_{45} & 0 \\ 0 & 0 & 0 & -s_{45} & s_{44} & 0 \\ s_{61} & -s_{61} & 0 & 0 & 0 & \frac{1}{2}(s_{11}-s_{12}) \end{pmatrix}$

(contd.)

Table 24-3 Continued

	622 **6mm** $\bar{6}$**m**2 6/**mmm**	$\begin{pmatrix} s_{11} & s_{12} & s_{13} & 0 & 0 & 0 \\ s_{12} & s_{11} & s_{13} & 0 & 0 & 0 \\ s_{31} & s_{31} & s_{33} & 0 & 0 & 0 \\ 0 & 0 & 0 & s_{44} & 0 & 0 \\ 0 & 0 & 0 & 0 & s_{44} & 0 \\ 0 & 0 & 0 & 0 & 0 & \frac{1}{2}(s_{11}-s_{12}) \end{pmatrix}$
Cubic	23 **m**3	$\begin{pmatrix} s_{11} & s_{12} & s_{13} & 0 & 0 & 0 \\ s_{13} & s_{11} & s_{12} & 0 & 0 & 0 \\ s_{12} & s_{13} & s_{11} & 0 & 0 & 0 \\ 0 & 0 & 0 & s_{44} & 0 & 0 \\ 0 & 0 & 0 & 0 & s_{44} & 0 \\ 0 & 0 & 0 & 0 & 0 & s_{44} \end{pmatrix}$
	432 **m**3**m** $\bar{4}$3**m**	$\begin{pmatrix} s_{11} & s_{12} & s_{12} & 0 & 0 & 0 \\ s_{12} & s_{11} & s_{12} & 0 & 0 & 0 \\ s_{12} & s_{12} & s_{11} & 0 & 0 & 0 \\ 0 & 0 & 0 & s_{44} & 0 & 0 \\ 0 & 0 & 0 & 0 & s_{44} & 0 \\ 0 & 0 & 0 & 0 & 0 & s_{44} \end{pmatrix}$
Isotropic		$\begin{pmatrix} s_{11} & s_{12} & s_{12} & 0 & 0 & 0 \\ s_{12} & s_{11} & s_{12} & 0 & 0 & 0 \\ s_{12} & s_{12} & s_{11} & 0 & 0 & 0 \\ 0 & 0 & 0 & \frac{1}{2}(s_{11}-s_{12}) & 0 & 0 \\ 0 & 0 & 0 & 0 & \frac{1}{2}(s_{11}-s_{12}) & 0 \\ 0 & 0 & 0 & 0 & 0 & \frac{1}{2}(s_{11}-s_{12}) \end{pmatrix}$

Source: Ref. 11

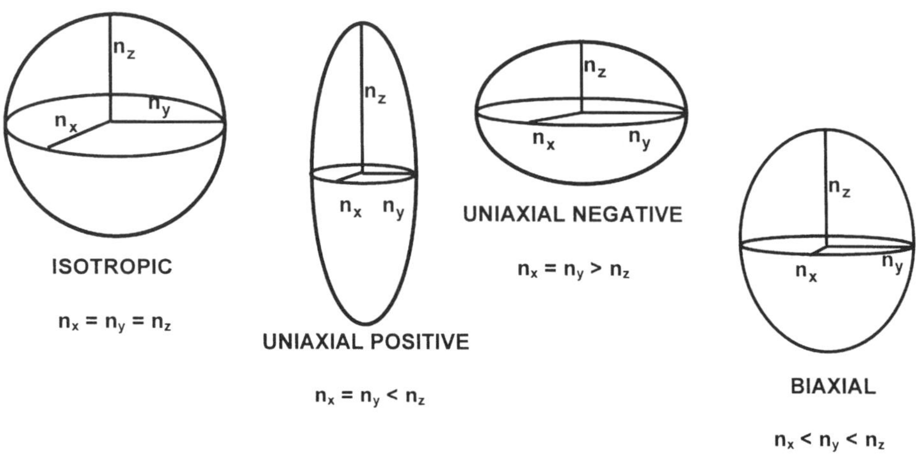

Figure 24-1 Index ellipsoids.

24.4.2 Natural Birefringence

Many materials have natural birefringence, i.e., they are uniaxial or biaxial in their natural (absence of applied fields) state. These materials are often used in passive devices such as polarizers and retarders. Calcite is one of the most important naturally birefringent materials for optics, and is used in a variety of well known polarizers, e.g., the Nichol, Wollaston, or Glan-Thompson prisms. As we shall see later, naturally isotropic materials can be made birefringent, and materials that have natural birefringence can be made to change that birefringence with the application of electromagnetic fields.

24.4.3 The Wave Surface

There are two additional methods of depicting the effect of crystal anisotropy on light. Neither is as satisfying or useful to this author as the index ellipsoid; however, both will be mentioned for the sake of completeness and in order to facilitate understanding of those references that use these models. They are most often used to explain birefringence, e.g., in the operation of calcite-based devices [12–14].

The first of these is called the wave surface. As a light wave from a point source expands through space, it forms a surface that represents the wave front. This surface consists of points having equal phase. At a particular instant in time, the wave surface is a representation of the velocity surface of a wave expanding in the medium; it is a measure of the distance through which the wave has expanded from some point over some time period. Because the wave will have expanded further (faster) when experiencing a low refractive index and expanded less (slower) when experiencing high index, the size of the wave surface is inversely proportional to the index.

To illustrate the use of the wave surface, consider a uniaxial crystal. Recall that we have defined the optic axis of a uniaxial crystal as the direction in which the speed of propagation is independent of polarization. The optic axes for positive and negative uniaxial crystals are shown on the index ellipsoids in Fig. 24-2, and the optic axes for a biaxial crystal are shown on the index ellipsoid in Fig. 24-3.

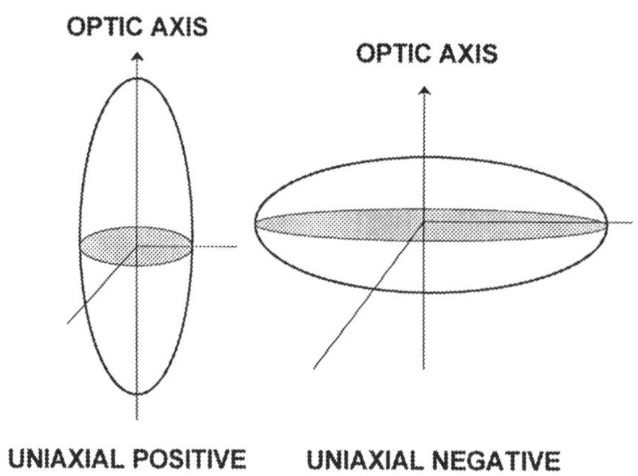

Figure 24-2 Optic axis on index ellipsoid for uniaxial positive and uniaxial negative crystals.

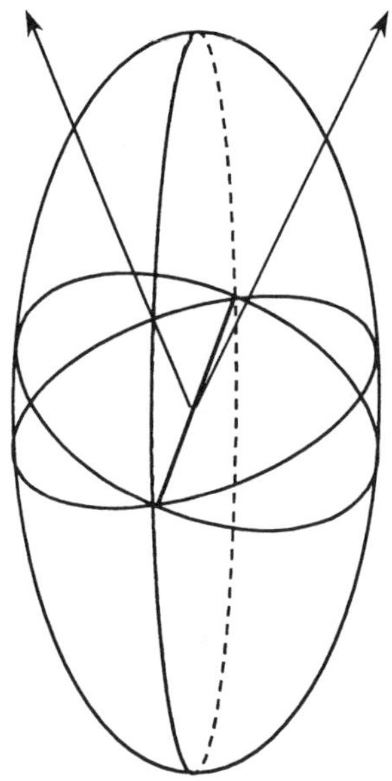

Figure 24-3 Optic axes on index ellipsoid for biaxial crystals.

The wave surfaces are now shown in Fig. 24-4 for both positive and negative uniaxial materials. The upper diagram for each pair shows the wave surface for polarization perpendicular to the optic axes (also perpendicular to the principal section through the ellipsoid), and the lower diagram shows the wave surface for polarization in the plane of the principal section. The index ellipsoid surfaces are shown for reference. Similarly, cross-sections of the wave surfaces for biaxial materials are shown in Fig. 24-5. In all cases, polarization perpendicular to the plane of the page is indicated with solid circles along the rays, whereas polarization parallel to the plane of the page is shown with short double-headed arrows along the rays.

24.4.4 The Wavevector Surface

A second method of depicting the effect of crystal anisotropy on light is the wavevector surface. The wavevector surface is a measure of the variation of the value of k, the wavevector, for different propagation directions and different polarizations. Recall that

$$\mathbf{k} = \frac{2\pi}{\lambda} = \frac{\omega \mathbf{n}}{\mathbf{c}} \tag{24-24}$$

so $\mathbf{k} \propto \mathbf{n}$. Wavevector surfaces for uniaxial crystals will then appear as shown in Fig 24-6. Compare these to the wave surfaces in Fig. 24-4.

Wavevector surfaces for biaxial crystals are more complicated. Cross-sections of the wavevector surface for a biaxial crystal where $\mathbf{n_x} < \mathbf{n_y} < \mathbf{n_z}$ are shown in Fig. 24-7. Compare these to the wave surfaces in Fig. 24-5.

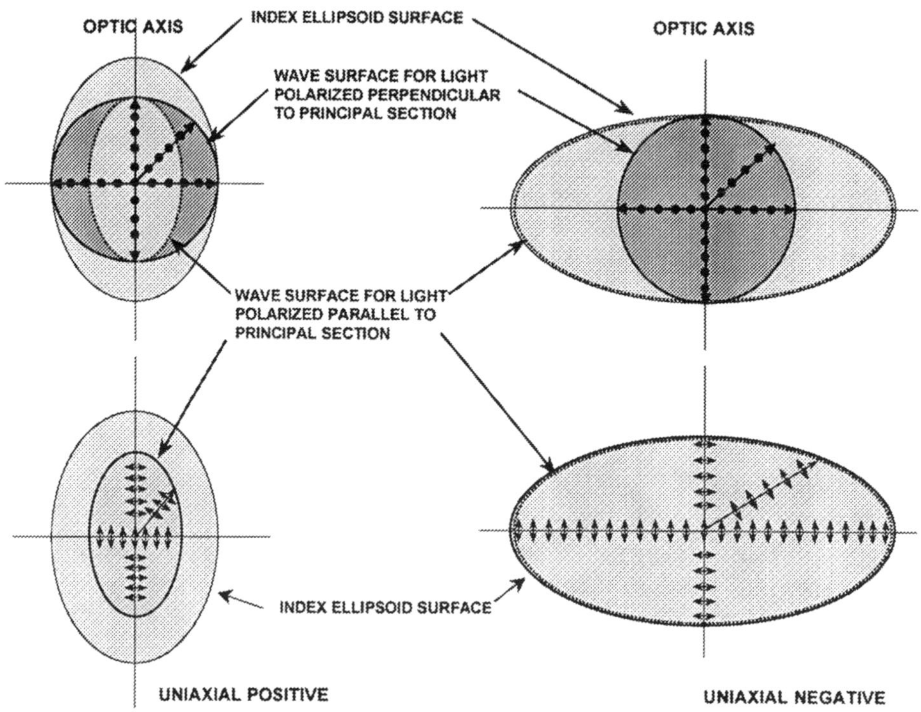

Figure 24-4 Wave surfaces for uniaxial positive and negative materials.

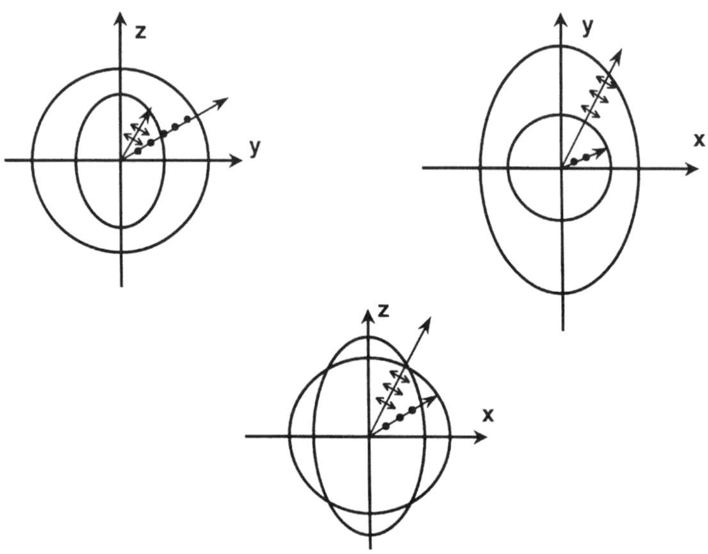

Figure 24-5 Wave surfaces for biaxial materials in principal planes.

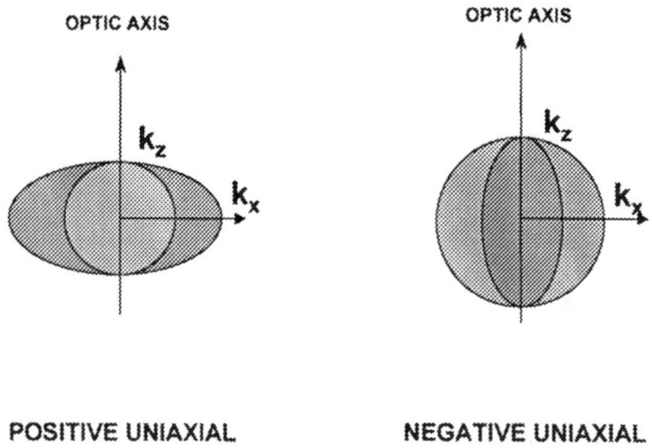

Figure 24-6 Wavevector surfaces for positive and negative uniaxial crystals.

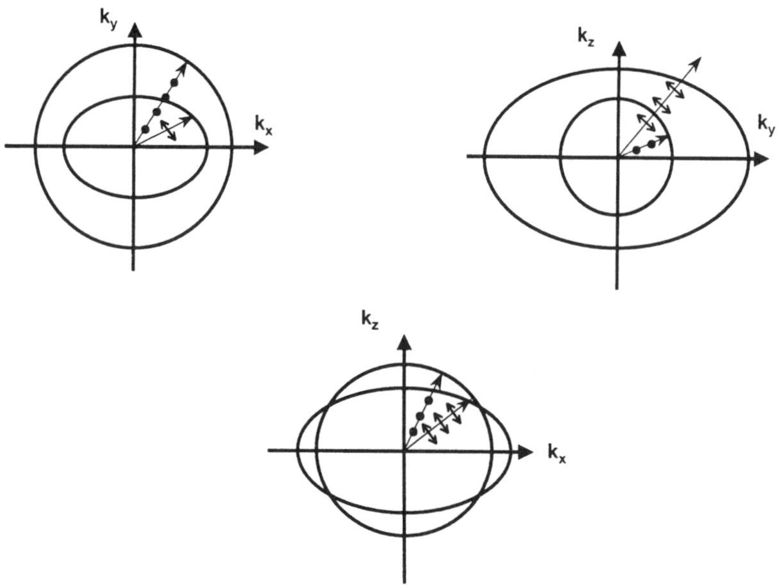

Figure 24-7 Wavevector surface cross sections for biaxial crystals.

24.5 APPLICATION OF ELECTRIC FIELDS: INDUCED BIREFRINGENCE AND POLARIZATION MODULATION

When fields are applied to materials, whether isotropic or anisotropic, birefringence can be induced or modified. This is the principle of a modulator; it is one of the most important optical devices, since it gives control over the phase and/or amplitude of light.

The alteration of the index ellipsoid of a crystal on application of an electric and/or magnetic field can be used to modulate the polarization state. The equation for the index ellipsoid of a crystal in an electric field is

$$\eta_{ij}(\mathbf{E})\mathbf{x}_i\mathbf{x}_j = 1 \qquad (24\text{-}25)$$

or

$$(\eta_{ij}(0) + \Delta\eta_{ij})\mathbf{x_i x_j} = 1 \tag{24-26}$$

This equation can be written as

$$\mathbf{x}^2\left(\frac{1}{\mathbf{n_x^2}} + \Delta\left(\frac{1}{\mathbf{n}}\right)_1^2\right) + \mathbf{y}^2\left(\frac{1}{\mathbf{n_y^2}} + \Delta\left(\frac{1}{\mathbf{n}}\right)_2^2\right) + \mathbf{z}^2\left(\frac{1}{\mathbf{n_z^2}} + \Delta\left(\frac{1}{\mathbf{n}}\right)_3^2\right)$$
$$+ 2\mathbf{yz}\left(\Delta\left(\frac{1}{\mathbf{n}}\right)_4^2\right) + 2\mathbf{xz}\left(\Delta\left(\frac{1}{\mathbf{n}}\right)_5^2\right) + 2\mathbf{xy}\left(\Delta\left(\frac{1}{\mathbf{n}}\right)_6^2\right) = 1 \tag{24-27}$$

or

$$\mathbf{x}^2\left(\frac{1}{\mathbf{n_x^2}} + \mathbf{r_{1k}E_k} + \mathbf{s_{1k}E_k^2} + 2\mathbf{s_{14}E_2E_3} + 2\mathbf{s_{15}E_3E_1} + 2\mathbf{s_{16}E_1E_2}\right)$$
$$+ \mathbf{y}^2\left(\frac{1}{\mathbf{n_y^2}} + \mathbf{r_{2k}E_k} + \mathbf{s_{2k}E_k^2} + 2\mathbf{s_{24}E_2E_3} + 2\mathbf{s_{25}E_3E_1} + 2\mathbf{s_{26}E_1E_2}\right)$$
$$+ \mathbf{z}^2\left(\frac{1}{\mathbf{n_z^2}} + \mathbf{r_{3k}E_k} + \mathbf{s_{3k}E_k^2} + 2\mathbf{s_{34}E_2E_3} + 2\mathbf{s_{35}E_3E_1} + 2\mathbf{s_{36}E_1E_2}\right)$$
$$+ 2\mathbf{yz}(\mathbf{r_{4k}E_k} + \mathbf{s_{4k}E_k^2} + 2\mathbf{s_{44}E_2E_3} + 2\mathbf{s_{45}E_3E_1} + 2\mathbf{s_{46}E_1E_2})$$
$$+ 2\mathbf{zx}(\mathbf{r_{5k}E_k} + \mathbf{s_{5k}E_k^2} + 2\mathbf{s_{54}E_2E_3} + 2\mathbf{s_{55}E_3E_1} + 2\mathbf{s_{56}E_1E_2})$$
$$+ 2\mathbf{xy}(\mathbf{r_{6k}E_k} + \mathbf{s_{6k}E_k^2} + 2\mathbf{s_{64}E_2E_3} + 2\mathbf{s_{65}E_3E_1} + 2\mathbf{s_{66}E_1E_2}) = 1 \tag{24-28}$$

where the $\mathbf{E_k}$ are components of the electric field along the principal axes and repeated indices are summed.

If the quadratic coefficients are assumed to be small and only the linear coefficients are retained, then

$$\Delta\left(\frac{1}{\mathbf{n}}\right)_1^2 = \sum_{k=1}^{3} \mathbf{r_{lk}E_k} \tag{24-29}$$

and $k = 1, 2, 3$ corresponds to the principal axes x, y, and z. The equation for the index ellipsoid becomes

$$\mathbf{x}^2\left(\frac{1}{\mathbf{n_x^2}} + \mathbf{r_{1k}E_k}\right) + \mathbf{y}^2\left(\frac{1}{\mathbf{n_y^2}} + \mathbf{r_{2k}E_k}\right) + \mathbf{z}^2\left(\frac{1}{\mathbf{n_z^2}} + \mathbf{r_{3k}E_k}\right)$$
$$+ 2\mathbf{yz}(\mathbf{r_{4k}E_k}) + 2\mathbf{zx}(\mathbf{r_{5k}E_k}) + 2\mathbf{xy}(\mathbf{r_{6k}E_k}) = 1 \tag{24-30}$$

Suppose we have a cubic crystal of point group $\bar{4}3\mathbf{m}$, the symmetry group of such common materials as GaAs. Suppose further that the field is in the z direction. Then, the index ellipsoid is

$$\frac{\mathbf{x}^2}{\mathbf{n}^2} + \frac{\mathbf{y}^2}{\mathbf{n}^2} + \frac{\mathbf{z}^2}{\mathbf{n}^2} + 2\mathbf{r_{41}E_z xy} = 1 \tag{24-31}$$

The applied electric field couples the x-polarized and y-polarized waves. If we make the coordinate transformation:

$$\mathbf{x} = \mathbf{x}'\cos 45^\circ - \mathbf{y}'\sin 45^\circ$$
$$\mathbf{y} = \mathbf{x}'\sin 45^\circ - \mathbf{y}'\cos 45^\circ \tag{24-32}$$

and substitute these equations into the equation for the ellipsoid, the new equation for the ellipsoid becomes

$$\mathbf{x}'^2\left(\frac{1}{\mathbf{n}^2} + \mathbf{r}_{41}\mathbf{E}_z\right) + \mathbf{y}'^2\left(\frac{1}{\mathbf{n}^2} - \mathbf{r}_{41}\mathbf{E}_z\right) + \frac{\mathbf{z}^2}{\mathbf{n}^2} = 1 \tag{24-33}$$

and we have eliminated the cross term. We want to obtain the new principal indices. The principal index will appear in Eq. (24-33) as $1/\mathbf{n}_{x'}^2$ and must be equal to the quantity in the first parenthesis of the equation for the ellipsoid, i.e.,

$$\frac{1}{\mathbf{n}_{x'}^2} = \frac{1}{\mathbf{n}^2} + \mathbf{r}_{41}\mathbf{E}_z \tag{24-34}$$

We can solve for $\mathbf{n}_{x'}$ so (24-34) becomes

$$\mathbf{n}_{x'} = \mathbf{n}(1 + \mathbf{n}^2\mathbf{r}_{41}\mathbf{E}_z)^{1/2} \tag{24-35}$$

We assume $\mathbf{n}^2\mathbf{r}_{41}\mathbf{E}_z \ll 1$ so that the term in parentheses in (24-35) is approximated by

$$(1 + \mathbf{n}^2\mathbf{r}_{41}\mathbf{E}_z)^{1/2} \cong \left(1 - \frac{1}{2}\mathbf{n}^2\mathbf{r}_{41}\mathbf{E}_z\right) \tag{24-36}$$

The equations for the new principal indices are

$$\mathbf{n}_{x'} = \mathbf{n} - \frac{1}{2}\mathbf{n}^3\mathbf{r}_{41}\mathbf{E}_z \tag{24–37}$$
$$\mathbf{n}_{y'} = \mathbf{n} + \frac{1}{2}\mathbf{n}^3\mathbf{r}_{41}\mathbf{E}_z$$
$$\mathbf{n}_{z'} = \mathbf{n}.$$

As a similar example for another important materials type, suppose we have a tetragonal (point group $\bar{4}2\mathbf{m}$) uniaxial crystal in a field along z. The index ellipsoid becomes

$$\frac{\mathbf{x}^2}{\mathbf{n}_o^2} + \frac{\mathbf{y}^2}{\mathbf{n}_o^2} + \frac{\mathbf{z}^2}{\mathbf{n}_e^2} + 2\mathbf{r}_{63}\mathbf{E}_z\mathbf{x}\mathbf{y} = 1 \tag{24-38}$$

A coordinate rotation can be done to obtain the major axes of the new ellipsoid. In the present example, this yields the new ellipsoid:

$$\left(\frac{1}{\mathbf{n}_o^2} + \mathbf{r}_{63}\mathbf{E}_z\right)\mathbf{x}'^2 + \left(\frac{1}{\mathbf{n}_o^2} - \mathbf{r}_{63}\mathbf{E}_z\right)\mathbf{y}'^2 + \left(\frac{\mathbf{z}^2}{\mathbf{n}_e^2}\right) = 1 \tag{24-39}$$

As in the first example, the new and old z axes are the same, but the new $\mathbf{x}'$ and $\mathbf{y}'$ axes are 45° from the original $\mathbf{x}$ and $\mathbf{y}$ axes (see Fig. 24-8).

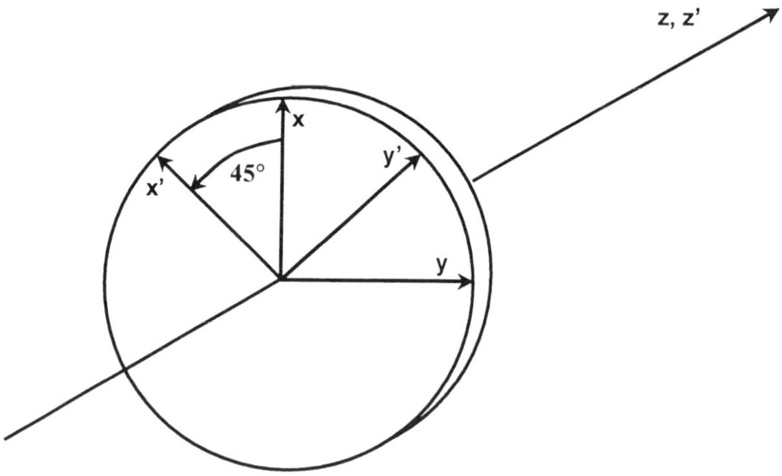

Figure 24-8 Rotated principal axes.

The refractive indices along the new x and y axes are

$$\mathbf{n}'_x = \mathbf{n}_o - \frac{1}{2}\mathbf{n}_o^3\mathbf{r}_{63}\mathbf{E}_z \tag{24-40}$$

$$\mathbf{n}'_y = \mathbf{n}_o + \frac{1}{2}\mathbf{n}_o^3\mathbf{r}_{63}\mathbf{E}_z$$

Note that the quantity $\mathbf{n}^3\mathbf{rE}$ in these examples determines the change in refractive index. Part of this product, $\mathbf{n}^3\mathbf{r}$, depends solely on inherent material properties, and is a figure of merit for electro-optical materials. Values for the linear and quadratic electro-optic coefficients for selected materials are given in Tables 24-4 and 24-5, along with values for $\mathbf{n}$ and, for linear materials, $\mathbf{n}^3\mathbf{r}$. While much of the information from these tables is from Yariv and Yeh [11], materials tables are also to be found in Kaminow [5,15]. Original sources listed in these references should be consulted on materials of particular interest. Additional information on many of the materials listed here, including tables of refractive index versus wavelength and dispersion formulas, can be found in Tropf et al. [16].

For light linearly polarized at 45°, the x and y components experience different refractive indices $\mathbf{n}'_x$ and $\mathbf{n}'_y$. The birefringence is defined as the index difference $\mathbf{n}'_y - \mathbf{n}'_x$. Since the phase velocities of the x and y components are different, there is a phase retardation Γ (in radians) between the x and y components of E given by

$$\Gamma = \frac{\omega}{c}(\mathbf{n}'_y - \mathbf{n}'_x)\mathrm{d} = \frac{2\pi}{\lambda}\mathbf{n}_o^3\mathbf{r}_{63}\mathbf{E}_z\mathrm{d} \tag{24-41}$$

where d is the path length of light in the crystal. The electric field of the incident light beam is

$$\bar{\mathbf{E}} = \frac{1}{\sqrt{2}}\mathbf{E}(\hat{\mathbf{x}} + \hat{\mathbf{y}}) \tag{24-42}$$

After transmission through the crystal, the electric field is

$$\frac{1}{\sqrt{2}}\mathbf{E}\left(\mathbf{e}^{i\Gamma/2}\hat{\mathbf{x}}' + \mathbf{e}^{-i\Gamma/2}\hat{\mathbf{y}}'\right) \tag{24-43}$$

Table 24-4 Linear Electro-optic Coefficients

Substance	Symmetry	Wavelength (μm)	Electrooptic Coefficients r_{lk} (10^{-12} m/V)	Indices of Refraction	n^3r (10^{-12} m/V)
CdTe	$\bar{4}3m$	1.0	$r_{41} = 4.5$	$n = 2.84$	103
		3.39	$r_{41} = 6.8$		
		10.6	$r_{41} = 6.8$	$n = 2.60$	120
		23.35	$r_{41} = 5.47$	$n = 2.58$	94
		27.95	$r_{41} = 5.04$	$n = 2.53$	82
GaAs	$\bar{4}3m$	0.9	$r_{41} = 1.1$	$n = 3.60$	51
		1.15	$r_{41} = 1.43$	$n = 3.43$	58
		3.39	$r_{41} = 1.24$	$n = 3.3$	45
		10.6	$r_{41} = 1.51$	$n = 3.3$	54
ZnSe	$\bar{4}3m$	0.548	$r_{41} = 2.0$	$n = 2.66$	
		0.633	r_{41}[a] $= 2.0$	$n = 2.60$	35
		10.6	$r_{41} = 2.2$	$n = 2.39$	
ZnTe	$\bar{4}3m$	0.589	$r_{41} = 4.51$	$n = 3.06$	
		0.616	$r_{41} = 4.27$	$n = 3.01$	
		0.633	$r_{41} = 4.04$	$n = 2.99$	108
			r_{41}[a] $= 4.3$		
		0.690	$r_{41} = 3.97$	$n = 2.93$	
		3.41	$r_{41} = 4.2$	$n = 2.70$	83
		10.6	$r_{41} = 3.9$	$n = 2.70$	77
$Bi_{12}SiO_{20}$	23	0.633	$r_{41} = 5.0$	$n = 2.54$	82
CdS	6mm	0.589	$r_{51} = 3.7$	$n_o = 2.501$	
				$n_e = 2.519$	
		0.633	$r_{51} = 1.6$	$n_o = 2.460$	
				$n_e = 2.477$	
		1.15	$r_{31} = 3.1$	$n_o = 2.320$	
			$r_{33} = 3.2$	$n_e = 2.336$	
			$r_{51} = 2.0$		
		3.39	$r_{13} = 3.5$	$n_o = 2.276$	
			$r_{33} = 2.9$	$n_e = 2.292$	
			$r_{51} = 2.0$		
		10.6	$r_{13} = 2.45$	$n_o = 2.226$	
			$r_{33} = 2.75$	$n_e = 2.239$	
			$r_{51} = 1.7$		
CdSe	6mm	3.39	r_{13}[a] $= 1.8$	$n_o = 2.452$	
			$r_{33} = 4.3$	$n_e = 2.471$	
PLZT[b]	∞m	0.546	$n_e^3 r_{33} - n_o^3 r_{13} = 2320$	$n_o = 2.55$	
($Pb_{0.814}La_{0.124}Zr_{0.4}Ti_{0.6}O_3$)					
$LiNbO_3$	3m	0.633	$r_{13} = 9.6$	$n_o = 2.286$	
			$r_{22} = 6.8$	$n_e = 2.200$	
			$r_{33} = 30.9$		
			$r_{51} = 32.6$		

(contd.)

Table 24-4 Continued

Substance	Symmetry	Wavelength (μm)	Electrooptic Coefficients r_{lk} (10^{-12} m/V)	Indices of Refraction	$\mathbf{n}^3\mathbf{r}$ (10^{-12} m/V)
		1.15	$r_{22} = 5.4$	$n_o = 2.229$	
				$n_e = 2.150$	
		3.39	$r_{22} = 3.1$	$n_o = 2.136$	
				$n_e = 2.073$	
$LiTaO_3$	**3m**	0.633	$r_{13} = 8.4$	$n_o = 2.176$	
			$r_{33} = 30.5$	$n_e = 2.180$	
			$r_{22} = -0.2$		
		3.39	$r_{33} = 27$	$n_o = 2.060$	
			$r_{13} = 4.5$	$n_e = 2.065$	
			$r_{51} = 15$		
			$r_{22} = 0.3$		
KDP (KH_2PO_4)	$\bar{4}$**2m**	0.546	$r_{41} = 8.77$	$n_o = 1.5115$	
			$r_{63} = 10.3$	$n_e = 1.4698$	
		0.633	$r_{41} = 8$	$n_o = 1.5074$	
			$r_{63} = 11$	$n_e = 1.4669$	
		3.39	$r_{63} = 9.7$		
			$n_o^3 r_{63} = 33$		
ADP ($NH_4H_2PO_4$)	$\bar{4}$**2m**	0.546	$r_{41} = 23.76$	$n_o = 1.5079$	
			$r_{63} = 8.56$	$n_e = 1.4683$	
		0.633	$r_{63} = 24.1$		
$RbHSeO_4$[c]		0.633			13,540
$BaTiO_3$	**4mm**	0.546	$r_{51} = 1640$	$n_o = 2.437$	
				$n_e = 2.365$	
KTN ($KTa_xNb_{1-x}O_3$)	**4mm**	0.633	$r_{51} = 8000$	$n_o = 2.318$	
				$n_e = 2.277$	
$AgGaS_2$	$\bar{4}$**2m**	0.633	$r_{41} = 4.0$	$n_o = 2.553$	
			$r_{63} = 3.0$	$n_e = 2.507$	

[a]These values are for clamped (high-frequency field) operation.
[b]PLZT is a compound of Pb, La, Zr, Ti, and O [17,18]. The concentration ratio of Zr to Ti is most important to its electro-optic properties. In this case, the ratio is 40 : 60.
[c]*Source:* Ref. 19.

If the path length and birefringence are selected such that $\Gamma = \pi$, the modulated crystal acts as a half-wave linear retarder and the transmitted light has field components:

$$\frac{1}{\sqrt{2}}\mathbf{E}\left(\mathbf{e}^{\mathrm{i}\pi/2}\hat{\mathbf{x}}' + \mathbf{e}^{-\mathrm{i}\pi/2}\hat{\mathbf{y}}'\right) = \frac{1}{\sqrt{2}}\mathbf{E}\left(\mathbf{e}^{i\pi/2}\hat{\mathbf{x}}' - \mathbf{e}^{\mathrm{i}\pi/2}\hat{\mathbf{y}}'\right)$$

$$= \mathbf{E}\frac{\mathbf{e}^{\mathrm{i}\pi/2}}{\sqrt{2}}\left(\hat{\mathbf{x}}' - \hat{\mathbf{y}}'\right) \tag{24-44}$$

The axis of linear polarization of the incident beam has been rotated by 90° by the phase retardation of π radians or one-half wavelength. The incident linear polarization state has been rotated into the orthogonal polarization state. An analyzer at the

Table 24-5 Quadratic Electro-optic Coefficients

Substance	Symmetry	Wavelength (μm)	Electro-optic Coefficients s_{ij} (10^{-18} m^2/V^2)	Index of Refraction	Temperature (°C)
$BaTiO_3$	**m3m**	0.633	$s_{11} - s_{12} = 2290$	$n = 2.42$	$T > T_c$ ($T_c = 120°C$)
PLZT[a]	∞**m**	0.550	$s_{33} - s_{13} = 26000/n^3$	$n = 2.450$	Room temperature
KH_2PO_4 (KDP)	$\bar{4}$**2m**	0.540	$n_e^3(s_{33} - s_{13}) = 31$ $n_o^3(s_{31} - s_{11}) = 13.5$ $n_o^3(s_{12} - s_{11}) = 8.9$ $n_o^3 s_{66} = 3.0$	$n_o = 1.5115$[b] $n_e = 1.4698$[b]	Room temperature
$NH_4H_2PO_4$ (ADP)	$\bar{4}$**2m**	0.540	$n_e^3(s_{33} - s_{13}) = 24$ $n_o^3(s_{31} - s_{11}) = 16.5$ $n_o^3(s_{12} - s_{11}) = 5.8$ $n_o^3 s_{66} = 2$	$n_o = 1.5266$[b] $n_e = 1.4808$[b]	Room temperature

[a]PLZT is a compound of Pb, La, Zr, Ti, and O [17,18]. The concentration ratio of Zr to Ti is most important to its electro-optic properties; in this case, the ratio is 65 : 35.
[b]At 0.546 μm.
Source: Ref. 11.

output end of the crystal aligned with the incident (or unmodulated) plane of polarization will block the modulated beam. For an arbitrary applied voltage producing a phase retardation of Γ the analyzer transmits a fractional intensity $\cos^2 \Gamma$. This is the principle of the Pockels cell.

Note that the form of the equations for the indices resulting from the application of a field is highly dependent on the direction of the field in the crystal. For example, Table 24-6 gives the electro-optical properties of cubic $\bar{4}3$**m** crystals when the field is perpendicular to three of the crystal planes. The new principal indices are obtained in general by solving an eigenvalue problem. For example, for a hexagonal material with a field perpendicular to the (111) plane, the index ellipsoid is

$$\left(\frac{1}{\mathbf{n}_o^2} + \frac{\mathbf{r}_{13}\mathbf{E}}{\sqrt{3}}\right)\mathbf{x}^2 + \left(\frac{1}{\mathbf{n}_o^2} + \frac{\mathbf{r}_{13}\mathbf{E}}{\sqrt{3}}\right)\mathbf{y}^2 + \left(\frac{1}{\mathbf{n}_e^2} + \frac{\mathbf{r}_{33}\mathbf{E}}{\sqrt{3}}\right)\mathbf{z}^2 + 2\mathbf{yzr}_{51}\frac{\mathbf{E}}{\sqrt{3}} + 2\mathbf{zxr}_{51}\frac{\mathbf{E}}{\sqrt{3}} = 1 \tag{24-45}$$

and the eigenvalue problem is

$$\begin{pmatrix} \frac{1}{\mathbf{n}_o^2} + \frac{\mathbf{r}_{13}\mathbf{E}}{\sqrt{3}} & 0 & \frac{2\mathbf{r}_{51}\mathbf{E}}{\sqrt{3}} \\ 0 & \frac{1}{\mathbf{n}_o^2} + \frac{\mathbf{r}_{13}\mathbf{E}}{\sqrt{3}} & \frac{2\mathbf{r}_{51}\mathbf{E}}{\sqrt{3}} \\ \frac{2\mathbf{r}_{51}\mathbf{E}}{\sqrt{3}} & \frac{2\mathbf{r}_{51}\mathbf{E}}{\sqrt{3}} & \frac{1}{\mathbf{n}_e^2} + \frac{\mathbf{r}_{33}\mathbf{E}}{\sqrt{3}} \end{pmatrix} \mathbf{V} = \frac{1}{\mathbf{n}'^2}\mathbf{V} \tag{24-46}$$

Table 24-6 Electro-optic Properties of Cubic $\bar{4}3\mathbf{m}$ Crystals

E Field Direction	Index Ellipsoid	Principal Indices
E perpendicular to (001) plane: $\mathbf{E_x} = \mathbf{E_y} = 0$ $\mathbf{E_z} = \mathbf{E}$	$\dfrac{\mathbf{x^2+y^2+z^2}}{\mathbf{n_o^2}} + 2\mathbf{r_{41}Exy} = 1$	$\mathbf{n'_x} = \mathbf{n_o} + \frac{1}{2}\mathbf{n_o^3 r_{41} E}$ $\mathbf{n'_y} = \mathbf{n_o} - \frac{1}{2}\mathbf{n_o^3 r_{41} E}$ $\mathbf{n'_z} = \mathbf{n_o}$
E perpendicular to (110) plane: $\mathbf{E_x} = \mathbf{E_y} = \mathbf{E}/\sqrt{2}$ $\mathbf{E_z} = 0$	$\dfrac{\mathbf{x^2+y^2+z^2}}{\mathbf{n_o^2}} + \sqrt{2}\mathbf{r_{41}E(yz+zx)} = 1$	$\mathbf{n'_x} = \mathbf{n_o} + \frac{1}{2}\mathbf{n_o^3 r_{41} E}$ $\mathbf{n'_y} = \mathbf{n_o} - \frac{1}{2}\mathbf{n_o^3 r_{41} E}$ $\mathbf{n'_z} = \mathbf{n_o}$
E perpendicular to (111) plane: $\mathbf{E_x} = \mathbf{E_y} = \mathbf{E_z} = \mathbf{E}/\sqrt{3}$	$\dfrac{\mathbf{x^2+y^2+z^2}}{\mathbf{n_o^2}} + \dfrac{2}{\sqrt{3}}\mathbf{r_{41}E(yz+zx+xy)} = 1$	$\mathbf{n'_x} = \mathbf{n_o} + \frac{1}{2\sqrt{3}}\mathbf{n_o^3 r_{41} E}$ $\mathbf{n'_y} = \mathbf{n_o} - \frac{1}{2\sqrt{3}}\mathbf{n_o^3 r_{41} E}$ $\mathbf{n'_z} = \mathbf{n_o} - \frac{1}{\sqrt{3}}\mathbf{n_o^3 r_{41} E}$

Source: Ref. 20.

The secular equation is then

$$\begin{pmatrix} \left(\dfrac{1}{\mathbf{n_o^2}} + \dfrac{\mathbf{r_{13}E}}{\sqrt{3}}\right) - \dfrac{1}{\mathbf{n'^2}} & 0 & \dfrac{2\mathbf{r_{51}E}}{\sqrt{3}} \\ 0 & \left(\dfrac{1}{\mathbf{n_o^2}} + \dfrac{\mathbf{r_{13}E}}{\sqrt{3}}\right) - \dfrac{1}{\mathbf{n'^2}} & \dfrac{2\mathbf{r_{51}E}}{\sqrt{3}} \\ \dfrac{2\mathbf{r_{51}E}}{\sqrt{3}} & \dfrac{2\mathbf{r_{51}E}}{\sqrt{3}} & \left(\dfrac{1}{\mathbf{n_o^2}} + \dfrac{\mathbf{r_{33}E}}{\sqrt{3}}\right) - \dfrac{1}{\mathbf{n'^2}} \end{pmatrix} = 0 \qquad (24\text{-}47)$$

and the roots of this equation are the new principal indices.

24.6 MAGNETO-OPTICS

When a magnetic field is applied to certain materials, the plane of incident linearly polarized light may be rotated in passage through the material. The magneto-optic effect linear with field strength is called the Faraday effect, and was discovered by Michael Faraday in 1845. A magneto-optic cell is illustrated in Fig. 24-9. The field is set up so that the field lines are along the direction of the optical beam propagation. A linear polarizer allows light of one polarization into the cell. A second linear polarizer is used to analyze the result.

The Faraday effect is governed by the equation:

$$\theta = \mathbf{VB}\mathrm{d} \qquad (24\text{-}48)$$

where **V** is the Verdet constant, θ is the rotation angle of the electric field vector of the linearly polarized light, **B** is the applied field, and d is the path length in the

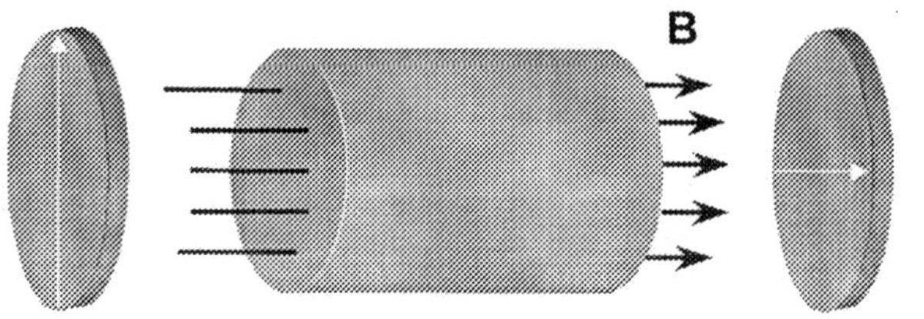

Figure 24-9 Illustration of a setup to observe the Faraday effect.

Table 24-7 Values of the Verdet Constant at $\lambda = 5893$ Å

Material	T (°C)	Verdet Constant (deg/G · mm)
Water[a]	20	2.18×10^{-5}
Air ($\lambda = 5780$ Å and 760 mm Hg)[b]	0	1.0×10^{-8}
NaCl[b]	16	6.0×10^{-5}
Quartz[b]	20	2.8×10^{-5}
CS_2[a]	20	7.05×10^{-5}
P[a]	33	2.21×10^{-4}
Glass, flint[a]	18	5.28×10^{-5}
Glass, Crown[a]	18	2.68×10^{-5}
Diamond[a]	20	2.0×10^{-5}

[a]*Source:* Ref. 11.
[b]*Source:* Ref. 10.

medium. The rotatory power ρ, defined in degrees per unit path length, is given by

$$\rho = \mathbf{VB} \tag{24-49}$$

A list of Verdet constants for some common materials is given in Table 24-7. The material that is often used in commercial magneto-optic-based devices is some formulation of iron garnet. Data tabulations for metals, glasses, and crystals, including many iron garnet compositions, can be found in Chen [21]. The magneto-optic effect is the basis for magneto-optic memory devices, optical isolators, and spatial light modulators [22,23].

Other magneto-optic effects in addition to the Faraday effect include the Cotton–Mouton effect, the Voigt effect, and the Kerr magneto-optic effect. The Cotton–Mouton effect is a quadratic magneto-optic effect observed in liquids. The Voigt effect is similar to the Cotton–Mouton effect but is observed in vapors. The Kerr magneto-optic effect is observed when linearly polarized light is reflected from the face of either pole of a magnet. The reflected light becomes elliptically polarized.

24.7 LIQUID CRYSTALS

Liquid crystals are a class of substances which demonstrate that the premise that matter exists only in solid, liquid, and vapor (and plasma) phases is a simplification. Fluids, or liquids, generally are defined as the phase of matter which cannot maintain any degree of order in response to a mechanical stress. The molecules of a liquid have random orientations and the liquid is isotropic. In the period 1888 to 1890 Reinitzer, and separately Lehmann, observed that certain crystals of organic compounds exhibit behavior between the crystalline and liquid states [24]. As the temperature is raised, these crystals change to a fluid substance that retains the anisotropic behavior of a crystal. This type of liquid crystal is now classified as thermotropic because the transition is effected by a temperature change, and the intermediate state is referred to as a mesophase [25]. There are three types of mesophases: smectic, nematic, and cholesteric. Smectic and nematic mesophases are often associated and occur in sequence as the temperature is raised. The term smectic derives from the Greek word for soap and is characterized by a material more viscous than the other mesophases. Nematic is from the Greek for thread and was named because the material exhibits a striated appearance (between crossed polaroids). The cholesteric mesophase is a property of the cholesterol esters, hence the name.

Figure 24-10a illustrates the arrangement of molecules in the nematic mesophase. Although the centers of gravity of the molecules have no long-range order as crystals do, there is order in the orientations of the molecules [26]. They tend to be oriented parallel to a common axis indicated by the unit vector $\hat{\mathbf{n}}$.

The direction of $\hat{\mathbf{n}}$ is arbitrary and is determined by some minor force such as the guiding effect of the walls of the container. There is no distinction between a positive and negative sign of $\hat{\mathbf{n}}$. If the molecules carry a dipole, there are equal numbers of dipoles pointing up as down. These molecules are not ferroelectric. The molecules are achiral, i.e., they have no handedness, and there is no positional order of the molecules within the fluid. Nematic liquid crystals are optically uniaxial.

The temperature range over which the nematic mesophase exists varies with the chemical composition and mixture of the organic compounds. The range is quite

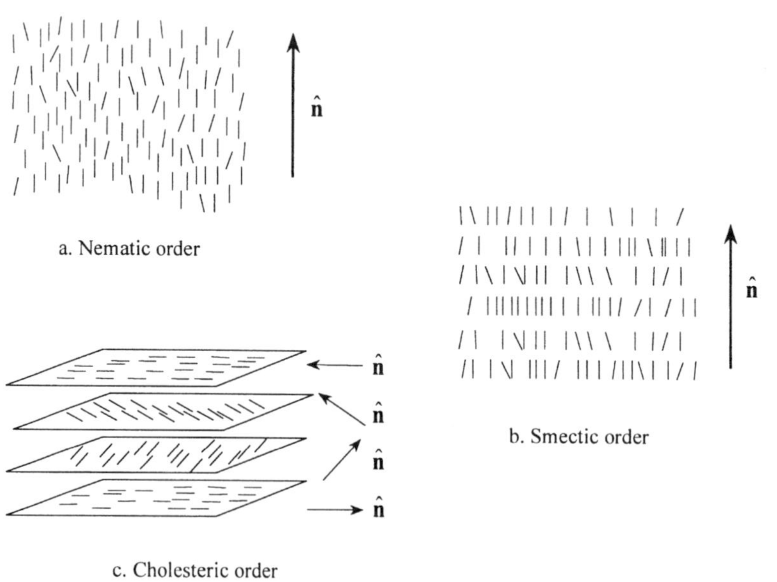

Figure 24-10 Schematic representation of liquid crystal order.

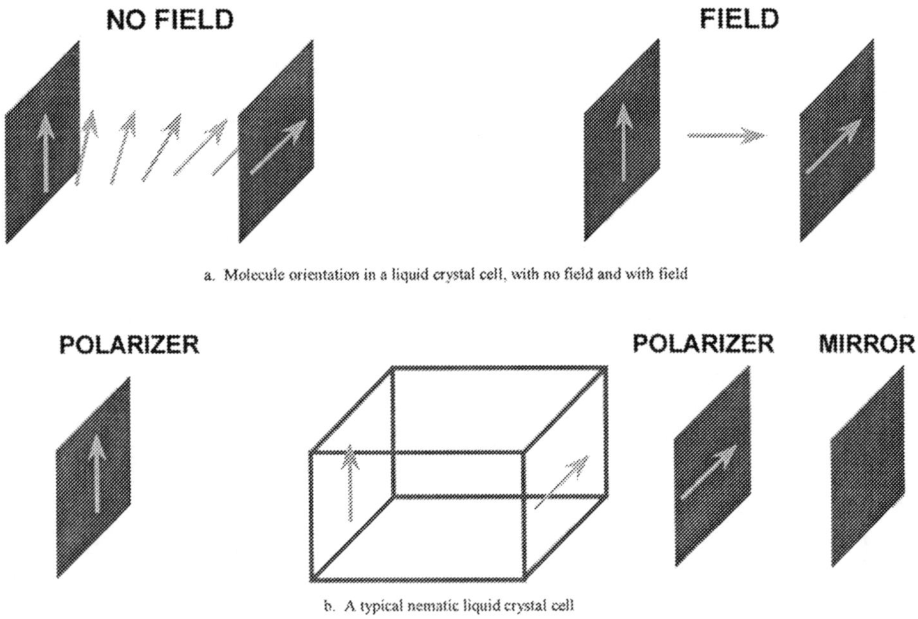

Figure 24-11 Liquid crystal cell operation.

wide; for example, in one study of ultraviolet imaging with a liquid crystal light valve, four different nematic liquid crystals were used [27]. Two of these were MBBA [*N*-(*p*-methoxybenzylidene)-*p*-(*n*-butylaniline)] with a nematic range of 17° to 43°C, and a proprietary material with a range of –20° to 51°C.

There are many known electro-optical effects involving nematic liquid crystals [25, 28, 29]. Two of the more important are field-induced birefringence, also called deformation of aligned phases, and the twisted nematic effect, also called the Schadt–Helfrich effect.

An example of a twisted nematic cell is shown in Fig. 24-11. Figure 24-11a shows the molecule orientation in a liquid crystal cell, without and with an applied field. The liquid crystal material is placed between two electrodes. The liquid crystals at the cell wall align themselves in some direction parallel to the wall as a result of very minor influences. A cotton swab lightly stroked in one direction over the interior surface of the wall prior to cell assembly is enough to produce alignment of the liquid crystal [30]. The molecules align themselves with the direction of the rubbing. The electrodes are placed at 90° to each other with respect to the direction of rubbing. The liquid crystal molecules twist from one cell wall to the other to match the alignments at the boundaries as illustrated, and light entering at one cell wall with its polarization vector aligned to the crystal axis will follow the twist and be rotated 90° by the time it exits the opposite cell wall. If the polarization vector is restricted with a polarizer on entry and an analyzer on exit, only the light with the 90° polarization twist will be passed through the cell. With a field applied between the cell walls, the molecules tend to orient themselves perpendicular to the cell walls, i.e., along the field lines. Some molecules next to the cell walls remain parallel to their original orientation, but most of the molecules in the center of the cell align themselves parallel to the electric field, destroying the twist. At the proper strength, the electric field will cause all the light to be blocked by the analyzer.

Figure 24-11b shows a twisted nematic cell as might be found in a digital watch display, gas pump, or calculator. Light enters from the left. A linear polarizer is the first element of this device and is aligned so that its axis is along the left-hand liquid crystal cell wall alignment direction. With no field, the polarization of the light twists with the liquid crystal twist, 90° to the original orientation, passes through a second polarizer with its axis aligned to the right-hand liquid crystal cell wall alignment direction, and is reflected from a mirror. The light polarization twists back the way it came and leaves the cell. Regions of this liquid crystal device that are not activated by the applied field are bright. If the field is now applied, the light does not change polarization as it passes through the liquid crystal and will be absorbed by the second polarizer. No light returns from the mirror, and the areas of the cell that have been activated by the applied field are dark.

A twisted nematic cell has a voltage threshold below which the polarization vector is not affected due to the internal elastic forces. A device 10 μm thick might have a threshold voltage of 3 V [25].

Another important nematic electro-optic effect is field-induced birefringence or deformation of aligned phases. As with the twisted nematic cell configuration, the liquid crystal cell is placed between crossed polarizers. However, now the molecular axes are made to align perpendicular to the cell walls and thus parallel to the direction of light propagation. By using annealed SnO_2 electrodes and materials of high purity, Schiekel and Fahrenschon [29] found that the molecules spontaneously align in this manner. Their cell worked well with 20 μm thick MBBA. The working material must be one having a negative dielectric anisotropy so that when an electric field is applied (normal to the cell electrodes) the molecules will tend to align themselves perpendicular to the electric field. The molecules at the cell walls tend to remain in their original orientation and the molecules within the central region will turn up to 90°; this is illustrated in Fig. 24-12.

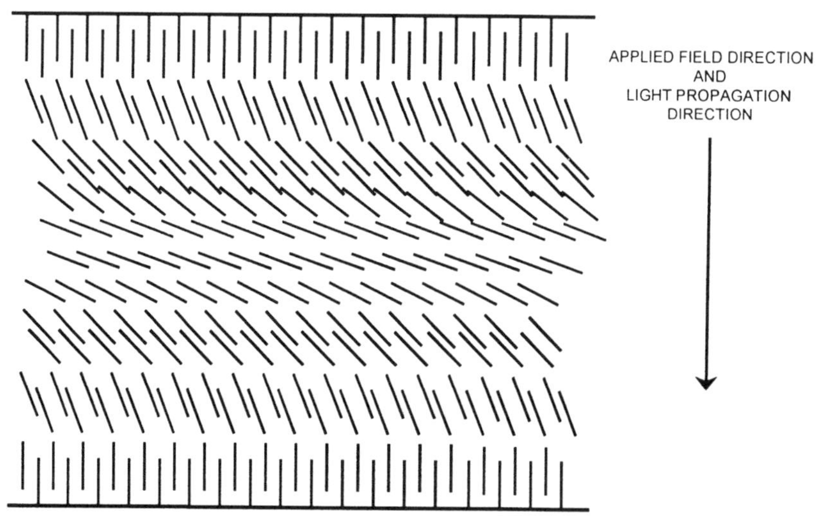

Figure 24-12 Deformation of liquid crystal due to applied voltage. (After Ref. 28.)

There is a threshold voltage typically in the 4–6 V range [25]. Above the threshold, the molecules begin to distort and become birefringent due to the anisotropy of the medium. Thus, with no field, no light exits the cell; at threshold voltage, light begins to be divided into ordinary and extraordinary beams, and some light will exit the analyzer. The birefringence can also be observed with positive dielectric anisotropy when the molecules are aligned parallel to the electrodes at no field and both electrodes have the same orientation for nematic alignment. As the applied voltage is increased, the light transmission increases for crossed polarizers [25]. The hybrid field-effect liquid crystal light valve relies on a combination of the twisted nematic effect (for the "off" state) and induced birefringence (for the "on" state) [31].

Smectic liquid crystals are more ordered than the nematics. The molecules are not only aligned, but they are also organized into layers, making a two-dimensional fluid. This is illustrated in Fig. 24-10b. There are three types of smectics: A, B, and C. Smectic A is optically uniaxial. Smectic C is optically biaxial. Smectic B is the most ordered, since there is order within layers. Smectic C, when chiral, is ferroelectric. Ferroelectric liquid crystals are known for their fast switching speed and bistability.

Cholesteric liquid crystal molecules are helical, and the fluid is chiral. There is no long range order, as in nematics, but the preferred orientation axis changes in direction through the extent of the liquid. Cholesteric order is illustrated in Fig. 24-10c.

For more information on liquid crystals and an extensive bibliography, see Wu [32,33], and Khoo and Wu [34].

24.8 MODULATION OF LIGHT

We have seen that light modulators are composed of an electro- or magneto-optical material on which an electromagnetic field is imposed. Electro-optical modulators may be operated in a longitudinal mode or in a transverse mode. In a longitudinal mode modulator, the electric field is imposed parallel to the light propagating through the material, and in a transverse mode modulator, the electric field is imposed perdendicular to the direction of light propagation. Either mode may be used if the entire wavefront of the light is to be modulated equally. The longitudinal mode is more likely to be used if a spatial pattern is to be imposed on the modulation. The mode used will depend on the material chosen for the modulator and the application.

Figure 24-13 shows the geometry of a longitudinal electro-optic modulator. The beam is normal to the face of the modulating material and parallel to the field imposed on the material. Electrodes of a material that is conductive yet transparent to the wavelength to be modulated are deposited on the faces through which the beam travels. This is the mode used for liquid crystal modulators.

Figure 24-14 shows the geometry of the transverse electro-optic modulator. The imposed field is perpendicular to the direction of light passing through the material. The electrodes do not need to be transparent to the beam. This is the mode used for modulators in laser beam cavities, e.g., a CdTe modulator in a CO_2 laser cavity.

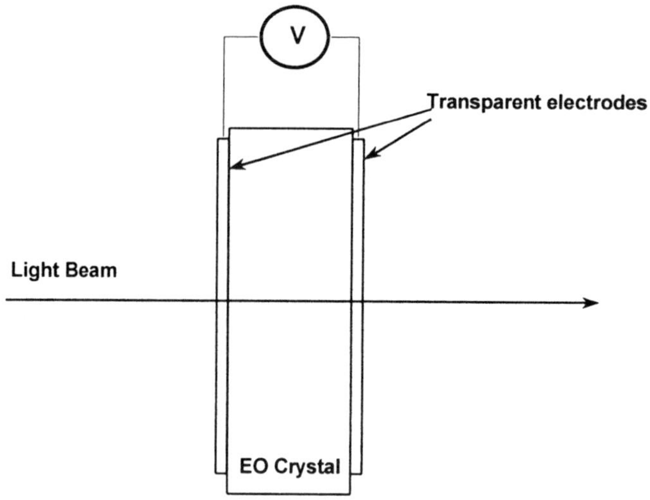

Figure 24-13 Longitudinal mode modulator.

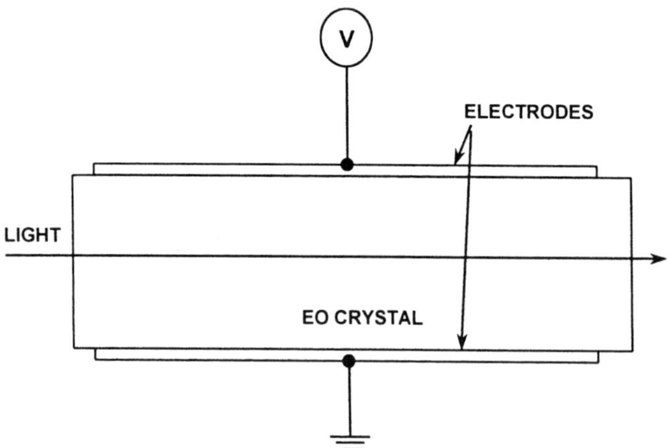

Figure 24-14 Transverse mode modulator.

24.9 CONCLUDING REMARKS

The origin of the electro-optic tensor, the form of that tensor for various crystal types, and the values of the tensor coefficients for specific materials have been discussed. The concepts of the index ellipsoid, the wave surface, and the wavevector surface were introduced. These are quantitative and qualitative models that aid in the understanding of the interaction of light with crystals. We have shown how the equation for the index ellipsoid is found when an external field is applied, and how expressions for the new principal indices of refraction are derived. Magneto-optics and liquid crystals were described. The introductory concepts of constructing an electro-optic modulator were discussed.

While the basics of electro- and magneto-optics in bulk materials have been covered, there is a large body of knowledge dealing with related topics which cannot be covered here. A more detailed description of electro-optic modulators is covered in Yariv and Yeh [11]. Information on spatial light modulators may be found in Efron [35]. Shen [36] describes the many aspects and applications of nonlinear optics, and current work in such areas as organic nonlinear materials can be found in SPIE Proceedings [37,38].

REFERENCES

1. Jackson, J. D., *Classical Electrodynamics*, 2nd ed., Wiley, New York, 1975.
2. Lorrain, P. and Corson, D., *Electromagnetic Fields and Waves*, 2nd ed., Freeman, New York, 1970.
3. Reitz, J. R. and Milford, F. J., *Foundations of Electromagnetic Theory*, 2nd ed., Addison-Wesley, Reading, MA, 1967.
4. Lovett, D. R., *Tensor Properties of Crystals*, Hilger, Bristol, UK, 1989.
5. Kaminow, I. P., *An Introduction to Electrooptic Devices*, Academic Press, New York, 1974.
6. Nye, J. F., *Physical Properties of Crystals: Their Representation by Tensors and Matrices*, Oxford University Press, Oxford, UK, 1985.
7. Senechal, M., *Crystalline Symmetries: An Informal Mathematical Introduction*, Hilger, Bristol, UK, 1990.
8. Wood, E. A., *Crystals and Light: An Introduction to Optical Crystallography*, Dover, New York, 1977.
9. Kittel, C., *Introduction to Solid State Physics*, Wiley, New York, 1971.
10. Hecht, E., *Optics*, Addison-Wesley, Reading, MA, 1987.
11. Yariv, A. and Yeh, P., *Optical Waves in Crystals*, Wiley, New York, 1984.
12. Jenkins, F. A. and White, H. E., *Fundamentals of Optics*, McGraw-Hill, New York, 1957.
13. Born, M. and Wolf, E., *Principles of Optics*, Pergamon Press, Oxford, UK, 1975.
14. Klein, M. V., *Optics*, Wiley, New York, 1970.
15. Kaminow, I. P., "Linear Electrooptic Materials," *CRC Handbook of Laser Science and Technology*, Vol. IV, *Optical Materials*, Part 2: *Properties,* Weber, M. J., ed., CRC Press, Cleveland, OH, 1986.
16. Tropf, W. J., Thomas, M. E., and Harris, T. J., "Properties of crystals and glasses," *Handbook of Optics, Volume II, Devices, Measurements, and Properties*, 2nd ed., McGraw-Hill, New York, 1995, Ch. 33.
17. Haertling, G. H. and Land, C. E. "Hot-pressed (Pb,La)(Zr,Ti)O_3 ferroelectric ceramics for electrooptic applications," *J. Am. Ceram. Soc.*, **54**, 1 (1971).
18. Land, C. E., "Optical information storage and spatial light modulation in PLZT ceramics," *Opt. Eng.*, **17**, 317 (1978).
19. Salvestrini, J. P., Fontana, M. D., Aillerie, M., and Czapla, Z., "New material with strong electro-optic effect: rubidium hydrogen selenate ($RbHSeO_4$)," *Appl. Phys. Lett.*, **64**, 1920 (1994).
20. Goldstein, D., *Polarization Modulation in Infrared Electrooptic Materials*, Ph.D. Dissertation, University of Alabama in Huntsville, Huntsville, AL, 1990.
21. Chen, D., "Data Tabulations," *CRC Handbook of Laser Science and Technology*, Vol. IV, *Optical Materials*, Part 2: *Properties*, Weber, M. J., ed., CRC Press, Cleveland, OH, 1986.
22. Ross, W. E., Psaltis, D., and Anderson, R. H., "Two-dimensional magneto-optic spatial light modulator for signal processing," *Opt. Eng.*, **22**, 485 (1983).

23. Ross, W. E., Snapp, K. M., and Anderson, R. H., "Fundamental characteristics of the Litton iron garnet magneto-optic spatial light modulator," Proc. SPIE, Vol. 388, *Advances in Optical Information Processing*, 1983.
24. Gray, G. W., *Molecular Structure and the Properties of Liquid Crystals*, Academic Press, New York, 1962.
25. Priestley, E. B., Wojtowicz, P. J., and Sheng, P., ed., *Introduction to Liquid Crystals*, Plenum Press, New York, 1974.
26. De Gennes, P. G., *The Physics of Liquid Crystals*, Oxford University Press, Oxford, UK, 1974.
27. Margerum, J. D., Nimoy, J., and Wong, S. Y., *Appl. Phys. Lett.*, **17**, 51 (1970).
28. Meier, G., Sackman, H., and Grabmaier, F., *Applications of Liquid Crystals*, Springer Verlag, Berlin, 1975.
29. Schiekel, M. F. and Fahrenschon, K., "Deformation of nematic liquid crystals with vertical orientation in electrical fields," *Appl. Phys. Lett.*, **19**, 391 (1971).
30. Kahn, F. J., "Electric-field-induced orientational deformation of nematic liquid crystals: tunable birefringence," *Appl. Phys. Lett.*, **20**, 199 (1972).
31. Bleha, W. P., Lipton, L. T., Wiener-Arnear, E., Grinberg, J., Reif, P. G., Casasent, D., Brown, H. B., and Markevitch, B. V., "Application of the liquid crystal light valve to real-time optical data processing," *Opt. Eng.*, **17**, 371 (1978).
32. Wu, S.-T., "Nematic liquid crystals for active optics," *Optical Materials, A Series of Advances*, Vol. 1, ed. S. Musikant, Marcel Dekker, New York, 1990.
33. Wu, S.-T., "Liquid crystals", *Handbook of Optics*, Vol. II, *Devices, Measurements, and Properties*, 2nd ed., McGraw Hill, New York, 1995, Ch. 14.
34. Khoo, I.-C. and Wu, S.-T., *Optics and Nonlinear Optics of Liquid Crystals*, World Scientific, River Edge, NJ, 1993.
35. Efron, U., ed., *Spatial Light Modulator Technology, Materials, Devices, and Applications*, Marcel Dekker, New York, 1995.
36. Shen, Y. R., *The Principles of Nonlinear Optics*, Wiley, New York, 1984.
37. Moehlmann, G. R., ed., *Nonlinear Optical Properties of Organic Materials IX*, Proc. SPIE, Vol. 2852 (1996).
38. Kuzyk, M. G., ed., *Nonlinear Optical Properties of Organic Materials X*, Proc. SPIE, Vol. 3147 (1997).

25

Optics of Metals

25.1 INTRODUCTION

We have been concerned with the propagation of light in nonconducting media. We now turn our attention to describing the interaction of light with conducting materials, namely, metals and semiconductors. Metals and semiconductors, *absorbing media*, are crystalline aggregates consisting of small crystals of random orientation. Unlike true crystals they do not have repetitive structures throughout their entire forms.

The phenomenon of conductivity is associated with the appearance of heat; it is very often called Joule heat. It is a thermodynamically irreversible process in which electromagnetic energy is transformed to heat. As a result, the optical field within a conductor is attenuated. The very high conductivity exhibited by metals and semiconductors causes them to be practically opaque. The phenomenon of conduction and strong absorption corresponds to high reflectivity so that metallic surfaces act as excellent mirrors. In fact, up to the latter part of the nineteenth century most large reflecting astronomical telescope mirrors were metallic. Eventually, metal mirrors were replaced with parabolic glass surfaces overcoated with silver, a material with a very high reflectivity. Unfortunately, silver oxidizes in a relatively short time with oxygen and sulfur compounds in the atmosphere and turns black. Consequently, silver-coated mirrors must be recoated nearly every other year or so, a difficult, time-consuming, expensive process. This problem was finally solved by Strong in the 1930s with his method of evaporating aluminum on to the surface of optical glass.

In the following sections we shall not deal with the theory of metals. Rather, we shall concentrate on the phenomenological description of the interaction of polarized light with metallic surfaces. Therefore, in Section 25.2 we develop Maxwell's equations for conducting media. We discover that for conducting media the refractive index becomes complex and has the form $\mathbf{n} = n(1-i\kappa)$ where n is the real refractive index and κ is the extinction coefficient. Furthermore, Fresnel's equations for reflection and transmission continue to be valid for conducting (absorbing) media. However, because of the rapid attenuation of the optical field within an absorbing medium, Fresnel's equations for transmission are inapplicable. Using the complex refractive index, we develop Fresnel's equations for reflection at normal incidence and describe them in terms of a quantity called the reflectivity. It is possible to develop Fresnel's reflection equations for non-normal incidence. However, the forms are very complicated and so approximate forms are derived for the s and p polarizations. It is rather remarkable that the phenomenon of conductivity may be taken into account simply by introducing a complex index of refraction. A complete understanding of the significance of n and κ can only be understood on the basis of the dispersion theory of metals. However, experience does show that large values of reflectivity correspond to large values of κ.

In Sections 25.3 and 25.4 we discuss the measurement of the optical constants n and κ. A number of methods have been developed over the past 100 years, nearly all of which are null-intensity methods. That is, n and κ are obtained from the null condition on the reflected intensity. The best-known null method is the *principle angle of incidence/principle azimuthal angle* method (Section 25.3). In this method a beam of light is incident on the sample and the incidence angle is varied until an incidence angle is reached where a phase shift of $\pi/2$ occurs. The incidence angle where this takes place is known as the *principle angle of incidence.* An additional phase shift of $\pi/2$ is now introduced into the reflected light with a quarter-wave retarder. The condition of the principal angle of incidence and the quarter-wave shift and the introduction of the quarter-wave retarder, as we shall see, creates linearly polarized light. Analyzing the phase-shifted reflected light with a polarizer that is rotated around its azimuthal angle leads to a null intensity (*at the principal azimuthal angle*) from which n and κ can be determined.

Classical null methods were developed long before the advent of quantitative detectors, digital voltmeters, and digital computers. Nulling methods are very valuable, but they have a serious drawback: the method requires a mechanical arm that must be rotated along with the azimuthal rotation of a Babinet–Soleil compensator and analyzer until a null intensity is found. In addition, a mechanical arm that yields scientifically useful readings is quite expensive. It is possible to overcome these drawbacks by reconsidering Fresnel's equations for reflection at an incidence angle of 45°. It is well known that Fresnel's equations for reflection simplify at normal incidence and at the Brewster angle for nonabsorbing (dielectric) materials. Less well known is that Fresnel's equations also simplify at an incidence angle of 45°. All of these simplifications were discussed in Chapter 8 assuming dielectric media. The simplifications at the incidence angle of 45° hold even for absorbing media. Therefore, in Section 25.4 we describe the measurement of an optically absorbing surface at an incidence angle of 45°. This method, called digital refractometry, overcomes the nulling problems and leads to equations to determine n and κ that can be solved on a digital computer by iteration.

25.2 MAXWELL'S EQUATIONS FOR ABSORBING MEDIA

We now solve Maxwell's equations for a homogeneous isotropic medium described by a dielectric constant ε, a permeability μ, and a conductivity σ. Using material equations (also called the constitutive relations):

$$\mathbf{D} = \varepsilon \mathbf{E} \tag{25-1a}$$

$$\mathbf{B} = \mu \mathbf{H} \tag{25-1b}$$

$$\mathbf{j} = \sigma \mathbf{E} \tag{25-1c}$$

Maxwell's equations become, in MKSA units,

$$\nabla \times \mathbf{H} - \varepsilon \frac{\partial \mathbf{E}}{\partial t} = \sigma \mathbf{E} \tag{25-2a}$$

$$\nabla \times \mathbf{E} + \mu \frac{\partial \mathbf{H}}{\partial t} = 0 \tag{25-2b}$$

$$\nabla \cdot \mathbf{E} = \frac{\rho}{\varepsilon} \tag{25-2c}$$

$$\nabla \cdot \mathbf{H} = 0 \tag{25-2d}$$

These equations describe the propagation of the optical field within and at the boundary of a conducting medium. To find the equation for the propagation of the field $\mathbf{E}$ we eliminate $\mathbf{H}$ between (25-2a) and (25-2b). We take the curl of (25-2b) and substitute (25-2a) into the resulting equation to obtain

$$\nabla \times (\nabla \times \mathbf{E}) + (\mu\varepsilon) \frac{\partial^2 \mathbf{E}}{\partial t^2} + \mu\sigma \frac{\partial \mathbf{E}}{\partial t} = 0 \tag{25-3}$$

Expanding the $\nabla \times (\nabla \times)$ operator, we find that (25-3) becomes

$$\nabla^2 \mathbf{E} = \mu\varepsilon \frac{\partial^2 \mathbf{E}}{\partial t^2} + \mu\sigma \frac{\partial \mathbf{E}}{\partial t} \tag{25-4}$$

Equation (25-4) is the familiar wave equation modified by an additional term. From our knowledge of differential equations the additional term described by $\partial \mathbf{E}/\partial t$ corresponds to damping or attenuation of a wave. Thus, (25-4) can be considered the damped or attenuated wave equation.

We proceed now with the solution of (25-4). If the field is strictly monochromatic and of angular frequency ω so that $\mathbf{E} \equiv \mathbf{E}(\mathbf{r}, t) = \mathbf{E}(\mathbf{r}) \exp(i\omega t)$, then substituting this form into (25-4) yields

$$\nabla^2 \mathbf{E}(\mathbf{r}) = (-\mu\varepsilon\omega^2)\mathbf{E}(\mathbf{r}) + (i\omega\mu\sigma)\mathbf{E}(\mathbf{r}) \tag{25-5}$$

which can be written as

$$\nabla^2 \mathbf{E}(\mathbf{r}) = (-\mu\omega^2)\left[\varepsilon - i\left(\frac{\sigma}{\omega}\right)\right]\mathbf{E}(\mathbf{r}) \tag{25-6}$$

In this form, (25-6) is identical to the wave equation except that now the dielectric constant is complex; thus,

$$\varepsilon = \varepsilon - i\left(\frac{\sigma}{\omega}\right) \tag{25-7}$$

where ε is the real dielectric constant.

The correspondence with nonconducting media is readily seen if ε is defined in terms of a complex refractive index $\mathbf{n}$ (we set $\mu = 1$ since we are not dealing with magnetic materials):

$$\varepsilon = \mathbf{n}^2 \tag{25-8}$$

We now express $\mathbf{n}$ in terms of the refractive index and the absorption of the medium. To find the form of $\mathbf{n}$ which describes both the refractive and absorbing behavior of a propagating field, we first consider the intensity $I(z)$ of the field after it has propagated a distance z. We know that the intensity is attenuated after a distance z has been traveled, so the intensity can be described by

$$I(z) = I_0 \exp(-\alpha z) \tag{25-9}$$

where α is the *attenuation* or *absorption coefficient*. We wish to relate α to κ, the extinction coefficient or attenuation index. We first note that $\mathbf{n}$ is a dimensionless quantity, whereas from (25-9) α has the dimensions of inverse length. We can express αz as a dimensionless parameter by assuming that after a distance equal to a wavelength λ the intensity has been reduced to

$$I(\lambda) = I_0 \exp(-4\pi\kappa) \tag{25-10}$$

Equating the arguments of the exponents in (25-9) and (25-10), we have

$$\alpha = \left(\frac{4\pi}{\lambda}\right)\kappa = 2k\kappa \tag{25-11}$$

where $k = 2\pi/\lambda$ is the wavenumber. Equation (25-9) can then be written as

$$I(z) = I_0 \exp\left[-\left(\frac{4\pi}{\lambda}\right)\kappa z\right] \tag{25-12}$$

From this result we can write the corresponding field $E(z)$ as

$$E(z) = E_0 \exp\left[-\left(\frac{2\pi}{\lambda}\right)\kappa z\right] \tag{25-13}$$

or

$$E(z) = E_0 \exp(-k\kappa z) \tag{25-14}$$

Thus, the field propagating in the z direction can be described by

$$E(z) = E_0 \exp(-k\kappa z)\exp[i(\omega t - kz)] \tag{25-15}$$

The argument of (25-15) can be written as

$$i\omega\left[t - \left(\frac{k}{\omega}\right)z + i\left(\frac{k\kappa}{\omega}\right)z\right] \tag{25-16a}$$

$$= i\omega\left[t - \frac{k}{\omega}\{1 - i\kappa\}z\right] \tag{25-16b}$$

But $k = \omega/v = \omega n/c$, so (25-16b) becomes

$$i\omega\left[t - \frac{n}{c}\{1 - i\kappa\}z\right] \tag{25-17a}$$

$$= i\omega\left[t - \left(\frac{\mathbf{n}}{c}\right)z\right] \tag{25-17b}$$

where

$$\mathbf{n} = n(1 - i\kappa) \tag{25-18}$$

Thus, the propagating field (25-15) can be written in the form:

$$\mathbf{E}(z) = E_0 \exp\left[i\omega\left(t - \left(\frac{\mathbf{n}}{c}\right)z\right)\right] \tag{25-19}$$

Equation (25-19) shows that conducting (i.e., absorbing) media lead to the same solutions as nonconducting media except that the real refractive index n is replaced by a complex refractive index $\mathbf{n}$. Equation (25-18) relates the complex refractive index to the real refractive index and the absorption behavior of the medium and will be used throughout the text.

From (25-7), (25-8), and (25-18) we can relate n and κ to σ. We have

$$\varepsilon = \mathbf{n}^2 = n^2(1 - i\kappa)^2 = \varepsilon - i\left(\frac{\sigma}{\omega}\right) \tag{25-20}$$

which leads immediately to

$$n^2(1-\kappa^2) = \varepsilon \tag{25-21a}$$

$$n^2\kappa = \frac{\sigma}{2\omega} = \frac{\sigma}{4\pi\nu} \tag{25-21b}$$

where $\nu = \omega/2\pi$

We solve these equations to obtain

$$n^2 = \frac{1}{2}\left[\sqrt{\varepsilon^2 + \left(\frac{\sigma}{4\pi\nu}\right)^2} + \varepsilon\right] \tag{25-22a}$$

$$n^2\kappa^2 = \frac{1}{2}\left[\sqrt{\varepsilon^2 + \left(\frac{\sigma}{4\pi\nu}\right)^2} - \varepsilon\right] \tag{25-22b}$$

Equation (25-22) is important because it enables us to relate the (measured) values of n and κ to the constants ε and σ of a metal or semiconductor. Because metals are opaque, it is not possible to measure these constants optically.

Since the wave equation for conducting media is identical to the wave equation for dielectrics, except for the appearance of a complex refractive index, we would expect the boundary conditions and all of its consequences to remain unchanged. This is indeed the case. Thus, Snell's law of refraction becomes

$$\sin\theta_i = \mathbf{n}\sin\theta_r \tag{25-23}$$

where the refractive index is now complex. Similarly, Fresnel's law of reflection and refraction continue to be valid. Since optical measurements cannot be made with Fresnel's refraction equations, only Fresnel's reflection equations are of practical interest. We recall these equations are given by

$$R_s = -\frac{\sin(\theta_i - \theta_r)}{\sin(\theta_i + \theta_r)}E_s \tag{25-24a}$$

$$R_p = \frac{\tan(\theta_i - \theta_r)}{\tan(\theta_i + \theta_r)}E_p \tag{25-24b}$$

In (25-24) θ_i is the angle of incidence and θ_r is the angle of refraction, and R_s, R_p, E_s, and E_p have their usual meanings.

We now derive the equations for the reflected intensity, using (25-24). We consider (25-24a) first. We expand the trigonometric sum and difference terms, substitute $\sin\theta_r = \mathbf{n}\sin\theta_r$ into the result, and find that

$$\frac{R_s}{E_s} = \left[\frac{\cos\theta_i - \mathbf{n}\cos\theta_r}{\cos\theta_i + \mathbf{n}\cos\theta_r}\right] \tag{25-25}$$

We first use (25-25) to obtain the reflectivity, that is, the normalized intensity at normal incidence. The reflectivity for the s polarization, (25-25) is defined to be

$$\mathscr{R}_s \equiv \left|\frac{R_s}{E_s}\right|^2 \tag{25-26}$$

At normal incidence $\theta_i = 0$, so from Snell's law, (25-23), $\theta_r = 0$ and (25-25) reduces to

$$\frac{R_s}{E_s} = \left[\frac{1-\mathbf{n}}{1+\mathbf{n}}\right] \tag{25-27}$$

Replacing $\mathbf{n}$ with the explicit form given by (25-18) yields

$$\frac{R_s}{E_s} = \left[\frac{(1-n)+in\kappa}{(1+n)-in\kappa}\right] \tag{25-28}$$

From the definition of the reflectivity (25-26) we then see that (25-28) yields

$$\mathscr{R}_s = \left[\frac{(n-1)^2+(n\kappa)^2}{(n+1)^2+(n\kappa)^2}\right] \tag{25-29}$$

We observe that for nonabsorbing media ($\kappa = 0$), (25-29) reduces to the well-known results for dielectrics. We also note that for this condition and for $n = 1$ the reflectivity is zero, as we would expect. In Fig. 25-1 a plot of (25-29) as a function of κ

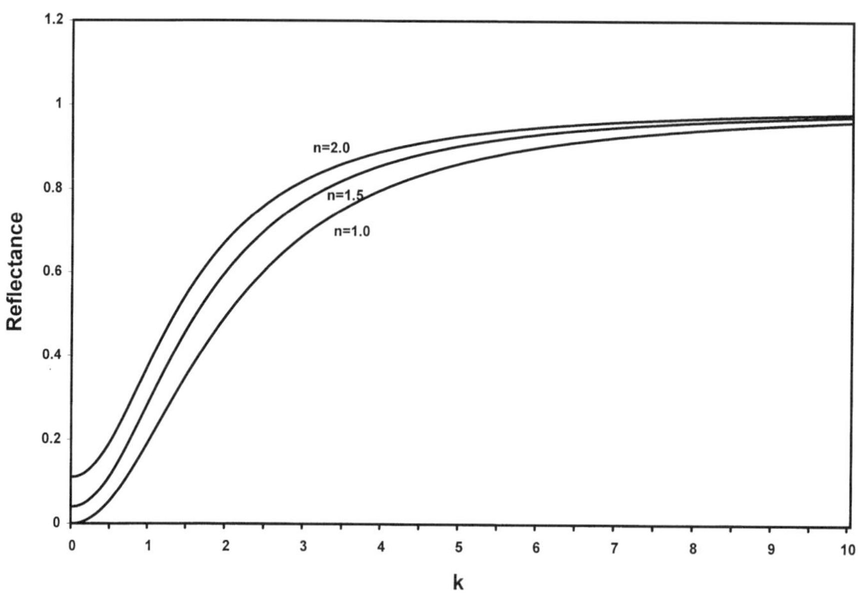

Figure 25-1 Plot of the reflectivity as a function of κ. The refractive indices are $n = 1.0$, 1.5, and 2.0, respectively.

is shown. We see that for absorbing media with increasing κ the reflectivity approaches unity. Thus, highly reflecting absorbing media (e.g., metals) are characterized by high values of κ.

In a similar manner we can find the reflectivity for normal incidence for the p polarization, (25-24b). Equation (25-24b) can be written as

$$\frac{R_p}{E_p} = \frac{\sin(\theta_i - \theta_r)\cos(\theta_i + \theta_r)}{\sin(\theta_i + \theta_r)\cos(\theta_i - \theta_r)} \tag{25-30}$$

At normal incidence the cosine factor in (25-30) is unity, and we are left with the same equation for the s polarization, (25-24a). Hence,

$$\mathscr{R}_p = \mathscr{R}_s \tag{25-31}$$

and for normal incidence the reflectivity is the same for the s and p polarizations.

We now derive the reflectivity equations for non-normal incidence. We again begin with (25-24a) or, more conveniently, its expanded form, (25-25)

$$\frac{R_s}{E_s} = \left[\frac{\cos\theta_i - \mathbf{n}\cos\theta_r}{\cos\theta_i + \mathbf{n}\cos\theta_r}\right] \tag{25-25}$$

Equation (25-25) is, of course, exact and can be used to obtain an exact expression for the reflectivity $\mathscr{R}_s$. However, the result is quite complicated. Therefore, we derive an approximate equation, much quoted in the literature, for $\mathscr{R}_s$ which is sufficiently close to the exact result. We replace the factor $\cos\theta_r$ by $(1 - \sin^2\theta_r)^{1/2}$ and use $\sin\theta_i = \mathbf{n}\sin\theta_r$. Then, (25-25) becomes

$$\frac{R_s}{E_s} = \left[\frac{\cos\theta_i - \sqrt{\mathbf{n}^2 - \sin^2\theta_i}}{\cos\theta_i + \sqrt{\mathbf{n}^2 - \sin^2\theta_i}}\right] \tag{25-32}$$

Equation (25-32) can be approximated by noting that $\mathbf{n}^2 \gg \sin^2\theta_i$, so (25-32) can be written as

$$\frac{R_s}{E_s} = \frac{\cos\theta_i - \mathbf{n}}{\cos\theta_i + \mathbf{n}} \tag{25-33}$$

We now substitute (25-18) into (25-33) and group the terms into real and imaginary parts:

$$\frac{R_s}{E_p} = \left[\frac{(\cos\theta_i - n) + in\kappa}{(\cos\theta_i + n) - in\kappa}\right] \tag{25-34}$$

The reflectivity $\mathscr{R}_s$ is then

$$\mathscr{R}_s = \left[\frac{(n - \cos\theta_i)^2 + (n\kappa)^2}{(n + \cos\theta_i)^2 + (n\kappa)^2}\right] \tag{25-35}$$

We now develop a similar, approximate, equation for $\mathscr{R}_p$. We first write (25-24b) as

$$\frac{R_s}{E_p} = \frac{\sin(\theta_i - \theta_r)\cos(\theta_i + \theta_r)}{\sin(\theta_i + \theta_r)\cos(\theta_i - \theta_r)} \tag{25-30}$$

The first factor is identical to (25-24a), so it can be replaced by its expanded form (25-25):

$$\frac{\sin(\theta_i - \theta_r)}{\sin(\theta_i + \theta_r)} = \left[\frac{\cos\theta_i - \mathbf{n}\cos\theta_r}{\cos\theta_i + \mathbf{n}\cos\theta_r}\right] \tag{25-36}$$

The second factor in (25-30) is now expanded, and again we use $\cos\theta_r = (1 - \sin^2\theta_r)^{1/2}$ and $\sin\theta_i = \mathbf{n}\sin\theta_r$:

$$\frac{\cos(\theta_i + \theta_r)}{\cos(\theta_i - \theta_r)} = \frac{(\cos\theta_i)\sqrt{\mathbf{n}^2 - \sin^2\theta_i} - \sin^2\theta_i}{(\cos\theta_i)\sqrt{\mathbf{n}^2 - \sin^2\theta_i} + \sin^2\theta_i} \tag{25-37}$$

Because $\mathbf{n}^2 \gg \sin^2\theta_i$, (25-37) can approximated as

$$\frac{\cos(\theta_i + \theta_r)}{\cos(\theta_i - \theta_r)} \cong \frac{\mathbf{n}\cos\theta_i - \sin^2\theta_i}{\mathbf{n}\cos\theta_i + \sin^2\theta_i} \tag{25-38}$$

We now multiply (25-36) by (25-38) to obtain

$$\frac{R_p}{E_p} = \left(\frac{\cos\theta_i - \mathbf{n}}{\cos\theta_i + \mathbf{n}}\right)\left(\frac{\mathbf{n}\cos\theta_i - \sin^2\theta_i}{\mathbf{n}\cos\theta_i + \sin^2\theta_i}\right) \tag{25-39}$$

Carrying out the multiplication in (25-39), we find that there is a $\sin^2\theta_i\cos\theta_i$ term. This term is always much smaller than the remaining terms and can be dropped. The remaining terms then lead to

$$\frac{R_p}{E_p} = \frac{-\mathbf{n}\cos\theta_i + 1}{\mathbf{n}\cos\theta_i + 1} \tag{25-40a}$$

or

$$\frac{R_p}{E_p} = -\frac{\cos\theta_i - 1/\mathbf{n}}{\cos\theta_i + 1/\mathbf{n}} \tag{25-40b}$$

Replacing $\mathbf{n}$ by $n(1 - i\kappa)$, grouping terms into real and imaginary parts, and ignoring the negative sign because it will vanish when we determine the reflectivity, gives

$$\frac{R_p}{E_p} = \frac{(n - 1/\cos\theta_i) - in\kappa}{(n + 1/\cos\theta_i) + in\kappa} \tag{25-41}$$

Multiplying (25-41) by its complex conjugate, we obtain the reflectivity $\mathcal{R}_p$:

$$\mathcal{R}_p = \frac{(n - 1/\cos\theta_i)^2 + (n\kappa)^2}{(n + 1/\cos\theta_i)^2 + (n\kappa)^2} \tag{25-42}$$

For convenience we write the equation for $\mathcal{R}_s$, (25-35), here also:

$$\mathcal{R}_s = \left[\frac{(n - \cos\theta_i)^2 + (n\kappa)^2}{(n + \cos\theta_i)^2 + (n\kappa)^2}\right] \tag{25-35}$$

In Figs. 25-2 through 25-5 plots are shown for the reflectivity as a function of the incidence angle θ_i of gold (Au), silver (Ag), copper (Cu), and platinum (Pt), using (25-35) and (25-39). The values for n and κ are taken from Wood's classic text *Physical Optics*.

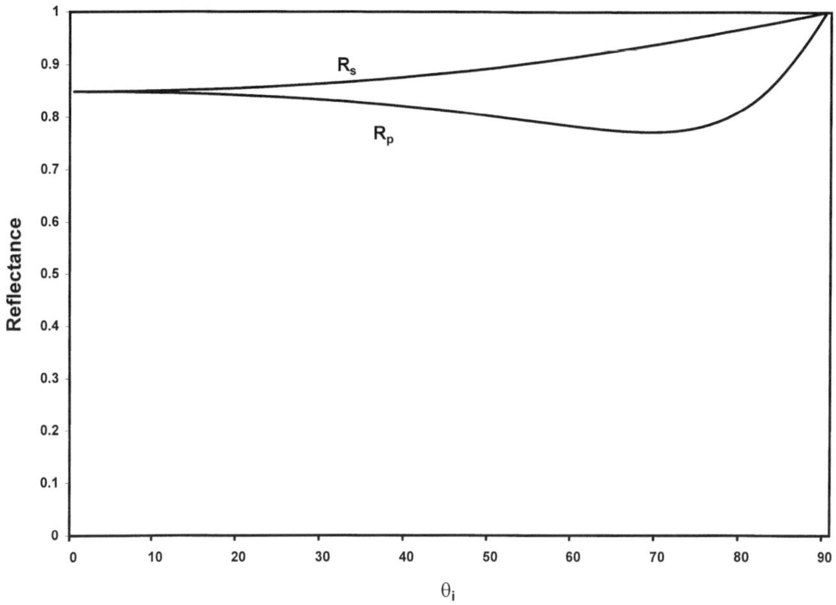

Figure 25-2 Reflectance of gold (Au) as a function of incidence angle. The refractive index and the extinction coefficient are 0.36 and 7.70 respectively. The normal reflectance value is 0.849.

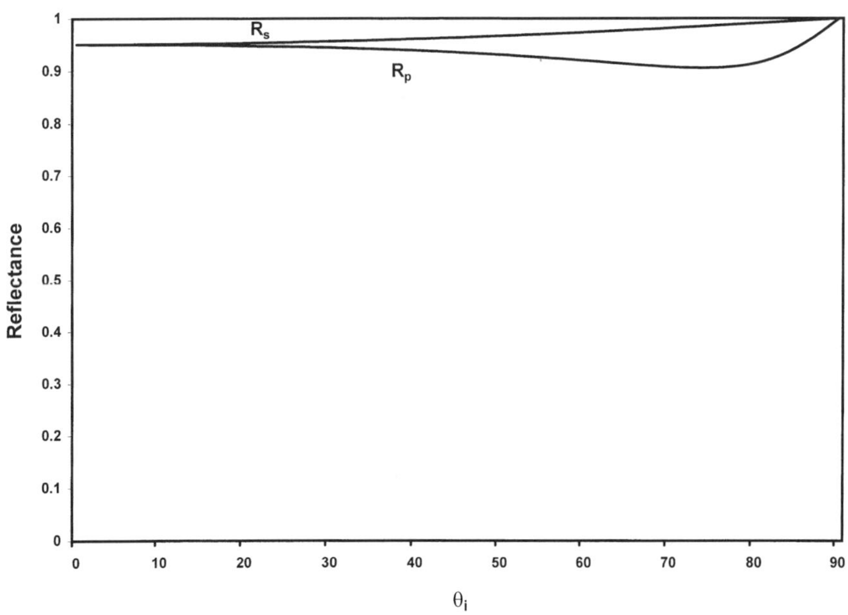

Figure 25-3 Reflectance of silver (Ag) as a function of incidence angle. The refractive index and the extinction coefficient are 0.18 and 20.2, respectively. The normal reflectance value is 0.951.

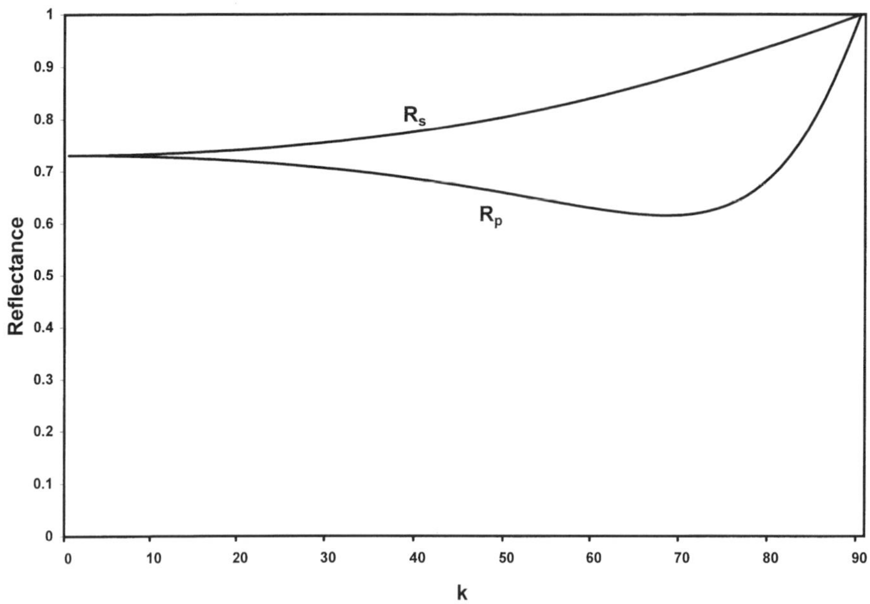

Figure 25-4 Reflectance of copper (Cu) as a function of incidence angle. The refractive index and the extinction coefficient are 0.64 and 4.08, respectively. The normal reflectance value is 0.731.

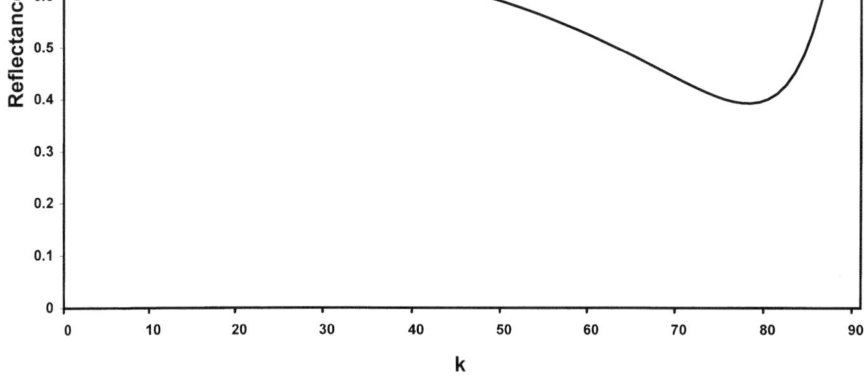

Figure 25-5 Reflectance of platinum (Pt) as a function of incidence angle. The refractive index and the extinction coefficient are 2.06 and 2.06, respectively. The normal reflectance value is 0.699.

In Figs. 25-2 through 25-5 we observe that the p reflectivity has a minimum value. This minimum is called the *pseudo-Brewster angle minimum* because, unlike the Brewster angle for dielectrics, the intensity does not go to zero for metals. Nevertheless, a technique based on this minimum has been used to determine n and κ. The interested reader is referred to the article by Potter.

Finally, we see that the refractive index can be less than unity for many metals. Born and Wolf have shown that this is a natural consequence of the simple classical theory of the electron and the dispersion theory. The theory provides a theoretical basis for the behavior of n and κ. Further details on the nature of metals and, in particular, the refractive index and the extinction coefficient (n and κ) as it appears in the dispersion theory of metals can be found in the reference texts by Born and Wolf and by Mott and Jones.

25.3 PRINCIPAL ANGLE OF INCIDENCE MEASUREMENT OF REFRACTIVE INDEX AND EXTINCTION COEFFICIENT OF OPTICALLY ABSORBING MATERIALS

In the previous section we saw that optically absorbing materials are characterized by a real refractive index n and an extinction coefficient κ. Because these constants describe the behavior and performance of optical materials such as metals and semiconductors, it is very important to know these "constants" over the entire optical spectrum.

Methods have been developed to measure the optical constants. One of the best known is the *principal angle of incidence* method. The basic idea is as follows. Incident $+45°$ linearly polarized light is reflected from an optically absorbing material. In general, the reflected light is elliptically polarized; the corresponding polarization ellipse is in nonstandard form. The angle of incidence of the incident beam is now changed until a phase shift of 90° is observed in the reflected beam. The incident angle where this takes place is called the *principal angle of incidence*. Its significance is that, at this angle, the polarization ellipse for the reflected beam is now in its standard form. From this condition relatively simple equations can then be found for n and κ. Because the polarization ellipse is now in its standard form, the orthogonal field components are parallel and perpendicular to the plane of incidence. The reflected beam is now passed through a quarter-wave retarder. The beam of light that emerges is linearly polarized with its azimuth angle at an unknown angle. The beam then passes through an analyzing polarizer that is rotated until a null intensity is found. The angle at which this null takes place is called the *principal azimuth angle*. From the measurement of the principal angle of incidence and the principal azimuth angle the optical constants n and κ can then be determined. In Fig. 25-6 we show the measurement configuration.

To derive the equations for n and κ, we begin with Fresnel's reflection equations for absorbing media:

$$R_s = -\frac{\sin(\theta_i - \theta_r)}{\sin(\theta_i + \theta_r)} E_s \tag{25-24a}$$

$$R_p = \frac{\tan(\theta_i - \theta_r)}{\tan(\theta_i + \theta_r)} E_p \tag{25-24b}$$

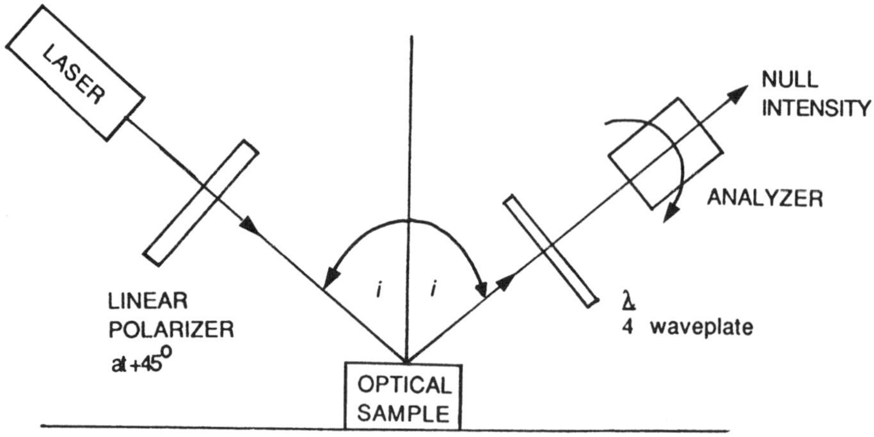

Figure 25-6 Measurement of the principal angle of incidence and the principal azimuth angle.

The angle θ_r is now complex, so the ratios R_p/E_p and R_s/E_s are also complex. Thus, the amplitude and phase change on being reflected from optically absorbing media. Incident polarized light will, in general, become elliptically polarized on reflection from an optically absorbing medium. We now let ϕ_p and ϕ_s be the phase changes and ρ_p and ρ_s the absolute values of the reflection coefficients r_p and r_s. Then, we can write

$$r_p = \frac{R_p}{E_p} = \rho_p \exp(i\phi_p) \tag{25-43a}$$

$$r_s = \frac{R_s}{E_s} = \rho_s \exp(i\phi_s) \tag{25-43b}$$

Equation (25-43) can be transformed to the Stokes parameters. The Stokes parameters for the incident beam are

$$S_0 = \cos\theta_i(E_s E_s^* + E_p E_p^*) \tag{25-44a}$$

$$S_1 = \cos\theta_i(E_s E_s^* - E_p E_p^*) \tag{25-44b}$$

$$S_2 = \cos\theta_i(E_s E_p^* + E_p E_s^*) \tag{25-44c}$$

$$S_3 = i\cos\theta_i(E_s E_p^* - E_p E_s^*) \tag{25-44d}$$

Similarly, the Stokes parameters for the reflected beam are defined as

$$S_0' = \cos\theta_i(R_s R_s^* + R_p R_p^*) \tag{25-45a}$$

$$S_1' = \cos\theta_i(R_s R_s^* - R_p R_p^*) \tag{25-45b}$$

$$S_2' = \cos\theta_i(R_s R_p^* + R_p R_s^*) \tag{25-45c}$$

$$S_3' = i\cos\theta_i(R_s R_p^* - R_p R_s^*) \tag{25-45d}$$

Substituting (25-43) into (25-45) and using (25-44) yields

$$\begin{pmatrix} S_0' \\ S_1' \\ S_2' \\ S_3' \end{pmatrix} = \frac{1}{2} \begin{pmatrix} \rho_s^2 + \rho_p^2 & \rho_s^2 - \rho_p^2 & 0 & 0 \\ \rho_s^2 - \rho_p^2 & \rho_s^2 + \rho_p^2 & 0 & 0 \\ 0 & 0 & 2\rho_s\rho_p \cos\Delta & 2\rho_s\rho_p \sin\Delta \\ 0 & 0 & -2\rho_s\rho_p \sin\Delta & 2\rho_s\rho_p \cos\Delta \end{pmatrix} \begin{pmatrix} S_0 \\ S_1 \\ S_2 \\ S_3 \end{pmatrix} \tag{25-46}$$

where $\Delta = \phi_s - \phi_p$.

We now allow the incident light to be +45° linearly polarized so that $E_p = E_s$. Furthermore, we introduce an azimuthal angle α (generally complex) for the reflected light, which is defined by

$$\tan\alpha = \frac{R_s}{R_p} = -\frac{\cos(\theta_i - \theta_r)}{\cos(\theta_i + \theta_r)} = P\exp(i\Delta) \tag{25-47}$$

where we have used (25-24), and P is real and we write it as

$$P = \tan\psi \tag{25-48a}$$

where ψ is called the azimuthal angle. From (25-43) we also see that

$$P = \frac{\rho_s}{\rho_p} \qquad \Delta = \phi_s - \phi_p \tag{25-48b}$$

We note that α is real in the following two cases:

1. For normal incidence ($\theta_i = 0$); then from (25-47) we see that $P = 1$ and $\Delta = \pi$.
2. For grazing incidence ($\theta_i = \pi/2$); then from (25-47) we see that $P = 1$ and $\Delta = 0$.

Between these two extreme values there exists an angle $\bar{\theta}_i$ called the *principal angle of incidence* for which $\Delta = \pi/2$. Let us now see the consequences of obtaining this condition. We first write (25-48b) as

$$\rho_s = P\rho_p \tag{25-49}$$

Substituting (25-49) into (25-46), we obtain the Stokes vector of the reflected light to be

$$\begin{pmatrix} S_0' \\ S_1' \\ S_2' \\ S_3' \end{pmatrix} = \frac{\rho_p^2}{2} \begin{pmatrix} 1 + P^2 & -(1 - P^2) & 0 & 0 \\ -(1 - P^2) & (1 + P^2) & 0 & 0 \\ 0 & 0 & 2P\cos\Delta & 2P\sin\Delta \\ 0 & 0 & -2P\sin\Delta & 2P\cos\Delta \end{pmatrix} \begin{pmatrix} S_0 \\ S_1 \\ S_2 \\ S_3 \end{pmatrix} \tag{25-50}$$

For incident +45° linearly polarized light, the Stokes vector is

$$\begin{pmatrix} S_0 \\ S_1 \\ S_2 \\ S_3 \end{pmatrix} = I_0 \begin{pmatrix} 1 \\ 0 \\ 1 \\ 0 \end{pmatrix} \tag{25-51}$$

Substituting (25-51) into (25-50), we find the Stokes vector of the reflected light to be

$$\begin{pmatrix} S'_0 \\ S'_1 \\ S'_2 \\ S'_3 \end{pmatrix} = \frac{\rho_p^2 I_0}{2} \begin{pmatrix} 1+P^2 \\ -(1-P^2) \\ 2P\cos\Delta \\ -2P\sin\Delta \end{pmatrix} \tag{25-52}$$

The ellipticity angle χ is

$$\chi = \frac{1}{2}\sin^{-1}\left(\frac{S'_3}{S'_0}\right) = \frac{1}{2}\sin^{-1}\left(\frac{-2P\sin\Delta}{1+P^2}\right) \tag{25-53a}$$

Similarly, the orientation angle ψ is

$$\psi = \frac{1}{2}\tan^{-1}\left(\frac{-2P\cos\Delta}{1-P^2}\right) \tag{25-53b}$$

We see that χ is greatest when $\Delta = \pi/2$ but then $\psi = 0$; i.e., the polarization ellipse corresponding to (25-52) is in its standard, nonrotated, form.

For $\Delta = \pi/2$ the Stokes vector, (25-52), becomes

$$\begin{pmatrix} S'_0 \\ S'_1 \\ S'_2 \\ S'_3 \end{pmatrix} = \frac{\rho_p^2 I_0}{2} \begin{pmatrix} 1+P^2 \\ -(1-P^2) \\ 0 \\ -2P \end{pmatrix} \tag{25-54}$$

and χ and ψ of the polarization ellipse corresponding to (25-54) are,

$$\chi = \frac{1}{2}\sin^{-1}\left(\frac{S'_3}{S'_0}\right) = \frac{1}{2}\sin^{-1}\left(\frac{-2P}{1+P^2}\right) \tag{25-55a}$$

$$\psi = \frac{1}{2}\tan^{-1}\left(\frac{S'_2}{S'_1}\right) = 0 \tag{25-55b}$$

We must now transform the elliptically polarized light described by the Stokes vector (25-54) to linearly polarized light. A quarter-wave retarder can be used to transform elliptically polarized light to linearly polarized light. The Mueller matrix for a quarter-wave retarder oriented at 0° is

$$M = \begin{pmatrix} 1 & 0 & 0 & 0 \\ 0 & 1 & 0 & 0 \\ 0 & 0 & 0 & -1 \\ 0 & 0 & 1 & 0 \end{pmatrix} \tag{25-56}$$

Multiplying (25-54) by (25-56) yields

$$\begin{pmatrix} S'_0 \\ S'_1 \\ S'_2 \\ S'_3 \end{pmatrix} = \frac{\rho_p^2 I_0}{2} \begin{pmatrix} 1+P^2 \\ -(1-P^2) \\ 2P \\ 0 \end{pmatrix} \tag{25-57}$$

which is the Stokes vector for linearly polarized light. The Mueller matrix for a linear polarizer at an angle β is

$$M = \frac{1}{2}\begin{pmatrix} 1 & \cos 2\beta & \sin 2\beta & 0 \\ \cos 2\beta & \cos^2 2\beta & \sin 2\beta \cos 2\beta & 0 \\ \sin 2\beta & \sin 2\beta \cos 2\beta & \sin^2 2\beta & 0 \\ 0 & 0 & 0 & 0 \end{pmatrix} \tag{25-58}$$

Multiplying (25-57) by (25-58), we obtain the intensity of the beam emerging from the analyzing polarizer:

$$I(\beta) = (1 + P^2) - (1 - P^2)\cos 2\beta + 2P \sin 2\beta \tag{25-59}$$

or

$$I(\beta) = A - B\cos 2\beta + C\sin 2\beta \tag{25-60a}$$

where

$$A = 1 + P^2 \tag{25-60b}$$

$$B = 1 - P^2 \tag{25-60c}$$

$$C = 2P \tag{25-60d}$$

Equation (25-60a) is now written as

$$I(\beta) = A\left[1 - \frac{B}{A}\cos 2\beta + \frac{C}{A}\sin 2\beta\right] \tag{25-61a}$$

We set

$$\cos\gamma = \frac{B}{A} \tag{25-61b}$$

$$\sin\gamma = \frac{C}{A} \tag{25-61c}$$

so that (25-61a) can now be written as

$$I(\beta) = A[1 - \cos(\gamma - 2\beta)] \tag{25-62a}$$

and

$$\gamma = \tan^{-1}\left(\frac{C}{B}\right) = \tan^{-1}\left(\frac{2P}{1 - P^2}\right) \tag{25-62b}$$

A null intensity for (25-62a) is obtained when

$$\beta = \frac{\gamma}{2} \tag{25-63a}$$

or

$$\gamma = 2\beta \tag{25-63b}$$

The azimuthal angle where the null intensity occurs is called the *principal azimuthal angle* $\bar{\psi}$.

We can relate (25-63b) to the *principal azimuthal angle* $\bar{\psi}$ as follows. We recall from (25-48a) that

$$P = \tan\psi \tag{25-48a}$$

Substituting (25-48a) into (25-62b), we find that

$$\gamma = \tan^{-1}\left(\frac{C}{B}\right) = \tan^{-1}[\tan 2\psi] = 2\beta \tag{25-64a}$$

or

$$\beta = \bar{\psi} \tag{25-64b}$$

The magnitude of P is then

$$P = \frac{\rho_s}{\rho_p} = \tan\bar{\psi} \tag{25-65}$$

It is possible to obtain the same results by irradiating the sample surface with circularly polarized light rather than linearly polarized light. We remove the quarter-wave retarder from the analyzing arm (see Fig. 25-1) and insert it between the $+45^\circ$ linear polarizer and the optical sample in the generating arm. The Stokes vector of the beam emerging from the linear polarizer in the generating arm is

$$\begin{pmatrix} S_0 \\ S_1 \\ S_2 \\ S_3 \end{pmatrix} = I_0 \begin{pmatrix} 1 \\ 0 \\ 1 \\ 0 \end{pmatrix} \tag{25-66}$$

Multiplying (25-66) by the Mueller matrix for a quarter-wave retarder oriented at 0°, (25-56), we obtain the Stokes vector for right circularly polarized light:

$$\begin{pmatrix} S_0 \\ S_1 \\ S_2 \\ S_3 \end{pmatrix} = I_0 \begin{pmatrix} 1 \\ 0 \\ 0 \\ 1 \end{pmatrix} \tag{25-67}$$

The Stokes vector (25-67) is now used in (25-50), whereupon the Stokes vector of the reflected beam is found to be

$$\begin{pmatrix} S'_0 \\ S'_1 \\ S'_2 \\ S'_3 \end{pmatrix} = \frac{\rho_p^2 I_0}{2} \begin{pmatrix} 1+P^2 \\ -(1-P^2) \\ 2P\sin\Delta \\ 2P\cos\Delta \end{pmatrix} \tag{25-68}$$

At the principal angle of incidence $\Delta = \pi/2$, so (25-68) reduces to

$$\begin{pmatrix} S'_0 \\ S'_1 \\ S'_2 \\ S'_3 \end{pmatrix} = \frac{\rho_p^2 I_0}{2} \begin{pmatrix} 1+P^2 \\ -(1-P^2) \\ 2P \\ 0 \end{pmatrix} \tag{25-69}$$

which is identical to the Stokes vector found in (25-57). Thus, the quarter-wave retarder can be inserted into either the generating or analyzing arm, because the phase shift of $\pi/2$ can be generated before or after the reflection from the optical surface.

We also point out that a quarter-wave retarder can also transform elliptically polarized light to linearly polarized light *if the polarization ellipse is in its standard form*. However, the orientation angle is different from 45°. To see this clearly, let us represent the Stokes vector of elliptically polarized light in its "standard" form; that is,

$$\begin{pmatrix} S_0 \\ S_1 \\ S_2 \\ S_3 \end{pmatrix} = \begin{pmatrix} A \\ B \\ 0 \\ D \end{pmatrix} \tag{25-70}$$

The Mueller matrix of a quarter-wave retarder is

$$M = \begin{pmatrix} 1 & 0 & 0 & 0 \\ 0 & 1 & 0 & 0 \\ 0 & 0 & 0 & -1 \\ 0 & 0 & 1 & 0 \end{pmatrix} \tag{25-56}$$

Multiplying (25-70) by (25-56) yields

$$\begin{pmatrix} S_0 \\ S_1 \\ S_2 \\ S_3 \end{pmatrix} = \begin{pmatrix} A \\ B \\ -D \\ 0 \end{pmatrix} \tag{25-71}$$

which is, of course, the Stokes vector for linearly polarized light. However, the polarization ellipse is now oriented at an angle ψ given by

$$\psi = \frac{1}{2}\tan^{-1}\left(\frac{-D}{B}\right) \tag{25-72}$$

We now relate the principle angle of incidence $\bar{\theta}_i$ ($\Delta = \pi/2$) and the principal azimuthal angle $\bar{\psi}$ to the optical constants n and κ. We recall that

$$\tan\alpha = \frac{R_s}{R_p} = \frac{-\cos(\theta_i - \theta_r)}{\cos(\theta_i + \theta_r)} = P\exp(i\Delta) \tag{25-47}$$

$$P = \frac{\rho_s}{\rho_p} \qquad \Delta = \phi_s - \phi_p \tag{25-48b}$$

We expand (25-47)

$$P\exp(i\Delta) = -\frac{\cos\theta_i\cos\theta_r + \sin\theta_i\sin\theta_r}{\cos\theta_i\cos\theta_r - \sin\theta_i\sin\theta_r} \tag{25-73a}$$

$$= \frac{\tan\theta_i\tan\theta_r + 1}{\tan\theta_i\tan\theta_r - 1} \tag{25-73b}$$

Solving (25-73b) for tan θ_i tan θ_r gives

$$\frac{1 + P\exp(i\Delta)}{1 - P\exp(i\Delta)} = -\tan\theta_i\tan\theta_r = -\frac{\tan\theta_i\sin\theta_i}{\sqrt{\mathbf{n}^2 - \sin^2\theta_i}} \tag{25-74}$$

At the principal angle of incidence $\Delta = \pi/2$. Furthermore, $\sin^2\theta_i \ll n^2$ and may be disregarded. Then, (25-74) becomes

$$\frac{1+iP}{1-iP} = \frac{-\sin\bar{\theta}_i \tan\bar{\theta}_i}{n(1-i\kappa)} \tag{25-75}$$

where $\bar{\theta}_i$ is the principal angle. Multiplying (25-75) by its complex conjugate leads immediately to

$$n\sqrt{1+\kappa^2} = \sin\bar{\theta}_i \tan\bar{\theta}_i \tag{25-76}$$

Equation (25-76) serves as a very useful check on n and κ.

We now establish the relations between n and κ and $\bar{\theta}_i$ and $\bar{\psi}$. First, we invert (25-75) and obtain

$$\frac{1-iP}{1+iP} = \frac{-n(1-i\kappa)}{\sin\bar{\theta}_i \tan\bar{\theta}_i} \tag{25-77}$$

Next, we replace P by $\tan\bar{\psi}$. Then, multiplying the numerator and the denominator of the left-hand side of (25-77) by $1 - i\tan\bar{\psi}$, we find that

$$\frac{n}{\sin\bar{\theta}_i \tan\bar{\theta}_i} = -\frac{1-\tan^2\bar{\psi}}{1+\tan^2\bar{\psi}} \tag{25-78a}$$

$$\frac{n\kappa}{\sin\bar{\theta}_i \tan\bar{\theta}_i} = \frac{-2\tan^2\bar{\psi}}{1+\tan^2\bar{\psi}} \tag{25-78b}$$

The right-hand sides of (25-78a) and (25-78b) reduce to $\cos 2\bar{\psi}$ and $\sin 2\bar{\psi}$, respectively. This leads immediately to

$$n = -\sin\bar{\theta}_i \tan\bar{\theta}_i \cos 2\bar{\psi} \tag{25-79a}$$

$$\kappa = \tan 2\bar{\psi} \tag{25-79b}$$

We can substitute (25-79a) and (25-79b) into (25-18) and find that

$$\mathbf{n} = -\sin\bar{\theta}_i \tan\bar{\theta}_i \exp(-i2\bar{\psi}) \tag{25-80}$$

Thus, by measuring the principal angle of incidence $\bar{\theta}_i$ and the principal azimuthal angle $\bar{\psi}$, we can determine n and κ from (25-79a) and (25-79b), respectively.

In the present formulation of relating n and κ to $\bar{\theta}_i$ and $\bar{\psi}$, the term $\sin^2\theta_i$ has been neglected. Interestingly, as pointed out by Wood, the inclusion of $\sin^2\theta_i$ leads to the same equations.

Further information on the principal angle of incidence method is given in the textbooks by Born and Wolf, Wood, and Longhurst. For example, Wood also describes the application of the method to the measurement of optical materials in the ultraviolet region of the optical spectrum.

25.4 MEASUREMENT OF REFRACTIVE INDEX AND EXTINCTION COEFFICIENT AT AN INCIDENT ANGLE OF 45°

In the previous section we saw that the principal angle of incidence method can be used to obtain the optical constants n and κ. We also pointed out that there is another method known as the pseudo-Brewster angle method; this method is described by Potter. The classical Brewster angle method, we recall, leads to a *null* intensity at the

Brewster angle for dielectrics. For absorbing optical materials, however, one can show that a *minimum* intensity is obtained instead; this is indicated in Figs. 25-2 through 25-5. From a measurement of the minimum intensity, n and κ can then be found. The Brewster angle method is very useful. However, a "wide" minimum is obtained, and this limits the accuracy of the results to only two or three decimal places at most.

All of these classical methods are based on a "null" or minimum intensity condition and reading of mechanical dials. Consequently, these methods can be called optomechanical; that is, only optics and mechanical components are used to determine n and κ. While these methods have long been the "standard" means for determining n and κ, they have a number of drawbacks. The first and most serious is that a mechanical arm must be used and moved to find the appropriate angle, e.g., the principal angle of incidence or the Brewster angle. Very often, apparatus to do this, such as a divided circle, is not readily available. Furthermore, a mechanical divided circle is quite expensive. Another drawback is that it is time consuming to move the mechanical arm and search for a null or minimum intensity. In addition, automating the movement of a mechanical arm is difficult and expensive. Finally, these measurement methods do not utilize to any significant extent the developments made in electronics and optical detectors in recent years.

Ideally, it would be preferable if n and κ could be measured without any mechanical movement whatsoever, especially, with respect to moving a mechanical arm. This can indeed be done by irradiating the optical surface at an incident angle of 45°. At this angle Fresnel's equations reduce to relatively simple forms, and the measurement of the reflected intensity can be easily made with an optical detector and a digital voltmeter. Mechanical fixed mounts are, of course, still necessary, but there are no major mechanical movements. Furthermore, the required mechanical and optical components are nearly always available in any modern optical laboratory. In addition, because the angles involved are 45° and the components are aligned perpendicular to each other, these measurements are easily carried out on an optical table. Finally, a digital voltmeter capable of reading to, say, $5\frac{1}{2}$ digits is relatively inexpensive. In this method, therefore, the optical constants are derived by using only quantitative detectors and reading on a digital voltmeter rather than a mechanical dial. In fact, the four Stokes parameters must be measured, but these measurements are made at settings of 0°, 45°, and 90°, which does not require searching for a null. Consequently, this measurement method can be called optoelectronic. It has been called digital refractometry. Therefore, we consider Fresnel's equations for reflection at an incident angle of 45°. From the measurement of the Stokes parameters of the reflected beam, n and κ can then be determined. We now derive the relations which relate the Stokes parameters to n and κ at an incident angle of 45°.

Figure 25-7 shows the incident orthogonal components E_p and E_s and the reflected field components R_p and R_s, respectively; p and s have their usual meanings. For absorbing optical materials Fresnel's reflection equations continue to hold, so

$$R_s = -\frac{\sin(\theta_i - \theta_r)}{\sin(\theta_i + \theta_r)} E_s \tag{25-81a}$$

$$R_p = \frac{\tan(\theta_i - \theta_r)}{\tan(\theta_i + \theta_r)} E_p \tag{25-81b}$$

In (25-81), θ_i is the angle of incidence and θ_r is the angle of refraction.

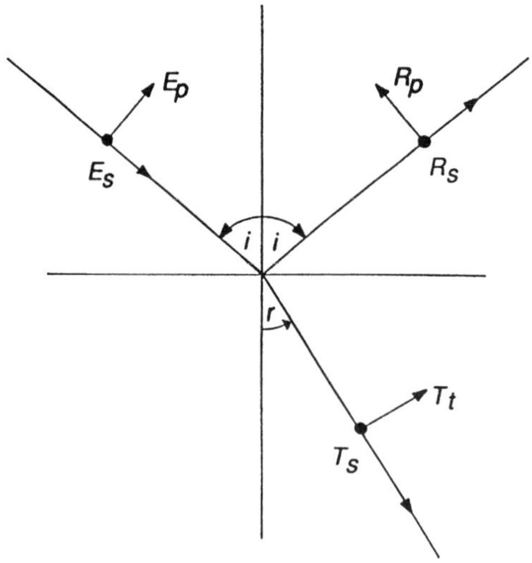

Figure 25-7 Optical field components for the incident and reflected fields.

Absorbing optical media are characterized by a complex refractive index $\mathbf{n}$ of the form:

$$\mathbf{n} = n(1 - i\kappa) \tag{25-18}$$

When $\theta_i = 45^\circ$, a relatively simple form of Fresnel's equations emerges, as we shall now show.

Snell's law of refraction continues to be valid for media described by (25-18), so we have

$$\sin\theta_r = \frac{\sin\theta_i}{\mathbf{n}} \tag{25-82}$$

Equation (25-82) can be expressed in terms of $\cos\theta_r$:

$$\cos\theta_r = \frac{\sqrt{\mathbf{n}^2 - \sin^2\theta_i}}{\mathbf{n}} \tag{25-83}$$

For an incident angle of 45°, (25-82) and (25-83) become, respectively,

$$\sin\theta_r = \frac{1}{\sqrt{2}\mathbf{n}} \tag{25-84a}$$

and

$$\cos\theta_r = \frac{\sqrt{2\mathbf{n}^2 - 1}}{\sqrt{2}\mathbf{n}} \tag{25-84b}$$

For an incident angle of 45°, (25-81) reduces to

$$R_s = -\left[\frac{\cos\theta_r - \sin\theta_r}{\cos\theta_r + \sin\theta_r}\right]E_s \tag{25-85a}$$

$$R_p = \left[\frac{\cos\theta_r - \sin\theta_r}{\cos\theta_r + \sin\theta_r}\right]^2 E_p \tag{25-85b}$$

Replacing the cosine and sine terms using (25-84), we can write (25-85) as

$$R_s = -\left[\frac{\sqrt{2\mathbf{n}^2 - 1} - 1}{\sqrt{2\mathbf{n}^2 - 1} + 1}\right]E_s \tag{25-86a}$$

In a similar manner the equation for R_p becomes

$$R_p = \left[\frac{\sqrt{2\mathbf{n}^2 - 1} - 1}{\sqrt{2\mathbf{n}^2 - 1} + 1}\right]^2 E_p \tag{25-86b}$$

We now set

$$\sqrt{2\mathbf{n}^2 - 1} = a - ib = A\exp(-i\phi) \tag{25-87}$$

Then, (25-86a) and (25-86b) can be written using (25-87) as

$$R_s = -\left[\frac{(a-1) - ib}{(a+1) - ib}\right]E_s \tag{25-88a}$$

$$R_p = \left[\frac{(a-1) - ib}{(a+1) - ib}\right]^2 E_p \tag{25-88b}$$

Equation (25-86) can be written also in terms of A and ϕ, as in (25-87). Straightforward substitution gives

$$R_s = \left[\frac{1 - A\exp(-i\phi)}{1 + A\exp(-i\phi)}\right]E_s \tag{25-89a}$$

$$R_p = \left[\frac{1 - A\exp(-i\phi)}{1 + A\exp(-i\phi)}\right]^2 E_p \tag{25-89b}$$

From (25-87) we have

$$A^2 = a^2 + b^2 \tag{25-90a}$$

$$\phi = \tan^{-1}\left(\frac{b}{a}\right) \tag{25-90b}$$

A and ϕ can also be expressed in terms of n and κ. We have from (25-87)

$$a - ib = \sqrt{2\mathbf{n}^2 - 1} \tag{25-91}$$

Substituting (25-18) into (25-91) and then squaring both sides of the equation leads to

$$a^2 - b^2 - i(2ab) = (2n^2 - 2n^2\kappa^2 - 1) - i(4n^2\kappa) \tag{25-92}$$

Equating real and imaginary terms we have

$$a^2 - b^2 = 2n^2 - 2n^2\kappa^2 - 1 \tag{25-93a}$$

$$ab = 2n^2\kappa \tag{25-93b}$$

We can also find an expression for a^2+b^2. We take the complex conjugate of (25-91):

$$a + ib = \sqrt{2n^2(1 + i\kappa)^2 - 1} \tag{25-94a}$$

We also have from (25-87) that

$$a - ib = \sqrt{2n^2(1 - i\kappa)^2 - 1} \tag{25-94b}$$

Multiplying (25-94a) by (25-94b) gives

$$A^2 = a^2 + b^2 = \sqrt{4n^4(1 + \kappa^2)^2 - 4n^2(1 - \kappa^2) + 1} \tag{25-95}$$

Adding and subtracting (25-93a) and (25-95) yields

$$a^2 = \frac{1}{2}\left[\sqrt{4n^4(1 + \kappa^2)^2 - 4n^2(1 - \kappa^2) + 1} + (2n^2 - 2n^2\kappa^2 - 1)\right] \tag{25-96a}$$

$$b^2 = \frac{1}{2}\left[\sqrt{4n^4(1 + \kappa^2)^2 - 4n^2(1 - \kappa^2) + 1} - (2n^2 - 2n^2\kappa^2 - 1)\right] \tag{25-96b}$$

Then, from (25-90b) and (25-96) we see that

$$\phi = \tan^{-1}\left[\frac{\sqrt{4n^4(1 + \kappa^2)^2 - 4n^2(1 - \kappa^2) + 1} - (2n^2 - 2n^2\kappa^2 - 1)}{\sqrt{4n^4(1 + \kappa^2)^2 - 4n^2(1 - \kappa^2) + 1} + (2n^2 - 2n^2\kappa^2 - 1)}\right]^{1/2} \tag{25-97}$$

For nonabsorbing materials $\kappa = 0$, so (25-95) and (25-97) reduce to

$$A^2 = a^2 = 2n^2 - 1 \qquad \text{and} \qquad \phi = 0 \tag{25-98}$$

as expected.

We must now transform the amplitude equations (25-89) to intensity equations, and from these derive the Stokes polarization parameters.

We defined the Stokes parameters of the incident and reflected beams in Eqs. (25-44) and (25-45).

Substituting (25-89) into (25-45) and using (25-44) yields

$$\begin{pmatrix} S'_0 \\ S'_1 \\ S'_2 \\ S'_3 \end{pmatrix} = \frac{1 - 2A\cos\phi + A^2}{(1 + 2A\cos\phi + A^2)^2}$$

$$\times \begin{pmatrix} 1 + A^2 & 2A\cos\phi & 0 & 0 \\ 2A\cos\phi & 1 + A^2 & 0 & 0 \\ 0 & 0 & 1 - A^2 & -2A\sin\phi \\ 0 & 0 & 2A\sin\phi & 1 - A^2 \end{pmatrix} \begin{pmatrix} S_0 \\ S_1 \\ S_2 \\ S_3 \end{pmatrix} \tag{25-99}$$

The 4 × 4 matrix is the Mueller matrix for optically absorbing materials at an incident angle of 45°. The presence of the off-diagonal terms in the upper and lower parts of the matrix shows that optically absorbing materials *simultaneously* change the amplitude and phase of the incident beam. To determine n and κ, we measure A and ϕ and solve (25-95) and (25-97) for n and κ by iteration. It is straightforward to show that (25-99) reduces to the equation for dielectrics by setting $\kappa = 0$; i.e.,

$$\begin{pmatrix} S'_0 \\ S'_1 \\ S'_2 \\ S'_3 \end{pmatrix} = \frac{1 - \sin 2\theta_r}{(1 + \sin 2\theta_r)^2} \begin{pmatrix} 1 & \sin 2\theta_r & 0 & 0 \\ \sin 2\theta_r & 1 & 0 & 0 \\ 0 & 0 & -\cos 2\theta_r & 0 \\ 0 & 0 & 0 & -\cos 2\theta_r \end{pmatrix} \begin{pmatrix} S_0 \\ S_1 \\ S_2 \\ S_3 \end{pmatrix} \tag{25-100}$$

We can derive an important relation between the intensity of an incident beam, I_0, and the orthogonal intensities of the reflected beam, I_s and I_p, respectively. Consider that we irradiate the surface of an optically absorbing material with a linear vertically polarized beam of intensity I_0; we call this the p polarized beam, and its Stokes vector is

$$\begin{pmatrix} S_0 \\ S_1 \\ S_2 \\ S_3 \end{pmatrix} = I_0 \begin{pmatrix} 1 \\ -1 \\ 0 \\ 0 \end{pmatrix} \tag{25-101}$$

Multiplying (25-99) out with (25-101) substituted for the incident Stokes vector gives

$$I_p = I_0 \frac{(1 + A^2 - 2A\cos\phi)^2}{(1 + A^2 + 2A\cos\phi)^2} \tag{25-102}$$

Next, we irradiate the optical surface with a linearly horizontally polarized beam; we call this the s polarized beam. Its Stokes vector is

$$\begin{pmatrix} S_0 \\ S_1 \\ S_2 \\ S_3 \end{pmatrix} = I_0 \begin{pmatrix} 1 \\ 1 \\ 0 \\ 0 \end{pmatrix} \tag{25-103}$$

Multiplying (25-99) out with (25-103) substituted for the incident Stokes vector yields

$$I_s = I_0 \frac{(1 + A^2 - 2A\cos\phi)}{(1 + A^2 + 2A\cos\phi)} \tag{25-104}$$

We eliminate the ratio factor between (25-102) and (25-104) and find that

$$I_s^2 = I_0 I_p \tag{25-105}$$

Equation (25-105) is a fundamental relation. It is the intensity analog of the amplitude relation:

$$R_s^2 = R_p \tag{25-106}$$

Equation (25-105) shows that it is incorrect to square (25-106) in order to obtain (25-105); the correct relation includes the source intensity I_0. From an experimental point of view (25-105) is very useful because it shows that by measuring the orthogonal intensities of the reflected beam the source intensity can be directly monitored or measured. Similarly, if I_0 is known, then (25-105) serves as a useful check on the measurement of I_s and I_p.

We now turn to the measurement of A and ϕ in (25-99).

Let us irradiate the optical surface with an optical beam of intensity I_0 which is right circularly polarized. The Stokes vector of the incident beam is then

$$\begin{pmatrix} S_0 \\ S_1 \\ S_2 \\ S_3 \end{pmatrix} = I_0 \begin{pmatrix} 1 \\ 0 \\ 0 \\ 1 \end{pmatrix} \tag{25-107}$$

Multiplying (25-99) out with (25-107) substituted for the incident Stokes vector, we find that the Stokes vector of the reflected beam is

$$\begin{pmatrix} S'_0 \\ S'_1 \\ S'_2 \\ S'_3 \end{pmatrix} = I_0 \frac{1 - 2A\cos\phi + A^2}{(1 + 2A\cos\phi + A^2)^2} \begin{pmatrix} 1 + A^2 \\ 2A\cos\phi \\ -2A\sin\phi \\ 1 - A^2 \end{pmatrix} \tag{25-108}$$

The reflected beam is elliptically polarized. We can determine the quantities A^2 and ϕ directly from measuring the Stokes parameters. Dividing S'_3 by S'_0 we find that

$$A^2 = \frac{S'_0 - S'_3}{S'_0 + S'_3} \tag{25-109a}$$

Dividing S'_2 by S'_1 gives

$$\phi = \tan^{-1}\left(\frac{-S'_2}{S'_1}\right) \tag{25-109b}$$

We also see that the ellipticity angle χ is

$$\chi = \frac{1}{2}\sin^{-1}\left(\frac{1-A^2}{1+A^2}\right) \tag{25-110a}$$

$$= \frac{1}{2}\sin^{-1}\left[\frac{1-\sqrt{4n^4(1+\kappa^2)^2 - 4n^2(1-\kappa^2)+1}}{1+\sqrt{4n^4(1+\kappa^2)^2 - 4n^2(1-\kappa^2)+1}}\right] \tag{25-110b}$$

The orientation angle ψ is, using (25-97),

$$\psi = \frac{-\phi}{2} \tag{25-111a}$$

$$= -\frac{1}{4}\tan^{-1}\left[\frac{\sqrt{4n^4(1+\kappa^2)^2 - 4n^2(1-\kappa^2)+1} - (2n^2 - 2n^2\kappa^2 - 1)}{\sqrt{4n^4(1+\kappa^2)^2 - 4n^2(1-\kappa^2)+1} + (2n^2 - 2n^2\kappa^2 - 1)}\right]^{1/2} \tag{25-111b}$$

For the condition where we have no absorption ($\kappa = 0$) χ and ψ become, respectively,

$$\chi = \frac{1}{2}\sin^{-1}\left(\frac{1-n^2}{n^2}\right) \tag{25-112a}$$

$$\psi = 0 \tag{25-112b}$$

as expected. To determine A^2 and ψ uniquely, we must measure all four Stokes parameters. In Part I, various methods for doing this were considered. Before we describe an experimental configuration for carrying out the measurement, we relate the above equations to another commonly used representation, the reflection coefficients representation.

The reflection coefficients are defined by Born and Wolf to be

$$r_s = \frac{R_s}{E_s} = \rho_s e^{i\gamma_s} \tag{25-113a}$$

$$r_p = \frac{R_p}{E_p} = \rho_p e^{i\gamma_p}; \qquad \gamma = \gamma_s - \gamma_p \tag{25-113b}$$

From the definitions of the Stokes parameters given by (25-44) and (25-45), the amplitude equations (25-113) are found to transform

$$\begin{pmatrix} S'_0 \\ S'_1 \\ S'_2 \\ S'_3 \end{pmatrix} = \frac{1}{2}\begin{pmatrix} \rho_s^2+\rho_p^2 & \rho_s^2-\rho_p^2 & 0 & 0 \\ \rho_s^2-\rho_p^2 & \rho_s^2+\rho_p^2 & 0 & 0 \\ 0 & 0 & 2\rho_s\rho_p\cos\gamma & 2\rho_s\rho_p\sin\gamma \\ 0 & 0 & -2\rho_s\rho_p\sin\gamma & 2\rho_s\rho_p\cos\gamma \end{pmatrix}\begin{pmatrix} S_0 \\ S_1 \\ S_2 \\ S_3 \end{pmatrix} \tag{25-114}$$

We can relate the coefficients in (25-114) to A and ψ in (25-108) by irradiating the surface with right circularly polarized light. The respective Stokes parameters of the reflected beam are

$$\frac{1}{2}\left[\rho_s^2 + \rho_p^2\right] = \left[\frac{1 + A^2 - 2A\cos\phi}{\left(1 + A^2 + 2A\cos\phi\right)^2}\right](1 + A^2) \tag{25-115a}$$

$$\frac{1}{2}\left[\rho_s^2 - \rho_p^2\right] = \left[\frac{1 + A^2 - 2A\cos\phi}{\left(1 + A^2 + 2A\cos\phi\right)^2}\right](2A\cos\phi) \tag{25-115b}$$

$$\rho_s\rho_p \sin\gamma = \left[\frac{1 + A^2 - 2A\cos\phi}{\left(1 + A^2 + 2A\cos\phi\right)^2}\right](-2A\sin\phi) \tag{25-115c}$$

$$\rho_s\rho_p \cos\gamma = \left[\frac{1 + A^2 - 2A\cos\phi}{\left(1 + A^2 + 2A\cos\phi\right)^2}\right](1 - A^2) \tag{25-115d}$$

Adding (25-115a) and (25-115b) gives

$$\rho_s^2 = \frac{I_s}{I_0} = \frac{1 + A^2 - 2A\cos\phi}{1 + A^2 + 2A\cos\phi} \tag{25-116a}$$

and subtracting (25-115b) from (25-115a) gives

$$\rho_p^2 = \frac{I_p}{I_0} = \frac{\left(1 + A^2 - 2A\cos\phi\right)^2}{\left(1 + A^2 + 2A\cos\phi\right)^2} \tag{25-116b}$$

The relation for γ in terms of A and ϕ is then obtained by dividing (25-115c) by (25-115d)

$$\tan\gamma = \left[\frac{-2A}{1 - A^2}\right]\sin\phi \tag{25-116c}$$

We see that the reflection coefficients in (25-116a) and (25-116b) are identical to the ratio of the orthogonal intensities, (25-102) and (25-104), of the reflected beam. We also see from (25-116a) and (25-116b) that

$$\rho_s^4 = \rho_p^2 \tag{25-117}$$

in agreement with our previous observations.

Figure 25-8 shows a block diagram of the experimental configuration for measuring n and κ.

In this measurement, a He–Ne laser is used as the optical source (6328 Å). The optical beam emerging from the laser is expanded and collimated; this creates a plane wave. In addition, an improved signal-to-noise ratio is obtained by chopping the beam. The frequency at which the beam is chopped is then used as a reference signal for a lock-in amplifier. The circular polarizer before the sample creates a circularly polarized beam which then irradiates the optical surface at an incident angle of 45°. The autocollimator is used to align the optical surface of the sample being measured to exactly 45°. The reflected beam is then analyzed by a circular polarizer in order to obtain the four Stokes parameters in accordance with the discussion in Part I. The

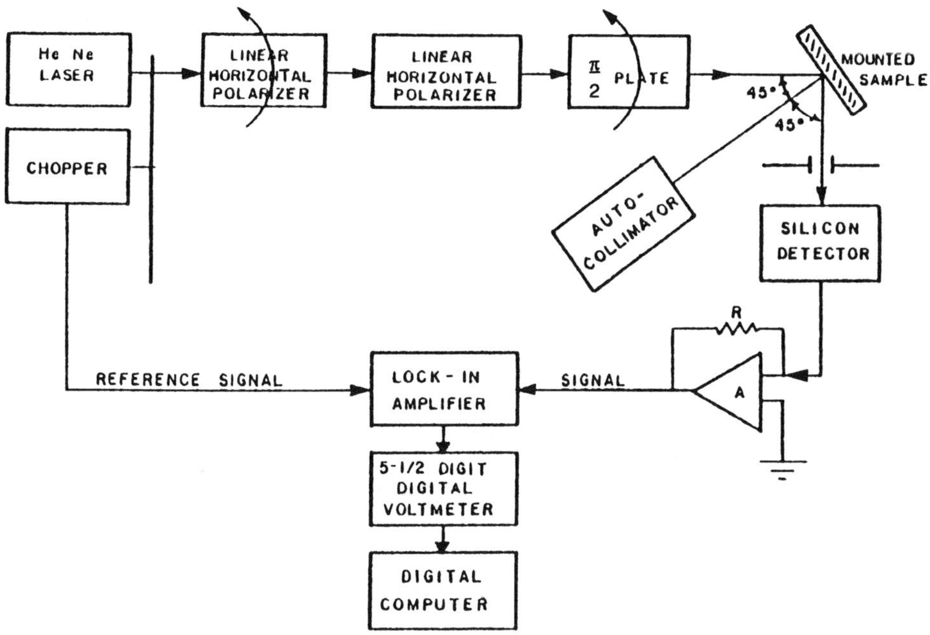

Figure 25-8 Experimental configuration for measuring the Stokes parameters and the optical constants n and κ of an optically absorbing material.

beam that emerges from the circular polarizer in the analyzing path is incident on a silicon detector. The chopped voltage signal is then fed to the lock-in amplifier along with the reference signal. The lock-in amplifier consists, essentially, of a phase-sensitive detector along with an RC network to smooth the output d.c. (analog) voltage. The d.c. voltage is then converted into a digital voltage by a digital voltmeter, e.g., a $5\frac{1}{2}$ digit voltmeter. The $5\frac{1}{2}$ means that the minimum voltage which can be displayed or "read" is five digits after the decimal point. The 1/2 term means that the number to the left of the decimal point can vary from 0 to 1 for an average of $(1 + 0)/2 = 1/2$. If we have a voltage greater than 1.99999 V, then the maximum displayed reading can only be read to four decimal places, e.g., 2.1732 V. The digital output is read by a digital computer, and the values of A^2 and ϕ are then calculated from (25-109a) and (25-109b), respectively.

The optical constants n and κ are calculated from the values of A^2 and ϕ in (25-95) and (25-97). For convenience we repeat the equations here:

$$A^2 = a^2 + b^2 = \sqrt{4n^4(1+\kappa^2)^2 - 4n^2(1-\kappa^2) + 1} \tag{25-95}$$

$$\phi = \tan^{-1}\left[\frac{\sqrt{4n^4(1+\kappa^2)^2 - 4n^2(1-\kappa^2) + 1} - (2n^2 - 2n^2\kappa^2 - 1)}{\sqrt{4n^4(1+\kappa^2)^2 - 4n^2(1-\kappa^2) + 1} + (2n^2 - 2n^2\kappa^2 - 1)}\right]^{1/2} \tag{25-97}$$

To determine n and κ, we first estimate these values. This is most easily done from the plots of (25-95) and (25-97) in Figs. 25-9 and 25-10.

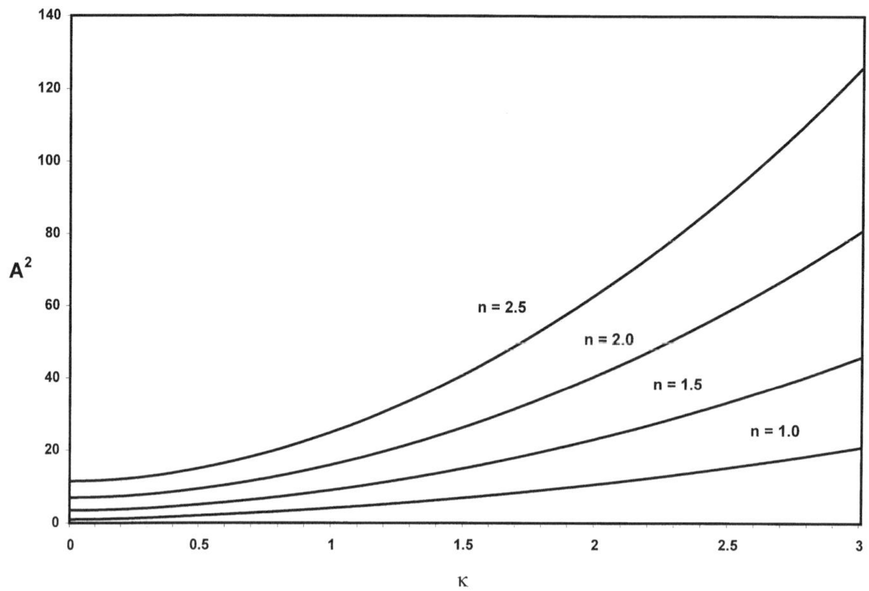

Figure 25-9 Plot of A^2 versus κ for varying values of n, Eq. (25-95).

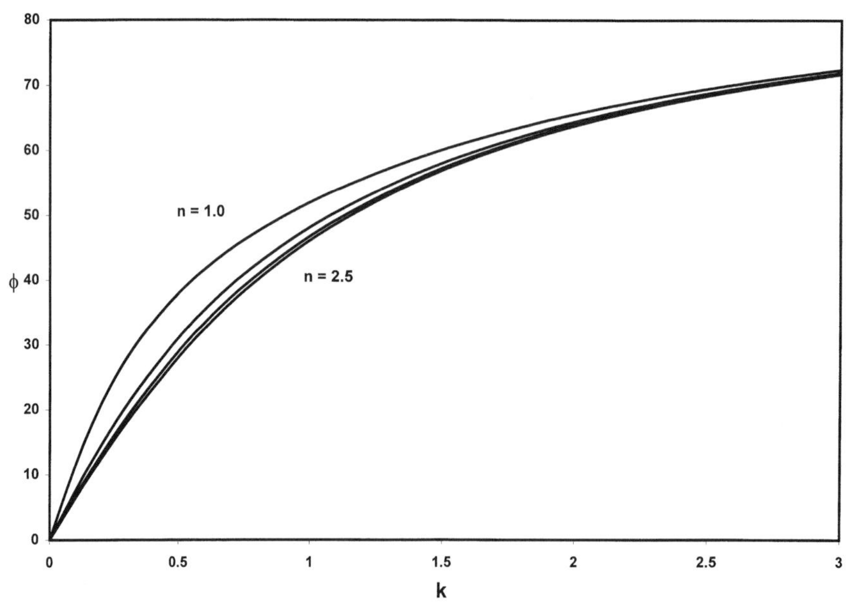

Figure 25-10 Plot of ϕ versus κ for varying values of n, Eq. (25-97).

Inspecting Figs. 25-9 and 25-10, we observe that small values of κ yield small values of A^2 and relatively small phase shifts ϕ. This information is very useful for determining the approximate value of the complex refractive index.

Let us now consider two examples of determining n and κ. In the first example, a sample is measured, and its normalized Stokes vector is

$$\begin{pmatrix} S_0 \\ S_1 \\ S_2 \\ S_3 \end{pmatrix} = \begin{pmatrix} 1.000 \\ 0.365 \\ -0.246 \\ -0.898 \end{pmatrix} \tag{25-118}$$

We find then from (25-118), (25-109a), and (25-109b) that

$$A^2 = 18.608 \tag{25-119a}$$

and

$$\phi = 33.979^\circ \tag{25-119b}$$

Using these values in (25-102) and (25-104), we then find that the ratio of the orthogonal intensities is

$$\frac{I_p}{I_s} = 0.837 \tag{25-120}$$

This value provides a final check on the measurement. To obtain a "seed" value for the complex refractive index, we construct Table 25-1, using (25-95) and (25-97).

Inspection of the table shows that, as the "blocks" of n increase, *the first entry of A^2 and ϕ in each block increase and decrease, respectively*. Thus, we need only match the pair of A^2 and ϕ that is closest to the actual value. In this case the desired values are $A^2 = 18.608$ and $\phi = 33.979^\circ$. The closest pair in the table which matches this is $A^2 = 15.046$ and $\phi = 30.182$, and the corresponding values of n and κ are 2.5 and 0.5. Thus, the "seed" complex refractive index is chosen to be

$$\mathbf{n} = 2.5(1 - i0.5) \tag{25-121}$$

Table 25-1 was constructed for small values of κ. If, for example, large values of A^2 and ϕ were found, this would indicate that a new table should be constructed from (25-95) and (25-97), starting with values of, say, $n = 0.5$ and $\kappa = 5.0$, etc.

We now iterate (25-95) and (25-97) around $n = 2.50$ and $\kappa = 0.50$, and we find that for $A^2 = 18.608$ and $\phi = 33.979^\circ$ the complex refractive index is represented by

$$\mathbf{n} = 2.6790(1 - i0.5745) \tag{25-122}$$

We can use this result to find the reflectivity of an optical beam at normal incidence. We recall from Section 25.2 that the reflectivity for any incident polarization is

$$R = \frac{(n-1)^2 + (n\kappa)^2}{(n+1)^2 + (n\kappa)^2} \tag{25-123}$$

Substituting the above values of n and κ into (25-123), we find that

$$R = 32.6\% \tag{25-124}$$

which shows that this optical material is a very poor reflector.

Table 25-1 Initial values for determining n and κ from (25-95) and (25-97)

n	κ	A^2	ϕ
0.5	0.5	0.800	68.954
0.5	1.0	1.414	67.815
0.5	1.5	2.211	68.254
0.5	2.0	3.202	68.954
1.0	0.5	2.062	42.393
1.0	1.0	4.123	57.961
1.0	1.5	6.946	64.113
1.0	2.0	10.630	67.007
1.5	0.5	5.088	33.833
1.5	1.0	9.055	54.216
1.5	1.5	15.038	62.774
1.5	2.0	23.114	66.465
2.0	0.5	9.434	31.280
2.0	1.0	16.032	52.752
2.0	1.5	26.401	62.244
2.0	2.0	40.608	66.256
2.5	0.5	15.046	30.182
2.5	1.0	25.020	50.052
2.5	1.5	41.020	61.987
2.5	2.0	63.105	66.156
3.0	0.5	21.915	29.608
3.0	1.0	36.014	51.667
3.0	1.5	58.892	61.844
3.0	2.0	90.604	66.701

We now consider another example. The normalized Stokes vector for this example is

$$\begin{pmatrix} S_0 \\ S_1 \\ S_2 \\ S_3 \end{pmatrix} = \begin{pmatrix} 1.000 \\ 0.053 \\ 0.160 \\ -0.986 \end{pmatrix} \tag{25-125}$$

We immediately find that

$$A^2 = 139.845 \tag{25-126a}$$

$$\phi = 71.672^\circ \tag{25-126b}$$

From Figs. 25-9 and 25-10 we see that the very large values of A^2 and ϕ indicate a relatively large value for κ. We again construct a seed table as before, and we find that the appropriate seed value for the complex refractive index is

$$\mathbf{n} = 0.5(1 - i15) \tag{25-127}$$

Proceeding as before, we obtain

$$\mathbf{n} = 0.65(1 - i12.78) \tag{25-128}$$

For this sample we find that the reflectivity R, (25-123), is

$$R = 96.4\% \tag{25-129}$$

This large value of R shows that this sample is an excellent reflector.

Thus, we see that Fresnel's equations for an incidence angle of 45° enable us to determine n and κ by taking advantage of all of the developments of modern electronics and computers. In particular, this method is readily automated. While the simplest measurement configuration has been shown in Fig. 25-8, more complicated ones, which simplify the measurements, such as a dual-beam configuration to measure I_s and I_p simultaneously, can be conceived.

The measurement of the refractive index and the extinction coefficient of materials is critical to the development of modern optical materials (e.g., fiber-optic glass, metals and metal alloys, and semiconductors). In this and previous sections we have dealt with determining the optical constants which are inherent to the material itself. In practice, this means that the material and, in particular, the optical surface must be free of any other substances resting on the surface (e.g., a thin film).

The problem of considering the effects of a thin film on an optical surface appears to have been first studied by Drude about 1890. He was probably initially interested in characterizing these thin films in terms of their optical properties. However, as he advanced in his investigations he came to realize that the subject was far from simple and required substantial effort. In fact, the fundamental equations could not be solved until the advent of digital computers. In order to determine n and κ for thin films as well as the substrates, he developed a method that has come to be known as ellipsometry. As time developed, further applications were found, e.g., the measurement of thin films deposited on optical lenses in order to improve their optical performance. In Chapter 29 we consider the fundamentals of ellipsometry.

REFERENCES

Books

1. Strong, J., *Procedures in Applied Optics*, Marcel Dekker, New York, 1989.
2. Mott, N. F. and Jones, H., *The Theory of the Properties of Metals and Alloys*, Dover, New York, 1958.
3. Born, M. and Wolf, E., *Principles of Optics*, 5th ed., Pergamon Press, New York, 1975.
4. Wood, R. W., *Physical Optics*, 3rd ed., Optical Society of America, Washington, DC, 1988.
5. Driscoll, W. G. and Vaughn, W., *Handbook of Optics*, McGraw-Hill, New York, 1978.
6. Longhurst, R. S., *Geometrical and Physical Optics*, 2nd ed., Wiley, New York, 1967.
7. Potter, R. F., *Optical Properties of Solids*, S. Nudclman, and S. S. Mitra, eds., Plenum Press, New York, 1969.
8. Humphreys-Owen, S. P. F., *Proc. Phys. Soc.* (London), *77*, 941 (1961).
9. Collett, E., *Opt. Commun.*, *63*, 217 (1987).

26

Polarization Optical Elements

26.1 INTRODUCTION

A polarization optical element is any optical element that modifies the state of polarization of a light beam. Polarizers, retarders, rotators, and depolarizers are all polarization optical elements, and we will discuss their properties in this chapter. The many references on polarization elements, and catalogs and specifications from manufacturers, are good sources of information. We include here a survey of elements, and brief descriptions so that the reader has at least a basic understanding of the range of available polarization elements.

26.2 POLARIZERS

A polarizer is an optical element that is designed to produce polarized light of a specific state independent of the incident state. The desired state might be linear, circular, or elliptically polarized, and an optical element designed to produce one of these states is a linear, circular, or elliptical polarizer. Polarization elements are based on polarization by absorption, refraction, and reflection. Since this list describes most of the things that can happen when light interacts with matter, the appearance of polarized light should not be surprising. We will cover polarization by all these methods in the following sections.

26.2.1 Absorption Polarizers: Polaroid

Polaroid is a material invented in 1928 by Edwin Land, who was then a 19-year-old student at Harvard University. (The generic name for Polaroid, sheet polarizer, applies to a polarizer whose thickness normal to the direction of propagation of light is much smaller than the width.) Land used aligned microcrystals of herapathite in a transparent medium of index similar to that of the crystalline material. Herapathite is a crystalline material discovered about 1852 by the English medical researcher William Bird Herapath. Herapath had been feeding quinine to dogs, and the substance that came to be known as herapathite crystallized out of the dogs' urine. Crystals of herapathite tend to be needle-shaped and the principal absorption axis is parallel to the long axis of the crystal. Land reduced crystals of herapathite to small size, aligned them, and placed them in a solution of cellulose acetate. This first absorption polarizer is known as J-sheet.

Sheet polarizer, operating on the principle of differential absorption along orthogonal axes, is also known as dichroic polarizer. This is because the unequal absorptions also happen to be spectrally dependent, i.e., linearly polarized light transmitted through a sample of Polaroid along one axis appears to be a different color from linearly polarized light transmitted along the orthogonal axis.

The types of sheet polarizer typically available are molecular polarizers, i.e., they consist of transparent polymers that contain molecules that have been aligned and stained with a dichroic dye. The absorption takes place along the long axis of the molecules, and the transmission axis is perpendicular to this. H-sheet, K-sheet, and L-sheet are of this type, with H-sheet being the most common. Sheet polarizers can be made in large sizes (several square feet) for both the visible and near infrared, and is an extremely important material, because, unlike calcite, it is inexpensive. Polaroid material can be laminated between glass plates and the performance of these polarizers is extremely good.

We now derive equations that describe sheet polarizer properties; the equations are equally applicable to any type of polarizer. Suppose we have a light source that is passed through an ideal polarizer with its transmission axis at some angle α from a reference. The output of the ideal polarizer then passes through a sheet polarizer with its transmission axis oriented at an angle θ with respect to a reference, as shown in Fig. 26-1. The Mueller matrix of this last polarizer is

$$M_{\text{pol}}(\theta) = \begin{pmatrix} A & B\cos 2\theta & B\sin 2\theta & 0 \\ B\cos 2\theta & A\cos^2 2\theta + C\sin^2 2\theta & (A-C)\sin 2\theta\cos 2\theta & 0 \\ B\sin 2\theta & (A-C)\sin 2\theta\cos 2\theta & A\sin^2 2\theta + C\cos^2 2\theta & 0 \\ 0 & 0 & 0 & C \end{pmatrix} \tag{26-1}$$

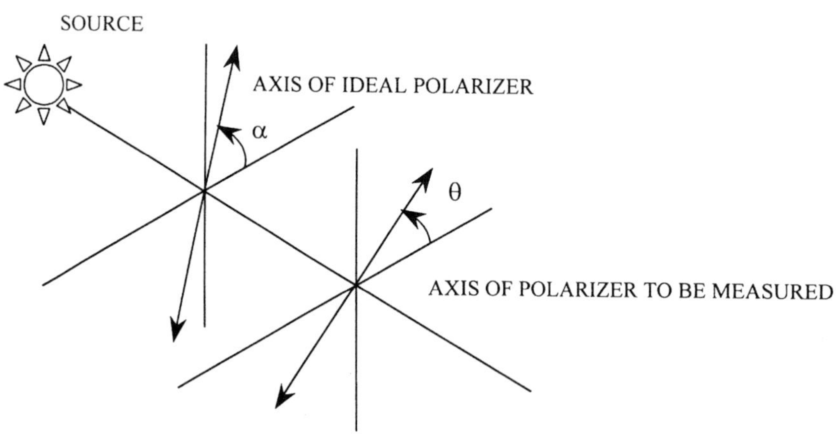

Figure 26-1 Measurement configuration for characterizing a single polarizer.

where

$$
\begin{aligned}
A &= \frac{p_x^2 + p_y^2}{2} \\
B &= \frac{p_x^2 - p_y^2}{2} \qquad (26\text{-}2) \\
C &= p_x p_y
\end{aligned}
$$

and where the quantities p_x and p_y are the absorption coefficients of the orthogonal optical axes, and $0 \leq p_x, p_y \leq 1$. The Stokes vector of the beam emerging from the ideal polarizer with its transmission axis at angle α is

$$
S = I_0 \begin{pmatrix} 1 \\ \cos 2\alpha \\ \sin 2\alpha \\ 0 \end{pmatrix} \qquad (26\text{-}3)
$$

where I_0 is the intensity of the beam. The light intensity emerging from the sheet polarizer is found from multiplying (26-3) by (26-1) where we obtain

$$
I(\theta, \alpha) = I_0[A + B\cos 2(\theta - \alpha)] \qquad (26\text{-}4)
$$

The maximum intensity occurs at $\theta = \alpha$ and is

$$
I_{\max} = I_0[A + B] = I_0 p_x^2 \qquad (26\text{-}5)
$$

The minimum intensity occurs at $\theta = \alpha + \pi/2$ and is

$$
I_{\min} = I_0[A - B] = I_0 p_y^2 \qquad (26\text{-}6)
$$

A linear polarizer has two transmittance parameters: the major principal transmittance k_1 and the minor principal transmittance k_2. The parameter k_1 is defined as the ratio of the transmitted intensity to the incident intensity when the incident beam is linearly polarized in that vibration azimuth which maximizes the transmittance. Similarly, the ratio obtained when the transmittance is a minimum is k_2. Thus,

$$
k_1 = \frac{I_{\max}}{I_0} = A + B = p_x^2 \qquad (26\text{-}7)
$$

$$
k_2 = \frac{I_{\min}}{I_0} = A - B = p_y^2 \qquad (26\text{-}8)
$$

The ratio k_1/k_2 is represented by R_t and is called the principal transmittance ratio. R_t of a high-quality polarizer may be as large as 10^5. The reciprocal of R_t is called the extinction ratio, and is often quoted as a figure of merit for polarizers. The extinction ratio should be a small number and the transmittance ratio a large number; if this is not the case, the term at hand is being misused.

Because the principal transmittance can vary over several orders of magnitude, it is common to express k_1 and k_2 in terms of logarithms. Specifically, k_1 and k_2 are defined in terms of the minor and major principal densities, d_1 and d_2:

$$d_1 = \log_{10}\left(\frac{1}{k_1}\right) \qquad d_2 = \log_{10}\left(\frac{1}{k_2}\right) \tag{26-9}$$

or

$$k_1 = 10^{-d_1} \qquad k_2 = 10^{-d_2} \tag{26-10}$$

Dividing k_1 by k_2 yields

$$R_t = 10^D \tag{26-11}$$

where $D = d_2 - d_1$ is the density difference or dichroitance.

The average of the principal transmittances is called the total transmittance k_t so that

$$k_t = \frac{k_1 + k_2}{2} = \frac{p_x^2 + p_y^2}{2} = A \tag{26-12}$$

The parameter k_t is the ratio of the transmitted intensity to incident beam intensity when the incident beam is unpolarized [multiply a Stokes vector for unpolarized light by the matrix in (26-1)]. Furthermore, we see that k_t is an intrinsic constant of the polarizer and does not depend on the polarization of the incident beam, as is the case with k_1 and k_2.

Figure 26-1 shows the measurement of k_1 and k_2 of a single polarizer. We assumed that we had a source of perfectly polarized light from an ideal (or nearly ideal) polarizer. Another way to determine k_1 and k_2 is to measure a pair of identical polarizers and use an unpolarized light source. This method requires an extremely good source of unpolarized light. It turns out to be surprisingly difficult to obtain a perfectly unpolarized light source. Nearly every optical source has some elliptical polarization associated with it, i.e., the emitted light is partially polarized to some degree. One reason this is so is because a reflection from nearly every type of surface, even one which is rough, creates polarized light. Assuming we can produce a light source that is sufficiently unpolarized as to lead to meaningful data, the parameters k_1 and k_2 can, in principle, be determined from a pair of identical polarizers. Figure 26-2 illustrates the experiment.

Let us assume we can align the polarization axes. From (26-1), the Stokes vector resulting from the passage of unpolarized light through the two aligned polarizers is

$$\begin{pmatrix} A & B & 0 & 0 \\ B & A & 0 & 0 \\ 0 & 0 & C & 0 \\ 0 & 0 & 0 & C \end{pmatrix}\begin{pmatrix} A & B & 0 & 0 \\ B & A & 0 & 0 \\ 0 & 0 & C & 0 \\ 0 & 0 & 0 & C \end{pmatrix}\begin{pmatrix} I_0 \\ 0 \\ 0 \\ 0 \end{pmatrix} = I_0\begin{pmatrix} A^2 + B^2 \\ 2AB \\ 0 \\ 0 \end{pmatrix} \tag{26-13}$$

The intensity for the beam emerging from the polarizer pair is

$$I(p) = I_0(A^2 + B^2) \tag{26-14}$$

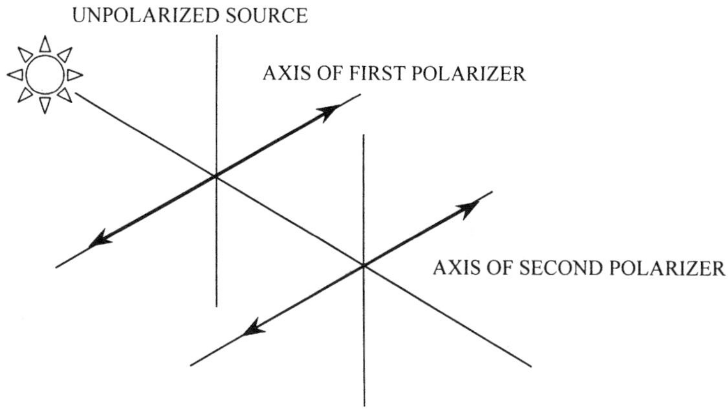

Figure 26-2 Measurement of k_1 and k_2 of identical polarizers.

This may be written:

$$I(p) = \frac{k_1^2 + k_2^2}{2} I_0 \tag{26-15}$$

We now rotate the polarizer closest to the unpolarized source through 90°. The Stokes vector of the beam emerging from the polarizer pair is now

$$\begin{pmatrix} A & B & 0 & 0 \\ B & A & 0 & 0 \\ 0 & 0 & C & 0 \\ 0 & 0 & 0 & C \end{pmatrix} \begin{pmatrix} A & -B & 0 & 0 \\ -B & A & 0 & 0 \\ 0 & 0 & C & 0 \\ 0 & 0 & 0 & C \end{pmatrix} \begin{pmatrix} I_0 \\ 0 \\ 0 \\ 0 \end{pmatrix} = I_0 \begin{pmatrix} A^2 - B^2 \\ 0 \\ 0 \\ 0 \end{pmatrix} \tag{26-16}$$

The intensity from the crossed pair, $I(s)$, is

$$I(s) = I_0\left(A^2 - B^2\right) \tag{26-17}$$

and this may be written:

$$I(s) = k_1 k_2 \tag{26-18}$$

Now let the ratio of intensities $I(p)/I_0$ when the polarizers are aligned be H_0 and let the ratio of intensities $I(s)/I_0$ when the polarizers are perpendicular be H_{90}. Then, we can write

$$H_0 = \frac{k_1^2 + k_2^2}{2} = \left(A^2 + B^2\right) \tag{26-19}$$

and

$$H_{90} = k_1 k_2 = \left(A^2 - B^2\right) \tag{26-20}$$

Multiplying (26-19) and (26-20) by 2 and adding gives

$$2H_0 + 2H_{90} = k_1^2 + 2k_1 k_2 + k_2^2 \tag{26-21}$$

Taking the square root, we have

$$\sqrt{2}(H_0 + H_{90})^{1/2} = k_1 + k_2 \tag{26-22}$$

Multiplying (26-19) and (26-20) by 2, subtracting, and taking the square root, we have

$$\sqrt{2}(H_0 - H_{90})^{1/2} = k_1 - k_2 \tag{26-23}$$

Now we solve for k_1 and k_2 by adding and subtracting (26-22) and (26-23):

$$k_1 = \frac{\sqrt{2}}{2}\left[(H_0 + H_{90})^{1/2} + (H_0 - H_{90})^{1/2}\right] \tag{26-24}$$

$$k_2 = \frac{\sqrt{2}}{2}\left[(H_0 + H_{90})^{1/2} - (H_0 - H_{90})^{1/2}\right] \tag{26-25}$$

The principal transmittance ratio can now be expressed in terms of H_0 and H_{90}:

$$R_t = \frac{k_1}{k_2} = \frac{\left[(H_0 + H_{90})^{1/2} + (H_0 - H_{90})^{1/2}\right]}{\left[(H_0 + H_{90})^{1/2} - (H_0 - H_{90})^{1/2}\right]} \tag{26-26}$$

Thus, if we have a perfect unpolarized light source and we can be assured of aligning the polarizers parallel and perpendicular to each other, we can determine the transmittance parameters k_1 and k_2 of a polarizer when they are arranged in a pair. However, as has been pointed out, it is very difficult to produce perfectly unpolarized light. It is much easier if a known high-quality polarizer is used to produce linearly polarized light and the measurement of k_1 and k_2 follows the measurement method illustrated in Fig. 26-1.

Suppose we cannot align the two polarizer axes perfectly. If one of the polarizers is rotated from the horizontal axis by angle θ, then we have the situation shown in Fig. 26-3.

The Stokes vector of the beam emerging from the first polarizer is

$$I_0 \begin{pmatrix} A & B & 0 & 0 \\ B & A & 0 & 0 \\ 0 & 0 & C & 0 \\ 0 & 0 & 0 & C \end{pmatrix} \begin{pmatrix} 1 \\ 0 \\ 0 \\ 0 \end{pmatrix} = I_0 \begin{pmatrix} A \\ B \\ 0 \\ 0 \end{pmatrix} \tag{26-27}$$

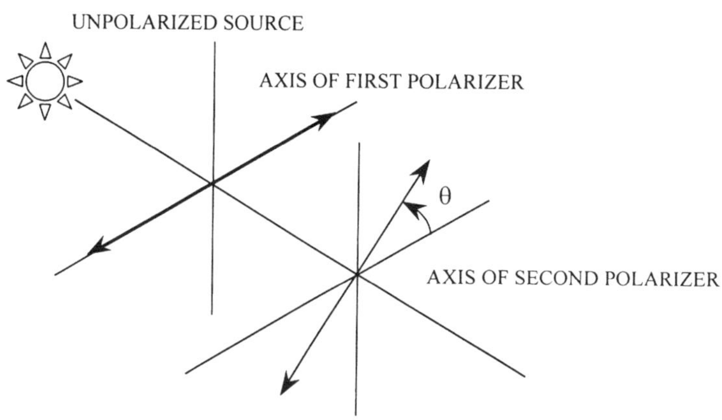

Figure 26-3 Nonaligned identical linear polarizers.

The second polarizer is represented by (26-1), and so the beam intensity emerging from the second polarizer is

$$I(\theta) = I_0\left[A^2 + B^2\cos 2\theta\right] \tag{26-28}$$

Using a trigonometric identity, this can be written as

$$I(\theta) = I_0\left[\left(A^2 - B^2\right) + 2B^2\cos^2\theta\right] \tag{26-29}$$

Equation (26-29) can be expressed in terms of H_0 and H_{90}, i.e.,

$$H(\theta) = \frac{I(\theta)}{I_0} = H_{90} + (H_0 - H_{90})\cos^2\theta \tag{26-30}$$

Equation (26-30) is a generalization of Malus' Law for nonideal polarizers. This relation is usually expressed for an ideal polarizer so that $A^2 = B^2 = 1/4$, $H_0 = 2A^2$, and $H_{90} = 0$ so that

$$H(\theta) = \frac{1}{2}\cos^2\theta \tag{26-31}$$

We now apply data to these results. In Fig. 26-4 the spectral curves of different types of Polaroid sheet are shown with the values of k_1 and k_2. In Table 26-1, values of H_0 and H_{90} are listed for the sheet Polaroids HN-22, HN-32, and HN-38 over the visible wavelength range. From this table we can construct Table 26-2 and determine the corresponding principal transmittances. We see from Table 26-2 that HN-22 has the largest principal transmittance ratio in comparison with HN-32 and HN-38, consequently it is the best Polaroid polarizer. Calcite polarizers typically have a principal transmittance ratio of 1×10^6 from 300 to 2000 nm. This is three times

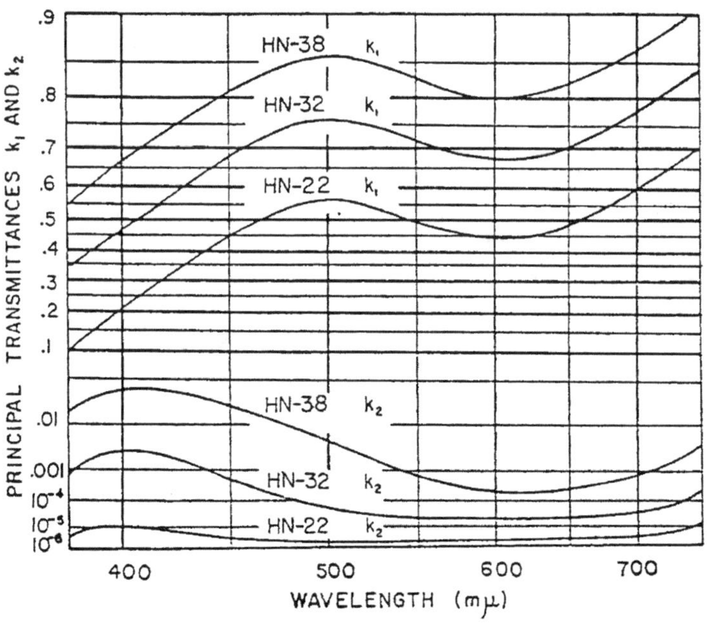

Figure 26-4 Curves of k_1 and k_2 for three grades of HN polarizer. (From Ref. 1.)

Table 26-1 Parallel-Pair H_0 and Crossed-Pair Transmittance H_{90} of HN Polarizers

Wavelength (nm)	HN-22		HN-32		HN-38	
	H_0	H_{90}	H_0	H_{90}	H_0	H_{90}
375	0.006	0.0000005	0.05	0.0003	0.15	0.01
400	0.02	0.0000002	0.11	0.002	0.22	0.03
450	0.10	0.000002	0.23	0.0003	0.33	0.02
500	0.15	0.000001	0.28	0.00004	0.37	0.004
550	0.12	0.000001	0.25	0.00001	0.34	0.0006
600	0.09	0.000001	0.22	0.00001	0.31	0.0002
650	0.11	0.000001	0.25	0.00001	0.34	0.0002
700	0.17	0.000002	0.30	0.000002	0.37	0.0006
750	0.24	0.000007	0.35	0.0002	0.41	0.004

Table 26-2 Principal Transmittances of HN-22, HN-32, and HN-38

Wavelength (nm)	R_t		
	HN-22	HN-32	HN-38
375	4.17×10^{-5}	3.00×10^{-3}	3.34×10^{-2}
400	5.00×10^{-5}	9.09×10^{-3}	6.85×10^{-2}
450	1.00×10^{-5}	6.52×10^{-4}	3.03×10^{-2}
500	3.33×10^{-6}	7.14×10^{-5}	5.41×10^{-3}
550	5.56×10^{-6}	2.00×10^{-5}	8.82×10^{-4}
600	4.55×10^{-6}	2.27×10^{-5}	3.23×10^{-4}
650	4.55×10^{-6}	2.00×10^{-5}	2.94×10^{-4}
700	5.88×10^{-6}	3.33×10^{-5}	8.11×10^{-4}
750	1.46×10^{-5}	2.86×10^{-4}	4.88×10^{-3}

better than that of Polaroid HN-22 at its best value. Nevertheless, in view of the lower cost of sheet polarizer, it is useful in many applications.

26.2.2 Absorption Polarizers: Polarcor

Polarcor is an absorption polarizer consisting of elongated silver particles in glass. This polarizer, developed commercially by Corning, has been produced with transmittance ratios of 10,000 in the near infrared. The polarizing ability of silver in glass was observed in the late 1960s [2], and polarizers with high transmittance ratios were developed in the late 1980s [3]. Because these polarizers depend on a resonance phenomenon, performance is strongly dependent on wavelength, but they can be engineered to operate with good performance over broad wavelength regions centered on near-infrared wavelengths from 800 to 1500 nm.

26.2.3 Wire-Grid Polarizers

A wire grid is a planar array of parallel wires. It is similar to the sheet polarizer in that the transmitted light is polarized perpendicularly to the wires. Light polarized parallel to the wires is reflected instead of absorbed as with the sheet polarizer. To be an effective polarizer, the wavelength should be longer than the spacing between the wires. For practical reasons, wire grids are usually placed on a substrate. Until relatively recently, they have been typically manufactured for the infrared region of the spectrum ($>2\,\mu m$) because the small grid spacing required for shorter wavelengths has been difficult to produce. Grid spacing for these infrared polarizers are normally $0.5\,\mu m$ or greater, although smaller spacings have been fabricated. With technological improvements in grid fabrication techniques, grids with wires of width $0.065\,\mu m$ or less have been produced. These grids are useful into the near infrared and visible [4,5]. Photomicrographs of wire-grid polarizers composed of $0.065\,\mu m$ aluminum wires are given in Fig. 26-5.

Since reflection loss and absorption reduce the transmittance ratio of wire grids, an antireflection coating is often applied to the substrate. The quality of this coating and its achromaticity are important factors in the overall performance of wire grids. Commercial wire grid polarizers have transmittance ratios of 20 to

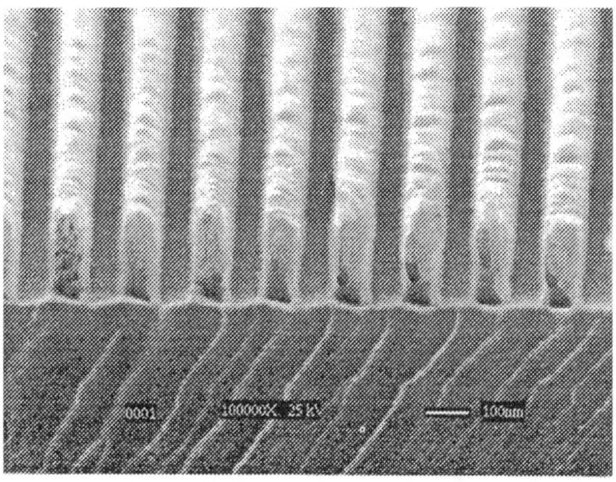

(a)

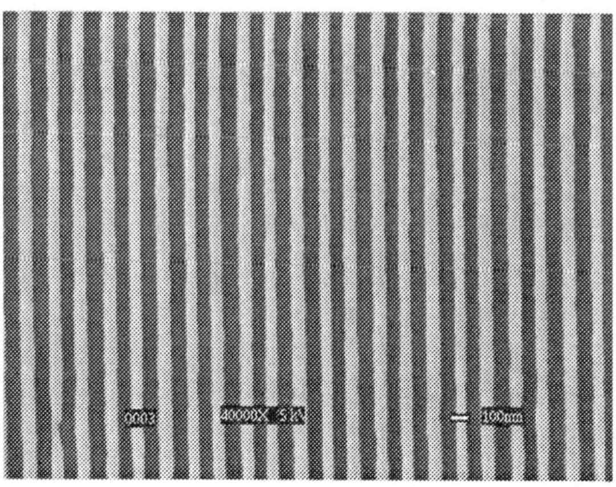

(b)

Figure 26-5 Photomicrographs of wire-grid polarizers. (a) Side view. (b) Top down view. (Courtesy of MOXTEK, Inc.)

10,000. More information on wire grids is given in Bennett and Bennett [6] and Bennett [7] and the cited patents [4,5,8].

26.2.4 Polarization by Refraction (Prism Polarizers)

Crystal prism polarizers use total internal reflection at internal interfaces to separate the polarized components. There are many designs of prism polarizers, and we will not cover all of these here. The reader should consult the excellent article by Bennett and Bennett [6] for a comprehensive treatment.

The basis of most prism polarizers is the use of a birefringent material, as described in Chapter 24. We illustrate the phenomenon of double refraction with the following example of the construction of a Nicol polarizing prism. We know that calcite has a large birefringence. (Calcite, the crystalline form of limestone, marble, and chalk, occurs naturally. It has not been grown artificially in anything but very small sizes. It can be used in prism polarizers for wavelengths from 0.25 to 2.7 μm.) If the propagation is not perpendicular to the direction of the optic axis, the ordinary and extraordinary rays separate. Each of these rays is linearly polarized. A Nicol prism is a polarizing prism constructed so that one of the linear polarized beams is rejected and the other is transmitted through the prism unaltered. It was the first polarizing prism ever constructed (1828). However, it is now obsolete and has been replaced by other prisms, such as the Glan–Thompson, Glan–Taylor, Rochon, and Wollaston prisms. These new designs have become more popular because they are optically superior; e.g., the light is nearly uniformly polarized over the field of view, whereas it is not for the Nicol prism. The Glan–Thompson type has the highest reported transmittance ratio [6].

In a Nicol prism a flawless piece of calcite is split so as to produce an elongated cleavage rhomb about three times as long as it is broad. The end faces, which naturally meet the edges at angles of $70^\circ 53'$, are ground so that the angles become 68° (this allows the field-of-view angle to be increased); apparently, this practice of "trimming" was started by Nicol himself. Figure 26-6 shows the construction of the Nicol prism. The calcite is sawed diagonally and at right angles to the ground and polished end faces. The halves are cemented together with Canada balsam, and the sides of the prism are covered with an opaque, light-absorbing coating. The refractive index of the Canada balsam is 1.54, a value intermediate to the ordinary ($n_o = 1.6584$) and extraordinary ($n_e = 1.4864$) refractive indices of the calcite. Its purpose is to deflect the ordinary ray (by total internal reflection) out of the prism and to allow the extraordinary ray to be transmitted through the prism.

We now compute the angles. The limiting angle for the ordinary ray is determined from Snell's law. At 5893 Å the critical angle θ_2 for total internal reflection at the calcite–balsam interface is obtained from

$$1.6583 \sin\theta_2 = 1.54 \sin 90^\circ \tag{26-32}$$

so that $\theta_2 = 68.28^\circ$. The cut is normal to the entrance face of the prism, so that the angle of refraction θ_{r_1} at the entrance face is $90^\circ - 68.28^\circ = 21.72^\circ$. The angle of incidence is then obtained from

$$\sin\theta_{i_1} = 1.6583 \sin 21.72^\circ \tag{26-33}$$

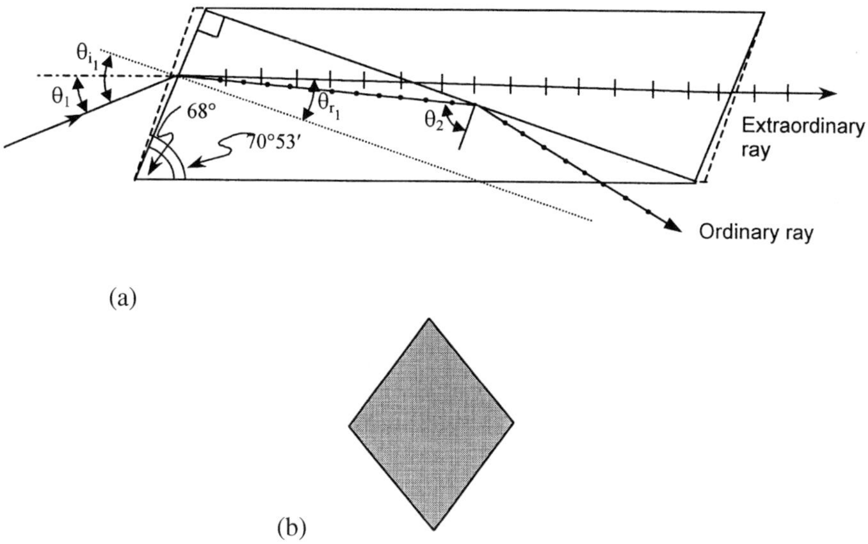

Figure 26-6 Diagram of a Nicol prism: (a) longitudinal section; (b) cross-section.

so that the angle of incidence is $\theta_{i_1} = 37.88^\circ$. Since the entrance face makes an angle of 68° with the longitudinal axis of the prism, the normal to the entrance face is $90^\circ - 68^\circ = 22^\circ$ with respect to the longitudinal axis. The limiting angle at which the ordinary ray is totally reflected at the balsam results in a half-field angle of $\theta_1 = 37.88^\circ - 22^\circ = 15.88^\circ$. A similar computation is required for the limiting angle for the extraordinary ray at which total reflection does not occur. The refractive index for the extraordinary ray is a function of the angle (let us call it ϕ) between the wave normal and the optic axis. Using the same procedure as before (but not shown in Fig. 26-6), we have $\theta_2' = 90^\circ - \theta_{r_1}'$, and the critical angle at the calcite/balsam interface is obtained from

$$\sin(90^\circ - \theta_{r_1}') = \cos\theta_{r_1}' = \frac{1.54}{n_\phi} \tag{26-34}$$

The index of refraction n_ϕ of the extraordinary wave traveling in a uniaxial crystal at an angle ϕ with the optic axis is given by

$$\frac{1}{n_\phi^2} = \frac{\sin^2\phi}{n_e^2} + \frac{\cos^2\phi}{n_o^2} \tag{26-35}$$

For our Nicol prism:

$$\phi = \theta_{r_1}' + 41^\circ 44' \tag{26-36}$$

and (26-35) becomes

$$\frac{\cos^2\theta_{r_1}'}{1.54^2} = \frac{\sin^2(\theta_{r_1}' + 41.73^\circ)}{n_e^2} + \frac{\cos^2(\theta_{r_1}' + 41.73^\circ)}{n_o^2} \tag{26-37}$$

This transcendental equation is easily solved using a computer, and θ'_{r_1} is found to be 7.44° and θ'_1 is 11.61°, using the values of the indices for $\lambda = 5893\text{Å}$. The semi field angle is 10.39° and the total field angle is 20.78°.

The cross-section of the Nicol prism is also shown in Fig. 26-6. Only the extraordinary ray emerges and the plane of vibration is parallel to the short diagonal of the rhombohedron, so that the direction of polarization is obvious. The corners of the prism are sometimes cut, making the direction of polarization more difficult to discern.

26.2.5 Polarization by Reflection

One has only to examine plots of the Fresnel equations, as described in Chapter 8, to see that polarization will almost always occur on reflection. Polarizers that depend on reflection are usually composed of plates oriented near the Brewster angle. Because sheet and prism polarizers do not operate in the infrared and ultraviolet, reflection polarizers are sometimes used in these regions. Brewster-angle polarizers are necessarily sensitive to incidence angle and are physically long devices because Brewster angles can be large, especially in the infrared where materials with high indices are used.

26.3 RETARDERS

A retarder is an optical element that produces a specific phase difference between two orthogonal components of incident polarized light. A retarder can be in prism form, called a rhomb, or it can be in the form of a window or plate, called a waveplate. Waveplates can be zero order, i.e., the net phase difference is actually the specified retardance, or multiorder, in which case the phase difference can be a multiple, sometimes large, of the specified retardance. Retarders are also sometimes called compensators, and can be made variable, e.g., the Babinet–Soleil compensator. Retarders may be designed for single wavelengths, or be designed to operate over larger spectral regions i.e., achromatic retarders.

26.3.1 Birefringent Retarders

The properties of isotropic, uniaxial, and biaxial optical materials were discussed in Chapter 24. We can obtain from that discussion that the phase retardation of linearly polarized light in going through a uniaxial crystal with its optic axis parallel to the faces of the crystal is

$$\Gamma = \frac{2\pi}{\lambda} d(n_e - n_o) \tag{26-38}$$

when the polarization is at an angle with the optic axis. The optical path difference experienced by the two components is $d(n_e - n_o)$ and the birefringence is $(n_e - n_o)$. These quantities are all positive for positive uniaxial materials, i.e., materials with $n_e > n_o$. The component of the light experiencing the refractive index n_e is parallel with the optic axis while the component experiencing the index n_o is perpendicular to the optic axis. The slow axis is the direction in the material in which light experiences the higher index n_e, i.e., for the positive uniaxial material, the direction of the optic axis. The fast axis is the direction in the material in which light experiences the lower index, n_o. It is the fast axis that is usually marked with a line on commercial waveplates. The foregoing discussion is the same for negative uniaxial material with the positions of n_e and n_o interchanged.

The most common commercial retarders are quarter wave and half wave, i.e., where there are $\pi/2$ and π net phase differences between components, respectively. The quarter-wave retarder produces circular polarization when the azimuth of the (linearly polarized) incident light is 45° to the fast axis. The half-wave retarder produces linearly polarized light rotated by an angle 2θ when the azimuth of the (linearly polarized) incident light is at an angle θ with respect to the fast axis of the half-wave retarder.

As we have seen above, the net retardance is an extensive property of the retarder; i.e., the retardance increases with path length through the retarder. When the net retardation for a retarder reaches the minimum net value desired for the element, that retarder is known as a single-order retarder (sometimes called a zero-order retarder). Although many materials have small birefringence, some (e.g., calcite) have large values of birefringence (see Table 26-3). Birefringence is, like index, a function of wavelength. A single-order retarder may not be possible because it would be too thin to be practical. A retarder called "first order" may be constructed by joining two pieces of material such that the fast axis of one piece is aligned with the slow axis of the other. The thicknesses of the pieces of material are adjusted so that the difference in the thicknesses of the two pieces is equal to the thickness of a single-order retarder. The retardation can be found from the equation

$$\Gamma = \frac{2\pi}{\lambda}(d_1 - d_2)(n_e - n_o) \tag{26-39}$$

where d_1 and d_2 are the thicknesses.

A multiple-order retarder is a retarder of thickness such that its net retardation is an integral number of wavelengths plus the desired fractional retardance, e.g., $5\lambda/4$, $3\lambda/2$, etc. Multiple-order retarders may be less expensive than single-order retarders, but they are sensitive to temperature and incidence angle.

Table 26-3 Birefringence for Optical Materials at 589.3 nm

Material	Birefringence ($n_e - n_o$)
Positive Uniaxial Crystals	
Ice (H_2O)	0.004
Quartz (SiO_2)	0.009
Zircon ($ZrSiO_4$)	0.045
Rutile (TiO_2)	0.287
Negative Uniaxial Crystals	
Beryl ($Be_3Al_2(SiO_3)_6$)	−0.006
Sodium nitrate ($NaNO_3$)	−0.248
Calcite ($CaCO_3$)	−0.172
Sapphire (Al_2O_3)	−0.008

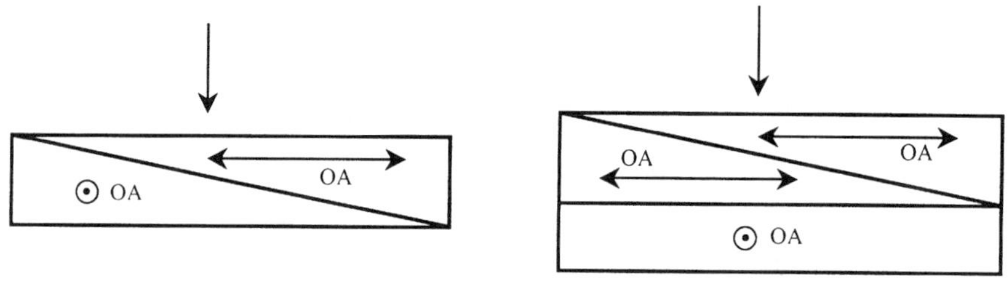

(a) (b)

Figure 26-7 Diagrams of (a) Babinet compensator, and (b) Soleil compensator where OA is the optic axis.

26.3.2 Variable Retarders

Retarders have been constructed of movable elements in order to produce variable retardance. Two of the most common designs based on movable wedges are the Babinet and Soleil (also variously called Babinet–Soleil, Soleil–Babinet, or Soleil–Bravais) compensators, shown in Fig. 26-7. The term compensator is used for these elements because they are often used to allow adjustable compensation of retardance originating in a sample under test.

The Babinet compensator, shown in Fig. 26-7a, consists of two wedges of a (uniaxial) birefringent material (e.g., quartz). The bottom wedge is fixed while the top wedge slides over the bottom by means of a micrometer. The optic axes of both wedges are parallel to the outer faces of the wedge pair, but are perpendicular to one another. At any particular location across the face of the Babinet compensator, the net retardation is

$$\Gamma = \frac{2\pi}{\lambda}(d_1 - d_2)(n_e - n_o) \tag{26-39}$$

where d_1 and d_2 are the thicknesses at that location. If monochromatic polarized light oriented at 45° to one of the optic axes is incident on the Babinet compensator, one component of the light becomes the extraordinary component and the other is the ordinary component in the first wedge. When the light enters the second wedge, the components exchange places, i.e., the extraordinary becomes the ordinary and vice versa. An analyzer whose azimuth is perpendicular to the original polarization can be placed behind the compensator to show the effect of the retardations. Everywhere where there is zero or a multiple of 2π phase difference there will be a dark band. When the upper wedge is translated, the bands shift. These bands indicate the disadvantage of the Babinet compensator—a desired retardance only occurs along these parallel bands.

The Soleil compensator, shown in Fig. 26-7b consists of two wedges with parallel optic axes followed by a plane parallel quartz prism with its optic axis perpendicular to the wedge axes. The top wedge is the only moving part again. The advantage of this design is that the retardance is uniform over the whole field where the wedges overlap.

Jerrard [9] gives a review of these and many other compensator designs.

26.3.3 Achromatic Retarders

The most common type of retarder is the waveplate, as described above, a plane parallel plate of birefringent material, with the crystal axis oriented perpendicular to the propagation direction of light. As the wavelength varies, the retardance of the zero-order waveplate must also vary, unless by coincidence the birefringence was linearly proportional to wavelength. Since this does not occur in practice, the waveplate is only approximately quarter wave (or whatever retardance it is designed for) for a small wavelength range. For higher order waveplates, $m = 3, 5, \ldots$, the effective wavelength range for quarter-wave retardance is even smaller.

The achromatic range of waveplates can be enlarged by assembling combinations of waveplates of birefringent materials [6]. This method has been common in the visible region; however, in the infrared the very properties required to construct such a device are the properties to be measured polarimetrically, and there are not an abundance of data available to readily design high-performance devices of this kind. Nevertheless, an infrared achromatic waveplate has been designed [10] using a combination of two plates. This retarder has a theoretical retardance variation of about 20° over the 3–11 μm range.

A second class of achromatic retardation elements is the total internal reflection prism. Here, a specific phase shift between the s and p components of light (linear retardance) occurs on total internal reflection. This retardance depends on the refractive index, which varies slowly with wavelength. However, since this retardance is independent of any thickness, unlike the waveplate, the variation of retardance with wavelength is greatly reduced relative to the waveplate. A common configuration for retarding prisms is the Fresnel rhomb, depicted in Fig. 26-8. This figure shows a Fresnel rhomb designed for the visible spectrum. The nearly achromatic behavior of this retarder is the desired property; however, the Fresnel rhomb has the disadvantages of being long with large beam offset. In an application where the retarder must be rotated, any beam offset is unacceptable. A quarter-wave Fresnel rhomb for the infrared, made of ZnSe and having a clear aperture of x in., has a beam offset of $1.7x$ in. and a length of $3.7x$ in.

Infrared Achromatic Retarder

Figure 26-9 shows a prism retarder that was designed for no beam deviation. This design includes two total internal reflections and an air–metal reflection. Similar prisms have been designed previously, but special design considerations for the infrared make this prism retarder unique. Previous designs for the visible have

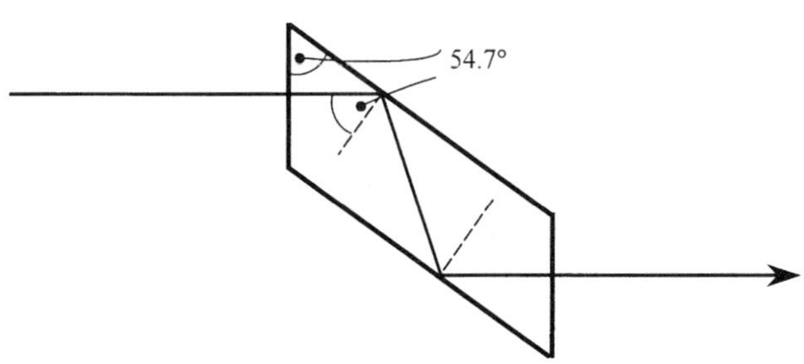

Figure 26-8 Fresnel rhomb.

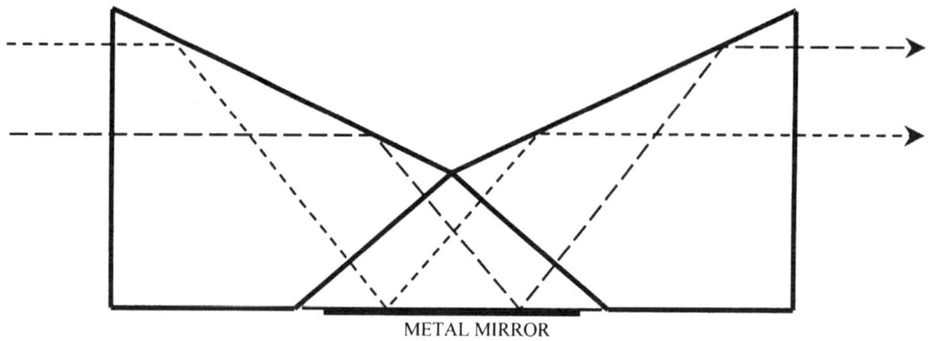

Figure 26-9 Infrared achromatic prism retarder design.

included a solid prism with similar shape to the retarder in Fig. 26-9, but with no air space [11], and a set of confronting rhombs called the double Fresnel rhomb. The latter design includes four total internal reflections. These designs are not appropriate for the infrared.

The prism design relies on the fact that there are substantial phase shifts between the s and p components of polarized light at the points of total internal reflection (TIR). The phase changes of s and p components on TIR are given by the formulas [12]:

$$\delta_s^{\text{prism}} = 2\tan^{-1}\frac{\left(n^2\sin^2\phi - 1\right)^{1/2}}{n\cos\phi} \tag{26-40}$$

and

$$\delta_p^{\text{prism}} = 2\tan^{-1}\frac{n\left(n^2\sin^2\phi - 1\right)^{1/2}}{\cos\phi} \tag{26-41}$$

where ϕ is the angle of incidence and n is the index of refraction of the prism material. The linear retardance associated with the TIR is the net phase shift between the two components

$$\Delta^{\text{prism}} = \delta_p^{\text{prism}} - \delta_s^{\text{prism}} \tag{26-42}$$

In addition there are phase shifts on reflection from the metal given by [6]

$$\delta_s^{\text{metal}} = \tan^{-1}\frac{2\eta_{0\text{s}}b}{\eta_{0\text{s}}^2 - \left(a^2 + b^2\right)} \tag{26-43}$$

$$\delta_p^{\text{metal}} = \tan^{-1}\frac{-2\eta_{0\text{p}}d}{c^2 + d^2 - \eta_{0\text{p}}^2} \tag{26-44}$$

where

$$\eta_{0\text{s}} = n_0\cos\theta_0 \tag{26-45}$$

$$\eta_{0\text{p}} = \frac{n_0}{\cos\theta_0} \tag{26-46}$$

$$a^2 + b^2 = \left[\left(n_1^2 - k_1^2 - n_0^2\sin^2\theta_0\right)^2 + 4n_1^2k_1^2\right]^{1/2} \tag{26-47}$$

$$c^2 + d^2 = \frac{\left(n_1^2 + k_1^2\right)^2}{\left(a^2 + b^2\right)} \tag{26-48}$$

$$b = \left[\frac{\left(a^2 + b^2\right)}{2} - \frac{\left(n_1^2 - k_1^2 - n_0^2\sin^2\theta_0\right)}{2}\right]^{1/2} \tag{26-49}$$

$$d = b\left(1 - \frac{n_0^2\sin^2\theta_0}{a^2 + b^2}\right) \tag{26-50}$$

and where n_0 is the refractive index of the incident medium, θ_0 is the angle of incidence, and n_1 and k_1 are respectively the index of refraction and extinction index for the metal mirror. The linear retardance associated with the metal mirror is the net phase shift between the s and p components:

$$\Delta^{\text{metal}} = \delta_p^{\text{metal}} - \delta_s^{\text{metal}} \tag{26-51}$$

The net retardance for the two TIRs and the metal reflection is then

$$\delta = 2\Delta^{\text{prism}} + \Delta^{\text{metal}} \tag{26-52}$$

The indices of refraction of materials that transmit well in the infrared are higher than indices of materials for the visible. Indices for infrared materials are generally greater than 2.0, where indices for materials for the visible are in the range 1.4–1.7. The higher indices for the infrared result in greater phase shifts between s and p components for a given incidence angle than would occur for the visible. Prism retarder designs for the infrared that have more than two TIRs soon become impractically large as the size of the clear aperture goes up or the desired retardance goes down. The length of a solid prism retarder of the shape of Fig. 26-9 is governed by the equation:

$$L = \frac{ad_a}{\tan(90^\circ - \theta)} \tag{26-53}$$

where d_a is the clear aperture and θ is the angle of incidence for the first TIR. The theoretical minimum length of the two-prism design for a clear aperture of 0.5 in. and a retardance of a quarter wave is 2.1 in. The minimum length for the same retardance and clear aperture in a three TIR design is 4.5 in.

Materials that are homogeneous (materials with natural birefringence are unacceptable) and good infrared transmitters must be used for such a device. Suitable materials include zinc selenide, zinc sulfide, germanium, arsenic trisulfide glass, and gallium arsenide. Metals that may be used for the mirror include gold, silver, copper, lead, or aluminum, with gold being preferable because of its excellent reflective properties in the infrared and its resistance to corrosion.

Beam angles at the entry and exit points of the two-prism arrangement are designed to be at normal incidence to minimize Fresnel diattenuation. Figure 26-10 shows the theoretical phase shift versus wavelength for this design. For zinc selenide prisms and a gold mirror at the angles shown, the retardation is very close to a quarter of a wavelength over the 3 to 14 μm band. (The angles were computed to give

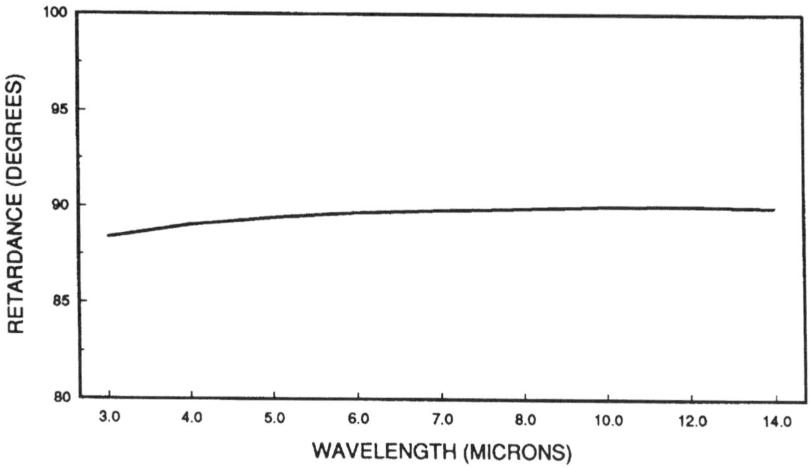

Figure 26-10 Theoretical retardance of achromatic prism retarder in the infrared.

Table 26-4 Numerical Data for Achromatic Retarder

Wavelength (μm)	ZnSe Index	Gold Index (n)	Gold Index (k)	Total Phase Shift
3	2.440	0.704	21.8	88.39
4	2.435	1.25	29.0	89.03
5	2.432	1.95	36.2	89.42
6	2.438	2.79	43.4	89.66
7	2.423	3.79	50.5	89.81
8	2.418	4.93	57.6	89.91
10	2.407	7.62	71.5	90.02
12	2.394	10.8	85.2	90.04
14	2.378	14.5	98.6	89.98

a retardance of 90° near 10 μm.) Table 26-4 gives numerical values of the phase shift along with indices for zinc selenide and gold. The indices for gold are from Ordal et al. [13] and the indices for ZnSe are from Wolfe and Zissis [14]. The requirement of a nearly achromatic retarder with no beam deviation is satisfied, although the disadvantage of the length of the device remains (the actual length is dependent on the clear aperture desired).

Achromatic Waveplate Retarders

As we have seen, waveplates are made of birefringent materials and the retardance is given by

$$\Gamma = \frac{2\pi}{\lambda}(n_e - n_0)d \tag{26-54}$$

The retardance is explicitly inversely proportional to wavelength. If the value of the birefringence:

$$\Delta n = (n_e - n_0) \tag{26-55}$$

for some material was directly proportional to wavelength then achromatic waveplates could be made from the material. This condition is not normally satisfied in nature.

Plates made up of two or three individual plates have been designed that are reasonably achromatic [6]. If we consider a plate made of two materials, *a* and *b*, having thicknesses d_a and d_b and wish to make the retardance equal at two wavelengths λ_1 and λ_2, we can write the equations:

$$N\lambda_1 = \Delta n_{1a} d_a + \Delta n_{1b} d_b \tag{26-56}$$

$$N\lambda_2 = \Delta n_{2a} d_a + \Delta n_{2b} d_b \tag{26-57}$$

where N is the retardance we require in waves, i.e., 1/4 , 1/2, etc., and the subscripts on the birefringence Δn designates the wavelength and material. Solving the equations for d_a and d_b we have

$$d_a = \frac{N(\lambda_1 \Delta n_{2b} - \lambda_2 \Delta n_{1b})}{\Delta n_{1a} \Delta n_{2b} - \Delta n_{1b} \Delta n_{2a}} \tag{26-58}$$

and

$$d_b = \frac{N(\lambda_2 \Delta n_{1a} - \lambda_1 \Delta n_{2a})}{\Delta n_{1a} \Delta n_{2b} - \Delta n_{1b} \Delta n_{2a}} \tag{26-59}$$

The optimization of the design is facilitated by changing the thickness of one of the plates and the ratio of the thicknesses [15]. There will generally be an extremum in the retardance function in the wavelength region of interest. A good achromatic design will have the extremum near the middle of the region. Changing the ratio of the thicknesses shifts the position of the extremum. Changing the thickness of one of the plates changes the overall retardance value.

There are important practical considerations for compound plate design. For example, it may not be possible to fabricate plates that are too thin, or they may result in warped elements; and plates that are thick will be more sensitive to angular variation of the incident light. Precision of alignment of the plates in a multiplate design is extremely important, and misalignments will result in oscillation of retardance. A compound waveplate for the infrared mentioned earlier is composed of two plates of CdS and CdSe with fast axes oriented perpendicularly [8]. This design calls for a CdS plate about 1.3 mm thick followed by a CdSe plate about 1 mm thick. The theoretical achromaticity over the 3–11 μm wavelength region is $90^\circ \pm 20^\circ$, although measurements indicate somewhat better performance [16]. The useful wavelength range of these achromatic waveplates is often determined by the design of the antireflection coatings.

26.4 ROTATORS

Rotation of the plane of polarization can occur through optical activity, the Faraday effect, and by the action of liquid crystals.

26.4.1 Optical Activity

Arago first observed optical activity in quartz in 1811. During propagation of light though a material, a rotation of the plane of polarization occurs that is proportional to the thickness of the material and also depends on wavelength. There are many substances that exhibit optical activity, notably quartz and sugar solutions (e.g., place a bottle of corn syrup between crossed polarizers!). Many organic molecules can exist as stereoisomers, i.e., a molecule of the same chemical formula is formed such that it either rotates light to the right or to the left. Since these molecules can have drastically different effects when taken internally, it has become important to distinguish and separate them when producing pharmaceuticals. Natural sugar is dextrorotatory, meaning it rotates to the right; amino acids are generally levorotatory, rotating to the left.

Optical activity can be explained in terms of left and right circularly polarized waves and the refractive indices that these waves experience. The rotatory power of an optically active medium is

$$\rho = \frac{\pi(n_L - n_R)}{\lambda} \tag{26-60}$$

in degrees per centimeter, where n_L is the index for left circularly polarized light, and n_R is the index for right circularly polarized light.

The rotation angle is

$$\delta = \frac{\pi(n_L - n_R)d}{\lambda} \tag{26-61}$$

Suppose we have a linearly polarized wave entering an optically active medium. The linearly polarized wave can be represented as a sum of circular components. Using the Jones formalism:

$$\begin{pmatrix} 1 \\ 0 \end{pmatrix} = \frac{1}{2}\begin{pmatrix} 1 \\ -i \end{pmatrix} + \frac{1}{2}\begin{pmatrix} 1 \\ i \end{pmatrix} \tag{26-62}$$

We have written the linear polarized light as a sum of left circular and right circular components. After traveling a distance d through the medium, the Jones vector will be

$$\frac{1}{2}\begin{pmatrix} 1 \\ -i \end{pmatrix} e^{i2\pi n_L d/\lambda} + \frac{1}{2}\begin{pmatrix} 1 \\ i \end{pmatrix} \mathrm{e}^{i2\pi n_R d/\lambda}$$
$$= \frac{1}{2} e^{i2\pi(n_R + n_L)d/2\lambda} \left\{ \begin{pmatrix} 1 \\ -i \end{pmatrix} e^{-i2\pi(n_R - n_L)d/2\lambda} + \begin{pmatrix} 1 \\ i \end{pmatrix} e^{i2\pi(n_R - n_L)d/2\lambda} \right\} \tag{26-63}$$

Let

$$\psi = \frac{2\pi(n_R + n_L)d}{2\lambda} \quad \text{and} \quad \delta = \frac{2\pi(n_L - n_R)d}{2\lambda} \tag{26-64}$$

Substituting these values into the right hand side of (26-64) gives

$$e^{i\psi}\left\{\frac{1}{2}\begin{pmatrix}1\\-i\end{pmatrix}e^{i\delta}+\frac{1}{2}\begin{pmatrix}1\\i\end{pmatrix}e^{-i\delta}\right\}=e^{i\psi}\left\{\begin{pmatrix}\frac{1}{2}(e^{i\delta}+e^{-i\delta})\\-\frac{1}{2}i(e^{i\delta}-e^{-i\delta})\end{pmatrix}\right\}=e^{i\psi}\begin{pmatrix}\cos\delta\\\sin\delta\end{pmatrix} \tag{26-65}$$

which is a linearly polarized wave whose polarization has been rotated by δ.

26.4.2 Faraday Rotation

The Faraday effect has been described in Chapter 24. Faraday rotation can be used as the basis for optical isolators. Consider a Faraday rotator between two polarizers that have their axes at 45°. Suppose that the Faraday rotator is such that it rotates the incident light by 45°. It should then pass through the second polarizer since the light polarization and the polarizer axis are aligned. Any light returning through the Faraday rotator is rotated an additional 45° and will be blocked by the first polarizer. In this way, very high isolation, up to 90 dB [17], is possible. Rotation in devices based on optical activity and liquid crystals retrace the rotation direction and cannot be used for isolation. Faraday rotation is the basis for spatial light modulators, optical memory, and optical crossbar switches.

26.4.3 Liquid Crystals

A basic description of liquid crystals has been given in Chapter 24. Liquid crystal cells of various types can be configured to act as polarization rotators. The rotation is electrically controllable, and may be continuous or binary. For a detailed treatment of liquid crystals, see Khoo and Wu [18].

26.5 DEPOLARIZERS

A depolarizer reduces the degree of polarization. We recall that the degree of polarization is given by

$$P=\frac{\sqrt{S_1^2+S_2^2+S_3^2}}{S_0} \tag{26-66}$$

An ideal depolarizer produces a beam of unpolarized light regardless of the initial polarization state, so that the goal of an ideal depolarizer is to reduce P to 0. The Mueller matrix for an ideal depolarizer is

$$\begin{pmatrix}1&0&0&0\\0&0&0&0\\0&0&0&0\\0&0&0&0\end{pmatrix} \tag{26-67}$$

A partial depolarizer (or pseudodepolarizer) reduces the degree of polarization. It could reduce one, two, or all three of the Stokes vector components by varying amounts, and there are many possibilities [19]. Examples of depolarizers in an everyday environment include waxed paper and projection screens. Integrating spheres have been shown to function as excellent depolarizers [20]. Commercial depolarizers are offered that are based on producing a variable phase shift across their apertures. These rely on obtaining a randomized mix of polarization states over the beam width. A small beam will defeat this depolarization scheme.

REFERENCES

1. Shurcliff, W. A., *Polarized Light—Production and Use*, Harvard University Press, 1962.
2. Stookey, S. D. and Araujo, R. J., "Selective polarzation of light due to absorption by small elongated silver particales in glass," Appl. Opt. *7*(5), 777–779, 1968.
3. Taylor M. and Bucher, G., "High contract polarizers for the near infrared," in *Polarization Considerations for Optical Systems II*, Proc. SPIE, Vol. 1166, R. A. Chipman, ed., 1989.
4. Perkins, R. T., Hansen, D. P., Gardner, E. W., Thorne, J. M., and Robbins, A. A., "Broadband wire grid polarizer for the visible spectrum," US Patent 6 122 103, Sept. 19, 2000.
5. Perkins, R. T., Gardner, E. W., and Hansen, D. P., "Imbedded wire grid polarizer for the visible spectrum," US Patent 6, 288 840, Sept. 11, 2001.
6. Bennett, J. M., and Bennett, H. E., "Polarization," in *Handbook of Optics*, W. G. Driscoll and W. Vaughan, eds., McGraw-Hill, New York, 1978.
7. Bennett, J. M., "Polarizers," in *Handbook of Optics*, 2nd ed., M. Bass, ed., McGraw-Hill, New York, 1995.
8. Chipman, R. A. and Chenault, D. B., "Infrared Achromatic Retarder," US Patent No. 4 961 634, Oct. 9, 1990.
9. Jerrard, H. G., "Optical compensators for measurement of elliptical polarization," *JOSA 38*, 35–59 1948.
10. Chenault, D. B. and Chipman, R. A., "Infrared spectropolarimetry," in *Polarization Considerations for Optical Systems II*, Proc. SPIE, Vol. 1166, R. A. Chipman, ed., 1989.
11. Clapham, P. B., Downs, M. J., and King, R. J., "Some applications of thin films to polarization devices," *Appl. Opt. 8*, 1965–1974, (1969).
12. Jenkins, F. A. and White, H. E., *Fundamentals of Optics*, McGraw-Hill, New York, 1957.
13. Ordal, M. A., Long, L. L., Bell, R. J., Bell, S. E., Bell, R. R., Alexander, R. W., Jr., and Ward, C. A., "Optical properties of the metals Al, Co, Cu, Au, Fe, Pb, Ni, Pd, Pt, Ag, Ti, and W in the infrared and far infrared," *Appl. Opt.*, *22*, 1099–1119 (1983).
14. Wolfe, W. L. and Zissis, G. J., *The Infrared Handbook*, Office of Naval Research, Washington, DC, 1978.
15. Chenault, David B., "Achromatic Retarder Design Study", Nichols Research Corporation Report No. NRC-TR-96-075, 1996.
16. Chenault, D. B., "Infrared Spectropolarimetry," Ph.D. Dissertation, University of Alabama, Huntsville, AL, 1992.
17. Saleh, B. E. A. and Teich M. C., *Fundamentals of Photonics*, John Wiley. New York, 1991.
18. Khoo, I-C. and Wu, S-T., *Optics and Nonlinear Optics of Liquid Crystals*, World Scientific, Singapore, 1993.
19. Chipman, R. A., "Depolarization," in *Polarization: Measurement, Analysis, and Remote Sensing II*, Proc. SPIE *3754*, D. H. Goldstein and D. B. Chenault, eds., 1999.
20. McClain, S. C., Bartlett, C. L., Pezzaniti, J. L., and Chipman, R. A., "Depolarization measurements of an integrating sphere," *Appl. Opt. 34*, 152–154 (1995).

27

Stokes Polarimetry

27.1 INTRODUCTION

In this chapter, we discuss methods of measuring (or creating) the Stokes vector, the real four-element entity that describes the state of polarization of a beam of light. The measurement process can be represented as

$$\mathbf{I} = \mathbf{A}\mathbf{S} \tag{27-1}$$

where $\mathbf{I}$ is the vector of flux measurements as made by the detector, $\mathbf{A}$ is a matrix whose dimensions depend on the number of measurements and whose elements depend on the optical system, and $\mathbf{S}$ is the incident Stokes vector. Since we want to determine the incident Stokes vector, we must invert Eq. (27-1) so that $\mathbf{S}$ is given by

$$\mathbf{S} = \mathbf{A}^{-1}\mathbf{I} \tag{27-2}$$

This system of equations is generated through a set of measurements and can be solved through Fourier or nonFourier techniques. Both solution methods will be discussed in this chapter.

A set of elements that analyzes a polarization state of incoming light is a polarization state analyzer (PSA). A set of elements that generates a polarization state is a polarization state generator (PSG). The PSA and PSG are functionally depicted in Fig. 27-1. All of the polarimeter types described in this chapter can be or have to be used with electronics and computers in order to automate the data collection process.

A Stokes polarimeter is complete if it measures all four elements of the Stokes vector, and incomplete if it measures less than four. We will describe several types of Stokes polarimeters in the remainder of the chapter. Rotating element polarimetry, oscillating element polarimetry, and phase modulation polarimetry are all methods that make a series of measurements over time to obtain the Stokes vector [1]. Other techniques, division of amplitude and division of wavefront polarimetry, described in the last section of the chapter, are designed to measure all four elements of the Stokes vector simultaneously.

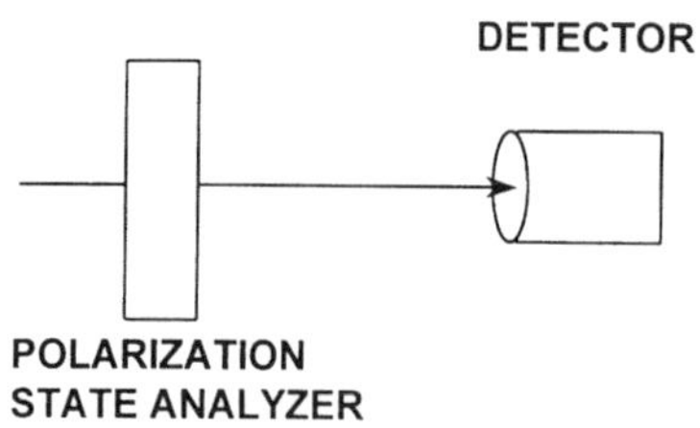

a. Polarization state analyzer with detector

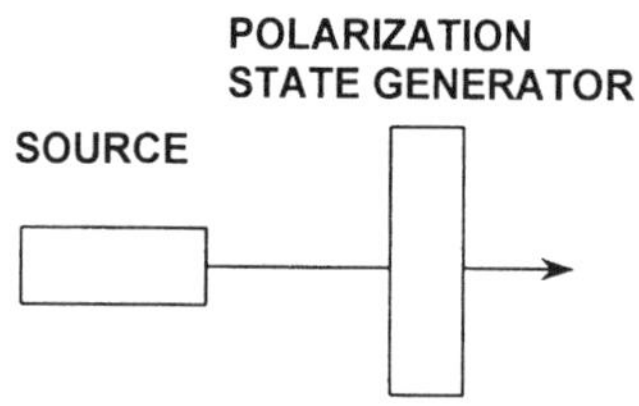

b. Polarization state generator with a source

Figure 27-1 Functional diagrams of Stokes polarimetry.

27.2 ROTATING ELEMENT POLARIMETRY

Stokes polarimeters that use rotating elements are shown in Fig. 27-2. The elements shown are all linear retarders and polarizers (analyzers). The measured Stokes elements are shown in the box to the right of each diagram, where the large black dots indicate the Stokes components that are measured.

27.2.1 Rotating Analyzer Polarimeter

Shown in Fig. 27-2a, the polarizer (analyzer) in this polarimeter rotates and produces a modulating signal at the detector, which is given by

$$I = \frac{a_0}{2} + \frac{a_2}{2}\cos 2\theta + \frac{b_2}{2}\sin 2\theta \tag{27-3}$$

where θ is the azimuthal angle of the polarizer. The coefficients a_0, a_2, and b_2 are the first three elements of the Stokes vector. At least three measurements must be made to determine the three measurable elements of the Stokes vector.

Equation (27-3) and subsequent expressions for the modulated signal in this chapter on Stokes polarimetry and in Chapter 28 on Mueller matrix polarimetry are all derived from algebraic equations representing these polarimetric systems. For example, for the rotating analyzer polarimeter, we have the equation:

$$\begin{pmatrix} S'_0 \\ S'_1 \\ S'_2 \\ S'_3 \end{pmatrix} = \frac{1}{2}\begin{pmatrix} 1 & \cos 2\theta & \sin 2\theta & 0 \\ \cos 2\theta & \cos^2 2\theta & \sin 2\theta\cos 2\theta & 0 \\ \sin 2\theta & \sin 2\theta\cos 2\theta & \sin^2 2\theta & 0 \\ 0 & 0 & 0 & 0 \end{pmatrix}\begin{pmatrix} S_0 \\ S_1 \\ S_2 \\ S_3 \end{pmatrix} \tag{27-4}$$

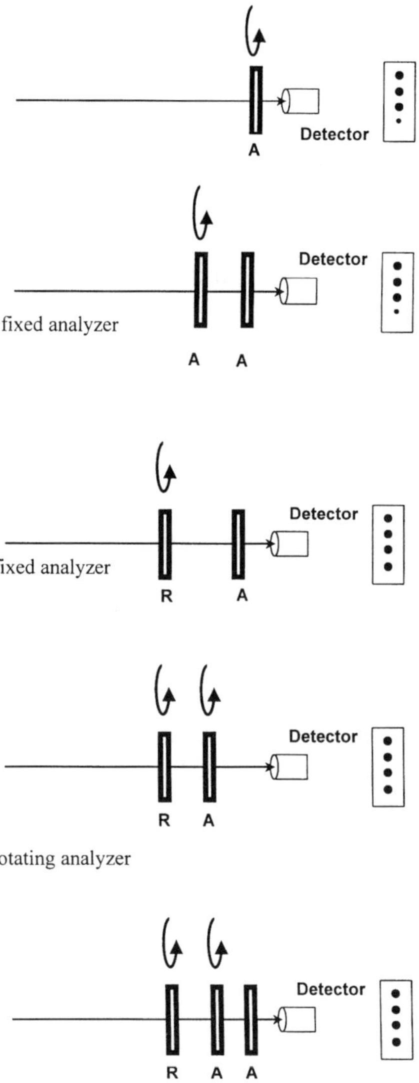

a. Rotating analyzer

b. Rotating analyzer plus fixed analyzer

c. Rotating retarder plus fixed analyzer

d. Rotating retarder and rotating analyzer

e. Rotating retarder, rotating analyzer, and fixed analyzer

Figure 27-2 Rotating element polarimeters. (After Ref. 1.)

where the input Stokes vector is multiplied by the Mueller matrix for a rotated ideal linear polarizer to obtain the (primed) output Stokes vector. We only need carry out the multiplication of the first row of the Mueller matrix with the input Stokes vector because we will be measuring the output signal $I = S_0'$. Thus,

$$I = \frac{S_0}{2} + \frac{S_1}{2}\cos 2\theta + \frac{S_2}{2}\sin 2\theta \tag{27-5}$$

Comparing this equation with (27-3), we have the correspondence:

$$\begin{aligned} S_0 &= a_0 \\ S_1 &= a_2 \\ S_2 &= b_2 \end{aligned} \tag{27-6}$$

The coefficients have been purposely written as a's and b's to represent the modulated signal as a Fourier series where the fundamental frequency of modulation and its harmonics are the angle θ and its multiples. We will continue to do this for the polarimeters described in this chapter and the next.

27.2.2 Rotating Analyzer and Fixed Analyzer Polarimeter

A fixed analyzer in front of the detector in this configuration shown in Fig. 27-2b means that the detector observes only one polarization, and any detector polarization sensitivity is made superfluous. A modulated signal composed of two frequencies is measured, and can be expressed as the Fourier series:

$$I = \frac{a_0}{4} + \frac{1}{4}\sum_{n=1}^{2}(a_{2n}\cos 2n\theta + b_{2n}\sin 2n\theta) \tag{27-7}$$

The first three elements of the Stokes vector are

$$\begin{aligned} S_0 &= a_0 - a_4 \\ S_1 &= \frac{2}{3}(a_2 - a_0 + 2a_4) \\ S_2 &= 0.4(2b_2 + b_4). \end{aligned} \tag{27-8}$$

27.2.3 Rotating Retarder and Fixed Analyzer Polarimeter

This is the basic complete Stokes polarimeter and is illustrated in Fig. 27-2c. The detector observes only a single polarization, and the modulated signal is again composed of two frequencies. The signal is again expressed as a Fourier series:

$$I = \frac{a_0}{2} + \frac{1}{2}\sum_{n=1}^{2}(a_{2n}\cos 2n\theta + b_{2n}\sin 2n\theta) \tag{27-9}$$

where now the angle θ is the azimuthal angle of the retarder. If the retarder is quarter wave, the Stokes vector is given in terms of the Fourier coefficients as

$$\begin{aligned} S_0 &= a_0 - a_4 \\ S_1 &= 2a_4 \\ S_2 &= 2b_4 \\ S_3 &= b_2 \end{aligned} \tag{27-10}$$

27.2.4 Rotating Retarder and Analyzer Polarimeter

Both elements rotate in this polarimeter of Fig. 27-2d. When the analyzer is rotated at three times the retarder angle and the retarder is quarter wave, the detected signal is given by

$$I = \frac{a_0}{2} + \frac{1}{2}\sum_{n=1}^{3}(a_{2n}\cos 2n\theta + b_{2n}\sin 2n\theta) \tag{27-11}$$

where θ is the rotation angle of the retarder. The Stokes vector is

$$\begin{aligned} S_0 &= a_0 \\ S_1 &= a_2 + a_6 \\ S_2 &= b_6 - b_2 \\ S_3 &= b_4 \end{aligned} \tag{27-12}$$

27.2.5 Rotating Retarder and Analyzer Plus Fixed Analyzer Polarimeter

This case, combining the previous two cases and shown in Fig. 27-2e, produces as many as nine harmonics in the detected signal when the analyzer is rotated by the factors 5/2, 5/3, or −3/2 times the retarder angle so that

$$I = \frac{a_0}{4} + \frac{1}{4}\sum_{\substack{n=1 \\ n\neq 9}}^{10}(a_n\cos n\theta + b_n\sin n\theta) \tag{27-13}$$

The Stokes vector is given in terms of the Fourier coefficients, when the rotation factor is 5/2 and the fixed analyzer is at 0°, as

$$\begin{aligned} S_0 &= a_0 - a_4 \\ S_1 &= 2a_1 \\ S_2 &= 2b_1 \\ S_3 &= b_3 \end{aligned} \tag{27-14}$$

27.3 OSCILLATING ELEMENT POLARIMETRY

Oscillating element polarimeters rotate the polarization of light using some electro- or magneto-optical device such as a Faraday cell or a liquid crystal cell (see Chapter 24). If, for example, the plane of polarization is rotated by an angle θ in a Faraday cell, this has the effect of having mechanically rotated all subsequent elements by an angle $-\theta$. The modulation is typically sinusoidal, which simulates an oscillating element, although a saw-tooth signal could be used to drive the modulation to result in an equivalent to a synchronous rotation of the element. The advantages of oscillating element polarimeters include operation at high frequencies, and the absence of moving parts to disturb alignment. A disadvantage, when the modulation is sinusoidal, is the additional complication in the signal content. The azimuthal angles are sinusoids, and the detected intensity now contains an infinite number of harmonics whose amplitudes depend on Bessel functions of the modulation

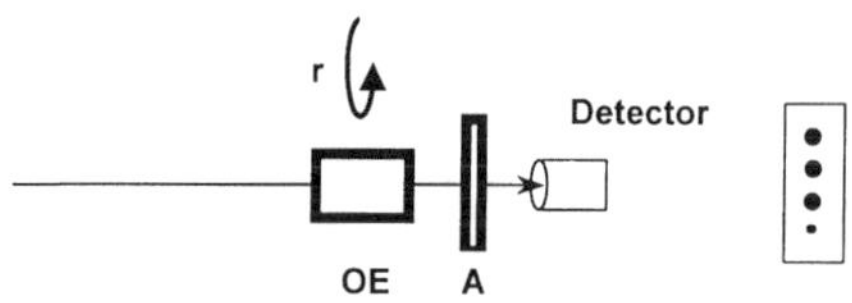

a. Oscillating analyzer polarimeter

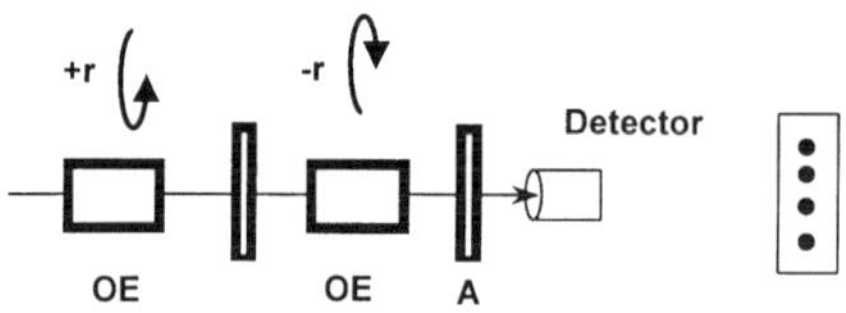

b. Oscillating retarder and fixed analyzer

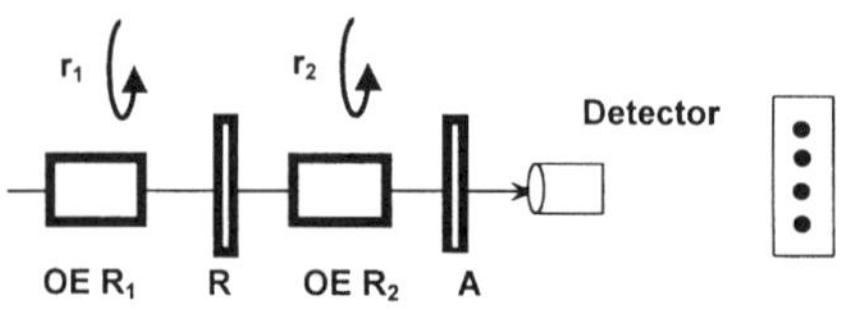

c. Oscillating retarder and analyzer

Figure 27-3 Oscillating element polarimeters. (After Ref. 1.)

amplitude. Oscillating element polarimeters derive harmonic content from the relationships (Bessel function expansions):

$$\sin(\theta \sin \omega t) = 2 \sum_{n=0}^{\infty} J_{2n+1}(\theta) \sin[(2n+1)\omega t] \tag{27-15}$$

$$\cos(\theta \sin \omega t) = J_0(\theta) + 2 \sum_{n=0}^{\infty} J_{2n}(\theta) \cos 2n\omega t \tag{27-16}$$

Experimentally, a lock-in amplifier is required for each detected frequency. Three oscillating element polarimeters are shown in Fig. 27-3 and we describe these polarimeters in the following subsections.

27.3.1 Oscillating Analyzer Polarimeter

The oscillating analyzer polarimeter (see Ref. 2) is shown in Fig. 27-3a. This polarimeter, like the rotating analyzer polarimeter, measures the first three components of the Stokes vector and hence is an incomplete polarimeter. The oscillating element produces an effective analyzer azimuth of

$$\theta = \theta_0 + \theta_1 \sin \omega t \tag{27-7}$$

where the azimuth θ_0 is determined by the mechanical azimuth of the fixed analyzer and/or a d.c. bias current in the Faraday cell, and θ_1 is the amplitude of the sinusoidal optical rotation produced by the Faraday cell. Substituting (27-17) into (27-3) we have

$$
\begin{aligned}
I = \frac{a_0}{2} &+ \frac{a_2 \cos 2\theta_0 + b_2 \sin 2\theta_0}{2} \cos(2\theta_1 \sin \omega t) \\
&+ \frac{-a_2 \sin 2\theta_0 + b_2 \cos 2\theta_0}{2} \sin(2\theta_1 \sin \omega t)
\end{aligned}
\tag{27-18}
$$

If we now use Eqs. (27-15) and (27-16) to replace $\cos(2\theta_1 \sin \omega t)$ and $\sin(2\theta_1 \sin \omega t)$, we have

$$
\begin{aligned}
I = \frac{a_0}{2} &+ \left(\frac{a_2}{2} \cos 2\theta_0 + \frac{b_2}{2} \sin 2\theta_0\right)[J_0(2\theta_1) + 2J_2(2\theta_1) \cos 2\omega t] \\
&+ \left(\frac{-a_2}{2} \sin 2\theta_0 + \frac{b_2}{2} \cos 2\theta_0\right)[2J_1(2\theta_1) \sin \omega t]
\end{aligned}
\tag{27-19}
$$

where we have neglected terms in frequency higher than 2ω.

The zero frequency (d.c.), fundamental, and second harmonic of the detected signal are then

$$
\begin{aligned}
I(0) &= \left[1 + J_0(2\theta_1)\left(\frac{a_2}{2} \cos 2\theta_0 + \frac{b_2}{2} \sin 2\theta_0\right)\right] \\
I(\omega) &= \left[2J_1(2\theta_1)\left(\frac{-a_2}{2} \sin 2\theta_0 + \frac{b_2}{2} \cos 2\theta_0\right)\right] \sin \omega t \\
I(2\omega) &= \left[2J_2(2\theta_1)\left(\frac{a_2}{2} \cos 2\theta_0 + \frac{b_2}{2} \sin 2\theta_0\right)\right] \cos 2\omega t
\end{aligned}
\tag{27-20}
$$

The d.c., fundamental, and second harmonic of the signal are detected synchronously, and the amplitude ratios are

$$
\begin{aligned}
\eta_\omega &= I(\omega)/I(0) \\
\eta_{2\omega} &= I(2\omega)/I(0)
\end{aligned}
\tag{27-21}
$$

and these are, using (28-20),

$$
\begin{aligned}
\eta_\omega &= \frac{2J_2(2\theta_1)((-a_2/2) \sin 2\theta_0 + (b_2/2) \cos 2\theta_0)}{1 + J_0(2\theta_1)((a_2/2) \cos 2\theta_0 + (b_2/2) \sin 2\theta_0)} \\
\eta_{2\omega} &= \frac{2J_2(2\theta_1)((a_2/2) \cos 2\theta_0 + (b_2/2) \sin 2\theta_0)}{1 + J_0(2\theta_1)((a_2/2) \cos 2\theta_0 + (b_2/2) \sin 2\theta_0)}
\end{aligned}
\tag{27-22}
$$

These last equations can be inverted to give the coefficients

$$
\begin{aligned}
a_2 &= \frac{\eta_\omega J_2(2\theta_1) \sin 2\theta_0 - \eta_{2\omega} J_1(2\theta_1) \cos 2\theta_0}{J_1(2\theta_1)[\eta_{2\omega} J_0(2\theta_1) - 2J_2(2\theta_1)]} \\
b_2 &= \frac{-\eta_\omega J_2(2\theta_1) \cos 2\theta_0 - \eta_{2\omega} J_1(2\theta_1) \sin 2\theta_0}{J_1(2\theta_1)[\eta_{2\omega} J_0(2\theta_1) - 2J_2(2\theta_1)]}
\end{aligned}
\tag{27-23}
$$

If $\theta = 0^\circ$ and $2\theta_1 = 137.8^\circ$, $J_0(2\theta_1) = 0$, and the Stokes vector is given by

$$\begin{aligned} S'_0 &= I_0 \\ S'_1 &= \frac{\eta_{2\omega}}{2J_2(2\theta_1)} \\ S'_2 &= \frac{\eta_\omega}{2J_1(2\theta_1)} \end{aligned} \tag{27-24}$$

where the primes indicate the output Stokes parameters.

27.3.2 Oscillating Retarder with Fixed Analyzer Polarimeter

This polarimeter, the equivalent of the rotating retarder polarimeter, is shown in Fig. 27-3b. As indicated in the figure, this is a complete Stokes polarimeter. A retarder is surrounded by two optical rotators with equal and opposite rotations. For example, a quarter-wave retarder might have a Faraday cell on one side and an identical Faraday cell on the other side but connected to an electrical signal source of opposite polarity. A light beam passing through a linear retarder of retardance δ with fast axis azimuth θ_R and a linear polarizer (analyzer) of azimuth θ_A results in an output intensity corresponding to the first Stokes parameter of the emergent light:

$$\begin{aligned} S'_0 = \frac{S_0}{2} &+ \frac{S_1}{2}[\cos 2\theta_R \cos(2\theta_A - 2\theta_R) - \sin 2\theta_R \sin(2\theta_A - 2\theta_R)\cos\delta] \\ &+ \frac{S_2}{2}[\sin 2\theta_R \cos(2\theta_A - 2\theta_R) + \cos 2\theta_R \sin(2\theta_A - 2\theta_R)\cos\delta] \\ &+ \frac{S_3}{2}[\sin(2\theta_A - 2\theta_R)\sin\delta] \end{aligned} \tag{27-25}$$

If we assume that $\delta = \pi/2$ and $\theta_A = 0$, I is the detected signal, and k is a proportionality constant, then we have

$$kI = \left(S_0 + \frac{1}{2}S_1\right) + \frac{1}{2}S_1 \cos 4\theta_R + \frac{1}{2}S_2 \sin 4\theta_R - S_3 \sin 2\theta_R \tag{27-26}$$

or

$$kI = \beta_0 + \beta_1 \cos 4\theta_R + \beta_2 \sin 4\theta_R - \beta_3 \sin 2\theta_R \tag{27-27}$$

where

$$\beta_0 = S_0 + \frac{1}{2}S_1 \tag{27-28a}$$

$$\beta_1 = \frac{1}{2}S_1 \tag{27-28b}$$

$$\beta_2 = \frac{1}{2}S_2 \tag{27-28c}$$

and

$$\beta_3 = -S_3 \tag{27-28d}$$

The two optical rotators on either side of the retarder effectively oscillate the retarder azimuth and we have

$$\theta_R = \theta_{R_0} + \theta_{R_1} \sin \omega t \tag{27-29}$$

where θ_{R_0} is the bias azimuth, and θ_{R_1} is the rotation amplitude. Using (27-29) in (27-27) and again making use of the Bessel function expansions, we can obtain the Fourier amplitudes of the detected signal as

$$kI_{\text{dc}} = \beta_0 + \beta_1[\cos 4\theta_{R_0} J_0(4\theta_{R_1})] + \beta_2[\sin 4\theta_{R_0} J_0(4\theta_{R_1})] + \beta_3[\sin 2\theta_{R_0} J_0(2\theta_{R_1})] \tag{27-30a}$$

$$kI_{\omega} = \beta_1[-2\sin 4\theta_{R_0} J_1(4\theta_{R_1})] + \beta_2[2\cos 4\theta_{R_0} J_1(4\theta_{R_1})] + \beta_3[2\cos 2\theta_{R_0} J_1(2\theta_{R_1})] \tag{27-30b}$$

$$kI_{2\omega} = \beta_1[2\cos 4\theta_{R_0} J_2(4\theta_{R_1})] + \beta_2[2\sin 4\theta_{R_0} J_2(4\theta_{R_1})] + \beta_3[2\sin 2\theta_{R_0} J_2(2\theta_{R_1})] \tag{27-30c}$$

$$kI_{3\omega} = \beta_1[-2\sin 4\theta_{R_0} J_3(4\theta_{R_1})] + \beta_2[2\cos 4\theta_{R_0} J_3(4\theta_{R_1})] + \beta_3[2\cos 2\theta_{R_0} J_3(2\theta_{R_1})] \tag{27-30d}$$

In vector-matrix form, the last three equations are

$$k\begin{pmatrix} I_{\omega} \\ I_{2\omega} \\ I_{3\omega} \end{pmatrix} = \begin{pmatrix} -2\sin 4\theta_{R_0} J_1(4\theta_{R_1}) & 2\cos 4\theta_{R_0} J_1(4\theta_{R_1}) & 2\cos 2\theta_{R_0} J_1(2\theta_{R_1}) \\ 2\cos 4\theta_{R_0} J_2(4\theta_{R_1}) & 2\sin 4\theta_{R_0} J_2(4\theta_{R_1}) & 2\sin 2\theta_{R_0} J_2(2\theta_{R_1}) \\ -2\sin 4\theta_{R_0} J_3(4\theta_{R_1}) & 2\cos 4\theta_{R_0} J_3(4\theta_{R_1}) & 2\cos 2\theta_{R_0} J_3(2\theta_{R_1}) \end{pmatrix} \begin{pmatrix} \beta_1 \\ \beta_2 \\ \beta_3 \end{pmatrix} \tag{27-31}$$

This equation can be solved for β_1, β_2, and β_3 by inverting the 3×3 matrix. Equation (27-30a) can then be used to find β_0, and (27-28) used to find the Stokes vector elements.

27.3.3 Oscillating Retarder and Analyzer Polarimeter

The oscillating retarder and analyzer polarimeter is the generalization of oscillating element designs [3]. This polarimeter is shown in Fig. 27-3c. A retarder is surrounded by two optical rotators as in the oscillating retarder and fixed analyzer polarimeter, but now the rotators produce rotations θ_{r_1} and θ_{r_2}. The retarder is oriented at some angle θ_R and the linear polarizer is oriented at some angle θ_P. With no optical rotators, the detected signal is given by

$$kI = S_0 + (S_1 \cos 2\theta_R + S_2 \sin 2\theta_R)\cos(2\theta_P - 2\theta_R) + S_3 \sin(2\theta_P - 2\theta_R) \tag{27-32}$$

Consider that the rotator R_2 in Fig. 27-3c is replaced by two equivalent rotators in series that have rotations $-r_1$ and $r_1 + r_2$. The sum of these is r_2 and we have not changed the resultant net rotation. The retarder is now surrounded by rotators with rotations r_1 and $-r_1$ and this is equivalent to the retarder in the new azimuth $\theta_R + r_1$. The rotator with rotation $r_1 + r_2$ rotates the polarizer azimuth to $\theta_P + r_1 + r_2$. If we replace the angles in (27-32) with the azimuthal angles resulting from the addition of the rotators, we have

$$\begin{aligned} kI = S_0 &+ S_1 \cos(2\theta_R + 2r_1)\cos(2\theta_P - 2\theta_R + 2r_2) \\ &+ S_2 \sin(2\theta_R + 2r_1)\cos(2\theta_P - 2\theta_R + 2r_2) \\ &+ S_3 \sin(2\theta_P - 2\theta_R + 2r_2) \end{aligned} \tag{27-33}$$

If we reference the angular coordinates to the azimuth of the polarizer, we can set $\theta_P = 0$ and rewrite (27-33) as

$$\begin{aligned} kI = S_0 &+ \frac{1}{2}S_1[\cos 4\theta_R \cos(2r_1 - 2r_2) - \sin 4\theta_R \sin(2r_1 - 2r_2) + \cos(2r_1 + 2r_2)] \\ &+ \frac{1}{2}S_2[\sin 4\theta_R \cos(2r_1 - 2r_2) + \cos 4\theta_R \sin(2r_1 - 2r_2) + \sin(2r_1 + 2r_2)] \\ &- S_3(\sin 2\theta_R \cos 2r_2 - \cos 2\theta_R \sin 2r_2) \end{aligned} \tag{27-34}$$

Now consider that the rotators are oscillated at the same frequency and are either in phase or out of phase by π, then the rotations produced are given by

$$r_1 = \theta_{r_1} \sin \omega t \tag{27-35a}$$

and

$$r_2 = \theta_{r_2} \sin \omega t \tag{27-35b}$$

We can now substitute the expressions of (27-35) into (27-34) and again use the Bessel function expansions of (27-15) and (27-16) to obtain the equation:

$$kI = MS_n \tag{27-36}$$

where

$$I = \begin{pmatrix} I_\omega \\ I_{2\omega} \\ I_{3\omega} \end{pmatrix} \qquad S_n = \begin{pmatrix} S_1 \\ S_2 \\ S_3 \end{pmatrix} \tag{27-37}$$

and

$$M = \begin{pmatrix} -\sin 4\theta_R J_1(2\theta_{r_1} - 2\theta_{r_2}) & \cos 4\theta_R J_2(2\theta_{r_1} - 2\theta_{r_2}) + J_2(2\theta_{r_1} + 2\theta_{r_2}) & 2\cos 2\theta_R J_1(2\theta_{r_2}) \\ \cos 4\theta_R J_2(2\theta_{r_1} - 2\theta_{r_2}) + J_2(2\theta_{r_1} + 2\theta_{r_2}) & \sin 4\theta_R J_2(2\theta_{r_1} - 2\theta_{r_2}) & -2\sin 2\theta_R J_2(2\theta_{r_2}) \\ -\sin 4\theta_R J_3(2\theta_{r_1} - 2\theta_{r_2}) & \cos 4\theta_R J_2(2\theta_{r_1} - 2\theta_{r_2}) + J_2(2\theta_{r_1} + 2\theta_{r_2}) & 2\cos 2\theta_R J_3(2\theta_{r_2}) \end{pmatrix} \tag{27-38}$$

The zero frequency term is given by

$$\begin{aligned} I_{\text{dc}} = S_0 &+ \frac{1}{2}S_1[\cos 4\theta_R J_0(2\theta_{r_1} - 2\theta_{r_2}) + J_0(2\theta_{r_1} + 2\theta_{r_2})] \\ &+ \frac{1}{2}S_2[\sin 4\theta_R J_0(2\theta_{r_1} - 2\theta_{r_2})] - S_3[\sin 2\theta_R J_0(2\theta_{r_2})] \end{aligned} \tag{27-39}$$

S_n is found by multiplying the signal vector I by the inverse of M and then S_0 is obtained from (27-39).

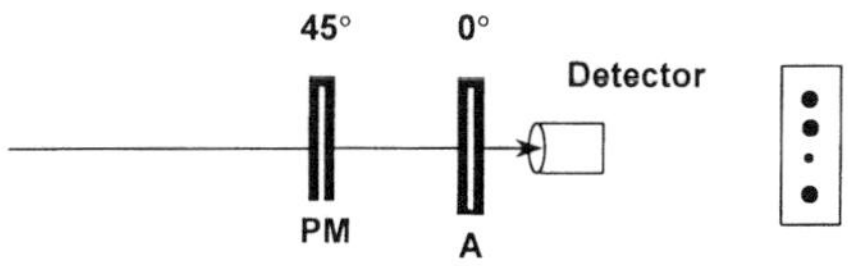

a. A single modulator with fixed analyzer

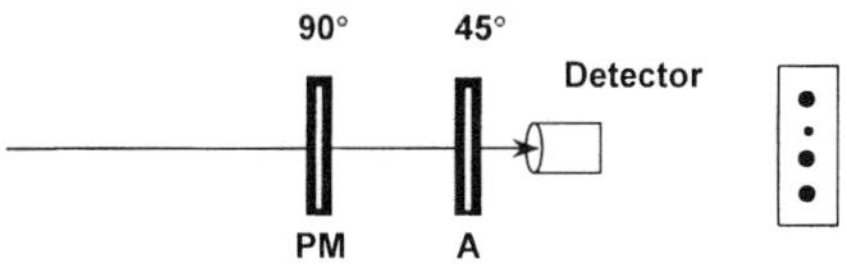

b. A single modulator with fixed analyzer

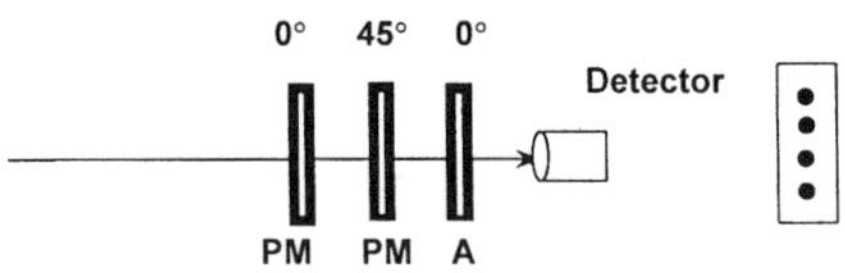

c. Dual modulators with fixed analyzer

Figure 27-4 Phase modulation polarimeters. (After Ref. 1.)

27.4 PHASE MODULATION POLARIMETRY

Phase modulation polarimeters are shown in Fig. 27-4. These polarimeters use devices that vary in retardance in response to an electrical signal. A common type of phase modulator is the photoelastic modulator (see Chapter 24).

27.4.1 Phase Modulator and Fixed Analyzer Polarimeter

This polarimeter, shown in Fig. 27-4a, uses a single modulator with a fixed linear analyzer. The axes of the modulator and analyzer are inclined at 45° to each other.

The detected signal is given by

$$I = \frac{S_0}{2} + \frac{1}{2}(S_1 \cos 2\theta_A + S_2 \sin 2\theta_A)\cos\Delta + S_3 \sin\Delta \tag{27-40}$$

where θ_A is the azimuthal angle of the analyzer and Δ is the retardance of the modulator. The modulator retardance is

$$\Delta = \delta \sin \omega t \tag{27-41}$$

where ω is the frequency of modulation and δ is the magnitude of the modulation. The detected intensity is given by

$$I = \frac{I_0}{2} + \frac{I_1}{2}\sin\omega t + \frac{I_2}{2}\cos 2\omega t \tag{27-42}$$

If $\delta = 137.8^\circ$ [$J_0(\delta) = 0$ and $\theta = 0^\circ$] the Stokes vector is given by

$$S_0 = I_0$$
$$S_1 = \frac{I_2}{2J_2(\delta)} \tag{27-43}$$
$$S_3 = \frac{I_1}{2J_1(\delta)}$$

If the polarimeter elements are both rotated by 45° (see Fig. 27-4b), we will measure the Stokes vector:

$$S_0 = I_0$$
$$S_2 = \frac{I_2}{2J_2(\delta)} \tag{27-44}$$
$$S_3 = \frac{I_1}{2J_1(\delta)}$$

27.4.2 Dual-Phase Modulator and Fixed Analyzer Polarimeter

The dual-phase modulator and fixed analyzer polarimeter is shown in Fig. 27-4c. The first modulator (closest to the analyzer) is aligned 45° to the analyzer and has time-varying retardation:

$$\Delta_1 = \delta_1 \sin \omega_1 t \tag{27-45}$$

The second modulator, aligned to the analyzer axis, has time-varying retardation:

$$\Delta_2 = \delta_2 \sin \omega_2 t \tag{27-46}$$

All four Stokes parameters can be measured with this system. The signal is

$$I = \frac{S_0}{2} + \frac{S_1 \cos \Delta_2}{2} + \frac{S_2 \sin \Delta_2 \sin \Delta_1}{2} - \frac{S_3 \sin \Delta_2 \cos \Delta_1}{2} \tag{27-47}$$

and if we demand that $\delta_1 = \delta_2 = 137.8^\circ$ then

$$I = \frac{I_0}{2} + \frac{I_1 \cos 2\omega_2 t}{2} \pm \frac{I_2 \cos(\omega_2 \pm \omega_1)}{2} + \frac{I_3 \sin(\omega_2 \pm 2\omega_1)t}{2} \tag{27-48}$$

and higher frequency terms. The Stokes vector is then given by

$$S_0 = I_0$$
$$S_1 = \frac{I_1}{2J_2(\delta_2)}$$
$$S_2 = \frac{I_2}{2J_1(\delta_1)J_1(\delta_2)} \tag{27-49}$$
$$S_3 = \frac{-I_3}{2J_2(\delta_1)J_1(\delta_2)}$$

27.5 TECHNIQUES IN SIMULTANEOUS MEASUREMENT OF STOKES VECTOR ELEMENTS

In the polarimetry techniques we have described in this chapter up to this point, all depend on a time sequential activity. That is, in rotating element polarimetry, polarizers and retarders are rotated and measurements are made at various angular positions of the elements; in oscillating element polarimetry, rotators are oscillated, and measurements are made at various points in the oscillation; in phase modulation polarimetry, measurements are made at various phase values in the modulation. We would like to be able to make all required measurements at the same time to ensure that time is not a factor in the result. In order to do this we can divide the wavefront spatially and make simultaneous measurements of different quantities at different points in space, or we can separate polarizations by dividing the amplitude of the wavefront. Polarimeters of these types generally have no moving parts.

27.5.1 Division of Wavefront Polarimetry

Wavefront division relies on analyzing different parts of the wavefront with separate polarization elements. This has been done using a pair of boresighted cameras that were flown on the space shuttle [4,5]. A linear polarizer was placed in front of each camera where the polarizers were orthogonal to each other. Chun et al. [6] have performed wavefront division polarimetry using a single infrared camera. Metal wire-grid polarizers were formed on a substrate using microlithography in the pattern shown in Fig. 27-5. This wire-grid array was placed in front of the detector array so that light from different parts of the object space pass through different polarization elements and on to different detectors. Each detector element of the infrared focal plane array has its own polarizer. These polarizers are linear polarizers at four different orientations, as shown in Fig. 27-5, and the pattern is repeated up to the size of the array. There are no circular components measured and thus this is an incomplete polarimeter.

Figure 27-5 Pattern of micropolarizers in a wavefront division polarimeter.

The advantage of this polarimetric measurement method is the simultaneous measurement of the Stokes vector elements available from the polarization element array. The reduction in resolution of the detector by the number of different polarization elements and the spatial displacement of information within the polarization element pattern are disadvantages.

27.5.2 Division of Amplitude Polarimetry

In amplitude division polarimetry, the energy in the entire wavefront of the incident beam is split and analyzed before passing to detectors. The detectors should be spatially registered so that any detector element is looking at the same point in space as all other detector elements. This method can employ as few as two detectors with analysis of two orthogonally polarized components of light, or it can measure the complete Stokes vector using four detectors. There are a number of variations of division of amplitude polarimetry and we will describe several.

Four-Channel Polarimeter Using Polarizing Beam Splitters

A diagram of a four-channel polarimeter [7] is shown in Fig. 27-6. This polarimeter uses three polarizing beam splitters and two retarders. Readings are made at four detectors. The input Stokes vector is determined from the four detector measurements and from use of a transfer Mueller matrix found during the calibration procedure. The polarizing beam splitters have transmissions of 80% and 20% for the

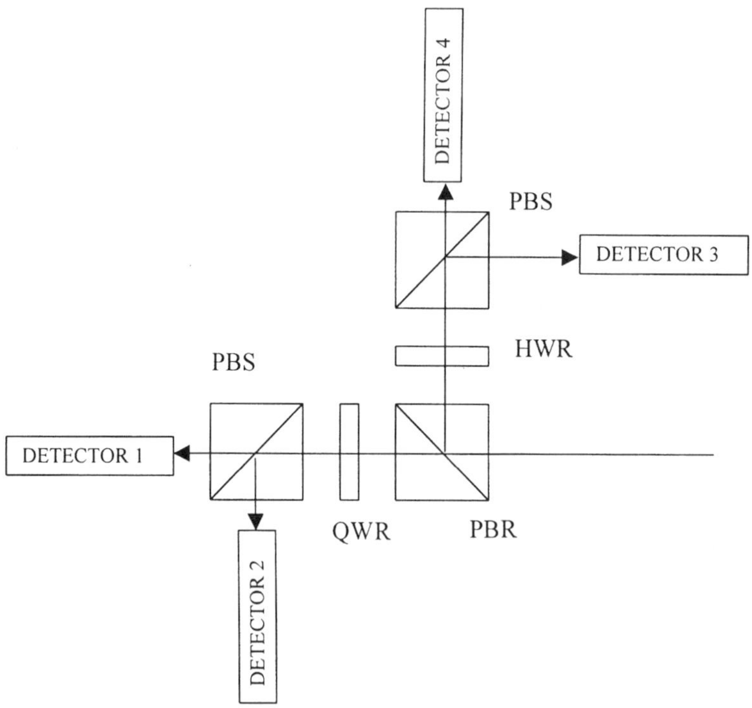

Figure 27-6 A four-channel polarimeter. PBS is a beam splitter, QWR is a quarter-wave retarder, and HWR is a half-wave retarder.

parallel and perpendicular components. A quarter-wave retarder before detectors 1 and 2 is oriented at 45° and the half-wave retarder before detectors 3 and 4 is oriented at 22.5°.

The advantage of this system is the simultaneous measurement of all four Stokes components for each point in object space. Care must be taken to ensure spatial registration of the detectors and equalization of detector response. Two-channel polarimeters [8] are substantially easier to construct.

Azzam's Four-Detector Photopolarimeter

Another type of amplitude division complete Stokes polarimeter is the four-detector photopolarimeter of Azzam [9,10]. A diagram of this polarimeter is shown in Fig. 27-7, and a photograph of a commercial version of this instrument is given in Fig. 27-8.

In this four-detector polarimeter, a light beam strikes four detectors in sequence, as shown in Fig. 27-7. Part of the light striking the first three is specularly reflected to the remaining detectors in the sequence, while the last detector absorbs substantially all the remaining light. The signal measured by each detector is proportional to the fraction of the light that it absorbs, and that fraction is a linear combination of the Stokes parameters. The light intensity measured by the detector is then linearly related to the input Stokes vector. The four detected signals are related to the input Stokes vector by

$$\mathbf{I} = \begin{pmatrix} i_0 \\ i_1 \\ i_2 \\ i_3 \end{pmatrix} = \mathbf{A}\begin{pmatrix} S_0 \\ S_1 \\ S_2 \\ S_3 \end{pmatrix} = \mathbf{AS} \tag{27-50}$$

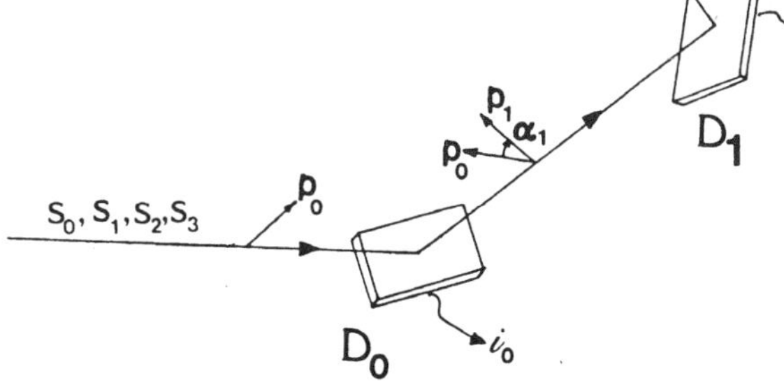

Figure 27-7 Optical diagram of the four-detector photopolarimeter. (From Ref. 9.)

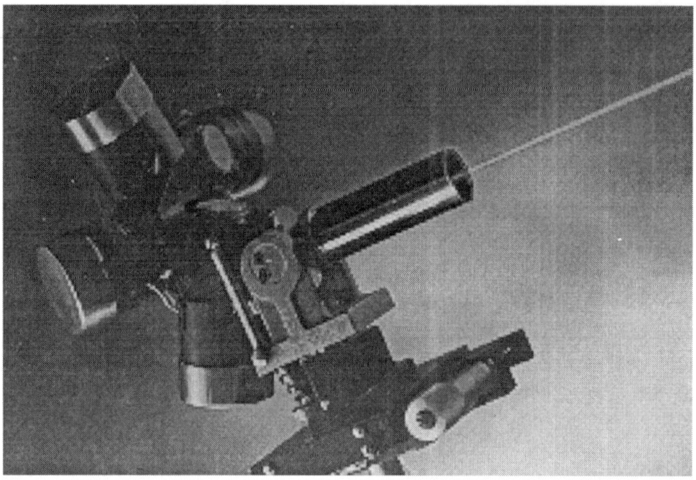

Figure 27-8 Photograph of a commercial four-detector photopolarimeter. (Courtesy of Gaertner Scientific Corp., Skokie, IL.)

where **A** is a Mueller matrix of the instrument. The input Stokes vector is then obtained from

$$\mathbf{S} = \mathbf{A}^{-1}\mathbf{I} \tag{27-51}$$

In order to determine the Stokes vector uniquely, the instrument matrix must be nonsingular. We now derive this instrument matrix.

The Stokes vectors of the light reflected from the surfaces of the photodetectors D_0, D_1, and D_2 are

$$\begin{aligned} \mathbf{S}^{(0)} &= \mathbf{M}_0\mathbf{S} \\ \mathbf{S}^{(1)} &= \mathbf{M}_1\mathbf{R}_1\mathbf{M}_0\mathbf{S} \\ \mathbf{S}^{(2)} &= \mathbf{M}_2\mathbf{R}_2\mathbf{M}_1\mathbf{R}_1\mathbf{M}_0\mathbf{S} \end{aligned} \tag{27-52}$$

where **S** is the input Stokes vector,

$$\mathbf{M}_l = r_l \begin{bmatrix} 1 & -\cos 2\psi_l & 0 & 0 \\ -\cos 2\psi_l & 1 & 0 & 0 \\ 0 & 0 & \sin 2\psi_l \cos \Delta_l & \sin 2\psi_l \sin \Delta_l \\ 0 & 0 & -\sin 2\psi_l \sin \Delta_l & \sin 2\psi_l \cos \Delta_l \end{bmatrix} \tag{27-53}$$

is the Mueller matrix of the lth detector, and

$$\mathbf{R}_l = \begin{bmatrix} 1 & 0 & 0 & 0 \\ 0 & \cos 2\alpha_l & \sin 2\alpha_l & 0 \\ 0 & -\sin 2\alpha_l & \cos 2\alpha_l & 0 \\ 0 & 0 & 0 & 1 \end{bmatrix} \tag{27-54}$$

is the rotation matrix describing the rotation of the plane of incidence between successive reflections; r_l is the reflectance of the lth detector for incident unpolarized or circularly polarized light and $\tan\psi_l e^{i\Delta_l} = r_{pl}/r_{sl}$ is the ratio of the complex reflection coefficients of the surface for polarizations parallel and perpendicular to the local plane of incidence.

Let us form a vector $\mathbf{L}$ composed of the first elements of the Stokes vectors $\mathbf{S}$, $\mathbf{S}^{(0)}$, $\mathbf{S}^{(1)}$, and $\mathbf{S}^{(2)}$, i.e., the elements that are proportional to the intensities. This can be accomplished by multiplying each of these Stokes vectors by the row vector:

$$\boldsymbol{\Gamma} = \begin{bmatrix} 1 & 0 & 0 & 0 \end{bmatrix} \tag{27-55}$$

so that we have

$$\mathbf{L} = \begin{bmatrix} S_0 \\ S_0^{(0)} \\ S_0^{(1)} \\ S_0^{(2)} \end{bmatrix} \tag{27-56}$$

This vector $\mathbf{L}$ is linearly related to the input Stokes vector by

$$\mathbf{L} = \mathbf{FS} \tag{27-57}$$

where $\mathbf{F}$ is given in terms of its rows by

$$\mathbf{F} = \begin{bmatrix} \mathbf{F}_0 \\ \mathbf{F}_1 \\ \mathbf{F}_2 \\ \mathbf{F}_3 \end{bmatrix} = \begin{bmatrix} \boldsymbol{\Gamma} \\ \boldsymbol{\Gamma}\mathbf{M}_0 \\ \boldsymbol{\Gamma}\mathbf{M}_1\mathbf{R}_1\mathbf{M}_0 \\ \boldsymbol{\Gamma}\mathbf{M}_2\mathbf{R}_2\mathbf{M}_1\mathbf{R}_1\mathbf{M}_0 \end{bmatrix} \tag{27-58}$$

The last three rows of this matrix are the first three rows of the matrices $\mathbf{M}_0$, $\mathbf{M}_1\mathbf{R}_1\mathbf{M}_0$, and $\mathbf{M}_2\mathbf{R}_2\mathbf{M}_1\mathbf{R}_1\mathbf{M}_0$. If we insert the appropriate forms of Eqs. (27-53) and (27-54) into (27-58) we obtain the matrix:

$$\mathbf{F} = \begin{bmatrix} 1 & 0 & 0 & 0 \\ f_{10} & f_{11} & 0 & 0 \\ f_{20} & f_{21} & f_{22} & f_{23} \\ f_{30} & f_{31} & f_{32} & f_{33} \end{bmatrix} \tag{27-59}$$

where

$$f_{10} = r_0$$

$$f_{11} = -r_0 \cos 2\psi_0$$

$$f_{20} = r_0 r_1 (1 + \cos 2\psi_0 \cos 2\psi_1 \cos 2\alpha_1)$$

$$f_{21} = -r_0 r_1 (\cos 2\psi_0 + \cos 2\psi_1 \cos 2\alpha_1)$$

$$f_{22} = -r_0 r_1 (\sin 2\psi_0 \cos \Delta_0 \cos 2\psi_1 \sin 2\alpha_1)$$

$$f_{23} = -r_0 r_1 (\sin 2\psi_0 \sin \Delta_0 \cos 2\psi_1 \sin 2\alpha_1)$$

$$
\begin{aligned}
f_{30} = r_0 r_1 r_2 (1 &+ \cos 2\psi_0 \cos 2\psi_1 \cos 2\alpha_1 + \cos 2\psi_1 \cos 2\psi_2 \cos 2\alpha_2 \\
&+ \cos 2\psi_0 \cos 2\psi_2 \cos 2\alpha_1 \cos 2\alpha_2 \\
&- \cos 2\psi_0 \sin 2\psi_1 \cos \Delta_1 \cos 2\psi_2 \sin 2\alpha_1 \sin 2\alpha_2)
\end{aligned}
$$

$$
\begin{aligned}
f_{31} = -r_0 r_1 r_2 (&\cos 2\psi_0 + \cos 2\psi_1 \cos 2\alpha_1 + \cos 2\psi_0 \cos 2\psi_1 \cos 2\psi_2 \cos 2\alpha_2 \\
&+ \cos 2\psi_2 \cos 2\alpha_1 \cos 2\alpha_2 - \sin 2\psi_1 \cos \Delta_1 \cos 2\psi_2 \sin 2\alpha_1 \sin 2\alpha_2)
\end{aligned}
$$

$$
\begin{aligned}
f_{32} = -r_0 r_1 r_2 (&\sin 2\psi_0 \cos \Delta_0 \cos 2\psi_1 \sin 2\alpha_1 + \sin 2\psi_0 \cos \Delta_0 \cos 2\psi_2 \sin 2\alpha_1 \cos 2\alpha_2 \\
&+ \sin 2\psi_0 \cos \Delta_0 \sin 2\psi_1 \cos \Delta_1 \cos 2\psi_2 \cos 2\alpha_1 \sin 2\alpha_2 \\
&- \sin 2\psi_0 \sin \Delta_0 \sin 2\psi_1 \sin \Delta_1 \cos 2\psi_2 \sin 2\alpha_2)
\end{aligned}
$$

$$
\begin{aligned}
f_{33} = -r_0 r_1 r_2 (&\sin 2\psi_0 \sin \Delta_0 \cos 2\psi_1 \sin 2\alpha_1 + \sin 2\psi_0 \sin \Delta_0 \cos 2\psi_2 \sin 2\alpha_1 \cos 2\alpha_2 \\
&+ \sin 2\psi_0 \cos \Delta_0 \sin 2\psi_1 \sin \Delta_1 \cos 2\psi_2 \sin 2\alpha_2 \\
&+ \sin 2\psi_0 \sin \Delta_0 \sin 2\psi_1 \cos \Delta_1 \cos 2\psi_2 \cos 2\alpha_1 \sin 2\alpha_2)
\end{aligned}
\tag{27-60}
$$

The signal from each of the four detectors is proportional to the light absorbed by it. The light absorbed is the difference between the incident flux and the reflected flux; thus, the signal from the first detector is the difference between the first two elements of the vector **L** (27-56) multiplied by a proportionality constant that is dependent on the detector responsivity; the signal from the second detector is proportional to the difference between the second and third elements of the vector **L**; the signal from the third detector is proportional to the difference between the third and fourth elements of the vector **L**; and since the last detector is assumed to absorb the remaining light, the signal from this detector is proportional to the remaining flux. The signal from each detector is then expressed as

$$
\begin{aligned}
i_0 &= k_0 (S_0 - S_0^{(0)}) \\
i_1 &= k_1 (S_0^{(0)} - S_0^{(1)}) \\
i_2 &= k_2 (S_0^{(1)} - S_0^{(2)}) \\
i_3 &= k_3 S_0^{(2)}
\end{aligned}
\tag{27-61}
$$

In matrix form, (27-61) can be expressed as

$$
\mathbf{I} = \mathbf{KDL} \tag{27-62}
$$

where **K** is the detector responsivity matrix, **L** is the vector in (27-56), and **D** is constructed so that it takes the difference between elements of the vector **L**, i.e.,

$$
\mathbf{K} = \begin{bmatrix} k_0 & 0 & 0 & 0 \\ 0 & k_1 & 0 & 0 \\ 0 & 0 & k_2 & 0 \\ 0 & 0 & 0 & k_3 \end{bmatrix} \tag{27-63}
$$

and

$$\mathbf{D} = \begin{bmatrix} 1 & -1 & 0 & 0 \\ 0 & 1 & -1 & 0 \\ 0 & 0 & 1 & -1 \\ 0 & 0 & 0 & 1 \end{bmatrix} \tag{27-64}$$

Substituting (27-57) into (27-62) we obtain

$$\mathbf{I} = \mathbf{KDFS} \tag{27-65}$$

and we observe in comparing (27-65) and (27-50) that the instrument matrix **A** is

$$\mathbf{A} = \mathbf{KDF} \tag{27-66}$$

We know **K**, **D**, and **F** from (27-59), (27-63), and (27-64), and we have found the instrument matrix.

In order to compute $\mathbf{A}^{-1}$, **A** must be nonsingular and its determinant must be nonzero. We find the determinant from

$$\det \mathbf{A} = (\det \mathbf{K})(\det \mathbf{D})(\det \mathbf{F}) \tag{27-67}$$

which becomes, when we make substitutions,

$$\det \mathbf{A} = -(k_0 k_1 k_2 k_3)(r_0^3 r_1^2 r_2)(\sin 2\alpha_1 \sin 2\alpha_2) \times (\sin^2 2\psi_0 \cos 2\psi_0 \sin 2\psi_1 \cos 2\psi_1 \cos 2\psi_2) \sin \Delta_1 \tag{27-68}$$

If any factor in this equation is zero, the determinant becomes zero. We can now make some observations about the conditions under which this can happen. The first term in parentheses is the product of the responsivities of the detectors. It is undesirable and unlikely that any of these are zero, but this might happen if a detector is not working. The next term in parentheses is a product of the reflectances of the first three detectors. If any of these are zero, light will not get to the fourth detector, and the system will not work. Again, this is a condition that is undesirable and unlikely. The third term in parentheses is a geometrical condition: these factors are nonzero as long as the planes of incidence of two successive reflections are not coincident or orthogonal. The detectors can be arranged so that this does not happen. The fourth term in parentheses vanishes when

$$\begin{aligned} \psi_0 &= 0, \frac{\pi}{2} \\ \psi_1 &= 0, \frac{\pi}{2} \\ \psi_0 &= \frac{\pi}{4} \\ \psi_1 &= \frac{\pi}{4} \\ \psi_2 &= \frac{\pi}{4} \end{aligned} \tag{27-69}$$

The first two conditions in (27-69) are equivalent to having the first two detectors as perfect linear polarizers. The last three conditions would require that the first three detectors reflect p and s polarizations equally or function as retarders. Since the detectors are designed to be absorbing elements and typical reflections from absorbing surfaces will not fulfill these conditions, they are unlikely. The last factor, $\sin \Delta_1$, is the sine of the differential reflection phase shift at the second detector. A phase shift of 0 or π is usually associated with Fresnel reflections from nonabsorbing dielectrics. Again, we have absorbing detectors and this condition is not fulfilled.

Further details of polarimeter optimization, light path choice, spectral performance, and calibration are given in Azzam [10]. A fiber-optic implementation of the four-detector polarimeter is described in Bouzid et al. [11], and a corner cube configuration version of the polarimeter is discussed in Liu and Azzam [12].

Division of Amplitude Polarimeters Using Gratings

A number of polarimeters based on division of amplitude using gratings have been proposed [13–16]. Diffraction gratings split a single incident light beam into multiple beams and introduce significant polarization [17]. Azzam has demonstrated a polarimeter based on conical diffraction [10]. This instrument is shown in Fig. 27-9. An incident beam strikes a metal diffraction grating at an oblique incidence angle ϕ. The grating is positioned such that the lines of the gratings are at some arbitrary angle α to the plane of incidence, and this is the condition for conical diffraction. With this geometry, the diffraction efficiency is dependent on all elements of the Stokes vector, and thus this instrument is a complete polarimeter. A linear detector is placed at the location of each diffracted order to be detected. When four detectors are used, the same relationships apply to the grating polarimeter as in the four-detector polarimeter; i.e., the signal is linearly related to the incident Stokes vector by

$$\mathbf{I} = \mathbf{AS} \tag{27-70}$$

and we again invert the instrument matrix $\mathbf{A}$ to obtain the Stokes vector as in (27-51), i.e.,

$$\mathbf{S} = \mathbf{A}^{-1}\mathbf{I} \tag{27-51}$$

The derivation of the instrument matrix for this polarimeter follows the calibration procedures established for the four-detector polarimeter.

A polarimeter using a grating in the normal spectroscopic orientation, i.e., in a planar diffraction condition, has been designed and constructed [14]. This polarimeter is illustrated in Fig. 27-10. Polarizers are placed in front of the detectors in this

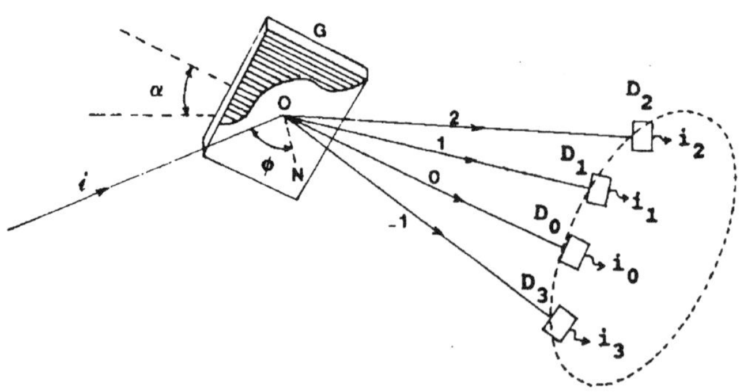

Figure 27-9 Photopolarimeter using conical diffraction. (From Ref. 13.)

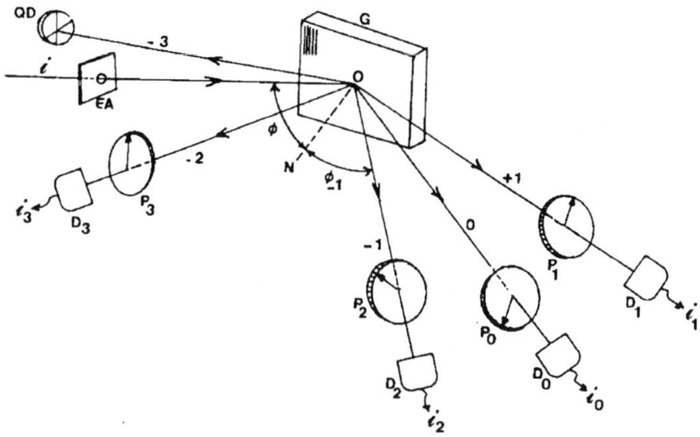

Figure 27-10 Photopolarimeter using planar diffraction. (From Ref. 14.)

design in order to make the instrument sensitive to all Stokes parameters. Four detectors are used in four diffracted orders. At least two of the diffracted beams must have polarizers in order for this polarimeter to be complete. An instrument matrix is determined through a calibration process.

A 16-beam grating-based polarimeter has also been designed and demonstrated [16]. A proposed polarimeter using transmission gratings and four linear detector arrays is designed to measure spectral and polarization information simultaneously [15].

Division of Amplitude Polarimeter Using a Parallel Slab

A wavefront may be divided in amplitude using the multiple reflections obtained in a planar dielectric slab [18]. Figure 27-11 shows a polarimeter based on a parallel plane slab of material of index $n_1(\lambda)$. A coating of metal of complex index $n_2 - ik_2$ is placed on the bottom surface of the slab. A light beam incident on the slab at angle ϕ undergoes multiple reflections in the slab, which results in a set of parallel and equally spaced outgoing beams. Linear polarizers are arranged in front of detectors in these beams with as many inclination angles of the transmission axes as there are detectors. The signal from the mth detector is then a linear combination of the elements of the Stokes vector, i.e.,

$$i_m = \sum_{j=0}^{3} a_{mj} S_j, \qquad m = 0, 1, 2, \ldots \tag{27-71}$$

where the mth vector $\mathbf{a}_m = [a_{m0} \quad a_{m1} \quad a_{m2} \quad a_{m3}]$ is the first row of the Mueller matrix of the mth light path. If we limit the detectors to four, the output signal vector is related to the input Stokes vector by the equation we have seen before for division of amplitude polarimeters:

$$\mathbf{I} = \mathbf{AS} \tag{27-70}$$

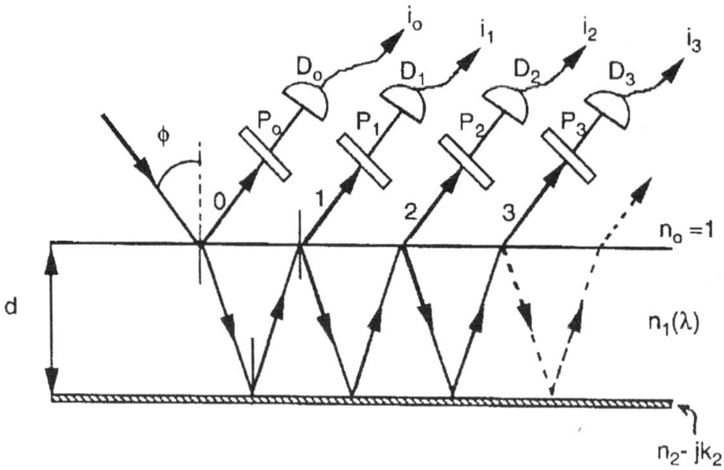

Figure 27-11 Parallel slab polarimeter. (From Ref. 18.)

The matrix **A** is the instrument matrix determined through calibration, and, as in previous division of amplitude examples, an unknown Stokes vector is found from the equation

$$\mathbf{S} = \mathbf{A}^{-1}\mathbf{I} \tag{27-51}$$

El-Saba et al. [18] show that for a slab of fused silica coated with a layer of silver and operated at 633 nm, the preferred angle of incidence for maximum energy in the beams and maximum value of the determinant of the instrument matrix is around 80°.

27.6 OPTIMIZATION OF POLARIMETERS

To this point we have not discussed specific polarization element angular settings. We have made reference to the use of quarter-wave retarders, primarily because we can construct a complete Stokes polarimeter using the readily available quarter-wave retarder and linear polarizer. We now ask the question, are there measurement angles and values of retardance that will result in a more efficient and/or better polarimeter?

This question was first addressed with regard to the angular positions of the quarter-wave retarder and linear polarizer in a rotating retarder and fixed analyzer polarimeter [19] and a rotating retarder, rotating analyzer polarimeter [20]. It was found in the first instance that angles of (−45°, 0°, 30°, 60°) or (−90°, −45°, 30°, 60°) resulted in the least sensitivity with regard to flux noise and rotation positional errors. In the second instance, if we let the rotation angle of the polarizer be θ and the rotation angle of the retarder be φ and define an α and β such that

$$\begin{aligned}\alpha &= 2\varphi \\ \beta &= 2(\theta - \varphi)\end{aligned} \tag{27-72}$$

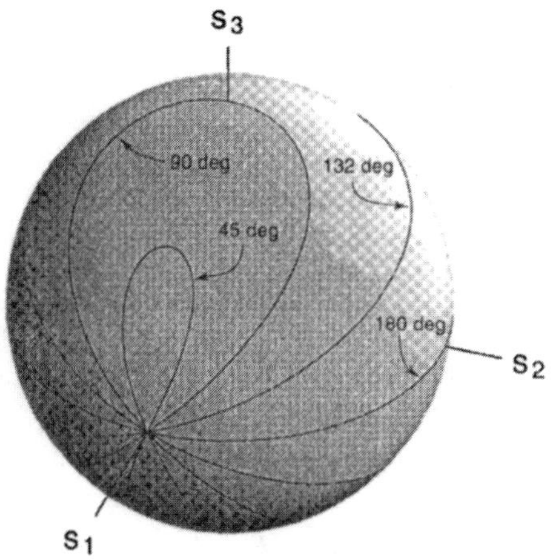

Figure 27-12 Locus of points on the Poincaré sphere for retardance values 45°, 90°, 132°, and 180° for a rotating retarder polarimeter. (From Ref. 21.)

then an optimal set of α and β is

$$[0^\circ, 90^\circ], \left[0^\circ, -\sin^{-1}\left(\frac{1}{3}\right)\right], \left[120^\circ, -\sin^{-1}\left(\frac{1}{3}\right)\right], \left[240^\circ, -\sin^{-1}\left(\frac{1}{3}\right)\right] \tag{27-73}$$

If we allow both the measurement angles and retardance to take part in the optimization process for a rotating retarder polarimeter, we find that the optimal value of retardance is 0.3661λ ($\approx 132^\circ$) and the optimal retarder positions are either ($\pm 15.12^\circ$, $\pm 51.69^\circ$) or ($\pm 74.88^\circ$, $\pm 38.31^\circ$) where these angle pairs are complements of each other [21,22]. These values were found through numerical optimization described in the cited references where the optimal values offer the best signal-to-noise performance and least sensitivity to element misalignment. Figure 27-12 shows the locus of points on the Poincaré sphere for values of retardance of 45°, 90°, 132°, and 180°. The figure indicates that better "global coverage" of the sphere is made possible by using the retardance of 132°.

Figure 27-13 reinforces this intuition where the intersection of the curve for the retardance value 132° with the four retarder positions ($\pm 15.12^\circ$, $\pm 51.69^\circ$) forms the corners of a regular tetrahedron inscribed in the Poincaré sphere, points as far apart as possible as one can make them on the surface of the sphere.

Figure 27-14 shows plots of a figure of merit for the rotating retarder fixed polarizer polarimeter versus number of measurements for the system with a quarter-wave retarder and an optimal retarder with both equally spaced angles and the optimal measurement angles. The results of this plot indicate that the optimal retarder with repeated optimal angles offers the best performance.

At this time, 132° retarders are not standard items from optical supply houses, and the improvement in performance gained by using these optimal elements may not be worth the cost and risk of ordering custom elements.

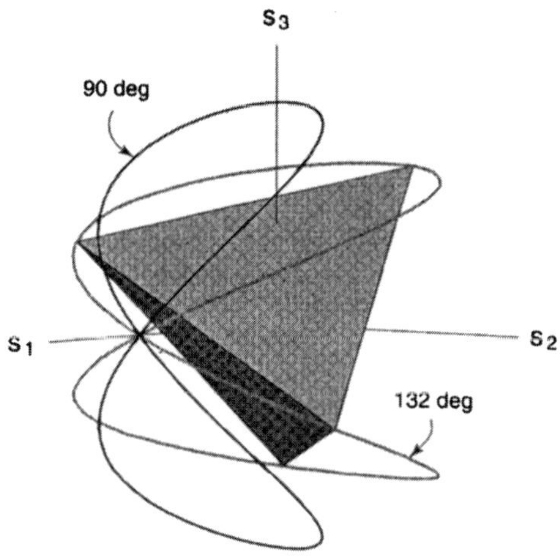

Figure 27-13 Curves for retardance values of 90° and 132° intersecting the retarder angles (±15.12°, ±51.69) to form the regular tetrahedron. (From Ref. 21.)

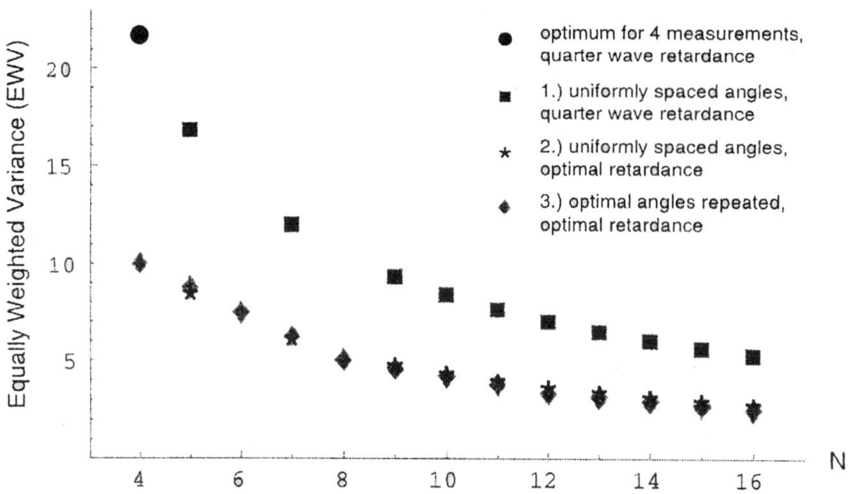

Figure 27-14 Plots of a figure of merit versus number of measurements for several measurement methods. (From Ref. 21.)

REFERENCES

1. Hauge, P. S., "Recent developments in instrumentation in ellipsometry," *Surface Sci.*, *96*, 108–140 (1980).
2. Azzam, R. M. A., "Oscillating-analyzer ellipsometer," *Rev. Sci. Instrum.*, *47*(5), 624–628 (1976).
3. Azzam, R. M. A., "Photopolarimeter using two modulated optical rotators," *Opt. Lett.*, *1*(5), 181–183 (1977).

4. Whitehead, V. S. and Coulson, K., "The space shuttle as a polarization observation platform," in *Polarization Considerations for Optical Systems II*, Proc. SPIE, Vol. 1166, San Diego, CA, Aug. 9–11, 1989; R. A. Chipman, ed.; SPIE, Bellingham, WA, 42–51, 1989.
5. Duggin, W. J., Israel, S. A., Whitehead, V. S., Myers, J. S., and Robertson, D. R., "Use of polarization methods in earth resources investigations," in *Polarization Considerations for Optical Systems II*, Proc SPIE, Vol. 1166, San Diego, CA, Aug. 9–11, 1989; R. A. Chipman, ed.; SPIE, Bellingham, WA, 11–22, 1989.
6. Chun, C. S. L., Fleming, D. L., Harvey, W. A., and Torok, E. J. "Polarization-sensitive infrared sensor for target discrimination," in *Polarization: Measurement, Analysis, and Remote Sensing*, Proc. SPIE, Vol. 3121, San Diego, CA, July 30–Aug. 1, 1997; D. H. Goldstein and R. A. Chipman, eds.; SPIE, Bellingham, WA, 55–62, 1997.
7. Gamiz, V. L., "Performance of a four channel polarimeter with low light level detection," Proc. SPIE, Vol. 3121, 35–46 (1997).
8. Wolff, L. B., "Polarization camera for computer vision with a beam splitter," *J. Opt. Soc. Am. A*, *11*(11), 2935–2945 (1994).
9. Azzam, R. M. A., "Arrangement of four photodetectors for measuring the state of polarization of light," *Opt. Lett.*, *10*, 309–311 (1985).
10. Azzam, R. M. A., Elminyawi, I. M., and El-Saba, A. M., "General analysis and optimization of the four-detector photopolarimeter," *J. Opt. Soc. Am. A*, *5*(5), 681–689 (1988).
11. Bouzid, A., Abushagur, M. A. G., El-Saba, A., and Azzam, R. M. A., "Fiber-optic four-detector polarimeter," *Opt. Comm.*, *118*, 329–334 (1995).
12. Liu, J. and Azzam, R. M. A. "Corner-cube four-detector photopolarimeter," *Optics & Laser Technol.*, *29*(5), 233–238 (1997).
13. Azzam, R. M. A., "Division-of-amplitude photopolarimeter based on conical diffraction from a metallic grating," *Appl. Opt.*, *31*(19), 3574–3576 (1992).
14. Azzam, R. M. A. and Giardina, K. A., "Photopolarimeter based on planar grating diffraction," *J. Opt. Soc. Am. A* *10*(6), 1190–1196 (1993).
15. Todorov, T. and Nikolova, L., "Spectrophotopolarimeter: fast simultaneous real-time measurement of light parameters," *Opt. Lett.* *17*(5), 358–359 (1992).
16. Cui, Y. and Azzam, R. M. A., "Calibration and testing of a sixteen-beam grating-based division-of-amplitude photopolarimeter," *Rev. Sci. Instrum.*, *66*(12), 5552–5558 (1995).
17. Bennett, J. M. and Bennett, H. E., "Polarization," in *Handbook of Optics*, W. G. Driscoll and W. Vaughan, eds., McGraw-Hill, New York, 1978.
18. El-Saba, A. M., Azzam, R. M. A., and Abushagur, M. A. G., "Parallel-slab division-of-amplitude photopolarimeter," *Opt. Lett.*, *21*, 1709–1711 (1996).
19. Ambirajan, A. and Look, D. C., "Optimum angles for a polarimeter: part I," *Opt. Eng.*, *34*, 1651–1655 (1995).
20. Ambirajan, A. and Look, D. C., "Optimum angles for a polarimeter: part II," *Opt. Eng.*, *34*, 1656–1659 (1995).
21. Sabatke, D. S., Descour, M. R., Dereniak, E. L., Sweatt, W. C., Kemme, S. A., and Phipps, G. S., "Optimization of retardance for a complete Stokes polarimeter," *Opt. Lett.*, *25*(11), 802–804 (2000).
22. Tyo, J. S., "Design of optimal polarimeters: maximization of signal-to-noise ratio and minimization of systematic error," *Appl. Opt.*, *41*(4), 619–630 (2002).

28

Mueller Matrix Polarimetry

28.1 INTRODUCTION

The real 4×4 matrix that completely describes the polarization properties of a material in reflection or transmission is measured in Mueller matrix polarimetry. A Mueller matrix polarimeter is complete if all 16 of the elements are measured, and incomplete otherwise. To be complete, a Mueller matrix polarimeter must have a complete polarization state analyzer (PSA) and a complete polarization state generator (PSG). Figure 28-1 is a conceptual diagram of a Mueller matrix polarimeter.

The equation we wish to solve in Mueller matrix polarimetry is

$$\begin{bmatrix} I \\ \bullet \\ \bullet \\ \bullet \end{bmatrix} = \mathbf{aMp} = \begin{bmatrix} a_1 & a_2 & a_3 & a_4 \\ \bullet & \bullet & \bullet & \bullet \\ \bullet & \bullet & \bullet & \bullet \\ \bullet & \bullet & \bullet & \bullet \end{bmatrix} \begin{bmatrix} m_{11} & m_{12} & m_{13} & m_{14} \\ m_{21} & m_{22} & m_{23} & m_{24} \\ m_{31} & m_{32} & m_{33} & m_{34} \\ m_{41} & m_{42} & m_{43} & m_{44} \end{bmatrix} \begin{bmatrix} p_1 \\ p_2 \\ p_3 \\ p_4 \end{bmatrix} \tag{28-1}$$

where $\mathbf{M}$ is the Mueller matrix to be measured, the vector $\mathbf{p}$ is the Stokes vector of the light entering the sample represented by $\mathbf{M}$, the vector $\mathbf{a}$ is the first row of the PSA Mueller matrix, and I is the signal from the detector. Note that the vector $\mathbf{p}$ is the product of the Stokes vector of the source and the Mueller matrix of the PSG, and only the first row of $\mathbf{a}$ is needed since the measured signal from the detector is the single value representing the first element of the output Stokes vector. We should measure at least 16 values of I with 16 settings of the PSG and PSA in order to obtain 16 equations in the 16 unknowns of the elements of the sample Mueller matrix. Very often more than 16 measurements are made so that the matrix elements are overdetermined. Measurement methods using Fourier or non-Fourier data-reduction techniques may be used.

In this chapter we shall discuss a small selection of Mueller matrix polarimeters that have found practical use. This will serve to illustrate the variation in method and serve as examples for those contemplating measurement of Mueller matrices. Hauge [1] gives a more complete review of various types of incomplete and complete Mueller matrix polarimeters. We review practical examples of rotating-element

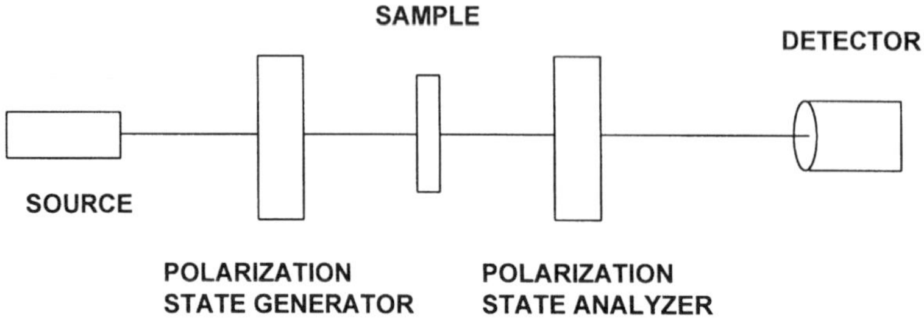

Figure 28-1 Conceptual diagram of a Mueller matrix polarimeter.

and phase-modulating polarimeters. Another type, the four-detector polarimeter, is also reviewed in this chapter.

28.1.1 Polarimeter Types

There are a number of different methods that have been devised to collect Mueller matrices. Many Mueller matrix polarimeters are either rotating-element polarimeters or phase-modulating polarimeters. Rotating-element polarimeters use mechanical rotation of polarizers or retarders to achieve the desired measurements. Phase modulating polarimeters use an electro-optical modulator to induce a time-varying retardation. Either of these polarimeter types may be complete or incomplete. Examples of different configurations of these two types are depicted in Figs. 28-2 and 28-3 and these show the Mueller matrix elements that are measured in each case (represented by the large dots).

28.1.2 Rotating Element Polarimeters

Figure 28-2a shows a rotating polarizer—rotating analyzer polarimeter. When the polarizer is rotated by an angle θ and the analyzer by angle 3θ synchronously, the Fourier series representing the normalized intensity has the form (I_0 is the source intensity):

$$\frac{I}{I_0} = \frac{a_0}{4} + \frac{1}{4}\sum_{k=1}^{4}(a_{2k}\cos 2k\theta + b_{2k}\sin 2k\theta) \tag{28-2}$$

The nine Fourier coefficients determine nine elements of the Mueller matrix:

$$M = \begin{bmatrix} a_0 & a_2 & b_2 & \cdot \\ a_6 & a_4 + a_8 & -b_4 + b_8 & \cdot \\ b_6 & b_4 + b_8 & a_4 - a_8 & \cdot \\ \cdot & \cdot & \cdot & \cdot \end{bmatrix} \tag{28-3}$$

A rotating polarizer–rotating compensator plus fixed analyzer polarimeter is shown in Fig. 28-2b. If the polarizer and retarder of this polarimeter are rotated synchronously in a 3 : 1 ratio, the normalized detected intensity can be expanded in the

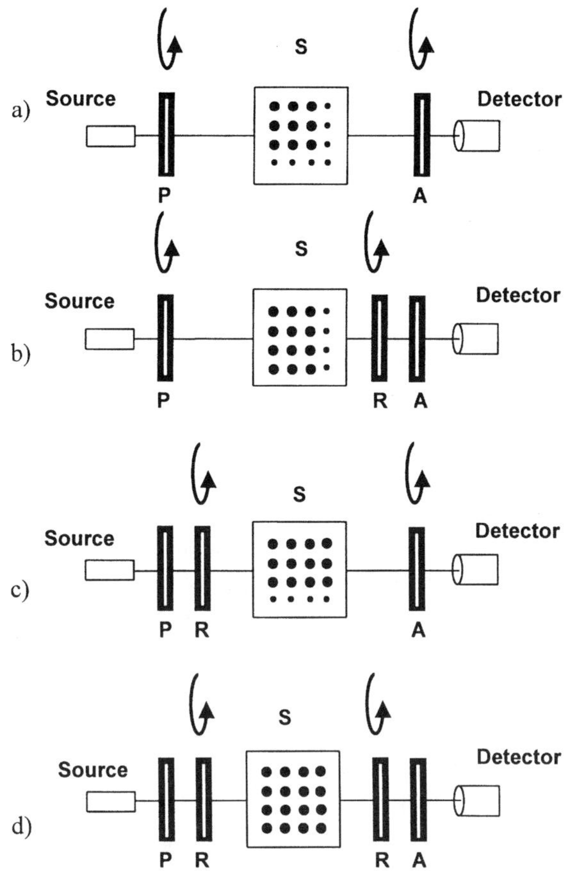

Figure 28-2 Rotating element polarimeters; P is a polarizer, A is an analyzer, R is a retarder, and S is the sample. Measured elements of the Mueller matrix are indicated by large dots. (After Ref. 1.)

Fourier series:

$$\frac{I}{I_0} = \frac{a_0}{4} + \frac{1}{4}\sum_{k=1}^{7}(a_{2k}\cos 2k\theta + b_{2k}\sin 2k\theta) \tag{28-4}$$

The 15 Fourier coefficients overdetermine the 12 elements of the Mueller matrix in the first three columns:

$$M = \begin{bmatrix} (a_0 - a_6) & (a_1 - a_5 - a_7) & (b_1 - b_5 + b_7) & \cdot \\ 2a_6 & 2(a_5 + a_7) & 2(b_7 - b_5) & \cdot \\ 2b_6 & 2(b_5 + b_7) & 2(a_5 - a_7) & \cdot \\ -2b_3 & -2b_2 & -2a_2 & \cdot \end{bmatrix} \tag{28-5}$$

The polarimeter in Fig. 28-2c determines the first three rows of the Mueller matrix. The last rotating-element polarimeter in Fig. 28-2d is the dual rotating-retarder polarimeter, and we will discuss this polarimeter in more detail in Section 28.2 below.

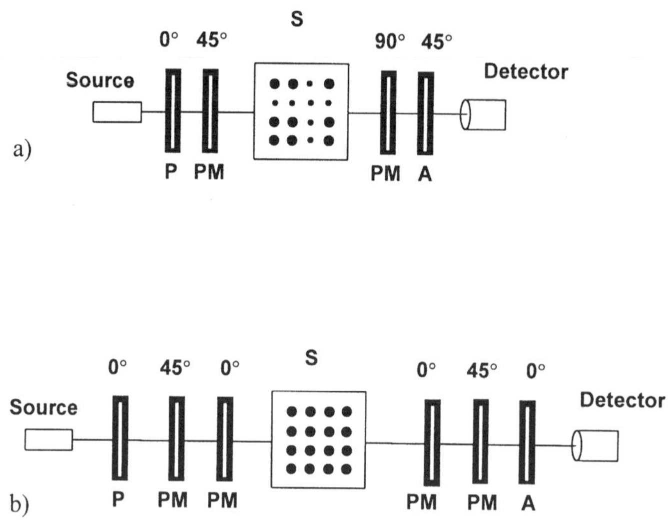

Figure 28-3 Phase-modulating polarimeters. A phase modulator/phase modulator polarimeter is shown in a); a dual-phase modulator polarimeter is shown in b). Measured elements of the Mueller matrix are indicated by large dots. P is a polarizer, A is an analyzer, PM is a phase modulator, and S is a sample. (After Ref. 1.)

28.1.3 Phase-Modulating Polarimeters

Two types of phase modulation polarimeters are shown in Fig. 28-3. One has a single modulator on either side of the sample, and the other has a double modulator on either side. We describe the double modulator case in more detail later in this chapter. For the single modulator case, it can be shown that the detected intensity, when the modulator axes are inclined at 45° to each other, as shown in Fig. 28-3a, is

$$\frac{I}{I_0} = \frac{1}{4}\begin{bmatrix} 1 & 0 & \cos\Delta_2 & \sin\Delta_2 \end{bmatrix} M \begin{bmatrix} 1 \\ \cos\Delta_1 \\ 0 \\ \sin\Delta_1 \end{bmatrix} \tag{28-6}$$

where I_0 is the source intensity and

$$\begin{aligned} \cos\Delta_i &= \cos(\delta_i \sin\omega_i t) \\ \sin\Delta_i &= \sin(\delta_i \sin\omega_i t) \end{aligned} \tag{28-7}$$

and the subscripts 1 and 2 identify the first and second modulators. The detected signal is then given by

$$\begin{aligned} \frac{I}{I_0} = \frac{1}{4}(M_{00} &+ M_{01}\cos\Delta_1 + M_{01}\sin\Delta_1 + M_{20}\cos\Delta_2 + M_{21}\cos\Delta_1\cos\Delta_2 \\ &+ M_{23}\sin\Delta_1\cos\Delta_2 + M_{30}\sin\Delta_2 + M_{31}\cos\Delta_1\sin\Delta_2 \\ &+ M_{33}\sin\Delta_1\sin\Delta_2). \end{aligned} \tag{28-8}$$

The frequencies ω and phases δ are chosen such that the nine matrix elements are measured by sequential or parallel phase-sensitive detection, i.e., lock-in amplifiers.

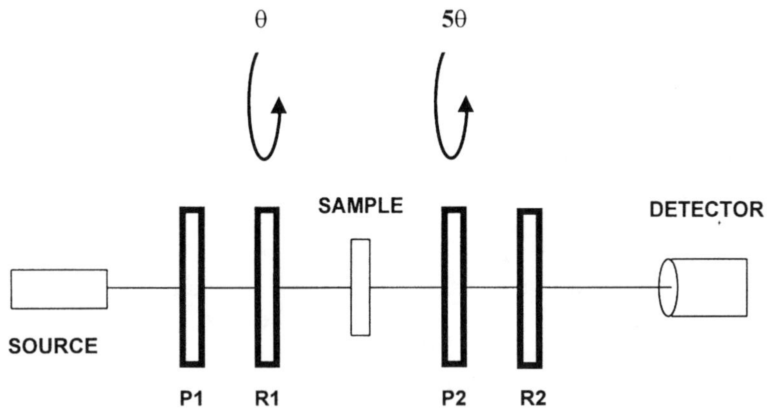

Figure 28-4 Dual rotating-retarder polarimeter. P1 and P2 are polarizers, R1 and R2 are retarders.

One type of complete Mueller matrix polarimeter is represented in Fig. 28-4. This is the dual rotating-retarder polarimeter [2]. It consists of a complete polarimeter as a PSG and a complete polarimeter as a PSA. The retarders are rotated and Fourier analysis is performed on the resulting modulated signal to obtain the Mueller matrix of the sample. In Section 28.2, we will examine this polarimeter in more detail. This dual rotating-retarder method has been implemented as a nonimaging laser polarimeter in order to examine electro-optical samples in transmission [3]. An imaging version of this polarimeter has been constructed to obtain highly resolved polarimetric images of liquid crystal televisions [4] and electro-optic modulators [5]. This same method has been used in the construction of spectropolarimeters to evaluate samples in transmission and reflection [6,7]. In Section 28.3, we will discuss other types of Mueller matrix polarimeters. The polarimetric methods that were discussed in the first part of this book were based on manual methods. The methods described here are all automated and typically depend on computers to collect and process the information.

28.2 DUAL ROTATING-RETARDER POLARIMETRY

This polarimeter configuration is based on a concept originally proposed by Azzam [2], elaborated on by Hauge [8], and by Goldstein [3], and has been used in spectropolarimetry as we shall see [6,7]. The technique has also been used with the sample in reflection to measure birefringence in the human eye at visible wavelengths [9–11]. We have shown in Fig. 28-1 a functional block diagram of a general Mueller matrix polarimeter. The polarimeter has five sections: the source, the polarizing optics, the sample, the analyzing optics, and the detector.

28.2.1 Polarimeter Description

The polarizing optics consist of a fixed linear polarizer and a quarter-wave retarder that rotates. The sample region is followed by the analyzing optics, which consist of a quarter-wave retarder that rotates followed by a fixed linear polarizer. This is shown in Fig. 28-4. One of the great advantages of this configuration is that the polarization sensitivity of the detector is not important because the orientation of the final polarizer is fixed.

The two retarders are rotated at different but harmonic rates, and this results in a modulation of the detected intensity. The Mueller matrix of the sample is found through a relationship between the Fourier coefficients of a series representing the modulation and the elements of the sample matrix.

The second retarder is rotated at least five times the rate of the first, and data might typically be collected for every 2° to 6° of rotation of the first retarder. The stages are stopped completely after each incremental rotation, and an intensity reading is recorded. The resulting data set is a modulated waveform, which is then processed according to the algorithms we shall describe shortly.

The polarizing elements in the polarimeter are required to be aligned with respect to a common axis to start the measurements (this would typically be the axis of the polarized laser or the axis of the first polarizer if an unpolarized source is used). This alignment is done manually to try to minimize orientation errors, and the residual orientation errors are removed through a computational compensation method that we will describe.

28.2.2 Mathematical Development: Obtaining the Mueller Matrix

This polarimeter measures a signal that is modulated by rotating the retarders. The elements of the Mueller matrix are encoded on the modulated signal. The output signal is then Fourier analyzed to determine the Mueller matrix elements. The second retarder is rotated at a rate of five times that of the first. This generates 12 harmonic frequencies in the Fourier spectrum of the modulated intensity.

The Mueller matrix for the system is

$$\mathbf{P}_2\mathbf{R}_2(\theta)\mathbf{M}\mathbf{R}_1(\theta)\mathbf{P}_1 \tag{28-9}$$

where $\mathbf{P}$ indicates a linear polarizer, $\mathbf{R}(\theta)$ indicates an orientation-dependent retarder, and $\mathbf{M}$ is the sample and the matrix quantity to be determined. Mueller matrices are then substituted for a linear retarder with quarter-wave retardation and a fast axis at θ and 5θ for $\mathbf{R}_1$ and $\mathbf{R}_2$, respectively; a horizontal linear polarizer for $\mathbf{P}_2$; a horizontal linear polarizer for $\mathbf{P}_1$; and a sample for $\mathbf{M}$. The detected intensity is given by

$$I = c\mathbf{AMP} \tag{28-10}$$

where $\mathbf{P} = \mathbf{R}_1\mathbf{P}_1\mathbf{S}$ is the Stokes vector of light leaving the polarizing source ($\mathbf{S}$ is the Stokes vector of the light from the source), $\mathbf{A} = \mathbf{P}_2\mathbf{R}_2$ is the Mueller matrix of the analyzing optics, $\mathbf{M}$ is the Mueller matrix of the sample, and c is a proportionality constant obtained from the absolute intensity. Explicitly,

$$I = c\sum_{i,j=1}^{4} a_i p_j m_{ij} \tag{28-11}$$

or

$$I = c\sum_{i,j=1}^{4} \mu_{ij} m_{ij} \tag{28-12}$$

where the a_i are the elements of the first row of **A**, the p_j are the elements of **P**, the m_{ij} are the elements of the Mueller matrix **M**, and where

$$\mu_{ij} = a_i p_j \tag{28-13}$$

The order of matrix multiplication can be changed as shown above in going from (28-10) to (28-11) because we are only measuring intensity, i.e., the first element of the Stokes vector. Only the first row of the matrix **A** is involved in the calculation:

$$\begin{bmatrix} a_1 & a_2 & a_3 & a_4 \\ \bullet & \bullet & \bullet & \bullet \\ \bullet & \bullet & \bullet & \bullet \\ \bullet & \bullet & \bullet & \bullet \end{bmatrix} \begin{bmatrix} m_{11} & m_{12} & m_{13} & m_{14} \\ m_{21} & m_{22} & m_{23} & m_{24} \\ m_{31} & m_{32} & m_{33} & m_{34} \\ m_{41} & m_{42} & m_{43} & m_{44} \end{bmatrix} \begin{bmatrix} p_1 \\ p_2 \\ p_3 \\ p_1 \end{bmatrix} = \begin{bmatrix} I \\ \bullet \\ \bullet \\ \bullet \end{bmatrix} \tag{28-14}$$

and multiplying through:

$$\begin{aligned} I = {} & a_1(m_{11}p_1 + m_{12}p_2 + m_{13}p_3 + m_{14}p_4) \\ & + a_2(m_{21}p_1 + m_{22}p_2 + m_{23}p_3 + m_{24}p_4) \\ & + a_3(m_{31}p_1 + m_{32}p_2 + m_{33}p_3 + m_{34}p_4) \\ & + a_4(m_{41}p_1 + m_{42}p_2 + m_{43}p_3 + m_{44}p_4) = \sum_{i,j=1}^{4} \mu_{ij} m_{ij} \end{aligned} \tag{28-15}$$

When the rotation ratio is $5:1$ the μ_{ij} are given by

$$\begin{aligned} \mu_{11} &= 1 \\ \mu_{12} &= \cos^2 2\theta \\ \mu_{13} &= \sin 2\theta \cos 2\theta \\ \mu_{14} &= \sin 2\theta \\ \mu_{21} &= \cos^2 10\theta \\ \mu_{22} &= \cos^2 2\theta \cos^2 10\theta \\ \mu_{23} &= \sin 2\theta \cos 2\theta \cos^2 10\theta \\ \mu_{24} &= \sin 2\theta \cos^2 10\theta \\ \mu_{31} &= \sin 10\theta \cos 10\theta \\ \mu_{32} &= \cos^2 2\theta \sin 10\theta \cos 10\theta \\ \mu_{33} &= \sin 2\theta \cos 2\theta \sin 10\theta \cos 10\theta \\ \mu_{34} &= \sin 2\theta \sin 10\theta \cos 10\theta \\ \mu_{41} &= -\sin 10\theta \\ \mu_{42} &= -\cos^2 2\theta \sin 10\theta \\ \mu_{43} &= -\sin 2\theta \cos 2\theta \sin 10\theta \\ \mu_{44} &= -\sin 2\theta \sin 10\theta \end{aligned} \tag{28-16}$$

These equations can be expanded in a Fourier series to yield the Fourier coefficients, which are functions of the Mueller matrix elements. The inversion of the these relations gives the Mueller matrix elements in terms of the Fourier coefficients:

$$
\begin{aligned}
m_{11} &= a_0 - a_2 + a_8 - a_{10} + a_{12} \\
m_{12} &= 2a_2 - 2a_8 - 2a_{12} \\
m_{13} &= 2b_2 + 2b_8 - 2b_{12} \\
m_{14} &= b_1 - 2b_{11} = b_1 + 2b_9 = b_1 + b_9 - b_{11} \\
m_{21} &= -2a_8 + 2a_{10} - 2a_{12} \\
m_{22} &= 4a_8 + 4a_{12} \\
m_{23} &= -4b_8 + 4b_{12} \\
m_{24} &= -4b_9 = 4b_{11} = 2(-b_9 + b_{11}) \\
m_{31} &= -2b_8 + 2b_{10} - 2b_{12} \\
m_{32} &= 4b_8 + 4b_{12} \\
m_{33} &= 4a_8 - 4a_{12} \\
m_{34} &= 4a_9 = -4a_{11} = 2(a_9 - a_{11}) \\
m_{41} &= 2b_3 - b_5 = -b_5 + 2b_7 = (b_3 - b_5 + b_7) \\
m_{42} &= -4b_3 = -4b_7 = -2(b_3 + b_7) \\
m_{43} &= -4a_3 = 4a_7 = 2(-a_3 + a_7) \\
m_{44} &= -2a_4 = 2a_6 = (a_6 - a_4)
\end{aligned}
\tag{28-17}
$$

The 5:1 rotation ratio is not the only ratio that can be used to determine Mueller matrix elements, but it is the lowest ratio in which the expressions for the Fourier coefficients may be inverted to give the Mueller matrix elements.

Intensity values in the form of voltages are measured as the retarders are incrementally advanced such that the first retarder is rotated through 180°. The Fourier coefficients must be obtained from the measured intensity values. There are several methods of formulating the solution to this problem.

If the problem is formulated as

$$xa = I, \tag{28-18}$$

where I is a vector of 36 intensity values, a is the set of Fourier coefficients, and x is a 26×25 matrix where each row is of the form:

$$(1 \quad \cos 2\theta \quad \cos 4\theta \quad \ldots \quad \cos 24\theta \quad \sin 2\theta \quad \sin 4\theta \quad \ldots \quad \sin 24\theta)$$

where the θ for each row represents the angle of the fast axis of the first retarder, then the solution is

$$a = (x^T x)^{-1} x^T I \tag{28-19}$$

(The minimum number of equations needed to solve for the coefficients uniquely is 25 so that the maximum rotation increment for the first retarder is 7.2°; for this example, 36 equations are obtained from 5° rotational increments through 180°.) This solution is equivalent to the least-squares solution [12]. In the least-squares formulation the expression for the instrument response is

$$I(\theta) = a_0 + \sum_{j=1}^{12} (a_j \cos 2j\theta + b_j \sin 2j\theta), \tag{28-20}$$

but the actual measurement $\Phi(\theta)$ may be different from this value due to noise and/or error. The sum of the square of these differences may be formed, i.e.,

$$\sum [\Phi(\theta_l) - I(\theta_l)]^2 = E(a_0, a_1, \ldots, a_{12}, b_1, \ldots, b_{12}) \tag{28-21}$$

where E is a function of the coefficients and l is the subscript of the retarder angle. The values of the coefficients can now be found by taking the partial derivative of E with respect to the coefficients and setting these equal to zero:

$$\frac{\partial E}{\partial a_k} = 0, \qquad \frac{\partial E}{\partial b_k} = 0. \tag{28-22}$$

The expression becomes, for the derivative with respect to a_l,

$$\sum_{l=0}^{35} \left[\Phi(\theta_l) - \left[a_0 + \sum_{j=1}^{12} (a_j \cos 2j\theta_l + b_j \sin 2j\theta_l \right] \right] \times (-2 \cos 2k\theta_l) = 0 \tag{28-23}$$

Solving this system of 36 equations in 25 unknowns will give the least-squares solution for the coefficients, which is identical to the solution obtained from (28-19).

28.2.3 Modulated Intensity Patterns

Simulated modulated intensity patterns for no sample and various examples of ideal polarization elements are given in Figs. 28-5 through 28-8. The abscissa represents measurement number in a sequence of 36 (corresponding to 5° increments over 180°) and the ordinate represents detector voltage, normalized to 0.5.

The quality of the measurement and the type of element in the sample position can be recognized by observation of the measured intensity modulation. For example, the pattern of a retarder with its fast axis aligned and one with its slow axis aligned are immediately recognizable and differentiated. Good measurements yield modulated intensity patterns that are essentially identical to the simulations.

28.2.4 Error Compensation

The true nature of the sample may be obscured by errors inherent in the polarimeter optical system. The Mueller matrix elements must be compensated for the known errors in retardance of the retarders and the errors caused by the inability to align the polarizing elements precisely. The fact that there are errors that cannot be eliminated through optical means leads to an error analysis and a compensation procedure to be implemented during polarimeter data processing.

A summary of an error analysis of a dual-rotating retarder Mueller matrix polarimeter is presented in this section. The derivation of the compensated Mueller matrix elements using the small-angle approximation is documented in detail [13],

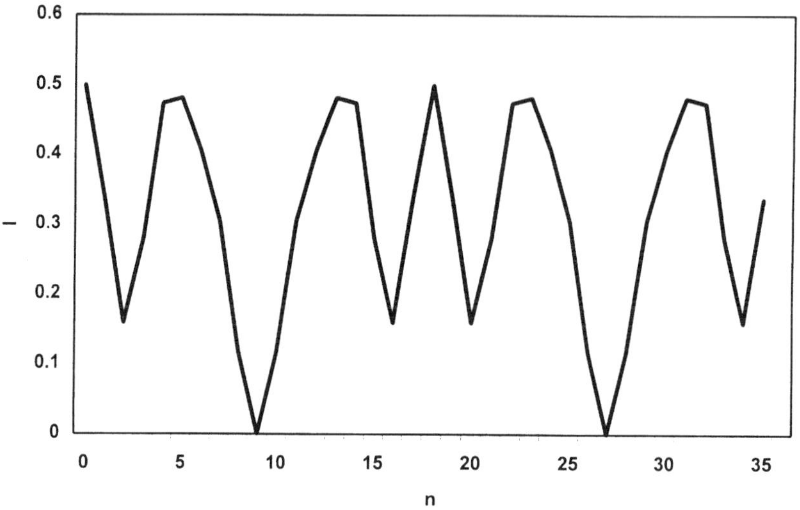

Figure 28-5 Modulated intensity for no sample.

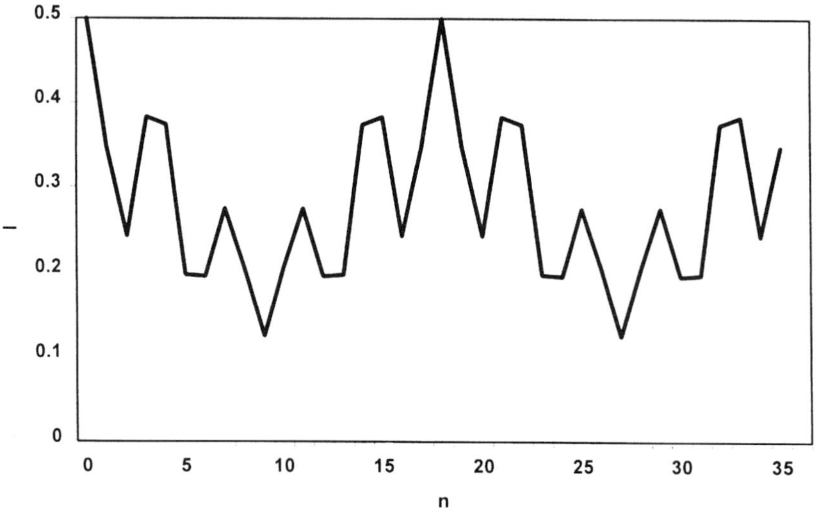

Figure 28-6 Modulated intensity for a linear horizontal polarizer.

and exact compensation equations for the Mueller matrix elements have been derived [14]. Errors in orientational alignment and errors caused by nonideal retardation elements are considered in these compensations. A compensation for imperfect retardation elements is then made possible with the equations derived, and the equations permit a calibration of the polarimeter for the azimuthal alignment of the polarization elements. A similar analysis was done earlier [8] for a dual rotating compensator ellipsometer; however, that analysis did not include errors in the last polarizer but did include errors caused by diattenuation in the retardation elements. Experimental experience with the polarimeter described here

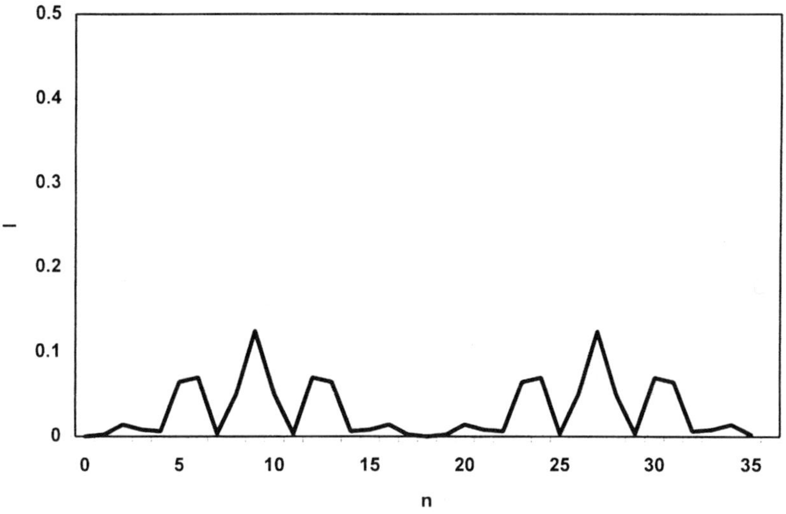

Figure 28-7 Modulated intensity for a linear vertical polarizer.

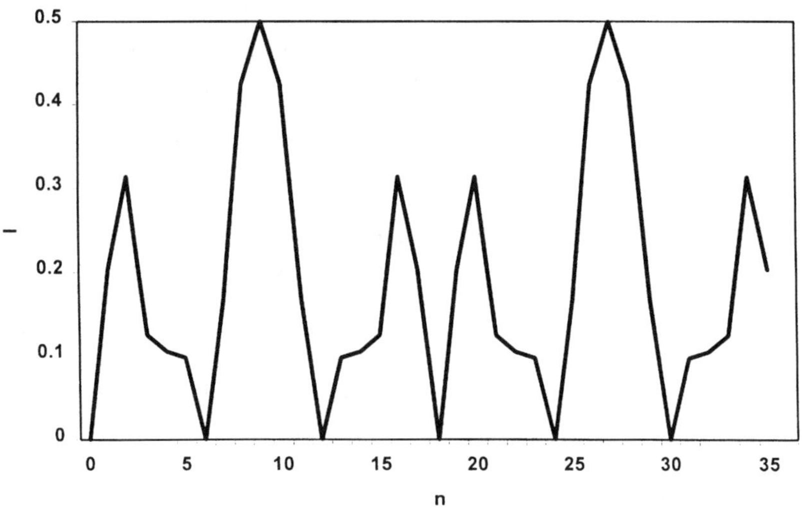

Figure 28-8 Modulated intensity for a half wave plate at 45°.

indicates that the deviation of the retarders from quarter wave is important compared with the diattenuation of the retarders [3].

In the error analysis, the effect of retardation associated with the polarizers and polarization associated with the retarders have not been included. It is also assumed that there are no angular errors associated with the stages that rotate the elements. It is only the relative orientations of the polarizers and retarders that are relevant, and the analysis is simplified by measuring all angles relative to the angle of the polarization from the first polarizer. The errors are illustrated in Fig. 28-9. The three

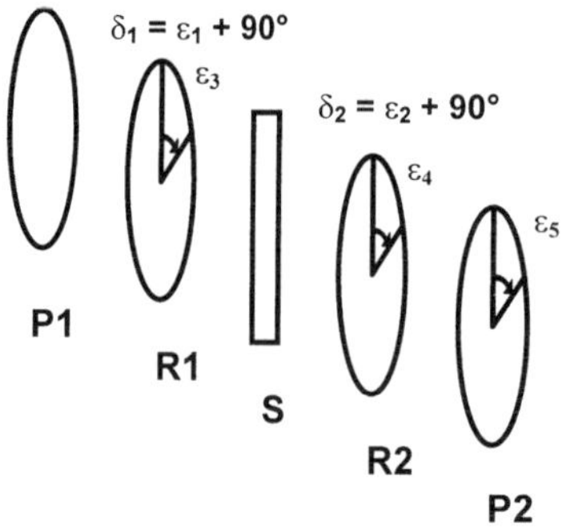

Figure 28-9 Significant error sources in the dual rotating-retarder polarimeter.

polarization elements have errors associated with their initial azimuthal alignment with respect to the first polarizer. These are shown as ε_3, ε_4, and ε_5 in Fig. 28-9. In addition, one or both retarders may have retardances that differ from quarter wave. These are shown as δ_1 and δ_2 where ε_1 and ε_2 are the deviations from quarter wave in Fig. 28-9. In general, both retarders will have different retardances and the three polarization elements will be slightly misaligned in azimuth.

The following calibration procedure is used. First, the polarimeter is operated with no sample and Fourier coefficients obtained from the measured modulated intensity. Second, using error-compensation equations with matrix elements of the identity matrix inserted for the Mueller matrix elements, errors in the element orientations and retardances are calculated. Third, in the routine use of the polarimeter, the systematic errors in the Fourier coefficients arising from the imperfections are compensated for by using the error-compensated equations with experimentally determined error values to obtain the error-compensated sample Mueller matrix elements as a function of measured Fourier coefficients.

With no sample in the polarimeter, the sample matrix is the identity matrix. Because all off-diagonal elements in the sample Mueller matrix are zero, all odd Fourier coefficients in (28-20) become zero. Because the diagonal elements equal one, the coefficients of the twelfth harmonic vanish also.

The Fourier coefficients are found to be functions of the errors, after we find the μ's as in (28-16) but this time as functions of errors. The Fourier coefficients are

$$
\begin{aligned}
a_0 &= \frac{1}{2}m_{11} + \frac{1}{4}\beta_3 m_{12} + \frac{1}{4}\beta_4 \cos 2\varepsilon_5 m_{21} + \frac{1}{8}\beta_3\beta_4 \cos 2\varepsilon_5 m_{22} \\
&\quad + \frac{1}{4}\beta_4 \sin 2\varepsilon_5 m_{31} + \frac{1}{8}\beta_3\beta_4 \sin 2\varepsilon_5 m_{32} \\
a_1 &= \frac{1}{2}\sin\delta_1 \sin 2\varepsilon_3 m_{14} + \frac{1}{4}\beta_4 \sin\delta_1 \sin 2\varepsilon_3 \cos 2\varepsilon_5 m_{24} \\
&\quad + \frac{1}{4}\beta_4 \sin\delta_1 \sin 2\varepsilon_3 \sin 2\varepsilon_5 m_{34}
\end{aligned}
$$

$$
\begin{aligned}
a_2 &= \frac{1}{4}\beta_1 \cos 4\varepsilon_3 m_{12} + \frac{1}{4}\beta_1 \sin 4\varepsilon_3 m_{13} + \frac{1}{8}\beta_1\beta_4 \cos 4\varepsilon_3 \cos 2\varepsilon_5 m_{22} \\
&\quad + \frac{1}{8}\beta_1\beta_4 \sin 4\varepsilon_3 \cos 2\varepsilon_5 m_{23} + \frac{1}{8}\beta_1\beta_4 \cos 4\varepsilon_3 \sin 2\varepsilon_5 m_{32} \\
&\quad + \frac{1}{8}\beta_1\beta_4 \sin 4\varepsilon_3 \sin 2\varepsilon_5 m_{33} \\
a_3 &= -\frac{1}{8}\beta_1 \sin\delta_2 \sin\alpha_3 m_{42} - \frac{1}{8}\beta_1 \sin\delta_2 \cos\alpha_3 m_{43} \\
a_4 &= -\frac{1}{4}\sin\delta_1 \sin\delta_2 \cos\alpha_1 m_{44} \\
a_5 &= \frac{1}{2}\sin\delta_2 \sin\alpha_5 m_{41} + \frac{1}{4}\beta_3 \sin\delta_2 \sin\alpha_5 m_{42} \\
a_6 &= \frac{1}{4}\sin\delta_1 \sin\delta_2 \cos\alpha_2 m_{44} \\
a_7 &= -\frac{1}{8}\beta_1 \sin\delta_2 \sin\alpha_4 m_{42} + \frac{1}{8}\beta_1 \sin\delta_2 \cos\alpha_4 m_{43} \\
a_8 &= \frac{1}{16}\beta_1\beta_2 \cos\alpha_9 (m_{22} + m_{33}) + \frac{1}{16}\beta_1\beta_2 \sin\alpha_9 (m_{32} - m_{23}) \\
a_9 &= \frac{1}{8}\beta_2 \sin\delta_1 \sin\alpha_6 m_{24} + \frac{1}{8}\beta_2 \sin\delta_1 \cos\alpha_6 m_{34} \\
a_{10} &= \frac{1}{4}\beta_2 \cos\alpha_{11} m_{21} + \frac{1}{8}\beta_2\beta_3 \cos\alpha_{11} m_{22} + \frac{1}{4}\beta_2 \sin\alpha_{11} m_{31} \qquad \text{(28-24)} \\
&\quad + \frac{1}{8}\beta_2\beta_3 \sin\alpha_{11} m_{32} \\
a_{11} &= -\frac{1}{8}\beta_2 \sin\delta_1 \sin\alpha_7 m_{24} - \frac{1}{8}\beta_2 \sin\delta_1 \cos\alpha_7 m_{34} \\
a_{12} &= \frac{1}{16}\beta_1\beta_2 \cos\alpha_{10} (m_{22} - m_{33}) + \frac{1}{16}\beta_1\beta_2 \sin\alpha_{10} (m_{23} + m_{32}) \\
b_0 &= 0 \\
b_1 &= \frac{1}{2}\sin\delta_1 \cos 2\varepsilon_3 m_{14} + \frac{1}{4}\beta_4 \sin\delta_1 \cos 2\varepsilon_3 \cos 2\varepsilon_5 m_{24} \\
&\quad + \frac{1}{4}\beta_4 \sin\delta_1 \cos 2\varepsilon_3 \sin 2\varepsilon_5 m_{34} \\
b_2 &= -\frac{1}{4}\beta_1 \sin 4\varepsilon_3 m_{12} + \frac{1}{4}\beta_1 \cos 4\varepsilon_3 m_{13} + \frac{1}{8}\beta_1\beta_4 \cos 4\varepsilon_3 \cos 2\varepsilon_5 m_{23} \\
&\quad - \frac{1}{4}\beta_1\beta_4 \sin 4\varepsilon_3 \cos 2\varepsilon_5 m_{22} \\
&\quad + \frac{1}{8}\beta_1\beta_4 \cos 4\varepsilon_3 \sin 2\varepsilon_5 m_{33} - \frac{1}{8}\beta_1\beta_4 \sin 4\varepsilon_3 \sin 2\varepsilon_5 m_{32} \\
b_3 &= -\frac{1}{8}\beta_1 \sin\delta_2 \cos\alpha_3 m_{42} + \frac{1}{8}\beta_1 \sin\delta_2 \sin\alpha_3 m_{43}
\end{aligned}
$$

$$
\begin{aligned}
b_4 &= \frac{1}{4}\sin\delta_1 \sin\delta_2 \sin\alpha_1 m_{44} \\
b_5 &= -\frac{1}{2}\sin\delta_2 \cos\alpha_5 m_{41} - \frac{1}{4}\beta_3 \sin\delta_2 \cos\alpha_5 m_{42} \\
b_6 &= -\frac{1}{4}\sin\delta_1 \sin\delta_2 \sin\alpha_2 m_{44} \\
b_7 &= -\frac{1}{8}\beta_1 \sin\delta_2 \cos\alpha_4 m_{42} - \frac{1}{8}\beta_1 \sin\delta_2 \sin\alpha_4 m_{43} \\
b_8 &= -\frac{1}{16}\beta_1\beta_2 \sin\alpha_9(m_{22} + m_{33}) - \frac{1}{16}\beta_1\beta_2 \cos\alpha_9(m_{23} - m_{32}) \\
b_9 &= -\frac{1}{8}\beta_2 \sin\delta_1 \cos\alpha_6 m_{24} + \frac{1}{8}\beta_2 \sin\delta_1 \sin\alpha_6 m_{34} \\
b_{10} &= -\frac{1}{4}\beta_2 \sin\alpha_{11} m_{21} - \frac{1}{8}\beta_2\beta_3 \sin\alpha_{11} m_{22} + \frac{1}{4}\beta_2 \cos\alpha_{11} m_{31} \\
&\quad + \frac{1}{8}\beta_2\beta_3 \cos\alpha_{11} m_{32} \\
b_{11} &= \frac{1}{8}\beta_2 \sin\delta_1 \cos\alpha_7 m_{24} - \frac{1}{8}\beta_2 \sin\delta_1 \sin\alpha_7 m_{34} \\
b_{12} &= -\frac{1}{16}\beta_1\beta_2 \sin\alpha_{10}(m_{22} - m_{33}) + \frac{1}{16}\beta_1\beta_2 \cos\alpha_{10}(m_{23} + m_{32})
\end{aligned}
$$

where

$$
\begin{aligned}
\beta_1 &= 1 - \cos\delta_1 \\
\beta_2 &= 1 - \cos\delta_2 \\
\beta_3 &= 1 + \cos\delta_1 \\
\beta_4 &= 1 + \cos\delta_2 \\
\alpha_1 &= 2\varepsilon_4 - 2\varepsilon_3 - 2\varepsilon_5 \\
\alpha_2 &= 2\varepsilon_4 + 2\varepsilon_3 - 2\varepsilon_5 \\
\alpha_3 &= 2\varepsilon_4 - 4\varepsilon_3 - 2\varepsilon_5 \\
\alpha_4 &= 2\varepsilon_4 + 4\varepsilon_3 - 2\varepsilon_5 \\
\alpha_5 &= 2\varepsilon_5 - 2\varepsilon_4 \\
\alpha_6 &= 2\varepsilon_5 - 4\varepsilon_4 + 2\varepsilon_3 \\
\alpha_7 &= 2\varepsilon_5 - 4\varepsilon_4 - 2\varepsilon_3 \\
\alpha_8 &= -2\varepsilon_5 + 4\varepsilon_4 - 2\varepsilon_3 = -\alpha_6 \\
\alpha_9 &= 4\varepsilon_4 - 4\varepsilon_3 - 2\varepsilon_5 \\
\alpha_{10} &= 4\varepsilon_4 + 2\varepsilon_3 - 2\varepsilon_5 \\
\alpha_{11} &= 4\varepsilon_4 - 2\varepsilon_5
\end{aligned}
\tag{28-25}
$$

These equations can be inverted for this case where there is no sample, so that the sample Mueller matrix is the identity matrix, and we then solve for the errors in terms of the Fourier coefficients. The equations yield the errors as

$$
\begin{aligned}
\varepsilon_3 &= \frac{1}{4}\tan^{-1}\left(\frac{b_8}{a_8}\right) - \frac{1}{4}\tan^{-1}\left(\frac{b_{10}}{a_{10}}\right) \\
\varepsilon_4 &= \frac{1}{2}\tan^{-1}\left(\frac{b_2}{a_2}\right) - \frac{1}{2}\tan^{-1}\left(\frac{b_6}{a_6}\right) + \frac{1}{4}\tan^{-1}\left(\frac{b_8}{a_8}\right) - \frac{1}{4}\tan^{-1}\left(\frac{b_{10}}{a_{10}}\right) \\
\varepsilon_5 &= \frac{1}{2}\tan^{-1}\left(\frac{b_2}{a_2}\right) + \frac{1}{2}\tan^{-1}\left(\frac{b_8}{a_8}\right) - \frac{1}{2}\tan^{-1}\left(\frac{b_{10}}{a_{10}}\right) \\
\delta_1 &= \cos^{-1}\left(\frac{a_{10}\cos\alpha_9 - a_8\cos\alpha_{11}}{a_{10}\cos\alpha_9 + a_8\cos\alpha_{11}}\right) \\
\delta_2 &= \cos^{-1}\left(\frac{a_2\cos\alpha_9 - a_8\cos(4\varepsilon_3 - 2\varepsilon_5)}{a_2\cos\alpha_9 + a_8\cos(4\varepsilon_3 - 2\varepsilon_5)}\right)
\end{aligned}
\tag{28-26}
$$

These values for the errors found from the calibration are now to be substituted back into the equations for the Mueller matrix elements by using measured values of the Fourier coefficients with a sample in place:

$$
\begin{aligned}
m_{44} &= \frac{4}{\sin\delta_1 \sin\delta_2}\left(-\frac{a_4}{\cos\alpha_1} + \frac{a_6}{\cos\alpha_2}\right) \\
m_{43} &= 8\frac{-a_3\cos\alpha_3 + b_3\sin\alpha_3 + a_7\cos\alpha_4 - b_7\sin\alpha_4}{\beta_1\sin\delta_2} \\
m_{42} &= -8\frac{a_3\sin\alpha_3 + b_3\cos\alpha_3 + a_7\sin\alpha_4 + b_7\cos\alpha_4}{\beta_1\sin\delta_2} \\
m_{41} &= \frac{-\beta_3 m_{42}}{2} - \frac{4b_5}{\cos\alpha_5\sin\delta_2} \\
m_{24} &= 8\frac{a_9\sin\alpha_6 - b_9\cos\alpha_6 - a_{11}\sin\alpha_7 + b_{11}\cos\alpha_7}{\beta_2\sin\delta_1} \\
m_{34} &= 8\frac{a_9\cos\alpha_6 + b_9\sin\alpha_6 - a_{11}\cos\alpha_7 - b_{11}\sin\alpha_7}{\beta_2\sin\delta_1} \\
m_{14} &= \frac{-\beta_4\cos 2\varepsilon_5 m_{24}}{2} + \frac{4b_1}{\cos 2\varepsilon_3\sin\delta_1} - \frac{\beta_4\sin 2\varepsilon_5 m_{34}}{2} \\
m_{22} &= 16\frac{a_8\cos\alpha_9 + a_{12}\cos\alpha_{10} - b_8\sin\alpha_9 - b_{12}\sin\alpha_{10}}{\beta_1\beta_2} \\
m_{33} &= 16\frac{a_8\cos\alpha_9 - a_{12}\cos\alpha_{10} - b_8\sin\alpha_9 + b_{12}\sin\alpha_{10}}{\beta_1\beta_2} \\
m_{23} &= 16\frac{-a_8\sin\alpha_9 + a_{12}\sin\alpha_{10} - b_8\cos\alpha_9 + b_{12}\cos\alpha_{10}}{\beta_1\beta_2}
\end{aligned}
\tag{28-27}
$$

$$m_{32} = 16\frac{a_8 \sin\alpha_9 + a_{12}\sin\alpha_{10} + b_8\cos\alpha_9 + b_{12}\cos\alpha_{10}}{\beta_1\beta_2}$$

$$m_{12} = \frac{16a_2\cos 4\varepsilon_3 - 16b_2\sin 4\varepsilon_3 - \beta_1\beta_4\cos 2\varepsilon_5 m_{22} - \beta_1\beta_4\sin 2\varepsilon_5 m_{32}}{2\beta_1}$$

$$m_{13} = \frac{16a_2\sin 4\varepsilon_3 + 16b_2\cos 4\varepsilon_3 - \beta_1\beta_4\cos 2\varepsilon_5 m_{23} - \beta_1\beta_4\sin 2\varepsilon_5 m_{33}}{2\beta_1}$$

$$m_{21} = \frac{16a_{10}\cos\alpha_{11} - 16b_{10}\sin\alpha_{11} - \beta_2\beta_3 m_{22}}{2\beta_2}$$

$$m_{31} = \frac{-(\beta_2\beta_3 m_{32} - 16b_{10}\cos\alpha_{11} - 16a_{10}\sin\alpha_{11})}{2\beta_2}$$

$$m_{11} = 4a_0 - \frac{1}{2}\beta_3 m_{12} - \frac{1}{2}\beta_4\cos 2\varepsilon_5 m_{21} - \frac{1}{4}\beta_3\beta_4\cos 2\varepsilon_5 m_{22} - \frac{1}{2}\beta_4\sin 2\varepsilon_5 m_{31} - \frac{1}{4}\beta_3\beta_4\sin 2\varepsilon_5 m_{32}$$

28.2.5 Optical Properties from the Mueller Matrix

One objective of Mueller matrix polarimetry might be to obtain electro- and magneto-optic coefficients of crystals. The coefficients are derived from the Mueller matrices measured as a function of applied field strength. The method by which this derivation is accomplished is briefly summarized here [15].

The application of an electric field across a crystal produces an index change. Principal indices are obtained by solving an eigenvalue problem (see Chapter 24). For example, for a $\bar{4}3m$ cubic material with index n_0 and with a field E perpendicular to the (110) plane, the index ellipsoid is

$$\frac{x^2 + y^2 + z^2}{n_0^2} + \sqrt{2}r_{41}E(yz + zx) = 1 \tag{28-28}$$

The eigenvalue problem is solved, and the roots of the secular equation are the new principal indices:

$$\begin{aligned} n'_x &= n_0 + \frac{1}{2}n_0^3 r_{41}E \\ n'_y &= n_0 - \frac{1}{2}n_0^3 r_{41}E \\ n'_z &= n_0 \end{aligned} \tag{28-29}$$

The principal indices of the $\bar{4}3m$ cubic material for an electric field applied transversely and longitudinally are given by Namba [16].

The phase retardation accumulated by polarized light in traversing a medium with anisotropic properties is given by

$$\Gamma = 2\pi(n_a - n_b)L/\lambda \tag{28-30}$$

where L is the medium thickness in the direction of propagation, λ is the wavelength of light, and n_a, n_b are the indices experienced in two orthogonal directions perpendicular to the direction of propagation. In the longitudinal mode of operation, the electric field and propagation direction are both along the z axis. The refractive indices experienced by the light are in the plane containing the x and y principal axes. If the light polarization and crystal are aligned such that the polarization is 45° from either principal axis, the phase retardation will be

$$\Gamma = 2\pi(n'_y - n'_x)L/\lambda \tag{28-31}$$

where n'_y, n'_x are the (new) principal indices with the field applied. (For crystals with natural birefringence and no electric field, these indices may just be the principal indices.)

The phase delays for light polarized at 45° to the principal axes of the $\bar{4}3m$ material can now be calculated. The phase retardation for the $\bar{4}$3m cubic material is

$$\Gamma_{\text{cubic}} = 2\pi n_0^3 r_{41} EL/\lambda \tag{28-32}$$

If the electric field is expressed in terms of electric potential and charge separation, i.e., $E = V/d$, then the phase retardation is

$$\Gamma_{\text{cubic}}^{\text{long}} = 2\pi n_0^3 r_{41} V/\lambda \tag{28-33}$$

because the charge separation d is equal to the optical path through the crystal.

The phase retardation for $\bar{4}3m$ cubic material in the transverse mode is also given by (28-26). In the transverse mode the charge separation is not the same as the optical path so that when E is given as V/d, the phase delay is given as

$$\Gamma_{\text{cubic}}^{\text{trans}} = 2\pi n_0^3 r_{41} VL/d\lambda \tag{28-34}$$

The cubic crystal described is expected to act as a linear retarder. The Mueller matrix formalism representation of a retarder with a fast axis at arbitrary orientation angle θ is

$$\begin{bmatrix} 1 & 0 & 0 & 0 \\ 0 & \cos^2 2\theta + \sin^2 2\theta \cos\delta & (1-\cos\delta)\sin 2\theta\cos 2\theta & -\sin 2\theta \sin\delta \\ 0 & (1-\cos\delta)\sin 2\theta \cos 2\theta & \sin^2 2\theta + \cos^2 2\theta\cos\delta & \cos 2\theta \sin\delta \\ 0 & \sin 2\theta \sin\delta & -\cos 2\theta\sin\delta & \cos\delta \end{bmatrix} \tag{28-35}$$

where the retardance is δ. If the retarder fast axis is assumed to be at 0°, the matrix becomes, substituting for δ the retardance of the crystal,

$$\begin{bmatrix} 1 & 0 & 0 & 0 \\ 0 & 1 & 0 & 0 \\ 0 & 0 & \cos\frac{2\pi}{\lambda} n^3 r_{41} V\frac{L}{d} & \sin\frac{2\pi}{\lambda} n^3 r_{41} V\frac{L}{d} \\ 0 & 0 & -\sin\frac{2\pi}{\lambda} n^3 r_{41} V\frac{L}{d} & \cos\frac{2\pi}{\lambda} n^3 r_{41} V\frac{L}{d} \end{bmatrix} \tag{28-36}$$

It is now clear that the electro-optic coefficient r_{42} can be obtained from the measured Mueller matrix.

Note that for purposes of obtaining the electro-optic coefficient experimentally, the fast axis of an electro-optic crystal acting as an ideal retarder can be at any orientation. The (4,4) matrix element of the matrix for a retarder with the fast axis at angle θ is independent of fast-axis orientation, and the fast-axis orientation can be eliminated elsewhere by adding the (2,2) and (3,3) matrix elements or squaring and adding elements in the fourth row and column. Given a measured Mueller matrix of a crystal, a known applied voltage, and a known refractive index, one can easily obtain the electro-optic coefficient r_{41}.

28.2.6 Measurements

As an example of a calibration measurement and compensation, the ideal and measured Mueller matrices for a calibration (no sample) are, respectively,

$$\begin{bmatrix} 1 & 0 & 0 & 0 \\ 0 & 1 & 0 & 0 \\ 0 & 0 & 1 & 0 \\ 0 & 0 & 0 & 1 \end{bmatrix}$$

$$\begin{bmatrix} 0.998 & 0.026 & 0.019 & -0.002 \\ 0.002 & 0.976 & -0.030 & 0.009 \\ 0.007 & 0.033 & 0.966 & -0.002 \\ 0.002 & -0.004 & -0.002 & 1.000 \end{bmatrix}$$

The measured results, normalized to unity, are given without any error compensation. The measured matrix is clearly recognizable as a noisy representation of the corresponding ideal matrix.

Error compensation may be demonstrated with the experimental calibration Mueller matrix. The source of the large error for the two middle elements of the diagonal is the retardance errors of the wave plates. Using calculated values for the errors and compensation by using the small-angle approximation error analysis as discussed above [13], one sees that the renormalized compensated Mueller matrix for no sample becomes

$$\begin{bmatrix} 0.997 & -0.006 & 0.004 & 0.002 \\ 0.007 & 1.000 & -0.007 & 0.009 \\ 0.008 & -0.007 & 0.990 & -0.003 \\ 0.003 & -0.006 & -0.007 & 0.998 \end{bmatrix}$$

Equations for the exact error compensation give slightly better results.

28.2.7 Spectropolarimetry

Spectropolarimetry is the measurement of both spectral and polarization information. A spectropolarimeter has been described [7] based on a Fourier transform infrared (FTIR) spectrometer with the dual rotating-retarder polarimeter described previously. An optical diagram of this instrument, based on a Nicolet 6000 FTIR spectrometer, is given in Fig. 28-10 and shows the complete polarimeter within the sample compartment. The spectrometer performs the normal spectral scanning, and after a scan period the dual rotating retarder changes to a new rotational position. This continues, as described in the previous section, until all polarization information is collected. The data are then reduced to produce a Mueller matrix for each

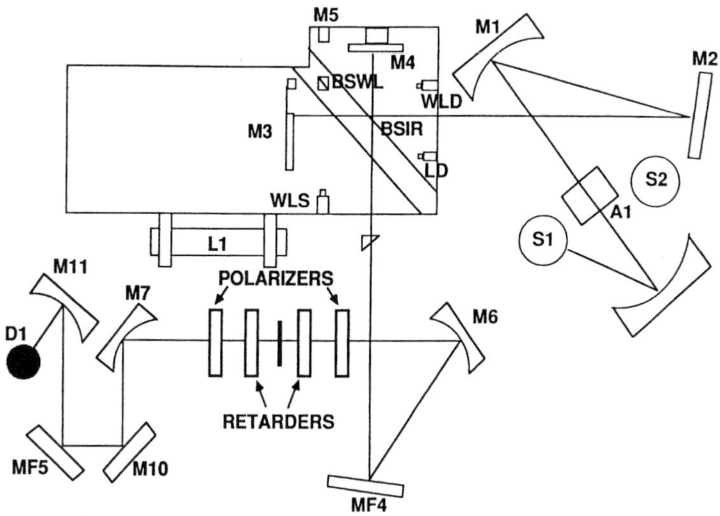

Figure 28-10 Optical diagram of a spectropolarimeter based on a Fourier-transform infrared spectrometer. L1 is a laser, S1 and S2 are sources, D1 is the detector, elements starting with M are mirrors, elements starting with BS are beam splitters, WLS and WLD are white light source and white light detector, LD is the laser detector, and BSIR is the infrared beam splitter.

wavelength of the FTIR spectrometer scan. This spectropolarimeter has been used to analyze polarization properties of optical samples in reflection and transmission.

Spectropolarimetry requires polarization elements that are achromatic across a spectral region of the data collection. Polarizers that are achromatic are generally more readily available than achromatic retarders. For the infrared (2–25 μm), wire-grid polarizers are achromatic over large ranges within this region, although their diattenuation performance is not generally as good as that of prism polarizers. Achromatic waveplates have been designed that are achromatic over wavelength ranges somewhat smaller than the polarizers, and these custom elements can be expensive and they have achromatic performance poorer than that of the polarizers. Fortunately, the compensation techniques described in the last section apply to this problem, and are used to great advantage to correct for the imperfect achromaticity of the retarders.

28.2.8 The Measurement Matrix Method

An alternative to the Fourier method described above is the measurement matrix method (see Ref. 17). Similar to (28-14), we have

$$\begin{bmatrix} a_{q,1} & a_{q,2} & a_{q,3} & a_{q,4} \\ \bullet & \bullet & \bullet & \bullet \\ \bullet & \bullet & \bullet & \bullet \\ \bullet & \bullet & \bullet & \bullet \end{bmatrix} \begin{bmatrix} m_{11} & m_{12} & m_{13} & m_{14} \\ m_{21} & m_{22} & m_{23} & m_{24} \\ m_{31} & m_{32} & m_{33} & m_{34} \\ m_{41} & m_{42} & m_{43} & m_{44} \end{bmatrix} \begin{bmatrix} p_{q,1} \\ p_{q,2} \\ p_{q,2} \\ p_{q,2} \end{bmatrix} = \begin{bmatrix} I_q \\ \bullet \\ \bullet \\ \bullet \end{bmatrix}$$

$$= \sum_{j=0}^{3} \sum_{k=0}^{3} a_{q,j} m_{j,k} s_{q,k} \tag{28-37}$$

for the qth measurement at the qth position of the PSG and PSA. We now write the Mueller matrix as a 16×1 vector:

$$\mathbf{M} = \begin{bmatrix} m_{00} & m_{01} & m_{02} & m_{03} & m_{10} & \ldots & m_{33} \end{bmatrix}^T \tag{28-38}$$

We also define a 16×1 measurement vector for the qth measurement as

$$\begin{aligned}\mathbf{W}_q &= \begin{bmatrix} w_{q,00} & w_{q,01} & w_{q,02} & w_{q,03} & w_{q,10} & \cdots & w_{q,33} \end{bmatrix}^T \\ &= \begin{bmatrix} a_{q,0}s_{q,0} & a_{q,0}s_{q,1} & a_{q,0}s_{q,2} & a_{q,0}s_{q,3} & a_{q,1}s_{q,0} & \cdots & a_{q,3}s_{q,3} \end{bmatrix}^T\end{aligned} \tag{28-39}$$

The qth measurement is then the dot product of $\mathbf{M}$ and $\mathbf{W}$:

$$I_q = \mathbf{W}_q \bullet \mathbf{M} = \begin{bmatrix} a_{q,0}s_{q,0} & a_{q,0}s_{q,1} & a_{q,0}s_{q,2} & a_{q,0}s_{q,3} & a_{q,1}s_{q,0} & \cdots & a_{q,3}s_{q,3} \end{bmatrix} \begin{bmatrix} m_{00} \\ m_{01} \\ m_{02} \\ m_{03} \\ m_{10} \\ \vdots \\ m_{33} \end{bmatrix} \tag{28-40}$$

We make a set of Q measurements so that we obtain a $Q\times 16$ matrix where the qth row is the measurement vector $\mathbf{W}_q$. The measurement equation relates the measurement vector $\mathbf{I}$ to the sample Mueller vector:

$$\mathbf{I} = \mathbf{W}\mathbf{M} = \begin{bmatrix} I_0 \\ I_1 \\ \vdots \\ I_{Q-1} \end{bmatrix} = \begin{bmatrix} w_{0,00} & w_{0,01} & \cdots & w_{0,33} \\ w_{1,00} & w_{1,01} & \cdots & w_{1,33} \\ \vdots & & & \\ w_{Q-1,00} & w_{Q-1,01} & \cdots & w_{Q-1,33} \end{bmatrix} \begin{bmatrix} m_{00} \\ m_{01} \\ \vdots \\ m_{33} \end{bmatrix} \tag{28-41}$$

If $\mathbf{W}$ contains 16 linearly independent columns, all 16 elements of the Mueller matrix can be determined. If $Q = 16$, then the matrix inverse is unique and the Mueller matrix elements are determined from the data-reduction equation:

$$\mathbf{M} = \mathbf{W}^{-1}\mathbf{P} \tag{28.42}$$

If more than 16 measurements are made, which is usually the case, $\mathbf{M}$ is overdetermined, although now $\mathbf{W}$ may not have a unique inverse. The optimal polarimetric data-reduction equation is equivalent to a least-squares solution.

28.3 OTHER MUELLER MATRIX POLARIMETRY METHODS

Other polarimetric methods have been used to obtain Mueller matrices. We describe three of them in this section.

28.3.1 Modulator-Based Mueller Matrix Polarimeter

Another class of polarimeters has been designed using electro-optical modulators. Thompson et al. [18] describe a polarimeter for scattering measurements, which uses four modulators. These modulators are Pockels cells made of potassium dideuterium

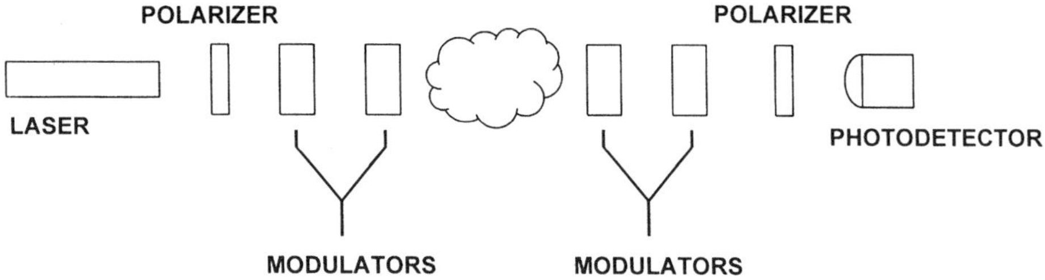

Figure 28-11 Functional diagram of the four-modulator polarimeter.

phosphate (KD*P). A functional diagram of this four-modulator polarimeter is shown in Fig. 28-11.

All elements of the Mueller matrix are measured simultaneously in this polarimeter. The polarizers are aligned and fixed in position. The four Pockels cells are driven at four different frequencies. The normalized Stokes vector after the first polarizer is

$$S_0 = \begin{bmatrix} 1 \\ 1 \\ 0 \\ 0 \end{bmatrix} \tag{28-43}$$

so that the Stokes vector at the detector is

$$S_f = I_0(P_2 M_4 M_3 F M_2 M_1) S_0 \tag{28-44}$$

where M_1, M_2, M_3, and M_4 are the modulator Mueller matrices, P_2 is the second polarizer matrix, I_0 is the initial intensity, and F is the sample matrix. The intensity at the detector is the first element of this vector and is given by

$$\begin{aligned} I_f = \frac{I_0}{2}(& f_{11} + f_{12}\cos\delta_1 + f_{13}\sin\delta_1\sin\delta_2 - f_{14}\sin\delta_1\cos\delta_2 + f_{21}\cos\delta_4 + f_{22}\cos\delta_1\cos\delta_4 \\ & + f_{23}\sin\delta_1\sin\delta_2\cos\delta_4 - f_{24}\sin\delta_1\cos\delta_2\cos\delta_4 + f_{31}\sin\delta_3\sin\delta_4 + f_{32}\cos\delta_1\sin\delta_3\sin\delta_4 \\ & + f_{33}\sin\delta_1\sin\delta_2\sin\delta_3\sin\delta_4 - f_{34}\sin\delta_1\cos\delta_2\sin\delta_3\sin\delta_4 + f_{41}\cos\delta_3\sin\delta_4 \\ & + f_{42}\cos\delta_1\cos\delta_3\sin\delta_4 + f_{43}\sin\delta_1\sin\delta_2\cos\delta_3\sin\delta_4 - f_{44}\sin\delta_1\cos\delta_2\cos\delta_3\sin\delta_4) \end{aligned} \tag{28-45}$$

where the f_{ij} are the elements of the sample matrix and δ_1, δ_2, δ_3, and δ_4 are the retardances of the four modulators. The retardances of the modulators are driven by oscillators at different frequencies so that they are

$$\delta_i = \delta_{oi}\cos\omega_i t \tag{28-46}$$

where δ_{oi} is the amplitude of the retardance of the ith retarder. The trigonometric functions in the oscillating retardances are expanded in terms of Bessel functions of the retardation amplitudes, these results are substituted into the expression for the intensity, and the Fourier expansion of the coefficients of the f_{ij} is taken. The primary frequencies at which each matrix element occurs are

$$\begin{array}{cccc} 0 & 2\omega_1 & \omega_1 \pm \omega_2 & \omega_1 \pm 2\omega_2 \\ 2\omega_4 & 2\omega_1 \pm 2\omega_4 & \omega_1 \pm \omega_2 \pm 2\omega_4 & \omega_1 \pm 2\omega_2 \pm 2_4 \\ \omega_3 \pm \omega_4 & 2\omega_1 \pm \omega_3 \pm \omega_4 & \omega_1 \pm \omega_2 \pm \omega_3 \pm \omega_4 & \omega_1 \pm 2\omega_2 \pm \omega_3 \pm \omega_4 \\ 2\omega_3 \pm \omega_4 & 2\omega_1 \pm 2\omega_3 \pm \omega_4 & \omega_1 \pm \omega_2 \pm 2\omega_3 \pm \omega_4 & \omega_1 \pm 2\omega_2 \pm 2\omega_3 \pm \omega_4 \end{array} \quad (28\text{-}47)$$

The modulation frequencies are chosen so that there are unique frequencies of signal corresponding to each matrix element. Lock-in amplifiers for these frequencies are used in the detector electronics.

Initial alignment of the modulators with the polarization direction is not perfect, and the foregoing analysis can be repeated with a constant retardation error for each modulator. This results in somewhat more complex expressions for the characteristic frequencies for the matrix elements. A calibration procedure minimizes the errors due to misalignment. An accuracy of 1% is said to be attainable with iterative calibration.

28.3.2 Mueller Matrix Scatterometer

The scatter of light reflected from a surface into the sphere surrounding the point of incidence is measured in order to understand reflection properties of the surface. The effect of polarization in the reflection process can be measured with a Mueller matrix scatterometer, described by Schiff et al [19]. The sample is mounted on a goniometer so that in-plane or out-of-plane measurements may be made. There are optics associated with the source (PSG) and receiver (PSA) that allow complete polarization control, shown in Fig. 28-12. The source optics consist of a linearly polarized laser source, a half-wave plate to control orientation of the linear polarization, and a quarter-wave retarder. The receiver optics consist of a quarter-wave retarder and a linear polarizer.

The power measured by the detector is given by

$$\begin{array}{ccccc} & & 1\times 4 & 4\times 4 & 4\times 1 \\ P_0 & = & [r] & [M] & [s]P_i \\ & & \text{Rec} & \text{Sample} & \text{Source} \end{array} \quad (28\text{-}48)$$

where P_i is the input power from the laser, vector s is the (normalized) source optics Stokes vector, M is the sample matrix, and r is basically the top row of the Mueller matrix for the receiving optics. In order to measure M, the source optics are set so that six Stokes vectors are produced corresponding to the normalized Stokes vectors for linear horizontal, linear vertical, $\pm 45^\circ$ linear, and right and left circularly polarized light, i.e., S_1, S_2, and S_3 are set to ± 1, one at a time. The PSA is set to these six polarization states for each of the six states of the PSG to produce 36 measurements. Expressing this in matrix form we have

$$\begin{array}{ccccc} 6\times 6 & & 6\times 4 & 4\times 4 & 4\times 6 \\ [P_0] & = & [R] & [M] & [S]P_i \\ & & \text{Rec} & \text{Sample} & \text{Source} \end{array} \quad (28\text{-}49)$$

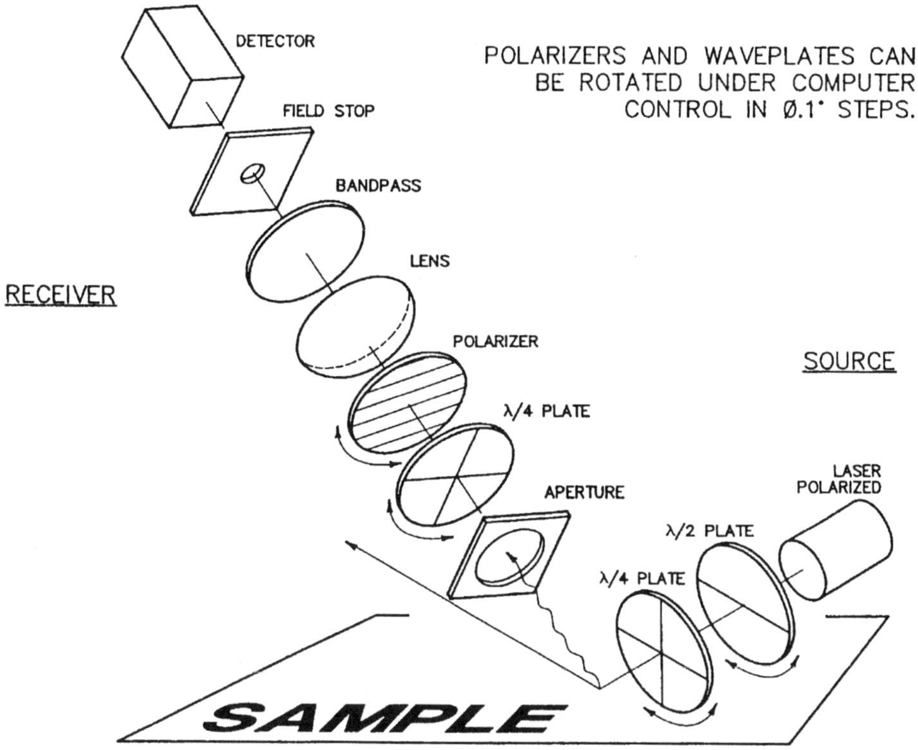

Figure 28-12 Diagram of a Mueller matrix scatterometer. (From Ref. 19.)

A calibration must be performed to compensate for errors, since there are multiple error sources that will not allow the production of ideal polarization states. The R and S matrices above give 48 unknowns. A measurement is made with no sample to give 36 values of P_0, and 12 more equations are obtained from the quadrature relations associated with the overdefinition of the Stokes vectors. This comprises a system of 48 equations and 48 unknowns. Solving these produces the matrices $[S]$ and $[R]$, and now measurements of P_0 can be made with a sample in place, and the matrix M can be calculated from

$$[M] = \left([R]^T[R]\right)^{-1}[R]^T[P_0][S]^T\left([S][S]^T\right)^{-1}\left(\frac{1}{P_i}\right) \tag{28-50}$$

28.3.3 Four-Detector Photopolarimeter

The four-detector photopolarimeter was described in Chapter 27. It is a complete Stokes polarimeter. A Mueller matrix polarimeter is constructed by using a four-detector photopolarimeter as the PSA and a conventional polarizer—quarter-wave retarder pair as a PSG. The polarizer is set at some fixed azimuth, and the output signal (a four-element vector) from the four-detector photopolarimeter is recorded as a function of the azimuth of the fast axis of the quarter-wave retarder. The signal is subject to Fourier analysis to yield a limited series whose vectorial coefficients determine the columns of the measured Mueller matrix.

Calibration of the instrument is required and takes place with no sample present. The optical elements are aligned so that light is directed straight through. The fast axis of the quarter-wave retarder is aligned with the fixed polarizer by adjusting it in small steps until S_3 from the four-detector photopolarimeter is 0. After the light passes through the quarter-wave plate, the Stokes vector is

$$\mathbf{S}(\theta) = \begin{bmatrix} 1 + g\cos 2\theta \\ (1-f) + g\cos 2\theta + f\cos 4\theta \\ g\sin 2\theta + f\sin 4\theta \\ \sin 2\theta \end{bmatrix}$$
$$= \mathbf{S}_0 + \mathbf{S}_{1c}\cos 2\theta + \mathbf{S}_{1s}\sin 2\theta + \mathbf{S}_{2c}\cos 4\theta + \mathbf{S}_{2s}\sin 4\theta \quad (28\text{-}51)$$

where θ is the retarder azimuth, and f and g are characteristic of the quarter wave retarder and where

$$\mathbf{S}_0 = \begin{bmatrix} 1 \\ (1-f) \\ 0 \\ 0 \end{bmatrix}, \quad \mathbf{S}_{1c} = \begin{bmatrix} g \\ g \\ 0 \\ 0 \end{bmatrix}, \quad \mathbf{S}_{1s} = \begin{bmatrix} 0 \\ 0 \\ g \\ 1 \end{bmatrix}, \quad \mathbf{S}_{2c} = \begin{bmatrix} 0 \\ f \\ 0 \\ 0 \end{bmatrix}, \quad \mathbf{S}_{2s} = \begin{bmatrix} 0 \\ 0 \\ f \\ 0 \end{bmatrix} \quad (28\text{-}52)$$

The values of f and g are determined by a rotating quarter-wave test [20]. The value of g is the diattenuation of the quarter-wave retarder, and $2f-1$ is the retardance error from quarter wave in radians.

The output vector of the four-detector polarimeter with a sample in position is

$$\mathbf{I}(\theta) = \mathbf{AMS}(\theta) \quad (28\text{-}53)$$

where $\mathbf{M}$ is the sample Mueller matrix and $\mathbf{A}$ is the instrument calibration matrix [20]. Using $\mathbf{S}(\theta)$ from (28-52) gives a Fourier series for $I(\theta)$ of the same composition as $\mathbf{S}(\theta)$ with vectorial coefficients given by

$$\begin{aligned} \mathbf{I}_0 &= \mathbf{A}[\mathbf{C}_{1M} + (1-f)\mathbf{C}_{2M}] \\ \mathbf{I}_{1c} &= g\mathbf{A}[\mathbf{C}_{1M} + \mathbf{C}_{2M}] \\ \mathbf{I}_{1s} &= \mathbf{A}[g\mathbf{C}_{3M} + \mathbf{C}_{4M}] \\ \mathbf{I}_{2c} &= f\mathbf{AC}_{2M} \\ \mathbf{I}_{2s} &= f\mathbf{AC}_{3M} \end{aligned} \quad (28\text{-}54)$$

where $\mathbf{C}_{1M}$, $\mathbf{C}_{2M}$, $\mathbf{C}_{3M}$, and $\mathbf{C}_{4M}$ are the columns of the Mueller matrix $\mathbf{M}$. These columns are then given by

$$\begin{aligned} \mathbf{C}_{2M} &= (1/f)\mathbf{A}^{-1}\mathbf{I}_{2c} \\ \mathbf{C}_{3M} &= (1/f)\mathbf{A}^{-1}\mathbf{I}_{2s} \\ \mathbf{C}_{1M} &= \mathbf{A}^{-1}\mathbf{I}_0 - (1-f)\mathbf{C}_{2M} \\ \mathbf{C}_{2M} &= \mathbf{A}^{-1}\mathbf{I}_{1s} - g\mathbf{C}_{3M} \end{aligned} \quad (28\text{-}55)$$

REFERENCES

1. Hauge, P. S., "Recent development in instrumentation in ellipsometry," *Surface Sci., 96*, 108–140 (1980).
2. Azzam, R. M. A., "Photopolarimetric measurement of the Mueller matrix by Fourier analysis of a single detected signal," *Opt. Lett. 2*, 148–150 (1978).
3. Goldstein, D. H., "Mueller matrix dual-rotating-retarder polarimeter," *Appl. Opt. 31*, 6676–6683 (1992).
4. Pezzaniti, J. L., McClain, S. C., Chipman, R. A., and Lu, S.-Y., "Depolarization in liquid-crystal televisions," *Opt. Lett. 18*(23), 2071–2073 (1993).
5. Sornsin, E. A., and Chipman, R. A., "Electro-optic light modulator characterization using Mueller matrix imaging," *Proc. SPIE, 3121*, 161–166, Aug. 1997.
6. Goldstein, D. H. and Chipman, R. A., "Infrared Spectropolarimeter," US Patent No. 5 045 701, September 3, 1991.
7. Goldstein, D. H., Chipman, R. A., and Chenault, D. B., "Infrared Spectropolarimetry," *Opt. Eng. 28*, 120–125 (1989).
8. Hauge, P. S., "Mueller matrix ellipsometry with imperfect compensators," *J. Opt. Soc. Am., 68*, 1519–1528 (1978).
9. klein Brink, H. B., "Birefringence of the human crystalline lens in vivo," *J. Opt. Soc. Am. A 8*, 1788–1793 (1991).
10. klein Brink, H. B. and van Blokland, G. J., "Birefringence of the human foveal area assessed in vivo with Mueller matrix ellipsometry," *J. Opt. Soc. Am. A. 5*, 49–57 (1988).
11. van Blokland, G. J., "Ellipsometry of the human retina in vivo: preservation of polarization," *J. Opt. Soc. Am. A, 2*, 72–75 (1985).
12. Strang, G., *Linear Algebra and Its Applications*, 2nd ed., Academic Press, New York, 1976, p.112.
13. Goldstein, D. H. and Chipman, R. A., "Error Analysis of Mueller Matrix Polarimeters," *J. Opt. Soc. Am. 7*(4), 693–700 (1990).
14. Chenault, D. B., Pezzaniti, J. L., and Chipman, R. A., "Mueller matrix algorithms," *Proc. SPIE 1746* 231–246 (1992).
15. Goldstein, D. H., Chipman, R. A., Chenault, D. B., and Hodgson, R. R. "Infrared material properties measurements with polarimetry and spectropolarimetry," *Proc. SPIE 1307*, pp. 448–462 (1990).
16. Namba, C. S., "Electro-optical effect of zincblende," *J. Opt. Soc. Am. 51*, 76–79 (1961).
17. Chipman, R. A. "Polarimetric impulse response", *Proc. SPIE, 1317, Polarimetry: Radar, Infrared, Visible, Ultraviolet, and X-Ray*, May 1990, pp. 223–241.
18. Thompson, R. C., Bottiger, J. R., and Fry, E. S., "Measurement of polarized light interactions via the Mueller matrix", *Appl. Opt., 19*(8) 1323–1332 (1980).
19. Schiff, T. C., Stover, J. C., Bjork, D. R., Swimley, B. D., Wilson, D. J., and Southwood, M. E. "Mueller matrix measurements with an out-of-plane polarimetric scatterometer," *Proc. SPIE 1746*, 295–306, (1992).
20. Azzam, R. M. A. and Lopez, A. G. "Accurate calibration of the four-detector photopolarimater with imperfect polarizing optical elements," *J. Opt. Soc. Am. A, 6*, 1513–1521 (1989).

29

Ellipsometry

29.1 INTRODUCTION

One of the most important applications of polarized light is the measurement of the complex refractive index and thickness of thin films. A field of optics has been developed to do this and has come to be known as ellipsometry. In its broadest sense ellipsometry is the art of measuring and analyzing the elliptical polarization of light; the name appears to have been given in 1944 by Alexandre Rothen, one of the pioneers in the field. However, the field of ellipsometry has become much more restrictive so that now it almost always applies to the measurement of the complex refractive index and thickness of thin films. In its most fundamental form it is an optical method for measuring the optical parameters of a thin film by analyzing the reflected polarized light. The optical parameters are the refractive index n, the extinction coefficient κ, and the thickness d of a thin film deposited on a substrate. The optical procedure for determining these parameters is done in a very particular manner, and it is this manner which has come to be known as ellipsometry. The fundamental concepts of ellipsometry are quite simple and straightforward. However, we shall see that this seeming simplicity is deceptive. Nevertheless, it is very elegant.

The fact that a thin film on a substrate could significantly change the measured characteristics of an optical material; e.g., a microthin coating of oil on water came, apparently, as a surprise to nineteenth-century optical physicists. The great Lord Rayleigh admitted as much when he was experimenting with the surface viscosity of liquids and said:

> Having proved that the superficial viscosity of water was due to a greasy contamination whose thickness might be much less than one-millionth of a millimetre, I too hastily concluded that films of such extraordinary tenuity were unlikely to be of optical importance until prompted by a remark of Sir G. Stokes, I made an actual estimate of the effect to be expected.

At about the time that Rayleigh was investigating the optical properties of light reflected from the surface of liquids, Drude was investigating the optical properties of light reflected from solids. In two fundamental articles published in 1889 and 1890 he laid the foundations for ellipsometry. As we have pointed out many times, at that time the only optical detector was the human eye, which has only a capability of "measuring" a null-intensity condition. Drude cleverly exploited this very limited quantitative condition of the human eye to determine the optical parameters of a thin film. He recognized that an optical material such as a metal behaves simultaneously as a polarizer and a phase shifter so that, in general, light reflected from the optical surface of a metal is elliptically polarized. Analysis shows that by adjusting the amplitude and the phase of the *incident* beam it is possible to transform the *reflected* elliptically polarized light to linearly polarized light. Drude did this by inserting a polarizer and a compensator (retarder) between the optical source and the sample.

By setting the compensator with its fast axis at 45° and rotating the polarizer through an angle P, the reflected elliptically polarized light could be transformed to linearly polarized light. The reflected linearly polarized light was then analyzed by another linear polarizer (the analyzer) by rotating it through an angle Q until a null intensity was observed. Analysis showed that these angles could be used to determine the ellipsometric parameters ψ and Δ, which described the change in amplitude and phase in the reflected wave. Further analysis based, e.g., on Fresnel's reflection equations could then relate ψ and Δ to n, κ, and d. The elegance of the method will become apparent when this analysis is presented in the following sections.

Ellipsometry can be used to determine the optical constants of a reflecting material or the optical constants and thickness of the film deposited on an optical substrate. It has a number of advantages over other methods for determining the optical constants. Among these are its applicability to the measurement of strongly absorbing materials, the simplicity of the measurement method, and the ease of the sample preparation. In addition, it is nondestructive and requires only a very small sample size. For studying the properties of surface films its directness, sensitivity, and simplicity are without parallel. Also, ellipsometry can be applied to the measurement of surface films whose thickness ranges from monatomic dimensions to micrometers. Throughout this range the index of refraction n of a film can be determined and, for absorbing film media, the extinction coefficient κ as well.

Ellipsometry can be conveniently divided into two parts. The first is the measurement technique for determining ψ and Δ. The second is the theory required to relate the optical parameters of the thin film to the measured values of ψ and Δ. Throughout this section we use the formalism of the Stokes parameters and the Mueller matrices to derive some important results. We begin by deriving the fundamental equation of ellipsometry, that is, the equation relating ψ and Δ to n, κ, and d.

29.2 FUNDAMENTAL EQUATION OF CLASSICAL ELLIPSOMETRY

In this section we derive an equation that relates the amplitude and phase of the incident and reflected beams from a thin film, the so-called ellipsometric parameters, to the complex refractive index and the thickness of the film. The equation is called the fundamental equation of ellipsometry. To derive this equation, we consider Fig. 29-1. In the figure E_p and E_s are the incident field components parallel (p)

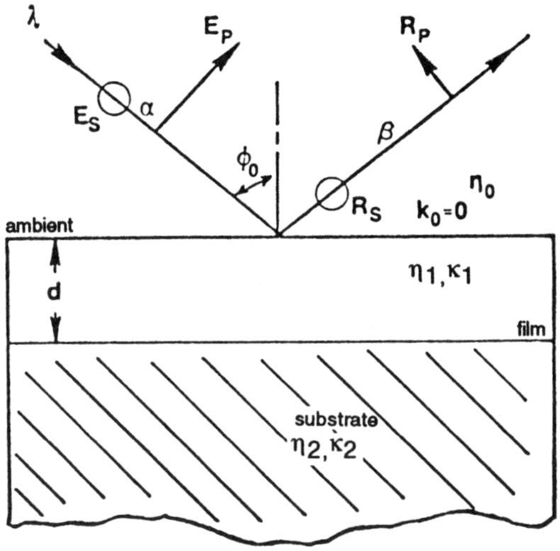

Figure 29-1 Reflection of an incident beam by an optical film of thickness d with a refractive index n_1 and an extinction coefficient κ_1.

and perpendicular (s) to the plane of paper. Similarly, R_p and R_s are the parallel and perpendicular reflected components, respectively. For the incident field components we can write

$$E_p = E_{0p}e^{i\alpha_p} \tag{29-1a}$$

$$E_s = E_{0s}e^{i\alpha_s} \tag{29-1b}$$

A similar pair of equations can also be written for the reflected field, namely,

$$R_p = R_{0p}e^{i\beta_p} \tag{29-2a}$$

$$R_s = R_{0s}e^{i\beta_s} \tag{29-2b}$$

In (29-1) and (29-2) the propagation factor, $\omega t - \kappa z$, has been suppressed. Measurements have shown that $R_{p,s}$ is directly related to $E_{p,s}$, and, in general, for optically absorbing materials the incident field will be attenuated and undergo a phase shift. In order to describe this behavior we introduce complex reflection coefficients, ρ_p and ρ_s:

$$R_p = \rho_p E_p \tag{29-3a}$$

$$R_s = \rho_s E_s \tag{29-3b}$$

or, in general,

$$\rho_m = \frac{R_m}{E_m} \qquad m = p, s \tag{29-4}$$

Substituting (29-1) and (29-2) into (29-4) then yields

$$\rho_m = \left(\frac{R_{0m}}{E_{0m}}\right) e^{i(\beta_m - \alpha_m)} \qquad m = p, s \tag{29-5}$$

We define a complex relative amplitude attenuation as

$$\rho = \frac{\rho_p}{\rho_s} = \left(\frac{R_{0p}/E_0 p}{E_{0s}/R_0 s}\right) e^{i(\beta - \alpha)} \tag{29-6}$$

where $\alpha = \alpha_p - \alpha_s$ and $\beta = \beta_p - \beta_s$. The quantities α and β describe the phase before and after reflection, respectively.

Traditionally, the factors in (29-6) are written in terms of the tangent of the angle ψ:

$$\tan\psi = \frac{R_{0p}/E_0 p}{E_{0s}/R_0 s} \tag{29-7a}$$

and a phase angle Δ:

$$\Delta = \beta - \alpha = (\beta_p - \beta_s) - (\alpha_p - \alpha_s) \tag{29-7b}$$

From (29-7) we can then express (29-6) as

$$\rho = \tan\psi e^{i\Delta} \tag{29-8}$$

Thus, ellipsometry involves the measurement of tan ψ, the change in the amplitude ratio and Δ, the change in phase. The quantities ψ and Δ are functions of the optical constants of the medium, the thin film and the substrate, the wavelength of light, the angle of incidence, and, for an optical film deposited on a substrate, its thickness. With these factors in mind we now express (29-8) as

$$\rho = \tan\psi e^{i\Delta} = f(n, \kappa, d) \tag{29-9}$$

Equation (29-9) is called *fundamental equation of ellipsometry*. Ideally, by measuring ψ and Δ the quantities n, κ, and d can be determined. In (29-9), ρ has been expressed in terms of a general functional form, $f(n, \kappa, d)$. Later, we derive the specific form of $f(n, \kappa, d)$ for a thin film deposited on a substrate.

Equation (29-9) shows that the basic problem of ellipsometry is to measure ψ and Δ and relate it to $f(n, \kappa, d)$. In the next section we develop the equations for measuring ψ and Δ. In the following section we relate these measurements to $f(n, \kappa, d)$. We shall soon see that the form of (29-9) is deceptively simple and that a considerable effort is needed to solve it.

29.3 CLASSICAL MEASUREMENT OF THE ELLIPSOMETRIC PARAMETERS PSI (ψ) AND DELTA (Δ)

In this section we describe the classical measurement of ψ and Δ in the fundamental equation of ellipsometry, (29-9). This is done by using a polarizer and compensator *before* the sample and a polarizer (analyzer) *after* the sample. The objective of the present analysis is to relate the angular settings on the polarizers and the compensator to ψ and Δ. Figure 29-2 shows the experimental configuration.

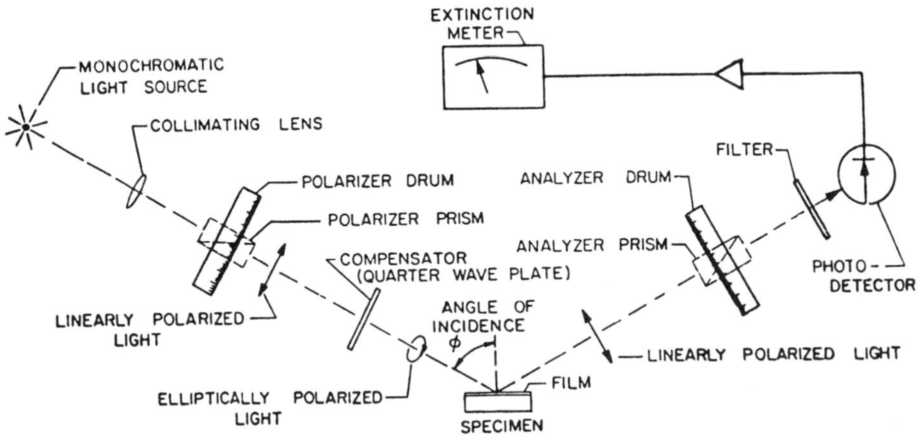

Figure 29-2 Experimental configuration to measure ψ and Δ of an optical sample. (Courtesy of Gaertner Scientific Corp., Skokie, IL.)

We first determine the Mueller matrix of the combination of the linear polarizer and the compensator in the "generating" arm. The linear polarizer can be rotated to any angle P. The compensator, on the other hand, has its fast axis fixed at 45°, but its phase ϕ can be varied from 0° to 360°. The Mueller matrices for the polarizer and compensator are then, respectively,

$$M_{\text{pol}}(P) = \frac{1}{2}\begin{pmatrix} 1 & \cos 2P & \sin 2P & 0 \\ \cos 2P & \cos^2 2P & \cos 2P \sin 2P & 0 \\ \sin 2P & \cos 2P \sin 2P & \sin^2 2P & 0 \\ 0 & 0 & 0 & 0 \end{pmatrix} \tag{29-10}$$

and

$$M_{\text{comp}}(+45^\circ) = \begin{pmatrix} 1 & 0 & 0 & 0 \\ 0 & \cos\phi & 0 & \sin\phi \\ 0 & 0 & 1 & 0 \\ 0 & -\sin\phi & 0 & \cos\phi \end{pmatrix} \tag{29-11}$$

The Mueller matrix for the polarizer–compensator combination, (29-10) and (29-11), is

$$M_{\text{PSG}} = M_{\text{comp}}(\phi) M_{\text{pol}}(P) \tag{29-12}$$

and so

$$M_{\text{PSG}} = \frac{1}{2}\begin{pmatrix} 1 & \cos 2P & \sin 2P & 0 \\ \cos\phi\cos 2P & \cos\phi\cos^2 2P & \cos\phi\cos 2P\sin 2P & 0 \\ \sin 2P & \cos 2P \sin 2P & \sin^2 2P & 0 \\ -\sin\phi\cos 2P & -\sin\phi\cos^2 2P & -\sin\phi\cos 2P\sin 2P & 0 \end{pmatrix} \tag{29-13}$$

where PSG stands for polarization state generator.

The Stokes vector of the beam incident on the polarizer–compensator combination is represented by its most general form:

$$S = \begin{pmatrix} S_0 \\ S_1 \\ S_2 \\ S_3 \end{pmatrix} \tag{29-14}$$

Multiplying (29-14) by (29-13), we obtain the Stokes vector of the beam incident on the samples:

$$S' = \begin{pmatrix} S'_0 \\ S'_1 \\ S'_2 \\ S'_3 \end{pmatrix} = \frac{1}{2}(S_0 + S_1 \cos 2P + S_2 \sin 2P) \begin{pmatrix} 1 \\ \cos\phi \cos 2P \\ \sin 2P \\ -\sin\phi \sin 2P \end{pmatrix} \tag{29-15}$$

which is a Stokes vector for elliptically polarized light. The orientation angle Ψ of the beam is

$$\tan 2\Psi = \tan \frac{\tan 2P}{\cos\phi} \tag{29-16a}$$

and, similarly, the ellipticity angle χ is

$$\sin 2\chi = -\sin\phi \cos 2P \tag{29-16b}$$

Thus, by varying P and ϕ we can generate any state of elliptically polarized light.

We now write (29-15) simply as

$$S = I_0 \begin{pmatrix} 1 \\ \cos\phi \cos 2P \\ \sin 2P \\ -\sin\phi \cos 2P \end{pmatrix} \tag{29-17}$$

and drop the primes on the Stokes vector (parameters).

The phase shift between the components emerging from the polarizer–compensator according to the relations derived in Section 29.2 is expressed in terms of an angle α. The Stokes parameters of the beam incident on the sample can then be written in terms of its field components as seen from (29-1) as

$$S_0 = E_s E_s^* + E_p E_p^* = E_{0s}^2 + E_{0p}^2 \tag{29-18a}$$

$$S_1 = E_s E_s^* - E_p E_p^* = E_{0s}^2 - E_{0p}^2 \tag{29-18b}$$

$$S_2 = E_s E_p^* + E_p E_s^* = 2E_{0s}E_{0p} \cos\alpha \tag{29-18c}$$

$$S_3 = i(E_s E_p^* - E_p E_s^*) = 2E_{0s}E_{0p} \sin\alpha \tag{29-18d}$$

The phase shift α is seen from (29-17) and (29-18) to be

$$\tan\alpha = \frac{\sin\alpha}{\cos\alpha} = \frac{S_3}{S_2} = \frac{-\sin\phi \cos 2P}{\sin 2P} \tag{29-19}$$

Now

$$\sin(2P - 90^\circ) = -\cos 2P \tag{29-20a}$$

$$\cos(2P - 90^\circ) = \sin 2P \tag{29-20b}$$

Substituting (29-20) into (29-19) then yields

$$\tan\alpha = \sin\phi \tan(2P - 90^\circ) \tag{29-21}$$

Thus, the phase α of the beam emerging from the polarizer–compensator combination can be varied by adjusting the phase shift ϕ of the compensator and the polarizer orientation angle P. In particular, if we have a quarter-wave retarder so that $\phi = 90^\circ$, then from (29-21) $\alpha = 2P - 90^\circ$. By rotating the polarizer angle from $P = 0$ ($\alpha = -90^\circ$) to $P = 90^\circ$ ($\alpha = 90^\circ$), the total phase change is 180°. In terms of the Stokes vector S, (29-17), for $\phi = 90^\circ$ we then have

$$S = I_0 \begin{pmatrix} 1 \\ 0 \\ \sin 2P \\ -\cos 2P \end{pmatrix} \tag{29-22}$$

Equation (29-22) is the Stokes vector for elliptically polarized light; its orientation angle Ψ is always 45°. However, according to (29-22) the ellipticity angles corresponding to $P = 0^\circ$, 45°, and 90°, are $\chi = -45^\circ$, 0°, and $+45^\circ$, and the respective Stokes vectors are $\{1, 0, 0, -1\}$, $\{1, 0, 1, 0\}$, and $\{1, 0, 0, +1\}$; these vectors correspond to left circularly polarized light, linear $+45^\circ$ polarized light, and right circularly polarized light, respectively. By rotating the polarizer from 0° to 90°, we can generate any state of elliptically polarized light ranging from left circularly polarized light to right circularly polarized light.

The ratio of the amplitudes E_p and E_s of the beam emerging from the polarizer–compensator can be defined in terms of an angle L, by

$$\tan L = \frac{E_p}{E_s} \tag{29-23a}$$

From (29-18a), (29-18b), and (29-17) we have

$$\frac{S_1}{S_0} = \frac{E_s E_s^* - E_p E_p^*}{E_s E_s^* + E_p E_p^*} = \cos\phi \cos 2P \tag{29-23b}$$

or

$$\frac{1 - (E_p/Es)\,(E_p^*/E_s^*)}{1 + (E_p/E_s)\,(E_p^* E_s^*)} = \cos\phi \cos 2P \tag{29-23c}$$

Because tan L is real, (29-23a) can be expressed as

$$(\tan L)^* = \frac{E_p^*}{E_s^*} = \tan L \tag{29-24}$$

Thus, (29-23c) can be written with the aid of (29-24) as

$$\frac{1-\tan^2 L}{1+\tan^2 L}=\frac{1-(\sin^2 L)/(\cos^2 L)}{1+(\sin^2 L)/(\cos^2 L)}=\cos^2 L-\sin^2 L=\cos\phi\cos 2P \tag{29-25}$$

or

$$\cos 2L=\cos\phi\cos 2P \tag{29-26}$$

We note that if S_1 is defined as the negative of (29-18b); that is,

$$S_1=E_pE_p^*-E_sE_s^* \tag{29-27}$$

then (29-26) becomes

$$\cos 2L=-\cos\phi\cos 2P \tag{29-28}$$

which is the form usually given in ellipsometry. Thus, again, by varying ϕ and P, the angle L can be selected. For circularly polarized light $E_s=E_p$, so $L=45^\circ$, (29-23), and $\cos 2L=0$. For linearly horizontally polarized light $E_p=0$, $L=0$, and $\cos 2L=1$. Finally, for linearly vertically polarized light, $E_s=0$, $L=90^\circ$, and $\cos 2L=-1$.

Equations (29-21) and (29-28) appear very often in ellipsometry and so are rewritten here together as the pair:

$$\tan\alpha=\sin\phi\tan(2P-90^\circ) \tag{29-29a}$$

$$\cos 2L=-\cos\phi\cos 2P \tag{29-29b}$$

We emphasize that (29-29a) and (29-29b) relate the amplitude and phase of the optical beam incident on the sample to the value of the compensator phase ϕ and the polarizer angle P, respectively.

The procedure for measuring ψ and Δ consists of rotating the generating polarizer and the analyzing polarizer until the reflected beam is extinguished. Because the compensator is fixed with its fast axis at 45°, only two polarizing elements rather than three must be adjusted. The Stokes vector of the reflected light is

$$S'=\begin{pmatrix} E'^2_{0s}+E^2_{0p} \\ E'^2_{0s}-E'^2_{0p} \\ 2E'_{0s}E'_{0p}\cos\beta \\ 2E'_{0s}E'_{0p}\sin\beta \end{pmatrix} \tag{29-30}$$

where β, using the notation in Section 29.2, is the phase associated with the *reflected* beam. To obtain linearly polarized light, $\sin\beta$ in (29-30) must be zero, so

$$\beta=0^\circ, 180^\circ \tag{29-31}$$

Thus, there are two values of β which lead to linearly polarized light. The Stokes vector S' in (29-30), using (29-31), then becomes

$$S'=\begin{pmatrix} E'^2_{0s}+E'^2_{0p} \\ E'^2_{0s}-E'^2_{0p} \\ \pm 2E'_{0s}E'_{0p} \\ 0 \end{pmatrix} \tag{29-32}$$

The condition on β then transforms (29-76) to

$$\Delta = \beta - \alpha = -\alpha \qquad (\beta = 0^\circ) \tag{29-33a}$$

or

$$\Delta = 180^\circ - \alpha \qquad (\beta = 180^\circ) \tag{29-33b}$$

The angles of the polarizer in the *generating* arm corresponding to (29-33a) and (29-33b) can be written as P_0 and P'_0 , respectively.
We have

$$\tan \alpha = \sin \phi \tan(2P_0 - 90^\circ) \tag{29-34a}$$

$$\cos 2L_0 = -\cos \phi \cos 2P_0 \tag{29-34b}$$

and

$$\tan \alpha' = \sin \phi \tan(2P'_0 - 270^\circ) \tag{29-35a}$$

$$\cos 2L'_0 = -\cos \phi \cos 2P'_0 \tag{29-35b}$$

The linearly polarized reflected beam will be extinguished when the *analyzer* angles corresponding to P_0 and P'_0 are A_0 and A'_0, respectively. This leads immediately to the following forms for $\tan \psi$, (29-7a):

$$\tan \psi = \frac{|R_p|}{|R_s|} = \frac{R_{0p}\, E_{0s}}{R_{0s}\, E_{0p}} \tag{29-36}$$

Substituting (29-23a) into (29-36), we have

$$\tan \psi = \frac{R_{0p}}{R_{0s}} \cot L_0 \tag{29-37}$$

where we have used the measurement value L_0. We also see that

$$\tan(-A_0) = \frac{R_{0p}}{R_{0s}} \tag{29-38}$$

(the angle $-A_0$ is opposite to P_0). Then, using (29-38), (29-37) becomes

$$\tan \psi = \cot L_0 \tan(-A_0) \tag{29-39}$$

for the polarizer-analyzer pair settings of P_0 and A_0. Similarly, for the pair P'_0 and A'_0 we have

$$\tan \psi = \cot L'_0 \tan A'_0 \tag{29-40}$$

From (29-34a) and (29-35a) we see that

$$P'_0 = P_0 \pm 90^\circ \tag{29-41a}$$

and

$$A'_0 = A_0 \pm 90^\circ \tag{29-41b}$$

Using (29-41) and setting (29-39) equal to (29-40) yields

$$\cot L'_0 = \tan L_0 \tag{29-42}$$

so that multiplying (29-39) and (29-40) gives

$$\tan^2\psi = \tan(A_0')\tan(-A_0) \tag{29-43}$$

Equation (29-43) shows that $\tan\psi$ can be determined by measuring A'_0 and A_0, the angular settings on the analyzer. Similarly, the phase shift Δ can be obtained from (29-33) and (29-34) or (29-35).

For the special case where $\phi = 90°$, a quarter-wave retarder, the equations relating ψ and Δ simplify. From (29-34a) and (29-34b) we have

$$\Delta = 2P_0 - 90° = 2P_0' - 270° \tag{29-44}$$

from (29-34b) and (29-35b):

$$L_0' = L_0 \tag{29-45}$$

and from (29-35) and (29-40):

$$-A_0 = A_0' \tag{29-46}$$

If a Babinet–Soleil compensator is used, then the phase shift ϕ can be set to 90° and A_0, A_0', P_0, and P'_0 can be used to give $\tan\psi$ and Δ, (29-43) and (29-44), respectively; that is,

$$\tan^2\psi = \tan^2 A_0 = \tan^2(-A_0') \tag{29-47a}$$

so

$$\psi = A_0 = -A_0' \tag{29-47b}$$

and

$$\Delta = 2P_0 - 90° = 2P_0' - 270° \tag{29-47c}$$

In order to select the correct equations for calculating Δ and ψ from a pair of extinction settings, it is necessary to establish whether the settings correspond to the condition $\Delta' = -\Delta$ or $\Delta' = \Delta + 180°$. This is accomplished by observing that, although Δ may have any value between 0° and 360°, ψ is limited to values between 0° and 90°. From this fact the sign of the analyzer extinction setting, according to $\psi = -A_0 = A_0'$, determines whether the setting corresponds to the primed or unprimed case.

The relations presented above describe the measurement formulation of ellipsometry. The formulation rests on the conditions required to obtain a null intensity; that is, linearly polarized light will be obtained for reflected light if $\sin\beta = 0°$ or 180°. From this condition one works backwards to find the corresponding values of P and A and then ψ and Δ.

There are other configurations and formulations of ellipsometry. One of the most interesting has been given by Holmes and Feucht. Their formulation is particularly valuable because it leads to a single expression for the complex reflectivity ρ in terms of the polarizer angles P and A; we now designate the analyzing polarizer angle by A. Moreover, it includes the "imperfections" of the compensator with its fast axis at an angle C. This formulation was used by F. L. McCrackin, one of the first researchers to use digital computers to solve the ellipsometric equations, in the early 1960s.

We recall that ρ of an optical surface is related to the ellipsometric parameters ψ and Δ by

$$\rho = \tan \psi e^{i\Delta} \tag{29-8}$$

We assume the same ellipsometric measurement configuration as before, namely, an ideal polarizer and a compensator in the generating arm and an ideal polarizer in the analyzing arm. The transmission axes of the polarizers are at P and A, respectively. The compensator is considered to be slightly absorbing, and its fast axis is at an angle C. Lastly, the beam incident on the generating polarizer is assumed to be linearly horizontally polarized with a unit amplitude. We use the Jones calculus to carry out the caluculations.

The Jones matrix for the incident beam is

$$J_{\text{inc}} = \begin{pmatrix} 1 \\ 0 \end{pmatrix} \tag{29-48}$$

The Jones matrix of a rotated linear polarizer is

$$\begin{aligned} J_{\text{pol}} &= \begin{pmatrix} \cos P & -\sin P \\ \sin P & \cos P \end{pmatrix} \begin{pmatrix} 1 & 0 \\ 0 & 1 \end{pmatrix} \begin{pmatrix} \cos P & \sin P \\ -\sin P & \cos P \end{pmatrix} \\ &= \begin{pmatrix} \cos^2 P & \sin P \cos P \\ \sin P \cos P & \sin^2 P \end{pmatrix} \end{aligned} \tag{29-49}$$

Multiplying (29-48) by (29-49) then gives

$$J = \cos P \begin{pmatrix} \cos P \\ \sin P \end{pmatrix} \tag{29-50}$$

The term $\cos P$ is an amplitude factor, which can be ignored, and so the Jones matrix of the beam incident on the compensator is

$$J = \begin{pmatrix} \cos P \\ \sin P \end{pmatrix} \tag{29-51}$$

The Jones matrix for an ideal compensator is

$$J_{\text{comp}} = \begin{pmatrix} e^{i\phi_x} & 0 \\ 0 & e^{i\phi_y} \end{pmatrix} \tag{29-52}$$

If there is also absorption along each of the axes, then the Jones matrix (29-52) can be rewritten as

$$J_{\text{comp}} = \begin{pmatrix} a_x e^{i\phi_x} & 0 \\ 0 & a_y e^{i\phi_y} \end{pmatrix} \tag{29-53}$$

where $0 \le a_{x,y} < 1$. We see that we can now write (29-53) as

$$J_{\text{comp}} = \begin{pmatrix} 1 & 0 \\ 0 & a_c \end{pmatrix} \tag{29-54}$$

where $a_c = (a_y/a_x)\exp(i\phi)$ and $\phi = \phi_y - \phi_x$; a_c is called the absorption ratio of the compensator; we have ignored the factor $a_x e^{i\phi_x}$ outside of the matrix in (29-54). The Jones matrix of the compensator, (29-54), with its fast axis rotated to an angle C is

$$J_{\text{comp}} = \begin{pmatrix} \cos C & -\sin C \\ \sin C & \cos C \end{pmatrix} \begin{pmatrix} 1 & 0 \\ 0 & a_c \end{pmatrix} \begin{pmatrix} \cos C & \sin C \\ -\sin C & \cos C \end{pmatrix}$$
$$= \begin{pmatrix} \cos^2 C + a_c \sin^2 C & (1 - a_c)\sin C \cos C \\ (1 - a_c)\sin C \cos C & \sin^2 C + a_c \cos^2 C \end{pmatrix} \tag{29-55}$$

Multiplying (29-51) by (29-55), the Jones matrix of the beam incident on the optical sample is

$$J = \begin{pmatrix} \cos C \cos(C - P) + a_c \sin C \sin(C - P) \\ \sin C \cos(C - P) - a_c \cos C \sin(C - P) \end{pmatrix} \tag{29-56}$$

We must now determine the Jones matrix of the optical sample. By definition, the reflected beam is related to the incident beam by

$$R_p = \rho_p E_p \tag{29-3a}$$

$$R_s = \rho_s E_s \tag{29-3b}$$

where ρ_p and ρ_s are the complex reflection coefficients for the parallel and perpendicular components, respectively. The Jones matrix of the sample is then seen from (29-3) to be

$$J_{\text{samp}} = \begin{pmatrix} \rho_p & 0 \\ 0 & \rho_s \end{pmatrix} \tag{29-57}$$

The complex relative amplitude attenuation ρ in (29-8) is defined by

$$\rho = \frac{\rho_p}{\rho_s} \tag{29-6}$$

so (29-57) can be written as

$$J_{\text{samp}} = \begin{pmatrix} \rho & 0 \\ 0 & 1 \end{pmatrix} \tag{29-58}$$

where we have ignored the factor ρ_s because it will drop out of our final equation, which is a ratio.

The Jones matrix of the beam incident on the analyzing polarizer is now seen from multiplying (29-56) by (29-58) to be

$$J = \begin{pmatrix} \rho[\cos C \cos(C - P) + a_c \sin C \sin(C - P)] \\ \sin C \cos(C - P) - a_c \cos C \sin(C - P) \end{pmatrix} = \begin{pmatrix} E_x \\ E_y \end{pmatrix} \tag{29-59}$$

Equation (29-59) shows that the reflected light is, in general, elliptically polarized. However, if C, P, and a_c are adjusted so that the reflected light is linearly polarized, then the azimuthal angle γ of the linearly polarized light is

$$\tan\gamma = \frac{E_y}{E_x} = \frac{\sin C \cos(C - P) - a_c \cos C \sin(C - P)}{\rho[\cos C \cos(C - P) + a_c \sin C \sin(C - P)]} \tag{29-60}$$

The linearly polarized light, (29-60), is now analyzed by the analyzer. We know that if the analyzer is rotated through 90° from γ we will obtain a null intensity. Thus, we have

$$A = \gamma + 90^\circ \tag{29-61a}$$

so

$$\gamma = A - 90^\circ \tag{29-61b}$$

Taking the tangent of both sides of (29-61b) yields

$$\tan\gamma = \frac{-1}{\tan A} \tag{29-62}$$

Solving now for ρ in (29-60), using (29-62) and factoring out $\cos C \cos(C-P)$ from numerator and denominator yields

$$\rho = \frac{\tan A[\tan C + a_c \tan(P - C)]}{a_c \tan C \tan(P - C) - 1} \tag{29-63}$$

where we have expressed (29-63) with the argument $P-C$ rather than $C-P$, as is customary in ellipsometry:

Equation (29-63) enables us to determine ρ from the reading A, P, C and a knowledge of a_c. As an example of (29-63), suppose that we use a perfect quarter-wave retarder so that $a_c = i(\sqrt{-1})$. Furthermore, suppose that P is measured to be 60°, $C = 30°$, and $A = 45°$. Substituting these values into (29-63), we find that

$$\rho = \frac{-\sqrt{3} - i2\sqrt{3}}{5} \tag{29-64}$$

Equating (29-64) to (29-8) we find that

$$\psi = \tan^{-1}\left(\sqrt{\frac{3}{5}}\right) = 37.8^\circ \tag{29-65a}$$

$$\Delta = \tan^{-1}(2) = 63.4^\circ \tag{29-65b}$$

Because (29-63) is so easy to use, it is probably the simplest way to determine the ellipsometric parameters ψ and Δ from ρ.

As we mentioned, other ellipsometric configuration can be conceived, e.g., placing the compensator in the analyzing arm. However, for a variety of reasons the most popular configuration is the one discussed here. Further information on the measurement of the ellipsometric parameters can be found in the references.

29.4 SOLUTION OF THE FUNDAMENTAL EQUATION OF ELLIPSOMETRY

We now turn to the problem of finding a specific form for $f(n, \kappa, d)$, the right-hand side of the ellipsometric equation, and then the solution of the fundamental equation of ellipsometry. The model proposed by Drude, and the one which has been used with great success, is that of a homogeneous thin film superposed on a substrate. An optical beam is then incident on the thin film and undergoes multiple reflections

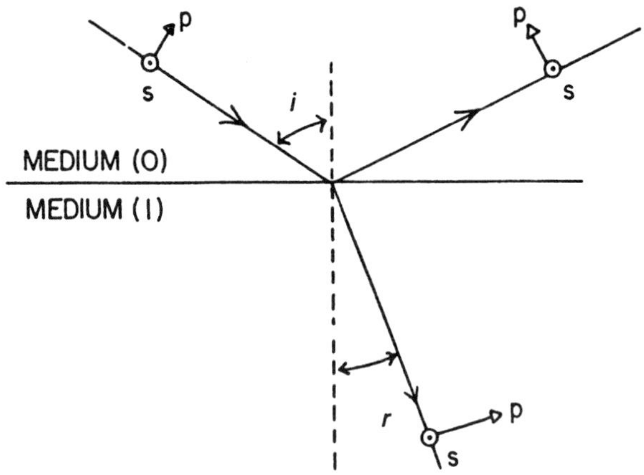

Figure 29-3 Oblique reflection and transmission of a plane wave at the planar interface between two semi-infinite media 1 and 2. (Modified from Azzam and Bashara.)

within the film. From a knowledge of the polarization state of the incident and reflected beams, the refractive index, extinction coefficient, and thickness of the film can be determined.

In order to solve this problem several related problems must be addressed. The first is to determine the relation between the refractive indices of two different media and the complex relative amplitude attenuation ρ. In Fig. 29-3 we show the oblique reflection and transmission of a plane wave incident on a boundary.

Fresnel's equations for the reflection coefficients r_p and r_s can be written as (see Chapter 8)

$$r_p = \frac{n_2 \cos\theta_i - n_1 \cos\theta_r}{n_2 \cos\theta_i + n_1 \cos\theta_r} \tag{29-66a}$$

$$r_s = \frac{n_1 \cos\theta_i - n_2 \cos\theta_r}{n_1 \cos\theta_i + n_2 \cos\theta_r} \tag{29-66b}$$

The complex relative amplitude attenuation ρ is defined to be

$$\rho = \frac{r_p}{r_s} \tag{29-67}$$

Substituting (29-66) into (29-67) gives

$$\rho = \frac{r_p}{r_s} = \frac{x\cos\theta_i - \cos\theta_r}{x\cos\theta_i + \cos\theta_r}\,\frac{\cos\theta_i + x\cos\theta_r}{\cos\theta_i - x\cos\theta_r} \tag{29-68}$$

where $x = n_2/n_1$. The refractive angle θ_r can be eliminated from (29-68) by using Snell's law, which we write as

$$\sin\theta_r = \frac{\sin\theta_i}{x} \tag{29-69}$$

so (29-68) can then be rewritten as

$$\rho = \frac{r_p}{r_s} = \left(\frac{x^2\cos\theta_i - \sqrt{x^2 - \sin^2\theta_i}}{x^2\cos\theta_i + \sqrt{x^2 - \sin^2\theta_i}}\right)\left(\frac{\cos\theta_i + \sqrt{x^2 - \sin^2\theta_i}}{\cos\theta_i - \sqrt{x^2 - \sin^2\theta_i}}\right) \tag{29-70}$$

We now set

$$a = \cos\theta_i \qquad \text{and} \qquad b = \sqrt{x^2 - \sin^2\theta_i} \tag{29-71}$$

and let

$$U = a^2x^2 - b^2 \tag{29-72a}$$

$$V = ab(1 - x^2) \tag{29-72b}$$

so (29-70) now becomes

$$\rho = \frac{U + V}{U - V} \tag{29-72c}$$

Setting $f = U/V$, we solve (29-72a) and (29-72b) for x^2 and find that

$$x^2 = \sin^2\theta_i\left[1 + \frac{\tan^2\theta_i}{f^2}\right] \tag{29-73}$$

Equation (29-72c) can be solved for f in terms of ρ, and we find that

$$f = \frac{1 + \rho}{1 - \rho} \tag{29-74}$$

Finally, from $x = n_2/n_1$ and (29-74) we see that (29-73) then becomes

$$\frac{n_2}{n_1} = \sin\theta_i\left[1 + \left(\frac{1 - \rho}{1 + \rho}\right)^2 \tan^2\theta_i\right]^{1/2} \tag{29-75}$$

which is the desired relation between n_2, n_1, and ρ.

A slightly different form of (29-75) can be written by writing $\tan\theta_i$ as $(\sin\theta_i/\cos\theta_i)$. A little bit of further algebra then leads to

$$\frac{n_2}{n_1} = \tan\theta_i\left[1 - \frac{4\rho}{(1 + \rho)^2}\sin^2\theta_i\right]^{1/2} \tag{29-76}$$

The elimination of the refractive angle θ_r is advantageous from a computational point of view because it is easier to evaluate (29-75) [or (29-76)] in terms of ρ rather than a complex angle.

We recall that, for materials with a real refractive index n, Fresnel's reflection coefficient at the Brewster angle θ_{i_B} is $r_p = 0$, so $\rho = 0$. We then see that (29-76) reduces to

$$\tan\theta_{i_B} = \frac{n_2}{n_1} \tag{29-77}$$

For a medium such as air, whose refractive index is practically equal to 1, (29-77) becomes

$$n = \tan \theta_{i_B} \tag{29-78}$$

which is the usual form of Brewster's law.

It is of interest to solve (29-76) for ρ and then investigate the behaviour of ρ as a function of the incidence angle θ_i. Solving (29-76) for ρ leads to a quadratic equation in ρ whose solution is

$$\rho = \frac{-(x+y) + \sqrt{(x+y)^2 - x^2}}{x} \tag{29-79a}$$

where (we set $n_2 = n$ and $n_1 = 1$)

$$x = \frac{n^2 - \tan^2 \theta_i}{2} \tag{29-79b}$$

$$y = \frac{\sin^4 \theta_i}{\cos^2 \theta_i} \tag{29-79c}$$

The positive value of the square root is chosen in (29-79a) because, as we shall see, this correctly describes the behavior of ρ. For an incidence angle of $\theta_i = 0$, (29-79) becomes

$$x = \frac{n^2}{2} \qquad y = 0 \qquad \rho = -1 \tag{29-80}$$

The negative value of ρ shows that at normal incidence there is a 180° phase shift between the incident and reflected waves.

The next angle of interest is the Brewster angle, where we find that

$$x = 0 \qquad y = \frac{n^4}{n^2 + 1} \qquad \rho = 0 \tag{29-81}$$

Finally, the determination of ρ at an incidence angle of $\theta_i = 90°$ can be found from the limiting value as $\theta_i \to 90°$. First, (29-79a) is written as

$$\rho = -\left(1 + \frac{y}{x}\right) + \sqrt{\left(1 + \frac{y}{x}\right)^2 - 1} \tag{29-82a}$$

For large values of θ_i we see that (29-79b) and (29-79c) can be written as

$$x \cong -\frac{\tan^2 \theta_i}{2} \tag{29-82b}$$

$$y = \frac{\sin^4 \theta_i}{\cos^2 \theta_i} \tag{29-82c}$$

so

$$\frac{y}{x} = -2 \sin^2 \theta_i \tag{29-82d}$$

Equation (29-82a) then becomes

$$\rho = -(1 - 2\sin^2\theta_i) + \sqrt{(1 - 2\sin^2\theta_i)^2 - 1} \tag{29-82e}$$

In the limit as $\theta_i \to 90^\circ$ we see that $\rho \to 1$, so we have

$$x \to -\infty \qquad y \to \infty \qquad \rho \to 1 \tag{29-83}$$

This behavior is confirmed in Figs. 29-4 and 29-5. In the first figure a plot is made of $\rho(\theta_i)$ versus θ_i. We see that $\rho = -1$ at $\theta_i = 0^\circ$, $\rho = 0$ at $\theta_i = \theta_{i_B}$ (the Brewster angle), and $\rho = 1$ at $\theta_i = 90^\circ$. Similarly, in Fig. 29.5 a plot is made of the absolute magnitude of $\rho(\theta_i)$.

In terms of measurable quantities the reflectances $\mathscr{R}_p$ and $\mathscr{R}_s$ are of practical importance and are defined by

$$\mathscr{R}_p = |r_p|^2 \tag{29-84a}$$

$$\mathscr{R}_s = |r_s|^2 \tag{29-84b}$$

which gives the fraction of the total intensity of an incident plane wave that appears in the reflected wave for the p and s polarizations.

At this point it is of interest to use (29-76) to determine the complex refractive index of a material for a specific angle of incidence. We see that, at an incidence angle of 45° and for $n_1 = 1$, (29-76) reduces to the simple form:

$$n_2^2 = \frac{1 + \rho^2}{(1 + \rho)^2} \tag{29-85}$$

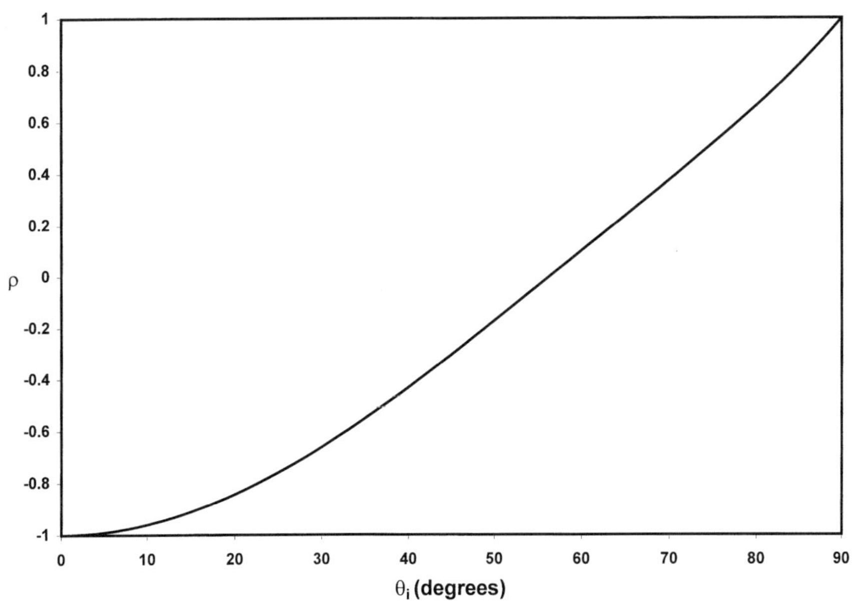

Figure 29-4 Plot of the complex relative amplitude attenuation ρ, (29-79a) as a function of incident angle θ_i.

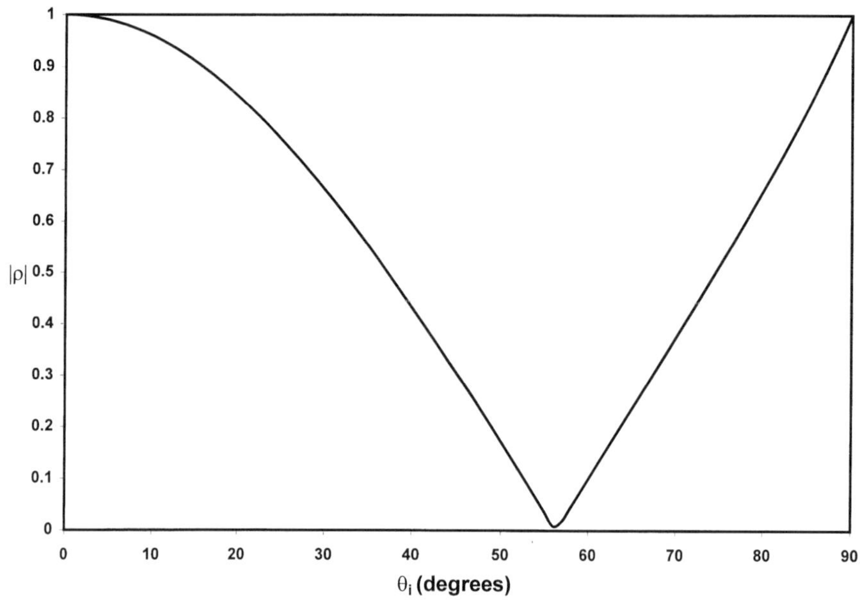

Figure 29-5 Plot of the absolute magnitude of ρ, (29-79a), as a function of θ_i.

The complex relative amplitude attenuation ρ can be written as

$$\rho = a + ib \tag{29-86}$$

Substituting (29-86) into (29-85) and grouping terms into real and imaginary parts yields

$$n_2^2 = \frac{(CE + DF) - i(CF - DE)}{E^2 + F^2} \tag{29-87a}$$

$$= A - iB \tag{29-87b}$$

where

$$A = \frac{CE + DF}{E^2 + F^2} \tag{29-87c}$$

$$B = \frac{CF - DE}{E^2 + F^2} \tag{29-87d}$$

and

$$C = 1 + a^2 - b^2 \tag{29-87e}$$

$$D = 2ab \tag{29-87f}$$

$$E = (1 + a)^2 - b^2 \tag{29-87g}$$

$$F = 2b(1 + a) \tag{29-87h}$$

We recall that n_2 is complex and defined in terms of its real refractive index n and extinction coefficient κ as

$$n_2 = n(1 - i\kappa) \tag{29-88}$$

We can now find n and κ in terms of A and B by equating the square of (29-88) to (29-87b):

$$n_2^2 = n^2(1 - i\kappa)^2 = A - iB \tag{29-89}$$

Expanding (29-89) and equating the real and imaginary parts yields

$$n^2 - n^2\kappa^2 = A \tag{29-90a}$$

$$2n^2\kappa = B \tag{29-90b}$$

Equation (29-90) then leads to a quadratic equation in n^2:

$$n^4 - An^2 - \frac{B^2}{4} = 0 \tag{29-91}$$

whose solution is

$$n^2 = \frac{A \pm \sqrt{A^2 + B^2}}{2} \tag{29-92a}$$

and for κ, (29-90b),

$$\kappa = \frac{B}{A \pm \sqrt{A^2 + B^2}} \tag{29-92b}$$

For real refractive indices, B must be zero so we must choose the positive sign in (29-92a) and we have

$$n^2 = A \tag{29-93a}$$

$$\kappa = 0 \tag{29-93b}$$

We can now consider a specific example. In Section 29.3 we saw that ellipsometric measurements on a material led to a value for ρ of

$$\rho = \frac{-\sqrt{3} - i2\sqrt{3}}{5} \tag{29-64}$$

From (29-86) to (29-92) the complex refractive index n_2 is then found to be

$$n_2 = 0.3953(1 - i0.4641) \tag{29-94}$$

Equation (29-76) is very important because, in practice, thin films are deposited on substrates and the complex refractive index of the substrate, written n_2, must be known in order to characterize the thin film.

In the problem described we have assumed that the incident beam propagates in medium 0 and is reflected and transmitted at the interface of medium 0 and 1. We can denote the reflection and transmission coefficients at the interface by r_{01} and t_{01}; by convention the order of the subscripts denotes that the beam is travelling *from* the medium represented by the first subscript (0) *to* the medium represented by the second subscript (1). If the incident beam is propagating in medium 1 and is reflected and transmitted at the interface of medium 0, then the reflection and transmission coefficients are denoted by r_{10} and t_{10}, respectively. It is necessary to know the relation between these coefficients. A direct way to do this is to interchange n_1 and n_2 in Fresnel's reflection and transmission equations. Another method, due to Stokes, is not only elegant but very novel and is given in Section 29.4.1. If the ambient medium is designated by 0 and the film by 1, then the following relations are found:

$$r_{10} = -r_{01} \tag{29-95a}$$

$$t_{01}t_{10} = 1 - r_{01}^2 \tag{29-95b}$$

With this background we can now consider a specific form for $f(n, \kappa, d)$, the thin film deposited on a substrate and very often called the ambient–film–substrate (AFS) system. This system is shown in Fig. 29-6. The film has parallel-plane boundaries of thickness d and is sandwiched between semi-infinite ambient and substrate media. The three media are all homogeneous and optically isotropic with complex refractive indices n_0, n_1, and n_2, respectively. In most cases the ambient medium is transparent and n_0 is real.

In the figure the incident beam is seen to undergo multiple reflections and transmissions at the interfaces between the ambient and the thin film and the thin film and the substrate. We know that there will be destructive or constructive interference for these multiple reflections. The interference will take place constructively if the phase shift between each of the adjacent beams from the thin film into the ambient medium differs by 2π radians. In order to proceed with the problem it is necessary to determine the relation between the phase shift between each of the adjacent beams and the film thickness. Figure 29-7 shows the geometry of the path difference between two adjacent beams.

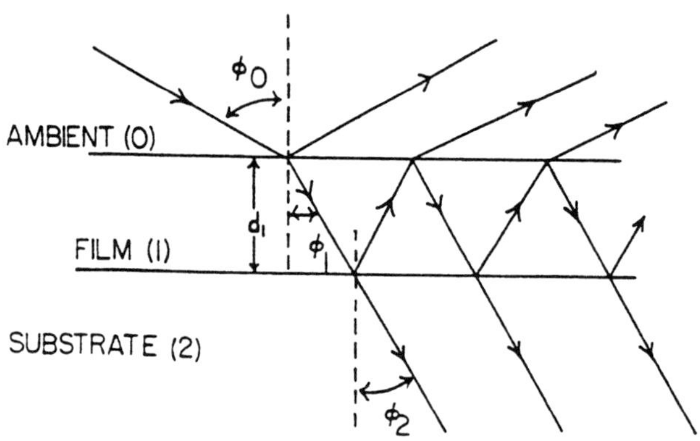

Figure 29-6 Oblique reflection and transmission of a plane wave by an ambient (0)–film (1)–substrate (2) system with parallel-plane boundaries. The film thickness is d, ϕ_0 is the angle of incidence in the ambient medium, and ϕ_1 and ϕ_2 at the angles of refraction in the film and the substrate, respectively. (From Azzam and Bashara.)

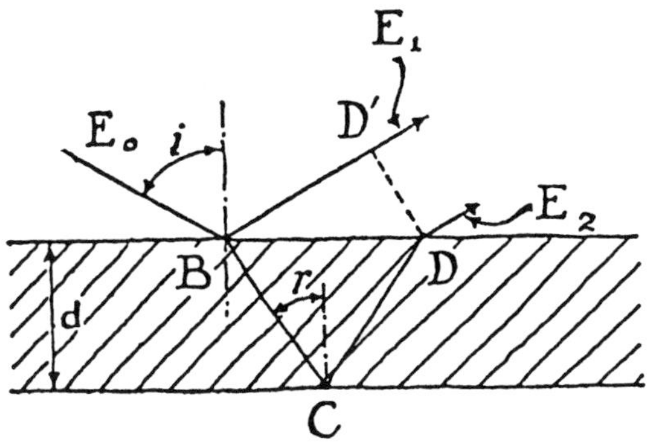

Figure 29-7 Geometry of the path difference between two adjacent beams on reflection at oblique incidence by front and back surfaces of a thin film. (From Strong, Ref. 6.)

In Fig. 29-7 the path lengths between the two adjacent beams are BD' and $BC+CD$, respectively. The optical path difference is Δl so the phase difference is $\Delta\phi = k\Delta l$ or,

$$\Delta\phi = k[n(BC + CD) - BD'] \tag{29-96}$$

where $k = 2\pi/\lambda$ and λ is the free-space wavelength of the incident light. We see that

$$BC = CD = \frac{d}{\cos\theta_r} \tag{29-97a}$$

$$BD' = BD\sin\theta_i = n(BD)\sin\theta_r \tag{29-97b}$$

and

$$BD = \frac{2d}{\cos\theta_r}\sin\theta_r \tag{29-97c}$$

Substituting (29-97) into (29-96) yields

$$\Delta\phi = \frac{4\pi nd}{\lambda}\cos\theta_r \tag{29-98}$$

In Fig. 29-6 we see that we replace θ_r by ϕ_1, so we have

$$\Delta\phi = \frac{4\pi nd}{\lambda}\cos\phi_1 \tag{29-99}$$

If $\Delta\phi = 2\pi$, then there is constructive interference between the adjacent beams; that is, the waves are in phase with one another. Similarly, if $\Delta\phi = \pi$, there is destructive interference, so the waves are completely out of phase with one another.

Equation (29-99) is readily expressed in terms of the incident angle ϕ_0. From Snell's law we see that (29-99) can be written as

$$\Delta\phi = \frac{4\pi d}{\lambda}\left(n_1^2 - n_0^2\sin^2\phi_0\right)^{1/2} \tag{29-100}$$

We must now add all the contributions of the beams contributing to the total reflected beam. For the moment we ignore the polarizations s and p; they will be restored later. If the incident field is written as E_0, then we see that the first four beams are

$$E_1 = r_{01} E_0 \tag{29-101a}$$

$$E_2 = t_{01} t_{10} r_{12} e^{-i\Delta\phi} E_0 \tag{29-101b}$$

$$E_3 = t_{01} t_{10} r_{10} r_{12}^2 e^{-i2\Delta\phi} E_0 \tag{29-101c}$$

$$E_4 = t_{01} t_{10} r_{10}^2 r_{12}^2 e^{-i3\Delta\phi} E_0 \tag{29-101d}$$

so the total field E is

$$\begin{aligned} E = {} & r_{01} E_0 + t_{01} t_{10} r_{12} e^{-i\Delta\phi} E_0 + t_{01} t_{10} r_{10} r_{12}^2 e^{-i2\Delta\phi} E_0 \\ & + t_{01} t_{10} r_{10}^2 r_{12}^2 e^{-i3\Delta\phi} E_0 \end{aligned} \tag{29-102}$$

We can write all the terms after the first term $r_{01} E_0$ for N beams as

$$t_{01} t_{10} r_{12} e^{-i\Delta\phi} \left[1 + r_0 r_{12} e^{-i\Delta\phi} + r_{10}^2 r_{12}^2 e^{-i2\Delta\phi} + \cdots + r_{10}^{N-1} r_{12}^{N-1} e^{-i(N-1)\Delta\phi} \right] \tag{29-103}$$

The terms within the brackets can be written as

$$S = 1 + x + x^2 + \cdots + x^{N-1} \tag{29-104a}$$

where

$$x = r_{10} r_{12} e^{-i\Delta\phi} \tag{29-104b}$$

Equation (29-104a) is a geometric sum. The solution is readily obtained by multiplying (29-104a) through by x:

$$xS = x + x^2 + x^3 + \cdots + x^N \tag{29-104c}$$

and then subtracting (29-104c) from (29-104a) to obtain

$$S = \frac{1 - x^N}{1 - x} \tag{29-105}$$

The factor x is always less than 1, so that for an infinite number of beams $N \to \infty$ and the limiting value of S in (29-105) is

$$S = \frac{1}{1 - x} \tag{29-106}$$

Thus, we see that (29-102) becomes

$$r = r_{01} + \frac{t_{01} t_{10} r_{12} e^{-i\Delta\phi}}{1 - r_{10} r_{12} e^{-i\Delta\phi}} \tag{29-107a}$$

or

$$r = \frac{r_{01} + r_{12} e^{-i\Delta\phi}}{1 + r_{01} r_{12} e^{-i\Delta\phi}} \tag{29-107b}$$

where $r = E/E_0$ and we have used Stokes' relations $r_{10} = -r_{01}$ and $t_{01} t_{10} = 1 - r_{01}^2$. We observe that Stokes' relations are extremely important because they not only enable us to determine the correct signs between the coefficients but they also allow us to express R in terms of r_{01} and r_{12} only, the reflection coefficients for the ambient–film (0–1) interface and the film–substrate (1–2) interface, respectively.

Equation (29-107b) is valid when the incident wave is linearly polarized either parallel (p) or perpendicular (s) to the plane of incidence. Thus, we may express the complex reflection coefficients as, adding the subscripts for the polarization components,

$$\rho_p = \frac{r_{01p} + r_{12p}e^{-i\Delta\phi}}{1 + r_{01p}r_{12p}e^{-i\Delta\phi}} \tag{29-108a}$$

$$\rho_s = \frac{r_{01s} + r_{12s}e^{-i\Delta\phi}}{1 + r_{01s}r_{12s}e^{-i\Delta\phi}} \tag{29-108b}$$

where $\Delta\phi$ is the same for the p and s polarizations and is given by (29-100). The Fresnel reflection coefficients at the 0–1 and 1–2 interfaces for the p and s polarizations are now

$$r_{01p} = \frac{n_1 \cos\phi_0 - n_0 \cos\phi_1}{n_1 \cos\phi_0 + n_0 \cos\phi_1} \tag{29-109a}$$

$$r_{12p} = \frac{n_2 \cos\phi_1 - n_1 \cos\phi_2}{n_2 \cos\phi_1 + n_1 \cos\phi_2} \tag{29-109b}$$

and

$$r_{01s} = \frac{n_0 \cos\phi_0 - n_1 \cos\phi_1}{n_0 \cos\phi_0 + n_1 \cos\phi_1} \tag{29-110a}$$

$$r_{12s} = \frac{n_1 \cos\phi_1 - n_2 \cos\phi_2}{n_1 \cos\phi_1 + n_2 \cos\phi_2} \tag{29-110b}$$

The three angles ϕ_0, ϕ_1, and ϕ_2 between the directions of propagation of the plane waves in media 0, 1, and 2, and the normal to the film boundaries are related by Snell's law:

$$n_0 \sin\phi_0 = n_1 \sin\phi_1 = n_2 \sin\phi_2 \tag{29-111}$$

Thus, between (29-109), (29-110) and (29-111) all the quantities can be found for determining the reflection coefficients r_{01p}, r_{12p} and r_{01s}, r_{12s}.

We can consider an example of the calculation of these coefficients. For simplicity, so that we can see how a calculation of this type is carried through, let us consider that we have media that are characterized only by *real* refractive indices, e.g., a thin-film dielectric deposited on a glass substrate. Let the ambient medium be represented by air, so the refractive index is $n_0 = 1$ and the film and substrate refractive indices are $n_1 = 1.5$ and $n_2 = 2.0$, respectively. Further, let the incident angle be $\phi_0 = 30°$. We then find from Snell's law (29-111) that

$$\phi_0 = 30° \tag{29-112a}$$

$$\phi_1 = 19.4712° \tag{29-112b}$$

$$\phi_2 = 14.4775° \tag{29-112c}$$

We now substitute these values along with the corresponding refractive indices into (29-109) and (29-110) and find that

$$r_{01p} = 0.5916 \tag{29-113a}$$

$$r_{12p} = 0.6682 \tag{29-113b}$$

$$r_{01s} = 0.2679 \tag{29-113c}$$

$$r_{12s} = 0.4776 \tag{29-113d}$$

Inspecting (29-108a) and (29-108b) we see that there are only two unknown quantities, the complex amplitude reflection coefficient ρ_p (or ρ_s) and $\Delta\phi$. Thus, if we measure either ρ_p or ρ_s, we can determine $\Delta\phi$; in practice we actually measure $|\rho_p|^2$ and $|\rho_s|^2$.

The usual problem is to determine the thickness of the thin film d, that is, to determine $\Delta\phi$. We can readily determine $\Delta\phi$ if all the coefficients are real. For example, we can rewrite (29-108a) as

$$\rho_p = \frac{a + be^{-i\Delta\phi}}{1 + abe^{-i\Delta\phi}} \tag{29-114a}$$

where

$$a = r_{01p} \qquad b = r_{12p} \tag{29-114b}$$

Multiplying (29-114a) by its complex conjugate then gives

$$|\rho_p|^2 = \frac{a^2 + b^2 + 2ab\cos\Delta\phi}{1 + a^2b^2 + 2ab\cos\Delta\phi} \tag{29-115}$$

Equation (29-115) is readily solved for $\Delta\phi$:

$$\Delta\phi = \cos^{-1}\left[\frac{(a^2 + b^2) - |\rho_p|^2(1 + a^2b^2)}{2ab\left(|\rho_p|^2 - 1\right)}\right] \tag{29-116}$$

Thus, by measuring $|\rho_p|^2$ and knowing a and b from (29-114b), we determine $\Delta\phi$ and, from (29-100), the film thickness d.

However, the above equations do not describe the fundamental equation of ellipsometry. To obtain this equation we must introduce ρ, which is equal to the ratio of ρ_p and ρ_s; that is, dividing (29-108a) by (29-108b), we have

$$\rho = \frac{\rho_p}{\rho_s} = \tan\psi e^{i\Delta} = \left(\frac{r_{01p} + r_{12p}e^{-i\Delta\phi}}{1 + r_{01p}r_{12p}e^{-i\Delta\phi}}\right)\left(\frac{1 + r_{01s}r_{12s}e^{-i\Delta\phi}}{r_{01s} + r_{12s}e^{-i\Delta\phi}}\right) \tag{29-117}$$

Equation (29-117) is *the fundamental equation of ellipsometry*. The right-hand side is the specific form of $f(n, \kappa, d)$. We now see, however, that $f(n, \kappa, d)$ is a very complicated function. In actuality it relates the measured ellipsometric angles ψ and Δ to the optical properties of a three-phase system, namely, the (complex) refractive indices of the ambient (n_0), the film (n_1), the substrate (n_2), the film thickness (d_1) for given values of the vacuum wavelength (λ) of the ellipsometer light beam, and the angle of incidence (ϕ_0) in the ambient; the subscript 1 on d indicates that it is the thickness of the film associated with the medium n_1. The equation can now be written symbolically as

$$\rho = \tan\psi e^{i\Delta} = f(n_0, n_1, n_2, d_1, \phi_0, \lambda) \tag{29-118}$$

Equation (29-118) may be broken into two real equations for ψ and Δ, namely,

$$\psi = \tan^{-1}\left|f(n_0, n_1, n_2, d_1, \phi_0, \lambda)\right| \tag{29-119a}$$

$$\Delta = \arg[f(n_0, n_1, n_2, d_1, \phi_0, \lambda)] \tag{29-119b}$$

where $|\rho|$ and $\arg\rho$ are the absolute value and argument (angle of the complex function ρ), respectively.

Azzam and Bashara have correctly stated that "although the function ρ may appear from (29-117) to be deceptively simple, it is, in reality, quite complicated and can be handled satisfactorily only by a digital computer". In fact, the solution of (29-117) had to wait until the development of digital computers in the 1950s and 1960s. Inspection of (29-118) shows that ρ is, in general, explicitly dependent on *nine* real arguments; the real and imaginary parts of the three complex refractive indices n_0, n_1, n_2, the film thickness d, the angle of incidence ϕ_0, and the wavelength. Not surprisingly, therefore, the solution of (29-117) must be obtained in a piecemeal fashion following the same development given above for real refractive indices (and, therefore, reflection coefficients). Here, however, the numerical solution is greatly complicated because the reflection coefficients are now complex. Fortunately, computer programs have been developed which enable the complex refractive indices to be determined as well.

In practice, the refractive indices of the ambient medium, thin film, and substrate are very often known, and the quantity of interest is the thickness of the film. The thickness of the thin film, d_1, can be found in the following way. We write (29-108a) and (29-108b) as

$$\rho_p = \frac{a + bX}{1 + abX} \tag{29-120a}$$

$$\rho_s = \frac{c + dX}{1 + cdX} \tag{29-120b}$$

where a, b, c, and d are the complex coefficients in (29-108a) and (29-108b) and

$$X = e^{-i\Delta\phi} \tag{29-120c}$$

From (29-117) we then have

$$\rho = \frac{(a + bX)(a + cdX)}{(1 + abX)(c + dX)} \tag{29-121a}$$

where

$$(a, b) = (r_{01p}, r_{12p}) \tag{29-121b}$$

$$(c, d) = (r_{01s}, r_{12s}) \tag{29-121c}$$

Carrying out the multiplication in (29-121a), we then find that

$$\rho = \frac{A + BX + CX^2}{D + EX + FX^2} \tag{29-122a}$$

where

$$A = r_{01p} \tag{29-122b}$$

$$B = r_{12p} + r_{01p}r_{01s}r_{12s} \tag{29-122c}$$

$$C = r_{12p}r_{01s}r_{12s} \tag{29-122d}$$

$$D = r_{01s} \tag{29-122e}$$

$$E = r_{12s} + r_{01p}r_{12p}r_{01s} \tag{29-122f}$$

$$F = r_{01p}r_{12p}r_{12s} \tag{29-122g}$$

Equation (29-122a) can now be written as a quadratic equation:

$$a_2X^2 + a_1X + a_0 = 0 \tag{29-123a}$$

where

$$a_2 = \rho F - C = r_{12p}r_{12s}(\rho r_{01p} - r_{01s}) \tag{29-123b}$$

$$a_1 = \rho E - B = \rho(r_{12s} + r_{01p}r_{12p}r_{01s}) - (r_{12p} + r_{01p}r_{01s}r_{12s}) \tag{29-123c}$$

$$a_0 = \rho D - A = \rho r_{01s} - r_{01p} \tag{29-123d}$$

The two solutions of (29-123a) are

$$X_1 = \frac{-a_1 + \sqrt{a_1^2 - 4a_2a_0}}{2a_2} \tag{29-124a}$$

$$X_2 = \frac{-a_1 - \sqrt{a_1^2 - 4a_2a_0}}{2a_2} \tag{29-124b}$$

Thus, we have found a formal solution to the problem. To solve for X_1 and X_2, we substitute the values of a_2, a_1, and a_0 from (29-123) into (29-124). The result is a complex number:

$$X_{1,2} = U \pm iV \tag{29-125}$$

We recall (from (29-120c) and (29-99)) that $X_{1,2}$ is

$$X_{1,2} = \exp\left(\frac{-4\pi i n_1 \cos\phi_1 d_1}{\lambda}\right) \tag{29-126}$$

Furthermore,

$$n_1 \cos\phi_1 = \left[n_1^2 - (n_0 \sin\phi_0)^2\right]^{1/2} \tag{29-127}$$

Substituting (29-127) into (29-126) yields

$$X_{1,2} = \exp\left(\frac{-4\pi i\left[n_1^2 - (n_0 \sin\phi_0)^2\right]^{1/2} d_1}{\lambda}\right) = U \pm iV \tag{29-128}$$

This can be rewritten still further by setting

$$D = \frac{\lambda}{2}\left[n_1^2 - (n_0 \sin\phi_0)^2\right]^{-1/2} \tag{29-129}$$

thus,

$$X_{1,2} = \exp\left[-i2\pi\left(\frac{d}{D}\right)\right] \tag{29-130}$$

where we have dropped the subscript 1 on d. Thus, we need only iterate d until $X_{1,2}$ is equal to the right-hand side of (29-128). In order to do this, however, the square root in (29-124) must first be converted to Cartesian form. We briefly review this process. We express the square root in (29-124) as

$$\sqrt{a + ib} = x + iy \tag{29-131a}$$

$$= ce \tag{29-131b}$$

We square both sides and equate the real and imaginary terms and find that

$$c^2 \cos 2\theta = a \tag{29-132a}$$

$$c^2 \sin 2\theta = b \tag{29-132b}$$

Squaring and adding both sides of (29-132) then leads to

$$c = (a^2 + b^2)^{1/4} \tag{29-133}$$

Next, we divide (29-132b) by (29-132a) to obtain

$$\tan 2\theta = \frac{b}{a} \tag{29-134}$$

Using the trigonometric identity:

$$\tan 2\theta = \frac{2\tan\theta}{1 - \tan^2\theta} \tag{29-135}$$

leads (29-134) to a quadratic equation of the form:

$$b\tan^2\theta + 2a\tan\theta - b = 0 \tag{29-136}$$

The solutions are found immediately to be

$$\tan\theta_1 = \frac{-a + \sqrt{a^2 + b^2}}{b} \tag{29-137a}$$

$$\tan\theta_2 = \frac{-a - \sqrt{a^2 + b^2}}{b} \tag{29-137b}$$

We restrict the angle θ to the positive quadrant, so we take the first solution. Constructing the familiar right triangle from (29-137a), we then find that

$$\sin\theta = \frac{-a + \sqrt{a^2 + b^2}}{\sqrt{2}\left[(a^2 + b^2) - a\sqrt{a^2 + b^2}\right]^{1/2}} \tag{29-138a}$$

$$\cos\theta = \frac{b}{\sqrt{2}\left[(a^2 + b^2) - a\sqrt{a^2 + b^2}\right]^{1/2}} \tag{29-138b}$$

As an example of the formulas, we consider expressing the following simple expression in Cartesian coordinates:

$$\sqrt{4 + 3i} = x + iy \tag{29-139}$$

We see that $a = 4$ and $b = 3$ and we readily find that

$$\sqrt{4 + 3i} = \sqrt{5}\left[\frac{3}{\sqrt{10}} + i\frac{1}{\sqrt{10}}\right] = \frac{1}{\sqrt{2}}[3 + i] \tag{29-140}$$

The equality is readily checked by squaring both sides of (29-140). The Cartesian form of the square root in (29-124) is now added (or subtracted) from $-a$ in the numerator. We now have Cartesian forms in the numerator and the denominator. We can then write

$$\begin{aligned} X_{1,2} &= \frac{m + in}{o + ip} \\ &= \frac{(mo + pn) + i(no - mp)}{o^2 + p^2} = U + iV \end{aligned} \tag{29-141}$$

We can express $U + iV$ in complex polar coordinates and write (29-141) as

$$\exp\left[-i2\pi\left(\frac{d}{D}\right)\right] = U + iV = A\exp(-i\alpha) \tag{29-142a}$$

where A and α are real quantities and we have

$$A = \sqrt{U^2 + V^2} \tag{29-142b}$$

$$\alpha = \tan\left(\frac{-V}{U}\right) \tag{29-142c}$$

Finally, we take the natural logarithm of both sides of (29-142a) and obtain

$$-i2\pi\left(\frac{d}{D}\right) = \ln A - i\alpha \tag{29-143}$$

so

$$d = \frac{D}{2\pi}[\alpha + i\ln A] \tag{29-144}$$

where

$$D = \frac{\lambda}{2}\left[n_1^2 - (n_0\sin\phi_0)^2\right]^{-1/2} \tag{29-129}$$

If n_1 is real, then (29-144) can be iterated by using a range of values from $d = 0$ to $d = D$ until the correct value is found. We also observe that if n_1 and n_0 are real there is no imaginary part because d must be real.

If n_1 is complex, (29-144) should first be squared and the result separated into its real and imaginary parts. When this is done, we find that

$$d^2\left[n^2(1-\kappa^2) - n_0^2 \sin^2 \phi_0\right] = \left(\frac{\lambda}{4\pi}\right)^2 \left[\alpha^2 - (\ln A)^2\right] \tag{29-145a}$$

$$d^2 n^2 \kappa = -\alpha \ln A \tag{29-145b}$$

If n and κ are known, then (29-145a) can be iterated until the solution is found. The result can then be checked by using (29-145b). However, if both d and the optical constants n and κ are not known, then both equations can be iterated by using a range of values for d, n, and κ until the equations are satisfied. It is clear that this process is tedious at best, but is readily carried out on a digital computer. One can see that it is a time-consuming process even to write a computer program in order to evaluate the appropriate ellipsometric equations presented here. Fortunately, computer programs have been written and are available from manufacturers of ellipsometers.

Archer has carried out a well-known computer solution to the evaluation of ψ and Δ for a transparent film on a substrate of a single crystal of silicon. He solved the above equations and made a Cartesian plot of ψ and Δ as shown in Fig. 29-8. The constants used in the evaluation were an angle of incidence of 70.00°, a wavelength of 5461 Å, and a complex index of refraction for silicon of 4.050 − 0.028i.

Each curve in Fig. 29-8 is the locus of points of increasing thickness for a film of fixed index of refraction. The arrows show the direction of increasing thickness, and the underlined numbers are the indices of refraction of the films. A thickness scale is marked off on each curve in 20° increments in δ. The phase shift is denoted by δ, which is measured in degrees and given by

$$\delta = \left(\frac{360^\circ}{\lambda}\right) d\left[n_1^2 - \sin^2 \phi\right]^{1/2} \tag{29-146}$$

and may be used to convert from degrees to Ångström units. The δ scales for all of the curves have a common origin at 0°, which is the point $(\bar{\Delta}, \bar{\psi})$ for a film-free silicon surface. The quantities Δ and ψ are cyclic functions of thickness, and the curves repeat periodically with every 180° change in δ. For a film index of refraction 1.5, for example, the period is 2430 Å.

A significant property of the dependence of Δ and ψ on the index of refraction of the film is that, for all practical cases, no two curves overlap or intersect. Consequently, each point in the plane corresponds to a unique value for the index of refraction of the film. Strictly speaking, curves for very low and very high indices of refraction do intersect, but the extreme values are seldom, if ever, encountered. Although it is an academic point, as the index of refraction becomes indefinitely large, the corresponding curve coincides with the curve marked 100. Only the position of the δ scale on the curve shifts with increasing index of refraction.

The property of uniqueness allows the determination of the thickness and index of refraction of an unknown transparent film from a single measurement of Δ and ψ.

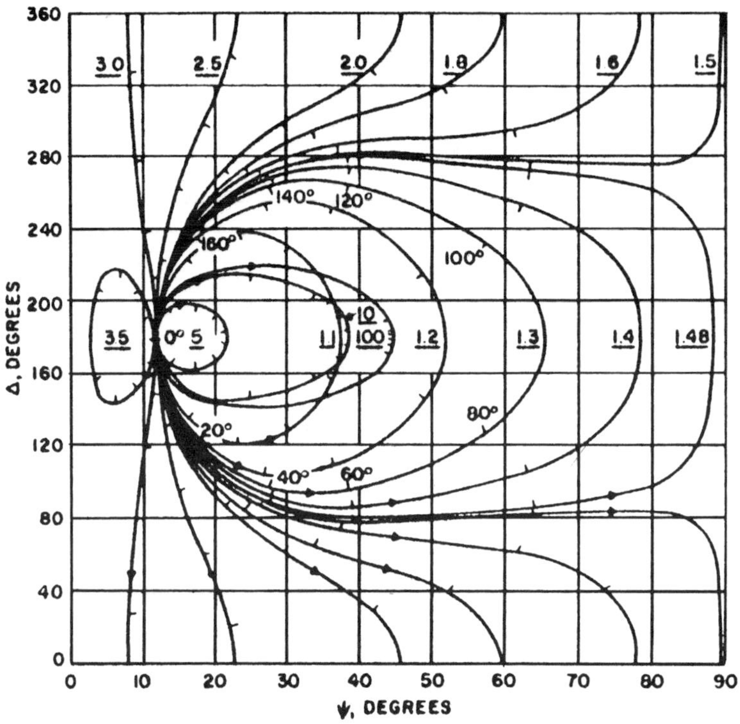

Figure 29-8 The dependence of Δ and ψ on the properties of transparent films on silicon. The parameter is the index of refraction of the film (underlined numbers). The thickness scale is marked off in 20° increments in δ. The thickness is given by $15.17\delta/(n^2 - 0.8830)^{1/2}$ Å. (From Archer.)

Figure 29-8 thus constitutes a nomogram for translating the measurement into thickness and index of refraction.

To summarize, in the precious section equations were developed to measure the ellipsometric parameters ψ and Δ; the measurement of these two parameters allows us to determine ρ. In this section the appropriate equations were solved to determine the thickness d and the optical constants n and κ from a knowledge of ρ. Specifically, this is accomplished by determining the complex reflectivities, (29-109) and (29-110) along with (29-111). With these values the quadratic equation for X (29-123a) and (29-120c) is solved, where a_2, a_1 and a_0 are given by (29-123a), (29-123b), and (29-123c): X is an exponential function for d, the thickness of the thin film, and by some further algebraic manipulation is determined by using either (29-144) or (29-145).

Ellipsometry has received wide attention for the past 40 years. The subject has been best described by Azzam and Bashara, and their text contains a wealth of information and knowledge as well as numerous references. In addition, they also treat in detail and with much mathematical skill the subject of polarized light, especially, as it relates to ellipsometry. Because of the wide range and applications of ellipsometry, the reader will find the references of great interest. The introduction to ellipsometry presented here should provide the interested reader with the background to read and understand the papers and books listed in the references.

29.4.1 Stokes' Treatment of Reflection and Refraction at an Interface

In the above discussion the reflection and transmission coefficients were used in the derivation of the equations of ellipsometry. In particular, it is necessary to know the reflection coefficients for a beam traveling from one medium to another, and vice versa. This problem appears to have first been treated by Stokes. In this section we derive these relations. A very clear discussion of this derivation has been given by Hecht and Zajac, and we follow their treatment closely.

Suppose we have an incident wave of amplitude E_{0i} incident on the planar interface separating two dielectric media as shown in Fig. 29-9. Since r and t are the fractional reflected and transmitted amplitudes, respectively (and where $n_i = n_1$ and $n_t = n_2$), we have $E_{0r} = rE_{0i}$ and $E_{0t} = tE_{0i}$. Fermat's principle also allows reversibility, that is, with the one proviso that there is no energy dissipation (absorption) a wave's direction of propagation can be reversed. In the language of physics one speaks of *time-reversal invariance*; i.e., if a process occurs, the reversed process can also occur.

In Fig. 29-9c two incident waves of amplitude $E_{0i}r$ and $E_{0i}t$ are shown. A portion of the wave whose amplitude is $E_{0i}t$ is both reflected and transmitted at the interface. Without making any assumptions let r' and t' be the amplitude reflection and transmission coefficients for a wave incident from below (i.e., $n_i = n_2$ and $n_t = n_1$). Consequently, the reflected portion is $E_{0i}tr'$, while that transmitted is $E_{0i}tt'$. Similarly, the incoming wave whose amplitude is $E_{0i}r$ splits into segments of amplitude $E_{0i}rr$ and $E_{0i}rt$. If the configuration of Fig. 29-9c is to be identical with that of Fig. 29-9b, we must have

$$E_0tt' + E_0rr = E_0 \tag{29-147}$$

$$E_0rt + E_0tr' = 0 \tag{29-148}$$

Hence,

$$tt' = 1 - r^2 \tag{29-149}$$

and

$$r' = -r \tag{29-150}$$

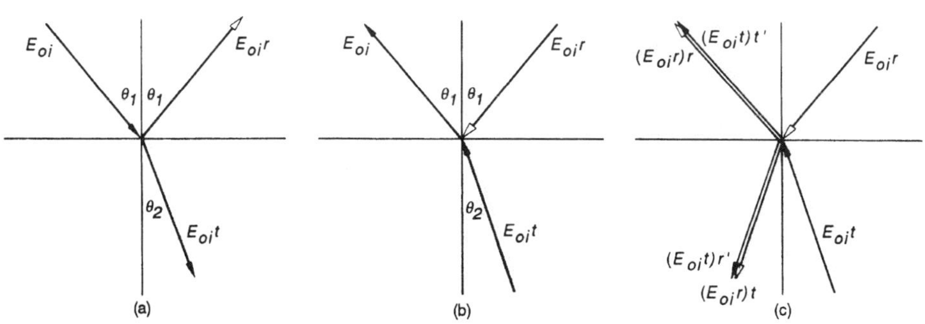

Figure 29-9 Reflection and refraction via the Stokes treatment. (From Hecht and Zajac, Ref. 8.)

which are Stokes' relations used in the main body of the text but written as $r' = r_{10}$, $r = r_{01}$ $t' = t_{10}$, and $t = t_{01}$. In their derivation Hecht and Zajac point out some other subtleties with respect to Stokes' treatment, and the reader is referred to their text for a further discussion.

29.5 FURTHER DEVELOPMENTS IN ELLIPSOMETRY: THE MUELLER MATRIX REPRESENTATION OF ψ AND Δ

The foundations of ellipsometry were developed primarily by P. Drude around 1890. At that time the optical sources were extremely limited with respect to their types and performance. Furthermore, it was only possible to measure ψ and Δ using the human eye as a detector, and this is only possible using a null-intensity condition. Thus, ellipsometry and its mathematical representation was developed under very restrictive conditions, namely, constant optical sources which allowed the settings and mechanical dial movements for the generating and analyzing polarizers to be moved relatively slowly until the null-intensity condition was found. In other words, classical ellipsometry can only be done under conditions in which the optical source and the sample (thin film) do not change and there is a considerable amount of time available to make the required measurements.

If we use optical sources of very short duration (e.g., pulsed lasers) or the sample is continually changing (e.g., the continuous deposition of an optical film on to a substrate), then clearly the classical formulation of the measurement process is inadequate. The concepts of representing the optical surface in terms of ψ and Δ are still, of course, valid, but a different procedure must be developed for measuring these quantities. Ideally, then, it would be useful to develop a formulation of ellipsometry which is valid regardless of the behavior of the optical source and the type of optical detector.

This can be done by reformulating the equations of ellipsometry in terms of the *ABCD* Mueller matrix and the Stokes polarization parameters. In this final section we develop this matrix and solve for ψ and Δ in terms of the Stokes parameters.

Consider that we have an optical beam incident on an optical surface. The Stokes vector of the incident beam is

$$S_0 = E_s E_s^* + E_p E_p^* \tag{29-18a}$$

$$S_1 = E_s E_s^* - E_p E_p^* \tag{29-18b}$$

$$S_2 = E_s E_p^* + E_p E_s^* \tag{29-18c}$$

$$S_3 = i(E_s E_p^* - E_p E_s^*) \tag{29-18d}$$

Similarly, the Stokes vector of the reflected beam is

$$S'_0 = R_s R_s^* + R_p R_p^* \tag{29-151a}$$

$$S'_1 = R_s R_s^* - R_p R_p^* \tag{29-151b}$$

$$S'_2 = R_s R_p^* + R_p R_s^* \tag{29-151c}$$

$$S'_3 = i(R_s R_p^* - R_p R_s^*) \tag{29-151d}$$

We saw earlier that the complex reflection coefficient are defined by

$$\rho_s = \frac{R_s}{E_s} \tag{29-152a}$$

$$\rho_p = \frac{R_p}{E_p} \tag{29-152b}$$

or

$$R_p = \rho_p E_p \tag{29-3a}$$

$$R_s = \rho_s E_s \tag{29-3b}$$

Substituting (29-3) into the equations for the reflected Stokes parameters, (29-151) yields

$$S_0' = (\rho_s \rho_s^*) E_s E_s^* + (\rho_p \rho_p^*) E_p E_p^* \tag{29-153a}$$

$$S_1' = (\rho_s \rho_s^*) E_s E_s^* - (\rho_p \rho_p^*) E_p E_p^* \tag{29-153b}$$

$$S_2' = (\rho_s \rho_p^*) E_s E_p^* + (\rho_p \rho_s^*) E_p E_s^* \tag{29-153c}$$

$$S_3' = i[(\rho_s \rho_p^*) E_s E_p^* - (\rho_p \rho_s^*) E_p E_s^*] \tag{29-153d}$$

We have (29-18) for the input Stokes vector and (29-153) for the output Stokes vector. The complete equation, with the resulting Mueller matrix, is

$$\begin{pmatrix} S_0' \\ S_1' \\ S_2' \\ S_3' \end{pmatrix} = \frac{1}{2} \begin{pmatrix} \rho_s\rho_s^* + \rho_p\rho_p^* & \rho_s\rho_s^* - \rho_p\rho_p^* & 0 & 0 \\ \rho_s\rho_s^* - \rho_p\rho_p^* & \rho_s\rho_s^* + \rho_p\rho_p^* & 0 & 0 \\ 0 & 0 & \rho_s\rho_p^* + \rho_p\rho_s^* & -i(\rho_s\rho_p^* - \rho_p\rho_s^*) \\ 0 & 0 & i(\rho_s\rho_p^* - \rho_p\rho_s^*) & \rho_s\rho_p^* + \rho_p\rho_s^* \end{pmatrix} \begin{pmatrix} S_0 \\ S_1 \\ S_2 \\ S_3 \end{pmatrix} \tag{29-154}$$

The matrix has the familiar form of the *ABCD* matrix.

We also saw that

$$\tan\psi = \frac{R_{0p}/R_{0s}}{E_{0p}/E_{0s}} \tag{29-155a}$$

$$\Delta = \beta - \alpha \tag{29-155b}$$

and

$$\rho = \frac{\rho_p}{\rho_s} = \tan\psi e^{i\Delta} \tag{29-155c}$$

The last relation can be written as

$$\rho_p = \rho_s \tan\psi e^{i\Delta} \tag{29-155d}$$

Substituting (29-155d) into (29-154), we then find that

$$\begin{pmatrix} S'_0 \\ S'_1 \\ S'_2 \\ S'_3 \end{pmatrix} = \frac{\rho_s \rho_s^*}{2} \begin{pmatrix} 1+\tan^2\psi & 1-\tan^2\psi & 0 & 0 \\ 1-\tan^2\psi & 1+\tan^2\psi & 0 & 0 \\ 0 & 0 & 2\tan\psi\cos\Delta & 2\tan\psi\sin\Delta \\ 0 & 0 & -2\tan\psi\sin\Delta & 2\tan\psi\cos\Delta \end{pmatrix} \begin{pmatrix} S_0 \\ S_1 \\ S_2 \\ S_3 \end{pmatrix} \tag{29-156}$$

Equation (29-156) represents ψ and Δ in terms of the *ABCD* Mueller matrix. The matrix can be used regardless of the duration of the optical source, that is, with both c.w. and pulsed optical sources. Because of this general formulation of ψ and Δ, (29-156) is of fundamental importance to ellipsometry. Equation (29-156) can be used to determine ψ and Δ using a specific polarization state of the incident beam. For example, consider an incident beam that is right circularly polarized so that its Stokes vector is

$$S = I_0 \begin{pmatrix} 1 \\ 0 \\ 0 \\ 1 \end{pmatrix} \tag{29-157}$$

Multiplication of (29-157) with (29-156) then yields the Stokes vector for the reflected beam:

$$S' = \begin{pmatrix} S'_0 \\ S'_1 \\ S'_2 \\ S'_3 \end{pmatrix} = \frac{\rho_s \rho_s^* I_0}{2} \begin{pmatrix} 1+\tan^2\psi \\ 1-\tan^2\psi \\ \tan\psi\sin\Delta \\ \tan\psi\cos\Delta \end{pmatrix} \tag{29-158}$$

Solving (29-158) for ψ and Δ in terms of the reflected Stokes parameters, S', we find that

$$\tan\psi = \left[\frac{S'_0 - S'_1}{S'_0 + S'_1}\right]^{1/2} \tag{29-159a}$$

$$\tan\Delta = \frac{S'_2}{S'_3} \tag{29-159b}$$

Thus, by measuring each of the four Stokes parameters of the reflected beam, we can determine ψ and Δ. In forming (29-159a) and (29-159b) we see that the factor $(\rho_s \rho_s^*)$ $I_0/2$ cancels out. Hence, we can simply drop the factor $\rho_s \rho_s^* I_0$, but we retain the 1/2 since this allows us to represent a polarizer and phase shifter (retarder) in their standard forms. Thus, the *ABCD* or Mueller matrix for ellipsometry is

$$M = \frac{1}{2} \begin{pmatrix} 1+\tan^2\psi & 1-\tan^2\psi & 0 & 0 \\ 1-\tan^2\psi & 1+\tan^2\psi & 0 & 0 \\ 0 & 0 & 2\tan\psi\cos\Delta & 2\tan\psi\sin\Delta \\ 0 & 0 & -2\tan\psi\sin\Delta & 2\tan\psi\cos\Delta \end{pmatrix} \tag{29-160}$$

The form of (29-160) for an ideal polarizer and an ideal compensator is easily found. For a perfect polarizer there is no phase shift, so $\Delta = 0$ and (29-160) is written as

$$M_{\text{pol}} = \frac{1}{2}\begin{pmatrix} 1+\tan^2\psi & 1-\tan^2\psi & 0 & 0 \\ 1-\tan^2\psi & 1+\tan^2\psi & 0 & 0 \\ 0 & 0 & 2\tan\psi & 0 \\ 0 & 0 & 0 & 2\tan\psi \end{pmatrix} \tag{29-161}$$

Equation (29-161) is another representation of a linear polarizer. As an example of (29-161), an ideal linear horizontal polarizer is described by

$$M_{\text{pol}} = \frac{1}{2}\begin{pmatrix} 1 & 1 & 0 & 0 \\ 1 & 1 & 0 & 0 \\ 0 & 0 & 0 & 0 \\ 0 & 0 & 0 & 0 \end{pmatrix} \tag{29-162}$$

Comparing (29-162) with (29-161) we see that $\tan\psi = 0$. According to the definition given by (29-155a), this is exactly what we would expect if there were no R_{0p} component but only an R_{0s} component. Similarly, the Mueller matrix for a perfect compensator is

$$M_{\text{comp}} = \begin{pmatrix} 1 & 0 & 0 & 0 \\ 0 & 1 & 0 & 0 \\ 0 & 0 & \cos\phi & \sin\phi \\ 0 & 0 & -\sin\phi & \cos\phi \end{pmatrix} \tag{29-163}$$

Comparing (29-163) with (29-160), we see that we must have

$$M = \begin{pmatrix} 1 & 0 & 0 & 0 \\ 0 & 1 & 0 & 0 \\ 0 & 0 & \cos\Delta & \sin\Delta \\ 0 & 0 & -\sin\Delta & \cos\Delta \end{pmatrix} \tag{29-164}$$

and $\tan^2\psi = 1$; (29-164) shows that the emerging beam is unattenuated and the magnitude of the reflected beam is unchanged from the incident beam. This, too, is the behavior expected of a perfect phase-shifting material. From (29-163) and (29-164) we see also that $\Delta = \phi$ as expected.

Let us now determine ψ and Δ in (29-160) by generating an elliptically polarized beam as before using a linear polarizer at angle P and a quarter-wave retarder fixed at $+45^\circ$. The Stokes vector of the beam incident on the sample is

$$S = I_0\begin{pmatrix} 1 \\ 0 \\ \sin 2P \\ -\cos 2P \end{pmatrix} \tag{29-165}$$

Multiplying (29-165) by (29-160), we find that the reflected Stokes vector is

$$S' = \frac{I_0}{2}\begin{pmatrix} 1+\tan^2\psi \\ 1-\tan^2\psi \\ 2\tan\psi\sin(2P-\Delta) \\ 2\tan\psi\cos(2P-\Delta) \end{pmatrix} \tag{29-166}$$

This is, of course, the Stokes vector of elliptically polarized light. In order for the reflected light to be linearly polarized, we must have

$$\cos(2P-\Delta) = 0 \tag{29-167}$$

Thus, (29-167) is satisfied if the generating linear polarizer is set to

$$2P_1 - \Delta = 90^\circ \tag{29-168a}$$

or

$$2P_2 - \Delta = -90^\circ \tag{29-168b}$$

Thus, solving (29-168a) and (29-168b) for Δ gives

$$\Delta = 2P_1 - 90^\circ \tag{29-169a}$$

$$\Delta = 2P_2 + 90^\circ \tag{29-169b}$$

so

$$\Delta = 2P_2 + 90^\circ = 2P_1 - 90^\circ \tag{29-170}$$

Equation (29-170) is recognized as the condition that was obtained before on the measurement of Δ when the problem was treated following the classical formulation in Section 29.3. Substracting (29-169b) from (29-169a), we then find that

$$P_2 = P_1 - 90^\circ \tag{29-171}$$

We note that for the condition (29-171) the reflected Stokes vector becomes

$$S' = \frac{I_0}{2}\begin{pmatrix} 1+\tan^2\psi \\ 1-\tan^2\psi \\ \pm 2\tan\psi \\ 0 \end{pmatrix} \tag{29-172}$$

where the $\pm$ sign refers to (29-169a) and (29-169b), respectively.

In order to find $\tan\psi$, or ψ, we now consider the null-intensity condition created by using an analyzing linear polarizer. The Mueller matrix of the analyzer is

$$M = \frac{1}{2}\begin{pmatrix} 1 & \cos 2Q & \sin 2Q & 0 \\ \cos 2Q & \cos^2 2Q & \cos 2Q\sin 2Q & 0 \\ \sin 2Q & \sin 2Q\cos 2Q & \sin^2 2Q & 0 \\ 0 & 0 & 0 & 0 \end{pmatrix} \tag{29-173}$$

We now assume that the angle P has been adjusted so that the reflected beam has become linearly polarized and is represented by (29-172). The intensity of the beam emerging from the analyzer is obtained by multiplying (29-172) by (29-173) and writing the first Stokes parameter as $I(\psi, Q)$:

$$I(\psi, Q) = \frac{I_0}{4}\left[\left(1 + \tan^2\psi\right) + \left(1 - \tan^2\psi\right)\cos 2Q \pm 2\tan\psi\sin 2Q\right] \quad (29\text{-}174)$$

where the + sign refers to the P_1 condition and the – sign refers to the P_2 condition, respectively. The null-intensity conditions for Q_1 and Q_2 corresponding to P_1 and P_2 are, respectively,

$$I(\psi_1, Q_1) = 0 = \left(1 + \tan^2\psi\right) + \left(1 - \tan^2\psi\right)\cos 2Q_1 + 2\tan\psi\sin 2Q_1 \quad (29\text{-}175a)$$

$$I(\psi_1, Q_2) = 0 = \left(1 + \tan^2\psi\right) + \left(1 - \tan^2\psi\right)\cos 2Q_2 - 2\tan\psi\sin 2Q_2 \quad (29\text{-}175b)$$

Subtracting (29-175b) from (29-175a) gives

$$\left(1 - \tan^2\psi\right)[\cos 2Q_1 - \cos 2Q_2] + 2\tan\psi[\sin 2Q_1 + \sin 2Q_2] = 0 \quad (29\text{-}176)$$

Equation (29-176) can only be satisfied if

$$\cos 2Q_1 - \cos 2Q_2 = 0 \quad (29\text{-}177a)$$

and

$$\sin 2Q_1 + \sin 2Q_2 = 0 \quad (29\text{-}177b)$$

Squaring (29-177a) and (29-177b) and adding the results yields

$$\cos 2Q_1 \cos 2Q_2 - \sin 2Q_1 \sin 2Q_2 = 1 \quad (29\text{-}178a)$$

or

$$\cos(2Q_1 + 2Q_2) = 1 \quad (29\text{-}178b)$$

Thus, we find that

$$Q_2 = -Q_1 \quad (29\text{-}179a)$$

$$Q_2 = 90^\circ - Q_1 \quad (29\text{-}179b)$$

which are exactly the conditions found earlier for the analyzer.

With a knowledge of Q_1 (or Q_2) we can now solve for $\tan\psi$ and ψ. We see that (29-175a) [or (29-175b)] can be rearranged as a quadratic equation:

$$(1 - \cos 2Q_1)\tan^2\psi + 2\sin 2Q_1\tan\psi + (1 + \cos 2Q_1) = 0 \quad (29\text{-}180)$$

Equation (29-180) can be solved to obtain

$$\tan\psi = \frac{-\sin 2Q_1}{1 - \cos 2Q_1} \quad (29\text{-}181a)$$

which reduces to

$$\tan\psi = -\cot Q_1 \quad (29\text{-}181b)$$

The tangent and cotangent functions in (29-181b) can be rewritten in terms of their sine and cosine functions; thus,

$$\cos(\psi - Q_1) = 0 \tag{29-182a}$$

and we finally have

$$\psi = 90^\circ - Q_1 = 270^\circ - Q_1 \tag{29-182b}$$

Equations (29-182a) and (29-170) are of fundamental importance, so they are rewritten here as the pair

$$\Delta = 2P_2 + 90^\circ = 2P_1 - 90^\circ \tag{29-170}$$

and

$$\psi = 90^\circ - Q_1 = 270^\circ - Q_1 \tag{29-182b}$$

Finally, in the foregoing analysis the angular settings on the polarizer and the compensator in the generating arm were made so that linearly polarized, rather than elliptically polarized, light was reflected from the optical sample.

Let us now assume that these adjustments are not carried out first, but that we wish to determine the conditions on the settings such that the intensity of the beam emerging from the analyzer is a minimum, which in this case is zero (null). The intensity of the beam is found by multiplying (29-166) and (29-173), so we have

$$\begin{aligned} I(\psi, \Delta, P, Q) = \frac{I_0}{4}[(1 + \tan^2\psi) + (1 - \tan^2\psi)\cos 2Q \\ + 2\tan\psi \sin(2P - \Delta)\sin 2Q] \end{aligned} \tag{29-183}$$

The minimum intensity is found from the conditions:

$$\frac{\partial I(\psi, \Delta, P, Q)}{\partial P} = 0 \tag{29-184a}$$

$$\frac{\partial I(\psi, \Delta, P, Q)}{\partial Q} = 0 \tag{29-184b}$$

Differentiating (29-183) according to (29-184a) leads immediately to

$$\cos(2P - \Delta) = 0 \tag{29-185}$$

which is exactly the same result we obtained in (29-167); that is,

$$2P - \Delta = 90^\circ, 270^\circ \tag{29-186}$$

Next, (29-183) is differentiated according to (29-184b) and we find that

$$\tan\psi = -\frac{(1 + \cos 2Q)}{\sin 2Q} = -\cot Q \tag{29-187}$$

which is identical to (29-181b)

We thus see that we can obtain all the previous conditions derived in Section 29.3 relating Δ and ψ to P and Q. we emphasize that with quantitative optical detectors the optical surface can be irradiated, for example, with right circularly polarized light, whereupon the measurement of all four Stokes parameters can then yield Δ and ψ, (29-159a) and (29-159b).

This concludes our discussion of ellipsometry. We see that the Stokes polarization parameters and the Mueller matrix allow us not only to obtain easily the formulas of classical ellipsometry, as was done in previous sections, but to reformulate the subject in a very general way, namely, representing an optical surface in terms of the *ABCD* (or Mueller) matrix.

REFERENCES

Papers

1. Drude, P., *Ann. Phys. Chem.*, *36*, 532, 865 (1889).
2. Drude, P., *Ann. Phys. Chem.*, *39*, 481, (1890).
3. Mueller, R. H., *Surf.* Sci., *56*, 19 (1976).
4. Hauge, P. S., *Surf. Sci.*, *96*, 108 (1980).
5. Winterbottom, A. B., in *Ellipsometry in the Measurement of Surfaces and Films*, Symp. Proc., Washington, DC, 1963, (E. Passaglia et al., eds.), N. B. S. Miscellaneous Publ. 256, Washington, DC, 1964, p. 97.
6. McCrackin, F. L., Passaglia, E., Stromberg, R. R. and Steinberg, H. L., *J. Res. Natl. Bur. Std.*, *67A*, 363 (1963).
7. Holmes, D. A. and Feucht, D. L., *J. Opt. Soc. Am.*, *57*, 283 (1967).
8. Aspnes, D. E., *J. Opt. Soc. Am.*, *6*, 639 (1974).
9. Aspnes, D. E. and Studna, A. A., *Appl. Opt.*, *10*, 1024 (1971).
10. Muller, R. S., *Surf. Sci.*, *16*, 34 (1979).
11. Archer, R. J., *J. Opt. Soc. Am.*, *52*, 970 (1962).
12. Azzam, R. M. A. and Bashara, N. M., *Appl. Opt.*, *11*, 2210 (1972).
13. Azzam, R. M. A. and Bashara, N. M., *Appl. Phys.*, *1*, 203 (1973).
14. Moritani, A., Okuda, Y., Kubo, H., and Nakai, J., *Appl. Opt.*, *22*, 2429 (1983).
15. Jellison, G. E., Jr. and Modine, F. A., *J. Opt. Soc. Am.* *72*, 1253 (1982).
16. Collett, E., *Surf, Sci.*, *96*, 156 (1980).
17. Jellison, G. E., Jr. and Lowndes D. H., *Appl. Opt.* *24*, 2948 (1985).

Books

1. Born, M. and Wolf, E., *Principles of Optics*, 3rd, ed., Pergamon Press, New York, 1965.
2. Azzam, R. M. A. and Bashara, N. M., *Ellipsometry and Polarized Light*, North-Holland, Amsterdam, 1977.
3. Heavens, O. S., *Optical Properties of Thins Solid Films*, Dover, New York, 1965.
4. Vasicek, A., *Optics of Thin Films*, North-Holland, Amsterdam, 1960.
5. Abeles, F., in *Progress in Optics*, Vol. II (E. Wolf, ed.), North-Holland, Amsterdam, 1963.
6. Strong, J., *Concepts of Classical Optics*, Freeman, San Francisco, 1959.
7. Stone, J. M., *Radiation and Optics*, McGraw-Hill, New York, 1963.
8. Hecht, E. and Zajac, A. *Optics*, Addison-Wesley, reading, MA, 1974.
9. Gerrard, A. and Burch, J. M., *Introduction to Matrix Methods in Optics*, Wiley, London, 1975.
10. Clarke, D. and Grainger, J. F., *Polarized Light and Optical Measurement*, Pergamon Press, Oxford, 1971.

Appendix A

Jones and Stokes Vectors

	Normalized Jones Vectors	Normalized Stokes Vectors	
Linear horizontally polarized light	$\begin{bmatrix} 1 \\ 0 \end{bmatrix}$	$\begin{bmatrix} 1 \\ 1 \\ 0 \\ 0 \end{bmatrix}$	(A.1)
Linear vertically polarized light	$\begin{bmatrix} 0 \\ 1 \end{bmatrix}$	$\begin{bmatrix} 1 \\ -1 \\ 0 \\ 0 \end{bmatrix}$	(A.2)
Linear 45° polarized light	$\frac{1}{\sqrt{2}}\begin{bmatrix} 1 \\ 1 \end{bmatrix}$	$\begin{bmatrix} 1 \\ 0 \\ 1 \\ 0 \end{bmatrix}$	(A.3)
Linear −45° polarized light	$\frac{1}{\sqrt{2}}\begin{bmatrix} 1 \\ -1 \end{bmatrix}$	$\begin{bmatrix} 1 \\ -1 \\ 0 \\ 0 \end{bmatrix}$	(A.4)
Right circularly polarized light	$\frac{1}{\sqrt{2}}\begin{bmatrix} 1 \\ -i \end{bmatrix}$	$\begin{bmatrix} 1 \\ 0 \\ 0 \\ 1 \end{bmatrix}$	(A.5)
Left circularly polarized light	$\frac{1}{\sqrt{2}}\begin{bmatrix} 1 \\ i \end{bmatrix}$	$\begin{bmatrix} 1 \\ 0 \\ 0 \\ -1 \end{bmatrix}$	(A.6)

Appendix B

Jones and Mueller Matrices

Jones matrix for free space
$$\begin{bmatrix} 1 & 0 \\ 0 & 1 \end{bmatrix} \tag{B.1}$$

Jones matrix for an isotropic absorbing material whose transmittance is p^2
$$\begin{bmatrix} p & 0 \\ 0 & p \end{bmatrix} \tag{B.2}$$

Jones matrix for linear polarizer at 0°
$$\begin{bmatrix} 1 & 0 \\ 0 & 0 \end{bmatrix} \tag{B.3}$$

Jones matrix for linear polarizer at 90°
$$\begin{bmatrix} 0 & 0 \\ 0 & 1 \end{bmatrix} \tag{B.4}$$

Jones matrix for linear polarizer at 45°
$$\frac{1}{2}\begin{bmatrix} 1 & 1 \\ 1 & 1 \end{bmatrix} \tag{B.5}$$

Jones matrix for a right circular polarizer
$$\frac{1}{2}\begin{bmatrix} 1 & i \\ -i & 1 \end{bmatrix} \tag{B.6}$$

Jones matrix for a left circular polarizer
$$\frac{1}{2}\begin{bmatrix} 1 & -i \\ i & 1 \end{bmatrix} \tag{B.7}$$

Jones matrix for a linear retarder at angle θ
$$\begin{bmatrix} \cos^2\theta & \cos\theta\sin\theta \\ \cos\theta\sin\theta & \sin^2\theta \end{bmatrix} \tag{B.8}$$

Jones matrix for linear retarder with fast axis at angle θ and retardation δ

$$\begin{bmatrix} e^{i\delta}\cos^2\theta + \sin^2\theta & (e^{i\delta}-1)\sin\theta\cos\theta \\ (e^{i\delta}-1)\sin\theta\cos\theta & e^{i\delta}\sin^2\theta + \cos^2\theta \end{bmatrix} \tag{B.9}$$

Jones matrix for quarter wave linear retarder with fast axis at 0°

$$\begin{bmatrix} e^{i\pi/4} & 0 \\ 0 & e^{-i\pi/4} \end{bmatrix} \tag{B.10}$$

Jones matrix for half-wave retarder with fast axis at 45° $\begin{bmatrix} 0 & 1 \\ 1 & 0 \end{bmatrix}$ (B.11)

Mueller matrix for free space

$$\begin{bmatrix} 1 & 0 & 0 & 0 \\ 0 & 1 & 0 & 0 \\ 0 & 0 & 1 & 0 \\ 0 & 0 & 0 & 1 \end{bmatrix} \tag{B.12}$$

Mueller matrix for an isotropic absorbing material whose transmittance is k

$$\begin{bmatrix} k & 0 & 0 & 0 \\ 0 & k & 0 & 0 \\ 0 & 0 & k & 0 \\ 0 & 0 & 0 & k \end{bmatrix} \tag{B.13}$$

Mueller matrix for a linear polarizer at angle θ

$$\frac{1}{2}\begin{bmatrix} 1 & \cos 2\theta & \sin 2\theta & 0 \\ \cos 2\theta & \cos^2 2\theta & \cos 2\theta \sin 2\theta & 0 \\ \sin 2\theta & \cos 2\theta \sin 2\theta & \sin^2 2\theta & 0 \\ 0 & 0 & 0 & 1 \end{bmatrix} \tag{B.14}$$

Mueller matrix for a horizontal linear polarizer

$$\frac{1}{2}\begin{bmatrix} 1 & 1 & 0 & 0 \\ 1 & 1 & 0 & 0 \\ 0 & 0 & 0 & 0 \\ 0 & 0 & 0 & 0 \end{bmatrix} \tag{B.15}$$

Mueller matrix for a vertical linear polarizer

$$\frac{1}{2}\begin{bmatrix} 1 & -1 & 0 & 0 \\ -1 & 1 & 0 & 0 \\ 0 & 0 & 0 & 0 \\ 0 & 0 & 0 & 0 \end{bmatrix} \tag{B.16}$$

Mueller matrix for a linear polarizer at 45°
$$\frac{1}{2}\begin{bmatrix} 1 & 0 & 1 & 0 \\ 0 & 0 & 0 & 0 \\ 1 & 0 & 1 & 0 \\ 0 & 0 & 0 & 0 \end{bmatrix} \tag{B.17}$$

Mueller matrix for a right circular polarizer
$$\frac{1}{2}\begin{bmatrix} 1 & 0 & 0 & 1 \\ 0 & 0 & 0 & 0 \\ 0 & 0 & 0 & 0 \\ 1 & 0 & 0 & 1 \end{bmatrix} \tag{B.18}$$

Mueller matrix for a left circular polarizer
$$\frac{1}{2}\begin{bmatrix} 1 & 0 & 0 & -1 \\ 0 & 0 & 0 & 0 \\ 0 & 0 & 0 & 0 \\ -1 & 0 & 0 & 1 \end{bmatrix} \tag{B.19}$$

Mueller matrix for a linear retarder with fast axis at angle θ and retardation δ
$$\begin{bmatrix} 1 & 0 & 0 & 0 \\ 0 & \cos^2 2\theta + \sin^2 2\theta\cos\delta & (1-\cos\delta)\sin 2\theta\cos 2\theta & -\sin 2\theta\sin\delta \\ 0 & (1-\cos\delta)\sin 2\theta\cos 2\theta & \sin^2 2\theta + \cos^2 2\theta\cos\delta & \cos 2\theta\sin\delta \\ 0 & \sin 2\theta\sin\delta & -\cos 2\theta\sin\delta & \cos\delta \end{bmatrix} \tag{B.20}$$

Linear quarter wave retarder with fast axis at 0°
$$\begin{bmatrix} 1 & 0 & 0 & 0 \\ 0 & 1 & 0 & 0 \\ 0 & 0 & 0 & 1 \\ 0 & 0 & -1 & 0 \end{bmatrix} \tag{B.21}$$

Linear half-wave retarder with fast axis at 45°
$$\begin{bmatrix} 1 & 0 & 0 & 0 \\ 0 & -1 & 0 & 0 \\ 0 & 0 & 1 & 0 \\ 0 & 0 & 0 & -1 \end{bmatrix} \tag{B.22}$$

Appendix C

Relationships Between the Jones and Mueller Matrix Elements

Mueller matrix elements in terms of Jones matrix elements:

$$m_{11} = (j_{11}j_{11}^* + j_{12}j_{12}^* + j_{21}j_{21}^* + j_{22}j_{22}^*)/2 \tag{C.1}$$

$$m_{21} = (j_{11}j_{11}^* + j_{21}j_{21}^* - j_{12}j_{12}^* - j_{22}j_{22}^*)/2 \tag{C.2}$$

$$m_{13} = (j_{12}j_{11}^* + j_{22}j_{21}^* + j_{11}j_{12}^* + j_{21}j_{22}^*)/2 \tag{C.3}$$

$$m_{14} = i(j_{12}j_{11}^* + j_{22}j_{21}^* - j_{11}j_{12}^* - j_{21}j_{22}^*)/2 \tag{C.4}$$

$$m_{21} = (j_{11}j_{11}^* + j_{12}j_{12}^* - j_{21}j_{21}^* - j_{22}j_{22}^*)/2 \tag{C.5}$$

$$m_{22} = (j_{11}j_{11}^* - j_{21}j_{21}^* - j_{12}j_{12}^* + j_{22}j_{22}^*)/2 \tag{C.6}$$

$$m_{23} = (j_{11}j_{12}^* + j_{12}j_{11}^* - j_{21}j_{22}^* - j_{22}j_{21}^*)/2 \tag{C.7}$$

$$m_{24} = i(j_{12}j_{11}^* + j_{21}j_{22}^* - j_{22}j_{21}^* - j_{11}j_{12}^*)/2 \tag{C.8}$$

$$m_{31} = (j_{11}j_{22}^* + j_{21}j_{11}^* + j_{12}j_{22}^* + j_{22}j_{12}^*)/2 \tag{C.9}$$

$$m_{32} = (j_{11}j_{21}^* + j_{21}j_{11}^* - j_{12}j_{22}^* - j_{22}j_{12}^*)/2 \tag{C.10}$$

$$m_{33} = (j_{11}j_{22}^* + j_{12}j_{21}^* + j_{21}j_{12}^* + j_{22}j_{11}^*)/2 \tag{C.11}$$

$$m_{34} = i(-j_{11}j_{22}^* + j_{12}j_{21}^* - j_{21}j_{12}^* + j_{22}j_{11}^*)/2 \tag{C.12}$$

$$m_{41} = i(j_{11}j_{21}^* + j_{12}j_{22}^* - j_{21}j_{11}^* - j_{22}j_{12}^*)/2 \tag{C.13}$$

$$m_{42} = i(j_{11}j_{21}^* - j_{12}j_{22}^* - j_{21}j_{11}^* + j_{22}j_{12}^*)/2 \tag{C.14}$$

$$m_{43} = i(j_{11}j_{22}^* + j_{12}j_{21}^* - j_{21}j_{12}^* - j_{22}j_{11}^*)/2 \tag{C.15}$$

$$m_{44} = (j_{11}j_{22}^* - j_{12}j_{21}^* - j_{21}j_{12}^* + j_{22}j_{11}^*)/2 \tag{C.16}$$

Expressing the Jones matrix elements in polar form, i.e. $j = re^{i\theta}$, the Jones matrix elements in terms of the Mueller matrix elements are:

$$r_{11} = [(m_{11} + m_{12} + m_{21} + m_{22})/2]^{1/2} \tag{C.17}$$

$$r_{12} = [(m_{11} - m_{12} + m_{21} - m_{22})/2]^{1/2} \tag{C.18}$$

$$r_{21} = [(m_{11} + m_{12} - m_{21} - m_{22})/2]^{1/2} \tag{C.19}$$

$$r_{22} = [(m_{11} - m_{12} - m_{21} + m_{22})/2]^{1/2} \tag{C.20}$$

$$\cos(\theta_{11} - \theta_{12}) = \frac{(m_{13} + m_{23})}{\left[(m_{11} + m_{21})^2 - (m_{12} + m_{22})^2\right]^{1/2}} \tag{C.21}$$

$$\sin(\theta_{11} - \theta_{12}) = \frac{(m_{14} + m_{24})}{\left[(m_{11} + m_{21})^2 - (m_{12} + m_{22})^2\right]^{1/2}} \tag{C.22}$$

$$\cos(\theta_{21} - \theta_{11}) = \frac{(m_{31} + m_{32})}{\left[(m_{11} + m_{12})^2 - (m_{21} + m_{22})^2\right]^{1/2}} \tag{C.23}$$

$$\sin(\theta_{21} - \theta_{11}) = \frac{(m_{41} + m_{42})}{\left[(m_{11} + m_{12})^2 - (m_{21} + m_{22})^2\right]^{1/2}} \tag{C.24}$$

$$\cos(\theta_{11} - \theta_{22}) = \frac{(m_{33} + m_{44})}{\left[(m_{11} + m_{22})^2 - (m_{21} + m_{12})^2\right]^{1/2}} \tag{C.25}$$

$$\sin(\theta_{22} - \theta_{11}) = \frac{(m_{43} - m_{34})}{\left[(m_{11} + m_{22})^2 - (m_{21} + m_{12})^2\right]^{1/2}} \tag{C.26}$$

Appendix D

Vector Representation of the Optical Field: Application to Optical Activity

We have emphasized the Stokes vector and Jones matrix formulation for polarized light. However, polarized light was first represented by another formulation introduced by Fresnel and called the vector representation for polarized light. This representation is still much used and for the sake of completeness we discuss it. This formulation was introduced by Fresnel to describe the remarkable phenomenon of optical activity in which the "plane of polarization" of a linearly polarized beam was rotated as the optical field propagated through an optically active medium. Fresnel's mathematical description of this phenomenon was a brilliant success. After we have discussed the vector representation we shall apply it to describe the propagation of light through an optically active medium.

For a plane wave propagating in the z direction the components of the optical field in the xy plane are

$$E_x(z,t) = E_{0x}\cos(kz - \omega t + \delta_x) \tag{D-1a}$$

$$E_y(z,t) = E_{0y}\cos\big(kz - \omega t + \delta_y\big) \tag{D-1b}$$

Eliminating the propagator $kz-\omega t$ between (D-1a) and (D-1b) yields

$$\frac{E_x^2(z,t)}{E_{0x}^2} + \frac{E_y^2(z,t)}{E_{0y}^2} - \frac{2E_x(z,t)E_y(z,t)\cos\delta}{E_{0x}E_{0y}} = \sin^2\delta \tag{D-2}$$

The Stokes vector corresponding to (D-1) is, of course,

$$S = \begin{pmatrix} E_{0x}^2 + E_{0y}^2 \\ E_{0x}^2 - E_{0y}^2 \\ 2E_{0x}E_{0y}\cos\delta \\ 2E_{0x}E_{0y}\sin\delta \end{pmatrix} \tag{D-3}$$

In the xy plane we construct the vector $\mathbf{E}(z,t)$:

$$\mathbf{E}(z,t) = E_x(z,t)\mathbf{i} + E_y(z,t)\mathbf{j} \tag{D-4}$$

where **i** and **j** are unit vectors in the x and y directions, respectively. Substituting (D-1) into (D-4) gives

$$\mathbf{E}(z,t) = E_{0x}\cos(kz-\omega t+\delta_x)\mathbf{i} + E_{0y}\cos(kz-\omega t+\delta_y)\mathbf{j} \tag{D-5}$$

We can also express the optical field in terms of complex quantities by writing

$$E_x(z,t) = E_{0x}\cos(kz-\omega t+\delta_x) = \text{Re}\{E_{0x}\exp[i(kz-\omega t+\delta_x)]\} \tag{D-6a}$$

$$E_y(z,t) = E_{0y}\cos(kz-\omega t+\delta_y) = \text{Re}\{E_{0y}\exp[i(kz-\omega t+\delta_y)]\} \tag{D-6b}$$

where Re{...} means the real part is to be taken. In complex quantities (D-5) can be written as

$$\mathbf{E}(z,t) = E_{0x}\exp(i\delta_x)\mathbf{i} + E_{0y}\exp(i\delta_y)\mathbf{j} \tag{D-7}$$

In (D-7) we have factored out and then suppressed the exponential propagator [exp$i(kz$-$\omega t)$], since it vanishes when the intensity is formed. Further, factoring out the term $\exp(i\delta_x)$ in (D-7), we can write

$$\mathbf{E}(z,t) = E_{0x}\mathbf{i} + E_{0y}\exp(i\delta)\mathbf{j} \tag{D-8}$$

where $\delta = \delta_y - \delta_x$.

The exponential propagator [exp$i(kz-\omega$t)] is now restored in (D-8) and the real part taken:

$$\mathbf{E}(z,t) = E_{0x}\cos(kz-\omega t)\mathbf{i} + E_{0y}\exp(kz-\omega t+\delta)\mathbf{j} \tag{D-9}$$

Equation (D-9) is the vector representation for elliptically polarized light. There are two special forms of (D-9). The first is for $\delta = 0°$ or $180°$, which leads to linearly polarized light at an angle ψ [see (D-2)]. If either E_{0y} or E_{0x} is zero, we have linear horizontally polarized light or linear vertically polarized light respectively. For linearly polarized light (D-9) reduces to

$$\mathbf{E}(z,t) = (E_{0x}\mathbf{i} \pm E_{0y}\mathbf{j})\cos(kz-\omega t) \tag{D-10}$$

where $\pm$ corresponds to $\delta = 0°$ and $180°$, respectively. The corresponding Stokes vector is seen from (D-3) to be

$$S = \begin{pmatrix} E_{0x}^2 + E_{0y}^2 \\ E_{0x}^2 - E_{0y}^2 \\ \pm 2E_{0x}E_{0y} \\ 0 \end{pmatrix} \tag{D-11}$$

The orientation angle ψ of the linearly polarized light is

$$\tan 2\psi = \frac{S_2}{S_1} = \frac{\pm 2E_{0x}E_{0y}}{E_{0x}^2 - E_{0y}^2} \tag{D-12}$$

From the well-known trigonometric half-angle formulas we readily find that

$$\tan\psi = \frac{E_{0y}}{E_{0x}} \tag{D-13}$$

which is exactly what we would expect from inspection of (D-10).

The other special form of (D-9) is for $\delta = -90^\circ$ or 90°, whereupon the polarization ellipse reduces to the standard form of an ellipse. This reduces further to the equation of a circle if $E_{0x} = E_{0y} = E_0$. For $\delta = -90^\circ$, (D-9) reduces to

$$\mathbf{E}(z, t) = E_0[\cos(kz - \omega t)\mathbf{i} + \sin(kz - \omega t)\mathbf{j}] \tag{D-14}$$

and for $\delta = 90^\circ$

$$\mathbf{E}(z, t) = E_0[\cos(kz - \omega t)\mathbf{i} - \sin(kz - \omega t)\mathbf{j}] \tag{D-15}$$

The behavior of (D-14) and (D-15) is readily seen by considering the equations at $z = 0$ and then allowing ωt to take on the values 0 to 2π radians in intervals of $\pi/2$. One readily sees that (D-14) describes a vector $\mathbf{E}(z, t)$ which rotates *clockwise* at an angular frequency of ω. Consequently, (D-14) is said to describe *left circularly polarized* light. Similarly, in (D-15), $\mathbf{E}(z, t)$ rotates *counterclockwise* as the wave propagates toward the viewer and, therefore, we have *right circularly polarized* light.

Equations (D-14) and (D-15) lead to a very interesting observation. If we label $\mathbf{E}(z, t)$ in (D-14) and (D-15) as $\mathbf{E}_l(z, t)$ and $\mathbf{E}_r(z, t)$, respectively, and add the two equations we see that

$$\mathbf{E}_l(z, t) + \mathbf{E}_r(z, t) = 2E_0 \cos(\omega t - kz)\mathbf{i} = E_x(z, t)\mathbf{i} \tag{D-16}$$

Thus, a linearly polarized wave can be synthesized from two oppositely polarized circular waves of equal amplitude. This property played a key role in enabling Fresnel to describe the propagation of a beam in an optically active medium. The vector representation introduced by Fresnel revealed for the first time the mathematical existence of circularly polarized light; before Fresnel no one suspected the possible existence of circularly polarized light. Before we conclude this section another important property of the vector formulation must be discussed.

Elliptically polarized light can be decomposed into two orthogonal polarized states (coherent decomposition). We consider the form of the polarization ellipse which can be represented in terms of (1) linearly $\pm 45^\circ$ polarized light and (2) right and left circularly polarized light, respectively. We decompose an elliptically polarized beam into linear $\pm 45^\circ$ states of arbitrary amplitudes A and B (real) and write (D-8) as

$$\mathbf{E}(z, t) = E_{0x}\mathbf{i} + E_{0y}\exp(i\delta)\mathbf{j} = A(\mathbf{i} + \mathbf{j}) + B(\mathbf{i} - \mathbf{j}) \tag{D-17a}$$

$$= (A + B)\mathbf{i} + (A - B)\mathbf{j} \tag{D-17b}$$

Taking the vector dot product of the left- and right-hand sides of (D-17) and equating terms yields

$$E_{0x} = A + B \tag{D-18a}$$

$$E_{0y}e^{i\delta} = A - B \tag{D-18b}$$

Because A and B are real quantities, the left-hand side of (D-18b) can be real only for $\delta = 0^\circ$ or 180°. Thus, (D-18) becomes

$$E_{0x} = A + B \tag{D-19a}$$

$$\pm E_{0y} = A - B \tag{D-19b}$$

which leads immediately to

$$A = \frac{E_{0x} \pm E_{0y}}{2} \tag{D-20a}$$

$$B = \frac{E_{0x} \mp E_{0y}}{2} \tag{D-20b}$$

We see that elliptically polarized light *cannot* be represented by linear $\pm 45°$ polarization states. The only state that can be represented in terms of $L \pm 45°$ light is linear horizontally polarized light. This is readily seen by writing

$$E_{0x}\mathbf{i} = \left(\frac{E_{0x}}{2}\right)\mathbf{i} + \left(\frac{E_{0x}}{2}\right)\mathbf{i} + \left(\frac{E_{0x}}{2}\right)\mathbf{j} - \left(\frac{E_{0x}}{2}\right)\mathbf{j} \tag{D-21a}$$

$$= \frac{E_{0x}}{2}[\mathbf{i} + \mathbf{j}] + \frac{E_{0x}}{2}[\mathbf{i} - \mathbf{j}] \tag{D-21b}$$

We see that the right-hand side of (D-21b) consists of linear $\pm 45°$ polarized components of equal amplitudes.

It is also possible to express linearly polarized light, $E_{0x}\mathbf{i}$, in terms of right and left circularly polarized light of *equal* amplitudes. We can write, using complex quantities,

$$E_{0x}\mathbf{i} = \left(\frac{E_{0x}}{2}\right)\mathbf{i} + \left(\frac{E_{0x}}{2}\right)\mathbf{i} + i\left(\frac{E_{0x}}{2}\right)\mathbf{j} - i\left(\frac{E_{0x}}{2}\right)\mathbf{j} \tag{D-22a}$$

$$= \frac{E_{0x}}{2}[\mathbf{i} + i\mathbf{j}] + \frac{E_{0x}}{2}[\mathbf{i} - i\mathbf{j}] \tag{D-22b}$$

We see that (D-22b) describes two oppositely circularly polarized beams of equal amplitudes.

We now represent elliptically polarized light in terms of right and left circularly polarized light of amplitudes (real) A and B. We express (D-8) as

$$\mathbf{E}(z, t) + E_{0x}\mathbf{i} + E_{0y}\exp(\mathbf{i}\delta)\mathbf{j} = A(\mathbf{i} + i\mathbf{j}) + B(\mathbf{i} - i\mathbf{j}) \tag{D-23a}$$

$$= (A + B)\mathbf{i} + i(A - B)\mathbf{j} \tag{D-23b}$$

We then find

$$E_{0x} = A + B \tag{D-24a}$$

$$E_{0y}e^{i\delta} = i(A - B) \tag{D-24b}$$

We see immediately that for $\delta = \pm 90°$, (D-24) becomes

$$E_{0x} = A + B \tag{D-25a}$$

$$\pm E_{0y} = (A - B) \tag{D-25b}$$

so (D-23b) then becomes

$$\mathbf{E}(z, t) = E_{0x}\mathbf{i} \pm iE_{0y}\mathbf{j} \tag{D-25c}$$

Equation (D-25c) is the vector representation of the *standard form of the polarization* ellipse. For convenience we only consider the + value of (D-25b) so the amplitudes (that is, the radii) of the circles are

$$A = \frac{E_{0x} \pm E_{0y}}{2} \tag{D-26a}$$

$$B = \frac{E_{0x} - E_{0y}}{2} \tag{D-26b}$$

The condition $\delta = \pm 90°$ restricts the polarization ellipse to the standard form of the ellipse [see (D-2)], namely,

$$\frac{E_x^2(z,t)}{E_{0x}^2} + \frac{E_y^2(z,t)}{E_{0y}^2} = 1 \tag{D-27}$$

Thus, only the nonrotated form of the polarization ellipse can be represented by right and left circularly polarized light of unequal amplitudes, A and B (D-26).

In Fig. D-1 we show elliptically polarized light as the superposition of the right (R) and left (L) circularly polarized light. We can determine the points where the circles (RCP) and (LCP) intersect the polarization ellipse. We write (D-27) as

$$\frac{x^2}{(A+B)^2} + \frac{y^2}{(A-B)^2} = 1 \tag{D-28}$$

and the RCP and LCP circles as

$$x^2 + y^2 = A^2 \tag{D-29a}$$

$$x^2 + y^2 = B^2 \qquad \text{(A-29b)}$$

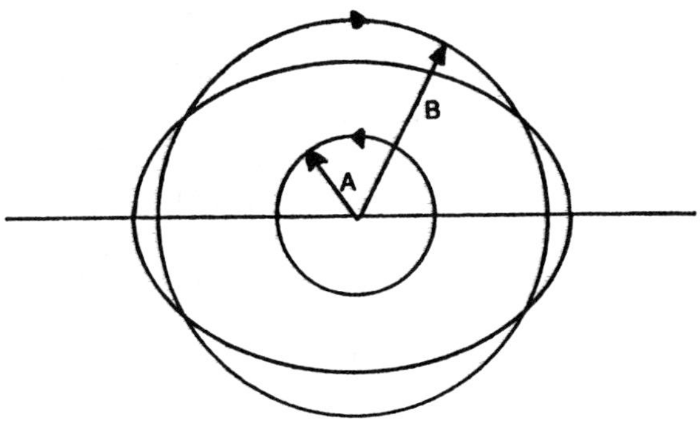

Figure D-1 Superposition of oppositely circularly polarized light of unequal amplitudes to form elliptically polarized light.

where we have set $E_x = x$ and $E_y = y$. Straightforward algebra shows the points of intersection (x_R, y_R) for the RCP circle are

$$x_R = \pm\frac{A+B}{2}\sqrt{\frac{2A-B}{A}} \tag{D-30a}$$

$$y_R = \pm\frac{A-B}{2}\sqrt{\frac{2A+B}{A}} \tag{D-30b}$$

and the points of intersection (x_L, y_L) for the LCP circle are

$$x_L = \pm\frac{A+B}{2}\sqrt{\frac{2B-A}{B}} \tag{D-31a}$$

$$y_L = \pm\frac{A-B}{2}\sqrt{\frac{2B+A}{B}} \tag{D-31b}$$

Equations (D-30) and (D-31) can be confirmed by squaring and adding (D-30a) and (D-30b) and, similarly, (D-31a) and (D-31b). We then find that

$$x_R^2 + y_R^2 = A^2 \tag{D-32a}$$

$$x_L^2 + y_L^2 = B^2 \tag{D-32b}$$

as expected.

As a numerical example of these results consider that we have an ellipse where $A = 3$ and $B = 1$. From (D-30) and (D-31) we then find that

$$x_R = \frac{\pm 2\sqrt{5}}{3} \tag{D-33a}$$

$$y_R = \pm\sqrt{5} \tag{D-33b}$$

and the points of intersection (x_L, y_L) for the LCP circle are

$$x_L = \pm 2i \tag{D-34a}$$

$$y_L = \pm\sqrt{5} \tag{D-34b}$$

Thus, as we can see from Fig. D-1, the RCP circle intersects the polarization ellipse, whereas the existence of the imaginary number in (D-34a) shows that there is no intersection for the LCP circle.

We now use these results to analyze the problem of the propagation of an optical beam through an optically active medium. Before we do this, however, we provide some historical and physical background to the phenomenon of optical activity.

Optical activity was discovered in 1811 by Arago, when he observed that the plane of vibration of a beam of linearly polarized light underwent a continuous rotation as it propagated along the optic axis of quartz. Shortly thereafter Biot (1774–1862) discovered this same effect in vaporous and liquid forms of various substances, such as the distilled oils of turpentine and lemon and solutions of sugar and camphor. Any material that causes the **E** field of an incident linear plane wave to appear to rotate is said to be *optically active*. Moreover, Biot discovered that the rotation could be left- or right-handed. If the plane of vibration appears to revolve *counterclockwise*, the substance is said to be *dextrorotatory* or *d-rotatory* (Latin *dextro*, right). On the other hand, if **E** rotates *clockwise* it is said to be *levorotatory* or *l-rotatory* (Latin *levo*, left).

The English astronomer and physicist Sir John Herschel (1792–1871), son of Sir William Herschel, the discoverer of the planet Uranus, recognized that the *d*-rotatory and *l*-rotatory behavior in quartz actually corresponded to two different crystallographic structures. Although the molecules are identical (SiO_2), crystal quartz can be either right-or left-handed, depending on the arrangement of these molecules. In fact, careful inspection shows that there are two forms of the crystals, and they are the same in all respects except that one is the mirror image of the other; they are said to be *enantiomorphs* of each other. All transparent enantiomorphic structures are optically active.

In 1825, Fresnel, without addressing himself to the actual mechanism of optical activity, proposed a remarkable solution. Since an incident linear wave can be represented as a superposition of R- and L-states, he suggested that these two forms of circularly polarized light propagate at different speeds in an optically active medium. An active material shows *circular birefringence*; i.e., it possesses two indices of refraction, one for the R-state (n_R) and one for the L-state (n_L). In propagating through an optically active medium, the two circular waves get out of phase and the resultant linear wave appears to rotate. We can see this behavior by considering this phenomenon analytically for an incident beam that is elliptically polarized; linearly polarized light is then a degenerate case.

In Fig. D-2 we show an incident elliptically polarized beam entering an optically active medium with field components E_x and E_y. After the beam has propagated through the medium the field components are E'_x and E'_y.

Fresnel suggested that in an optically active medium a right circularly polarized beam propagates with a wavenumber k_R and a left circularly polarized beam propagates with a different wavenumber k_L. In order to treat this problem analytically we consider the decomposition of $E_x(z, t)$ and $E_y(z, t)$ separately. Furthermore, we suppress the factor ωt in the equations because the time variation plays no role in the final equations.

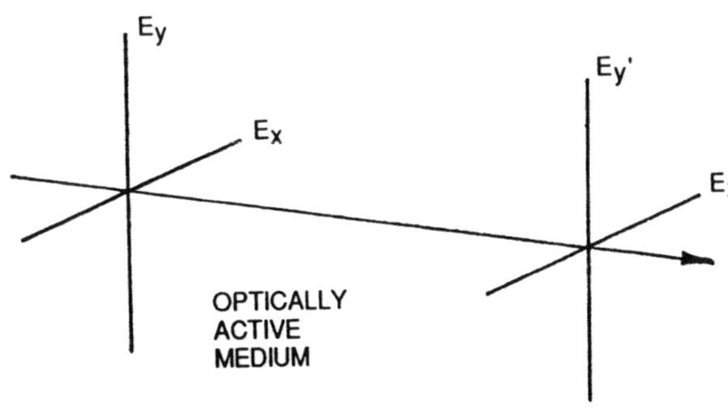

Figure D-2 Field components of an incident elliptically polarized beam propagating through an optically active medium.

For the $E_x(z)$ component we can write this in terms of circular components as

$$\mathbf{E}_{Rx}(z) = \frac{E_x}{2}[\cos(k_R z)\mathbf{i} - \sin(k_R z)\mathbf{j}] \tag{D-35a}$$

$$\mathbf{E}_{Lx}(z) = \frac{E_x}{2}[\cos(k_L z)\mathbf{i} + \sin(k_L z)\mathbf{j}] \tag{D-35b}$$

Adding (D-35a) and (D-35b) we see that, at $z=0$,

$$\mathbf{E}_{Rx}(0) + \mathbf{E}_{Lx}(0) = E_x\mathbf{i} \tag{D-36}$$

which shows that (D-35) represents the x component of the incident field. Similarly, for the $E_y(z)$ component we can write

$$\mathbf{E}_{Ry}(z) = \frac{E_y}{2}[\sin(k_R z)\mathbf{i} + \cos(k_R z)\mathbf{j}] \tag{D-37a}$$

$$\mathbf{E}_{Ly}(z) = \frac{E_y}{2}[-\sin(k_L z)\mathbf{i} + \cos(k_L z)\mathbf{j}] \tag{D-37b}$$

Adding (D-37a) and (D-37b) we see that, at $z=0$,

$$\mathbf{E}_{Ry}(0) + \mathbf{E}_{Ly}(0) = E_{yx}\mathbf{j} \tag{D-38}$$

so (D-37) corresponds to the y component of the incident field. The total field $\mathbf{E}'(z)$ in the optically active medium is

$$\mathbf{E}'(z) = E'_x\mathbf{i} + E'_y\mathbf{j} + = \mathbf{E}_{Rx} + \mathbf{E}_{Lx} + \mathbf{E}_{Ry} + \mathbf{E}_{Ly} \tag{D-39}$$

Substituting (D-35) and (D-37) into (D-39) we have

$$\begin{aligned}\mathbf{E}'(z) = \mathbf{i}&\left[\frac{E_x}{2}[\cos k_R z + \cos k_L z] + \frac{E_y}{2}[\sin k_R z + \sin k_L z]\right] \\ &+ \mathbf{j}\left[\frac{-E_x}{2}[\sin k_R z - \sin k_L z] + \frac{E_y}{2}[\cos k_R z + \sin k_L z]\right]\end{aligned} \tag{D-40}$$

Hence, we see that

$$E'_x(z) = \frac{E_x}{2}[\cos k_R z + \cos k_L z] + \frac{E_y}{2}[\sin k_R z + \sin k_L z] \tag{D-41a}$$

$$E'_y(z) = -\frac{E_x}{2}[\sin k_R z - \sin k_L z] + \frac{E_y}{2}[\cos k_R z + \cos k_L z] \tag{D-41b}$$

Equations (D-41a) and (D-41b) can be simplified by rewriting the terms:

$$\cos k_R z + \cos k_L z \tag{D-42a}$$

$$\sin k_R z - \sin k_L z \tag{D-42b}$$

Let

$$a = \frac{(k_R + k_L)z}{2} \tag{D-43a}$$

$$b = \frac{(k_R - k_L)z}{2} \tag{D-43b}$$

so

$$k_R z = a + b \tag{D-44a}$$

$$k_L z = a - b \tag{D-44b}$$

and (D-42) then becomes

$$\cos k_R z + \cos k_L z = \cos(a+b) + \cos(a-b) \tag{D-45a}$$

$$\sin k_R z - \sin k_L z = \sin(a+b) - \sin(a-b) \tag{D-45b}$$

Using the familiar sum and difference formulas for the cosine and sine terms of the right-hand sides of (D-45a) and (D-45b) along with (D-43), we find that

$$\cos k_R z + \cos k_L z = 2\cos\left[\frac{(k_R + k_L)z}{2}\right]\cos\left[\frac{(k_R - k_L)z}{2}\right] \tag{D-46a}$$

$$\sin k_R z - \sin k_L z = 2\cos\left[\frac{(k_R + k_L)z}{2}\right]\sin\left[\frac{(k_R - k_L)z}{2}\right] \tag{D-46b}$$

The term $\cos(k_R + k_L)z/2$ in (D-46a) and (D-46b) plays no role in the final equations and can be dropped. Substituting the remaining cosine and sine term in (D-46) into (D-41), we finally obtain

$$E'_x(z) = \frac{E_x}{2}\cos\frac{(k_R - k_L)z}{2} + \frac{E_y}{2}\sin\frac{(k_R - k_L)z}{2} \tag{D-47a}$$

$$E'_y(z) = -\frac{E_x}{2}\sin\frac{(k_R - k_L)z}{2} + \frac{E_y}{2}\cos\frac{(k_R - k_L)z}{2} \tag{D-47b}$$

We see that (D-47) are the equations for rotation of E_x and E_y. We can write (D-47) in terms of the Stokes vector and the Mueller matrix as

$$\begin{pmatrix} S'_0 \\ S'_1 \\ S'_2 \\ S'_3 \end{pmatrix} = \begin{pmatrix} 1 & 0 & 0 & 0 \\ 0 & \cos 2\beta & \sin 2\beta & 0 \\ 0 & -\sin 2\beta & \cos 2\beta & 0 \\ 0 & 0 & 0 & 0 \end{pmatrix} \begin{pmatrix} S_0 \\ S_1 \\ S_2 \\ S_3 \end{pmatrix} \tag{D-48a}$$

where

$$\beta = \frac{(k_R - k_L)z}{2} \tag{D-48b}$$

The angle of rotation β can be expressed in terms of the refractive indices n_R and n_L of the medium and the wavelength λ of the incident beam by writing

$$k_R = k_0 n_R = \frac{2\pi n_R}{\lambda} \tag{D-49a}$$

$$k_L = k_0 n_L = \frac{2\pi n_L}{\lambda} \tag{D-49b}$$

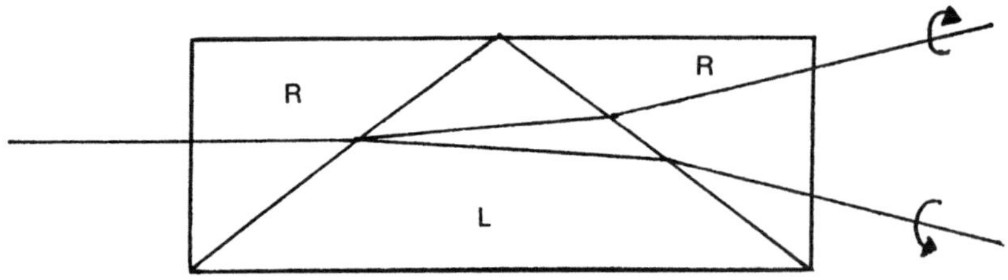

Figure D-3 Fresnel's construction of a composite prism consisting of R-quartz and L-quartz to demonstrate optical activity and the existence of circularly polarized light.

and $k_0 = 2\pi/\lambda$. If $n_R \leq n_L$ the medium is *d*-rotatory, and if $n_R \geq n_L$ the medium is *l*-rotatory. Substituting (D-49) into (D-48), we then have

$$\beta = \frac{\pi(n_R - n_L)z}{\lambda} \tag{D-50}$$

The quantity β/d is called the *specific rotatory power*. For quartz it is found to be 21.7°/mm for sodium light, from which it follows that $|n_R - n_L| = 7.1 \times 10^{-5}$. Thus, the small difference in the refractive indices shows that at an optical interface the two oppositely circularly polarized beams will be very difficult to separate. Fresnel was able to show the existence of the circular components and separate them by an ingenious construction of a composite prism consisting of R- and L-quartz, as shown in Fig. D-3. He reasoned that since the two component traveled with different velocities they should be refracted by different amounts at an oblique interface. In the prism the separation is increased at each interface. This occurs because the right-handed circular component is faster in the R-quartz and slower in the L-quartz. The reverse is true for the left-handed component. The former component is bent down and the latter up, the angular separation increasing at each oblique interface. If the two images of a linearly polarized source are observed through the compound prism and then examined with a linear polarizer the respective intensities are unaltered when the polarizer is rotated. Thus, the beams must be circularly polarized.

The subject of optical activity is extremely important. In the field of biochemistry a remarkable behavior is observed. When organic molecules are synthesized in the laboratory, an equal number of *d*- and *l*-isomers are produced, with the result that the mixture is optically inactive. One might expect in nature that equal amounts of *d*- and *l*-stereoisomers would exist. This is by no means the case. Natural sugar (sucrose, $C_{12}H_{22}O_6$) always appears in the *d*-rotatory form, regardless of where it is grown or whether it is extracted from sugar cane or sugar beets. Moreover the sugar dextrose of *d*-glucose ($C_6H_{12}O_{11}$) is the most important carbohydrate in human metabolism. Evidently, living cells can distinguish in a manner not yet fully understood between *l*- and *d*-molecules.

One of the earliest applications of optical activity was in the sugar industry, where the angle of rotation was used as a measure of the quality of the sugar (saccharimetry). In recent years optical activity has become very important in other branches of chemistry. For example, the artificial sweetener aspartame and the pain reducer ibuprofen are optically active. In the pharmaceutical industry it has been estimated that approximately 500 out of the nearly 1300 commonly used drugs are optically active. The difference between the *l*- and *d*-forms can, it is believed, lead to very undesirable consequences. For example, it is believed that the optically active sedative drug thalidomide when given in the *l*-form acts as a sedative, but the *d*-form is the cause of birth defects.

Interest in optical activity has increased greatly in recent years. Several sources are listed in the references. Of special interest is the stimulating article by Applequist, which describes the early investigations of optical activity by Biot, Fresnel, and Pasteur, as well as recent investigations, and provides a long list of related references.

REFERENCES

1. Hecht, E. and Zajac, A., *Optics*, Addison-Wesley, Reading, MA, 1979.
2. Wood, R. W., *Physical Optics*, 3rd ed., Optical Society of America, Washington, DC, 1988.
3. Callewaert, D. M. and Genuyea, J., *Basic Chemistry*, Worth, New York, 1980.
4. Applequist, J., *American Scientist*, *75*, 59 (1987).

Bibliography

Books primarily about polarization

Azzam, R. M. A and Bashara, N. M., *Ellipsometry and Polarized light*, North-Holland, Amsterdam, 1977.

Brosseau, C., *Fundamentals of Polarized light: A Statistical Optics Approach*, Wiley, New York, 1998.

Clarke, D. and Grainger, J. F., *Polarized Light and Optical Measurement*, Pergamon Press, Oxford, 1971.

Huard, S., *Polarization of Light*, Wiley, New York, 1997.

Shurcliff, W., *Polarized Light*, Oxford University Press, London, 1962.

Collected papers

Swindell, W., *Polarized Light in Optics*, Dowden, Hutchinson, & Ross, Stroudsburg, PA, 1975.

Optics books with material on polarization

Born, M. and Wolf, E., *Principles of Optics*, Pergamon Press, New York, 1980.

O'Neill, E. L., *Introduction to Statistical Optics*, Addison-Wesley, Reading, MA, 1963.

Saleh, B. and Teich, M., *Fundamentals of Photonics*, Wiley-Interscience, New York, 1991.

van de Hulst, H. C., *Light Scattering by Small Particles*, Dover, New York, 1981.

Handbook chapters

Azzam, R. M. A "Ellipsometry", Ch. 27 in Vol. 2, *Handbook of Optics*, 2nd ed., M. Bass, ed., McGraw-Hill, New York, 1994.

Bennett, J. M. and Bennett, H. E., "Polarization", Ch. 10 in *Handbook of Optics*, W. G. Driscoll and W.Vaughan, Eds., McGraw-Hill, New York, 1980.

Bennett, J. M., "Polarizers", Ch. 3 in Vol. 2, *Handbook of Optics*, 2nd ed., M. Bass, ed., McGraw-Hill, New York, 1994.

Chipman, R. A., "Polarimetry", Ch. 22 in Vol. 2, *Handbook of Optics*, 2nd ed., M. Bass, ed., McGraw-Hill, New York, 1994.

Introductory optics texts

Hecht, E., *Optics*, Addison-Wesley, Reading, MA, 1987.

Jenkins, F. A. and White, H. E., *Fundamentals of Optics*, McGraw-Hill, New York, 1976.

Polarization mathematics

Gerrard, A. and Burch, J. M., *Introduction to Matrix Methods in Optics*, Wiley, London, 1975.

Theocaris, P. S. and Gdoutos, E. E., *Matrix Theory of Photoelasticity*, Springer-Verlag, Berlin, 1979.

Miscellaneous

Konnen, G. P., *Polarized Light in Nature*, Cambridge University Press, Cambridge, 1985.

Yariv, A. and Yeh, P., *Optical Waves in Crystals*, Wiley, New York, 1984.

SPIE Proceedings

Volume 4819, *Polarization Measurement, Analysis, and Applications V*, eds. D. H. Goldstein and D. B. Chenault, July 2002.

Volume 4481, *Polarization Analysis, Measurement, and Remote Sensing IV*, eds. D. H. Goldstein, D. B. Chenault, W. G. Egan, and M. J. Duggin, Aug. 2001.

Volume 4133, *Polarization Analysis, Measurement, and Remote Sensing III*, eds. D. B. Chenault, M. J. Duggin, W. G. Egan, and D. H. Goldstein, Aug. 2000.

Volume 3754, *Polarization: Measurement, Analysis, and Remote Sensing II*, eds. D. H. Goldstein and D. B. Chenault, July 1999.

Volume 3121, *Polarization: Measurement, Analysis, and Remote Sensing*, eds, D. H. Goldstein and D. B. Chenault, Aug. 1997.

Volume 2873, *International Symposium on Polarization Analysis and Applications to Device Technology*, eds. T. Yoshizawa and H. Yokota, Aug. 1996.

Volume 2265, *Polarization Analysis and Measurement II*, eds. D. H. Goldstein and D. B. Chenault, Aug. 1994.

Volume 1747, *Polarization and Remote Sensing*, ed. W. G. Egan, July 1990.

Volume 1746, *Polarization Analysis and Measurement*, eds. D. H. Goldstein and R. A. Chipman, July 1992.

Volume 1317, *Polarimetry: Radar, Infrared, Visible, Ultraviolet, and X-Ray*, eds. R. A. Chipman and J. W. Morris, Oct. 1990.

Volume 1166, *Polarization Considerations in Optical Systems II*, ed. R. A. Chipman, Aug. 1989.

Volume 891, *Polarization Considerations in Optical Systems*, ed. R. A. Chipman, Jan. 1988.

MS23, *Selected Papers on Polarization*, ed. B. H. Billings, Nov. 1990.

Index

ABCD polarization matrix, 187–194
Absorbing media (*see* Optics of metals)
Ampere's law, 136, 301
Analyzer, 67
Arago, Dominique Francois, 2, 14, 21, 403, 411, 530
Attenuated total reflection, 142
Auxiliary angle, 29, 40
Azzam, R.M.A., 547, 552, 563, 609, 614

Babinet–Soleil compensator, 95, 102, 225, 226
Bartholinus, Erasmus, 1, 445, 447, 452
Bessel function expansion, 538
Birefringence
  induced, 463–470
  natural, 460
Bohr, Niels, 418
Bohr's correspondence principle, 433
Brewster's angle, 2, 140, 154–155, 199, 202–203, 205
Brewster, David, 2, 133, 155
de Broglie, Louis, 418

Calcite, 1, 69, 452, 460, 520
Canada balsam, 520
Cayley–Hamilton theorem, 182
Čerenkov effect, 382–393
  critical angle, 387
  interference, 386–387
  Stokes vector, 393
  wake, 387
Čerenkov, P.A., 383
Chandrasekhar's phase matrix, 195
Chandrasekhar, Subrahmanyan, 32, 47, 194, 420
Circularly polarized light
  left handed, 30, 37, 39
  right handed, 30, 37, 39
Cloude, S., 168
Coherency matrix, 168–172
Compensator (*see* Retarder)
Compton scattering, 420
Conical diffraction, 552
Continuity equation, 302
Corpuscular theory of light, 1, 11
Coulomb's law, 301
Critical angle, 140
Crystals
  biaxial, 453, 457
  centrosymmetric, 453
  isotropic, 457
  lattice type, 451
  Miller indices, 453
  point group, 451–452
  symmetry, 451–452
  uniaxial, 453, 457, 522
  unit cell, 453
C-vector, 171
Cycloid, 339
Cyclotron frequency, 336, 348

Decomposition of the Mueller matrix, Lu–Chipman, 175
Degree of polarization, 48–49, 167
  for quantum systems, 426
Density matrix, 421–427
Depolarization, 175, 181–184
Depolarizer, 66, 531–532
  ideal, 531
Dextrorotatory, 530
Diattenuation, 175–182, 237
Diattenuator, 66, 237
Dichroism, 237
Dielectric constant, 135, 448
Dielectric plates, 187–209
Dielectric tensor, 405, 449–450
Dipole radiation field, 322
Dirac matrices, 168–169
Displacement current, 301
Drude, 509, 585–586, 597, 616

Eigenpolarizations, 176
Eigenvalues, 190
Eigenvectors, 190
Electric displacement, 448
Electric polarization, 448
Electric susceptibility, 448
Electron area, 396
Electron radius, 348, 395, 397
Electro-optic modulator
clamped, 451
unclamped, 451
Electro-optic tensor coefficients
linear (Pockels), 451, 467
quadratic (Kerr), 451, 469
Ellipsometry
ABCD form of the Mueller matrix for ellipsometry, 616–618
ambient-film-substrate (AFS) system, 604
complex refractive index and film thickness measurement, 585
determination of the complex refractive index, 601–603
ellipsometric parameters psi (ψ) and delta ($\triangle$) measurement, 586, 588–597
Mueller matrix representation, 616–623
fundamental equation of, 586–588
historical background, 585–586
Holmes and Feucht's formulation, 594
measurement formulation of, 594
model, homogeneous thin film on substrate, 597
Mueller matrix in terms of reflectivity coefficients, 617
multiple reflections within a thin film, 604–607
null intensity condition, 621
solution of the fundamental equation of, 597–614
use of computer to obtain solution, 609, 613
Elliptically polarized light, 83–86
Enantiomorphic, 403

Fano, U., 420
Faraday cell, 537, 538, 540
Faraday effect, 351, 404, 471, 529
Faraday rotation in plasma, 414–416
Mueller matrix for, 416
[Faraday effect]
Faraday rotation in transparent medium, 411–414
Mueller matrix for, 413
Faraday's law, 301, 351
Faraday, Michael, 351, 411
Faraday rotation, 403–405, 531
45° polarized light, 26, 36, 38
Fourier transform, 7–8
Fresnel–Arago interference laws (*see* Interference laws)
Fresnel, Augustin Jean, 2, 21
Fresnel rhomb, 161–162, 525
Fresnel's equations for reflection and transmission, 134–145, 598, 607

Generator, 67
Glan–Thompson prism, 69
Glass, BK7, 161
Green's function, 305–306

Harmonic oscillator, 8–10
Heisenberg, Werner, 48, 364, 418
Herapath, William Bird, 511
Herapathite, 511
Herschel–Stokes experiment, 292–294
Hertz, Heinrich, 2, 298
Hipparchus, 241
Horizontally polarized light, 25, 36, 38
Huygens, Christian, 1, 11, 21, 447

Iceland spar, 445, 447
Impermeability tensor, 450
Index ellipsoid, 453–459, 463–470, 574
Index of refraction (*see* Refractive index)
Integrating sphere, 532
Interference laws
statements, 275–276
summary, 294–295

Jones calculus (*see* Jones formalism)
Jones formalism, 165, 211–239
Jones matrix, 166, 216–239
for circular polarizer, 235
for diattenuator, 237
for elliptical polarizer, 236–237
for neutral density filter, 218
for polarizer, 217
relationship to Mueller matrix, 166
for retarder, 219, 238
for rotator, 217

Jones, R. Clark, 165, 212
Jones vector, 165, 212–216
  for circularly polarized light, 214
  for linear horizontally polarized light, 213
  for linear −45° polarized light, 214
  for linear +45° polarized light, 214
  for linear vertically polarized light, 214
  orthogonal/orthonormal property, 214

Kent and Lawson, method of, 223–224
Kerr effect, 404, 471
Kronecker delta, 215, 278
Kronecker product, 166

Land, Edwin, 446, 511–512
Laplacian, 5
Larmor, J., 355
Larmor precession frequency, 354–355
Larmor's radiation formula, 312, 316
Larmor's theorem, 357
Law of cosines, 246, 252–258
Law of sines, 252–258
Levorotatory, 530
Liquid crystal electro-optic effects
  field-induced birefringence, 473
  twisted nematic effect, 473–475
Liquid crystals, 471–475, 531
  cholesteric, 472
  mesophase, 472
  nematic, 472
  smectic, 472
    ferroelectric, 475
  thermotropic, 472
Lorentz force equation, 332, 353
Lorentz, H., 319, 352
Lorentz–Zeeman effect (*see* Zeeman effect)
Lu–Chipman Decomposition, 175

Magnetic continuity law, 301
Magneto-optics (*see also* Faraday effect; Kerr effect; Verdet constant)
  Cotton–Mouton effect, 471
  magneto-optic cell, 470
  Voigt effect, 471
Malus, 1, 133
Malus' law, 517
Mass spectrometer, 337
Maxwell, James Clerk, 2, 297
Maxwell's equations, 135, 301, 312, 405
Maxwell's theory of the electrodynamic field, 297
  constitutive relations, 302
  Hertz's confirmation, 298
  solution of radiation equation by Green's function, 305–306
  source of the optical field, 305
  transverse field, 304
  wave equation and the speed of light, 303–305
McCrackin, F. L., 594
Measurement of the Mueller matrices
  for a polarizer, 117–124
  for a retarder, 124–130
  for a rotator, 130–132
Metals (*see* Optics of metals)
Modal matrix, 191
Modulator, 463–475
  longitudinal mode, 475–476
  transverse mode, 475–476
Molecular rotation, 404
Motion of charge in electromagnetic field
  constant electric field, 332–334
  constant magnetic field, 335–337
  crossed electric and magnetic field, 338–346
Mueller formalism, 165
Mueller, Hans, 66, 166
Mueller matrices
  for reflection and transmission
    Brewster angle, 154–155
    45° incidence, 155–156
    normal incidence, 152
    in terms of the Fresnel coefficients, 204–209
    total internal reflection, 158–163
Mueller matrix, 65–86
  ABCD polarization matrix, 188–194
  constraints on, 167
  decomposition, Lu–Chipman, 175
  depolarizer, ideal, 531
  eigenvector and eigenvalue analysis, 168
    examples, 171–175
  general linear polarizer, 68, 70
  half-wave retarder, 74, 79
  horizontal polarizer, 69
  mathematics of, 165–185
  neutral density filter, 68
  null, 71
  physically realizable, 165, 167, 168–175
  polarizer, rotated, 78, 512

[Mueller matrix]
quarter-wave retarder, 73
rotated, 82
relationship to Jones matrix, 166
retarder, 73
anisotropic absorbing, 90
rotated, 79, 575
rotator, 75–76
vertical polarizer, 69
von Neumann, John, 419
Newton, Isaac, 1, 11, 21
Null methods, 480

Oersted, H., 351
Optical activity, 403–411, 530–531
Mueller matrix, 408
natural rotation, 403
orientation angle of polarization ellipse, 409
Optical crossbar switch, 531
Optical isolator, 531
Optic axis, 453, 522
Optics of metals
absorbing media, 479
aluminum, evaporating, 479
astronomical mirrors, 479
attenuation coefficient, 482
complex refractive index, 480, 483
conductivity, 480, 483
extinction coefficient, 482
field propagation, 482
Fresnel's law of reflection, 483
Fresnel's reflection equations for absorbing media, 489–490, 497
Joule heat, 479
Maxwell's equations, 480–481
Mueller matrix, 491, 501
oxidation of silver, 479
principal angle of incidence, 480
method of measuring refractive index and extinction coefficient, 489–496
principal azimuthal angle, 480, 494
pseudo-Brewster angle, 489, 496
reflection coefficients, 503–504
reflectivity
for non-normal incidence, 485–489
for normal incidence, s and p components, 484–485
refractive index and extinction coefficient measurement at 45°, 496–509
[Optics of metals]
Snell's law for absorbing media, 483
Stokes parameters, 490, 504
Stokes vector, 491–495, 502
wave equation, 481

Park, III, N. G., 66
Parseval's theorem, 389
Pauli spin matrices, 62, 168, 427
Permeability, relative, 135
Permittivity,
of free space, 135, 448
relative, 135
Perrin, Francis, 419
Petal plot, 356
Photomultiplier tube, 102
Planck, 418
Plane, equation of, 10
Poincaré, Henri, 42, 242, 419
Poincaré sphere, 42, 241–273, 555
Pockels cells, 578–579
Polar decomposition, 175, 182
Polarimeter
optimization of, 554–556
Stokes
complete, 533
incomplete, 533
Polarimetry
division of amplitude, 533
four-channel, 546
four-detector photopolarimeter, 547–552
using gratings, 552–553
using parallel slab, 553–554
division of wavefront, 533, 545
oscillating element, 533, 537
analyzer, 538–540
retarder and analyzer, 541–542
retarder with fixed analyzer, 540–541
phase modulation, 533, 543
dual phase modulator and fixed analyzer, 544
phase modulator and fixed analyzer, 543–544
rotating element, 533, 534
analyzer, 534–536
analyzer and fixed analyzer, 536
retarder and analyzer, 537
retarder and analyzer plus fixed analyzer, 537
retarder and fixed analyzer, 536
Stokes, 533–557

[Polarimetry]
  Mueller matrix, 559–583
    complete, 559
    dual rotating retarder, 563–574
    four-detector photopolarimeter, 581–582
    incomplete, 559
    measurement matrix method, 577–578
    modulator-based, 578–580
    phase-modulating, 562–563
    rotating element, 560–561
    scatterometer, 580–581
Polarizance, 178–179
Polarization
  p, 139
  s, 137
  TE, 137
  TM, 139
Polarization ellipse, 21–30, 23, 222, 244
  ellipticity angle, 27, 35, 39, 41
  degenerate states, 26
  orientation angle, 27, 34, 39, 41
Polarization state analyzer (PSA), 533–534
Polarization state generator (PSG), 533–534
Polarizer, 44, 66, 511
  absorption, 511–520
  Brewster angle, 522
  calcite, 517
  densities, major and minor principal, 514
  dichroic, 512
  dichroitance, 514
  extinction ratio, 513
  Glan–Thompson, 460
  major principal transmittance, 513
  minor principal transmittance, 513
  molecular, 512
  pile of plates, 187, 197
  polarcor, 518
  principal transmittance ratio, 513
  prism, 520–522
    Glan–Taylor, 520
    Glan–Thompson, 520
    Nicol, 460, 520–522
    Rochon, 520
    Wollaston, 460, 520
  total transmittance, 514
  wire grid, 518–520
Polaroid, 69, 446, 511–518
  H-sheet, 512, 517–518
  J-sheet, 512
  K-sheet, 512
  L-sheet, 512
Poynting vector, 144–145, 312–313, 370
Pseudodepolarizer, 531
Pseudopolarizer, 126
Pseudorotator, 81, 127, 220, 414

Quantum mechanics, 418
Quartz, 403, 530

Radiation
  emitted by charge moving in circle, 325–328
  emitted by charge moving in constant electric field, 347
  emitted by charge moving in constant magnetic field
    nonrelativistic, 348–349
    relativistic, 376–382
  emitted by charge moving in ellipse, 32–329
  emitted by linear oscillating charge,
    nonrelativistic, 320–322
    relativistic, 369–375
  emitted by randomly oriented charges, 322–325
Radiation field equations
  nonrelativistic, 307–308
  relativistic, 369–400
Rayleigh, Lord, 585
Rayleigh scattering, 400
Reflection and transmission of a wave at an interface, 15–19
Reflection, internal, 140
Refractive index, 448
  extraordinary, 452
  ordinary, 452
  principal, 450–453
Refractive index measurement, 480
Refractometry, 480
Retardance, 175, 178–182
Retarder, 44, 66, 522–529
  achromatic, 525–529
    Fresnel rhomb, 525
    infrared, 525–528
    waveplate, 528–529
  fast axis, 522–523
  half wave, 523
  multiorder, 522
  quarter-wave, 46, 523
  rhomb, 522
  slow axis, 522

[Retarder]
  variable, 524
    Babinet–Soleil, 522, 524
  zero order, 522
Rose diagrams, 360
Rose equation, 361
Rotary power, 403–404
Rotator, 66
  Faraday, 531
  liquid crystal, 531
  optically active, 530–531
Rothen, Alexandre, 585

Saccharimeter, 403
Saccharimetry, 403
Scattering, 393–400
  differential cross section, 397
  Rayleigh, 398–400
  Thomson, 395–397
  total cross section, 397
Schrödinger, Erwin, 418
Schwarz's inequality, 60
Secular equation, 190
Selection rule, 436
Sheet polarizer, 511–512
Snell's law, 140
Sodium, D lines, 351–352
Soleillet, Paul, 419
Sommerfeld, A., 383
Spectropolarimetry, 576–577
Spherical triangle, 85–86, 246–248
Stereoisomer, 530
Stokes, George Gabriel, 31, 34, 419
Stokes intensity formula, 46
Stokes parameters, 31–62
  additional properties of, 49–59
  additive property, 52
  for circularly polarized light, 37
  classical measurement of, 43–46
  complex representation, 35
  derivation of, 32–34
  incoherent superposition, 49–55
  for linearly polarized light, 36
  for quantum systems, 425–442
    for particle in infinite potential well, 435–437
    for rigid rotator, 438–442
  relation to Poynting vector, 312–315
  for relativistic motion, 371
  for unpolarized and partially polarized light, 47–49
  and Wolf's coherency matrix, 60–62
Stokes vector, 37–42
  for circularly polarized light, 39
  for linearly polarized light, 38
  for radiation emitted by accelerating charges, 345
    moving in constant electric field, 347
    moving in constant magnetic field, 348
  unpolarized light, 147
Stokes vector measurement
  circular polarizer method, 91–95
  classical method, 87–91
  Fourier analysis method, 96–101
  Kent and Lawson method, 102–108
  null intensity method, 95–96
Sugar, 530
Sunglasses, polarizing, 149
Susceptibility, electric, 135
Susceptibility tensor, 449–450
Synchrotron radiation, 376–382

Target coherency matrix, 168–169
Tests for
  completely polarized light, 109
    circular, 110, 112
    linear, 109–110, 112
  elliptically polarized light, 109–115
  unpolarized light, 109
Thomson, J.J., 298
  discovery of the electron, 298, 319
Thomson scattering, 393–397
Trochoid, 339, 341

Unpolarized light, 48
  mathematical statements for, 276–278

Verdet constant, 404, 413, 471
Verdet, Emil, 47, 419
Vertically polarized light, 25, 36, 38

Wave equation, 3–12
  classical, 305
Waveplate (*see* Retarder)
Wave surface, 460–461
Wave theory of light, 2
Wavevector surface, 461–463
Wolf, E., 364
Wolf's coherency matrix, 59–62

Young, Thomas, 1, 3, 11, 21
  two-pinhole interference experiment, 1, 11–14, 278–292

Zeeman effect,
  differential equations of motion, 354
  discovery of the splitting of spectral lines, 319
  historical introduction, 351–352
  Larmor precession frequency, 354–355
  longitudinal, 364
  Lorentz force equation, 353
[Zeeman effect]
  motion of a bound charge in a constant magnetic field, 352–361
  petal plots, 356
  position equations, 356
  Stokes vector
    longitudinal Zeeman effect, 364
    transverse Zeeman effect, 365
  transverse, 365
Zeeman, P., 351–352